Introduction to the Magnetic Properties of Solids

Introduction to the Magnetic Properties of Solids

A. S. CHAKRAVARTY
Saha Institute of Nuclear Physics
Calcutta, India

A Wiley-Interscience Publication
JOHN WILEY & SONS
New York · Brisbane · Chichester · Toronto

Library of Congress Cataloging in Publication Data:

Chakravarty, Ardhendu Sekhar, 1931-
Introduction to the magnetic properties of solids.

"A Wiley-Interscience publication."
Includes index.
1. Solids—Magnetic properties. I. Title.
QC176.8.M3C47 530.4'1 80-12793
ISBN 0-471-07737-2

Printed in the United States of America

10 9 8 7 6 5 4 3 2 1

To my sons Sugato and Sujoy
who have always overestimated me

To my wife Chua
who has always underestimated me

and to my friends and colleagues
who have their own assessments

Preface

Magnetism in solids, owing to its importance in modern technology, continues to attract the attention of physicists and chemists. However, although the phenomena of ferromagnetism and antiferromagnetism in magnetic solids are qualitatively understood through the application of the molecular field theory, the search for satisfactory microscopic theories for a variety of magnetic structures in solids still continues, and scientists are working toward the development of a suitable theoretical model that will take into account the correlations in the motions of the electrons in an effective and satisfactory way to explain the critical point phenomenon. Our present knowledge is based on the one-electron approximation, and only the future can say whether we can do better than this. Many of the basic fundamental questions remain to be answered, but phenomenological models like the simple molecular field theories of Weiss and Stoner, the generalized Heisenberg model advocated by Van Vleck, and the band-theoretic approach of Slater go a long way toward an understanding of magnetism in solids. Furthermore, the recent correlated effective field (CEF) theory and the microscopic theories also shed much light on the magnetism of the ordered phase in magnetic solids. It should be borne in mind, however, that in theoretical physics the differences between theories stem not from conceptual disagreement over the phenomenon of magnetism in solids but rather from different philosophies of theoretical approximations.

During the last three decades research on the magnetism of solids has been extensive, and to understand and appreciate fully this work it is of the utmost importance to trace the theoretical developments in a chronological fashion. Indeed, this was the principal motivation for writing this book, and in so doing I have attempted to present a broad picture of the subject, including areas currently under investigation. This book analyzes chronologically the development of the different theoretical models, as well as their successes and failures, from the time of Weiss to the present Green functional approach. It deals, on the one hand, with magnetism in dilute magnetic systems, which can be understood by the application of the ligand field theory and, on the other, with condensed magnetic systems, where the exchange interaction is strong enough to give rise to some kind of order leading to ferromagnetism, antiferromagnetism, or other types of magnetic structures. Quantum field-theoretic techniques are currently used to explain the magnetization in the condensed systems.

The ligand field theory, in general, adequately explains the magnetic properties of the iron group of inorganic complexes. Such success cannot be claimed, however, in the case of compounds of the palladium and platinum groups, which owe their magnetic properties to the outermost $4d$ and $5d$ electrons, where the difficulty in preparing single crystals has inhibited experimental work. The failure of empirical investigation has, in turn, discouraged any extensive theoretical work. It is known, however, that compounds of the palladium and platinum groups generally become antiferromagnetic below the Neel temperature, and a basic theoretical understanding of these materials below this temperature is achieved through the molecular field theory, the spin-wave theory of Bloch, and the Green functional approach. The treatment of this area has been somewhat simplified in order to provide an understanding of the basic ideas without becoming involved with too much mathematical detail. Above the Neel temperature these materials become paramagnetic, but the electronic magnetic moment is greatly reduced because of the bonding effect, which is primarily due to the fact that the $4d$ and $5d$ electrons are less tightly bonded to the metal atom than their sister $3d$ electrons. As a result, the molecular orbital formation occurs readily, and the ligand field parameter, $10Dq$, increases considerably in magnitude. Furthermore, the problem of spin-orbit coupling becomes much more important, and therefore to treat it as a small perturbation is no longer permissible. The appropriate theoretical model for such compounds above the Neel temperature is the theory of intermediate coupling. I have dealt with this aspect in some detail since there has been considerable theoretical and experimental work on these materials during the last two decades.

Chapters 1 to 4 are introductory and deal with the atomic spectra, the electronic structures, and the magnetic effects of atoms and ions. The concept of symmetry and group theory (a familiarity with which, as presented by Cotton in his *Chemical Applications of the Group Theory*, is assumed) outlined in Chapter 5 is essential for an understanding of the ligand field theory. Chapter 6 is devoted to the crystal field or the ionic theory and explains what happens when the transition metal ion is embedded inside the static crystalline electric field of surrounding ligands having different symmetries. The crystal field theory is applied to the iron group of complexes in Chapter 7, and some worked examples illustrate the method of calculation. Chapter 8 deals with the systems whose ground state is orbitally nondegenerate, having only spin degeneracy. These are treated by the spin Hamiltonian method. The strong-field coupling scheme, developed principally by Tanabe, Sugano, and Kamimura, is outlined in Chapter 9. This coupling scheme describes the situation where the strength of the crystal field is comparable in magnitude to the Coulomb interaction between the d electrons, both being greater than the spin-orbit interaction. The compounds of the palladium and platinum groups are dealt with in Chapter 10, where the intermediate coupling scheme is applied to explain the magnetic properties of these materials. Chapter 11 is devoted to the molecular orbital theory, which successfully explains the nature of chemical bonds in the transition metal compounds, the reduction of the orbital moment, the spin-orbit coupling constant, and the Racah parameters, and, above all, provides a semiempirical method to calculate the $10Dq$ parameter. In the same chapter I have also dealt with the modified Wolfsberg -Helmholz (W-H) method, which qualitatively explains the optical properties of the transition metal complexes, especially the $10Dq$ value. The optical properties of ML_6 complexes are discussed in Chapter 12, using the vibronic (vibrational-electronic) model to explain the intensities of the optical absorption spectra. The vibronic model also explains the spin-allowed and spin-forbidden transitions between the crystal field states of the transition metal complexes. The Jahn-Teller effect, which is important for both the optical and the magnetic properties of solids, is also discussed in this chapter.

The first part of the book, therefore, deals with the paramagnetic properties of dilute magnetic systems. From Chapter 13 onward, however, the discussion centers on condensed magnetic solids. Chapters 13 and 14 are devoted to the molecular fields in ferromagnetic and antiferromagnetic solids, respectively, while Chapter 15 deals with magnetism in amorphous solids. The latter topic leads to a very important and fascinating class of materials known as spin glass, which has recently attracted considerable interest. However, our understanding of amorphous solids is still very unsatisfactory, and the discussion of these substances is therefore rather limited. Chapter 16 deals with cooperative phenomena and the second quantization formalism leading to the spin-wave theory of Bloch, while Chapters 17 and 18 are devoted to the quantum field-theoretic approach to the many-body aspects of magnetism. The detailed application of the Green function formalism for an understanding of ordered magnetic systems is treated in a simplified manner in the concluding chapters (Chapters 17 and 18).

Inevitably, as a result of confining the discussion of magnetism in solids to just one volume, some omissions and simplifications have proved necessary. Rare earths, already dealt with by Abragam and Bleaney, together with the relaxation mechanism in solids, have been excluded, and certain phenomena such as magnetic resonance have not been given detailed treatment.

It is my pleasure to acknowledge the great help, as well as constructive criticism, I received from three of my students: Dr. V. P. Desai, Dr. I. Chatterjee, and especially Mr. Subrata Basu. My thanks are also due to many of my colleagues, here and abroad, notably Professor J. W. Richardson, Professor R. E. Watson, Dr. A. D. Liehr, Professor C. J. Ballhausen, Professor M. H. L. Pryce, Professor R. Bersohn, Professor A. Bose, Professor D. K. Ghosh, Dr. P. Rudra, Dr. P. Ghosh, and Dr. R. K. Moitra, for helping me in various ways. I am grateful to Mr. David Blagbrough of the British Council, Calcutta, Professor C. K. Majumder, Professor B. Dutta Roy, Dr. A. K. Sengupta, and Mr. Subrata Basu for critically reading the manuscript and suggesting many improvements. It is also important to mention the name of my son, Sugato, without whose constant persuasion and nagging this book would never have been completed. I also wish to acknowledge with gratitude the help and cooperation I received from Dr. Theodore P. Hoffman, the Chemistry Editor, and from Mrs. Valda Aldzeris, the Editorial Manager of John Wiley & Sons, for an excellent editing of my manuscript.

My sincere thanks are due to Professor J. H. Van Vleck of Harvard University for his encouragement to write this book.

Lastly, it is a pleasure to thank Professor A. K. Saha and Professor D. N. Kundu of Saha Institute of Nuclear Physics for providing me with facilities to work unhindered and for their enthusiastic support and encouragement.

A. S. CHAKRAVARTY

Calcutta, India
August 1980

Contents

Introduction to the Magnetic Properties of Solids

1

Introduction

The elements from titanium to copper in the first group, zirconium to silver in the second group, and hafnium to gold in the third group of the periodic table have chemical properties that change gradually within the group and are therefore known as the transition elements. They are also known as *d*-block elements on account of their unfilled *d* shells. The transition elements are characterized by colored compounds, by an ability to form both complex and simple compounds because of the high polarizability of their ions, and by having different oxidation states. The compounds of the transition elements have wide-ranging magnetic properties, and it is now accepted that the distinctive magnetic and optical properties of their ions are essentially due to the unfilled *d* shells.

The atomic structure of the transition elements can be expressed in the general form

$$[X]kd^n(k+1)s^{1 \text{ or } 2}$$

where X is an inert gas configuration as follows:

$$\begin{aligned} X &= \text{Ar}, \quad k=3 \text{ for first group elements} \\ &= \text{Kr}, \quad k=4 \text{ for second group elements} \\ &= \text{Xe}, \quad k=5 \text{ for third group elements} \end{aligned}$$

It is well known that the transition metals form complexes or compounds with other elements of the periodic table. In transition metal chemistry it is usual to distinguish between compounds and complex ions. However, the electronic properties of a metal ion are determined by its nearest neighbors, and it makes little difference whether these are part of a binary solid or of a discrete complex ion, since the optical absorption spectrum of the "compound" MnF_2, for example, is very similar to that of the "complex ion" $Mn(H_2O)_6^{2+}$ in aqueous solution. Two extreme classes of transition metal compounds are conveniently distinguished, the metallic and the nonmetallic.

The former class includes the alloys, interstitial hydrides, borides, carbides, and the metals themselves, while the latter is represented by inorganic salts such as Tutton salts. From a theoretical standpoint the essential distinction between the two classes is that in a nonmetallic solid the *d* electrons of the partly filled shells may be assigned individually to particular metal atoms. Each metal atom or ion has its own set of *d* electrons localized near it, and they have very little interaction with the electrons belonging to neighboring metal atoms. In a metallic compound the *d* electrons are owned collectively by all the metal atoms and cannot be separated into nearly noninteracting sets. In this book we are concerned with both types of compounds.

The structures of a great number of transition metal compounds have been determined by X-ray crystallographic methods. It has been found from structural analysis of the transition metal compounds that the most common arrangement of the nearest neighbors about the metal ion is that of a more or less distorted octahedron. Thus the anhydrous fluorides of Mn^{2+}, Fe^{2+}, Co^{2+}, and Ni^{2+} crystallize in the rutile structure, each metal atom being surrounded by an almost regular octahedron of fluoride ions. The hydrated ions $Mn(H_2O)_6^{2+}$, $Fe(H_2O)_6^{2+}$, $Co(H_2O)_6^{2+}$, and $Ni(H_2O)_6^{2+}$ in crystals all have regular octahedral symmetry, though with some slight distortion.

Two other important types of stereochemical arrangement must also be mentioned. The first is tetrahedral coordination of the metal ion, which is fairly common particularly among the Co^{2+} compounds, for example, $CoCl_4^{2-}$, $ReCl_4^-$, and the blue form of $CoCl_2(NH_3)_2$; the second is found in the planar complexes, in which the metal is surrounded by a square of four molecules or ions, which are very common for metal ions with d^8 configuration. Other stereochemical arrangements are also found but are not as common as the types mentioned above.

The more distant environment of the metal ion, namely, that of the central octahedral, tetrahedral, or planar group, is extremely variable. Fortunately, groups other than the nearest neighbors are relatively unimportant in determining the chemical and physical properties of the ion. Consequently, we can often neglect all but the nearest neighbors of the metal ion, and these are usually referred to as the ligands.

It has been observed from spectroscopic and magnetic data that the properties of ions in solids of known structure and those of the corresponding ions in solution are almost identical. This proves that the inner coordination group is usually maintained almost unchanged in solution. For example, the optical spectra of the hydrated transition metal ions in aqueous solution are almost identical with those of the same hydrated ions in crystalline solids. It will be apparent in due course that this near-identity usually implies the same stereochemistry in the two situations.

The simplest possible picture of the electronic structure of the transition metal compounds is the purely ionic one. In the ionic model the formal valencies of different ions in the structure are interpreted literally to imply the presence of the corresponding ions. The only role of the ligands, in this model, has been to produce a crystalline field that splits the various orbitals of the metal atom. Such a point of view is equivalent to regarding a complex molecule as held together by purely ionic forces, replacing any inherent bonding by point charges attracting and repulsing each other. But we know from Earnshaw's theorem of electrostatics that no system of charges can be in stable equilibrium while at rest. Hence a modification in our concept of bonding is needed to save the situation. The essential modification is provided by the valence bond theory, as developed by Pauling. Here an attempt is made to distinguish between "ionic" compounds, which are held together by electrostatic forces, as in the alkali halide crystals, and "covalent" compounds, which are held together by directed bonds. Conceptually this theory has proved so attractive to chemists that it has been the basis of most recent chemical thinking on the subject. However, it has not proved fruitful in the field of quantitative calculation. The ionic theory, on the other hand, has been adapted for use in a more quantitative way. Largely through the influence of Bethe and Van Vleck, a detailed understanding of the magnetic properties of the divalent and trivalent ions in certain types of environment was built up. This theory, which is known as the crystal field theory, was restricted in its range of application since it failed to include the "chemical" interactions that exist in most metal compounds of high valence. However, within its range, by applying the perturbation theory to the quantum-mechanical description of the free ions, it achieved notable quantitative success. To incorporate the concept of chemical interactions in an effective way the crystal field theory was suitably modified to yield the ligand field theory, also known as the molecular orbital (MO) theory. The MO theory has been quite successful in explaining experimental observations on the covalent compounds of the transition metals that demanded an explanation in terms of electron sharing between atoms. The same theory acknowledges the presence of the ligands to a much greater extent than does the crystal field theory. Hence the ligand field method differs from the crystal field procedure in that the structural unit for the wavefunction is the whole complex ion, rather than the single central atom. To construct suitable wavefunctions, one has to use the so-called linear combinations of atomic orbitals (LCAO) method. In this connection the use of the symmetry considerations proves to be extremely useful, as we shall see later in this book.

As discussed earlier, the ions of the transition metals are formed by the removal of the outermost electrons and therefore contain only d electrons in the outermost shell. These open shell electrons are responsible for the physical and chemical properties of the transition metal complexes. The inner shell electrons comprising the closed shell core have a zero resultant orbital and spin angular momentum and therefore play no part in determining the magnetic and optical properties of these complexes. Thus the question of what happens to the free metal ions when they form complexes is essentially related to what happens to the wavefunctions and the energy levels of the outermost d electrons. In the case of free ions, the d^n electronic configuration corresponds to a number of *LS* terms in the Russell-Saunders coupling scheme. These terms again undergo splitting by the spin-orbit interaction, and the energies of the multiplets are functions of the Racah parameters (A, B, and C) and the spin-orbit parameter (ζ). These free-ion parameters are usually reduced in complexes, a phenomenon known as the nephelauxetic (cloud expanding) effect. The latter is determined by the nephelauxetic ratio, which measures the ratio of these parameters in complexes to those in free ions. The origin is either the "central field" covalency or the "symmetry-restricted" covalency. In the first case the nuclear charge of the central ion is screened by the ligand electrons; in the second case the central d electron is delocalized on to the ligands. From this one can conclude that the metal-ligand bond in the complexes is not purely ionic but also has some covalent character. The covalency effect arises from the electron transfer either from the metal to the ligand or from the ligand to the metal. Of the three groups of transition metal ions, the compounds of the first transition series are known to be less covalent than those of the second and third transition series because the $3d$ wavefunctions are less extended than the $4d$ or $5d$ wavefunctions. The metal-ligand overlap is therefore small.

An interesting property of the transition metals is their ability to exist in different valence states and therefore to form a variety of complexes. These complexes have beautiful colors, which can be explained with the help of the ligand field theory. The magnetic anisotropy and the reduction of the magnetic moment can also be explained by this theory. Although the ligand field theory is old in origin, it is only recently that its importance has been fully recognized. During the last two decades the ligand field theory has successfully explained the optical and magnetic properties of the transition metal ions in complexes and has also provided an understanding of the nature of bonding in inorganic complexes of the transition metal ions.

The basic idea of the ligand field theory is the same as that of the crystal field theory, which was introduced in 1929 by Becquerel[1] and formulated with the aid of group theory by Bethe[2] in the same year. The idea of the crystal field is that the electrons in the metal ion experience an electrostatic field produced by the charges on the surrounding ions, called ligands. The crystal field theory assumes no transfer of electrons, either from the metal to the ligand or from the ligand to the metal, and is therefore more successful in explaining the properties of the ionic compounds but is not satisfactory for the covalent compounds. Apart from Becquerel and Bethe, other pioneers of the crystal field theory were Van Vleck, Penney, Schlapp, Jordahl, and Gorter. Van Vleck[3] first applied the crystal field theory to explain the "spin-only" paramagnetism of the salts of the iron group, where he considered the quenching of the orbital angular momentum to be due to the crystalline electric field. This theory, through the work of Schlapp and Penney[4] and of Jordahl,[5] was also able to predict cases of small deviations from the spin-only formula and showed that the variations of magnetic anisotropy and susceptibility with temperature can be reasonably explained.

These properties were shown to be due to the removal by the crystal field of the orbital degeneracy of the electronic levels of the free ions. Gorter[6] showed that the crystal field of a regular tetrahedron would produce the same energy levels as were produced by a regular octahedron, but with the energy level structure inverted. Treatment in terms of the crystal field theory is a valid approximation only if the electronic wavefunctions of the ion do not overlap significantly with the charge distribution of the surrounding ions. If the overlap with the surrounding ligands becomes significant, the ligand electrons must also be included in the system. The wavefunctions are then approximated by considering the mixing between the orbitals of the central metal ion and those of the surrounding ligands. The crystal field theory is improved therefore by introducing this covalency effect via the orbital reduction factor. The improved theory is known as the ligand field theory. It should be remembered that both the crystal field theory and the ligand field theory are semiempirical in nature and are thus in the parametrized form, the parameters being estimated by comparing them with the experimentally observed magnetic and optical properties.

Let us now consider the ordered magnetic systems. By the use of new experimental techniques, strong magnetism has been discovered in many ionic solids such as the transition metals (Fe, Co, and Ni) and the rare earths. Antiferromagnetism has been discovered in a number of materials that form a well-defined class. In addition, other types of magnetic structures are also in existence. The known magnetic structures now form a wide and continuous range. Starting with the strong positive magnetism of ferromagnetic materials, in which the magnetic ions are spontaneously ordered in the same direction, we find the more complex orderings of ferrimagnetism and antiferromagnetism, together with helical and modulated structures. The theoretical basis of the different types of behavior of the ordered magnetic systems, namely, ferromagnetism and antiferromagnetism, is examined through chronological analysis of the work of earlier investigators.

Experimental observations permit the characterization of several classes of magnetically ordered solids. A sample of pure or alloyed Fe, Ni, or Co, for instance, shows strong ferromagnetic properties that are induced by an applied field and are found to be generally related to the shape and magnetic history of the sample, as well as to the field strength. However, when the applied magnetic field is withdrawn, the ferromagnetic sample retains its magnetism. This residual magnetism is a characteristic property of ferromagnetic materials, the magnitude of which can be revealed experimentally using single crystals.

A ferromagnetic crystal is spontaneously magnetized to a high degree at every point, but different regions of the crystal are magnetized in different, although crystallographically equivalent, directions. For example, an Fe crystal is cubic, that is, there are six equivalent preferred directions corresponding to $\pm x$-, $\pm y$-, and $\pm z$-directions of the Cartesian coordinates, parallel to the edges of the cubic elementary cell. A demagnetized Fe crystal, therefore, is subdivided into regions or "domains," each spontaneously magnetized in one or another of the six equivalent preferred directions listed above. If now a very small magnetic field is applied along one of these six directions, the magnetization will develop in this direction at the expense of the others. Thus, in the ideal case, the magnetization developed would correspond to that of a single domain. That a single crystal can be magnetized to saturation in its preferred direction was first proposed by Weiss[7] and subsequently confirmed by direct observation.

There are thus two distinct aspects to ferromagnetic behavior: (1) the occurrence and magnitude of the spontaneous magnetization, and (2) the way in which this spontaneous magnetization is distributed directionally in the form of domains. This book, which attempts to answer why a particular metal is ferromagnetic and others are not, is concerned with the first aspect only.

In accordance with accepted practice, ferromagnetism and ferrimagnetism have been grouped together, for although within each unit cell of a ferrimagnetic crystal there may be ionic moments in two or more directions, each unit cell has a nonzero resultant magnetic moment. Therefore a ferrimagnetic substance also possesses spontaneous magnetization. Viewed from the macroscopic standpoint of domain structures, ferromagnetic and ferrimagnetic materials are quite similar. On the other hand, antiferromagnetic materials, while possessing characteristic directions and domain structures, have no resultant magnetism.

We are therefore primarily concerned with the nature and degree of the spontaneous order within a domain, which involves three principal considerations: (1) the pattern of order within each unit cell, (2) the extent to which the order is disturbed by thermal agitation, and (3) the nature of the elementary magnetic moments contributing to the order.

It is important to note that the magnetic order of a ferromagnetic sample can be obtained from the statistical-mechanical relation between the entropy and the degree of order. Anomalous variations of specific heat with temperature, close to the critical temperature at which the magnetic order is destroyed by thermal agitation, reveal magnetic structural changes of various kinds.

Another powerful technique that demonstrates the magnetic (spin) structures of magnetic materials is neutron diffraction. If a monoenergetic beam of slow neutrons (having a wavelength of about 1 Å) is directed toward the magnetic material in a particular direction, a diffraction pattern very similar to that of X-rays is produced by the scattering of the neutrons at the atomic nuclei in a crystal. If the crystal contains ions with permanent magnetic moments, there is additional scattering, comparable in strength to nuclear scattering, due to the magnetic interactions between the ions and the neutrons. Now magnetic scattering modifies the diffraction pattern according to the kind of magnetic order present. When the ions in different magnetic sublattices are in sites that are also crystallographically different, the associated diffraction lines are always present, but the magnetic ordering changes their relative intensities. By examining which lines in the diffraction pattern are strengthened or reduced by the onset of the magnetic order, the pattern of the order can be deduced. In addition, the magnitudes of the moments at different lattice sites and the degree of order can also be obtained from the intensities observed. A large number of materials have been investigated in this way, the details of which are outside the scope of this book.

In addition to considering the magnetic properties of the nonmetallic solids, we shall also deal with the above-mentioned aspects of ordered magnetic structures and shall see that the Heisenberg[8] exchange interaction is responsible for magnetism. In ordered magnetic solids we are therefore concerned with the diagonalization of the Hamiltonian of a many-body system in order to understand the magnetism developed in such complicated magnetic structures.

Before concluding this introduction, a few words regarding the notations used in this book may be helpful. Gaussian units are used throughout. The symbol k represents the Boltzmann constant, whose magnitude is $1.3708 \pm 0.0014 \times 10^{-6}$ erg/°C, and it necessarily appears along with T, the absolute temperature; $\mu_B = eh/4\pi mc$ is the Bohr magneton, whose value is 0.9174×10^{-20} emu, where e and m represent the charge and mass of an electron and have magnitudes of 4.77×10^{-10} esu and 9.04×10^{-28} g, respectively. $\mathcal{H}$ represents the Hamiltonian (to be distinguished from the magnetic field **H**). Also, μ_{eff} is the effective Bohr magneton number, defined in terms of the susceptibility by the relation

$$\mu_{\text{eff}} = \sqrt{\frac{3kT\chi}{N\mu_B^2}}$$

where χ is the magnetic susceptibility per unit volume, and χ_{mol} would be the susceptibility per gram molecule. In addition $\hbar = h/2\pi$, where h is Planck's constant; $\hbar = 1.0544 \times 10^{-27}$ erg sec. N is the number of molecules per unit volume. Lastly, quantities in boldface type are vectors.

REFERENCES

1. J. Becquerel, *Z. Phys.* **58**, 205 (1929).
2. H. Bethe, *Ann. Phys.* **3**, 135 (1929).
3. J. H. Van Vleck, *Theory of Magnetic and Electrical Susceptibilities*, Oxford University Press, London, England, 1932; *J. Chem. Phys.* **3**, 803 (1935); *ibid* **3**, 807 (1935); *J. Chem. Phys.* **41**, 67 (1937); J. H. Van Vleck and W. G. Penney, *Phil Mag.* **17**, 96 (1934); *J. Chem. Phys.* **7**, 61 (1939); *Phys. Rev.* **57**, 426 (1940).
4. R. Schlapp and W. G. Penney, *Phys. Rev.* **42**, 666 (1932); W. G. Penney and R. Schlapp, *Phys. Rev.* **41**, 194 (1932).
5. O. M. Jordahl, *Phys. Rev.* **45**, 87 (1934).
6. C. J. Gorter, *Phys. Rev.* **42**, 437 (1932).
7. P. Weiss, *J. Phys.* **6**, 667 (1907); *ibid.* **1**, 166 (1930).
8. W. Heisenberg, *Z. Phys.*, **38**, 411 (1926).

2

Angular Momenta and Related Matters

2.1 THE HAMILTONIAN FOR AN ATOMIC SYSTEM

To understand the behavior of a metal ion in the environment imposed upon it in a compound, it is necessary, first of all, to understand the electronic structures of free atoms and ions. This is, of course, true for compounds in which one can presume that the environment represents a rather small distrubance on the motions of the electrons of the metal ion. However, it is also very useful to be able to refer to the atomic structures of free atoms even when the disturbance is large. This is especially true for the complexes or compounds in which we are interested in this book, namely, those in which the unpaired metal electrons are localized on or at least near their parent metal ions. In this section we therefore describe the features of the theory of atomic structure that are most relevant to our later treatment of the metal compounds.

Before proceeding further, we shall digress slightly to explain Dirac's[1] "bra" and "ket" notations, which we shall use throughout this book, simply because they are mathematically simple and easy to visualize.

Whenever we have a set of vectors in any mathematical theory, we can always set up a second set of vectors, known as the "dual" vectors. Suppose we have a number n that is a linear function of a "ket" vector $|A\rangle$, that is, to each ket vector $|A\rangle$ there corresponds one number n, so that the number corresponding to $|A\rangle+|A'\rangle$ is the sum of the numbers corresponding to $|A\rangle$ and $|A'\rangle$, and the number corresponding to $c|A\rangle$ is c times the number corresponding to $|A\rangle$, c being any numerical factor.

Knowing a complete set of ket vectors $|A\rangle$, we can derive its dual set, known as the "bra" vectors and denoted by $\langle A|$, the bra symbol being the mirror image of the ket symbol. The scalar product of a bra vector $\langle B|$ and a ket vector $|A\rangle$ is written as $\langle B|A\rangle$, and it denotes a number. A bra vector is considered to be completely defined when its scalar product with every ket vector is given, so that, if $\langle P|A\rangle=0$ for all $|A\rangle$, then $\langle P|=0$.

Now, on account of the one-to-one correspondence between bra and ket vectors, any state of a dynamical system may be specified by a bra vector just as well as by a ket vector. Given any two ket vectors $|A\rangle$ and $|B\rangle$, we can construct from them a number $\langle B|A\rangle$ by taking the scalar product of the first with the conjugate imaginary of the second, and vice versa. We shall assume that these two numbers are always equal, that is, $\langle B|A\rangle = \overline{\langle A|B\rangle}$. Putting $|B\rangle = |A\rangle$ here, we find that the number $\langle A|A\rangle$ must be real. We make the further assumption that $\langle A|A\rangle > 0$, except when $|A\rangle = 0$. We shall call a bra and a ket vector orthogonal if their scalar product is zero, and two bras or two kets will be called orthogonal if the scalar product of one with the conjugate imaginary of the other is zero. Also, we shall say that the two states of our dynamical system are orthogonal if the vectors corresponding to these two states are orthogonal. The length of a bra vector $\langle A|$ or of the conjugate imaginary ket vector $|A\rangle$ is defined as the square root of the positive number $\langle A|A\rangle$. When we are given a state and we wish to set up a bra or ket vector to correspond to it, only the direction of the vector is given; the vector itself is undetermined to the extent of an arbitrary numerical factor. It is often convenient to choose this numerical factor so that the vector is of length unity. This procedure is called normalization, and the vector so chosen is said to be normalized. The vector is not completely determined even then, however, since one can still multiply it by any number of modulus unity, that is, by any number $e^{i\alpha}$, where α is real, without changing its length. We shall call such a number a phase factor.

We shall not give here the details of the algebra of bra and ket vectors, for which the reader is referred to Dirac's *Principles of Quantum Mechanics*,[1] but shall mention only, in short, the general physical interpretation of any observable, O. If we take any O and any state $|A\rangle$, say, corresponding to the vector $|A\rangle$, we can form the number $\langle A|O|A\rangle$, which is necessarily real and also is uniquely determined when $\langle A|$ is normalized, since if we multiply $\langle A|$ by the numerical factor $e^{i\alpha}$, α being some real number, we must multiply $|A\rangle$ by $e^{-i\alpha}$ and $\langle A|O|A\rangle$ will thus remain unaltered. $\langle A|O|A\rangle$ is the average value of the observable O for the state $|A\rangle$. We can therefore make the general assumption that, if the observable O for the system in the state corresponding to $|A\rangle$ is measured a large number of times, the average of all the results obtained will be $\langle A|O|A\rangle$, provided that $|A\rangle$ is normalized.

An atom or an ion consists of a relatively massive, positively charged nucleus together with a number of electrons. The electrons are held near the nucleus by their electrostatic attraction to the latter and to some extent apart from each other by their mutual electrostatic repulsions. Because of its relatively large mass it is a good approximation to regard the nucleus as being at rest. This means, then, that essentially we have a classical Hamiltonian:

$$\mathcal{H} = \sum_{k=1}^{n} \left(\frac{1}{2m} \mathbf{p}_\kappa^2 - \frac{Ze^2}{\mathbf{r}_\kappa} \right) + \sum_{k<\lambda}^{n} \frac{e^2}{\mathbf{r}_{\kappa\lambda}} \tag{2.1}$$

for the system. In (2.1), $\mathbf{p}_\kappa$ is the momentum vector of the κth electron and $\mathbf{r}_\kappa$ its distance from the nucleus, m the mass of the electron, $-e$ its charge, $+Ze$ the charge on the nucleus, and $\mathbf{r}_{\kappa\lambda}$ the distance from the κth to the λth electron. There are n electrons, and the whole system has a charge $(Z-n)e$, so that for a neutral atom $n=Z$ and for a positive ion $n<Z$. Note that (2.1) is not a complete and accurate Hamiltonian for the system, however. The effect of regarding the nucleus as being at rest is an approximation and can be taken into account by interpreting m in (2.1) as the reduced mass $\mu M/(\mu+M)$, where μ is the true mass of an electron and M the mass of the nucleus. There are other ways in which (2.1) is incomplete, which will prove important in our theory. The electron has a magnetic moment and so, quite often, does the nucleus. This introduces into (2.1) extra terms that represent magnetic interactions between the various particles of the system. But these magnetic interactions are usually very small compared to the effect of the environment of a metal ion, which is again smaller than the electrostatic interactions arising from (2.1). Hence the electronic structures of atoms or ions having the Hamiltonian (2.1) are especially important.

In quantum theory the classical Hamiltonian $\mathcal{H}$ is taken over directly, but $\mathbf{p}_\kappa$ and $\mathbf{r}_\kappa$ are now regarded as operators, the components of which do not all commute. The commutation rule of the coordinates and the momenta is given by

$$\mathbf{q}_\kappa \mathbf{p}_{\lambda q'} - \mathbf{p}_{\lambda q'} \mathbf{q}_\kappa = i\hbar\, \delta_{qq'} \delta_{\kappa\lambda} \tag{2.2}$$

where $\delta_{ab}=0$ unless $a=b$, when $\delta_{ab}=1$. In (2.2), $\mathbf{p}_{\kappa q}$ represents the qth component of $\mathbf{p}_\kappa$, and similarly for others. The equations of motion in Schrödinger's form is

$$i\hbar \frac{d}{dt} |x\rangle = \mathcal{H} |x\rangle \tag{2.3}$$

and if we express the ket $|x\rangle$ in terms of the coordinates of the electrons and the time, $\mathbf{p}_\kappa$ may be equated to the differential operator

$$\mathbf{p}_\kappa = -i\hbar \nabla_\kappa = -i\hbar \left(\frac{\partial}{\partial x_\kappa}, \frac{\partial}{\partial y_\kappa}, \frac{\partial}{\partial z_\kappa} \right) \tag{2.4}$$

It can be readily seen that these $\mathbf{p}_\kappa$ satisfy (2.2). Equation (2.3) then becomes Schrödinger's wave equation for the system

$$\mathcal{H}(-i\hbar\nabla_\kappa, \mathbf{r}_\kappa) |\phi(\mathbf{r}_\kappa, t)\rangle = i\hbar \frac{\partial}{\partial t} |\phi(\mathbf{r}_\kappa, t)\rangle \tag{2.5}$$

But the Hamiltonian $\mathcal{H}$ represents the total energy of the system. Since it does not involve the time explicitly, we may find the stationary states that are eigenfunctions of $\mathcal{H}$. We can write, therefore,

$$|\phi(\mathbf{r}_\kappa, t)\rangle = |e^{-iEt/\hbar} \psi(\mathbf{r}_\kappa)\rangle \tag{2.6}$$

and

$$\mathcal{H} |\psi\rangle = E |\psi\rangle \tag{2.7}$$

when ψ is independent of time, and ϕ then satisfies (2.5). we are interested in these stationary states for the atomic system. We now have the wave equation given by (2.7), where

$$\mathcal{H} = \sum_{\kappa=1}^{n}\left(-\frac{\hbar^2}{2m}\nabla_\kappa^2 - \frac{Ze^2}{r_\kappa}\right) + \sum_{\kappa<\lambda}^{n}\frac{e^2}{r_{\kappa\lambda}} \tag{2.8}$$

It is well known that the electronic systems can be represented only by the solutions of (2.8) that are fully antisymmetric with respect to an interchange of electrons; this is a consequence of Fermi-Dirac statistics, which the electrons obey. It is an experimental fact that an electron has a magnetic moment and also that it possesses spin; the spins are associated with different orientations of the magnetic moment. The extra terms in the Hamiltonian due to the magnetic moment of the electron are usually small, as remarked earlier. Because of Fermi-Dirac statistics it turns out that the existence of two independent states for the electron has a very great influence on the energies of many-electron systems, even though the terms in the Hamiltonian that involve the spin are small. We shall consider the algebraic techniques for dealing with the spin and then discuss its influence on the electronic energy of an atomic system. Before doing so, however, we must discuss the properties of the orbital angular momentum.

2.2 ORBITAL ANGULAR MOMENTUM

In quantum mechanics there are always two important constants of motion for an isolated system. The first is the total energy, which is represented by our Hamiltonian, and the second is the total angular momentum. The angular momentum commutes with the Hamiltonian and therefore can be taken as a constant of motion. Hence it is very useful to examine the properties, the eigenstates, and the eigenvalues of the angular momentum. There are, however, two kinds of angular momentum which are constants of motion in quantum theory: the total angular momentum and the spin angular momentum. We shall learn later that, if magnetic interactions between the particles are included, only the sum of the orbital and the spin momentum is a constant of motion. This is known as the total angular momentum.

The orbital angular momentum of a particle about the origin is defined as the vector $\mathbf{l}=\mathbf{r}\times\mathbf{p}$. Using the well-known commutation relation, we find that

$$\begin{aligned}\mathbf{l}_x\mathbf{l}_y - \mathbf{l}_y\mathbf{l}_x &= i\hbar\mathbf{l}_z\\ \mathbf{l}_y\mathbf{l}_z - \mathbf{l}_z\mathbf{l}_y &= i\hbar\mathbf{l}_x\\ \mathbf{l}_z\mathbf{l}_x - \mathbf{l}_x\mathbf{l}_z &= i\hbar\mathbf{l}_y\end{aligned} \tag{2.9}$$

Introducing a notation for the commutator of two quantities, a and b:

$$[\mathbf{a},\mathbf{b}] = \mathbf{ab} - \mathbf{ba} \tag{2.10}$$

we rewrite (2.9) as

$$[\mathbf{l}_x, \mathbf{l}_y] = i\hbar \mathbf{l}_z \tag{2.11}$$

and similarly for the cyclic permutations of x, y, z.

For n particles the total angular momentum about the origin is defined as the sum

$$\mathbf{L} = \sum_{k=1}^{n} \mathbf{l}_k = \sum_{k=1}^{n} \mathbf{r}_k \times \mathbf{p}_k \tag{2.12}$$

of the angular momentum vectors of the individual particles. Using the commutation relations, we can show that

$$[\mathbf{L}_x, \mathbf{L}_y] = i\hbar \mathbf{L}_z \tag{2.13}$$

Thus any result we prove for **L** remains valid on replacing **L** by **l**, and vice versa.

The orbital angular momentum **L** for an n-particle system is very important in the theory of atomic structure simply because its components commute with the Hamiltonian (2.8). It can be proved easily; therefore we do not prove it here. However, the set $\mathcal{H}$, L_x, L_y, and L_z do not form a commuting set of observables because $\mathbf{L}_x$, $\mathbf{L}_y$ and $\mathbf{L}_z$ do not commute with each other. But it can easily be shown that $\mathbf{L}_z$ and $\mathbf{L}^2$ do commute, and also $\mathbf{L}^2$ and $\mathcal{H}$. Thus $\mathcal{H}$, $\mathbf{L}^2$, and $\mathbf{L}_z$ (or $\mathbf{L}_x$ or $\mathbf{L}_y$) form a commuting set of observables. Now in quantum mechanics a set of commuting observables possesses a complete set of simultaneous eigenkets; this means that we can expand an arbitrary ket as a sum over functions each of which is simultaneously an eigenket of $\mathcal{H}$, $\mathbf{L}^2$, and $\mathbf{L}_z$. This particular set of simultaneous eigenkets plays an important role in the theory of atomic structure.

We shall be interested later in more general types of angular momenta. A general angular momentum is defined to be any real vector having commutation relation (2.9). We now deduce various properties of the simultaneous eigenkets of $\mathcal{H}$, $\mathbf{L}^2$, and $\mathbf{L}_z$ using only the commutation relations. This means that our results will also be true for general angular momenta which commute with $\mathcal{H}$.

Let us write $|E\beta L_z'\rangle$ for a simultaneous eigenket, where E, β, and L_z' are the eigenvalues of $\mathcal{H}$, $\mathbf{L}^2$, and $\mathbf{L}_z$, respectively, and define the shift operator

$$\mathbf{L}^+ = \mathbf{L}_x + i\mathbf{L}_y$$

and its complex conjugate

$$\mathbf{L}^- = \mathbf{L}_x - i\mathbf{L}_y$$

Now, using relations of type (2.13), we can show that

$$[\mathbf{L}_z, \mathbf{L}^\pm] = \pm\hbar \mathbf{L}^\pm \tag{2.14}$$

and hence

$$\mathbf{L}_z\mathbf{L}^{\pm}|E\beta L_z'\rangle = (L_z' \pm \hbar)\mathbf{L}^{\pm}|E\beta L_z'\rangle$$

This justifies the name "shift operator," to denote this shifting of one simultaneous eigenket of $\mathbf{L}^2$ and $\mathbf{L}_z$ into another having the same eigenvalue for $\mathbf{L}^2$ but a different eigenvalue for $\mathbf{L}_z$. Repeating the shift operations n times, we can show that

$$\mathbf{L}_z(\mathbf{L}^{\pm})^n|E\beta L_z'\rangle = (L_z' \pm n\hbar)(\mathbf{L}^{\pm})^n|E\beta L'_z\rangle \tag{2.15}$$

Note that (2.14) or (2.15) represents two distinct equations in one, of which we take the upper or the lower sign throughout. Without going into details of quantum mechanics, which can be found in any standard textbook on this subject, such as Landau and Lifshitz,[2] we write the eigenvalue equations that are important for our purpose:

$$\left.\begin{aligned} \alpha|\alpha' E L M_L\rangle &= \alpha'|\alpha' E L M_L\rangle \\ \mathcal{H}|\alpha' E L M_L\rangle &= E|\alpha' E L M_L\rangle \\ \mathbf{L}^2|\alpha' E L M_L\rangle &= L(L+1)\hbar^2|\alpha' E L M_L\rangle \\ \mathbf{L}_z|\alpha' E L M_L\rangle &= M_L\hbar|\alpha' E L M_L\rangle \end{aligned}\right\} \tag{2.16}$$

where α stands for a set $\alpha_1, \alpha_2, \ldots, \alpha_n$, of commuting observables. It can also be shown that L is an integer. For a given α', E, and L, M_L takes all values from $-\mathrm{L}$ to $+L$ in steps of unity. Using

$$\mathbf{L}^{\mp}\mathbf{L}^{\pm} = \mathbf{L}^2 - \mathbf{L}_z^2 \mp \hbar\mathbf{L}_z \tag{2.17}$$

which can easily be derived from (2.13), and the value $L(L+1)\hbar^2$ for β, we have

$$\mathbf{L}^{\pm}|\alpha' E L M_L\rangle = \hbar\left[(L \mp M_L)(L \pm M_L + 1)\right]^{1/2}|\alpha' E L M_L \pm 1\rangle \tag{2.18}$$

for all M_L. Here, of course, we define

$$|\alpha' E L M_L\rangle = 0$$

for $|M|>L$.

2.3 SPIN ANGULAR MOMENTUM

We have already seen that all orbital angular momenta actually have $\mathbf{L}$ integral. Angular momenta that take nonintegral values do exist, however, and we now describe them. There are good experimental and theoretical reasons for supposing that electrons and also protons and neutrons each possess an intrinsic angular momentum **s** of magnitude $+\frac{1}{2}\hbar$, that is, $\mathbf{s}^2$ has the eigenvalue $\frac{3}{4}\hbar^2$. This means that there are just two independent internal states for the electron, corresponding to the eigenvalues $\pm\frac{1}{2}\hbar$ of the component of **s** along some specified axis. The momentum **s** is called the spin angular momentum. According to Dirac's theory of the electron,[1] the electron has the spin **s** and a magnetic moment $\boldsymbol{\mu}$ associated with spin, which are related to one another by

$$\boldsymbol{\mu} = -\frac{e}{mc}\mathbf{s} \tag{2.19}$$

The experimental value of the magnetic moment is in very good agreement with the value $e\hbar/2mc$ deduced from (2.19).

Now, using (2.16) and (2.18), which hold for any angular momentum, we can write explicitly the results of the operation of **s** on the eigenkets for a single electron. Writing γ, $\mathbf{s}^2$, $\mathbf{s}_z$ for a complete commuting set of observables for an electron, and operating by $\mathbf{s}^{\pm}$ and $\mathbf{s}_z$, we have

$$\begin{aligned}
&\mathbf{s}_z|\gamma'\tfrac{1}{2}\tfrac{1}{2}\rangle = \tfrac{1}{2}\hbar|\gamma'\tfrac{1}{2}\tfrac{1}{2}\rangle && \mathbf{s}_z|\gamma'\tfrac{1}{2}-\tfrac{1}{2}\rangle = -\tfrac{1}{2}\hbar|\gamma'\tfrac{1}{2}-\tfrac{1}{2}\rangle \\
&\mathbf{s}^+|\gamma'\tfrac{1}{2}\tfrac{1}{2}\rangle = 0 && \mathbf{s}^+|\gamma'\tfrac{1}{2}-\tfrac{1}{2}\rangle = \hbar|\gamma'\tfrac{1}{2}\tfrac{1}{2}\rangle \\
&\mathbf{s}^-|\gamma'\tfrac{1}{2}\tfrac{1}{2}\rangle = \hbar|\gamma'\tfrac{1}{2}-\tfrac{1}{2}\rangle && \mathbf{s}^-|\gamma'\tfrac{1}{2}-\tfrac{1}{2}\rangle = 0
\end{aligned} \tag{2.20}$$

From this it follows that

$$\begin{aligned}
\mathbf{s}_x|\gamma'\tfrac{1}{2}\tfrac{1}{2}\rangle &= \tfrac{1}{2}\hbar|\gamma'\tfrac{1}{2}-\tfrac{1}{2}\rangle \\
\mathbf{s}_x|\gamma'\tfrac{1}{2}-\tfrac{1}{2}\rangle &= \tfrac{1}{2}\hbar|\gamma'\tfrac{1}{2}\tfrac{1}{2}\rangle \\
\mathbf{s}_y|\gamma'\tfrac{1}{2}\tfrac{1}{2}\rangle &= \tfrac{1}{2}i\hbar|\gamma'\tfrac{1}{2}-\tfrac{1}{2}\rangle \\
\mathbf{s}_y|\gamma'\tfrac{1}{2}-\tfrac{1}{2}\rangle &= -\tfrac{1}{2}i\hbar|\gamma'\tfrac{1}{2}\tfrac{1}{2}\rangle
\end{aligned}$$

Because of the factor $\frac{1}{2}\hbar$ in the equations for $\mathbf{s}_x$, $\mathbf{s}_y$, and $\mathbf{s}_z$, it is convenient to consider, with **s**, the associated vector $\boldsymbol{\sigma}$ defined by

$$\mathbf{s} = \tfrac{1}{2}\hbar\boldsymbol{\sigma} \tag{2.21}$$

from which it follows that

$$\boldsymbol{\sigma}_x^2 = \boldsymbol{\sigma}_y^2 = \boldsymbol{\sigma}_z^2 = 1 \tag{2.22}$$

In a similar manner one finds that any pair of components of $\boldsymbol{\sigma}$ anticommute, that is,

$$\begin{aligned}
\boldsymbol{\sigma}_x\boldsymbol{\sigma}_y + \boldsymbol{\sigma}_y\boldsymbol{\sigma}_x &= 0 \\
\boldsymbol{\sigma}_y\boldsymbol{\sigma}_z + \boldsymbol{\sigma}_z\boldsymbol{\sigma}_y &= 0 \\
\boldsymbol{\sigma}_z\boldsymbol{\sigma}_x + \boldsymbol{\sigma}_x\boldsymbol{\sigma}_z &= 0
\end{aligned} \tag{2.23}$$

The total spin of a system of n particles, each of spin $\frac{1}{2}\hbar$, is given by

$$\mathbf{S} = \sum_{k=1}^{n} \mathbf{s}_k \tag{2.24}$$

and the process of vector addition shows that $\mathbf{S}$ has only integral values for **n** even and half-odd integral values for n odd. Clearly the largest possible value of $\mathbf{S}_z$ is $\frac{1}{2}n\hbar$ for an n-electron system. The matrix operators of the components of the spin momentum are

$$\mathbf{S}_z^{\text{op}} = \frac{\hbar}{2}\begin{pmatrix} 1 & 0 \\ 0 & -1 \end{pmatrix}; \quad \mathbf{S}_x^{\text{op}} = \frac{\hbar}{2}\begin{pmatrix} 0 & 1 \\ 1 & 0 \end{pmatrix}; \quad \mathbf{S}_y^{\text{op}} = \frac{\hbar}{2}\begin{pmatrix} 0 & -i \\ i & 0 \end{pmatrix} \tag{2.25}$$

The commutation properties of pairs of operators are of interest in connection with ferromagnetism and antiferromagnetism, to be dealt with later. These properties can be easily found by using (2.25). They are as follows:

$$\begin{aligned} \mathbf{s}_x\mathbf{s}_y - \mathbf{s}_y\mathbf{s}_x &= i\hbar\mathbf{s}_z \\ \mathbf{s}_y\mathbf{s}_z - \mathbf{s}_z\mathbf{s}_y &= i\hbar\mathbf{s}_x \\ \mathbf{s}_z\mathbf{s}_x - \mathbf{s}_x\mathbf{s}_z &= i\hbar\mathbf{s}_y \end{aligned} \tag{2.26}$$

These are of exactly the same form as the corresponding relations for the orbital angular momentum operators given in Section 2.2. Using (2.25), it can be shown that

$$\mathbf{S}^2 = s_x^2 + s_y^2 + s_z^2 = \tfrac{3}{4}\hbar^2\begin{pmatrix} 1 & 0 \\ 0 & 1 \end{pmatrix} \tag{2.27}$$

where

$$\begin{pmatrix} 1 & 0 \\ 0 & 1 \end{pmatrix}$$

is the identity matrix. It can also be shown that

$$\mathbf{s}_x^2 = \mathbf{s}_y^2 = \mathbf{s}_z^2 = \frac{\hbar^2}{4}\begin{pmatrix} 1 & 0 \\ 0 & 1 \end{pmatrix}$$

so that the square of each of the components has an eigenvalue equal to $(\hbar/2)^2$ in any spin state.

So far we have considered only the spin functions for states quantized in the z-direction. If the electron has a definite value of spin momentum in some other direction, these momenta can be described by spin functions in the Pauli formalism. The states written as

$$\chi_\alpha = \begin{pmatrix} 1 \\ 0 \end{pmatrix}$$

have the significance that the z-component of the spin of an electron in the state

$$\chi_\alpha(\sigma) = 1 \qquad \text{for} \quad \sigma = 1$$

has the value $+\frac{1}{2}\hbar$ (corresponding to $\sigma = 1$) with relative probability 1 (i.e., certainty), and in the state

$$\chi_\alpha(\sigma) = 0 \qquad \text{for} \quad \sigma = -1$$

has the value $-\frac{1}{2}\hbar$ (corresponding to $\sigma = -1$) with relative probability 0. For a state quantized in a direction other than the z-axis we should expect the relative probabilities for both components of the spin function to be between 0 and 1. If the direction of quantization has the direction cosines l, m, and n (only two of which are independent since $l^2 + m^2 + n^2 = 1$), the corresponding spin function $\chi_{lm}(\sigma)$ must satisfy

$$\mathbf{s}_{lm}\chi_{lm}(\sigma) = \pm\tfrac{1}{2}\hbar\chi_{lm}(\sigma) \tag{2.28}$$

where $\mathbf{s}_{lm}$ is the operator for the spin component in the (l,m,n)-direction. Then $\mathbf{s}_{lm}$ is given by

$$\mathbf{s}_{lm} = l\mathbf{s}_x + m\mathbf{s}_y + n\mathbf{s}_z$$

and, using (2.25), we have

$$\mathbf{s}_{lm} = \frac{\hbar}{2}\begin{pmatrix} n & l - im \\ l + im & -n \end{pmatrix} \tag{2.29}$$

We can now find the column matrices that satisfy (2.28) and (2.29). They are, for $s_{lm} = \frac{1}{2}\hbar$,

$$\chi_{lm} = N_+\begin{pmatrix} n+1 \\ l+im \end{pmatrix}$$

and, for $s_{lm} = -\frac{1}{2}\hbar$,

$$\chi_{lm} = N_-\begin{pmatrix} n-1 \\ l+im \end{pmatrix} \tag{2.30}$$

where N_+ and N_- are the normalization constants and can be shown to be $2^{-1/2}(1 \pm n)^{1/2}$. From the above, it follows that any arbitrary spin function can be expanded in terms of χ_α and χ_β. Thus, using (2.30) for $s_{lm} = +\frac{1}{2}\hbar$, we have

$$
\begin{aligned}
N_+^{-1}\chi_{lm} &= \begin{pmatrix} n+1 \\ l+im \end{pmatrix} \\
&= (n+1)\begin{pmatrix} 1 \\ 0 \end{pmatrix} + (l+im)\begin{pmatrix} 0 \\ 1 \end{pmatrix} \\
&= (n+1)\chi_\alpha + (l+im)\chi_\beta
\end{aligned} \tag{2.31}
$$

and a similar expression can be obtained $s_{lm} = -\frac{1}{2}\hbar$.

So far, in discussing the form of the spin functions, we have neglected the spin-orbit interaction. When the spin-orbit interaction is included, we can no longer factorize the total wave function into independent spatial and spin parts. In such cases the total state of an electron can be described by

$$
\psi(\mathbf{r}) = \begin{pmatrix} \psi_+(\mathbf{r}) \\ \psi_-(\mathbf{r}) \end{pmatrix} \tag{2.32}
$$

In terms of the spin coordinate $\boldsymbol{\sigma}$, the state of the electron can be written as $\psi(\mathbf{r}, \boldsymbol{\sigma})$, such that

$$
\begin{aligned}
\psi(\mathbf{r}, +1) &= \psi_+(\mathbf{r}) \\
\psi(\mathbf{r}, -1) &= \psi_-(\mathbf{r})
\end{aligned}
$$

When the spin-orbit interaction is neglected, the general wave function given above reduces to

$$
\begin{aligned}
\psi_+(\mathbf{r}) &= a\psi(\mathbf{r}) \\
\psi_-(\mathbf{r}) &= b\psi(\mathbf{r})
\end{aligned}
$$

where a and b are just numbers, so that

$$
\psi(\mathbf{r}, \boldsymbol{\sigma}) = \psi(\mathbf{r})\chi(\boldsymbol{\sigma}) \tag{2.33}
$$

2.4 ADDITION OF ANGULAR MOMENTA

We now consider ions containing more than one electron. The magnetic properties of an ion originate from the orbital and spin momenta of its electrons. The relative orientations of the orbital and spin moments of the electrons in a stationary state are determined by the interactions between the electrons. The most important interaction is the Coulomb interaction, which couples the orbital motions of the electrons and indirectly, through the Pauli exclusion requirement, their spin momenta as well. The next most important interaction is the spin-orbit interactions, which magnetically couples the spin and the orbital motions, particularly the orbital motion of each electron to its own spin. Another important influence for an ion in a crystal is the electric field due to the neighboring ions. In determining the magnetism of the transition metal complexes in which we are interested, we are required to minimize the Coulomb energy first. This minimization determines which one-electron states should be occupied thereby fixing the resultant orbital and spin momenta **L** and **S**. The spin-orbit coupling and the crystal field effects are then considered.

We now consider the resultant orbital momentum of two electrons:

$$\mathbf{L}=\mathbf{l}_1+\mathbf{l}_2$$

The commutation properties can be easily shown to be

$$\begin{aligned}\mathbf{L}_x\mathbf{L}_y-\mathbf{L}_y\mathbf{L}_x&=i\hbar\mathbf{L}_z\\ \mathbf{L}_y\mathbf{L}_z-\mathbf{L}_z\mathbf{L}_y&=i\hbar\mathbf{L}_x\\ \mathbf{L}_z\mathbf{L}_x-\mathbf{L}_x\mathbf{L}_z&=i\hbar\mathbf{L}_y\end{aligned}\tag{2.34}$$

for the resultant orbital momentum, **L**, of any number of electrons. Similarly, for the spin momentum **S** of any number of electrons, the commutation relations are

$$\begin{aligned}\mathbf{S}_x\mathbf{S}_y-\mathbf{S}_y\mathbf{S}_x&=i\hbar\mathbf{S}_z\\ \mathbf{S}_y\mathbf{S}_z-\mathbf{S}_z\mathbf{S}_y&=i\hbar\mathbf{S}_x\\ \mathbf{S}_z\mathbf{S}_x-\mathbf{S}_x\mathbf{S}_z&=i\hbar\mathbf{S}_y\end{aligned}\tag{2.35}$$

These have exactly the same form as the expressions for a single electron given earlier.

Note that, since the square of the magnitude of **L**, defined as

$$\mathbf{L}^2=\mathbf{L}_x^2+\mathbf{L}_y^2+\mathbf{L}_z^2\tag{2.36}$$

commutes with the Hamiltonian

$$\mathcal{H}=-\frac{\hbar^2}{2m}\sum_k\nabla_k^2+\sum_k V_k(\mathbf{r}_k)+\sum_{jk}V_{jk}(\mathbf{r}_j,\mathbf{r}_k)\tag{2.37}$$

where the first term represents the kinetic energies of the electrons, the second is the potential energy of the electrons in the field of the nucleus, and the last is the interelectron potential energies, $\mathbf{L}^2$ can have a definite value in a stationary state of an isolated ion. The permissible values are

$$\mathbf{L}^2=L(L+1)\hbar^2\tag{2.38}$$

where L is a quantum number taking integral values only. For each permissible value of $\mathbf{L}$, say L', there are $(2L'+1)$ distinct states; these can be distinguished by different values of the quantum number M_L, which is the z-component of $\mathbf{L}$. The $(2L'+1)$ values of M_L are

$$M_L = L', L'-1, \ldots, -(L'-1), -L' \tag{2.39}$$

that is,

$$M_L \leqslant L'.$$

The general arguments presented above for L_z apply equally well to S_z if $\mathcal{H}$ [equation (2.37)] contains no spin-dependent terms. There is, however, a quantative difference in the results because the z-component of a single electron's spin momentum takes half-integral values, that is, $\pm\frac{1}{2}\hbar$. Therefore we have

$$S_z = M_S\hbar \tag{2.40}$$

where the quantum number M_S must take half-integral values when the number of electrons in the ion is odd and integral values when the number of electrons is even. Moreover, though a highly excited ion can have an arbitrarily large orbital momentum, S_z cannot exceed $\frac{1}{2}n\hbar$, where n is the number of electrons in an ion. Similarly, as in the case of $\mathbf{L}^2$, we have for $\mathbf{S}^2$

$$\mathbf{S}^2 = S(S+1)\hbar^2 \tag{2.41}$$

where S, like M_S, is an integer for n even and a half-integer for n odd, and S cannot exceed $n/2$. For M_S there are $(2S'+1)$ values, given by

$$M_S = S, (S-1), \ldots, -(S-1), -S \tag{2.42}$$

The total number of states having the particular values L' and S', for the quantum numbers L and S, is thus $(2L'+1)(2S'+1)$.

Now the total angular momentum of a system is simply defined as the sum

$$\mathbf{J} = \mathbf{L} + \mathbf{S} \tag{2.43}$$

of the orbital and the spin angular momenta. Like $\mathbf{S}$, $\mathbf{J}$ has integral or half-integral eigenvalues according to whether the number of electrons in the system is even or odd. Here we assume, however, that the nucleus has no spin, although this is not always the case. But since the effects of the nuclear spin are small compared to those of the electrons, we neglect them.

2.5 WIGNER COEFFICIENTS

We now consider the addition of angular momenta in some detail. Let us write $\mathbf{j}_1$ and $\mathbf{j}_2$ for two commuting angular momenta and $\mathbf{j}$ for their sum. We now know that, if we take the $(2j_1+1)(2j_2+1)$ products $|j_1m_1\rangle|j_2m_2\rangle$, which we may write as $|j_1j_2m_1m_2\rangle$, suitable linear combinations of these give each of the basic states for $j=|j_1-j_2|,\ldots,j_1+j_2$, just once. Hence the coefficients in these linear combinations must be completely determined by j, j_1, j_2, m, m_1, and m_2, apart from arbitrary phase factors common to all states with the same **j**. Here m is the quantum number corresponding to j_z. Wigner has given an explicit but rather complicated formula for these coefficients. We shall not derive the formula but just discuss it.

We are to find the coefficients in the expansion

$$|j_1 j_2 j m\rangle = \sum_{m_1, m_2} |j_1 j_2 m_1 m_2\rangle\langle j_1 j_2 m_1 m_2 | j_1 j_2 j m\rangle \tag{2.44}$$

and, as j_1, j_2 are common to all kets in (2.44), we abbreviate the relation to

$$|jm\rangle = \sum_{m_1, m_2} |m_1 m_2\rangle\langle m_1 m_2 | jm\rangle \tag{2.45}$$

where $|m_1 m_2\rangle$ is an eigenstate of j_z with $j_z'=(m_1+m_2)\hbar$, and therefore $\langle m_1 m_2|jm\rangle=0$ unless $m_1+m_2=m$. Thus $\delta(m, m_1+m_2)$ is a factor of $\langle m_1 m_2|jm\rangle$.

Wigner's coefficients are obtained from (2.44) by multiplying through by $\langle j_1 j_2 m_1 m_2|$. They are, omitting the details,

$$\langle j_1 j_2 m_1 m_2 | j_1 j_2 j m\rangle = \delta(m, m_1+m_2)$$

$$\cdot\sqrt{\frac{(j+j_1-j_2)!(j-j_1+j_2)!(j_1+j_2-j)!(j+m)!(j-m)!(2j+1)}{(j+j_1+j_2+1)!(j_1-m_1)!(j_1+m_1)!(j_2-m_2)!(j_2+m_2)!}}$$

$$\cdot\sum_r \frac{(-1)^{r+j_2+m_2}(j+j_2+m_1-r)!(j_1-m_1+r)!}{(j-j_1+j_2-r)!(j+m-r)!r!(r+j_1-j_2-m)!} \tag{2.46}$$

where the summation is over all values of r such that all factorials occurring are of nonnegative integers $(0!=1)$. These Wigner coefficients are real and satisfy

$$\langle j_1 j_2 jm | j_1 j_2 m_1 m_2\rangle = \langle j_1 j_2 m_1 m_2 | j_1 j_2 j m\rangle \tag{2.47}$$

Equation (2.46) is fundamental to the theory of atomic structure. However, the formula is so complicated that it is more convenient to use tables that show the result of adding any j_1 and $j_2=0, \frac{1}{2}, 1, \frac{3}{2}, \cdots$ in order to obtain these transformation coefficients. The nonvanishing Wigner (or Clebsch-Gordan) coefficients $\langle j_1 j_2 m_1 m_2|j_1 j_2 j m\rangle$ for $j_2=\frac{1}{2}$, 1, and $\frac{3}{2}$ are presented in Tables 2.1, 2.2, and 2.3, respectively. In Section 2.8, we shall work out one important example with the help of these tables in order to show their usefulness.

Table 2.1

$\langle j_1 \tfrac{1}{2} m_1 m_2 | j_1 \tfrac{1}{2} j m \rangle$

j	$m_2=\frac{1}{2}$	$m_2=-\frac{1}{2}$
$j_1+\frac{1}{2}$	$\sqrt{\dfrac{j_1+m+\frac{1}{2}}{2j_1+1}}$	$\sqrt{\dfrac{j_1-m+\frac{1}{2}}{2j_1+\frac{1}{2}}}$
$j_1-\frac{1}{2}$	$-\sqrt{\dfrac{j_1-m+\frac{1}{2}}{2j_1+1}}$	$\sqrt{\dfrac{j_1+m+\frac{1}{2}}{2j_1+1}}$

Table 2.2

$\langle j_1 1 m_1 m_2 | j_1 1 j m \rangle$

j	$m_2=1$	$m_2=0$	$m_2=-1$
j_1+1	$\sqrt{\dfrac{(j_1+m)(j_1+m+1)}{(2j_1+1)(2j_1+2)}}$	$\sqrt{\dfrac{(j_1-m+1)(j_1+m+1)}{(2j_1+1)(j_1+1)}}$	$\sqrt{\dfrac{(j_1-m)(j_1-m+1)}{(2j_1+1)(2j_1+2)}}$
j_1	$-\sqrt{\dfrac{(j_1+m)(j_1-m+1)}{2j_1(j_1+1)}}$	$\dfrac{m}{\sqrt{j_1(j_1+1)}}$	$\sqrt{\dfrac{(j_1-m)(j_1+m+1)}{2j_1(j_1+1)}}$
j_1-1	$\sqrt{\dfrac{(j_1-m)(j_1-m+1)}{2j_1(2j_1+1)}}$	$-\sqrt{\dfrac{(j_1-m)(j_1+m)}{j_1(2j_1+1)}}$	$\sqrt{\dfrac{(j_1+m+1)(j_1+m)}{2j_1(2j_1+1)}}$

Table 2.3

$\langle j_1 \tfrac{3}{2} m_1 m_2 | j_1 \tfrac{3}{2} j m \rangle$

j	$m_2=\frac{3}{2}$	$m_2=\frac{1}{2}$
$j_1+\frac{3}{2}$	$\sqrt{\dfrac{(j_1+m-\frac{1}{2})(j_1+m+\frac{1}{2})(j_1+m+\frac{3}{2})}{(2j_1+1)(2j_1+2)(2j_1+3)}}$	$\sqrt{\dfrac{3(j_1+m+\frac{1}{2})(j_1+m+\frac{3}{2})(j_1-m+\frac{3}{2})}{(2j_1+1)(2j_1+2)(2j_1+3)}}$
$j_1+\frac{1}{2}$	$-\sqrt{\dfrac{3(j_1+m-\frac{1}{2})(j_1+m+\frac{1}{2})(j_1-m+\frac{3}{2})}{2j_1(2j_1+1)(2j_1+3)}}$	$-(j_1-3m+\frac{3}{2})\sqrt{\dfrac{j_1+m+\frac{1}{2}}{2j_1(2j_1+1)(2j_1+3)}}$
$j_1-\frac{1}{2}$	$\sqrt{\dfrac{3(j_1+m-\frac{1}{2})(j_1-m+\frac{1}{2})(j_1-m+\frac{3}{2})}{(2j_1-1)(2j_1+1)(2j_1+2)}}$	$-(j_1+3m-\frac{1}{2})\sqrt{\dfrac{j_1-m+\frac{1}{2}}{(2j_1-1)(2j_1+1)(2j_1+2)}}$
$j_1-\frac{3}{2}$	$-\sqrt{\dfrac{(j_1-m-\frac{1}{2})(j_1-m+\frac{1}{2})(j_1-m+\frac{3}{2})}{2j_1(2j_1-1)(2j_1+1)}}$	$\sqrt{\dfrac{3(j_1+m-\frac{1}{2})(j_1-m-\frac{1}{2})(j_1-m+\frac{1}{2})}{2j_1(2j_1-1)(2j_1+1)}}$

Table 2.3 continued

$\langle j_1 \tfrac{3}{2} m_1 m_2 | j_1 \tfrac{3}{2} j m \rangle$

j	$m_2=-\frac{1}{2}$	$m_2=-\frac{3}{2}$
$j_1+\frac{3}{2}$	$\sqrt{\frac{3(j_1+m+\frac{3}{2})(j_1-m+\frac{1}{2})(j_1-m+\frac{3}{2})}{(2j_1+1)(2j_1+2)(2j_1+3)}}$	$\sqrt{\frac{(j_1-m-\frac{1}{2})(j_1-m+\frac{1}{2})(j_1-m+\frac{3}{2})}{(2j_1+1)(2j_1+2)(2j_1+3)}}$
$j_1+\frac{1}{2}$	$(j_1+3m+\frac{3}{2})\sqrt{\frac{j_1-m+\frac{1}{2}}{2j_1(2j_1+1)(2j_1+3)}}$	$\sqrt{\frac{3(j_1+m+\frac{3}{2})(j_1-m-\frac{1}{2})(j_1-m+\frac{1}{2})}{2j_1(2j_1+1)(2j_1+3)}}$
$j_1-\frac{1}{2}$	$-(j_1-2m-\frac{1}{2})\sqrt{\frac{j_1+m+\frac{1}{2}}{(2j_1-1)(2j_1+1)(2j_1+2)}}$	$\sqrt{\frac{3(j_1+m+\frac{1}{2})(j_1+m+\frac{3}{2})(j_1-m-\frac{1}{2})}{(2j_1-1)(2j_1+1)(2j_1+2)}}$
$j_1-\frac{3}{2}$	$-\sqrt{\frac{3(j_1+m-\frac{1}{2})(j_1+m+\frac{1}{2})(j_1-m-\frac{1}{2})}{2j_1(2j_1-1)(2j_1+1)}}$	$\sqrt{\frac{(j_1+m-\frac{1}{2})(j_1+m+\frac{1}{2})(j_1+m+\frac{3}{2})}{2j_1(2j_1-1)(2j_1+1)}}$

2.6 MANY-ELECTRON WAVEFUNCTIONS

To determine which state is of least energy for a particular ion containing N electrons, we have to construct all the possible wavefunctions satisfying the Pauli exclusion principle. We begin by writing a wavefunction Ψ of the spatial and spin coordinates of all N electrons, given by

$$\Psi(\mathbf{r}_1,\mathbf{r}_2,\mathbf{r}_3,\ldots,\mathbf{r}_N;\, \boldsymbol{\sigma}_1,\boldsymbol{\sigma}_2,\boldsymbol{\sigma}_3,\ldots,\boldsymbol{\sigma}_N) \tag{2.48}$$

We can obtain the stationary states by solving the Schrödinger equation:

$$\mathcal{H}\Psi_n = E_n\Psi_n \tag{2.49}$$

where $\mathcal{H}$ is the time-independent Hamiltonian of the system and is the total energy of the system in the nth stationary state, represented by Ψ_n. With the spin-orbit coupling neglected, $\mathcal{H}$ is a summation over one-electron terms and is given by

$$\mathcal{H} = -\frac{\hbar^2}{2m}\sum \nabla_i^2 - \sum_i \frac{Ze^2}{r_i} + \sum \frac{e^2}{r_{ij}} \tag{2.50}$$

For the simplest case of all, the H atom, $N=1$, and the exact solution of the Schrödinger equation may be obtained in a straightforward manner. For $N \geqslant 2$ immediate difficulty is encountered from the presence of the cross terms in the potential energy, e^2/r_{ij}, which couple the motions of the electrons. With He-like atoms, however, the problem is still tractable by taking the wavefunction to include the variable r_{ij}. For all other systems, however, it has been necessary to resort to simplifications. The technique has been to replace the sum of all interelectronic interactions by a smoothed potential term $V_i(\mathbf{r})$, that is, the ith electron is taken to move in an averaged potential field of the remaining electrons and to move independently of them, as regards relative motion. It is then the difference between these effective potentials $V_i(\mathbf{r})$ and the exact expression for Coulomb interactions that is considered by the methods of perturbation theory. The best possible one-electron potential functions, that is, those that give the closest approximation to the exact wavefunction of the ion, are those that minimize the computed energy because, according to quantum mechanics, the energy computed for a stationary state of any system is a minimum against any small changes made in the normalized wavefunction away from the correct, exact form. Iterative procedures are used in practice to find these best "Hartree-Fock" potentials and the resulting wavefunctions, numerically, by successive approximation.[3, 4] It is important to remember here that our understanding of most atomic and solid state systems is based on "one-electron approximations" in which the true state of motion is approximated as closely as practicable by the product of a set of independent one-electron functions.

We now write the total Hamiltonian, breaking it into two parts:

$$\mathcal{H} = \sum_i \mathcal{H}_i + \sum_{i<j} \mathcal{H}_{ij} \tag{2.51}$$

where the first term on the right-hand side represents the one-electron Hamiltonian, and the second term represents the Coulomb interaction between the electrons, to be treated as a perturbation. The one-electron part of the Hamiltonian (2.51) is given by

$$\sum_i \mathcal{H}_i = -\frac{\hbar^2}{2m} \sum \nabla_i^2 - \sum \frac{Ze^2}{r_i} + \sum_i V_i(\mathbf{r}) \tag{2.52}$$

where $V_i(\mathbf{r})$ are the Hartree-Fock potentials.

The approximate Schrödinger equation, to which the approximate wavefunction is to be a solution, is

$$\sum \mathcal{H}_i \Psi(\mathbf{r}_1, \mathbf{r}_2, \ldots, \mathbf{r}_N : \boldsymbol{\sigma}_1, \boldsymbol{\sigma}_2, \ldots, \boldsymbol{\sigma}_N) = E\Psi(\mathbf{r}_1, \mathbf{r}_2, \ldots, \mathbf{r}_N ; \boldsymbol{\sigma}_1, \boldsymbol{\sigma}_2, \ldots, \boldsymbol{\sigma}_N) \tag{2.53}$$

This is satisfied by any product of the one-electron functions found as solutions to the one-electron Schrödinger equation:

$$\mathcal{H}_i \Psi(\mathbf{r}_i, \boldsymbol{\sigma}_i) = E_i \Psi(\mathbf{r}_i, \boldsymbol{\sigma}_i) \tag{2.54}$$

where E_i is the part of the total energy of the system that can be associated with the ith electron or, more precisely, with the electron that is described by $\Psi(\mathbf{r}_i, \boldsymbol{\sigma}_i)$. The precise form to be chosen for $\mathcal{H}$ will be discussed in Chapter 11 on molecular orbital theory.

Without going into further detail at this point, we may state that the solution of these equations yields a set of one-electron wavefunctions, similar in general to the wavefunctions obtained from the solution of the H-atom problem. The spin and the angular parts of the total wavefunctions are identical in both cases and are expressible in analytic form. The radial part, however, must be obtained numerically by the self-consistent field (SCF) method of Hartree and Fock. Each function Ψ is characterized by a set of four quantum numbers: n, l, m_l, and m_s. The configuration of a particular atomic state is described by giving the number of electrons the specified n and l-values. Using Fe^{2+} as an example, we can indicate its configuration by

$$1s^2, \qquad 2s^2 2p^6, \qquad 3s^2 3p^6, \qquad 3d^6$$

In this case, only six electrons are available to be distributed among the ten allowed $3d$ one-electron functions, meaning that there exist several Ψ or atomic states corresponding to this configuration. The total number, however, is severely restricted by the Pauli exclusion principle, that is, the antisymmetry requirement upon Ψ. Each of these states may be described in terms of the total angular momentum of the whole atom. If we let $\mathbf{l}_i$ and $\mathbf{s}_i$ equal the orbital and the spin angular momentum, respectively, of the electron, the two vectors $\mathbf{L}$ and $\mathbf{S}$ are defined as the vector sums of $\mathbf{l}_i$ and $\mathbf{s}_i$, respectively. Since the vector sums over the closed shells of electrons vanish, we are left with only the open $3d$ shell for further consideration of the problem.

In the approximation of the Hamiltonian operator so far considered, all these various Ψ's or atomic states have equal energies. This degeneracy is lifted, however, when the perturbation of $\mathcal{H}_{ij}$ is included. This perturbation is computed for each state. As a general rule (Hund's rule) it is found that, of all the allowed states, those with the largest S are lowest in energy; of all these, the one with the largest L is the lowest. Even at this point, there remains degeneracy in the total angular momentum, $\mathbf{J} = \mathbf{L} + \mathbf{S}$, arising from the possibility of several relative orientations of $\mathbf{L}$ with respect to $\mathbf{S}$. This last degeneracy is lifted when the spin-orbit coupling is taken into consideration.

Thus the state of an atom may be described in terms of the total energy, the total spin momentum, the total orbital momentum, and the total angular momentum. These quantities are expressed explicitly in terms of the individual energies and momenta of the one-electron wavefunctions, taken in antisymmetrized product form, corresponding to the original spherically symmetric atomic problem.

We now return to the question of how to construct a total wavefunction for an N-electron ion with the knowledge that one-electron functions can be (and have been) found by the Hartree-Fock method for a large number of isolated ions.[5] There is, in principle, an infinite set of solutions to each one-electron equation of the kind shown in (2.54). With these solutions it is certainly possible, in principle, to construct an N-electron wavefunction having as high an accuracy as desired because such a set is a "complete set" from which any function can be formed by linear combination. However, it is essential to find a wavefunction of great accuracy by using as few of the one-electron functions as possible. If the energies of the electrons in the states represented by the one-electron functions are widely separated, the most reasonable choice with which to construct a wavefunction for the ground state would be the N one-electron functions of least energy. Generally, however, some of the one-electron states may be degenerate in energy, in which case there is some difficulty in making a suitable choice.

Let us suppose that we have selected N one-electron functions whose energies are well below the rest. The N electrons can then be assigned to these states alone to give a good approximation to the ionic ground state. Although the simple product of these N functions, one for each electron,

$$\Phi = \psi_1(\mathbf{r}_1\sigma_1)\psi_2(\mathbf{r}_2\sigma_2)\cdots\psi_N(\mathbf{r}_N\sigma_N) \tag{2.55}$$

satisfies the approximate Schrödinger equation (2.53), it is not an acceptable ionic wavefunction because it fails to satisfy the fundamental requirement that the wavefunction for any system of electrons must be antisymmetrical to the interchange of the spatial and spin coordinates, $\mathbf{r}_a\sigma_a$ and $\mathbf{r}_b\sigma_b$, of any two of them. To make the wavefunction given above satisfy the antisymmetry requirement we must write it in the form of a determinant:

$$\begin{vmatrix} \psi_1(\mathbf{r}_1\sigma_1) & \psi_2(\mathbf{r}_1\sigma_1) & \cdots & \psi_N(\mathbf{r}_1\sigma_1) \\ \psi_1(\mathbf{r}_2\sigma_2) & \psi_2(\mathbf{r}_2\sigma_2) & \cdots & \psi_N(\mathbf{r}_2\sigma_2) \\ \vdots & & & \\ \psi_1(\mathbf{r}_N\sigma_N) & \psi_2(\mathbf{r}_N\sigma_N) & \cdots & \psi_N(\mathbf{r}_N\sigma_N) \end{vmatrix} \tag{2.56}$$

It can be shown that this is the only antisymmetrical function which can be constructed from the given N functions. A great difficulty will arise, however, when the one-electron states are arranged in degenerate groups. Those of least energy are progressively filled when making an assignment, but it may happen that the N electrons are insufficient in number to completely fill the highest degenerate group involved. No unique choice of N functions can then be justified. In this case a linear combination of the determinants is required. We shall encounter this question again in Chapter 13 when we deal with the band theory of ferromagnetism.

2.7 THE RUSSELL-SAUNDERS COUPLING SCHEME

Using the Hamiltonian $\mathcal{H}$ of (2.1), we saw that the total orbital angular momentum $\mathbf{L}$ commutes with $\mathcal{H}$. Since $\mathcal{H}$ does not involve the spin coordinates, the total spin angular momentum $\mathbf{S}$ also commutes with $\mathcal{H}$. Thus $\mathcal{H}$,

$\mathbf{L}^2$, $\mathbf{L}_z$, $\mathbf{S}^2$, and $\mathbf{S}_z$ form a set of five commuting observables, which means that the states of an atomic system can be classified as simultaneous eigenkets of these five observables. In practice, however, this is not generally possible because the Schrödinger equation can be solved exactly only for a single-electron system (the H atom). Instead, it is relatively easy to find the eigenkets of $\mathbf{L}^2$, $\mathbf{L}_z$, $\mathbf{S}^2$, and $\mathbf{S}_z$. It is often useful to use a complete set of kets that are simultaneous eigenkets of the observables listed above and therefore are approximate eigenkets of $\mathcal{H}$. Any simultaneous eigenket of $\mathbf{L}^2$, $\mathbf{L}_z$, $\mathbf{S}^2$, $\mathbf{S}_z$, and $\mathcal{H}$ can then be expanded in terms of our complete set; and if all the coefficients except one are negligible, the corresponding member of our complete set is an approximate eigenket of $\mathcal{H}$. Such a definition of an approximate eigenket of $\mathcal{H}$ can be justified by comparing with the experimental data.

Let us take a complete set of simultaneous eigenkets of $\mathbf{L}^2$, $\mathbf{L}_z$, $\mathbf{S}^2$, and $\mathbf{S}_z$. Our earlier discussion on the addition of angular momenta shows that there is a complete set of eigenkets of $\mathbf{S}^2$, $\mathbf{L}^2$, $\mathbf{J}^2$, and $\mathbf{J}_z$, where $\mathbf{J}=\mathbf{S}+\mathbf{L}$. These are associated with our first set through Wigner's formula (2.46). Any ket from either complete set is a simultaneous eigenket of $\mathbf{S}^2$ and $\mathbf{L}^2$. It is customary to call any such complete set of kets, each of which is a simultaneous eigenket of $\mathbf{S}^2$ and $\mathbf{L}^2$, a Russell-Saunders (or *LS*) coupling scheme. The two particular *LS* coupling schemes mentioned above are called, respectively, the SLM_SM_L and $SLJM$ schemes. Their eigenkets are written as $|\alpha' sLM_SM_L\rangle$ and $|\alpha' SLJM\rangle$, respectively, where α' represents additional observables needed to specify the complete set. Also, $\mathbf{S}$ corresponds to $\mathbf{j}_1$, and $\mathbf{L}$ to $\mathbf{j}_2$, in (2.46).

We have already seen that S may have a large effect on the allowed energies because of the antisymmetry of the wavefunction, and $\mathbf{L}$ also has a large effect on energy. $\mathbf{S}_z$ and $\mathbf{L}_z$, however, have no effect since both $\mathbf{S}$ and $\mathbf{L}$ are symmetrical functions of the electronic coordinates, which means that any permutation of the electrons leave them unchanged. As a consequence, if we have a ket $|\alpha' SLM_SM_L\rangle$ satisfying the antisymmetry requirement, all kets obtained by operating with the shift operators $\mathbf{S}^{\pm}$ and $\mathbf{L}^{\pm}$ will also satisfy it. Therefore the kets $|\alpha' SLM_SM_L\rangle$ can be grouped into sets, and each set contains $(2S+1)(2L+1)$ individual states, one for each allowed pair $(S_z' L_z')$. Such a set is called a term and is denoted by ${}^{2S+1}L$. When we wish to indicate the individual states of a term, we write the M_S and M_L values after the term symbol in that order; thus we write $|{}^{2S+1}L, M_S, M_L\rangle$. It is easy to show that all states belonging to a given term have the same first-order energy, that is,

$$\langle \alpha' SLM_SM_L|\mathcal{H}|\alpha' SLM_SM_L\rangle$$

is the same for all M_S, M_L when α', S, and L are given. The superscript $(2S+1)$ is called the multiplicity of the term, and we say that the term is a singlet, doublet, triplet,... according as $S=0, \frac{1}{2}, 1, \ldots$. When $L \geqslant S$, the multiplicity is equal to the number of different J values occurring for the corresponding kets in the $SLJM$ scheme. In practice, one usually knows the numerical value of L and then writes ${}^{2S+1}X$, where

$$X=S,P,D,F,G,\ldots \qquad \text{when } L=0,1,2,3,4,\ldots$$

The term that has the lowest energy is called the ground term. For example, the ground term of the neutral Fe atom is known to be ${}^5D$, which means that it has $S=L=2$.

2.8 WAVEFUNCTIONS OF THE nd^3 CONFIGURATION—AN EXAMPLE

With the help of an example we shall now explain how to determine the possible (SLM_SM_L) states of an atom (ion) for a given electronic configuration specified by a set of single-electron states. We choose the nd^3 configuration for this purpose.

In the nd^3 system, all three electrons have the same values of n and $l(l=2)$. Thus, according to Pauli's exclusion principle, it is necessary that no two electrons have the same values of m_l and m_s. We write $m_s=\frac{1}{2}$ as a plus superscript and $m_s=-\frac{1}{2}$ as a minus superscript over the associated m_l value. Thus the notation $\left(\overset{+}{2}\ \overset{-}{2}\ \overset{+}{1}\right)$ denotes a three-electron determinantal state involving the three single-electron states: $(n, l=2, m_l=+2, m_s=\frac{1}{2})$, $(n, l=2, m_l=+2, m_s=-\frac{1}{2})$, and $(n, l=2, m_l=+1, m_s=\frac{1}{2})$. We also have $M_L=\Sigma m_l$ and $M_S=\Sigma m_s$.

The possible determinantal states (microstates) for the d^3 configuration are constructed without violating the Pauli exclusion principle and are listed in Table 2.4, where they are arranged into several groups, each corresponding to a particular (M_S, M_L) pair. The table is shown only for positive values of M_L and can be easily extended for negative values of M_L by reversing the sign of each single-electron l value.

The total number of determinantal states is thus found to be 120, which is quite as expected, since the total number of ways in which two electrons can be placed into any two of the ten subtenths of the d level is just ${}^{10}C_2=120$.

When only one microstate corresponds to a particular (M_S, M_L) pair, say, (M_S', M_L'), that microstate will be eigenfunctions of $\mathbf{L}^2$ and $\mathbf{S}^2$ as well, and the corresponding values of L and S will be M_L' and M_S', respectively, provided that we have no microstate for $M_L>M_L'$, $M_S\geqslant M_S'$ or for $M_L\geqslant M_L'$, $M_S>M_S'$.

Thus from Table 2.4 we find that

$$|{}^2H;\tfrac{1}{2},5\rangle=\left(\overset{+}{2}\ \overset{-}{2}\ \overset{+}{1}\right) \tag{2.57}$$

$$|{}^4F;\tfrac{3}{2},3\rangle=\left(\overset{+}{2}\ \overset{+}{1}\ \overset{+}{0}\right) \tag{2.58}$$

From (2.57) we get

$$\mathbf{L}^-|{}^2H;\tfrac{1}{2},5\rangle=\mathbf{L}^-\left(\overset{+}{2}\ \overset{-}{2}\ \overset{+}{1}\right).$$

Table 2.4
Microstates of d^3 configuration

M_L \ M_S	$\frac{3}{2}$	$\frac{1}{2}$	$-\frac{1}{2}$	$-\frac{3}{2}$
6				
5		$(\overset{+}{2}\,\overset{-}{2}\,\overset{+}{1})$	$(\overset{+}{2}\,\overset{-}{2}\,\overset{-}{1})$	
4		$(\overset{+}{2}\,\overset{-}{2}\,\overset{+}{0})$ $(\overset{+}{2}\,\overset{+}{1}\,\overset{-}{1})$	$(\overset{+}{2}\,\overset{-}{2}\,\overset{-}{0})$ $(\overset{-}{2}\,\overset{+}{1}\,\overset{-}{1})$	
3	$(\overset{+}{2}\,\overset{+}{1}\,\overset{+}{0})$	$(\overset{+}{2}\,\overset{+}{1}\,\overset{-}{0})(\overset{+}{2}\,\overset{-}{1}\,\overset{+}{0})$ $(\overset{-}{2}\,\overset{+}{1}\,\overset{+}{0})(\overset{+}{2}\,\overset{-}{2}\,\overset{+}{-1})$	$(\overset{-}{2}\,\overset{-}{1}\,\overset{+}{0})(\overset{-}{2}\,\overset{+}{1}\,\overset{-}{0})$ $(\overset{+}{2}\,\overset{-}{1}\,\overset{-}{0})(\overset{+}{2}\,\overset{-}{2}\,\overset{-}{-1})$	$(\overset{-}{2}\,\overset{-}{1}\,\overset{-}{0})$
2	$(\overset{+}{2}\,\overset{+}{1}\,\overset{+}{-1})$	$(\overset{+}{2}\,\overset{+}{1}\,\overset{-}{-1})(\overset{+}{2}\,\overset{-}{1}\,\overset{+}{-1})$ $(\overset{-}{2}\,\overset{+}{1}\,\overset{+}{-1})(\overset{+}{1}\,\overset{-}{1}\,\overset{+}{0})$ $(\overset{+}{2}\,\overset{+}{0}\,\overset{-}{0})(\overset{+}{2}\,\overset{-}{2}\,\overset{+}{-2})$	$(\overset{-}{2}\,\overset{-}{1}\,\overset{+}{-1})(\overset{-}{2}\,\overset{+}{1}\,\overset{-}{-1})$ $(\overset{+}{2}\,\overset{-}{1}\,\overset{-}{-1})(\overset{-}{1}\,\overset{+}{1}\,\overset{-}{0})$ $(\overset{-}{2}\,\overset{-}{0}\,\overset{+}{0})(\overset{+}{2}\,\overset{-}{2}\,\overset{-}{-2})$	$(\overset{-}{2}\,\overset{-}{1}\,\overset{-}{-1})$

1	$(\overset{+}{2}\ \overset{+}{0}\ -\overset{+}{1})$ $(\overset{+}{1}\ \overset{+}{2}\ -\overset{+}{2})$	$(\overset{+}{2}\ \overset{+}{0}\ -\overset{-}{1})(\overset{+}{2}\ \overset{-}{0}\ -\overset{+}{1})(\overset{-}{2}\ \overset{+}{0}\ -\overset{+}{1})$ $(\overset{+}{1}\ \overset{-}{1}\ -\overset{+}{1})(\overset{+}{1}\ \overset{-}{0}\ \overset{+}{0})(\overset{+}{1}\ \overset{+}{2}\ -\overset{-}{2})$ $(\overset{+}{1}\ \overset{-}{2}\ -\overset{+}{2})(\overset{-}{1}\ \overset{+}{2}\ -\overset{+}{2})$	$(\overset{-}{2}\ \overset{-}{0}\ -\overset{+}{1})(\overset{-}{2}\ \overset{+}{0}\ -\overset{-}{1})(\overset{+}{2}\ \overset{-}{0}\ -\overset{-}{1})$ $(\overset{-}{1}\ \overset{+}{1}\ -\overset{-}{1})(\overset{-}{1}\ \overset{+}{0}\ \overset{-}{0})(\overset{-}{1}\ \overset{-}{2}\ -\overset{+}{2})$ $(\overset{-}{1}\ \overset{+}{2}\ -\overset{-}{2})(\overset{+}{1}\ \overset{-}{2}\ -\overset{-}{2})$	$(\overset{-}{2}\ \overset{-}{0}\ -\overset{-}{1})$ $(\overset{-}{1}\ \overset{-}{2}\ -\overset{-}{2})$
0	$(\overset{+}{1}\ \overset{+}{0}\ -\overset{+}{1})$ $(\overset{+}{2}\ \overset{+}{0}\ -\overset{+}{2})$	$(\overset{+}{1}\ \overset{+}{0}\ -\overset{-}{1})(\overset{+}{1}\ \overset{-}{0}\ -\overset{+}{1})(\overset{-}{1}\ \overset{+}{0}\ -\overset{+}{1})$ $(\overset{+}{2}\ \overset{+}{0}\ -\overset{-}{2})(\overset{+}{2}\ \overset{-}{0}\ -\overset{+}{2})(\overset{+}{2}\ \overset{-}{0}\ -\overset{+}{2})$ $(\overset{+}{1}\ \overset{-}{1}\ -\overset{+}{2})(\overset{+}{2}\ -\overset{+}{1}\ -\overset{-}{1})$	$(\overset{-}{1}\ \overset{-}{0}\ -\overset{+}{1})(\overset{-}{1}\ \overset{+}{0}\ -\overset{-}{1})(\overset{+}{1}\ \overset{-}{0}\ -\overset{-}{1})$ $(\overset{-}{2}\ \overset{-}{0}\ -\overset{+}{2})(\overset{-}{2}\ \overset{+}{0}\ -\overset{-}{2})(\overset{+}{2}\ \overset{-}{0}\ -\overset{-}{2})$ $(\overset{-}{1}\ \overset{+}{1}\ -\overset{-}{2})(\overset{-}{2}\ -\overset{-}{1}\ -\overset{+}{1})$	$(\overset{-}{1}\ \overset{-}{0}\ -\overset{-}{1})$ $(\overset{-}{2}\ \overset{-}{0}\ -\overset{-}{2})$

or

$$\sqrt{10}\,|{}^2H;\tfrac{1}{2},4\rangle=\sqrt{4}\left(\overset{+}{1}\ \overset{-}{2}\ \overset{+}{1}\right)+\sqrt{4}\left(\overset{+}{2}\ \overset{-}{1}\ \overset{+}{1}\right)+\sqrt{6}\left(\overset{+}{2}\ \overset{-}{2}\ \overset{+}{0}\right)$$

The first microstate on the right-hand side violates Pauli's principle and therefore is vanishing. Thus

$$\begin{aligned}|{}^2H;\tfrac{1}{2},4\rangle&=\sqrt{\tfrac{5}{2}}\left(\overset{+}{2}\ \overset{-}{1}\ \overset{+}{1}\right)+\sqrt{\tfrac{3}{5}}\left(\overset{+}{2}\ \overset{-}{2}\ \overset{+}{0}\right)\\&=-\sqrt{\tfrac{2}{5}}\left(\overset{+}{2}\ \overset{+}{1}\ \overset{-}{1}\right)+\sqrt{\tfrac{3}{5}}\left(\overset{+}{2}\ \overset{-}{2}\ \overset{+}{0}\right)\end{aligned}\tag{2.59}$$

Because the wavefunction $|{}^2G;\frac{1}{2},4\rangle$ is to be made up of the microstates $\left(\overset{+}{2}\ \overset{+}{1}\ \overset{-}{1}\right)$ and $\left(\overset{+}{2}\ \overset{-}{2}\ \overset{+}{0}\right)$, we must take the normalized combination of these microstates that is orthogonal to $|{}^2H;\frac{1}{2},4\rangle$. Consequently,

$$|{}^2G;\tfrac{1}{2},4\rangle=\sqrt{\tfrac{3}{5}}\left(\overset{+}{2}\ \overset{+}{1}\ \overset{-}{1}\right)+\sqrt{\tfrac{2}{5}}\left(\overset{+}{2}\ \overset{-}{2}\ \overset{+}{0}\right)\tag{2.60}$$

Operating with $\mathbf{L}^-$ upon (2.59) and (2.60), we get

$$\begin{aligned}|{}^2H;\tfrac{1}{2},3\rangle=&\sqrt{\tfrac{8}{15}}\left(\overset{+}{2}\ \overset{-}{1}\ \overset{+}{0}\right)-\sqrt{\tfrac{2}{15}}\left(\overset{+}{2}\ \overset{+}{1}\ \overset{-}{0}\right)\\&-\sqrt{\tfrac{2}{15}}\left(\overset{-}{2}\ \overset{+}{1}\ \overset{+}{0}\right)+\frac{1}{\sqrt{5}}\left(\overset{+}{2}\ \overset{-}{2}\ \overset{+}{-1}\right)\end{aligned}\tag{2.61}$$

and

$$\begin{aligned}|{}^2G;\tfrac{1}{2},3\rangle=&-\frac{1}{2\sqrt{5}}\left(\overset{+}{2}\ \overset{-}{1}\ \overset{+}{0}\right)+\frac{3}{2\sqrt{5}}\left(\overset{+}{2}\ \overset{+}{1}\ \overset{-}{0}\right)\\&-\frac{1}{\sqrt{5}}\left(\overset{-}{2}\ \overset{+}{1}\ \overset{+}{0}\right)+\sqrt{\tfrac{3}{10}}\left(\overset{+}{2}\ \overset{-}{2}\ \overset{+}{-1}\right)\end{aligned}\tag{2.62}$$

Operating with $\mathbf{L}^-$ and $\mathbf{S}^-$ upon (2.58), we get

$$|{}^4F;\tfrac{3}{2},2\rangle=\left(\overset{+}{2}\ \overset{+}{1}\ \overset{+}{-1}\right)\tag{2.63}$$

$$|{}^4F;\tfrac{1}{2},3\rangle=\frac{1}{\sqrt{3}}\left(\overset{+}{2}\ \overset{-}{1}\ \overset{+}{0}\right)+\frac{1}{\sqrt{3}}\left(\overset{+}{2}\ \overset{+}{1}\ \overset{-}{0}\right)+\frac{1}{\sqrt{3}}\left(\overset{-}{2}\ \overset{+}{1}\ \overset{+}{0}\right)\tag{2.64}$$

Now the state $|{}^2F;\frac{1}{2},3\rangle$ is to be built by using the four microstates $\left(\overset{+}{2}\ \overset{-}{1}\ \overset{+}{0}\right)$, $\left(\overset{+}{2}\ \overset{+}{1}\ \overset{-}{0}\right)$, $\left(\overset{-}{2}\ \overset{+}{1}\ \overset{+}{0}\right)$, and $\left(\overset{+}{2}\ \overset{-}{2}\ \overset{+}{-1}\right)$ and we must take the normalized linear combination of these microstates that is orthogonal to the states $|{}^2H;\frac{1}{2},3\rangle$, $|{}^2G;\frac{1}{2},3\rangle$, and $|{}^4F;\frac{1}{2},3\rangle$ as given by (2.61), (2.62), and (2.64), respectively. Then we find that

$$\begin{aligned}|{}^2F;\tfrac{1}{2},3\rangle=&\frac{1}{2\sqrt{3}}\left(\overset{+}{2}\ \overset{-}{1}\ \overset{+}{0}\right)+\frac{1}{2\sqrt{3}}\left(\overset{+}{2}\ \overset{+}{1}\ \overset{-}{0}\right)\\&-\frac{1}{\sqrt{3}}\left(\overset{-}{2}\ \overset{+}{1}\ \overset{+}{0}\right)-\frac{1}{\sqrt{2}}\left(\overset{+}{2}\ \overset{-}{2}\ \overset{+}{-1}\right)\end{aligned}\tag{2.65}$$

Operating with L^- upon (2.63), we have

$$|^4F;\tfrac{3}{2},1\rangle=\sqrt{\tfrac{3}{5}}\left(\overset{+}{2}\ \overset{+}{0}\ \overset{+}{-1}\right)-\sqrt{\tfrac{2}{5}}\left(\overset{+}{1}\ \overset{+}{2}\ \overset{+}{-2}\right) \tag{2.66}$$

The state $|^4P;\tfrac{3}{2},1\rangle$ is then constructed as a normalized linear combination of the microstates $\left(\overset{+}{2}\ \overset{+}{0}\ \overset{+}{-1}\right)$ and $\left(\overset{+}{1}\ \overset{+}{2}\ \overset{+}{-2}\right)$ so that it is orthogonal to the state $|^4F;\tfrac{3}{2},1\rangle$. Then

$$|^4P;\tfrac{3}{2},1\rangle=\sqrt{\tfrac{2}{5}}\left(\overset{+}{2}\ \overset{+}{0}\ \overset{+}{-1}\right)+\sqrt{\tfrac{3}{5}}\left(\overset{+}{1}\ \overset{+}{2}\ \overset{+}{-2}\right) \tag{2.67}$$

We note that only two microstates correspond to ($M_S=\tfrac{3}{2}, M_L=0$), and we already have two states $|^4F;\tfrac{3}{2},0\rangle$ and $|^4P;\tfrac{3}{2},0\rangle$ [to be obtained from (2.66) and (2.67), respectively, by operating with $\mathbf{L}^-$]. Therefore it follows that we cannot have the state $|^4S;\tfrac{3}{2},0\rangle$.

Operating with $\mathbf{L}^-$ upon (2.61), (2.62), and (2.65), we get

$$\begin{aligned}|^2H;\tfrac{1}{2},2\rangle=&-\frac{1}{\sqrt{30}}\left(\overset{+}{2}\ \overset{+}{1}\ \overset{-}{-1}\right)+\sqrt{\tfrac{3}{10}}\left(\overset{+}{2}\ \overset{-}{1}\ \overset{+}{-1}\right)\\&-\sqrt{\tfrac{2}{15}}\left(\overset{-}{2}\ \overset{+}{1}\ \overset{+}{-1}\right)+\frac{1}{\sqrt{5}}\left(\overset{+}{1}\ \overset{-}{1}\ \overset{+}{0}\right)\\&-\sqrt{\tfrac{3}{10}}\left(\overset{+}{2}\ \overset{+}{0}\ \overset{-}{0}\right)+\frac{1}{\sqrt{30}}\left(\overset{+}{2}\ \overset{-}{2}\ \overset{+}{-2}\right)\end{aligned} \tag{2.68}$$

$$\begin{aligned}|^2G;\tfrac{1}{2},2\rangle=&\tfrac{3}{2}\sqrt{\tfrac{3}{35}}\left(\overset{+}{2}\ \overset{+}{1}\ \overset{-}{-1}\right)+\tfrac{1}{2}\sqrt{\tfrac{3}{35}}\left(\overset{+}{2}\ \overset{-}{1}\ \overset{+}{-1}\right)\\&+2\sqrt{\tfrac{3}{35}}\left(\overset{+}{2}\ \overset{+}{0}\ \overset{-}{0}\right)-2\sqrt{\tfrac{3}{35}}\left(\overset{-}{2}\ \overset{+}{1}\ \overset{+}{-1}\right)\\&+\frac{1}{\sqrt{70}}\left(\overset{+}{1}\ \overset{-}{1}\ \overset{+}{0}\right)+\sqrt{\tfrac{3}{35}}\left(\overset{+}{2}\ \overset{-}{2}\ \overset{+}{-2}\right)\end{aligned} \tag{2.69}$$

$$\begin{aligned}|^2F;\tfrac{1}{2},2\rangle=&\frac{1}{2\sqrt{3}}\left(\overset{+}{2}\ \overset{+}{1}\ \overset{-}{-1}\right)-\frac{1}{2\sqrt{3}}\left(\overset{+}{2}\ \overset{-}{1}\ \overset{+}{-1}\right)\\&+\frac{1}{\sqrt{2}}\left(\overset{+}{1}\ \overset{-}{1}\ \overset{+}{0}\right)-\frac{1}{2\sqrt{3}}\left(\overset{+}{2}\ \overset{+}{0}\ \overset{-}{0}\right)\\&-\sqrt{\tfrac{2}{3}}\left(\overset{+}{2}\ \overset{-}{2}\ \overset{+}{-2}\right)\end{aligned} \tag{2.70}$$

Now we have to construct the state $|^2D;\tfrac{1}{2},2\rangle$ as a normalized linear combination of the six microstates $\left(\overset{+}{2}\ \overset{+}{1}\ \overset{-}{-1}\right)$, $\left(\overset{+}{2}\ \overset{-}{1}\ \overset{+}{-1}\right)$, $\left(\overset{-}{2}\ \overset{+}{1}\ \overset{+}{-1}\right)$, $\left(\overset{+}{1}\ \overset{-}{1}\ \overset{+}{0}\right)$, $\left(\overset{+}{2}\ \overset{+}{0}\ \overset{-}{0}\right)$, and $\left(\overset{+}{2}\ \overset{-}{2}\ \overset{+}{-2}\right)$, that it is orthogonal to the states $|^2H;\tfrac{1}{2},2\rangle$, $|^2G;\tfrac{1}{2},2\rangle$, and $|^2F;\tfrac{1}{2},2\rangle$, given by (2.68), (2.69), and (2.70), respectively. The number of coefficients in this combination is six, whereas the number of conditions (the normalization condition and the orthogonality conditions) is only four. Hence two coefficients in the combination can be chosen arbitrarily, that is, *two* linear independent combinations can be constructed, and we choose to make them orthogonal to each other. Thus we get *two* (independent) $|^2D;\frac{1}{2},2\rangle$ states, which we denote by $|^2_+D;\frac{1}{2},2\rangle$ and $|^2_-D;\frac{1}{2},2\rangle$. This construction can be done in an infinite number of ways; we give just one possibility:

$$|^2_+D;\tfrac{1}{2},2\rangle=\frac{1}{\sqrt{3}}\left(\overset{+}{2}\ \overset{+}{1}\ \overset{-}{-1}\right)+\frac{1}{\sqrt{3}}\left(\overset{+}{2}\ \overset{-}{1}\ \overset{+}{-1}\right)+\frac{1}{\sqrt{3}}\left(\overset{-}{2}\ \overset{+}{1}\ \overset{+}{-1}\right) \tag{2.71}$$

and

$$\begin{aligned}|^2_-D;\tfrac{1}{2},2\rangle=a\Bigg[&\left(\overset{+}{2}\ \overset{+}{1}\ \overset{-}{-1}\right)+\left(\overset{+}{2}\ \overset{-}{1}\ \overset{+}{-1}\right)-2\left(\overset{-}{2}\ \overset{+}{1}\ \overset{+}{-1}\right)\\&-\left(\frac{24\sqrt{3}}{6\sqrt{2}+5}\right)\left(\overset{+}{1}\ \overset{-}{1}\ \overset{+}{0}\right)-\left(\frac{12\sqrt{2}}{6\sqrt{2}+5}\right)\left(\overset{+}{2}\ \overset{+}{0}\ \overset{-}{0}\right)\\&-\left(\frac{30}{6\sqrt{2}+5}\right)\left(\overset{+}{2}\ \overset{-}{2}\ \overset{+}{-2}\right)\Bigg]\end{aligned} \tag{2.72}$$

where a is the appropriate factor that makes the state normalized.

Similarly, by operating with $\mathbf{L}^-$ upon (2.68) to (2.72), we would obtain the states $|^2H;\frac{1}{2},1\rangle$, $|^2G;\frac{1}{2},1\rangle$, $|^2F;\frac{1}{2},1\rangle$, $|^2_+D;\frac{1}{2},1\rangle$, and $|^2_-D;\frac{1}{2},1\rangle$, respectively. Also, by operating with $\mathbf{S}^-$ upon (2.66) and (2.67), we would obtain the states $|^4F;\frac{1}{2},1\rangle$ and $|^4P;\frac{1}{2},1\rangle$. (We do not give here, however, the complete expressions for these states. Then we can also construct the state $|^2P;\frac{1}{2},1\rangle$ by taking a proper normalized linear combination of the eight microstates corresponding to $(M_S=\frac{1}{2},M_L=1)$ such that it is orthogonal to the seven states mentioned above.

Finally we note that only eight microstates correspond to $(M_S=\frac{1}{2},M_L=0)$, and we already have eight states: $|^2H;\frac{1}{2},0\rangle$, $|^2G;\frac{1}{2},0\rangle$, $|^2F;\frac{1}{2},0\rangle$, $|^2_+D;\frac{1}{2},0\rangle$, $|^2_-D;\frac{1}{2},0\rangle$, $|^4F;\frac{1}{2},0\rangle$, $|^4P;\frac{1}{2},0\rangle$, and $|^2P;\frac{1}{2},0\rangle$.

Thus for the d^3 configuration we have completed the main task in the construction of states in the (SLM_SM_L) representation. Once we have these states, the remaining one can be readily obtained by successive applications of $\mathbf{L}^-$ and $\mathbf{S}^-$.

Following the procedure outlined above for the d^3 configuration, one can construct the (SLM_SM_L) states as linear combinations of determinantal microstates for any electronic configuration. Table 2.5 lists the microstates as well as the wavefunctions of the d^2 configuration, to be verified by the reader.

Table 2.5
Microstates of d^2 configuration

M_L \ M_S	1	0	−1
4		$(\overset{+}{2}\ \overset{-}{2})$	
3	$(\overset{+}{2}\ \overset{+}{1})$	$(\overset{+}{2}\ \overset{-}{1})(\overset{-}{2}\ \overset{+}{1})$	$(\overset{-}{2}\ \overset{-}{1})$
2	$(\overset{+}{2}\ \overset{+}{0})$	$(\overset{+}{2}\ \overset{-}{0})(\overset{+}{2}\ \overset{+}{0})(\overset{+}{1}\ \overset{-}{1})$	$(\overset{-}{2}\ \overset{-}{0})$
1	$(\overset{+}{2}\ \overset{+}{-1})(\overset{+}{1}\ \overset{+}{0})$	$(\overset{+}{2}\ \overset{-}{-1})(\overset{-}{2}\ \overset{+}{-1})$ $(\overset{+}{1}\ \overset{-}{0})(\overset{-}{1}\ \overset{+}{0})$	$(\overset{-}{2}\ \overset{-}{-1})(\overset{-}{1}\ \overset{-}{0})$
0	$(\overset{+}{2}\ \overset{+}{-2})$ $(\overset{+}{1}\ \overset{+}{-1})$	$(\overset{+}{2}\ \overset{-}{-2})(\overset{-}{2}\ \overset{+}{-2})$ $(\overset{+}{1}\ \overset{-}{-1})(\overset{-}{1}\ \overset{+}{-1})$ $(\overset{-}{0}\ \overset{+}{0})$	$(\overset{-}{2}\ \overset{-}{-2})$ $(\overset{-}{1}\ \overset{-}{-1})$

The principal (SLM_SM_L) states are as follows:

$$|{}^1G;0,4\rangle = (\overset{+}{2}\ \overset{-}{2})$$

$$|{}^1G;0,3\rangle = \frac{1}{\sqrt{2}}(\overset{+}{2}\ \overset{-}{1}) - \frac{1}{\sqrt{2}}(\overset{-}{2}\ \overset{+}{1})$$

$$|{}^1G;0,2\rangle = \sqrt{\tfrac{3}{14}}\,(\overset{+}{2}\ \overset{-}{0}) + \sqrt{\tfrac{8}{14}}\,(\overset{+}{1}\ \overset{-}{1}) - \sqrt{\tfrac{3}{14}}\,(\overset{-}{2}\ \overset{+}{0})$$

$$|{}^1G;0,1\rangle = \frac{1}{\sqrt{14}}(\overset{+}{2}\ \overset{-}{-1}) - \frac{1}{\sqrt{14}}(\overset{-}{2}\ \overset{+}{-1})$$
$$+\sqrt{\tfrac{6}{14}}\,(\overset{+}{1}\ \overset{-}{0}) - \sqrt{\tfrac{6}{14}}\,(\overset{-}{1}\ \overset{+}{0})$$

$$|{}^1G;0,0\rangle = \frac{1}{\sqrt{70}}(\overset{+}{2}\ \overset{-}{-2}) - \frac{1}{\sqrt{70}}(\overset{-}{2}\ \overset{+}{-2}) + \frac{4}{\sqrt{70}}(\overset{+}{1}\ \overset{-}{-1})$$
$$-\frac{4}{\sqrt{70}}(\overset{-}{1}\ \overset{+}{-1}) + \frac{6}{\sqrt{70}}(\overset{+}{0}\ \overset{-}{0})$$

$$|{}^3F;1,3\rangle = (\overset{+}{2}\ \overset{+}{1})$$

$$|{}^3F;1,2\rangle = (\overset{+}{2}\ \overset{+}{0})$$

$$|{}^3F;1,1\rangle = \sqrt{\tfrac{3}{5}}\,(\overset{+}{2}\ \overset{+}{-1}) + \sqrt{\tfrac{2}{5}}\,(\overset{+}{1}\ \overset{+}{0})$$

$$|{}^3F;1,0\rangle = \frac{1}{\sqrt{5}}(\overset{+}{2}\ \overset{+}{-2}) + \frac{2}{\sqrt{5}}(\overset{+}{1}\ \overset{+}{-1})$$

$$|^3F;0,3\rangle=\frac{1}{\sqrt{2}}\left(\overset{+}{2}\ \overset{-}{1}\right)+\frac{1}{\sqrt{2}}\left(\overset{-}{2}\ \overset{+}{1}\right)$$

$$|8u3F;0,2\rangle=\frac{1}{\sqrt{2}}\left(\overset{+}{2}\ \overset{-}{0}\right)+\frac{1}{\sqrt{2}}\left(\overset{-}{2}\ \overset{+}{0}\right)$$

$$|^3F;0,1\rangle=\sqrt{\tfrac{3}{10}}\left(\overset{+}{2}\ -\overset{-}{1}\right)+\sqrt{\tfrac{3}{10}}\left(\overset{-}{2}\ -\overset{+}{1}\right)+\frac{1}{\sqrt{5}}\left(\overset{+}{1}\ \overset{-}{0}\right)$$
$$+\frac{1}{\sqrt{5}}\left(\overset{-}{1}\ \overset{+}{0}\right)$$

$$|^3F;0,0\rangle=\frac{1}{\sqrt{10}}\left(\overset{+}{2}\ -\overset{-}{2}\right)+\frac{1}{\sqrt{10}}\left(\overset{-}{2}\ -\overset{+}{2}\right)+\frac{2}{\sqrt{10}}\left(\overset{+}{1}\ -\overset{-}{1}\right)$$
$$+\frac{2}{\sqrt{10}}\left(\overset{-}{1}\ -\overset{+}{1}\right)$$

$$|^3P;1,1\rangle=\sqrt{\tfrac{2}{5}}\left(\overset{+}{2}\ -\overset{+}{1}\right)-\sqrt{\tfrac{3}{5}}\left(\overset{+}{1}\ \overset{+}{0}\right)$$

$$|^3P;1,0\rangle=\frac{2}{\sqrt{5}}\left(\overset{+}{2}\ -\overset{+}{2}\right)-\frac{1}{\sqrt{5}}\left(\overset{+}{1}\ -\overset{+}{1}\right)$$

$$|^3P;0,1\rangle=\frac{1}{\sqrt{5}}\left(\overset{+}{2}\ -\overset{-}{1}\right)+\frac{1}{\sqrt{5}}\left(\overset{-}{2}\ -\overset{+}{1}\right)$$
$$-\sqrt{\tfrac{3}{10}}\left(\overset{+}{1}\ \overset{-}{0}\right)-\sqrt{\tfrac{3}{10}}\left(\overset{-}{1}\ \overset{+}{0}\right)$$

$$|^3P;0,0\rangle=\sqrt{\tfrac{2}{5}}\left(\overset{+}{2}\ -\overset{-}{2}\right)+\sqrt{\tfrac{2}{5}}\left(\overset{-}{2}\ -\overset{+}{2}\right)$$
$$-\frac{1}{\sqrt{10}}\left(\overset{+}{1}\ -\overset{-}{1}\right)-\frac{1}{\sqrt{10}}\left(\overset{-}{1}\ -\overset{+}{1}\right)$$

$$|^1D;0,2\rangle=\sqrt{\tfrac{2}{7}}\left(\overset{+}{2}\ \overset{-}{0}\right)-\sqrt{\tfrac{2}{7}}\left(\overset{-}{2}\ \overset{+}{0}\right)-\sqrt{\tfrac{3}{7}}\left(\overset{+}{1}\ \overset{-}{1}\right)$$

$$|^1D;0,1\rangle=\sqrt{\tfrac{3}{7}}\left(\overset{+}{2}\ -\overset{-}{1}\right)-\sqrt{\tfrac{3}{7}}\left(\overset{-}{2}\ -\overset{+}{1}\right)$$
$$-\frac{1}{\sqrt{14}}\left(\overset{+}{1}\ \overset{-}{0}\right)+\frac{1}{\sqrt{14}}\left(\overset{-}{1}\ \overset{+}{0}\right)$$

$$|^1D;0,0\rangle=\sqrt{\tfrac{2}{7}}\left(\overset{+}{2}\ -\overset{-}{2}\right)-\sqrt{\tfrac{2}{7}}\left(\overset{-}{2}\ -\overset{+}{2}\right)+\frac{1}{\sqrt{14}}\left(\overset{+}{1}\ -\overset{-}{1}\right)$$
$$-\frac{1}{\sqrt{14}}\left(\overset{-}{1}\ -\overset{+}{1}\right)-\sqrt{\tfrac{2}{7}}\left(\overset{+}{0}\ \overset{-}{0}\right)$$

Finally, we calculate for the d^2 and d^3 configurations the Wigner coefficients that are necessary to construct the states in the $(SL\,JM_J)$ representation as a linear combination of the (SLM_SML) states. The states in the two representations for configurations d^2 and d^3 are listed in Table 2.6.

Table 2.6

States in (SLM_SM_L) and $(SLJM_J)$ representations of d^2 and d^3 configurations

Configuration	States in (SLM_SM_L) representation	States in $(SLJM_J)$ representation
d^2	$\|^1S;0,0\rangle$	$\|^1S_0,0\rangle$
	$\|^3P;M_S,M_L\rangle$	$\|^3P_2,M_J\rangle,\|^3P_1,M_J\rangle,\|^3P_0,0\rangle$
	$\|^1D;0,M_L\rangle$	$\|^1D_2,M_J\rangle$
	$\|^3F;M_S,M_L\rangle$	$\|^3F_4,M_J\rangle,\|^3F_3,M_J\rangle,\|^3F_2,M_J\rangle$
	$\|^1G;0,M_L\rangle$	$\|^1G_4,M_J\rangle$
d^3	$\|^2H;M_S,M_L\rangle$	$\|^2H_{11/2},M_J\rangle,\|^2H_{9/2},M_J\rangle$
	$\|^4F;M_S,M_L\rangle$	$\|^4F_{9/2},M_J\rangle,\|^4F_{7/2},M_J\rangle,\|^4F_{5/2},M_J\rangle$ $\|^4F_{3/2},M_J\rangle$
	$\|^2G;M_S,M_L\rangle$	$\|^2G_{9/2},M_J\rangle,\|^2G_{7/2},M_J\rangle$
	$\|^2F;M_S,M_L\rangle$	$\|^2F_{7/2},M_J\rangle,\|^2F_{5/2},M_J\rangle$
	$\|^4P;M_S,M_L\rangle$	$\|^4P_{5/2},M_J\rangle,\|^4P_{3/2},M_J\rangle,\|^4P_{1/2},M_J\rangle$
	$\|^2_\pm D;M_S,M_L\rangle$	$\|^2_\pm D_{5/2},M_J\rangle,\|^2_\pm D_{3/2},M_J\rangle$
	$\|^2P;\frac{1}{2},1\rangle$	$\|^2P_{3/2},M_J\rangle,\|^2P_{1/2},M_J\rangle$

Let us assume that $j_1=S$, $j_2=L$, $j=J$, $m_1=M_S$, $m_2=M_L$, and $m=M_J$, in the general notations $|j_1j_2m_1m_2\rangle$ and $|j_1j_1jm\rangle$ introduced earlier. Then we write, for example,

$$|^4F;M_S,M_L\rangle=|\tfrac{3}{2}3\,M_S\,M_L\rangle$$

and

$$|^4F_{7/2},M_J\rangle=|\tfrac{3}{2}3\tfrac{7}{2}M_J\rangle$$

Let us construct the $(SLJM_J)$ states for the d^2 configuration. According to (2.44), we have, for example,

$$\begin{aligned}|^3P_2,0\rangle &= \sum_{m_1,m_2}\langle 1\,1\,m_1\,m_2|1\,1\,2\,0\rangle|1\,1\,m_1\,m_2\rangle\\ &=\langle 1\,1\,0\,0|1\,1\,2\,0\rangle|1\,1\,0\,0\rangle+\langle 1\,1\,1-1|1\,1\,2\,0\rangle|1\,1\,1-1\rangle\\ &\quad+\langle 1\,1-1\,1|1\,1\,2\,0\rangle|1\,1-1\,1\rangle\end{aligned}$$

Now, using Table 2.1, we get the three Wigner coefficients:

$$\langle 1\,1\,0\,0|1\,1\,2\,0\rangle=\sqrt{\frac{(1-0+1)(1+0+1)}{(2+1)(1+1)}}=\sqrt{\frac{2}{3}}$$

$$\langle 1\,1\,1-1|1\,1\,2\,0\rangle=\sqrt{\frac{(1-0)(1-0+1)}{(2+1)(2+2)}}=\frac{1}{\sqrt{6}}$$

$$\langle 1\,1-1\,1|1\,1\,2\,0\rangle=\sqrt{\frac{(1+0)(1+0+1)}{(2+1)(2+2)}}=\frac{1}{\sqrt{6}}$$

Then we have

$$|^3P_2,0\rangle=\sqrt{\tfrac{2}{3}}\ |1100\rangle+\frac{1}{\sqrt{6}}|111-1\rangle+\frac{1}{\sqrt{6}}|11-11\rangle$$
$$\equiv\sqrt{\tfrac{2}{3}}\ |^3P;0,0\rangle+\frac{1}{\sqrt{6}}|^3P;1,-1\rangle+\frac{1}{\sqrt{6}}|^3P;-1,1\rangle$$

As another example we have

$$|^1D_2,2\rangle=\sum_{m_1,m_2}\langle 02m_1m_2|0222\rangle|02m_1m_2\rangle$$
$$=\langle 0202|0222\rangle|0202\rangle$$

Now, using the general result

$$\langle 0j_20m_2|0j_2jm\rangle=\delta(jj_2)\,\delta(mm_2) \tag{2.73}$$

we have

$$|^1D_2,2\rangle=|0202\rangle\equiv|^1D;0,2\rangle$$

In this way one can work out other $(SLJM_J)$ states for the d^2 configuration by evaluating the necessary Wigner coefficients by means of Table 2.1 (when $S=1$) or by means of (2.73) (when $S=0$). We now write the results, which can easily be verified:

$$|^1S_0,0\rangle=|^1S;0,0\rangle$$

$$|^3P_2,2\rangle=|^3P;1,1\rangle$$
$$|^3P_2,1\rangle=\frac{1}{\sqrt{2}}|^3P;0,1\rangle+\frac{1}{\sqrt{2}}|^3P;1,0\rangle$$
$$|^3P_2,0\rangle=\sqrt{\tfrac{2}{3}}\ |^3P;0,0\rangle+\frac{1}{\sqrt{6}}|^3P;1,-1\rangle+\frac{1}{\sqrt{6}}|^3P;-1,1\rangle$$
$$|^3P_1,1\rangle=\frac{1}{\sqrt{2}}|^3P;1,0\rangle-\frac{1}{\sqrt{2}}|^3p;0,1\rangle$$
$$|^3P_1,0\rangle=\frac{1}{\sqrt{2}}|^3P;1,-1\rangle-\frac{1}{\sqrt{2}}|^3P;-1,1\rangle$$
$$|^3P_0,0\rangle=-\frac{1}{\sqrt{2}}|^3P;0,0\rangle+\tfrac{1}{2}|^3P;1-1\rangle+\tfrac{1}{2}|^3P;-1,1\rangle$$

$$|^1D_2,2\rangle=|^1D;0,2\rangle$$
$$|^1D_2,1\rangle=|^1D;0,1\rangle$$
$$|^1D_2,0\rangle=|^1D;0,0\rangle$$

$$|^3F_4,4\rangle = |^3F;1,3\rangle$$

$$|^3F_4,3\rangle = \tfrac{1}{2}|^3F;0,3\rangle + \frac{\sqrt{3}}{2}|^3F;1,2\rangle$$

$$|^3F_4,2\rangle = \sqrt{\tfrac{3}{7}}\,|^3F;0,2\rangle + \sqrt{\tfrac{15}{28}}\,|^3F;1,1\rangle + \frac{1}{\sqrt{28}}|^3F;-1,3\rangle$$

$$|^3F_4,1\rangle = \sqrt{\tfrac{15}{28}}\,|^3F;0,1\rangle + \sqrt{\tfrac{5}{14}}\,|^3F;1,0\rangle + \sqrt{\tfrac{3}{28}}\,|^3F;-1,2\rangle$$

$$|^3F_4,0\rangle = \sqrt{\tfrac{4}{7}}\,|^3F;0,0\rangle + \sqrt{\tfrac{3}{14}}\,|^3F;-1,1\rangle + \sqrt{\tfrac{3}{14}}\,|^3F;1,-1\rangle$$

$$|^3F_3,3\rangle = \frac{\sqrt{3}}{2}|^3F;0,3\rangle - \tfrac{1}{2}|^3F;1,2\rangle$$

$$|^3F_3,2\rangle = \frac{1}{\sqrt{3}}|^3F;0,2\rangle - \sqrt{\tfrac{5}{12}}\,|^3F;1,1\rangle + \tfrac{1}{2}|^3F;-1,3\rangle$$

$$|^3F_3,1\rangle = \frac{1}{2\sqrt{3}}|^3F;0,1\rangle - \frac{1}{\sqrt{2}}|^3F;1,0\rangle + \tfrac{1}{2}\sqrt{\tfrac{5}{3}}\,|^3F;-1,2\rangle$$

$$|^3F_3,0\rangle = \frac{1}{\sqrt{2}}|^3F;-1,1\rangle - \frac{1}{\sqrt{2}}|^3F;1,-1\rangle$$

$$|^3F_2,2\rangle = -\sqrt{\tfrac{5}{21}}\,|^3F;0,2\rangle + \frac{1}{\sqrt{21}}|^3F;1,1\rangle + \sqrt{\tfrac{5}{7}}\,|^3F;-1,3\rangle$$

$$|^3F_2,1\rangle = -2\sqrt{\tfrac{2}{21}}\,|^3F;0,1\rangle + \frac{1}{\sqrt{7}}|^3F;1,0\rangle + \sqrt{\tfrac{10}{21}}\,|^3F;-1,2\rangle$$

$$|^3F_2,0\rangle = -\sqrt{\tfrac{3}{7}}\,|^3F;0,0\rangle + \sqrt{\tfrac{2}{7}}\,|^3F;1,-1\rangle + \sqrt{\tfrac{2}{7}}\,|^3F;-1,1\rangle$$

$$|^1G_4,M_J\rangle = |^1G;0,M_J\rangle$$

We have explicitly given the $(SLJM_J)$ states only for positive values of M_J. By reversing the signs of M_S and M_L in the associated (SLM_SM_L) states, we would obtain the corresponding states with negative values of M_J.

For the d^3 configuration two values of S can appear: $S=\frac{1}{2}$ and $S=\frac{3}{2}$. Hence in the first case we would require Wigner coefficients of the form

$$\langle \tfrac{1}{2}L\,M_S\,M_L|\tfrac{1}{2}LJ\,M_J\rangle \equiv \langle L\tfrac{1}{2}\,M_L\,M_S|L\tfrac{1}{2}JM_J\rangle$$

which are tabulated in Table 2.2, and in the second case the Wigner coefficients must have the form

$$\langle \tfrac{3}{2}L\,M_S\,M_L|\tfrac{3}{2}LJ\,M_J\rangle \equiv \langle L\tfrac{3}{2}M_LM_S|L\tfrac{3}{2}J\,M_J\rangle$$

which are given in Table 2.3.

As an example, for $S=\frac{1}{2}$ let us consider the state $|^2F_{7/2},\frac{1}{2}\rangle$. We have

$$\begin{aligned}|^2F_{7/2},\tfrac{3}{2}\rangle &= \sum_{m_1,m_2} \langle \tfrac{1}{2}3\,m_1\,m_2|\tfrac{1}{2}3\tfrac{7}{2}\tfrac{3}{2}\rangle|\tfrac{1}{2}3\,m_1\,m_2\rangle \\ &= \langle \tfrac{1}{2}3\tfrac{1}{2}1|\tfrac{1}{2}3\tfrac{7}{2}\tfrac{3}{2}\rangle|\tfrac{1}{2}3\tfrac{1}{2}1\rangle \\ &\quad + \langle \tfrac{1}{2}3-\tfrac{1}{2}2|\tfrac{1}{2}3\tfrac{7}{2}\tfrac{3}{2}\rangle|\tfrac{1}{2}3-\tfrac{1}{2}2\rangle\end{aligned}$$

Now, using Table 2.2, we get the two Wigner coefficients:

$$\begin{aligned}\langle \tfrac{1}{2}3\tfrac{1}{2}1|\tfrac{1}{2}3\tfrac{7}{2}\tfrac{3}{2}\rangle &\equiv \langle 3\tfrac{1}{2}1\tfrac{1}{2}|3\tfrac{1}{2}\tfrac{7}{2}\tfrac{3}{2}\rangle \\ &= \sqrt{\frac{3+\frac{3}{2}+\frac{1}{2}}{2.3+1}} = \sqrt{\frac{5}{7}} \\ \langle \tfrac{1}{2}3-\tfrac{1}{2}2|\tfrac{1}{2}3\tfrac{7}{2}\tfrac{3}{2}\rangle &\equiv \langle 3\tfrac{1}{2}2-\tfrac{1}{2}|3\tfrac{1}{2}\tfrac{7}{2}\tfrac{3}{2}\rangle \\ &= \sqrt{\frac{3-\frac{3}{2}+\frac{1}{2}}{2.3+1}} = \sqrt{\frac{2}{7}}\end{aligned}$$

Then

$$\begin{aligned}|^2F_{7/2},\tfrac{3}{2}\rangle &= \sqrt{\tfrac{5}{7}}\;|\tfrac{1}{2}3\tfrac{1}{2}1\rangle + \sqrt{\tfrac{2}{7}}\;|\tfrac{1}{2}3-\tfrac{1}{2}2\rangle \\ &= \sqrt{\tfrac{5}{7}}\;|^2F;\tfrac{1}{2},1\rangle + \sqrt{\tfrac{2}{7}}\;|^2F;-\tfrac{1}{2},2\rangle.\end{aligned}$$

Next, as an example, for $S=\frac{3}{2}$ let us consider the state $|^4F_{5/2},\frac{3}{2}\rangle$. We have, for this case,

$$\begin{aligned}|^4F_{5/2},\tfrac{3}{2}\rangle &= \sum_{m_1,m_2} \langle \tfrac{3}{2}3\,m_1\,m_2|\tfrac{3}{2}3\tfrac{5}{2}\tfrac{3}{2}\rangle|\tfrac{3}{2}3\,m_1\,m_2\rangle \\ &= \langle \tfrac{3}{2}3\tfrac{3}{2}0|\tfrac{3}{2}3\tfrac{5}{2}\tfrac{3}{2}\rangle|\tfrac{3}{2}3\tfrac{3}{2}0\rangle \\ &\quad + \langle \tfrac{3}{2}3\tfrac{1}{2}1|\tfrac{3}{2}3\tfrac{5}{2}\tfrac{3}{2}\rangle|\tfrac{3}{2}3\tfrac{1}{2}1\rangle \\ &\quad + \langle \tfrac{3}{2}3-\tfrac{1}{2}2|\tfrac{3}{2}3\tfrac{5}{2}\tfrac{3}{2}\rangle|\tfrac{3}{2}3-\tfrac{1}{2}2\rangle \\ &\quad + \langle \tfrac{3}{2}3-\tfrac{3}{2}4|\tfrac{3}{2}3\tfrac{5}{2}\tfrac{3}{2}\rangle|\tfrac{3}{2}3-\tfrac{3}{2}4\rangle\end{aligned}$$

Now, using Table 2.3, we calculate the four Wigner coefficients, for example,

$$\begin{aligned}\langle \tfrac{3}{2}3\tfrac{3}{2}0|\tfrac{3}{2}3\tfrac{5}{2}\tfrac{3}{2}\rangle &\equiv \langle 3\tfrac{3}{2}0\tfrac{3}{2}|3\tfrac{3}{2}\tfrac{5}{2}\tfrac{3}{2}\rangle \\ &= \sqrt{\frac{3\left(3+\frac{3}{2}-\frac{1}{2}\right)\left(3-\frac{3}{2}+\frac{1}{2}\right)\left(3-\frac{3}{2}+\frac{3}{2}\right)}{(2\times3-1)(2\times3+1)(2\times3+2)}} \\ &= \frac{3}{\sqrt{35}}\end{aligned}$$

Similarly,

$$\langle \tfrac{3}{2}3\tfrac{1}{2}1|\tfrac{3}{2}3\tfrac{5}{2}\tfrac{3}{2}\rangle = -\tfrac{1}{2}\sqrt{\tfrac{7}{5}}$$

$$\langle \tfrac{3}{2}3-\tfrac{1}{2}2|\tfrac{3}{2}3\tfrac{5}{2}\tfrac{3}{2}\rangle = \frac{1}{\sqrt{14}}$$

and

$$\langle \tfrac{3}{2}3-\tfrac{3}{2}4|\tfrac{3}{2}3\tfrac{5}{2}\tfrac{3}{2}\rangle = \tfrac{3}{2}\frac{1}{\sqrt{7}}$$

Then we have

$$\begin{aligned}|^4F_{5/2},\tfrac{3}{2}\rangle &= \frac{3}{\sqrt{35}}|\tfrac{3}{2}3\tfrac{3}{2}0\rangle - \tfrac{1}{2}\sqrt{\tfrac{7}{5}}\,|\tfrac{3}{2}3\tfrac{1}{2}1\rangle \\ &\quad + \frac{1}{\sqrt{14}}|\tfrac{3}{2}3-\tfrac{1}{2}2\rangle + \tfrac{3}{2}\frac{1}{\sqrt{7}}|\tfrac{3}{2}3-\tfrac{3}{2}4\rangle \\ &= \frac{3}{\sqrt{35}}|^4F;\tfrac{3}{2},0\rangle - \tfrac{1}{2}\sqrt{\tfrac{7}{5}}\,|^4F;\tfrac{1}{2},1\rangle \\ &\quad + \frac{1}{\sqrt{14}}|^4F;-\tfrac{1}{2},2\rangle + \tfrac{3}{2}\frac{1}{\sqrt{7}}|^4F;-\tfrac{3}{2},4\rangle\end{aligned}$$

REFERENCES

1. P. A. M. Dirac, *Principles of Quantum Mechanics*, Oxford University Press, London, England, 1958.
2. L. D. Landau and E. M. Lifshitz, *Quantum Mechanics—Nonrelativistic Theory*, Pergamon Press, New York, 1958.
3. F. Seitz, *The Modern Theory of Solids*, McGraw-Hill, New York, 1940.
4. J. C. Slater, *Rev. Mod. Phys.* **35**, 484 (1963).
5. E. Clementi, *Tables of Atomic Functions*, IBM Corp, New York, 1965.
6. E. P. Wigner, *Group Theory and its Application to the Quantum Mechanics of Atomic Spectra*, Academic Press, New York, 1959.

3

Electronic Structures of Atoms and Ions

In this chapter we describe the electronic structures of atoms, making use of the various basic techniques developed in Chapter 2. We begin by considering a H-like atom where there is just one electron moving in the field of a single nucleus of charge Ze. This is the only type of atom for which it has been possible to obtain closed expressions for energies and the eigenfunctions of the stationary states.

Let us assume that the single electron moves in a field of spherical symmetry. The theory in the present form will also apply to other atoms that may be considered, to a good approximation, to have all their electrons, except one, coupled together to form an inert core. Then the last electron moves in the field of the nucleus and this inert core. The inert core is spherically symmetrical, and so the Schrödinger equation for the outer electron is given by

$$-\frac{\hbar^2}{2mr^2}\left(r\frac{\partial^2}{\partial r^2}r+\frac{1}{\sin\theta}\frac{\partial}{\partial\theta}\sin\theta\frac{\partial}{\partial\theta}+\frac{1}{\sin^2\theta}\frac{\partial^2}{\partial\phi^2}\right)\Psi+V\Psi=E\Psi \qquad (3.1)$$

where we have used the spherical polar coordinates

$$\begin{aligned} x&=r\sin\theta\cos\phi \\ y&=r\sin\theta\sin\phi \\ z&=r\cos\theta \end{aligned} \qquad (3.2)$$

Since solutions to (3.1) for a central field of force are derived in most textbooks on elementary quantum mechanics, we do not present a systematic derivation here. The reader should consult Condon and Shortley[1] for details.

The solutions of (3.1) are of the form

$$\Psi = R(r)\Theta(\theta)\Phi(\phi) \tag{3.3}$$

where $\Theta(\theta)$ and $\Phi(\phi)$ are normalized functions given by

$$\begin{aligned}\Theta_{lm} &= (-1)^m \sqrt{\frac{(2l+1)(l-m)!}{2(l+m)!}}\ P_l^m(\cos\theta), \qquad m \geqslant 0 \\ &= \sqrt{\frac{(2l+1)(l-|m|)!}{2(l+|m|)!}}\ P_l^m(\cos\theta), \qquad m \leqslant 0 \\ \Phi_m &= \frac{1}{\sqrt{2\pi}}\, e^{im\phi}, \qquad m = 0, \pm 1, \pm 2, \pm 3, \ldots\end{aligned} \tag{3.4}$$

and

$$Y_l^m(\theta, \phi) = \theta_{lm}\phi_m$$

The Y_l^m, thus defined, have their phases connected by the shift operators $\mathbf{L}^{\pm}$. This is the reason for the factor $(-1)^m$ for $m \geqslant 0$ in (3.4), which means that $\theta_{lm} = (-1)^m \theta_{l-m}$. The explicit forms of some of the θ_{lm}'s, in which we are interested are shown in Table 3.1 for all $l \leqslant 3$.

Table 3.1

The functions θ_{lm} for $l \leqslant 3$

l	θ_{lm}
0	$\theta_{00} = \frac{1}{\sqrt{2}}$
1	$\theta_{10} = \sqrt{\frac{3}{2}}\ \cos\theta;\ \theta_{1\pm1} = \mp\frac{\sqrt{3}}{2}\sin\theta$
2	$\theta_{20} = \frac{\sqrt{5}}{2\sqrt{2}}(3\cos^2\theta - 1)$
	$\theta_{2\pm1} = \mp\frac{\sqrt{15}}{2}\sin\theta\cos\theta$
	$\theta_{2\pm2} = \frac{\sqrt{15}}{4}\sin^2\theta$
3	$\theta_{30} = \frac{\sqrt{7}}{2\sqrt{2}}(2\cos^3\theta - 3\sin^2\theta\cos\theta)$
	$\theta_{3\pm1} = \mp\frac{\sqrt{21}}{4\sqrt{2}}\sin\theta(5\cos^2\theta - 1)$
	$\theta_{3\pm2} = \frac{\sqrt{105}}{4}\sin^2\theta\cos\theta$
	$\theta_{3\pm3} = \mp\frac{\sqrt{35}}{4\sqrt{2}}\sin^3\theta$

So far we have dealt with the angular parts of the wavefunction. Now the radial part $R(r)$ has to be taken care of. Inserting $\Theta(\theta)$ and $\Phi(\phi)$ into the Schrödinger equation (3.1), we have

$$\frac{1}{r^2}\frac{d}{dr}\left(r^2\frac{dR}{dr}\right)+\left[\frac{2m}{\hbar^2}(E-V)-\frac{l(l+1)}{r^2}\right]R=0 \tag{3.5}$$

for the radial function $R(r)$. Although we cannot solve (3.5) in general, for the special case of one electron in a Coulomb field ($V=-Ze^2/r$) it is possible to obtain a closed solution.

We now consider the states for which $l=0$. The simplest solution of (3.5), in this case, can be found to be

$$R_1(r)=e^{-Zr/a_0} \tag{3.6}$$

where a_0 is the Bohr radius and is 0.529×10^{-8} cm^{-1}; this sets the scale of atomic dimensions. The next simplest solution for $l=0$ is

$$R_2(r)=\left(2-\frac{Zr}{a_0}\right)e^{-Zr/2a_0} \tag{3.7}$$

This solution also decays exponentially, but twice as slowly because the index contains $r/2$ instead of r. There is also an r-dependence in the first factor of (3.7), and clearly $R_2(r)$ has a spherical nodal surface, $R(r)=0$, at $r=2a_0/Z$. The energy of the atom can be shown to be

$$E=-\frac{me^4Z^2}{2\hbar^2(l+t)^2},\qquad t=1,2,3,\ldots \tag{3.8}$$

It is customary to write $n=l+t$, where n is called the principal quantum number and l the orbital quantum number. Then the energies of the bound states are given, as a series of discrete levels, by (3.8). For a given value of n, l can take any integral value from 0 up to $(n-1)$. For each l there are $(2l+1)$ possible values for l_z and two possible values for s_z. In all, we have a degeneracy of $2n^2$ for the energy level corresponding to the principal quantum number n. This complete independence of E on l is a consequence of the particular form assumed for $V(r)$. For other central fields, however, E depends on l as well as n, as is usually assumed in atomic structure calculations, where each electron moves in a effective central field representing the Coulomb effects of the nucleus and the other electrons.

It is customary to use lower-case letters for the angular momentum operators and eigenvalues for single electrons. Thus the spatial eigenfunctions of the H-like atom may be written in the sequence

n	1	2	3	4	
l	0	0, 1	0, 1, 2	0, 1, 2, 3	...
Labels	$1s$	$2s, 2p$	$3s, 3p, 3d$	$4s, 4p, 4d, 4f$	...

of increasing energy.

We now come to the main object of our present interest, the many-electron atom. Here also we use the approximate Hamiltonian $\mathcal{H}$ given in (2.1), but we now add to $\mathcal{H}$ the so-called spin-orbit coupling term. The spin-orbit coupling is a magnetic interaction between the magnetic moment of the electron and the electric field of the nucleus. We shall describe its effects in detail in the next chapter. However, for almost all the atoms and ions that concern us in this book, the main effect of the spin-orbit coupling is to remove the degeneracy of the various levels belonging to a term. It gives rise to energies that are usually small compared to the separation between the terms, and so our eigenstates normally belong to levels of a term to a rather good approximation.

As discussed earlier, it is not possible to solve the wave equation associated with the $\mathcal{H}$ of (2.1) exactly if $n > 1$, where n is the number of electrons. Hence we have to use an approximation in order to obtain any result. It will be realized later that the broad qualitative behavior of ions in compounds depends very significantly on the nature of the ground term and of the lowest lying excited terms, that is, on the L and S values, and not as much on other features of the eigenfunctions of the ion. We now develop an approximate theory of the electrostatic energies of many-electron atoms.

We have already described one method of writing the wave-functions for an n-electron system that satisfy the antisymmetry requirement. We take an orthonormal set of one-electron functions and then build the determinantal functions from them according to (2.56). In the central field approximation we now utilize this method in a particular form. Here we assume that each electron moves in an average field due to the nucleus and all the other electrons and, furthermore, that this average field is spherically symmetric. Hence its eigenfunctions will be of the one-electron type already considered, meaning that the eigenfunctions are products of a radial function and an angular function taken to be one of the $Y_l^m(\theta,\phi)$ of (3.4). Before proceeding further, we emphasize that the nature of our assumption is that we consider the appropriate basic one-electron functions as if each electron moved independently in a spherically symmetric field. It is unnecessary to find a detailed justification for this average field because the field does not appear in our calculations and we calculate the approximate energies using the Hamiltonian (2.1). However, it is important to remember that the average central fields do appear in certain specific methods, for example, in the SCF methods for obtaining the radial functions, as we shall see later in this chapter.

First we obtain approximate electrostatic energies and the eigenkets of the terms and then, in the next chapter, consider the spin-orbit coupling as a perturbation on these. In the central field approximation we search for eigenfunctions that are finite sums of determinants of one-electron functions. We also require that our one-electron functions be solutions of a central field problem. This means that the wavefunction can be written as

$$\Psi = \sum c_a \sum_{\lambda} (n!)^{-1/2}(-1)^{\lambda} P_{\lambda}\{\phi_1^a(1)\phi_2^a(2)\cdots\phi_n^a(n)\} \tag{3.9}$$

where each ϕ_i^a is of the form

$$\phi_i^a(j) = R_{n_i l_i}(j) Y_{l^i}^{m^i}(\theta,\phi)(j)\tau_{s^i}(j) \tag{3.10}$$

where the τ_s^i are the spin functions and $s^i = \pm\frac{1}{2}$, corresponding to $m_{s^i} = \pm\frac{1}{2}$ along Oz. $R(j)$ is the radial function and is assumed to be real; P_λ is the permutation operator that permutes the electrons, that is, the arguments in (3.9) and not the functions. Clearly in a determinant only $2(2l+1)$ one-electron functions with a given n and l can occur; otherwise the determinant goes to zero. For a fixed n and l, R_{nl} is, of course, the same for all m and s in (3.10). We shall discuss its computation in a later section, but for the present we assume it to be known for all relevant nl values.

We now specify the determinants in (3.9) by giving the nl values occurring in each term. If we specify the number of times each nl value occurs and then take all the determinants for these numbers, we say that we have a configuration of electrons and write it as

$$(n_1l_1)^{a_1}(n_2l_2)^{a_2}\cdots(n_rl_r)^{a_r}$$

where a_i is an integer giving the number of electrons with quantum numbers n_il_i. Thus we write $1s^22s^22p^1$, for example, as a configuration. Any sum (3.9) over such determinants is said to belong to a configuration. Since all the operators **S**, **L**, and **J** clearly turn a function of a configuration into another of the same configuration, we can break up any configuration into terms and levels.

Next we determine which terms occur for a given configuration. Let us consider the configuration $3d4d$, where the first d electron has $n=3$ and the second has $n=4$. By the vector coupling procedure,[1] the total L value is given by $L=l+l'$ to $|l-l'|$, in steps of unity, and in this case L takes the values 4, 3, 2, 1, and 0. The total S value, similarly, is given by s_1+s_2 to $|s_1-s_2|$, in steps of unity, and in this case $s=1$ and 0. Thus this configuration contains the following terms:

$$^1S,\ ^3S;\quad ^1P,\ ^3P;\quad ^1D,\ ^3D;\quad ^1F,\ ^3F;\quad ^1G,\ ^3G$$

This procedure is quite general, and we can always obtain the terms of any configuration consisting of two or more inequivalent electrons by applying the so-called vector coupling procedure to both orbital and spin functions. In practice, the allowed terms of a general configuration are found by adding one electron at a time. Now, what happens when the electrons are equivalent? The matter is no longer as simple, since the Pauli exclusion principle now eliminates certain terms.[1] A convenient method of obtaining the allowed terms is to classify the determinantal functions by their M_S, M_L values. Let us take the configuration $3d^2$ as an example. The highest possible M_L value is

$M_L=4$, and this can be uniquely represented by

$$|3,2,\tfrac{1}{2},2;\ 3,2,-\tfrac{1}{2},2\rangle$$

The M_S value is obviously $M_S=0$. Thus with $M_L=4$ only the ^1G term occurs and not ^3G. Next we consider $M_L=3$. There are four independent functions now. These are obtained by giving the four possible pairs of values to m_s and $m_{s'}$ in $|3,2,m_s,2;\ 3,2,m_{s'},1\rangle$. The four M_S values are 1, 0, 0, and -1; therefore we have one triplet and one singlet with a $M_L=3$ component. The singlet is obviously the ^1G found above, and the triplet must be ^3F. In this way we can find all the allowed terms of $3d^2$. The determinantal functions with given M_S, M_L values for the configurations $3d^2$ and $3d^3$ are listed in Tables 3.2*a* and 3.2*b*. It can be seen from these tables that we have written down only the part of the M_S, M_L table for which $M_S \geqslant 0$ and $M_L \geqslant 0$. We also notice from Table 3.2*b* that there are two ^2D terms in a d^3 configuration.

The allowed terms for all configurations of equivalent p or d electrons can be easily obtained by following the procedure described above and are given in Condon and Shortley,[1] but for the sake of completeness we list them in Table 3.3.

Table 3.2*a*

Part of M_S, M_L table for $3d^2$

M_L \ M_S	$+1$	0	-1	Term
4	0	1	0	^1G
3	1	2	1	^3F
2	1	3	1	^1D
1	2	4	2	^3P
0	2	5	2	^1S

Table 3.2*b*

Part of M_S, M_L table for $3d^3$

M_L \ M_S	$\frac{3}{2}$	$\frac{1}{2}$	Terms
5	0	1	^2H
4	0	2	^2G
3	1	4	$^4F, {}^2F$
2	1	6	$^2D, {}^2D$
1	2	8	$^4P, {}^2P$
0	2	8	—

Table 3.3

Allowed terms for p and d configurations[1]

Configurations	Singlets	Doublets	Triplets	Quartets	Quintets	Sextets
p^1,p^5	—	P	—	—	—	—
p^2,p^4	SD	—	P	—	—	—
p^3	—	PD	—	S	—	—
d^1,d^9	—	D	—	—	—	—
d^2,d^8	SDG	—	PF	—	—	—
d^3,d^7	—	$PDDFGH$	—	PF	—	—
d^4,d^6	$SSDDFGGI$	—	$PPDFFGH$	—	D	—
d^5	—	$SPDDD$ $FFGGHI$	—	$PDFG$	—	S

When we have a configuration that has the maximum number of electrons with given nl, we say that the configuration contains a closed shell of electrons. A closed shell configuration contains only the ^1S term, and the addition of any number of closed shells to a configuration makes no difference to the M_S, M_L table. Hence the allowed terms of a configuration consisting of closed shells and partly filled shells are the same as those of the corresponding configuration obtained by omitting the closed shells. For example, the ground configuration of the Cr^{3+} ion is $1s^2,2s^22p^6,3s^23p^6,3d^3$, and therefore its allowed terms are the same as those for $3d^3$, that is, ^2H, ^2G, ^2F, ^2D, ^2D, ^2P, ^4P, and ^4F.

Having found the allowed terms of the d^n configurations, we now have to determine the first-order electrostatic energies of these terms. To obtain the energies we need to diagonalize the matrix of the Hamiltonian $\mathcal{H}$ given by (2.1) within the configuration. Since $\mathcal{H}$ commutes with $\mathbf{L}^2$ and $\mathbf{S}^2$, there will only be nondiagonal elements between terms of the same type. Thus, when d^n is expressed in the form of LS terms, $\mathcal{H}$ is already diagonal for d^1, d^9, d^2, d^8, and d^{10} but is not completely so for the other configurations. To remove the degeneracy we have to solve a secular equation to find the electrostatic energies; for example, to find the individual energies of *two* ^2D terms of d^3 we have to solve a quadratic equation, and to find the *three* ^2D terms of d^5 we have to solve a cubic equation. By using the diagonal sum rule[1] and the Slater-Condon parameters we can find the energies of all the terms of the d^n configurations. We now do this, of course very briefly, since they can be found in detail in Condon and Shortley.[1]

3.1 SLATER-CONDON PARAMETERS

To derive the term energies we have to evaluate the matrix elements of (2.51) between the functions Ψ_j defined in (2.56) in the determinantal form. For a given configuration we write them as Ψ_j, say, which is given by

$$\Psi_j = \sum_{\mu} (n!)^{-1/2}(-1)^{\mu} P_{\mu}\{\phi_1^j(1)\phi_2^j(2)\cdots\phi_n^j(n)\} \tag{3.11}$$

where P_μ is the permutation operator. We now write the Hamiltonian $\mathcal{H}$ in the form

$$\mathcal{H} = \sum_{\kappa=1}^{n} U(\kappa) + \sum_{\kappa<\lambda}^{n} V(\kappa\lambda) \tag{3.12}$$

where the first term on the right-hand side is a one-electron operator, given by

$$U(\kappa) = \frac{1}{2m}\nabla_\kappa^2 - \frac{Ze^2}{\mathbf{r}_\kappa} \tag{3.13}$$

and the second term is

$$V(\kappa\lambda) = \frac{e^2}{r_{\kappa\lambda}} \tag{3.14}$$

We now find the matrix elements of $\mathcal{H}$ between the states Ψ_i and Ψ_j which are of the form (3.11). We have two cases. If $\Psi_i = \Psi_j$, then, because of orthonormality, only terms with the ϕ^i in the same order as the ϕ^j will be non-vanishing. Thus, after some manipulation, we get

$$U_{ii} = \sum_{\kappa=1}^{n} \langle \phi_\kappa | U(\kappa) | \phi_\kappa \rangle \tag{3.15}$$

where the κth term of (3.15) is the contribution to U_{ii} from the κth orbital and not from the κth electron.

If Ψ_i differs from Ψ_j in just one function $\phi_\lambda^i, \phi_\lambda^j$, then also U_{ij} will have at most one nonvanishing term, namely,

$$U_{ij} = (-1)^{\mu} \langle \phi_\lambda^i | U | \phi_{\lambda'}^j \rangle \tag{3.16}$$

where μ is the number of permutations needed to put the functions of Ψ_i in the same order as those of Ψ_j, and therefore we omit the parameter of integration. As we have already remarked, U_{ij} is zero if Ψ_i differs from Ψ_j in more than one constituent one-electron function.

Similarly, the two-electron matrix elements V_{ii} and V_{ij} are given by

$$V_{ii} = \sum_{\kappa<\lambda}^{n} [C(\kappa,\lambda) - J(\kappa,\lambda)] \tag{3.17}$$

and

$$V_{ij}=(-1)^{\mu}\sum_{\lambda\neq\kappa}\left[\langle\phi_\kappa^i\phi_\lambda|V|\phi_\kappa^j\phi_\lambda\rangle-\langle\phi_\lambda\phi_\kappa^i|V|\phi_\kappa^j\phi_\lambda\rangle\right] \tag{3.18}$$

where we omit the superscripts i and j on ϕ_λ because $\phi_\lambda^i=\phi_\lambda^j$ for $\lambda\neq\kappa$. It may be argued, as we did with U_{ij}, that when Ψ_i and Ψ_j belong to the same configuration, ϕ_κ^i and ϕ_κ^j have the same nl values. Equation 3.18 represents the case where Ψ_i differs from Ψ_j in just one function, $\phi_\kappa^i, \phi_\kappa^j$, say. When Ψ_i and Ψ_j differ in two functions, $\phi_\kappa^i, \phi_\lambda^i$ as against $\phi_\kappa^j\phi_\lambda^j$, say, we have

$$V_{ij}=(-1)^{\mu}\left[\langle\phi_\kappa^i\phi_\lambda^i|V|\phi_\kappa^j\phi_\lambda^j\rangle-\langle\phi_\lambda^i\phi_\kappa^i|V|\phi_\kappa^j\phi_\lambda^j\rangle\right] \tag{3.19}$$

In (3.17) to (3.19) the integrals C and J are known as the Coulomb and the exchange integrals.

We now have to evaluate the two-electron matrix elements occurring in (3.17), (3.18), and (3.19), in order to obtain the energies of the terms belonging to a configuration. It is evident that they are all of the form

$$\langle\phi_i\phi_j|V|\phi_m\phi_l\rangle \tag{3.20}$$

It can be shown that V can be expanded as a series of products of one-electron operators and is given by

$$V=\frac{e^2}{r_>}\sum_{k=0}^{\infty}\frac{r_<^k}{r_>^k}P_k(\cos\omega) \tag{3.21}$$

where $P_k(\cos\omega)$ are the Legendre polynomials.[1] By the well-known spherical harmonic addition theorem, it can be shown that

$$P_l(\cos\omega)=\frac{4\pi}{2l+1}\sum_{m=-l}^{+l}Y_l^m(1)\bar{Y}_l^m(2) \tag{3.22}$$

We now use

$$\phi_i(j)=R_{n^il^i}(j)\,Y_{l^i}^{m_i}(j)\,\tau_{s_i}(j) \tag{3.23}$$

for the ϕ and results (3.21) and (3.22) to evaluate (3.20); thus

$$\begin{aligned}\langle\phi_i\phi_j|V|\phi_m\phi_l\rangle&=\left\langle\phi_i\phi_j\left|\sum_{k=0}^{\infty}\frac{e^2r_<^k}{r_>^{k+1}}P_k(\cos\omega)\right|\phi_m\phi_l\right\rangle\\&=\sum_{k=0}^{\infty}\left\langle R_{n^il^i}R_{n^jl^j}\left|\frac{e^2r_<^k}{r_>^{k+1}}\right|R_{n^ml^m}R_{n^tl^t}\right\rangle\\&\quad\cdot\langle Y_{l^ip^i}Y_{l^jp^j}|P_k(\cos\omega)|Y_{l^mp^m}Y_{l^tp^t}\rangle\delta(s^i,s^m)\delta(s^js^t)\\&=\sum_{k=0}^{\infty}R^k(n^il^i,n^jl^j,n^ml^m,n^tl^t)A_k\,\delta(s^i,s^m)\delta(s^j,s^t)\end{aligned} \tag{3.24}$$

where A_k involves functions of angle only and can be simplified to give

$$A_k = c^k(l^ip^i, l^mp^m)c^k(l^tp^t, l^jp^j)\delta(p^i+p^j, p^m+p^t) \tag{3.25}$$

where

$$c^k(lp, l'p') = \left(\frac{2}{2k+1}\right)^{1/2}\langle \theta_{lp}\theta_{k,p-p'}\theta_{l'p'}\rangle \tag{3.26}$$

From (3.26) it follows that

$$c^k(lp, l'p') = (-1)^{p-p'}c^k(l'p', lp)$$

The c^k are, of course, just numbers. A general formula for the c^k has been worked out by Gaunt,[2] but it is very complicated. Hence in practice one works out the c^k once and for all and then uses tables for them. We list the c^k values for s, p, and d electrons in Table 3.4.[1] They are zero unless k, l^i, l^m satisfy the following conditions:

$$\begin{gathered} k + l^i + l^m \text{ even} \\ |l^i - l^m| \leqslant k \leqslant l^i + l^m \end{gathered} \tag{3.27}$$

From (3.24) we have

$$\begin{aligned} \langle \phi_i\phi_j|V|\phi_m\phi_t\rangle = {} & \delta(s^is^m)\delta(s^js^t)\delta(p^i+p^j, p^m+p^t) \\ & \cdot \sum_{k=0}^{\infty} R^k(n^il^i, n^jl^j, n^ml^m, n^tp^t)c^k(l^ip^i, l^mp^m)c^k(l^t, p^t, l^jp^j) \end{aligned} \tag{3.28}$$

Table 3.4

Values of $c^k(l^ip^i, l^mp^m)^a$

l^i	l^m	p^i	p^m	c^0	$5c^2$	
s	s	0	0	1	0	
p	p	± 1	± 1	1	-1	
		± 1	0	0	$+\sqrt{3}$	
		0	0	1	$+2$	
		± 1	∓ 1	0	$-\sqrt{6}$	
		p^i	p^m	c^0	$7c^2$	$21c^4$
d	d	± 2	± 2	1	-2	$+1$
		± 2	± 1	0	$+\sqrt{6}$	$-\sqrt{5}$
		± 2	0	0	-2	$+\sqrt{15}$
		± 1	± 1	1	$+1$	-4
		± 1	0	0	$+1$	$+\sqrt{30}$
		0	0	1	$+2$	$+6$
		± 2	∓ 2	0	0	$+\sqrt{70}$
		± 2	∓ 1	0	0	$-\sqrt{35}$
		± 1	∓ 1	0	$-\sqrt{6}$	$-\sqrt{40}$

[a]See Ref. 1.

where the sum is, in fact, finite because of (3.27).

Corresponding to C and J introduced in (3.17) to (3.19), we have

$$C(\phi_i,\phi_j)=\sum_{k=0}^{\infty} a^k(l^ip^i,l^jp^j)F^k(n^il^i,n^jp^j)$$
$$J(\phi_i,\phi_j)=\delta(s^i,s^j)\sum_{k=0}^{\infty} b^k(l^ip^i,l^jp^j)G^k(n^il^i,n^jp^j) \quad (3.29)$$

where

$$a^k(l^ip^i,l^jp^j)=c^k(l^ip^i,l^ip^i)c^k(l^jp^j,l^jp^j)$$
$$b^k(l^ip^i,l^jp^j)=\left[c^k(l^ip^i,l^jp^j)\right]^2 \quad (3.30)$$

When dealing with a configuration of equivalent electrons, $n^i=n^j$ and $l^i=l^j$, and we have $F^k=G^k$. Thus, we need use only the F^k's. Since a^k and b^k are fractions, it is convenient to define new parameters:

$$F_k=\frac{1}{d_k^2}F^k \quad (3.31)$$

where d_k is the common denominator occurring in Table 3.4 for a given set of c^k's. Thus for p electrons we write

$$F_0=F^0$$
$$F_2=\tfrac{1}{25}F^2 \quad (3.32)$$

and for d electrons

$$F_0=F^0$$
$$F_2=\tfrac{1}{49}F^2 \quad (3.32a)$$
$$F_4=\tfrac{1}{441}F^4$$

We have now reached our goal: we can express the energies of the terms of any configuration as functions of the Slater-Condon parameters F_k (or F^k). For most configurations there are usually considerably more terms than parameters. Hence one can either calculate the R^k (see Section 3.2) with some assumed central field or can obtain them empirically by fitting the observed spectral data.

For the d^n configurations we now simply write the electrostatic energies in terms of the Racah parameters A, B, and C. These are related to the

Slater-Condon integrals (also known as parameters) F_k as follows:

$$\begin{aligned} A &= F_0 - 49F_4 \\ B &= F_2 - 5F_4 \\ C &= 35F_4 \end{aligned} \tag{3.33}$$

where $F_k = F^k / D_k$ (for d electrons, $D_0 = 1$, $D_2 = 49$, and $D_4 = 441$). The F^k are defined by

$$F^k(nl;n'l') = e^2 \int_0^\infty \int_0^\infty \frac{r_<^k}{r_>^{k+1}} \left[R_{nl}(1) R_{n'l'}(2) \right]^2 r_1^2 r_2^2 \, dr_1 \, dr_2 \tag{3.34}$$

where R_{nl} is the radial part of the wavefunction, which in the case of an SCF function can be written as a linear combination of Slater-type orbitals (STO):

$$R_{nl}(r) = \sum_{i=1}^{m} c_i r^{n_i} e^{-\zeta_i r} \tag{3.35}$$

The Racah parameters of the second ($4d^n$) and third ($5d^n$) series of the transition metal ions have been calculated by De, Desai, and Chakravarty[3] and are given, along with those of the $3d^n$ series, in Appendix III. For the $4d^n$ configurations the HFSCF wavefunctions of Watson[4] and of Richardson, Blackman, and Ranochak[5] are used, whereas for $5d^n$ configurations Burn's[6] wavefunctions are used since no SCF wavefunctions are available.

Since the Racah parameters are used extensively, the Coulomb and the exchange integrals for the d electrons expressed in terms of them are given in Table 3.5. The Coulomb and the exchange integrals are defined as

$$\begin{aligned} C(a,b) &= \left\langle a(1)b(2) \left| \frac{1}{r_{12}} \right| a(1)b(2) \right\rangle \\ J(a,b) &= \left\langle a(1)b(2) \left| \frac{1}{r_{12}} \right| b(1)a(2) \right\rangle \end{aligned} \tag{3.36}$$

Table 3.5

Coulomb and exchange integrals between d electrons expressed in terms of the Racah parameters

m_l values		$C(a,b)$	$J(a,b)$
± 2	± 2	$A+4B+2C$	$A+4B+2C$
± 2	∓ 2	$A+4B+2C$	$2C$
± 2	± 1	$A-2B+C$	$6B+C$
± 2	∓ 1	$A-2B+C$	C
± 2	0	$A-4B+C$	$4B+C$
± 1	± 1	$A+B+2C$	$A+B+2C$
± 1	∓ 1	$A+B+2C$	$6B+2C$
± 1	0	$A+2B+C$	$B+C$
0	0	$A+4B+3C$	$A+4B+3C$

Using Tables 3.5 and 3.2*a*, we can find the energies of the terms, for example, of the d^2 and d^3 configurations. Thus the first-order electrostatic energies for d^2 are

$$\left.\begin{aligned} E(^1S) &= A+14B+7C \\ E(^1D) &= A-3B+2C \\ E(^1G) &= A+4B+2c \\ E(^3P) &= A+7B \\ E(^3F) &= A-8B \end{aligned}\right\} \qquad (3.37)$$

and, similarly, for the d^3 configuration,

$$\left.\begin{aligned} E(^4F) &= 3A-15B \\ E(^4P) &= 3A \\ E(^2H) &= 3A-6B+3C = E(^2P) \\ E(^2G) &= 3A-11B+3C \\ E(^2F) &= 3A+9B+3C \\ E(^2D) &= (3A+5B+5C) \pm (193B^2+8BC+4C^2)^{1/2} \end{aligned}\right\} \qquad (3.38)$$

The Racah parameters *A*, *B*, and *C* are positive, and therefore we can predict that the term ^3F of the d^2 configuration and the term ^4F of the d^3 configuration should lie lowest, in accordance with Hund's rule. It is possible by somewhat similar methods to determine the first-order energies of the remaining terms of the configurations. For the d^4 configuration we mention only the energies of the quintet and triplet terms, and for the d^5 configuration only the energies of the sextet and quartet terms. They are, for the d^4 configuration,

$$\left.\begin{aligned} E(^5D) &= 6A-12B \\ E(^3H) &= 6A-17B+4C \\ E(^3G) &= 6A-12B+4C \\ E(^3F) &= 6A-5B+\tfrac{11}{2}C \pm (68B^2+4BC+C^2)^{1/2} \\ E(^3D) &= 6A-5B+4C \end{aligned}\right\} \qquad (3.39)$$

Similarly, for the d^5 configuration,

$$\left.\begin{aligned} E(^6S) &= 10A-35B \\ E(^4G) &= 10A-25B+5C \\ E(^4F) &= 10A-13B+7C \\ E(^4D) &= 10A-18B+5C \\ E(^4P) &= 10A-28B+7C \end{aligned}\right\} \qquad (3.40)$$

The states of highest spin are very important for our purpose because they include the ground term; and since the separation between the terms is usually very large compared to kT, the unit of thermal energy, the ground term is normally the only one that is appreciably populated. These states have been given here from one of Racah's famous papers,[7] and for future details the reader should consult these publications. In regard to the energies of the terms of d^{10-n} configurations ($n \leqslant 5$), one should remember that they are the same as those of d^n configurations for $n \leqslant 5$. The electron configurations and the ground state terms for ions of interest in magnetism are listed in Appendix II.

Without going into details, we have practically covered the basic results in the theory of electrostatic energies of ions in the d^n configuration. We have derived these energies by first-order perturbation theory and have removed the accidental degeneracy that may be present (in some cases) in a configuration if the electrostatic interaction is neglected altogether. But in removing the accidental degeneracy we have always remained within the configuration.

We can make a further correction to these energies, however, by taking account of the configuration interaction. In this method one simply takes account of the matrix elements between the corresponding terms of a finite number of neighboring configurations and diagonalizes $\mathcal{H}$ within them. Moreover, if it is assumed that our determinantal functions form a complete set, then by taking enough configurations we can approach as closely as we wish the exact eigenfunctions of $\mathcal{H}$. But this method has its own limitations. Restricting ourselves to one configuration, we have a smaller number of parameters, the F_k's, which we determine by making a best fit with the experimental energies of the atomic configurations. But more configurations mean more parameters, so that one has to calculate a large number of parameters, if not all, or restrict the calculation considering only the interactions between a very few configurations. Such calculations have been done by Hopgood and his co-workers[8] but are not, unfortunately, very useful in applications to ions in compounds of interest to us. Empirical fitting is therefore the only alternative way to obtain these parameters from experiments than to derive these completely theoretically.

For the $3d^n$ series a considerable collection of the magnitudes of B and C has been obtained by fitting the experimental data. These, along with values for the $4d^n$ and $5d^n$ series, are given in Appendix III.

3.2 RADIAL FUNCTIONS AND THE HARTREE-FOCK PROCEDURE

Theoretically, the parameters describing the electrostatic interactions in atoms can be obtained by taking suitable radial functions R_{nl} for the one-electron orbitals and then working the F_k out directly from their definitions by using (3.34). This method therefore requires a prior knowledge of the radial functions. By using the Hartree-Fock self-consistent field (HFSCF) method we can find the most accurate radial functions R_{nl}. This HFSCF method furnishes, by an iterative procedure, the best approximate solutions to a simple type of Schrödinger equation.

We now elaborate this method. We have to find the determinantal function that is the best approximate solution. To obtain the best single-determinantal solutions of the Schrödinger equation

$$\mathcal{H}\Psi=\left[\sum_{i=1}^{n}U(i)+\sum_{i<j}^{n}V(ij)\right]\Psi=E\Psi \tag{3.41}$$

where

$$\Psi=(n!)^{-1/2}\Sigma(-1)^{\mu}P_{\mu}\{\phi_1(1)\phi_2(2)\cdots\phi_n(n)\} \tag{3.42}$$

it has been shown by Hartree[8] that we can find Ψ by minimizing

$$E=\langle\Psi|\mathcal{H}|\Psi\rangle \tag{3.43}$$

subject to

$$\langle\phi_i\phi_j\rangle=\delta_{ij}$$

Then, because the real or the imaginary parts of ϕ_i can be varied separately, we can show that

$$\begin{aligned}0&=\langle\delta\Psi|\mathcal{H}|\Psi\rangle\\&=\sum_i\langle\delta\phi_i|U(i)|\phi_i\rangle+\sum_{i\neq j}\langle\delta\phi_i\phi_j|V(ij)|\phi_i\phi_j\rangle-\sum_{i\neq j}\langle\delta\phi_i\phi_j|V(ij)|\phi_j\phi_i\rangle\end{aligned}$$

$$\langle\delta\bar{\phi}_i\phi_j\rangle=0 \tag{3.44}$$

By using the Lagrange method of undetermined multipliers in (3.44), we can show that

$$U|\phi_i\rangle+\sum_{j\neq i}\langle\phi_j|V|\phi_j\rangle|\phi_i\rangle-\sum_{j\neq i}\langle\phi_j|V|\phi_i\rangle|\phi_j\rangle=\sum_j\epsilon_{ij}|\phi_j\rangle \tag{3.45}$$

Equations (3.45) are known as the Hartree-Fock equations and can be solved by an iterative procedure to give the best possible solutions of (3.41) in the form (3.42). Now, in practice, two simplifications can be made. The coefficients of $|\phi_i\rangle$ and $|\phi_j\rangle$ in (3.45) are not necessarily spherically symmetrical. This lack of spherical symmetry does not make a great difference to the accuracy of the solutions, however, and therefore at each stage of iteration one averages these coefficients over the angular space. The other simplification is to neglect the exchange term in (3.45), in which case $\epsilon_{ij}=0$ for $i\neq j$, and we have the Hartree equations

$$U|\phi_i\rangle+\sum_{j\neq i}\left\langle\phi_j\left|\frac{e^2}{r_{ij}}\right|\phi_j\right\rangle|\phi_i\rangle=\epsilon_{ii}|\phi_i\rangle \tag{3.46}$$

In (3.46) we simply solve the one-electron problem for ϕ_i in the field of the nucleus and the average field due to all the other electrons. The energies and the eigenfunctions determined by (3.46) do not appear to differ greatly from those determined by (3.45). Both methods give the best possible absolute energies for atoms that differ from the experimental ones by about 0.5 eV per electron. Relative energies within and between configurations are usually considerably better.

Shortly we shall discuss the method of the self-consistent field which enables us to find the best possible radial wavefunctions for many-electron atoms. However, for many ordinary purposes it is sufficient to use the Slater-type orbitals (STOs) with the parameters ζ_i chosen according to the well-known Slater rules. These Slater orbitals are analytic functions resembling the exact solutions of the H-atom problem, but specified by parameters that depend on the nuclear charge (Z), the screening by other electrons in the atom (σ), and the effective principal quantum number (n^*). Slater's nodeless radial functions are of the form

$$R_{nl}(r) = Nr^{n^*-1}e^{-\zeta r/a_0} \tag{3.47}$$

where

$$\zeta = \frac{Z-\sigma}{n^*}, \qquad a_0 = \frac{\hbar^2}{me^2} \tag{3.48}$$

Equations 3.47 are the solutions of the one-electron problem with a potential energy

$$V(r) = -\frac{(Z-\sigma)e^2}{r} + \frac{n^*(n^*-1)\hbar^2}{2mr^2} \tag{3.49}$$

Slater gave the following rules for determining n^* and σ: $n^* = n$ for $n = 1,2,3$, but $n^* = 3.7$, 4.0, 4.2, respectively, for $n = 4,5,6$; Z and σ can be determined by dividing the electrons into groups:

$1s$; $2s, 2p$; $3s, 3p$; $3d$; $4s, 4p$; $4d$; $4f$; $5s, 5p$; $5d$

and taking contributions to σ for a given orbital according to the following prescription:

1. None from an orbital in a group outside the one considered.
2. 0.35 from each other occupied orbital in the group (0.30 for $1s$)
3. 1.00 from each orbital inside a d or f one. For s or p ones, 0.85 for each with n less by 1 and 1.00 for each orbital further in.

For example, let us take the Mn^{2+} ion, which has the configuration

$$1s^2,\quad 2s^22p^6,\quad 3s^23p^6,\quad 3d^5$$

with ^6S as the ground state. The radial orbitals (unnormalized) determined by Slater's rules given above are as follows:

$$\begin{aligned} \phi_{1s} &= e^{-24.7r/a_0}, & \phi_{2s} &= re^{-10.43r/a_0} \\ \phi_{2p} &= re^{-10.43r/a_0}, & \phi_{3s} &= r^2e^{-4.58r/a_0} \\ \phi_{3p} &= r^2e^{-4.58r/a_0}, & \phi_{3d} &= r^2e^{-1.87r/a_0} \end{aligned} \tag{3.50}$$

For a good discussion of Slater's functions for the transition metal ions, the reader should consult Brown.[11] For $1s$, $2s$, $2p$, $3s$, and $3p$ functions, the Slater orbitals do not provide too poor a description of atoms, although they appear to be somewhat better for the inner shells than for the valence shells. For negative ions and for the $3d$ valence shell of the transition metal ions, however, the Slater orbitals seem to be decidedly inadequate.

Let us now consider the Hartree-Fock self-consistent field (HFSCF) method, which is the best method for finding the radial wavefunctions of many-electron atoms. Within the framework of a configurational wavefunction, which means using wavefunctions composed of one-electron orbitals, the HFSCF provides us with the best possible set of radial functions for the orbitals. The phrase "best possible" is used in the sense of the variation theorem, and we seek the set of radial functions that minimizes the energy of the wavefunction for the configuration. Although this can strictly be done only if each function is expressed in infinite parametric form, it has been found that a linear combination of just a few STOs is quite adequate to reproduce energies within 0.001% of the true "Hartree-Fock minimum." We therefore try to obtain radial functions of the form

$$R_{nl}(r) = \sum_i c_i \,\mathrm{STO}(n_i \zeta_i) \tag{3.51}$$

Thus the problem of finding the best possible functions is reduced to finding the linear parameters c_i and the exponential parameters ζ_i for each orbital. The exponents ζ_i and the "principal quantum numbers" n_i are nonlinear parameters and are usually not varied or are varied only after the minimum energy for a particular set has been determined, in order to see whether a better set can be found. The technique of solving for the best c_i for a fixed basis set is due to Roothaan,[12, 13] and we describe it now.

In the framework of the self-consistent field theory, therefore, the energy obtained converges to the Hartree-Fock limit for sufficiently large and well-chosen basis sets. This theory presumably then gives the lowest possible energy for the single-configuration, one-electron orbital wavefunction. We shall, in the following, deal with the Roothaan-Hartree-Fock (RHF) equations very briefly and without proof. Details are given in Refs. 12 and 13. Let us begin with the closed shells.

With each orbital, ψ_i, described as a linear combination of basis functions ϕ_p, given by

$$\psi_i = \sum_{p=1}^{m} c_{ip}\phi_p \tag{3.52}$$

the ground state of the system is written as a single Slater determinant:

$$\Psi = |\psi_1(1)\bar{\psi}_1(2)\psi_2(3)\bar{\psi}_2(4)\cdots\psi_n(2n-1)\bar{\psi}_n(2n)| \tag{3.53}$$

Now, using the well-known Hamiltonian

$$\mathcal{H} = \sum_{i=1}^{2n} H(i) + \sum_{i=1}^{2n-1} \sum_{\substack{j=i+1 \\ i<j}}^{2n} \frac{e^2}{r_{ij}} \tag{3.54}$$

one can write the expression for the total energy of the system (without proof) as

$$E = 2\sum_{i=1}^{n} H_i + \sum_{i=1}^{n}\sum_{j=1}^{n}(2C_{ij} - J_{ij}) \tag{3.55}$$

where

$$\begin{aligned} H_i &= \langle\psi_i(1)|H(1)|\psi_i(1)\rangle \\ C_{ij} &= \left\langle \psi_i(1)\psi_j(2) \left| \frac{e^2}{r_{12}} \right| \psi_i(1)\psi_j(2) \right\rangle \\ J_{ij} &= \left\langle \psi_i(1)\psi_j(2) \left| \frac{e^2}{r_{12}} \right| \psi_j(1)\psi_i(2) \right\rangle \end{aligned} \tag{3.56}$$

Then Roothaan has shown that the ψ_i's are the solutions to the pseudo-eigenvalue equations

$$F\psi_i = \epsilon_i\psi_i \tag{3.57}$$

where F, known as the "Fock operator," is given by

$$F = H + \sum_{j=1}^{n}(2C_j - J_j) \tag{3.58}$$

and

$$\begin{aligned}\langle\psi_i|C_j|\psi_i\rangle &= C_{ij}\\ \langle\psi_i|J_j|\psi_i\rangle &= J_{ij}\end{aligned} \tag{3.59}$$

The orbital energies are then given by

$$\epsilon_i = H_i + \sum_{j=1}^{n} (2C_{ij} - J_{ij}) \tag{3.60}$$

Therefore

$$E = \sum_{i=1}^{n} (H_i + \epsilon_i) \tag{3.61}$$

Equation (3.57) is a pseudo-eigenvalue equation in the sense that both F and ψ_i are determined by the coefficients c_{ip}. The Fock operator F can be found only by trial and error by achieving self-consistency. We shall encounter this operator again in Section 11.7. The process of determining F is essentially iterative, meaning that initially an arbitrary set is chosen and F determined, and then this is used to calculate a set of orbitals and the orbital energies. These new orbitals are then used to find a new F and so on till convergence is achieved. There is no guarantee, however, that this process will converge; in fact, divergences have been found, though rarely. Since well-developed procedures are available,[13, 14] this does not pose any real problem. We shall not proceed with this aspect any further but instead proceed to the open shell equation, again very briefly.

It can easily be imagined that the Roothaan-Hartree-Fock equations for an open shell are much more complicated than those for a closed shell; therefore we shall not attempt to derive them as before. The conceptual physical difficulty arises from the different exchange potentials as seen by the up-spin and down-spin electrons. Mathematically this difficulty manifests itself as off-diagonal Lagrange multipliers which are very hard to eliminate.

One might think that this difficulty could be avoided by simply defining two sets of orbitals, one for the up-spin electrons and the other for the down-spins, but the trouble with this approach, which is known as the unrestricted Hartree-Fock (UHF) approach, is that the resulting Slater determinant is not an eigenfunction of $\mathbf{S}^2$. Although we can, in principle, resolve this difficulty by means of the projection operators, generalization of this method is again very difficult. In the RHF approach we can write the basic determinant of highest possible spin as

$$\begin{aligned}\Psi = |\psi_1^\alpha(1)\psi_1^\beta(2)\psi_2^\alpha(3)\psi_2^\beta(4)\cdots\psi_{n_c}^\alpha(2n_c-1)\psi_{n_c}^\beta(2n_c)\\ \cdot\psi_{n_c+1}^\alpha(2n_c+1)\cdots\psi_{n_c+n_0}^\alpha(2n_c+n_0)|\end{aligned} \tag{3.62}$$

It is evident from (3.62) that the first n_c orbitals are doubly occupied and thus form the closed shell part of the system. The next open shell orbitals, n_0, all contain electrons with α (up) spin only. This determinant is an eigenfunction of $\mathbf{S}^2$, with $S=M_S=\frac{1}{2}n_0$.

The energy of an arbitrary open shell determinant, following Roothaan, is given by

$$E=\left[2\sum_{k=1}^{n_c} H_k+\sum_{k=1}^{n_c}\sum_{l=1}^{n_c}(2C_{kl}-J_{kl})\right]+\left[\sum_{k=1}^{n_c}\sum_{m=n_c+1}^{n_0}(2C_{km}-J_{km})\right]$$
$$+\left[\sum_{m=n_c+1}^{n_0} H_m+\sum_{m=n_c+1}^{n_0}\sum_{n=n_c+1}^{n_0}(C-J'_{mn})\right] \tag{3.63}$$

where subscripts k and l refer to the closed shells, and m and n to the open shells. The first part of (3.63) represents the energy of the closed shell, and the third part the energy of the open shell; the second part gives the interaction between the closed and the open shells. Note that the determinant need not have α spin for all electrons in the open shell, but $J'_{mn}=0$ when m and n contain electrons with opposite spins.

For a large number of systems, Roothaan shows that it is possible to rewrite (3.63) in terms of the fractional occupancy of the open shell:

$$E=2\sum_{k=1}^{n_c} H_k+\sum_{k=1}^{n_c}\sum_{l=1}^{n_c}(2C_{kl}-J_{kl})+2\sum_{k=1}^{n_c}\sum_{m=1}^{n_c}(2C_{km}-J_{km})$$
$$+f\left[2\sum_{m=1}^{n_s} H_m+f\sum_{m=1}^{n_s}\sum_{n=1}^{n_s}(2aC_{mn}-bJ_{mn})\right] \tag{3.64}$$

where f is the fractional occupancy $(0<f<1)$ of the open shell, and a and b are constants to be determined. Now the sums over m and n extend, not over the number of electrons in the open shell, but rather over the number of degenerate orbitals (see Section 6.4) belonging to an irreducible representation. Thus $n_s=3$ for the t_{2g} shell and $n_s=2$ for the e_g shell in octahedral symmetry, and $n_s=1$ for any spatially nondegenerate shell. For the t_{2g}^4 configuration, for example, $f=\frac{2}{3}$, and for e_g^2, $f=\frac{3}{4}$. The constants a and b depend on the specific state of a given configuration, and $J_{mn}\neq 0$.

We prefer not to go into any more details of the calculation of the Fock operators for the closed and open shells here, but we shall again return to this subject when we deal with the molecular orbital theory in Chapter 11. Perhaps, one or two remarks would not be out of place at this point. The Fock operators take a particularly simple form if the irreducible representation to which the orbitals of the open shell belong is different from all the irreducible representations to which the closed shell orbitals belong. Typical examples are the atomic ions with open $2p$ or open $3d$ shells, but not ions with open $2s$, $3p$, or $4d$ shells. In such cases the orthogonality of the closed and open shells is automatic.

Although very few molecular calculations have actually been done using the Roothaan open shell equations, HFSCF wavefunctions for the transition metal ions that have been obtained by Watson,[4,9] Richardson and his co-workers,[5] and also Clementi[10] are currently available. At present these SCF wavefunctions are being used[3] very extensively.

REFERENCES

1. E. U. Condon and G. H. Shortley, *The Theory of Atomic Spectra*, Cambridge University Press, London, England, 1953.
2. J. A. Gaunt, *Phil. Trans. A*, **228**, 151.
3. Ibha De, V. P. Desai, and A. S. Chakravarty, *Ind. J. Phys.* **48**, 1133 (1974).
4. R. E. Watson, private communication (1971).
5. J. W. Richardson, M. J. Blackman, and J. E. Ranochak, *J. Chem. Phys.* **58**, 3010 (1973).
6. G. Burns, *J. Chem. Phys.* **41**, 1521 (1964).
7. G. Racah, *Phys. Rev.* **61**, 186; *ibid.* **62**, 438 (1942).
8. D. R. Hartree, *The Calculation of Atomic Structures*, John Wiley and Sons, New York, 1957.
9. R. E. Watson, *Phys. Rev.* **117**, 742 (1960).
10. E. Clementi, *Tables of Atomic Functions*, IBM Corp., New York, 1965.
11. D. A. Brown, *J. Chem. Phys.* **28**, 67 (1958).
12. C. C. J. Roothaan, *Rev. Mod. Phys.* **23**, 69 (1951); *ibid.* **32**, 179 (1960).
13. C. C. J. Roothaan and P. S. Bagus, *Methods in Computational Physics*, B. Alder (Ed.), Academic Press, New York, 1963.
14. R. McWeeny, *Proc. Roy. Soc. (London) A*, **235**, 490 (1956).

4

Magnetic Effects in Atomic Structure

4.1 SPIN-ORBIT COUPLING

In Chapter 3 we used the concept of electron spin. We have seen that the spin of the electron influences the energies and the allowed eigenstates, but there were no terms in the Hamiltonian showing an explicit dependence on it. In a more exact theory the spin-orbit interaction term has to be included in the Hamiltonian, though its effect is often small. In the present chapter we deal with spin-orbit coupling and see that it arises physically because the electron possesses a magnetic moment. There are two different ways of introducing the spin and the magnetic moment of the electron into the theory. The first one is the phenomenological way, where recourse is made to experiment to show, first, that the electron has two independent internal states (the orbital and the spin) and, second, that it has a magnetic moment. The other way is to proceed via Dirac's linear relativistic equation for the electron and to deduce both these properties from that equation. The latter approach is more fundamental but more difficult mathematically. We shall restrict ourselves to the first alternative and describe the specific magnetic effects in atomic structure from a rather empirical basis.

Experimentally the electron has a magnetic moment of magnitude $e\hbar/2mc$; and since the electron has a negative charge, it is natural to expect that, if this moment arises from rotation of the negative charge, the magnetic moment will lie in a direction directly opposed to the vector representing the rotation. Thus the magnetic moment $\boldsymbol{\mu}$ will be given by $-(e/mc)\mathbf{s}$. This is in agreement with experiment and also with the Dirac theory of the electron.

The actual value of the spin-orbit coupling constant cannot be obtained by purely classical arguments, however. Solution of the Dirac equation for the relativistic electron in a central field adds to the nonrelativistic Hamiltonian a term that can be interpreted as an interaction between the spin of the electron and its orbital motion. This is the spin-orbit interaction. The added term to the Hamiltonian is

$$\mathcal{H}_{so} = -\frac{e}{2m^2c^2}\frac{1}{r}\frac{dV(r)}{dr}\mathbf{l}\cdot\mathbf{s} \tag{4.1}$$

where $V(r)$ is the spherically symmetric potential for the electron. Classically this interaction may be viewed as the interaction of the magnetic moment of an electron spin with the magnetic field induced by the motion of the nucleus around the electron. The nucleus is seen from the coordinate system fixed on the electron.

For a nucleus of charge $Ze, V(r)$ is equal to Ze/r when no other electrons are present. Substituting this in (4.1), we can write the spin-orbit interaction as

$$\begin{aligned}\mathcal{H}_{so} &= \frac{Ze^2}{2m^2c^2}\frac{1}{r^3}\mathbf{l}\cdot\mathbf{s} \\ &= \zeta(r)\mathbf{l}\cdot\mathbf{s}\end{aligned} \tag{4.2}$$

Our discussion so far has referred to only one electron in a central field. For n electrons we use the same formula for each one individually and then sum over the electrons. We therefore have, for n electrons,

$$\mathcal{H}_{so} = \sum_{i=1}^{n} \zeta(r_i)\mathbf{l}_i\cdot\mathbf{s}_i \tag{4.3}$$

This, then, represents the spin-orbit interaction for an n-electron system. We now want to determine certain selection rules for $\mathcal{H}_{so}$. It can be shown by quantum mechanics that the total angular momentum vector $\mathbf{J}$ commutes rigorously with (4.3). However, since $\mathbf{L}$ and $\mathbf{S}$ are usually good quantum members, we would like to find out what selection rules are satisfied by $\mathcal{H}_{so}$ in a Russell-Saunders coupling scheme. The best way to do this is to take a SLM_SM_L scheme and expand an arbitrary matrix element of $\mathcal{H}_{so}$ as a sum of the products of the matrix elements of $\zeta(r_i)\mathbf{l}_i$ and $\mathbf{s}_i$ separately. Thus

$$\begin{aligned}&\langle\alpha''SLM_SM_L|\mathcal{H}_{so}|\alpha'S'L'M_S'M_L'\rangle \\ &\qquad = \sum_{i=1}^{n}\sum_{\alpha'''}\langle\alpha''SLM_SM_L|\zeta(r_i)\mathbf{l}_i|\alpha'''SL'M_SM_L'\rangle \\ &\qquad\quad \cdot\langle\alpha'''SL'M_SM_L'|\mathbf{s}_i|\alpha'S'L'M_S'M_L'\rangle\end{aligned} \tag{4.4}$$

Using the well-known selection rules for vectors with respect to the angular momenta, we can show that the required selection rules are

$$\begin{aligned}&\Delta L = 0, \pm 1, \qquad \Delta S = 0, \pm 1 \\ &\Delta J = 0, \qquad\quad\; \Delta M = 0\end{aligned} \tag{4.5}$$

for the spin-orbit coupling matrix elements in the LS coupling scheme.

In the form of $\mathcal{H}_{so}$ given in (4.3) we have not included the mutual magnetic interactions between the spin magnetic moments of different electrons and also those arising out of the magnetic moments associated with the orbital motions of the electrons. Such terms are represented by $\mathbf{l}_i \cdot \mathbf{l}_j$, $\mathbf{l}_i \cdot \mathbf{s}_j$, and $\mathbf{s}_i \cdot \mathbf{s}_j (i \neq j)$. These terms are almost always much smaller than the spin-orbit coupling energy in the atoms of interest to us.

We now show that, within a single Russell-Saunders term, the matrix elements of $\mathcal{H}_{so}$ assume a simple form. In this case we have, from (4.4),

$$\begin{aligned}\langle \alpha' S L M_S M_L | \mathcal{H}_{so} | \alpha' S L M_S' M_L' \rangle \\ = \sum_{i=1}^{n} \sum_{\alpha''} \langle \alpha' S L M_S M_L | \zeta(r_i) \mathbf{l}_i | \alpha'' S L M_S M_L' \rangle \\ \cdot \langle \alpha'' S L M_S M_L' | \mathbf{s}_i | \alpha' S L M_S' M_L' \rangle \end{aligned} \tag{4.6}$$

We notice in (4.6) that the first of the two matrix elements in any one of the terms of (4.6) is independent of M_S and the second of M_L. Our matrix elements of $\mathcal{H}_{so}$ then become

$$\begin{aligned}\sum_{i=1}^{n} \sum_{\alpha''} \gamma_L(\alpha', \alpha'', S L) \gamma_S(\alpha'', \alpha', S L) \cdot \langle \alpha' S L M_S M_L | \mathbf{L} | \alpha' S L M_S M_L' \rangle \\ \cdot \langle \alpha' S L M_S M_L' | \mathbf{S} | \alpha' S L M_S' M_L' \rangle = \lambda \langle \alpha' S L M_S M_L | \mathbf{L} \cdot \mathbf{S} | \alpha' S L M_S' M_L' \rangle\end{aligned} \tag{4.7}$$

where

$$\lambda = \lambda(\alpha', \alpha'', S L) = \sum_{i=1}^{n} \sum_{\alpha''} \gamma_L(\alpha', \alpha'', S L) \gamma_S(\alpha'', \alpha', S L)$$

is just a number independent of M_S, M_L, M_S', M_L'. We now transform back to the *SLJM* scheme. For this we expand a typical ket, $|\alpha' S L J M\rangle$, in that scheme in terms of the kets of the SLM_SM_L scheme. Then we have

$$|\alpha' S L J M\rangle = \sum |\alpha' S L M_S M_L\rangle \langle \alpha' S L M_S M_L | \alpha' S L J M\rangle$$

Now, using (4.6) and (4.7) and then transforming back to the *SLJM* scheme, we get

$$\begin{aligned}\langle \alpha' S L J M | \mathcal{H}_{so} | \alpha' S L J M \rangle = \lambda \langle \alpha' S L J M | \mathbf{L} \cdot \mathbf{S} | \alpha' S L J M \rangle \\ = \tfrac{1}{2} \lambda [J(J+1) - L(L+1) - S(S+1)]\end{aligned} \tag{4.8}$$

since

$$\mathbf{L} \cdot \mathbf{S} = \tfrac{1}{2}(\mathbf{J}^2 - \mathbf{L}^2 - \mathbf{S}^2)$$

It should be noted here that λ is a constant that depends only on the particular term we have chosen, and (4.8) gives us the complete matrix of $\mathcal{H}_{so}$ within a Russell-Saunders term. $\mathcal{H}_{so}$ is diagonal in J and M. The difference in energy between the level with J and the level with $(J-1)$ is seen to be λJ. This is known as the Landé interval rule, which states that the separation between two adjacent levels of a term is proportional to the higher J value.

We thus see that, within a given LS term, the matrix elements of $\mathcal{H}_{so}$ are proportional to those of $\mathbf{L}\cdot\mathbf{S}$. We also notice that the matrix elements in (4.7) are nonvanishing only when

$$M_L + M_S = M_L' + M_S' = M_J \tag{4.9}$$

This is, of course, true for the diagonal terms, but we also notice that $\mathbf{L}\cdot\mathbf{S}$ is able to couple terms together, provided that they have the same value of J. However, if one is interested in calculating the off-diagonal terms in $\mathbf{L}$ and $\mathbf{S}$, it is necessary to use the operator $\Sigma_i \zeta(r_i)\mathbf{l}_i\cdot\mathbf{s}_i$. Usually we ignore the admixtures of other L and S terms and consider only the diagonal terms.

We have seen that, if Russell-Saunders coupling holds in an atom, the diagonal matrix elements of $\mathcal{H}_{so}$ are proportional to those of $\mathbf{L}\cdot\mathbf{S}$. The constant of proportionality λ determines the splitting of a multiplet. Condon and Shortley[1] have shown that the diagonal matrix elements of the operator $\Sigma_i \zeta(r_i)\mathbf{l}_i\cdot\mathbf{s}_i$ are equal to those of $\lambda\mathbf{L}\cdot\mathbf{S}$, where, for the ground state of an ion,

$$\lambda = \pm \frac{\zeta_{nl}}{2S} \tag{4.10}$$

The plus and minus signs hold for ions with less than half-filled and more than half-filled shells, respectively. If LS coupling is a good approximation, the Landé interval rule holds and λ can then be determined directly from the experimental splitting of the ground multiplet. However, it was shown by Blume and Watson[2] that the residual spin-other orbit terms also have diagonal elements proportional to those of $\mathbf{L}\cdot\mathbf{S}$. These terms accordingly contribute to λ, although not to ζ_{nl}. We shall not go into these details here; the reader should refer to Blume and Watson.

Let us emphasize here that, in calculating the energies of levels in a configuration, we diagonalize the electrostatic energy first and then apply the spin-orbit energy as a perturbation. This is possible only when the spin-orbit energy is small compared to the electrostatic energy. This condition holds for $3d^n$ transition metal ions but not for second $(4d^n)$ and third $(5d^n)$ transition metal ions, where the spin-orbit interaction energy exceeds the energy of the electrostatic interaction. In such cases one has to diagonalize the spin-orbit energy first and then consider the electrostatic energy by applying the first-order perturbation theory. The latter procedure is called the jj-coupling scheme, and configurations for which one can obtain a reasonable agreement with experiment by using the procedure are said to obey the jj coupling. It should be noted, however, that, if we diagonalize both the electrostatic energy and the spin-orbit energy within a configuration, we obtain the same result regardless of which of the first-order functions we start from. In this case we say that we are making calculations in the intermediate coupling scheme. We shall discuss this coupling scheme in Section 10.3, in connection with compounds of the palladium and platinum groups.

In practice, however, it involves too much labor to diagonalize both the electrostatic and the spin-orbit energy even within one configuration. Hence we often resort to second-order corrections to the energy due to the spin-orbit interaction. As an example, let us take the Fe^{2+} ion. Its configuration is $3d^5$, outside the closed shells, and the ground term is ^6S. This term, of course, cannot be split by spin-orbit coupling to any order. However, it is possible for off-diagonal matrix elements of the spin-orbit coupling energy to mix into the ground state one or more of the excited levels of d^5. Therefore the ground state will not be exactly ^6S but will be given by

$$\psi = a_0\psi(^6S) + \sum_i a_i\phi_i \tag{4.11}$$

where a_0, a_i are numbers, and the ϕ_i are the excited levels. We use the perturbation theory to determine a_0, a_i, and ϕ_i.

We apply the selection rules to determine, in the LS scheme, which levels of d^5 can have nonvanishing matrix elements with ^6S. Since $\Delta S = 0, \pm 1$ and $\Delta L = 0, \pm 1$, the only possible state is ^4P. Finally, $\Delta J = 0$, which means that $^4P(J = \frac{5}{2})$ is the only level of d^5 that has a nonvanishing matrix element with ^6S. Then we have

$$\langle ^6S, M_S 0 | \mathcal{H}_{so} | ^4P_{5/2} M_J \rangle = \mu \delta_{M_S M_J} \tag{4.12}$$

where μ has to be determined. The energy difference between ^6S and ^4P for the d^5 configuration was given by (3.40) and is

$$\begin{aligned} E &= E(^4P_{5/2}) - E(^6S) \\ &= 10A - 28B + 7C - 10A + 35B \\ &= 7B + 7C \end{aligned} \tag{4.13}$$

since

$$\lambda(^4P) = \lambda(^6S) = 0$$

Calculating the matrix element given in (4.12) in a straightforward way, we can show that $\mu = \sqrt{5}\,\zeta$. Therefore the lowest level of d^5 is given approximately by

$$\psi = \psi(^6S, M0) - \frac{\sqrt{5}\,\zeta}{7(B + C)}\psi(^4P_{5/2}M) \tag{4.14}$$

The parameter ζ can be obtained by fitting the observed spectra with the theoretical formulas. These values for the transition metal ions are given in Appendix IV, where some indication of the methods of estimation is provided, together with references. It is also instructive to plot these values of ζ against the atomic numbers, and this has been done for the transition metal ions of interest in Figure 4.1. We see that the parameters for the palladium group are 2.5 to 3 times as large as those for the iron group, and those for the platinum group are between 7 and 10 times as large. It can be seen from the plot[5] that the general variation in all three series is very similar. It is also evident that in the second and third series the spin-orbit coupling should be taken into account very carefully in fitting the theoretical formulas to the available experimental data in order to get the energies.

Blume and his co-workers[2, 3] have performed the HFSCF calculations of the spin-orbit interactions in the transition metal ions, and these agree with

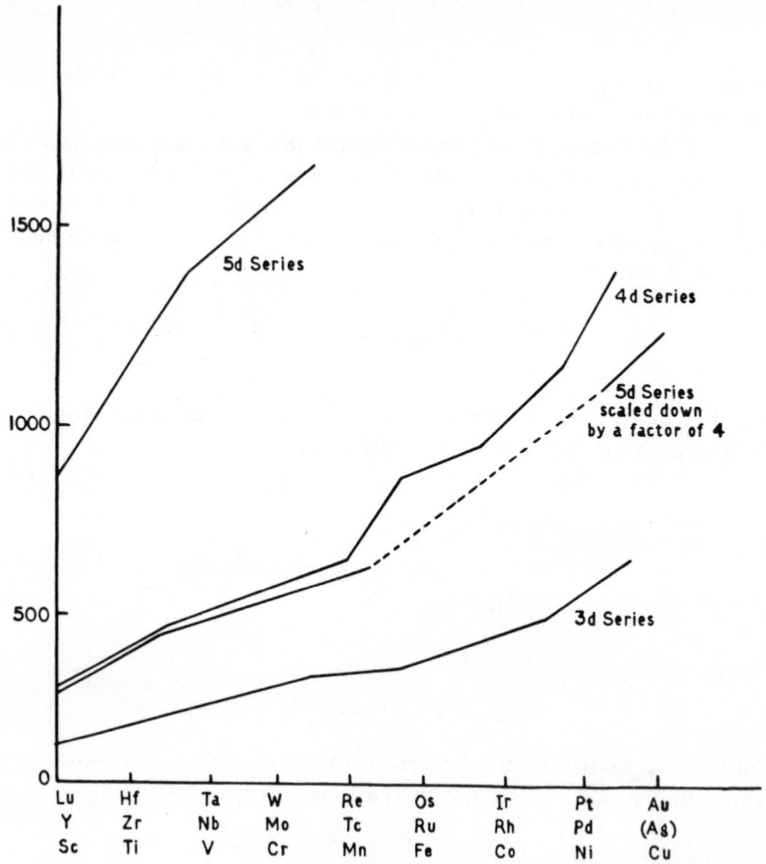

Fig. 4.1 The spin-orbit coupling constant ζ for the d electrons in neutral atoms. (Taken from J. S. Griffith.[5])

the experimental results reported by Dunn.[4] From the calculations and the experimental observations, the spin-orbit parameters for the three different series of transition metal ions are found to be in the ratio

$$\zeta_{3d}:\zeta_{4d}:\zeta_{5d}=1:2.5:5.0 \tag{4.15}$$

This is true, however, only where the diagonal matrix elements of $\mathcal{H}_{so}$ are concerned. For the nondiagonal matrix elements, which tend to break down the *LS* coupling by mixing the different terms of a configuration, we have the intermediate coupling scheme, and the eigenstates are no longer the *LS* states but contain mixtures of L and S giving the same total J and M_J, however, since the spin-orbit interaction commutes with J^2 and J_Z.

To treat intermediate coupling, it is most convenient to write the matrix elements of the electrostatic and the spin-orbit interactions between different terms of a configuration in the $(LSJM_J)$ representation. This simplifies the matrix to be diagonalized since $\mathcal{H}_{so}$ is diagonal in the (JM_J) representation, whereas in the (M_LM_S) representation it connects many M_LM_S states that add up to a given total M_J. The electrostatic term energies $\epsilon(LS)$, which are diagonal in the (LSM_LM_S) representation, are still diagonal in the $(LSJM_J)$ representation as they depend on L and S only. In the $(LSJM_J)$ representation it can be shown that the matrix element is given by[2, 3]

$$\langle L'S'JM_J|\mathcal{H}_{so}|LSJM_J\rangle=\sum_{\substack{M_L,M_S,\\ M_L',M_S'}}\zeta\begin{bmatrix} L' & S' & J\\ M_L' & M_S' & M_J\end{bmatrix}\begin{bmatrix} L & S & J\\ M_L & M_S & M_J\end{bmatrix}\cdot\sum_i\langle L'S'M_L'M_S'|\mathbf{l}_i\cdot\mathbf{s}_i|LSM_LM_S\rangle \tag{4.16}$$

where the square brackets denote the Clebsch-Gordan coefficients.

4.2 ATOMS IN EXTERNAL MAGNETIC FIELDS

We now discuss an atom in a constant external magnetic field **H**, described by the vector potential $\mathbf{A}=\frac{1}{2}(\mathbf{H}\times\mathbf{r})$. We shall neglect small effects such as the nuclear hyperfine structure in this section. Then the main modifications to our nonrelativistic Hamiltonian (2.1) are the inclusion of the spin-orbit coupling energy and the energy of orientation of the spin magnetic moment in the external magnetic field. Next, in order of importance, would be the magnetic interactions between the orbital and the spin magnetic moments of pairs of electrons. In the absence of the external field, the modified Hamiltonian is given by

$$\mathcal{H}_0=\sum_{i=1}^{n}\left[\frac{1}{2m}\mathbf{p}_i^2-\frac{Ze^2}{r_i}+\zeta(r_i)\mathbf{l}_i\cdot\mathbf{s}_i\right]+\sum_{i<j}\frac{e^2}{r_{ij}} \tag{4.17}$$

In the presence of a magnetic field, (4.17) becomes

$$\mathcal{H}_0+\mathcal{H}_1=\mathcal{H}_0+\sum_{i=1}^{n}\left[\frac{1}{2m}\left(\mathbf{p}_i+\frac{e}{c}\mathbf{A}_i\right)^2-\frac{1}{2m}\mathbf{p}_i^2\right.$$
$$\left.+\frac{e}{mc}\mathbf{S}_i\cdot\mathbf{H}+\frac{e}{c}\zeta(r_i)\mathbf{r}_i\times\mathbf{A}_i\cdot\mathbf{S}_i\right] \tag{4.18}$$

The last term of (4.18) is quite negligible compared to the other terms depending on **H**; the ratio between them can be shown to be of the order of $2.5\times10^{-5}n^{-2}$ for an electron in an nl orbital of hydrogen. With this observation, remembering that $\operatorname{div}\mathbf{A}=0$, we obtain our expression for $\mathcal{H}_1$:

$$\mathcal{H}_1=\sum_{i=1}^{n}\left[\frac{e}{mc}\mathbf{A}_i\cdot\mathbf{p}_i+\frac{e^2}{2mc^2}\mathbf{A}_i^2+\frac{e}{mc}\mathbf{H}\cdot\mathbf{S}_i\right] \tag{4.19}$$

Now, using $\mathbf{A}=\frac{1}{2}\mathbf{H}\times\mathbf{r}$ and the Bohr magneton, $\mu_{\mathrm{B}}=e\hbar/2mc$, we find that

$$\mathcal{H}_1=\sum_{i=1}^{n}\left[\frac{\mathbf{e}}{2mc}\mathbf{H}\times\mathbf{r}_i\cdot\mathbf{p}_i+\frac{e^2}{8mc^2}|\mathbf{H}\times\mathbf{r}_i|^2+\frac{e}{mc}\mathbf{H}\cdot\mathbf{S}_i\right]$$
$$=\sum_i\mu_{\mathrm{B}}(\mathbf{l}_i+2\mathbf{s}_i)\cdot\mathbf{H}+\frac{e^2}{8mc^2}\sum_i(x_i^2+y_i^2)\cdot\mathbf{H}^2 \tag{4.20}$$

The first term of (4.20) is called the paramagnetic part of $\mathcal{H}_1$ and is zero for atoms in ^1S states. The second term, which represents the diamagnetic part, is very small and, for atoms not in ^1S states, is negligible compared to the paramagnetic part. Since we are not interested in diamagnetism, we shall not deal with the second term of (4.20).

We now concentrate on the paramagnetic part of (4.20). Each component J of a fine structure multiplet is $(2J+1)$-fold degenerate, and the degeneracy is lifted by the application of the magnetic field. The energy operator associated with the Zeeman effect is

$$\mathcal{H}_m=\sum_i\mu_{\mathrm{B}}(\mathbf{l}_i+2\mathbf{s}_i)\cdot\mathbf{H}$$
$$=\mu_{\mathrm{B}}(\mathbf{L}+2\mathbf{S})\cdot\mathbf{H} \tag{4.21}$$

where **L** and **S** are measured in units of $\hbar$. Here μ_{B} is the Bohr magneton, $e\hbar/2mc$, **H** is the applied magnetic field, and the factor 2 before **s** is due to the magnetic anomaly of the spin. In weak magnetic fields we regard H_z (magnetic field in the z-direction) as a perturbation that is small (energies of the order of 1 cm^{-1}) compared to the separation between the levels of a term but nevertheless makes the main contribution to the energy, given by

$$E_m=\langle\psi|\,\mu_{\mathrm{B}}\mathbf{H}(2\mathbf{J}-\mathbf{L})|\psi\rangle \tag{4.22}$$

Now, assuming **H** parallel to the z-axis, we have

$$E_m = 2\mu_B HM - \mu_B H\langle\psi|\mathbf{L}_z|\psi\rangle \tag{4.23}$$

for a state $\psi = |\alpha' SLJM\rangle$. Since $\mathbf{L}_z$ commutes with J_z, $\langle\psi|\mathbf{L}_z|\psi\rangle$ is diagonal with respect to J_z (but not with respect to J). By the replacement theorem[5] the matrix element $\langle\psi|\mathbf{L}_z|\psi\rangle$ is proportional to M; its actual value can be derived by using Wigner's formula[1] and is

$$\langle\psi|\mathbf{L}_z|\psi\rangle = \frac{M[J(J+1)+L(L+1)-S(S+1)]}{2J(J+1)} \tag{4.24}$$

By substituting (4.24) in (4.23), we get, finally,

$$E_m = g\mu_B HM \tag{4.25}$$

where g is the Landé factor, after Landé, who discovered the formula empirically before quantum mechanics was able to predict it, and is given by

$$g = 1 + \frac{J(J+1)+S(S+1)-L(L+1)}{2J(J+1)} \tag{4.26}$$

It should be noted that in deriving (4.25) we neglected the matrix elements of $\mathbf{L}_z$ between states of different J. In other words, we supposed the matrix elements to be small compared to the multiplet splittings between the levels. This condition is satisfied in practice for most atoms, even for macroscopically strong magnetic fields.

4.2.1 Paramagnetism Due to Ions of the Transition Elements

A paramagnetic magnetization is in the same direction as the applied field and arises from a partial alignment of the permanent atomic or electronic dipole moments. When the spins of the electrons within an ion are coupled together, as in closed shells, the ground state becomes an ^1S state and diamagnetism always results. When the grouping of the electrons is incomplete within the ions or within the chemical bonds, however, or when the grouping is easily broken by the applied magnetic field, paramagnetism appears. For the transition metal ions the paramagnetism arises because of the incomplete d shells, whereas for the rare earths it is due to the $4f$ electrons. Electrons in the complete d or f shells have nonvanishing angular momentum and consequently magnetic moments.

The transition elements appear in groups in the periodic table (see Appendix I), each group having a different shell incompletely filled. First there are the iron group transition elements, from titanium to nickel. Free atoms of these elements have incomplete $3d$ shells. Outside these $3d$ electrons there are, in the neutral free atom, two $4s$ electrons. The next group is the rare earth or the lanthanide group, whose elements have incomplete $4f$ shells surrounded in a neutral atom by seven or eight electrons in the $n=5$ shell. Then there are three more transition groups, typified by palladium, platinum, and actinium; these have incomplete $4d$, $5d$, and $5f$ (or perhaps $6d$) shells, respectively.

To arrive at the paramagnetism of the independent ions let us consider only the orienting influence of the applied magnetic field on an individual ion containing an incomplete $3d$ shell. To derive the expression for the paramagnetic susceptibility we need to know the unperturbed energy levels and the nature of the thermal equilibrium distribution of the ions over these states. Each state of a free ion is characterized by a quantum number $\mathbf{J}$, which measures the magnitude of its total angular momentum. It has been already seen that $J(=L+S)$ must be an integer for an ion containing an even number of electrons and a half-integer for an odd number.

For a given value of J, there are $(2J+1)$ such states, each characterized by a different value of the magnetic quantum number M_J, which may take values from $-J$ to $+J$, increasing in steps of unity. These states differ in that, when a magnetic field is applied, they have different components of the permanent ionic dipole moment in the field direction. This component, μ_H, is thus determined by M_J and is given by

$$\mu_H = g\mu_B M_J \tag{4.27}$$

where μ_B is the Bohr magneton, and g is the ion's g-factor, which is fixed by L, S, and J according to the Landé formula (4.26). In addition to the internal energy, which is determined by the value of J, an ion in a magnetic field $\mathbf{H}$ therefore has a magnetic potential energy, given by

$$E_H = -g\mu_B M_J H \tag{4.28}$$

The energy levels are thus split by a magnetic field as a consequence of (4.28). The magnetic potential energy E_H derived above was obtained by neglecting the fact that the electronic charge cloud of the ion becomes slightly distorted by the applied magnetic field, which adds small, higher order terms to the magnetic energy. This gives rise to diamagnetism, which is usually (not always) much smaller than the paramagnetism with which we are concerned here. Distortion of the charge cloud by the magnetic field also gives rise to polarization, which normally is unimportant. This aspect will be treated in the next section.

We now distribute the ions over the M_J states, and for this we use Maxwell-Boltzmann statistics since the electrons of the ions are localized and the ions are in thermal equilibrium. The probability P_i of finding a given ion in the state i, which depends on the energy of the state E_i, is given by

$$P_i = Ae^{-E_i/kT} \tag{4.29}$$

where T is the absolute temperature, k the Boltzmann constant, and A a constant such that the sum $\Sigma_i P_i = 1$ (since the ion must be in one of its permissible states). Thus

$$A = \frac{1}{\sum_i e^{-E_i/kT}}$$

and therefore

$$P_i = \frac{e^{-E_i/kT}}{\sum_i e^{-E_i/kT}} \tag{4.30}$$

Equation 4.30 shows that, for a system in thermal equilibrium, the probability of finding a particular ion in a state i is smaller, the greater the energy of the state. It also shows that the higher the temperature, the higher is the relative probability of finding an ion in a given excited state.

If there are N similar ions per unit volume contributing to the induced paramagnetic magnetization **M**, and the ion in state i possesses a moment in the field direction, denoted by $\boldsymbol{\mu}_i$, then the magnetization **M** is given by

$$\mathbf{M} = \sum_i \boldsymbol{\mu}_i N P_i$$

or

$$M = \frac{\sum_i N g_i \mu_B M_{J_i} e^{-E_i/kT}}{\sum_i e^{-E_i/kT}} \tag{4.31}$$

where the summation is over all the possible states i, and E_i is given by (4.28). Since the probability of the paramagnetic ion being in an excited J state is small (such a state is usually situated at about 10,000 cm^{-1} from the ground state), we restrict ourselves to the ground state value of J, for all of which states the value of g is the same. Thus we have

$$E_i = -g\mu_B M_{J_i} H$$

and therefore (4.31) becomes

$$M = \frac{N g \mu_B \sum_{-J}^{+J} M_{J_i} e^{g\mu_B M_{J_i} H/kT}}{\sum_{-J}^{+J} e^{g\mu_B M_{J_i} H/kT}} \tag{4.32}$$

where M_{J_i} ranges from $-J$ to $+J$, in steps of unity. Putting $g\mu_B H/kT = x$ and performing the summation, we get

$$M = N\mu_m B_J\left(\frac{\mu_m H}{kT}\right) \tag{4.33}$$

where $\mu_m = g\mu_B J$, which is the maximum possible permissible component of the ionic dipole moment in the field direction, and $B_J(x)$, known as the Brillouin function of the variable x, is given by

$$B_J(x) = \frac{2J+1}{2J}\coth\left(\frac{2J+1}{2J}x\right) - \frac{1}{2J}\coth\frac{x}{2J} \tag{4.34}$$

For $x \ll 1$ we have

$$\coth x = \frac{1}{x} + \frac{x}{3} - \frac{x^3}{45} + \cdots$$

whence the susceptibility $\chi = M/H$ is given from (4.33) and (4.34) by

$$\chi = \frac{NJ(J+1)g^2\mu_B^2}{3kT} = \frac{C}{T}, \quad \text{say} \tag{4.35}$$

This dependence of χ upon T is known as Curie's law, which is obeyed by many magnetic materials.

If the splittings of the J states are not much greater than kT, the distribution of ions over the excited states also must be considered. If, in addition to the above, we also include the slight distortion of the ionic wavefunctions by the field, causing additional polarization that was neglected before, the expression for the total susceptibility is

$$\chi = \sum_i N_i\left[\frac{g_i^2\mu_B^2}{3kT}J_i(J_i+1) + \alpha_i\right] \tag{4.36}$$

where α_i is the polarizability due to the small field-induced distortion of the ith wavefunction. Explicitly, α_i is given[6] by

$$\alpha_i = \sum_{j \neq i} \frac{|\langle\psi_i|\,\mu_B(\mathbf{L}+2\mathbf{S})|\psi_j\rangle|^2}{E_j - E_i} \tag{4.37}$$

This term is known as the Van Vleck[6] temperature-independent second-order contribution to susceptibility.

We now consider three situations: (1) the multiplet separations narrow compared to kT, (2) the multiplet separations wide compared to kT, and (3) the multiplet separations comparable to kT. In the first case, where the polarization terms would seem at first sight to be overwhelmingly dominant because of small denominators, a simple Curie law behavior is regained. This remarkable result is a consequence of the fact that the admixture of a higher state into lower state causes a paramagnetic polarization, but the reverse admixture causes an equal diamagnetic polarization, as a result of the difference in sign of the denominator in (4.37). When summing over levels that are separated by energies much less than kT and are consequently equally populated, the result is almost a complete cancellation of the polarization terms, and the residual contribution, being determined by small population differences, is inversely proportional to temperature and is paramagnetic. In fact, Van Vleck[6] shows that in this case (4.36) reduces to the following Curie law:

$$\chi=\frac{N\mu_{\mathrm{B}}^2}{3kT}\left[L(L+1)+4S(S+1)\right] \tag{4.38}$$

where L and S refer to the ground state. In the second case only the ground state will be occupied significantly, and we will have the Curie law susceptibility derived above and, in addition, a small Van Vleck temperature-independent susceptibility, given by $N\alpha_g$, which is essentially paramagnetic because $E_j>E_i$. For the third case (4.36) has to be calculated in detail, and a complex temperature dependence of χ would be anticipated. Such calculations for the $3d^1$, $3d^2$, $3d^6$, and $3d^7$ complexes have been worked out in detail in Chapter 7.

It should be noted that an expression the same as (4.38) can be obtained by following the procedure leading to (4.35) if we calculate the orbital and spin contributions separately instead of using J, the total angular momentum quantum number. The spin and the orbital motions will then contribute separately, and by analogy with (4.35) we should get

$$\chi=\frac{N\mu_{\mathrm{B}}^2}{3kT}\left[g_L^2L(L+1)+g_S^2S(S+1)\right] \tag{4.39}$$

where $g_L=1$ and $g_S=2$, as is appropriate for the "orbital-only" and "spin-only" moments. In actual crystalline complexes of the transition metal ions, the crystal field is very much greater in magnitude than the spin-orbit coupling, as we shall see clearly in the following chapters. Hence **J** ceases to be a good quantum number, and there is partial quenching of the orbital angular momentum in the transition metal complexes, unlike the lanthanide group of complexes, where **J** remains a good quantum number since in these salts $\lambda\mathbf{L}\cdot\mathbf{S}\gg V_{\text{crys}}$. We shall not, of course, deal with the rare earths in this book.

4.2.2 Magnetic Susceptibilities

We give below a brief derivation of the basic formula, due to Van Vleck[6] for calculating the magnetic susceptibility, since the measurement and interpretation of the magnetic susceptibilities of the complexes and compounds of the transition metal ions constitute the subject in which we are mainly interested. Before Van Vleck's classic contribution to this field magnetochemists were usually content to measure the magnetic susceptibility of a complex at one temperature and then convert the result to the "number of Bohr magnetons" or "number of unpaired spins" by means of Curie's formula:

$$\chi = \frac{N\mu_{\text{eff}}^2}{3kT} \tag{4.40}$$

where

$$\mu_{\text{eff}}^2 = \mu_{\text{B}}^2 g_S^2 S(S+1) \tag{4.41}$$

During the last four decades there has been a multitude of magnetic measurements in this area, and these have revealed that the Curie law is not as universal as one would have liked. We shall deal with these specific cases in later chapters.

If we now assume that the energy of the level E_n of the atom or ion in the magnetic field **H** can be expanded as a power series:

$$E_n = E_n^0 + HE_{nm}^{(1)} + H^2E_{nm}^{(2)} + \cdots \tag{4.42}$$

where n and m specify the quantum numbers and E_n^0 is the energy in zero applied field, then the magnetic moment $\boldsymbol{\mu}$ in the field direction is given by

$$\begin{aligned} \mu_{nm} &= -\frac{\partial E_{nm}}{\partial H} \\ &= -E_{nm}^{(1)} - 2HE_{nm}^{(2)} - \cdots \end{aligned} \tag{4.43}$$

The moment per gram-ion is then obtained by averaging over all the states, weighted with the Boltzmann factor, giving

$$M = N\frac{\sum_{n,m} \mu_{nm} e^{-E_n/kT}}{\sum_n e^{-E_n/kT}} \tag{4.44}$$

whence χ, the susceptibility, is

$$\chi = \frac{\mathbf{M}}{\mathbf{H}} = -\frac{N}{H}\frac{\sum_{n,m} \mu_{nm} e^{-E_n/kT}}{\sum_n e^{-E_n/kT}} \tag{4.45}$$

where N is the Avogadro number. Using

$$e^{-E_n/kT} \simeq e^{-E_n^0/dT}\left(1-\frac{HE_{nm}^{(1)}}{kT}\right)$$

and (4.30), and supposing that the material possesses no mean residual moment in the absence of the field, that is,

$$\sum_{n,m} -E_{nm}^{(1)} e^{-E_n^0/kT}=0$$

we get

$$\chi=-\frac{N}{H}\frac{\sum_{n,m}\left(-E_{nm}^{(1)}-2HE_{nm}^{(2)}\right)\left(1-H\frac{E_{nm}^{(1)}}{kT}\right)e^{-E_n^0/kT}}{\sum_n e^{-E_n^0/kT}\left(1-\frac{HE_{nm}^{(1)}}{kT}\right)}$$

$$=N\frac{\sum_{n,m}\left[\frac{\{E_{nm}^{(1)}\}^2}{kT}-2E_{nm}^{(2)}\right]e^{-E_n^0/kT}}{\sum_n e^{-E_n^0/kT}} \tag{4.46}$$

In (4.46) we have retained only the part of the expansion of χ that is independent of **H**. This is usually an excellent approximation except at very low temperatures and high fields.

In (4.46) we have

$$E^{(1)}=\langle\psi_0|\,\mu_B(L_z+2S_z)|\psi_0\rangle$$

$$E^{(2)}=\sum_n\frac{|\langle\psi_n|\,\mu_B(L_z+2S_z)|\psi_0\rangle|^2}{E_0-E_n} \tag{4.47}$$

where L_z+2S_z is the component of $\mathbf{L}+2\mathbf{S}$ parallel to the external field **H**. For an octahedral field, therefore, χ, referred to one axis, can be written as,

$$\chi=\frac{N\mu_B^2}{3}\frac{\sum\left\{\frac{|\langle\psi_0|\mathbf{L}+2\mathbf{S}|\psi_0\rangle|^2}{kT}+2\sum_n\frac{|\langle\psi_n|\mathbf{L}+2\mathbf{S}|\psi_0\rangle|^2}{E_n-E_0}\right\}e^{-E_0/kT}}{\sum e^{-E_0/kT}} \tag{4.48}$$

In the particular case where the ground term is orbitally nondegenerate, it is an eigenstate of the total spin, and we neglect the matrix elements of **L** between the ground state and the excited states. Equation (4.48) then becomes

$$\chi = \frac{N\mu_B^2}{kT}\frac{\sum |\langle\psi_0|2S_z|\psi_0\rangle|^2}{2S+1} = \frac{N\mu_B^2 g_S^2}{(2S+1)kT}\sum_{n=-S}^{+S} n^2$$

$$= \frac{N\mu_B^2}{3kT} g_S^2 S(S+1) \tag{4.49}$$

Equation (4.49) is called the spin-only formula for the susceptibility. This is the termperature-dependent part of the susceptibility calculated under the assumption that only the ground state is populated. Experimentally, the effective magnetic moment $\boldsymbol{\mu}_{\text{eff}}$ is obtained from the observed susceptibilities according to the Curie law:

$$\chi = \frac{N\mu_{\text{eff}}^2}{3kT} \tag{4.50}$$

By comparison with (4.48) we see that

$$\mu_{\text{eff}}^2 = \mu_B^2 g_S^2 S(S+1) \tag{4.51}$$

REFERENCES

1. E. U. Condon and G. H. Shortley, *The Theory of Atomic Spectra*, 2nd ed., Cambridge University Press, London, England, 1953.
2. M. Blume and R. E. Watson, *Proc. Roy. Soc. A*, **271**, 565 (1963).
3. M. Blume, A. J. Freeman, and R. E. Watson, *Phys. Rev.* **134**, A320 (1964).
4. T. M. Dunn, *Trans. Faraday Soc.* **57**, 1441 (1961).
5. J. S. Griffith, *The Theory of Transition Metal Ions*, Cambridge University Press, London, England, 1961.
6. J. H. Van Vleck, *The Theory of Electric and Magnetic Susceptibilities*, Oxford University Press, London, England, 1932.

5

Symmetry and Group Theory

The remarkable success of the ligand field theory in explaining a variety of properties of the transition metal ions in complexes or compounds can be traced to exploitation of the symmetry properties of the Hamiltonian. The fact that the surrounding of the ion has certain symmetry enables us, with the help of group theory, to draw some conclusions without explicit calculations. In the ligand field theory, group theory plays an important role as far as the splitting of the energy levels is concerned. It shows how the levels in the higher group are resolved when the symmetry is lowered. However, group theory can give only a qualitative picture of the splitting and the angular behavior of the wavefunctions. It can also help in finding the allowed terms and allowed matrix elements of a given operator in a ligand field of known symmetry.

We will summarize here a number of results from the theory of group representations. The first is the very important property of the eigenstates of a physical system. These states must form the basis for irreducible representations of the group G of symmetry operations that commute with the Hamiltonian of the system. Before discussing the group representations we deal, very briefly, with the basic preliminaries of group theory.[8]

5.1 BASIC DEFINITIONS AND THEOREMS

A group is a collection of elements that are interrelated according to certain rules. Here we shall be concerned entirely with the groups formed by the sets of symmetry operations that may be carried out on molecules. For any set of elements to form a mathematical group, the following conditions or rules must be satisfied.

1. The product of any two elements in the group and the square of each element must be an element in the group.

If A and B are two elements of a group, we can combine them by simply writing AB or BA. Now the question arises, does it make any difference whether we write AB or BA? In ordinary algebra it does not, and we say that the multiplication is commutative. But in group theory the commutative law does not, in general, hold good. Thus AB may give C while BA may give D, where C and D are two more elements in the group. There are some groups, however, in which multiplication is commutative, and these are called Abelian groups. Since multiplication is not, in general, commutative, it is convenient to state whether an element B is to be multiplied by A in the sense AB or BA. In the first case we can say that B is left-multiplied by A, and in the second case is right-multiplied by A.

2. There is a unique element in the group that must commute with all the others and leave them unaltered.

   This element is designated by E and is known as the identity element. Symbolically we define it by writing

$$EX = XE = E$$

   where X is any element of the group.

3. The associative law of multiplication must hold, meaning that

$$A(BC) = (AB)C$$

   In general, this must hold for the continued product of any number of elements, that is,

$$\begin{aligned}(AB)(CD)(EF)(GH) &= (A)(BC)(D)(EF)(GH)\\ &= (AB)(C)(DE)(FG)(H)\\ &= \cdots\end{aligned}$$

4. Every element must have a reciprocal that is also an element of the group. The element R is the reciprocal of S if

$$RS = SR = E$$

   where E is the identity element. Also, E is its own reciprocal.

   In this connection we state, without proof, the following theorem, which is sometimes very useful:

   The reciprocal of a product of two or more elements is equal to the product of the reciprocals in reverse order. This means that

$$(ABC\cdots XYZ)^{-1} = Z^{-1}Y^{-1}X^{-1}\cdots C^{-1}B^{-1}A^{-1}$$

5.2 GROUPS, SUBGROUPS, CLASSES, AND SYMMETRY OPERATIONS

5.2.1 Some Examples of Groups

Groups may be either finite or infinite, that is, they may contain a limited or an unlimited number of elements. The symmetry groups with which we are concerned here are mostly finite. The number of elements in a finite group is called its order, and its conventional symbol is h.

As an example of a finite group let us consider the symmetrical figure formed by three points at the corners of an equilateral triangle, as shown in Figure 5.1. The symmetry is known as C_{3v} symmetry. The operations that send this figure into itself are as follows:

1. The identity operation E.
2. Operation A, which is a reflection in the yz plane.
3. Operation B, which is a reflection in the plane passing through point b and perpendicular to the line joining a and c.
4. Operation C, which is a reflection in the plane passing through c and perpendicular to the line joining a and b.
5. Operation D, a clockwise rotation through $2\pi/3$.
6. Operation F, anticlockwise rotation through $2\pi/3$.

Other symmetry operations are possible, but each is equivalent to one of the operations given above.

The successive application of any two of the operations listed above is equivalent to some single operation. Thus $DD = F$, $DA = C$, and so on. If we work out all the possible products of two operations, we obtain a multiplication table (Table 5.1), where the operation that is to be applied to the figure first is written across the top of the table. The set of operations E, A, B, C, D, F

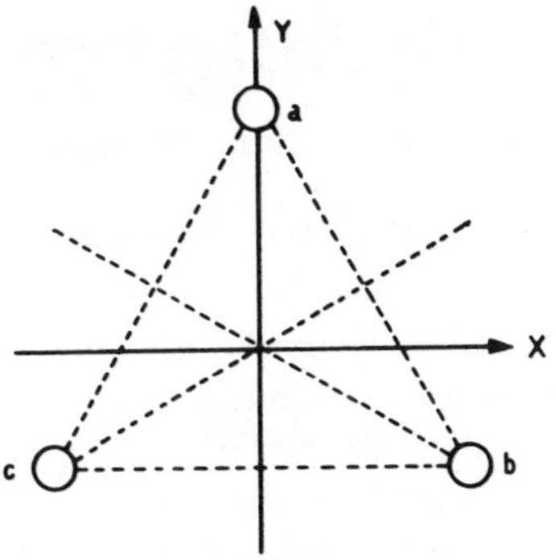

Fig. 5.1 C_{3v} symmetry.

Table 5.1

Multiplication table for the group of C_{3v} symmetry

	E	A	B	C	D	F
E	E	A	B	C	D	F
A	A	E	D	F	B	C
B	B	F	E	D	C	A
C	C	D	F	E	A	B
D	D	C	A	B	F	E
F	F	B	C	A	E	D

forms a group, and the table is known as the multiplication table for this group. The number of operations in the group is called the order of the group; here $h=6$.

We notice that in Table 5.1 the commutative law of multiplication does not hold good: $AB=D$ but $BA=F$, so that $AB \neq BA$.

From the multiplication table we notice that E is either on the diagonal or is placed symmetrically with respect to it. This is so because either an element is its own reciprocal (as for E, A, B, and C) or, since D is reciprocal to F, F is reciprocal to D. The identity element occurs only once in each column or row because each element can have only one reciprocal.

In Table 5.1 each column or row contains each element once and only once. This rule is true in general and is known as the rearrangement theorem. More formally, we can state that in the sequence

$$EA_k, A_2A_k, \ldots, A_hA_k$$

each group element A_i appears exactly once. The elements are merely rearranged by multiplying each by A_k, where A_k is any arbitrary element in the group. We do not give the proof here but refer the reader to any textbook on group theory.[8]

5.2.2 The Cyclic Groups

For any group element R, one can form the sequence

$$R, R^2, R^3, \ldots, R^n = E$$

This is called the period of R since the sequence would simply repeat this period over and over if it were extended. This group is also assumed to be finite. The integer n is called the order of R. This period clearly forms a group, although it need not exhaust all the elements of the group with which we started. This group is a cyclic group of order n. If it is only part of a larger group, it is referred to as a cyclic subgroup. Although the concept of the cyclic subgroup is introduced here, it should be remembered that subgroups need not be cyclic. Any subset of elements within a group that in itself forms a group is called a subgroup of the larger group. We also see that all cyclic groups must be Abelian.

In our example of the C_{3v} symmetric group, the period of D is

$$D, \qquad D^2 = F, \qquad D^3 = DF = E$$

Thus D is of order 3, and E, D, F form a cyclic subgroup of our entire group of order 6.

5.2.3 Subgroups

We find from Table 5.1 that within this group of order 6 there are smaller groups. Thus E is itself a group of order 1. There are groups of order 2, that is, E, A; E, B; E, C, and a group of order 3: E, D, F. These smaller groups that may be found within a larger group are called subgroups.

Are there any restriction on the nature of subgroups? Yes, there are. The orders of the subgroups must all be factors of the order of the main group. In this connection we state, without proof, the following theorem, which can be proved easily:

The order of any subgroup g, of a group of order h, must be a divisor of h. In other words, $h/g = k$, where k is an integer.

5.2.4 Classes

The elements of a group may be separated into smaller sets called classes. Before defining a class we must consider an operation known as the similarity transformation.

If A and R are two elements of a group, $R^{-1}AR$ will also be some element of the group, say B. We have therefore

$$B = R^{-1}AR$$

This relation says that B is a similarity transform of A by R. We also say that A and B are conjugate to each other. The following properties of conjugate elements are important:

1. Every element is conjugate to itself, meaning that, if we choose any particular element of A, it must be possible to find at least one element, X, such that $A = X^{-1}AX$.
2. If A is conjugate to B, then B is conjugate to A.
3. If A is conjugate to B and C, then B and C are conjugate to each other.

The proofs of these are easy, and we leave them as an exercise.

A class may be defined as a complete set of elements that are conjugate to each other. To determine the classes within a particular group we begin with one element and work out all of its transforms, using all the elements in the group, including itself, then take a second element that is not one of those found to be conjugate to the first and determine all its transforms, and so on until all the elements in the group have been placed in one class or another.

From the group of C_{3v} symmetry that we have been considering so long, it can be shown that E is itself a class of order 1; A, B, C are members of a class of order 3; and D, F are members of a class of order 2. Thus the classes have orders 1, 2, and 3, which are all factors of the group of order 6. For the classes we state, without proof, the following theorem:

The order of all classes must be integral factors of the order of the group.

5.2.5 Molecular Symmetry and the Symmetry Groups

It is found in reality that some molecules are more symmetrical than others or that some molecules have high symmetry whereas others have low symmetry or no symmetry at all. But what exactly do we mean by these statements? To clarify our ideas regarding molecular symmetry we must develop some rigid mathematical criteria of symmetry. Let us first consider the kinds of symmetry elements that a molecule may have and the symmetry operations generated by them. We then show that a complete set of symmetry operations (not elements) constitutes a mathematical group.

A symmetry operation is a movement of a body such that, after the movement has been carried out, every point of the body is coincident with an equivalent point of the body in its original orientation. In other words, the effect of the symmetry operation is to take the body into an equivalent configuration. A symmetry element is, however, a geometrical entity, such as a line, a plane, or a point, with respect to which one or more symmetry operations may be performed.

Symmetry elements and symmetry operations are closely interconnected because the operation can be defined only with respect to the element and, at the same time, the existence of a symmetry element can be demonstrated only by showing that the appropriate symmetry operations exist.

To specify molecular symmetry, four kinds of symmetry elements and operations are required. The symmetry elements are (1) reflection plane, (2) center of symmetry or center of inversion, (3) proper rotation axis, and (4) improper rotation axis. They generate four symmetry operations: reflections in the plane, inversion of all atoms through the center, one or more rotations about the axis, and one or more repetitions of this sequence: rotation followed by reflection in a plane perpendicular to the rotation axis.

5.2.6 Symmetry Planes and Reflections

A symmetry plane must pass through the body. Atoms lying in the plane constitute a special case since the operation of reflecting through the plane does not move them at all. Hence any planar molecule is bound to have at least one symmetry plane that is its molecular plane. It follows from what has been said above that there is a restriction on the numbers of the various kinds of atoms in a molecule having planar symmetry. All atoms of a given species that do not lie on the plane of symmetry must occur in even numbers because each one must have a twin on the other side of the plane. There is, of course, no restriction on the number of atoms lying on the symmetry plane. Moreover, if there is only one atom of a given species in a molecule, it must lie in each and every symmetry plane the molecule may possess. The standard symbol for a plane of symmetry is σ. It is important to note that one symmetry plane generates one symmetry operation, and the effect of applying the same reflection operation twice is to bring all atoms into their original positions. Thus $\sigma^2 = E$. Also, $\sigma^n = E$ when n is even and $\sigma^n = \sigma$ when n is odd.

It is easy to verify that the NH_3 molecule has *three* symmetry planes, a regular tetrahedral molecule has *six* planes of symmetry, and a regular octahedron possesses a total of *nine* symmetry planes.

5.2.7 The Inversion Center

If a molecule can be brought into an equivalent configuration by changing the coordinates (x, y, z) of every atom into $(-x, -y, -z)$, where the origin of coordinates lies at some point within the molecule, the point at which the origin lies is said to be a center of inversion. The standard symbol for it is i. Like a plane, the center of inversion is an element that generates only one operation. Some example of molecules having centers of inversion are octahedral (XY_6) molecules, planar (XY_4) molecules, and benzene. Tetrahedral molecules, however, have no center of inversion.

5.2.8 Proper Axis and Proper Rotations

A line drawn perpendicular to the plane of an equilateral triangle (C_{3v} symmetry, which we have been considering) and intersecting it at its geometrical center is a proper axis of rotation for that triangle. Rotating the triangle by $2\pi/3$ about this axis brings the triangle into an equivalent configuration. The general symbol for a proper axis of rotation is C_n, where n denotes the order of the axis. By "order" we mean the value of n in $2\pi/n$ such that $2\pi/n$ gives an equivalent configuration. A proper axis of order n generates n operations. The symbol C_n^m represents a rotation by $m \times 2\pi/n$. Let us consider the operation C_4^2. This is simply a rotation by $2 \times 2\pi/4 = 2\pi/2$ and can therefore be written as C_2. Many such correlations can be discovered in different symmetric groups.

5.2.9 Improper Axis and Improper Rotations

An improper rotation may be thought of as taking place in two successive steps: first a proper rotation and then a reflection through a plane perpendicular to the axis of rotation. The axis about which this occurs is called an axis of improper rotation and is denoted by the symbol S_n.

We are concerned chiefly with the following kinds of molecular symmetry in connection with the transition metal compounds. We now give their symmetry elements and operations, which can easily be verified.

5.3 IMPORTANT POINT GROUPS

5.3.1 The Point Group O_h

A regular octahedron has the following symmetry elements and operations (Figure 5.2):

E, the identity
$8C_3$ axes (one is shown in the figure)
There are eight threefold axes corresponding to four body diagonals about which we can rotate the molecule by $\pm 2\pi/3$
$3C_2$ axes (the coordinate axes in the figure)
$6C_4$ axes (the coordinate axes each with a rotation by $\pm\pi/2$)
$6C_2'$ axes (the six axes bisecting the coordinate axes)

To these symmetry operations must be added the inversion i. Hence altogether there are 48 elements in O_h, whereas in O the number of elements is 24. However, the inversion does not give anything new.

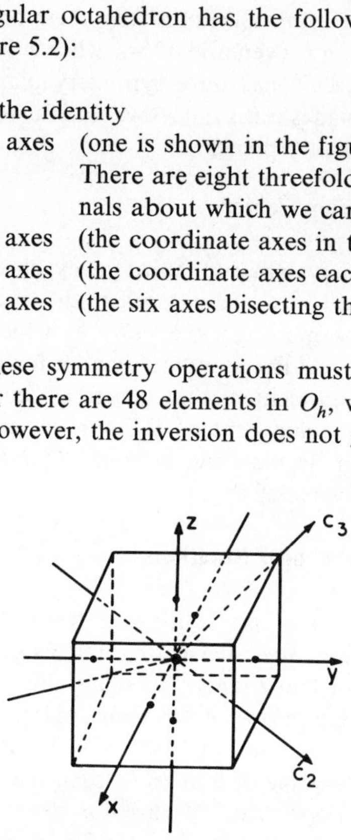

Fig. 5.2 The point group O_h. Regular octahedron (ML_6).

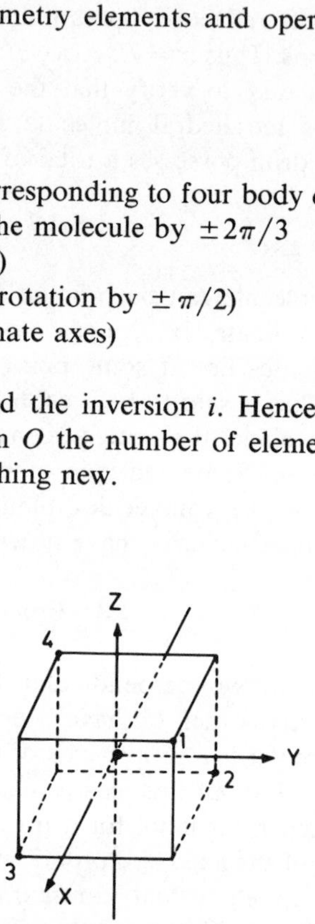

Fig. 5.3 The point group T_d. Regular tetrahedron (ML_4).

5.3.2 The Point Group T_d

A tetrahedral molecule possesses the following symmetry elements and operations (Figure 5.3):

E, the identity
$8C_3$, as in O_h

$3C_2$, as in O_h
$6S_4 = 6C_4\sigma_h$, the C_4 axis as in O_h
$6\sigma_d = 6C_2'i$, the C_2' axis as in O_h

The number of elements is 24.

5.3.3 The Point Group D_{4h}

This point group is characterized as follows (Figure 5.4):

E, the identity
$2C_4$ axes
$1C_2$ axis
$2C_2'$ axes
$2C_2''$ axes

In addition, there is the inversion operator i operating on all the symmetry operations listed above.

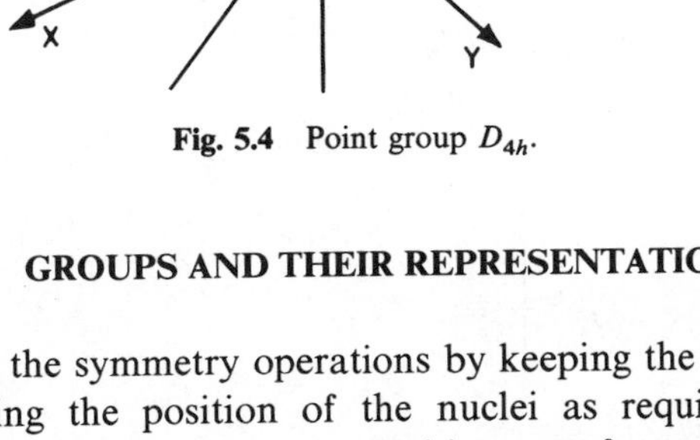

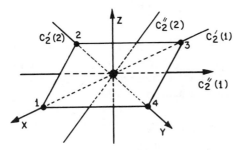

Fig. 5.4 Point group D_{4h}.

5.4 GROUPS AND THEIR REPRESENTATIONS

We shall describe the symmetry operations by keeping the coordinate system fixed and changing the position of the nuclei as required by the given symmetry operation. This is known as position transformation.

Symmetry operations like rotations and reflections are conveniently described mathematically by using matrix language. Let us study the effect of such operations on a single vector. The vector is of magnitude r and is situated in the xy plane. If the vector is rotated by an angle θ, this rotation will carry the point (x_1y_1) defined by the vector to the point (x_2y_2) as given in Figure 5.5. From Figure 5.5, we have, for the old coordinates,

$$x_1 = r\cos\phi$$
$$y_1 = r\sin\phi \tag{5.1}$$

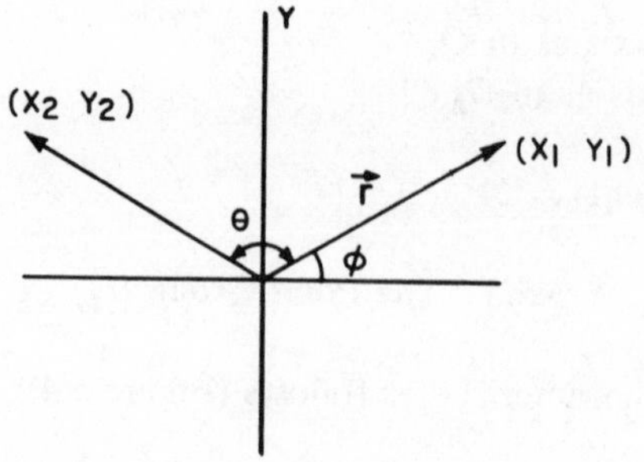

Fig. 5.5 Rotation of a vector by an angle θ.

and, for the new coordinates,

$$x_2 = r\cos(\theta+\phi) = x_1\cos\theta - y_1\sin\theta$$

$$y_2 = r\sin(\theta+\phi) = x_1\sin\theta + y_1\cos\theta$$

Using matrix language, we denote this rotation by

$$\begin{pmatrix} x_2 \\ y_2 \end{pmatrix} = \begin{pmatrix} \cos\theta & -\sin\theta \\ \sin\theta & \cos\theta \end{pmatrix}\begin{pmatrix} x_1 \\ y_1 \end{pmatrix} \tag{5.2}$$

This matrix of transformation is unitary, as can readily be verified. This is generally true for matrices representing rotations, reflections, and inversions. If the vector is defined in n-dimensional space and its components are $x_1, x_2, x_3, \ldots, x_n$, the transformed coordinates $x_1', x_2', x_3', \ldots, x_n'$ are given by

$$x_j' = \sum_k a_{jk} x_k, \qquad k, j = 1, 2, \ldots, n \tag{5.3}$$

where

$$a_{jk} = \begin{bmatrix} a_{11} & a_{12} & \cdots & a_{1n} \\ a_{21} & a_{22} & \cdots & a_{2n} \\ \vdots & \vdots & & \vdots \\ a_{n1} & a_{n2} & \cdots & a_{nn} \end{bmatrix} \tag{5.4}$$

If this transformation is followed by another:

$$x_i'' = \sum_j b_{ij} x_j' \tag{5.5}$$

the combined result of the two transformations can be described by a single transformation:

$$x_i'' = \sum_k c_{ik} x_k \tag{5.6}$$

Hence

$$x_i'' = \sum_j b_{ij} x_j' = \sum_j \sum_k b_{ij} a_{jk} x_k = \sum_k c_{ik} x_k$$

Thus

$$c_{ik} = \sum_j b_{ij} a_{jk} \tag{5.7}$$

Two groups P and P', are said to be isomorphic if, to each element $a,b,c,\ldots$ of P, there corresponds an element $a',b',c',\ldots$ of P' such that, if $ab=c$, then also $a'b'=c'$ for all products. It is important to note that the isomorphic groups have the same multiplication table. If the elements of, say, group P are square matrices, the matrices are called a representation of group P. The order of the matrices is called the dimension (or degree) of the representation.

We shall now find the representations of the point group of C_{3v} symmetry we were dealing with earlier.

Two one-dimensional representations can be written easily; they are as follows:

	E	D	F	A	B	C
A_1	1	1	1	1	1	1
A_2	1	1	1	−1	−1	−1

A two-dimensional representation can be found by using the matrix of transformation given earlier. For the identity element E we obtain from (5.2)

$$\begin{pmatrix} x_2 \\ y_2 \end{pmatrix} = \begin{pmatrix} \cos\theta & -\sin\theta \\ \sin\theta & \cos\theta \end{pmatrix} \begin{pmatrix} x_1 \\ y_1 \end{pmatrix}$$

With $\theta = 0$

$$E = \begin{pmatrix} 1 & 0 \\ 0 & 1 \end{pmatrix}$$

For D, with $\theta = -2\pi/3$, we have

$$D = \begin{bmatrix} -\dfrac{1}{2} & \dfrac{\sqrt{3}}{2} \\ -\dfrac{\sqrt{3}}{2} & -\dfrac{1}{2} \end{bmatrix}$$

For F, with $\theta=2\pi/3$, we have

$$F=\begin{bmatrix} -\frac{1}{2} & -\frac{\sqrt{3}}{2} \\ \frac{\sqrt{3}}{2} & -\frac{1}{2} \end{bmatrix}$$

The matrix for A can be easily found from

$$x_2=x_1+0\cdot y_1$$
$$y_2=y_1+0\cdot x_1$$

and is therefore given by

$$A=\begin{pmatrix} -1 & 0 \\ 0 & 1 \end{pmatrix}$$

The matrices for B and C can be found by using the multiplication table (Table 5.1) and the matrices given above. Notice that $B=FA$ and $C=DA$.

We have therefore the following set of matrices for C_{3v} symmetry for a two-dimensional representation, which we denote by E (not to be confused with the identity element):

$$\overset{E}{\begin{pmatrix} 1 & 0 \\ 0 & 1 \end{pmatrix}} \overset{D}{\begin{bmatrix} -\frac{1}{2} & \frac{\sqrt{3}}{2} \\ -\frac{\sqrt{3}}{2} & -\frac{1}{2} \end{bmatrix}} \overset{F}{\begin{bmatrix} -\frac{1}{2} & -\frac{\sqrt{3}}{2} \\ \frac{\sqrt{3}}{2} & -\frac{1}{2} \end{bmatrix}} \overset{A}{\begin{pmatrix} -1 & 0 \\ 0 & 1 \end{pmatrix}}$$

$$\cdot \begin{bmatrix} \frac{1}{2} & -\frac{\sqrt{3}}{2} \\ -\frac{\sqrt{3}}{2} & -\frac{1}{2} \end{bmatrix} \overset{C}{\begin{bmatrix} \frac{1}{2} & \frac{\sqrt{3}}{2} \\ \frac{\sqrt{3}}{2} & -\frac{1}{2} \end{bmatrix}} \quad (5.8)$$

Many other representations can also be found. For example, we might consider the three Cartesian coordinates of each nucleus. The matrices of transformation under the symmetry operations that correspond to them form a representation.

However, it is important to note that new representations can always be found from a given one by similarity transformations. In many cases it is possible to find a similarity transformation that transforms all the matrices

$a',b',c',\ldots$ into the form

$$a'' = \gamma' a' \gamma = \left\{\begin{array}{cccc} \boxed{a''_1} & & & \\ & \boxed{a''_2} & & \\ & & \boxed{a''_3} & \\ & & & \ddots \end{array}\right\} \tag{5.9}$$

$$b'' = \gamma' b' \gamma = \left\{\begin{array}{cccc} \boxed{b''_1} & & & \\ & \boxed{b''_2} & & \\ & & \boxed{b''_3} & \\ & & & \ddots \end{array}\right\} \tag{5.10}$$

and so on, where $a''_i,\ldots,b''_i,\cdots$ are square matrices and there are only zeros outside the squares. If a matrix yields to such a "reduction," the representation consisting of the original matrices $a',b',c',\cdots$ is called a reducible representation. It is possible that some of these representations can be further reduced by similarity transformation. However, in a few steps we can always reduce the original representation completely into representations that cannot be reduced any further by any similarity transformation. The latter representations are said to be irreducible representations, and every point group has a definite number of them. The representations A_1, A_2, and E for the group C_{3v}, which we have been considering, are all irreducible, and we shall see below that no others exist. It can be proved from group theory that the number of irreducible representations of a particular group is equal to the number of classes contained in the group. In general, any reducible representation is a linear combination of the irreducible representations and can be written as $\Gamma=\Sigma_i a_i\Gamma_i$. The a_i are positive integers or zero, that is, a given irreducible representation may occur in a reducible representation once, more than once, or not at all.

It is of great interest in spectroscopy to know how many times a given irreducible representation occurs in a reducible one. To determine this, we first make some general observations on unitary matrices.

If $\Gamma_i(R)$ is the matrix corresponding to an operation R and if $\Gamma_i(R)_{mn}$ is the mnth component of the matrix, we can show that

$$\sum_R \Gamma_i(R)_{mn}\Gamma_j(R)^*_{m'n'} = \frac{h}{l_i}\delta_{ij}\delta_{mm'}\delta_{nn'} \tag{5.11}$$

The meaning of (5.11) is obvious, and we do not prove it here. The proof can be obtained in any standard book on group theory.[8] Equation (5.11) is generally valid for all nonequivalent irreducible representations of any group. By "nonequivalent" we mean that the representations differ by more than a similarity transformation.

If there are two orthogonal vectors, A and B, in an n-dimensional space with complex components $A_1, A_2, A_3, \ldots, A_n$ and $B_1, B_2, B_3, \ldots, B_n$, respectively, we have

$$A_1B_1^* + A_2B_2^* + A_3B_3^* + \cdots + A_nB_n^* = 0$$

Thus only n independent mutually orthogonal vectors can be constructed in an n-dimensional space.

Similarly, we can show that

$$\Gamma_i(R_1)_{mn}, \Gamma_i(R_2)_{mn}, \ldots, \Gamma_i(R_h)_{mn}$$

can be considered as the components of an h-dimensional vector, where h is the order of the group. Such a vector is orthogonal to any other such vector obtained by another choice of m and n in the same irreducible representation. If the dimension of a given irreducible representation is l, l^2 vectors are obtainable from it. Since the number of independent orthogonal vectors in an h-dimensional space is h, we have

$$\sum_i l_i^2 = h \tag{5.12}$$

For the group C_{3v}

$$l_1 = 1, \qquad l_2 = 1, \qquad \text{and} \qquad l_3 = 2$$

for A_1, A_2, and E representations, respectively. Since h is 6, we have obtained all the irreducible representations and there are no others.

5.5 CHARACTERS OF REPRESENTATION

The trace, that is, the sum of the diagonal elements of a matrix of representation, is called the character in group theory, and it is the most important quantity in practical applications. For the ith irreducible representation and operation R, it is equal to

$$\chi_i(R) = \sum_m \Gamma_i(R)_{mm} \tag{5.13}$$

For one-dimensional representations the character is, of course, the same as the matrix itself ($+1$ and -1). It can easily be shown that the character remains unchanged by a similarity transformation. Therefore elements belonging to the same class have the same character. The character table for the group C_{3v} is then as follows:

C_{3v}	E	$2C_3$	$3\sigma_v$
A_1	1	1	1
A_2	1	1	-1
E	2	-1	0

If we now apply the orthogonality relation to the diagonal elements only, we obtain

$$\sum_R \Gamma_i(R)_{mm}\Gamma_j(R)^*_{m'm'} = \frac{h}{l_j}\delta_{ij}\delta_{mm'} \tag{5.14}$$

where $l_j \leqslant l_i$. We now sum over m and m' and observe that $\delta_{mm'}$ can be equal to unity l_j times. Thus

$$\sum_R \chi_i(R)\chi_j^*(R) = h\delta_{ij} \tag{5.15}$$

which means that the characters of the different irreducible representations form an orthogonal set of vectors as their matrices themselves.

The sum over R is taken over all operations belonging to a group. It can easily be transformed into a sum over classes if we remember that operations belonging to the same class have the same character. When this is done, there are only as many irreducible representations as there are classes.

For a reducible representation and for a given operation

$$\chi(R) = \sum_j a_j\chi_j(R) \tag{5.16}$$

since the reduction is obtained by a similarity transformation which does not change the character. Here a_j is the number of times the irreducible representation occurs in the reducible representation. Since $\chi(R)$ is just a linear combination of the $\chi_j(R)$, we have

$$\sum_R \chi(R)\chi_i^*(R) = \sum_R \sum_j a_j\chi_j(R)\chi_i^*(R) = ha_i \tag{5.17}$$

Thus the number of times that Γ_i occurs in a reducible representation Γ is equal to

$$a_i = \frac{1}{h}\sum_R \chi(R)\chi_i^*(R) \tag{5.18}$$

In terms of classes, we can write

$$a_i = \frac{1}{h} \sum_k N_k \chi(C_k) \chi_i^*(C_k) \tag{5.19}$$

where N_k is the number of elements in the class C_k and the sum runs over all classes in a group.

This is the most important relation from the viewpoint of the spectroscopist. It tells the number of times an irreducible representation occurs in a reducible one when only the characters are known.

We can encounter reducible representations in a variety of circumstances. For example, the *d*-functions provide a five-dimensional representation of O_h. Using (5.17), we see that the irreducible representations E_g (twofold degenerate) and T_{2g} (three-fold degenerate) are contained in O_h. The atomic orbitals used to form the molecular orbitals and the displacement coordinates of the nuclei in a molecule having a certain symmetry are examples of the reducible representation. We shall have occasion to deal with these situations later.

The character tables for the point groups important for our purpose can be found by following the procedure outlined above, and these, along with the basis functions, are given in Appendix V.

If an atom in the crystal possesses holohedral symmetry, it must be a symmetry center of the whole crystal. Then its symmetry operations can be classified into two categories of equal numbers of elements; the pure rotations, on the one hand, and rotations with reflection, on the other. The pure rotations form an invariant subgroup of the group. From every class of pure rotations a class of reflections or rotations with reflections arises by multiplication with the inversion at the nucleus.

Let us assume that C is a class of the invariant subgroup consisting of the pure rotations, that is, if X is any element of the invariant subgroup, then

$$X^{-1}CX = C$$

Now let I be the inversion that naturally commutes with every rotation ($I^2 = E$). It then follows that C is also a class in the entire group, since

$$(IX)^{-1}C(IX) = C$$

Also, IC forms a class of the group because

$$X^{-1}ICX = IC$$

Accordingly the group consists of precisely *twice* as many elements in precisely *twice* as many classes as the invariant subgroup in question, and therefore possesses twice as many representations as the latter, namely, (1) the positive representations, in which the inversion is represented by the unit matrix and the inversion class IC has the same character as the rotation class C, and (2) the negative representations, in which the inversion is represented by the negative unit matrix and multiplication of a class C by the inversion results in multiplying the character by -1. By comparison with the definition of Wigner and von Neumann[11] we see that, in a position of holohedral symmetry, a positive term of the free atom breaks up into nothing but positive crystal terms, and a negative term into nothing but negative crystal terms. Beyond this, one need not bother with the inversion but has only to carry out the reduction of the spherical rotation group, taken as representations of the group consisting of pure rotations of the crystal around the nucleus of the atom, which we call the crystal rotation group, for example, the cubic, tetragonal, and trigonal rotation groups. If we denote the inversion operation by C_i, then, for the point groups in which we are chiefly interested,

$$D_{3d} = D_3 \times C_i$$
$$D_{4h} = D_4 \times C_i$$
$$O_h = O \times C_i$$

Since the characters remain the same whether or not the inversion is included, we simply give, in Appendix V, the characters of the point groups we are interested in without considering the inversion.

It is also worth remembering here that representations for which the characters remain unaltered during the inversion are called even (g) representations and those that change sign during this operation are called odd (u) representations. The letter g or u is added as a subscript to the name of the representation. In many cases one need not bother with the "complete" character table because it is possible to tell beforehand whether odd or even representations will be obtained. For example, since all d orbitals are even under reflection at the center of symmetry, we can get only even representations by using the d orbitals as our basis functions. It will be seen later that the five d orbitals taken as basis functions span the two irreducible representations t_{2g} and e_g in the O_h point group.

5.6 DIRECT PRODUCT REPRESENTATIONS

We use the symbol G to denote a general finite group, and R to denote any general element of G. Also, Γ_i^a is the ith irreducible representation of G. The superscript a is used to distinguish irreducible representations of the same symmetry. The one-dimensional representation obtained by representing each element by 1 is denoted by A_1, and $\lambda(\Gamma)$ is used for the dimension of an irreducible representation Γ.

Now let us suppose f_γ^Γ to be a set of basis functions belonging to the irreducible representation Γ of group G. This basis set may sometimes be written as $|\Gamma\gamma\rangle$. The effect of any element R on f_γ^Γ is given by

$$Rf_\gamma^\Gamma = \sum_{\gamma'} D^\Gamma(R)_{\gamma'\gamma} f_{\gamma'}^\Gamma \tag{5.20}$$

where $D^{\Gamma}(R)$ is the matrix representation of the element R and the irreducible representation Γ. Just like the functions f_{γ}^{Γ}, it is also possible to find operators O_{γ}^{Γ} having the same transformation properties, that is,

$$RO_{\gamma}^{\Gamma}=\sum_{\gamma'} D^{\Gamma}(R)_{\gamma'\gamma}O_{\gamma'}^{\Gamma} \tag{5.21}$$

If $f_{\gamma_1}^{\Gamma_1}$ and $g_{\gamma_2}^{\Gamma_2}$ are bases for two irreducible representations Γ_1 and Γ_2, respectively, the product function $f_{\gamma_1}^{\Gamma_1}g_{\gamma_2}^{\Gamma_2}$ spans, a $[\lambda(\Gamma_1)\times\lambda(\Gamma_2)]$-dimensional representation of the group G. The product function is then transformed as

$$\begin{aligned} Rf_{\gamma_1}^{\Gamma_1}g_{\gamma_2}^{\Gamma_2} &= \sum_{\gamma_2', \gamma_2'} D_{\gamma_1'\gamma_1}^{\Gamma_1}(R)D_{\gamma_2'\gamma_2}^{\Gamma_2}(R)f_{\gamma_1'}^{\Gamma_1}g_{\gamma_2'}^{\Gamma_2} \\ &= \sum_{\gamma_1', \gamma_2'} D^{\Gamma_1\otimes\Gamma_2}(R)_{\gamma_1\gamma_2', \gamma_1'\gamma_2}f_{\gamma_1'}^{\Gamma_1}g_{\gamma_2'}^{\Gamma_2} \end{aligned} \tag{5.22}$$

where $D^{\Gamma_1\otimes\Gamma_2}(R)$ is the direct product (outer product) of the matrices $D^{\Gamma_1}(R)$ and $D^{\Gamma_2}(R)$. In general, this matrix is reducible and can be written as a direct sum of irreducible representations by unitary transformation, that is,

$$U^{+}D^{\Gamma_1\otimes\Gamma_2}(R)U=\begin{bmatrix} D^{\Gamma_i}(R) & 0 & 0 & \cdots & 0 \\ 0 & D^{\Gamma_j}(R) & 0 & \cdots & 0 \\ \vdots & & \ddots & & \vdots \\ & & & \ddots & \\ 0 & 0 & \cdots & \cdots & D^{\Gamma_k}(R) \end{bmatrix} \tag{5.23}$$

From the definition of characters of a representation, we have

$$\chi(\Gamma_1\otimes\Gamma_2)=\chi(\Gamma_1)\cdot\chi(\Gamma_2) \tag{5.24}$$

Then, using (5.17), one can determine how many times an irreducible representation occurs in the direct product $\Gamma_1\otimes\Gamma_2$. Unlike the situation for the three-dimensional rotation group R_3, in the direct product of two irreducible representations of a finite group a representation may be repeated because all the finite groups are not simply reducible.

The matrix U is independent of R and gives the so-called vector coupling coefficients. It is possible to find a linear combination of the products $f_{\gamma_1}^{\Gamma_1}g_{\gamma_2}^{\Gamma_2}$ such that they transform as $\phi_{\gamma_3}^{\Gamma_3}$, the γ_3th component of the basis for the irreducible representation Γ_3, provided that Γ_3 occurs in $\Gamma_1\otimes\Gamma_2$. When Γ_3 is repeated, one can find as many independent basis sets $|\Gamma_3\gamma_3\rangle$ as there are repetitions. Hence, if Γ_3 is repeated a_{Γ_3} times, we can find a_{Γ_3} independent basis sets and can write for the sth set

$$\phi_{\gamma_3}^{\Gamma_3}(s)=\sum_{\gamma_1\gamma_2}\langle\Gamma_1\gamma_1\Gamma_2\gamma_2|\Gamma_3\gamma_3\, s\rangle f_{\gamma_1}^{\Gamma_1}g_{\gamma_2}^{\Gamma_2} \tag{5.25}$$

where s can take a_{Γ_3} values $1,2,\ldots,a_{\Gamma_3}$. The numbers $\langle\Gamma_1\gamma_1\Gamma_2\gamma_2|\Gamma_3\gamma_3 s\rangle$ are known as the vector coupling coefficients or the Clebsch-Gordan coefficients, by analogy with the coupling of two angular momenta in the case of R_3, as given below:

$$|j_3m_3\rangle = \sum_{m_1m_2} \langle j_1m_1j_2m_2|j_3m_3\rangle|j_1m_1\rangle|j_2m_2\rangle \tag{5.26}$$

We write

$$|\Gamma_1\gamma_1\Gamma_2\gamma_2|\Gamma_3\gamma_3 s\rangle = \langle\Gamma_3\gamma_3 s|\Gamma_1\gamma_1\Gamma_2\gamma_2\rangle^* \tag{5.27}$$

or

$$U = U^*$$

Since U is unitary, the orthonormality conditions are satisfied by the rows and columns of U. Then we have

$$\begin{aligned}&\sum_{\gamma_1\gamma_2} \langle\Gamma_3\gamma_3 s|\Gamma_1\gamma_1\Gamma_2\gamma_2\rangle\langle\Gamma_1\gamma_1\Gamma_2\gamma_2|\Gamma_3'\gamma_3' s\rangle = \delta_{\Gamma_3\Gamma_3'}\delta_{\gamma_3\gamma_3'}\delta_{ss'}\\ &\sum_{\Gamma_3\gamma_3 s} \langle\Gamma_1\gamma_1\Gamma_2\gamma_2|\Gamma_3\gamma_3 s\rangle\langle\Gamma_3\gamma_3 s|\Gamma_1\gamma_1'\Gamma_2\gamma_2'\rangle = \delta_{\gamma_1\gamma_1'}\delta_{\gamma_2\gamma_2'}\end{aligned} \tag{5.28}$$

where $\delta_{ab}=1$ if $a=b$ and $\delta_{ab}=0$ if $a\neq b$. Using (5.28), we at once get the relation inverse to (5.25), that is,

$$f_{\gamma_1}^{\Gamma_1} g_{\gamma_2}^{\Gamma_2} = \sum_{\Gamma_3\gamma_3 s} \langle\Gamma_3\gamma_3 s|\Gamma_1\gamma_1\Gamma_2\gamma_2\rangle f_{\gamma_3}^{\Gamma_3}(s) \tag{5.29}$$

The phases of the angular momentum coupling coefficients can be so chosen as to make all of them real, but this is not always true for finite groups. In the case of cubic groups, if we take the real basis functions quantized along a fourfold axis of symmetry, then all the vector coupling coefficients are real. However, with the cubic bases quantized along the threefold (trigonal) axes, the coefficients are complex. These coefficients for the octahedral group were first tabulated by Tanabe and Sugano[2] in a paper. Griffith[3] has provided a rather extensive table of these coefficients. In a later work Griffith[4] has given a standard convention. In this connection we should also mention the extensive tabulations by Koster and his co-workers[5] and by Polo[6]. Although different phase conventions are used in the literature, it is highly desirable to follow the same phase convention throughout. We shall use Tanabe and Sugano's convention[2] consistently.

One important application of the direct product representation is in evaluating molecular integrals of the form

$$\int \phi(\Gamma_\gamma^k)\, O(\Gamma_\beta^j)\, \phi(\Gamma_\alpha^i)\, d\tau$$

where $\phi(\Gamma_\gamma^k)$, the operator $O(\Gamma_\beta^j)$, and $\phi(\Gamma_\alpha^i)$ belong to the irreducible representations Γ_γ, Γ_β, and Γ_α, respectively. Here $\phi(\Gamma_\gamma^k)$ and $\phi(\Gamma_\alpha^i)$ have been chosen in such a way as to form bases for the irreducible representations of the point group of the molecule. The integral above is nonvanishing only if it is invariant under all symmetry operations of the group. Then, unless $\Gamma_\alpha \otimes \Gamma_\beta$ contains the irreducible representation Γ_γ, the integral vanishes. In other words, the direct product $\Gamma_\gamma \otimes \Gamma_\beta \otimes \Gamma_\alpha$ must contain at least one completely invariant term. Since this product will, in general, be reducible,

$$\Gamma_\gamma \otimes \Gamma_\beta \otimes \Gamma_\alpha = \sum_i a_i \Gamma_i$$

where Γ_i are the irreducible representations of the group. Our integral will then be nonvanishing only if the totally symmetric representation Γ_1 is found in the expansion given above.

If $O(\Gamma_\beta^j)$ represents the Hamiltonian of the molecule, we know it to be invariant under all symmetry operations of the molecule. The integral given above is then nonvanishing only when $\Gamma_\gamma^k = \Gamma_\alpha^i$. The Hamiltonian is an operator expression for the energy of the molecule, and hence the energy of the molecule cannot change in either sign or magnitude as a result of a symmetry operation. We conclude therefore that $\phi(\Gamma_\gamma^k)$ and $\phi(\Gamma_\alpha^i)$ must belong to the same irreducible representation if the integral is to be nonvanishing.

5.7 THE WIGNER-ECKART THEOREM

From (5.20) and (5.21) we see that, just like the functions f_γ^Γ, the operators O_γ^Γ obey the same transformation properties. Such operators are known as irreducible tensor operators. When the operators are expressed in terms of irreducible tensor operators, the matrix elements of such operators can be separated into two parts, one part depending only on the symmetry aspects of the matrix element and the other depending on the physical aspects. This is the result of the Wigner-Eckart theorem.[4] Originally the theorem was proved for the rotation group R_3 and later was extended to the finite groups by Koster.[7] This theorem provides a generalization of the matrix element from scalar to higher rank tensor operators.[8] We now give a straightforward derivation of the Wigner-Eckart theorem by using the group-theoretical concepts of the representation theory and tensor algebra.[9,10]

Let $O_{\gamma_2}^{\Gamma_2}$ be an operator transforming as the γ_2th row of the irreducible representation Γ_2, or, in other words, $O_{\gamma_2}^{\Gamma_2}$ is the γ_2th component of an irreducible tensor operator O^{Γ_2} corresponding to the irreducible representation Γ_2 of the group in question.

We want to find a convenient expression for the matrix element

$$\langle f_{\gamma_1}^{\Gamma_1} | O_{\gamma_2}^{\Gamma_2} | f_{\gamma_3}^{\Gamma_3} \rangle \equiv \int f_{\gamma_1}^{\Gamma_1 *} O_{\gamma_2}^{\Gamma_2} f_{\gamma_3}^{\Gamma_3} \, d\tau \tag{5.30}$$

According to (5.29), we can write

$$O^{\Gamma_2}_{\gamma_2} f^{\Gamma_3}_{\gamma_3} = \sum_{\Gamma_k \gamma_k s} \langle \Gamma_k \gamma_k s | \Gamma_2 \gamma_2 \Gamma_3 \gamma_3 \rangle g^{\Gamma_k}_{\gamma_k}(s) \tag{5.31}$$

from which we obtain

$$\int f^{\Gamma_1^*}_{\gamma_1} O^{\Gamma_2}_{\gamma_2} f^{\Gamma_3}_{\gamma_3}\, d\tau = \sum_{\Gamma_k \gamma_k s} \langle \Gamma_k \gamma_k s | \Gamma_2 \gamma_2 \Gamma_3 \gamma_3 \rangle \int f^{\Gamma_1^*}_{\gamma_1} g^{\Gamma_k}_{\gamma_k}(s)\, d\tau \tag{5.32}$$

Since the integration is carried over the whole configuration space, $\int f^{\Gamma_1^*}_{\gamma_1} g^{\Gamma_k}_{\gamma_k}(s)\, d\tau$ will be invariant under any group operation R. Then we have

$$\int f^{\Gamma_1}_{\gamma_1} g^{\Gamma_k}_{\gamma_k}(s)\, d\tau = \int R f^{\Gamma_1^*}_{\gamma_1} R g^{\Gamma_k}_{\gamma_k}(s)\, d\tau$$

$$= \sum_{\gamma_1' \gamma_k'} D^{\Gamma_1^*}_{\gamma_1' \gamma_1}(R) D^{\Gamma_k}_{\gamma_k' \gamma_k}(R) \int f^{\Gamma_1^*}_{\gamma_1 \gamma_1'} g^{\Gamma_k}_{\gamma_k'}(s)\, d\tau$$

where we have used (5.20). Now, summing over R and dividing by h, we get from the above, on using (5.11),

$$\int f^{\Gamma_1^*}_{\gamma_1} g^{\Gamma_k}_{\gamma_k}(s)\, d\tau = \frac{1}{\lambda(\Gamma_1)} \delta_{\Gamma_1 \Gamma_k} \delta_{\gamma_1 \gamma_k} \sum_{\gamma_1'} \int f^{\Gamma_1^*}_{\gamma_1'} g^{\Gamma_1}_{\gamma_1'}(s)\, d\tau$$

$\lambda(\Gamma_1)$ being the dimension of the representation Γ_1.

Using this result in (5.32), we obtain

$$\int f^{\Gamma_1^*}_{\gamma_1} O^{\Gamma_2}_{\gamma_2} f^{\Gamma_3}_{\gamma_3}\, d\tau = \sum_s \langle \Gamma_1 \gamma_1 s | \Gamma_2 \gamma_2 \Gamma_3 \gamma_3 \rangle \cdot \frac{1}{\lambda(\Gamma_1)}$$

$$\cdot \sum_{\lambda_1'} \int f^{\Gamma_1^*}_{\gamma_1'} g^{\Gamma_1}_{\gamma_1'}(s)\, d\tau \tag{5.33}$$

The inverse of (5.31) is, according to (5.25),

$$g^{\Gamma_1}_{\gamma_1'}(s) = \sum_{\gamma_2' \gamma_3'} \langle \Gamma_2 \gamma_2' \Gamma_3 \gamma_3' | \Gamma_1 \gamma_1' s \rangle O^{\Gamma_2}_{\gamma_2'} f^{\Gamma_3}_{\gamma_3'}$$

Then we have

$$\sum_{\gamma_1'} \int f^{\Gamma_1^*}_{\gamma_1'} g^{\Gamma_1}_{\gamma_1'}(s)\, d\tau = \sum_{\gamma_1' \gamma_2' \gamma_3'} \langle \Gamma_2 \gamma_2' \Gamma_3 \gamma_3' | \Gamma_1 \gamma_1' s \rangle$$

$$\cdot \int f^{\Gamma_1^*}_{\gamma_1'} O^{\Gamma_2}_{\gamma_2'} f^{\Gamma_3}_{\gamma_3'}\, d\tau$$

which thus depends only on $\Gamma_1, \Gamma_2, \Gamma_3, s$ as well as on the three basis sets f^{Γ_1}, O^{Γ_2}, f^{Γ_3}, and it is usual to write

$$\sum_{\gamma_1'} \int f^{\Gamma_1^*} g_{\gamma_1'}^{\Gamma_1}(s)\, d\tau = \sqrt{\lambda(\Gamma_1)}\, \langle \Gamma_1 \| O^{\Gamma_2} \| \Gamma_3 \rangle^s$$

Then (5.33) becomes

$$\int f_{\gamma_1}^{\Gamma_1^*} O_{\gamma_2}^{\Gamma_2} f_{\gamma_3}^{\Gamma_3}\, d\tau = \sum_s \frac{\langle \Gamma_1 \gamma_1 s | \Gamma_2 \gamma_2 \Gamma_3 \gamma_3 \rangle}{\sqrt{\lambda(\Gamma_1)}} \langle \Gamma_1 \| O^{\Gamma_2} \| \Gamma_3 \rangle^s \tag{5.34}$$

Here $\langle \Gamma_1 \langle O^{\Gamma_2} \| \Gamma_3 \rangle^s$ is called the reduced matrix element depending on $\Gamma_1, \Gamma_2, \Gamma_3, s$, as well as on the basis sets $f^{\Gamma_1}, O^{\Gamma_2}, f^{\Gamma_3}$; $\langle \Gamma_1 \| O^{\Gamma_2} \langle \Gamma_3 \rangle^s$ represents the physical part of the matrix element since it depends on the complete nature of the basis sets $f^{\Gamma_1}, O^{\Gamma_2}, f^{\Gamma_3}$ that are involved in a particular problem. On the other hand, the first factor of (5.34) is the geometrical part of the matrix element, since it depends only on the symmetry aspects of the problem.

If we can evaluate the integral on the left-hand side of (5.34) for a_{Γ_1} sets of values of γ_1, γ_2, γ_3 and thereby obtain a_{Γ_1} independent linear equations in the a_{Γ_1} quantities $\langle \Gamma_1 \| O^{\Gamma_2} \| \Gamma_3 \rangle^s$ $(s = 1, 2, \ldots, a_{\Gamma_1})$, then, by solving these equations, we can find the values of these reduced matrices.

Relation (5.34) is generally known as the *Wigner-Eckart Theorem*. It is difficult to overestimate the usefulness of the theorem in all problems where the Hamiltonian of a physical system has sufficiently high symmetry.

As remarked earlier the theorem was first proved for the full rotation group R_3, which is the group of all rotations in three dimensions and hence is a continuous group, since all the group elements may be obtained by continuous variation of three parameters (e.g., the three Eulerian angles) specifying an arbitrary rotation.

A continuous group contains an infinite number of elements, that is, it is of infinite order. We note that R_3 is the rotational symmetry group for an atom in free space. There are an infinite number of classes of R_3, all rotations through the same angle (about whatever axis) being in the same class because two different rotation axes can always be made coincident by another rotation of the group.

Let **J** be the usual angular momentum operator associated with the rotation space of R_3. Let ψ be any function in this space, that is, with each point in the space we associate a value of ψ. Let us denote by $R_\alpha(\theta)$ a rotation by an angle θ about the α-axis $(\alpha = x, y, z)$. From the theory of angular momentum it is well known that under the operation $R_\alpha(\theta)$ the function ψ will be transformed to

$$\psi' \equiv R_\alpha(\theta)\psi = e^{-(i\theta/\hbar)\mathbf{J}_\alpha}\psi \tag{5.35}$$

Now we consider the case of an arbitrary rotation, which can always be specified by three Eulerian angles, as explained below.

Starting with an (x,y,z) coordinate system, we first apply the rotation $R_z(\alpha)$ and obtain a new coordinate system (x',y',z'). Then we apply the rotation $R_{y'}(\beta)$, and label the resulting axes as x'',y'',z''. Finally the rotation $R_{z''}(\gamma)$ is applied, and we obtain the final axes, x''',y''',z'''. The whole operation, consisting of these three successive steps, is denoted by $R(\alpha,\beta,\gamma)$:

$$R(\alpha,\beta,\gamma) \equiv R_{z''}(\gamma)R_{y'}(\beta)R_z(\alpha)$$

It is not difficult to show geometrically[8] that

$$R_{z''}(\gamma)R_{y'}(\beta)R_z(\alpha) \equiv R_z(\alpha)R_y(\beta)R_z(\gamma)$$

Then we have

$$R(\alpha,\beta,\gamma) \equiv R_z(\alpha)R_y(\beta)R_z(\gamma) \tag{5.36}$$

Hence, under the general rotation, the function ψ will be transformed to

$$R(\alpha,\beta,\gamma)\psi \equiv e^{-(i\alpha/\hbar)\mathbf{J}_z}e^{-(i\beta/\hbar)\mathbf{J}_y}e^{-(i\gamma/\hbar)\mathbf{J}_z}\psi \tag{5.37}$$

where we have used the basic result (5.35).

As we saw in Chapter 2, $\mathbf{J}_z$ and $\mathbf{J}^2$ can have a simultaneous eigenfunction (vector), $|Njm\rangle$, such that

$$\begin{aligned} \mathbf{J}^2|Njm\rangle &= \hbar^2 j(j+1)|Njm\rangle \\ \mathbf{J}_z|Njm\rangle &= \hbar m|Njm\rangle \end{aligned} \tag{5.38}$$

(Here N is any other set of necessary quantum numbers not relevant to the angular momentum.)

Also, by using the commutation relations for the components of $\mathbf{J}$, we obtain the following relations:

$$\mathbf{J}_+|Njm\rangle = \hbar\left[(j+m+1)(j-m)\right]^{1/2}|Njm+1\rangle \tag{5.39}$$

$$\mathbf{J}_-|Njm\rangle = \hbar\left[(j-m+1)(j+m)\right]^{1/2}|Njm-1\rangle \tag{5.40}$$

where

$$\begin{aligned} \mathbf{J}_+ &= \mathbf{J}_x + i\mathbf{J}_y \\ \mathbf{J}_- &= \mathbf{J}_x - i\mathbf{J}_y \end{aligned} \tag{5.41}$$

Replacing ψ by $|Njm\rangle$ in (5.37), we get (using 5.38)

$$\begin{aligned} R(\alpha,\beta,\gamma)|Njm\rangle &= e^{-(i\gamma/\hbar)m}e^{-(i\alpha/\hbar)\mathbf{J}_z}e^{-(i\beta/\hbar)\mathbf{J}_y}|Njm\rangle \\ &= e^{-i\gamma m}e^{-(i\alpha/\hbar)\mathbf{J}_z}\sum_{N'j'm'}|N'j'm'\rangle\langle N'j'm'|e^{-(i\beta/\hbar)\mathbf{J}_y}|Njm\rangle \end{aligned}$$

Now, the above matrix element vanishes unless $N'=N$ and $j'=j$ because, by operating with $e^{-(i\beta/\hbar)\mathbf{J}_y}$ upon $|Njm\rangle$, only states with different values of m will be generated. Then we have

$$R(\alpha,\beta,\gamma)|Njm\rangle = e^{-i\gamma m}\sum_{m'=-j}^{j} e^{-(i\alpha/\hbar)\mathbf{J}_z}|Njm'\rangle\langle Njm'|e^{-(i\beta/\hbar)\mathbf{J}_y}|Njm\rangle$$

$$= e^{-i\gamma m}\sum_{m'=-j}^{j} |Njm'\rangle e^{-i\alpha m'}\langle Njm'|e^{-(i\beta/\hbar)\mathbf{J}_y}|Njm\rangle$$

Thus we can write

$$R(\alpha,\beta,\gamma)|Njm\rangle = \sum_{m'=-j}^{j} D^{(j)}(\alpha,\beta,\gamma)_{m'm}|Njm'\rangle \tag{5.42}$$

where

$$D^{(j)}(\alpha,\beta,\gamma)_{m'm} = e^{-i(\gamma m+\alpha m')}\langle Njm'|e^{-(i\beta/\hbar)\mathbf{J}_y}|Njm\rangle \tag{5.43}$$

Equation (5.42) implies that the $(2j+1)$ functions (vectors) $|Njm\rangle$ (for given values of N,j) form a basis set for a $(2j+1)$-dimensional representation, $D^{(j)}$, of the group R_3, the matrix of the representation corresponding to the operation $R(\alpha,\beta,\gamma)$ being given by (5.43). The problem of finding these representation matrices is actually that of finding the elements $\langle Njm'|e^{-i\beta/\hbar \mathbf{J}_y}|Njm\rangle$. These have been determined for arbitrary integral or half-integral values of j by an indirect procedure due to H. Weyl,[8] which we do not discuss here. It can be further shown that these representations $D^{(j)}(j=0,\frac{1}{2},1,\frac{3}{2},\cdots)$ are all irreducible.

For rotation by an angle α about the z-axis ($\beta=0,\gamma=0$) we have from (5.43)

$$D^{(j)}(\alpha,0,0)_{m'm} = e^{-i\alpha m'}\langle Njm'|Njm\rangle = e^{-i\alpha m'}\delta_{mm'} \tag{5.44}$$

Hence the character is

$$\chi^{(j)}(\alpha) = \sum_{m=-j}^{j} D^{(j)}(\alpha,0,0)_{mm} = \sum_{m=-j}^{j} e^{-i\alpha m}$$

$$= \frac{\sin\left(j+\frac{1}{2}\right)\alpha}{\sin(\alpha/2)} \tag{5.45}$$

We have seen that all rotations by α are in the same class, regardless of axis. Hence this result is true for any rotation by an angle α. From (5.42) and (5.44) we have

$$R(\alpha,0,0)|Njm\rangle = e^{-i\alpha m}|Njm\rangle \tag{5.46}$$

In particular, for a rotation by 2π about the z-axis, we have

$$R(2\pi,0,0)|Njm\rangle = e^{-i2\pi m}|Njm\rangle \tag{5.47}$$

Now, for an integral value of j, m will be an integer (positive or negative), and in that case (5.47) gives

$$R(2\pi,0,0)|Njm\rangle = |Njm\rangle$$

In other words, under this rotation by 2π the function $|Njm\rangle$ remains unchanged, as is the usual physical situation, since a rotation by 2π should leave everything unchanged.

But for a half-integral value of j,m is also a half-integer (positive or negative), and then (5.47) gives

$$R(2\pi,0,0)|Njm\rangle = -|Njm\rangle$$

We also note that

$$R(4\pi,0,0)|Njm\rangle = |Njm\rangle$$

Thus, for a half-integral j, the phase of $|Njm\rangle$ remains unchanged, not under a rotation by 2π, but only under a rotation by 4π. Obviously this is an unnatural situation.

We tackle this strange situation by introducing the fiction that the points of the rotation space are taken into itself, not under a rotation by 2π, but only after a rotation by 4π. Then the rotation by 2π does not represent the identity element of the group but is given by the rotation through 4π. Such a full rotation group, in which the rotation through 4π is taken to be the identity element, will be denoted by R_3'. The same fiction can be introduced for any point group G of finite dimension by adding a new element R to the group, where R represents a rotation by 2π and has the property $R \neq E$ but $R^2 = E$. The new group is denoted by G' and is called the double group associated with the ordinary group G. We need not discuss the double groups any further here. Their nature and character tables will be described in some detail in Chapter 10, and there we shall have occasion to apply them to actual problems of interest.

We should mention here that it was Bethe[1] who first introduced the fictitious idea leading to the construction of double groups from ordinary groups. However, the mathematical interpretation of this idea became possible later.[4]

For integral values of j the coordinate representation of $|Njm\rangle$ can be given in terms of spherical harmonics $Y_j^m(\theta,\phi)$:

$$\langle r,\theta,\phi|Njm\rangle = f(N,r)\,Y_j^m(\theta,\phi) \tag{5.48}$$

where r,θ,ϕ are the usual spherical coordinates.

This can be easily verified by observing that (5.38), (5.39), and (5.40) remain valid on replacing $|Njm\rangle$ by $\langle r,\theta,\phi|Njm\rangle$ as given in (5.48) and employing the *coordinate-represented forms* of the operators $\mathbf{J}^2, \mathbf{J}_z, \mathbf{J}_x, \mathbf{J}_y$ as given below (in terms of spherical coordinates):

$$\left.\begin{aligned}
\mathbf{J}^2 &= -\hbar^2\left[\frac{1}{\sin\theta}\frac{\partial}{\partial\theta}\left(\sin\theta\frac{\partial}{\partial\theta}\right)+\frac{1}{\sin^2\theta}\frac{\partial^2}{\partial\phi^2}\right]\\
\mathbf{J}_z &= -i\hbar\frac{\partial}{\partial\phi}\\
\mathbf{J}_x &= i\hbar\left(\sin\phi\frac{\partial}{\partial\theta}+\cot\theta\cos\phi\frac{\partial}{\partial\phi}\right)\\
\mathbf{J}_y &= i\hbar\left(-\cos\phi\frac{\partial}{\partial\theta}+\cot\theta\sin\phi\frac{\partial}{\partial\phi}\right)
\end{aligned}\right\} \tag{5.49}$$

In the present case of the R_3 group we can also define the irreducible tensor operators in the same manner as before. Let $\mathbf{T}^{(j)}$ be such a tensor operator corresponding to the irreducible representation $D^{(j)}$ of dimension $(2j+1)$. The $(2j+1)$ components of $\mathbf{T}^{(j)}$ are $\mathbf{T}^{(j)}_m$ $(m=j,j-1,\ldots,-j)$ so that, under a rotation $R(\alpha,\beta,\gamma)$, $\mathbf{T}^{(j)}_m$ will be transformed to

$$R(\alpha,\beta,\gamma)\mathbf{T}^{(j)}_m = \sum_{m'=-j}^{j} D^{(j)}(\alpha,\beta,\gamma)_{m'm}\mathbf{T}^{(j)}_{m'} \tag{5.50}$$

the transformation matrix being given by (5.43). In the present case the Wigner-Eckart theorem (5.34) is obviously given by

$$\langle N_1 j_1 m_1|\mathbf{T}^{(j_2)}_{m_2}|N_3 j_3 m_3\rangle = \frac{\langle j_1 m_1|j_2 m_2 j_3 m_3\rangle}{\sqrt{2j_1+1}}\langle N_1 j_1\|\mathbf{T}^{(j_2)}\|N_3 j_3\rangle \tag{5.51}$$

where

$$\langle j_1 m_1|j_2 m_2 j_3 m_3\rangle \equiv \langle j_2 m_2 j_3 m_3|j_1 m_1\rangle^* = \langle j_2 m_2 j_3 m_3|j_1 m_1\rangle$$

which are the usual Wigner (or Clebsch-Gordan) coefficients and are real, so that there is no difference between $\langle j_1 m_1|j_2 m_2 j_3 m_3\rangle$ and $\langle j_2 m_2 j_3 m_3|j_1 m_1\rangle$. These coefficients were introduced in Chapter 2. We note that, unlike the general equation (5.34), there is no summation on the right-hand side of the present equation (5.51). This is so because in the direct product, $D^{(j_2)}\otimes D^{(j_3)}$, the representation $D^{(j_1)}$ appears only once with

$$j_1 = j_2+j_3, j_2+j_3-1,\ldots,|j_2-j_3|.$$

Thus

$$D^{j_2}\otimes D^{(j_3)} = D^{(j_2+j_3)} + D^{(j_2+j_3-1)} + \cdots + D^{(|j_2-j_3|)}$$

The coefficient $\langle j_1m_1|j_2m_2j_3m_3\rangle$ and hence the matrix element (5.51) will vanish unless the following two conditions are satisfied:

$$j_1 = j_2 + j_3, j_2 + j_3 - 1, \ldots, |j_2 - j_3|$$

and (5.52)

$$m_1 = m_2 + m_3$$

This is commonly known as the selection rule for the survival of the matrix element of a tensor operator (in the R_3 group) between two simultaneous eigenvectors of the operators $\mathbf{J}^2$ and $\mathbf{J}_z$.

We note that conditions (5.52) can also be written as

$$\begin{aligned} |\Delta j| &\equiv |j_1 - j_3| \leqslant j_2 \\ \Delta m &\equiv m_1 - m_3 = m_2 \end{aligned} \tag{5.53}$$

Let us consider a rotation through an infinitesimal angle $\delta\theta$ about the z-axis. The operator representing this rotation is, according to (5.35)

$$R_z(\delta\theta) = e^{-(i\delta\theta/\hbar)\mathbf{J}_z} \simeq \left(1 - \frac{i\,\delta\theta}{\hbar}\mathbf{J}_z\right) \tag{5.54}$$

since $\delta\theta$ is infinitesimal. The inverse of this operation is

$$R_z(\delta\theta)^{-1} = e^{(i\delta\theta/\hbar)\mathbf{J}_z} \simeq \left(1 + \frac{i\,\delta\theta}{\hbar}\mathbf{J}_z\right) \tag{5.55}$$

Under this rotation, $R_z(\delta\theta)$, the irreducible tensor operator $\mathbf{T}_m^{(j)}$ is transformed to

$$R_z(\delta\theta)\mathbf{T}_m^{(j)}R_z(\delta\theta)^{-1} = \sum_{m'} D^{(j)}[R_z(\delta\theta)]_{m'm}\mathbf{T}_{m'}^{(j)} \tag{5.56}$$

The left-hand side of this equation is [on using (5.54) and (5.55)]

$$\begin{aligned} \text{L.H.S.} &= \left(1 - i\frac{\delta\theta}{\hbar}\mathbf{J}_z\right)\mathbf{T}_m^{(j)}\left(1 + i\frac{\delta\theta}{\hbar}\mathbf{J}_z\right) \\ &= \mathbf{T}_m^{(j)} - i\frac{\delta\theta}{\hbar}\left[\mathbf{J}_z, \mathbf{T}_m^{(j)}\right] \end{aligned} \tag{5.57}$$

where we have omitted the term with $(\delta\theta)^2$, since $\delta\theta$ is infinitesimally small. According to (5.42), we have for the present case

$$R_z(\delta\theta)|Njm\rangle = \sum_{m'} D^{(j)}[R_z(\delta\theta)]_{m'm}|Njm'\rangle$$

From this we get

$$\langle Njm''|R_z(\delta\theta)|Njm\rangle = \sum_{m'} D^{(j)}[R_z(\delta\theta)]_{m'm}\delta_{m'm''}$$
$$= D^{(j)}[R_z(\delta\theta)]_{m''m}$$

or, by using (5.54),

$$D^{(j)}[R_z(\delta\theta)]_{m''m} = \delta_{m''m} - i\frac{\delta\theta}{\hbar}\langle Njm''|\mathbf{J}_z|Njm\rangle$$

Then the right-hand side of (5.56) is given by

$$\text{R.H.S.} = \mathbf{T}_m^{(j)} - i\frac{\delta\theta}{\hbar}\sum_{m'}\langle Njm'|\mathbf{J}_z|Njm\rangle\mathbf{T}_{m'}^{(j)} \tag{5.58}$$

Using (5.57) and (5.58) in (5.56), we get

$$[\mathbf{J}_z, \mathbf{T}_m^{(j)}] = \sum_{m'}\langle Njm'|\mathbf{J}_z|Njm\rangle\mathbf{T}_m^{(j)} \tag{5.59}$$

and, similarly, we have

$$[\mathbf{J}_x, \mathbf{T}_m^{(j)}] = \sum_{m'}\langle Njm'|\mathbf{J}_x|Njm\rangle\mathbf{T}_{m'}^{(j)}$$
$$[\mathbf{J}_y, \mathbf{T}_m^{(j)}] = \sum_{m'}\langle Njm'|\mathbf{J}_y|Njm\rangle\mathbf{T}_{m'}^{(j)} \tag{5.60}$$

Equation (5.59) at once gives

$$[\mathbf{J}_z\, \mathbf{T}_m^{(j)}] = m\hbar\mathbf{T}_m^{(j)} \tag{5.61}$$

Multiplying the second equation in (5.60) by i and then adding it to the first equation, we obtain

$$[\mathbf{J}_+, \mathbf{T}_m^{(j)}] = \sum_{m'}\langle Njm'|\mathbf{J}_+|Njm\rangle\mathbf{T}_{m'}^{(j)}$$

Similarly, on subtracting instead of adding, we get

$$[\mathbf{J}_-, \mathbf{T}_m^{(j)}] = \sum_{m'}\langle Njm'|\mathbf{J}_-|Njm\rangle\mathbf{T}_m^{(j)}$$

Then, because of (5.39) and (5.40), these equations reduce to

$$[\mathbf{J}_+, \mathbf{T}_m(j)] = \hbar[(j+m+1)(j-m)]^{1/2}\mathbf{T}_{m+1}^{(j)} \tag{5.62}$$

and

$$[\mathbf{J}_-, \mathbf{T}_m(j)] = \hbar[(j-m+1)(j+m)]^{1/2}\mathbf{T}_{m-1}^{(j)} \tag{5.63}$$

This set of three commutation relations—(5.61), (5.62), (5.63),—satisfied by the component operators $\mathbf{T}_m^{(j)}$ can be regarded as a definition for an irreducible tensor operator, $\mathbf{T}^{(j)}$. This definition in terms of the angular momentum operator is due to Racah and, in fact, is exactly equivalent to the usual definition given by the transformation equation (5.50).

Racah's definition is very convenient to use in many cases. Let us give a simple example. Using the usual commutation relations for the components of $\mathbf{J}$, we can easily see that the three operators

$$\mathbf{J}_{+1} = -\frac{1}{\sqrt{2}}\mathbf{J}_+$$

$$\mathbf{J}_0 = \mathbf{J}_z$$

$$\mathbf{J}_{-1} = \frac{1}{\sqrt{2}}\mathbf{J}_-$$

satisfy relations (5.59), (5.60), and (5.61) for $j=1$ and $m=1$, 0, and -1, and hence constitute a tensor operator, $\mathbf{J}^{(1)}$.

Relation (5.51) is readily applied in spectroscopy to find the selection rules for radiative transitions through electric- or magnetic-multipole interactions. The 2^l electric- or magnetic-multipole operator is a $(2l+1)$-component irreducible tensor operator. However, it should be remembered that the electric-multipole operator has parity $(-1)^l$, whereas the magnetic-multipole operator has parity $(-1)^{l+1}$ under inversion. This parity consideration may lead to additional restrictions on the selection rules if the quantum states of the system also have parity (even or odd), depending on one or more of the quantum numbers involved in the selection rules.

As an illustration, let us consider a very simple case—the radiative transitions of a one-electron atom through electric- or magnetic-dipole interaction. We first consider the electric-dipole transition. This dipole interaction is proportional to

$$\mathbf{E}\cdot\boldsymbol{\mu} = \mu(\mathbf{E}_x \sin\theta\cos\phi + \mathbf{E}_y \sin\theta\sin\phi + \mathbf{E}_z\cos\theta)$$

where $\boldsymbol{\mu}$ is the electric-dipole moment operator (vector) for the atom, and $\mathbf{E}$ represents the amplitude of the electric vector of the radiation. In terms of spherical harmonics, we write

$$\mathbf{E}\cdot\boldsymbol{\mu} \sim a_1 Y_1^1(\theta,\phi) + a_0 Y_1^0(\theta,\phi) + a_{-1} Y_1^{-1}(\theta,\phi) \tag{5.64}$$

The probability of transition between two atomic states (without considering the spin-orbit interaction), $|n\,l\,m_l\,m_s\rangle$ and $|n'\,l'\,m_l'\,m_s\rangle$, will be proportional to $|n\,l\,m_l\,m_s|\mathbf{E}\cdot\boldsymbol{\mu}|n'\,l'\,m_l'\,m_s\rangle|^2$. We note that the spin quantum number m_s is the same for the initial and final states, since the electric-dipole moment operator does not act on the spin.

Using (5.64) and (5.51), we then see that $\langle n\,l\,m_l\,m_s|\mathbf{E}\cdot\boldsymbol{\mu}|n'\,l'\,m_l'\,m_s\rangle$ contains terms proportional to the Clebsch-Gordan coefficients $\langle l\,m_l|1\,1\,l'\,m_l'\rangle$, $\langle l\,m_l|1\,0\,l'\,m_l'\rangle$, and $\langle l\,m_l|1-1\,l'\,m_l'\rangle$, and one of them will be nonvanishing only if

$$\Delta l \equiv l' - l = 0, \pm 1$$

and (5.65)

$$\Delta m_l \equiv m_l' - m_l = 0, \pm 1$$

Now the parity of $\mathbf{E}\cdot\boldsymbol{\mu}$ is -1, that is, odd, and hence its matrix element will vanish unless the parities of the initial and final states are different. Therefore for the nonvanishing of the matrix element we further require that both l and l' cannot be even or odd. Imposing this restriction upon (5.65), we obtain the actual selection rules for electric-dipole transitions:

$$\Delta l \equiv l' - l = \pm 1$$
$$\Delta m_l \equiv m_l' - m_l = 0, \pm 1 \tag{5.66}$$

Next, for magnetic-dipole transitions we first obtain rules (5.65) by the same considerations as before. But the magnetic-dipole moment operator has even parity under inversion, and hence we must impose the further restriction that both l and l' must be even or odd, so that the actual selection rule becomes

$$\Delta l = 0$$
$$\Delta m_l = 0, \pm 1 \tag{5.67}$$

Later we shall have many occasions to apply (5.34) as, for example, in the optical absorption analysis of the d^N complexes, which will be developed in Chapter 12.

REFERENCES

1. H. Bethe, *Ann. Phys* **3**, 135 (1929).
2. S. Sugano, Y. Tanabe, and H. Kamimura, *Multiplets of Transition-Metal Ions in Crystals*, Academic Press, New York, 1970.
3. J. S. Griffith, *The Theory of Transition Metal Ions*, Cambridge University Press, London, England, 1961.
4. J. S. Griffith, *The Irreducible Tensor Methods of Molecular Symmetry Groups*, Prentice-Hall, Englewood Cliffs, N.J., 1962.

5. G. F. Koster, J. O. Dimock, R. G. Wheeler, and H. Statz, *The Properties of the Thirty-two Point Groups*, M.I.T. Press, Cambridge, Mass., 1963.

6. S. R. Polo, *Studies on Crystal Field Theory*, RCA Report (1962).

7. G. F. Koster, *Phys. Rev.* **109**, 227 (1958).

8. M. Tinkham, *Group Theory and Quantum Mechanics*, McGraw-Hill, London, 1964.

9. B. R. Judd, *Operator Techniques in Atomic Spectroscopy*, McGraw-Hill, New York, 1963.

10. M. E. Rose, *Elementary Theory of Angular Momentum*, John Wiley and Sons, New York, 1957.

11. E. Wigner and J. von Neumann, *Z. Phys.* **49**, 73 (1928).

6

The Crystal Field Theory

6.1 FORMAL DEVELOPMENTS

The crystal field theory treats an inorganic complex as if such a compound could be regarded as an "ionic" molecule. It is then evident that the central metal atom in the complex is subjected to an electric field originating from the surrounding atoms or molecules, very similar to what would take place if the atom were located in a little cavity inside a crystalline lattice. Such a "crystalline field" would, of course, destroy the spherical symmetry of the free atom, and the consequences of this situation are dealt with by the crystal field theory. This model considers an isolated molecule and treats the electrons of the central metal ion as if they were subjected to an electric field originating from the surrounding molecules, called ligands.

The electrons on the ligands are not allowed to overlap or mix with the electrons of the metal ion. Thus the role played by the ligands is rather limited, and they are supposed only to provide a constant electric potential possessing the symmetry of the arrangement of the ligand nuclei in which the electrons of the metal ion can move.[1] It is important to realize that the word "ionic" as used in the crystal field theory is not to be interpreted in any way as representing the actual reality; rather, it serves as a convenient starting point to understand the reality.

One can formulate the quantitative treatment for such a model easily.[8] The Hamiltonian for the electrons of the metal ion consists of two terms:

$$\mathcal{H} = \mathcal{H}_F + V \tag{6.1}$$

where $\mathcal{H}_F$ is the Hamiltonian for the free ion, and V is the crystal potential provided by its ligands. We suppose that the eigenfunctions and the eigenvalues of $\mathcal{H}_F$ are known. Accordingly the potential V is regarded as a perturbation that determines the electronic motions and term values of the metal ion in the complex.

Our complete Hamiltonian is therefore given by

$$\mathcal{H} = -\frac{\hbar^2}{2m}\sum_i \nabla_i^2 - \sum_i \frac{Ze^2}{r_i} + \frac{1}{2}\sum_{i \neq j} \frac{e^2}{r_{ij}} + \sum_i \zeta(r_i)\mathbf{l}_i \cdot \mathbf{s}_i \tag{6.2}$$

Except for the term in V, (6.2) is exactly the Hamiltonian for the free ion. Before perturbing the eigenvalues of the free ion with V, it is very important to know how V compares, in order of magnitude, with the two other perturbing quantities in (6.2), the interelectronic repulsion term $\Sigma_{i \neq j} e^2/r_{ij}$ and the spin-orbit term $\zeta(r_i)\mathbf{l}_i \cdot \mathbf{s}_i$. It is found that three cases can be realized in the compounds of the transition metal ions:

$$\begin{aligned} &\text{Case I:} && \frac{e^2}{r_{ij}} > V_c > \zeta \mathbf{l}\cdot\mathbf{s} \\ &\text{Case II:} && V_c \geqslant \frac{e^2}{r_{ij}} > \zeta \mathbf{l}\cdot\mathbf{s} \\ &\text{Case III:} && V_c \geqslant \zeta \mathbf{l}\cdot\mathbf{s} > \frac{e^2}{r_{ij}} \end{aligned} \tag{6.3}$$

Here V_c represents the cubic part of the crystal field, which is the dominant term. Case I represents the situation in the complexes of the rare earths; Case II, the complexes of the first transition group; and Case III, the complexes of the palladium and platinum groups. It is customary to call the second case the weak crystal field case and the third case the strong crystal field case. There is, of course, no real distinction between the three cases; actually they overlap. The reason we distinguish between them is simply to specify our starting point.

The general situation in the complexes of the transition metal ions is summarised in Table 6.1. Whereas Cases I and II have been widely studied[2,11] in the last two decades, as will be apparent in the next chapter, Case III, which represents the palladium and platinum groups of complexes, has not been investigated satisfactorily as yet. This case will be dealt with adequately in Chapter 10.

Table 6.1

Rough comparison of the Coulomb, Crystal field, and Spin-orbit energies of Fe, Pd and, Pt group of Ions

Ions	Coulomb energy (cm^{-1})[a]	Crystal field energy (cm^{-1})	Spin-orbit energy (cm^{-1})
Fe group	$10–40\times10^3$	$10–20\times10^3$	100–800
Pd group	$5–20\times10^3$	$10–40\times10^3$	400–2000
Pt group	$5–20\times10^3$	$20–40\times10^3$	800–4000

[a]The second column represents the difference in energy between the ground term and the first excited term.

Since the unperturbed functions, being the eigenfunctions of the Hamiltonian

$$\mathcal{H} = -\frac{\hbar^2}{2m}\sum_i \nabla_i^2 - \sum_i \frac{Ze^2}{r_i} \tag{6.4}$$

are solutions to the problem with full spherical symmetry, it is convenient to expand V in a series of normalized spherical harmonics:

$$V = \sum_{i,j}\sum_{l,m} Y_l^m(\theta_i\phi_i)R_{nl}(r_i) \tag{6.5}$$

the first summation being over the i electrons of the metal ion and j ligands. It is now important to know the symmetry of the ligands around the metal ion because V must transform as the totally symmetric representation in the symmetry group of the molecule. This is due to the fact that the Hamiltonian for the molecule must remain totally symmetric under all symmetry operations. The explicit forms of V for various symmetries of the ligands will be derived in Sections 6.5 to 6.8.

6.2 SINGLE *d* ELECTRON IN CUBIC FIELD

To show the development of the crystal field theory and also to illustrate the application of the Wigner-Eckart theorem, let us consider a transition metal ion with a single electron in its d shell whose ligands are disposed at the apices of an octahedron, the ion being at the center of it. The ionic complex $[Ti(H_2O)_6]^{3+}$ is a good example. If we assume, as in the electrostatic approach,[1] that the ligands at the corners of the octahedron can be regarded as point negative charges, the d electron experiences over and above the spherically symmetric central field due to its nucleus, the electrostatic field due to the charge distribution. If we write the additional potential due to this charge distribution as V_c, the Hamiltonian for the d electron can be written as

$$\mathcal{H} = \mathcal{H}_F + V_c \tag{6.6}$$

It can easily be seen that, since a regular octahedron remains invariant under the operations of the cubic group O_h, so do V_c and also $\mathcal{H}$. We now derive the energies of the d electron in the presence of V_c by the well-known perturbation theory. However, we use a method quite different from the usual procedures. Our purpose is twofold: we do not want to simply repeat the arguments so well known,[4] and we would like to use this example to illustrate the fact that exploitation of the symmetry properties of the Hamiltonian of a physical problem leads to a complete resolution of the problem into two distinct parts, one relating to the symmetry aspects of the problem and the other depending on the physical aspects. The important point is that the part related to the symmetry aspects can be resolved without solving the part depending on the physical aspects. This explains the success of the ligand field theory, where some key parameters are treated semiempirically. The calculation of the parameters, of course, involves the solution of the physical aspects. It should be noted that what has been said above is nothing but a restatement of the Wigner-Eckart theorem (Section 5.7).

If we consider V_c as an operator, operating on the d functions $|d_m\rangle$, the energy levels in the presence of V_c are given by the roots of the matrix

$$\langle d_m | V_c | D_{m'} \rangle$$

where m ranges from -2 to $+2$ since $l=2$ for a d state. Instead of using the pure d functions, we may as well use a linear combination of them. By applying relation (5.19) to the rotation group R_3 and its subgroup O_h (see Appendix V), the d functions spanning the $j=2$ irreducible representation of R_3 can be decomposed into the irreducible representations of O_h as

$$\begin{array}{c} R_3 \quad O_h \\ D^2 \rightarrow T_{2g}, E_g \end{array}$$

The proper linear combinations of the d functions that span the T_{2g} and E_g representations are given by

$$\begin{aligned} |T_{2g}\xi\rangle &= \frac{i}{\sqrt{2}}(d_1 + d_{-1}) \\ |T_{2g}\eta\rangle &= -\frac{1}{\sqrt{2}}(d_1 - d_{-1}) \\ |T_{2g}\zeta\rangle &= -\frac{i}{\sqrt{2}}(d_2 - d_{-2}) \end{aligned} \tag{6.7}$$

and

$$\begin{aligned} |E_g\theta\rangle &= d_0 \\ |E_g\epsilon\rangle &= \frac{1}{\sqrt{2}}(d_2 + d_{-2}) \end{aligned} \tag{6.8}$$

where ξ, η, ζ are the three partners of the T_{2g} representation, and θ, ϵ those of the E_g representation, and

$$d_m = R_{nd}(r)\, Y_2^m(\theta, \phi) \tag{6.9}$$

$R_{nd}(r)$ being the radial part of the d functions of the free-ion Hamiltonian $\mathcal{H}_F$, obtained by solving the appropriate Schrodinger equation. The Y_l^m are the usual spherical harmonics. The zero-order functions (6.7) and (6.8) are so chosen that the ξ, η, ζ, etc., orbitals are all real. This choice makes the vector coupling coefficients real. The nonzero matrix elements of V_c can now be written using the Wigner-Eckart theorem:

$$\begin{aligned} \langle T_{2g}\alpha|V_c|T_{2g}\beta\rangle &= \frac{1}{\sqrt{3}}\langle T_{2g}\|V_c\|T_{2g}\rangle\delta_{\alpha\beta} \\ \langle E_g\alpha|V_c|E_g\beta\rangle &= \frac{1}{\sqrt{2}}\langle E_g\|V_c\|E_g\rangle\delta_{\alpha\beta} \\ \langle E_g\alpha|V_c|T_{2g}\beta\rangle &= 0 \end{aligned} \tag{6.10}$$

Note that here V_c transforms as the A_1 representation in O_h, and not as the $j=0$ representation of R_3.

Thus the energy levels of the d electron in a cubic field depend on two constants, that is, the two reduced matrix elements $\langle T_{2g}\|V_c\|T_{2g}\rangle$ and $\langle E_g\|V_c\|E_g\rangle$. However, on account of the center of gravity rule (see Appendix X), there is only one constant. By convention this is defined as

$$10Dq = E(e_g) - E(t_{2g}) \tag{6.11}$$

where we have written t_{2g} and e_g instead of T_{2g} and E_g, respectively, for a single electron.

The calculation of the actual energies depends on the evaluation of the reduced matrix elements, which we shall discuss in Section 6.5. In the next section we shall use the group-theoretical methods developed in Chapter 5 to find the term splittings in crystals, essentially following a method due to Bethe.[1]

6.3 TERM SPLITTING IN CRYSTALS

6.3.1 Splitting of Orbital Degeneracy

In the case of free ions, an electron is exposed to the spherically symmetric potential and the Hamiltonian is invariant under any rotation around the nucleus. An aggregate of such rotations around the arbitrary axis of direction forms a group called the continuous rotation group, in which the number of elements is infinite. Since rotations by the same angle belong to the same class, whatever may be the direction of the rotation axes, the number of classes is also infinite.

As discussed earlier, any atom or ion in the nl state has the wavefunctions

$$\phi_{nlm}(r,\theta,\phi) = R_{nl}(r)\,Y_l^m(\theta,\phi) \tag{6.12}$$

where $m = -l, -l+1, \ldots, l-1, l$. These form the bases of $(2l+1)$-dimensional irreducible representation $D^{(l)}$ of the continuous rotation group. To find the character of this group, let us consider an irreducible representation R_α for a rotation α around the z-axis. Then

$$R_\alpha \phi_{nlm} = \sum_{m'=-l}^{l} \phi_{nlm'} D^l_{m'm} \tag{6.13}$$

where $D^l_{m'm}$ is the matrix element of the rotation matrix $D^l(\alpha)$. Now, it can be shown that

$$R_\alpha Y_l^m(\theta,\phi) = e^{-im\alpha} Y_l^m(\theta,\phi) \tag{6.14}$$

and the rotation matrix $D^l(\alpha)$ takes the form

$$D^l(\alpha) = \begin{bmatrix} e^{il\alpha} & 0 & 0 & \cdots & & 0 \\ 0 & e^{i(l-1)\alpha} & 0 & & & 0 \\ 0 & 0 & \cdot & & & \\ \vdots & \vdots & & \ddots & & \vdots \\ & & & & \ddots & \\ 0 & 0 & \cdot & \cdots & & e^{-il\alpha} \end{bmatrix} \tag{6.15}$$

from which the character is calculated as

$$\chi^l(\alpha) = \sum_{m=-l}^{l} e^{-im\alpha} = \frac{\sin\left(l+\frac{1}{2}\right)\alpha}{\sin(\alpha/2)}, \qquad (\alpha \neq 0) \tag{6.16}$$

Note that (6.16) is inapplicable if $\alpha = 0$. However, it is obvious that in this case each diagonal element is equal to 1 and the character is, in general, equal to $(2l+1)$; in the present case $\chi(E)=5$, for d electrons.

In the presence of cubic symmetry, however, the Hamiltonian is no longer invariant under any rotation about any axis but is invariant under specific rotations that are the symmetry operations of the cubic group which is a subgroup of the continuous rotation group R_3. In the cubic group the representation D^l is reducible, in general; therefore we have

$$D^l = \sum_k c_{kl} D^k \tag{6.17}$$

where D^k is the kth irreducible representation of the cubic group and c_{kl} is the number of times that D^k appears.

To see the splitting of a term in a crystal of a particular symmetry let us start with the d wavefunctions under a crystal field of O_h symmetry. There are five distinct $3d$ oribital states, and the wavefunctions for each can be written as a product of the radial and angular factors that are given by (6.12). The radial factor is the same for the five $3d$ states, but $Y_2^m(\theta,\phi)$ is different for each, since m takes values from $-2,\ldots,+2$. For an isolated ion these states are degenerate, and any linear combination of these functions gives an equally satisfactory description of the stationary states of the ion. This is so because the high symmetry of the environment leaves the corresponding states, which differ only in the orientations of the orbital momentum, indistinguishable. Now, when the ion is placed in the crystal field of O_h symmetry, these five d wavefunctions form the basis for a representation of the octahedral group O_h, and we have to determine the manner in which the set of d orbitals is split. To determine the representation for which the set of d functions forms a basis, we must first determine the elements of the matrices that express the effect on the set of wavefunctions of each of the symmetry operations in the group. The characters associated with such symmetry operations have already been given by (6.16), which is sufficient to find the splitting of a single d electron in a crystal field of any symmetry.

But the results for a single electron in various types of orbitals apply also to the behavior of terms arising from groups of electrons. For example, just as a single d electron in a free atom has a wavefunction which belongs to a fivefold degenerate set corresponding to the five values that m may take, so a D state arising from a group of electrons has a completely analogous fivefold degeneracy because of the five values that the quantum number M may take. Moreover, the splitting of a D term will be just the same as the splitting of a set of one-electron d oribitals. Exactly the same relationship exists between f orbitals and F states, p orbitals and P states, and so on.

Now consider an atomic term with the orbital angular momentum L. The $(2L+1)$ spherical harmonics Y_L^M, which are degenerate in isotropic space, form the basis for a representation that, following Wigner, we denote by D_L for representations of the rotation group. From (6.16) we get

$$\chi(C_2)=\chi(\pi)=(-1)^L \tag{6.18}$$

$$\chi(C_3)=\chi\left(\frac{2\pi}{3}\right)=\begin{cases} 1 & L=0,3,\ldots \\ 0 & L=1,4,\ldots \\ -1 & L=2,5,\ldots \end{cases} \tag{6.19}$$

$$\chi(C_4)=\chi\left(\frac{\pi}{2}\right)=\begin{cases} 1 & L=0,1,4,5,\ldots \\ -1 & L=2,3,6,7,\ldots \end{cases} \tag{6.20}$$

Since the spherical harmonic of order L forms a basis for representing the group of all rotations, it certainly does so for the finite group of rotations O. In fact, we expect that we may be able to reduce the representation into smaller irreducible ones for this smaller group. To do this formally we merely need the character of D_L, together with the decomposition formula given earlier. We now tabulate the characters of the first few of these reducible representations of O (Table 6.2). Then

$$D_L=\sum_i a_i\Gamma_i \tag{6.21}$$

Table 6.2

Characters of Reducible Representations of O

O	E	$8C_3$	$3C_2$	$6C_2'$	$6C_4$
D_0	1	1	1	1	1
D_1	3	0	−1	−1	1
D_2	5	−1	1	1	−1
D_3	7	1	−1	−1	−1
D_4	9	0	1	1	1

where

$$a_i = (24)^{-1}\sum_k N_k \chi_i(C_k)\chi_L(C_k)$$

where N_k is the number of elements in class C_k.

For the simpler cases we can decompose the representation by inspection by noting, from the irreducible representation table (Appendix V), what rows need to be added together to produce the desired row in the D_L table. Let us list the first few cases:

$D_0 = A_1$ This does not split since it is one dimensional.

$D_1 = T_1$ Thus a P state ($L = 1$) is not split by a cubic field.

$D_2 = E_g + T_{2g}$ This means that D_2 must split since there are no five-dimensional irreducible representations of O. The actual decomposition shows that a D state ($L = 2$) is split into a twofold (E_g) and a threefold (T_{2g}) degenerate levels in O.

$D_3 = A_2 + T_1 + T_2$ Thus an F state ($L = 3$) is split into a nondegenerate and two threefold degenerate states.

$D_4 = A_1 + E + T_1 + T_2$ Thus a G state ($L = 4$) splits into a nondegenerate state, a doubly degenerate state, and two triply degenerate states in O symmetry.

These examples show that the degrees of residual degeneracy of an atom in a crystal field of given symmetry may be calculated almost trivially by using the group theory. Moreover, the various irreducible representations completely determine the symmetry or the transformation properties of the various sets of degenerate eigenfunctions.

6.3.2 Additional Splitting in Field of Lower Symmetry

In most crystals there are at least small departures from cubic symmetry at the lattice site of a magnetic ion. We could handle such cases by a method exactly similar to that employed above for group O, but using the appropriate smaller group. However, we obtain much more insight into the final level structure by considering the problem in steps, first working out the splitting of the free-ion terms into the irreducible representations of the octahedral group, and then working out the additional smaller splitting of these representations under the crystal field of lower symmetry.

As an example, we have shown that an $L=2$ state splits into E and T_2 under group O. Now let us assume that the octahedron of ions producing the crystal field is distorted by pulling out one of the four threefold axes. This is equivalent to a trigonal distortion and reduces the rotational symmetry to group D_3, which is a subgroup of O. To find the additional splitting of E and T_2 levels we need only decompose those representations which may be reducible with respect to subgroup D_3, though irreducible under O. To effect this reduction, we write the characters of the various irreducible representations of O, treated as representations of D_3. This can be done if we note by geometrical inspection that the C_2's of D_3, in this case, are the same as the $6C_2'$ type of twofold rotations of O. In this way we obtain Table 6.3. Comparing the two halves of the table, we see that the one-and two-dimensional representations A_1, A_2, and E are also irreducible under D_3. Hence the corresponding energy levels do not split any further under D_3. Next, we note that T_1 and T_2 must split because there are no three-dimensional representations of D_3. Inspection of the table or use of the decomposition formula (6.21) shows that

$$\begin{aligned} T_1 &\to E + A_2 \\ T_2 &\to E + A_1 \end{aligned} \tag{6.22}$$

Thus each triply degenerate level in a cubic field splits into a doubly degenerate and a nondegenerate level under a trigonal distortion along one of the four threefold axes of O.

The results that we have obtained are shown schematically in Figure 6.1.

Now the question arises, What is the order of these energy levels? The answer is that it is impossible to know the order of these levels without a priori knowledge of at least the perturbing fields. Such information can be

Table 6.3

Comparison of irreducible representations of D_3 and O

Irreducible representations of:		E	$2C_3$	$3C_2$
D_3	A_1	1	1	1
	A_2	1	1	−1
	E	2	−1	0
O	A_1	1	1	1
	A_2	1	1	−1
	E	2	−1	0
	T_1	3	0	−1
	T_2	3	0	1

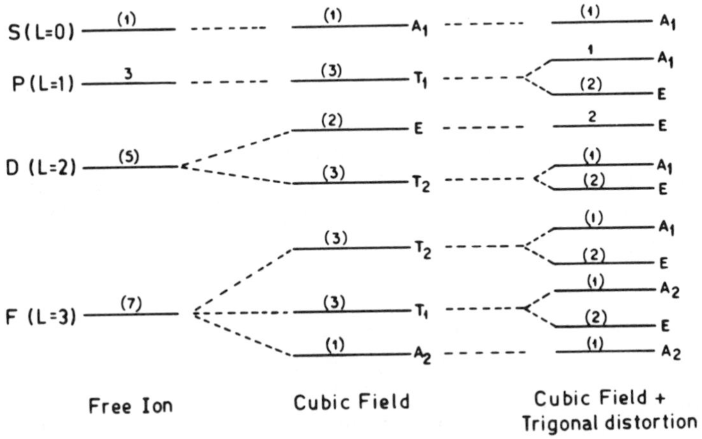

Fig. 6.1 Crystal field splittings (schematic) of the orbital angular momentum states ($L=0$ to 3). The numbers in the parentheses denote the degeneracies of the various energy levels.

obtained from a rough estimate of the crystal field potential, from the experimental absorption spectra of such crystalline complexes, or from the observed magnetic properties. We shall deal with these aspects in greater detail in Chapter 7.

6.3.3 Spin-Orbit Splitting

So far only the orbital degeneracy of an electron has been considered. If now we consider the spin of the electron, the d orbital becomes tenfold degenerate, and to study the splitting of this level it is necessary to consider the crystal double group representation. In this representation the space and spin rotations take place simultaneously, so that

$$R^{so} = R^s R^o \tag{6.23}$$

where R^o is the rotation in the coordinate space, and R^s is the rotation in the spin space. The spin rotation group is treated in exactly the same way as the space rotation, and hence the characters are given by

$$\chi^s(\alpha) = \frac{\sin\left(s+\frac{1}{2}\right)\alpha}{\sin(\alpha/2)} \tag{6.24}$$

The only difference here is that, in $\chi^l(\alpha), l$ is always an integer, whereas, for the spin, s can be a half-integer as well. For half-integral s the characters are, in general, double valued. The states which an electron in cubic symmetry can assume, when the spin-orbit interaction is also included, can be classified according to the irreducible representations of the cubic double group G'. The d state ($l=2$ and $s=\frac{1}{2}$) under the spin-orbit interaction is split into a sixfold $^2D_{5/2}$ and a fourfold $^2D_{3/2}$ term, and the required characters for the representations are obtained from (6.24) by putting $s=5/2$ and $3/2$, respectively. Comparing these characters with those of the cubic double group G' (Appendix V), we find that, for $j=3/2$, the representation is irreducible and corresponds to Γ_8, whereas for $j=5/2$ it is reducible and splits into two irreducible representations of G'. Hence

$$\Gamma(j=5/2)=\Gamma_7+\Gamma_8$$
$$\Gamma(j=3/2)=\Gamma_8 \tag{6.25}$$

We shall have occasion to use the double group representations when we consider the compounds of the palladium and platinum groups in Chapter 10.

6.4 THE LIGAND FIELD THEORY

As discussed earlier, the ligand field is the electric field originated at the site of the transition metal ion by its nearest neighbors. The ligand field theory is a natural extension of the crystal field theory, where the metal-ligand overlap or the covalency, as it is called, is usually taken into account in a semiempirical way (via the orbital reduction factor). The ligands are, however, to the first approximation, still represented by point charges or point dipoles. Some authors,[5] in studying the transition metal complexes or compounds, have used the terminology of the ligand field theory instead of the crystal field theory. In their opinion one should consider all other crystal components, in addition to the ones under immediate consideration, as contributing to this field. While it is meaningful to speak of crystallized complexes, this is no longer justified in the case of an isolated dissolved complex ion. Here the fields of the ligands coordinated about the central ion (i.e., the ligand field) must be considered.

In the ligand field theory one is interested in the changes in the electronic system of the central metal ion under the influence of the electric field produced by the ligands. The free-ion (LS) terms undergo stark splitting in this field. The Hamiltonian for the d electrons in the ligand field is given by (6.2). The eigenfunctions and the eigenvalues of the free ions were discussed earlier and are assumed to be known. The free ionic wavefunctions in the (LSM_LM_S) representation are the starting wavefunctions for the perturbation calculation for the crystal field designated by V, which is given by

$$V=\sum_i V(r_i)=\sum_{m,n} A_n^m r^n P_n^m(\cos\theta)e^{im\phi} \tag{6.26}$$

where

$$A_n^m = \sum_k \frac{e^2 Z_k}{r_k^{n+1}} \frac{(n-|m|)!}{(n+|m|)!} P_n^m(\cos\theta_k) e^{-im\phi_k}$$

in the point charge model.

The ligand field potential $V(r_i)$ is assumed to satisfy Laplace's equation and to have the local site symmetry of the metal ion.

In many complexes of the transition metal ions it has been observed that the effect of the ligand field, denoted by V, is comparable to $\mathcal{H}_F$, although both $\mathcal{H}_F$ and V are greater than $\mathcal{H}_{so}$. In such cases it is better to consider first the effect of V on the free-ion wavefunctions. This gives the crystal field wavefunctions which are linear combinations of the free-ion wavefunctions. One then considers the e^2/γ_{ij} interaction on these resulting wavefunctions which splits the ionic configuration into terms. This situation is very similar to that of the free-ion case. Now, however, $\mathcal{H}_{so}$ is considered as a perturbation. If the ligand field is very strong, Hund's rule may break down and consequently the ionic ground state may have an anomalously low spin. The cyanide complexes of the iron group fall in this category.

We shall treat the weak-field coupling scheme first [Case II of (6.3)] because the complexes of the iron group can be adequately described by this scheme. The strong-field and the intermediate coupling schemes, which are appropriate for the palladium and platinum groups, will be dealt with in Chapters 9 and 10, respectively.

6.5 LIGAND FIELD OF O_h SYMMETRY

Placing the metal atom at the center of a cube, we can coordinate the ligands with the metal atom in various ways, producing different crystal field symmetries. For example, when the ligands are placed at the centers of the cube faces (there will be six of them), a ligand field of octahedral symmetry (O_h) is produced. The fourfold coordination of the ligands gives a tetrahedral ligand field, denoted by T_d (Figure 5.3). Similarly, when eight identical ligand atoms are situated at the corners of the cube, we get a ligand field of cubic symmetry, denoted by O. The important point groups that are essential for our purpose were discussed in Chapter 5, and we do not treat them further.

We now derive the ligand field potential of O_h symmetry in the point charge model. The potential energy due to the interaction of two charges, e and eZ_k, situated at P and P_k and separated by a distance R_k, is given by e^2Z_k/R_k (Figure 6.2). The coordinates of P and P_k are (r,θ,ϕ) and (r_k,θ_k,ϕ_k), respectively. Then

$$R_k^2 = r^2 + r_k^2 - 2rr_k\cos\gamma$$

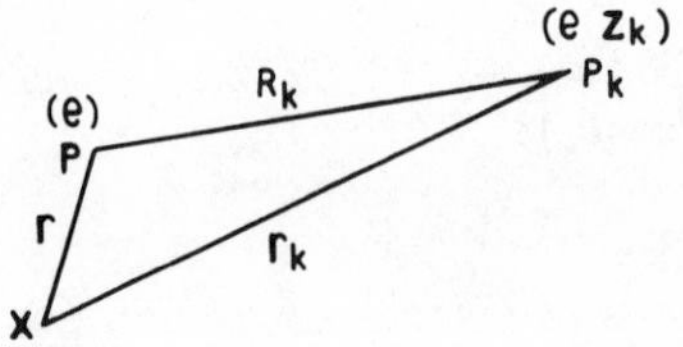

Fig. 6.2 The metal atom is situated at X, the ligand atom of charge eZ_k at P_k and the electron belonging to X at a distance r from X.

or

$$\frac{1}{R_k}=\frac{1}{r_k}\left\{1+\frac{r}{r_k}\left(\frac{r}{r_k}-2\cos\gamma\right)\right\}^{-1/2} \tag{6.27}$$

If $r_k > r$, (6.27) may be expanded as

$$\begin{aligned}\frac{1}{R_k}&=\frac{1}{r_k}\left\{1-\frac{1}{2}\frac{r^2}{r_k^2}+\frac{r}{r_k}\cos\gamma+\frac{3}{4}\cdot\frac{1}{2}\frac{r^2}{r_k^2}4\cos^2\gamma+\cdots\right\}\\&=\frac{1}{r_k}\left\{1+\frac{r}{r_k}\cos\gamma+\frac{1}{2}\frac{r^2}{r_k^2}(3\cos^2\gamma-1)+\cdots\right\}\\&=\frac{1}{r_k}\sum_{n=0}^{\infty}\frac{r^n}{r_k^{n+1}}P_n(\cos\gamma)\end{aligned} \tag{6.28}$$

where $P_n(\cos\gamma)$ are the Legendre polynomials. From the addition theorem of Legendre polynomials,[4] we have

$$P_n(\cos\gamma)=\sum_{m=-n}^{n}\frac{(n-|m|)!}{(n+|m|)!}P_n^m(\cos\theta)P_n^m(\cos\theta_k)e^{im(\phi-\phi_k)} \tag{6.29}$$

where the angular parts of the polar coordinates of P and P_k are related to the angle γ. Hence, if we have k point charges interacting with the electron at P, the potential is given by

$$V(r)=\sum_k\frac{Z_ke^2}{R_k} \tag{6.30}$$

If $r_k > r$, for all point charges eZ_k, we may replace $1/R_k$ by (6.28); then, substituting for $P_n(\cos\gamma)$ from (6.29), we obtain

$$V(r)=\sum_k\sum_{n=0}^{\infty}\frac{r^n}{r_k^{n+1}}\sum_{m=-n}^{n}\frac{(n-|m|)!}{(n+|m|)!}P_n^m(\cos\theta)P_n^m(\cos\theta_k)e^{im(\phi-\phi_k)} \tag{6.31}$$

The relation between the Legendre polynomial and the spherical harmonics is given by

$$Y_n^m(\theta,\phi)=(-1)^n\left[\frac{1}{4\pi}\frac{(2n+1)(n-|m|)!}{(n+|m|)!}\right]P_n^m(\cos\theta)e^{im\phi} \tag{6.32}$$

Therefore $V(r)$, in terms of spherical harmonics, is

$$V(r)=\sum_{k}\sum_{n=0}^{\infty}\frac{4\pi}{2n+1}\frac{e^2Z_k}{r_k^{n+1}}\cdot r^n\sum_{m=-n}^{n}Y_n^m(\theta,\phi)\,Y_n^{m*}(\theta_k,\phi_k) \tag{6.33}$$

where $Y_n^{m*}(\theta_k,\phi_k)$ is the complex conjugate of $Y_n^m(\theta_k,\phi_k)$ and is equal to $(-1)^nY_n^{-m}(\theta_k,\phi_k)$. Thus

$$V(r)=\sum_{n=0}^{\infty}\sum_{m=-n}^{n}A_n^m r^n Y_n^m(\theta,\phi)$$

where

$$A_n^m=\sum_{k}\frac{4\pi}{2n+1}\frac{e^2Z_k}{r_k^{n+1}}Y_n^{m*}(\theta_k,\phi_k) \tag{6.34}$$

Since we are interested only in the d electrons, in order to derive the ligand field potential for the six-coordinated cube (the O_h symmetry), we consider only $n=0$ and $n=4$. The higher terms in the potential expansion are neglected as they have no matrix elements between the states of d electrons. For O_h symmetry, therefore, the summation over k is from 1 to 6, and the polar coordinates for the six ligands are $(a,0,0)$, $(a,\pi/2,0)$, $(a,\pi/2,\pi/2)$, $(a,\pi/2,\pi)$, $(a,\pi/2,3\pi/2)$, $(a,\pi,0)$.

Hence, assuming the charges on the ligands to be the same, we have, for $n=0$, $m=0$,

$$A_0^0=\sqrt{4\pi}\,\frac{6Ze^2}{a}$$

for $n=4$, $m=0$,

$$A_4^0=\frac{7}{2}\sqrt{\frac{4\pi}{9}}\,\frac{Ze^2}{a^5}$$

and for $n=4$, $m=\pm4$

$$A_4^4+A_4^{-4}=\sqrt{\frac{35}{8}}\sqrt{\frac{4\pi}{9}}\,\frac{Ze^2}{a^5}$$

Thus

$$V(r)=\sqrt{4\pi}\,\frac{6Ze^2}{a}+\frac{7}{2}\sqrt{\frac{4\pi}{9}}\,\frac{Ze^2}{a^5}r^4\left[Y_4^0+\sqrt{\frac{5}{14}}\left(Y_4^4+Y_4^{-4}\right)\right] \tag{6.35}$$

The first term of (6.35) represents the potential energy of the electron located at the central ion, and this corresponds to a constant shift of the energy levels, whereas the second term lifts the degeneracy of the level. Thus the cubic field potential for O_h symmetry is

$$V_c(r) = A_4^0 r^4 \left[Y_4^0 + \sqrt{\frac{5}{14}} \left(Y_4^4 + Y_4^{-4} \right) \right] \tag{6.36}$$

where

$$A_4^0 = \frac{7}{2} \sqrt{\frac{4\pi}{9}} \frac{Ze^2}{a^5}$$

Similarly, for the eight-coordinated and the four-coordinated ligands, the perturbing potential has the same form, but the ligand parameters are

$$A_4^0 = \frac{28}{9} \sqrt{\frac{4\pi}{9}} \frac{Ze^2}{a^5} \qquad \text{for eight coordinations}$$

and

$$A_4^0 = \frac{14}{9} \sqrt{\frac{4\pi}{9}} \frac{Ze^2}{a^5} \qquad \text{for four coordinations}$$

These are the forms of the potentials when (0 0 1) is the z-axis. However, if (1 1 1) is the z-axis, the perturbing potential takes the form

$$V_c(r) = A_4^0 r^4 \left[Y_4^0 + \sqrt{\frac{10}{7}} \left(Y_4^3 + Y_4^{-3} \right) \right] \tag{6.37}$$

If we use Cartesian coordinates instead, (6.36) becomes

$$V_c(r) = D\left(x^4 + y^4 + z^4 - \frac{3}{5} r^4 \right) \tag{6.38}$$

where

$$D = \frac{5}{2} A_4^0 = \frac{35}{4} \sqrt{\frac{4\pi}{9}} \frac{Ze^2}{a^5}$$

In principle, it is possible to calculate the constant D, provided that we know the charge distribution around the cation, but usually D is taken as a semiempirical parameter to be determined from experiment. For sixfold coordination (O_h), D is positive, whereas it is negative for fourfold (T_d) and eightfold (O and D_{2d}) coordinations. Another constant, q, arises from (6.38) when one calculates the matrix element of this equation between the d-state wavefunctions; this is given by

$$q = \frac{2}{105}\int_0^\infty R^2(r) r^4 \cdot r^2\, dr = \frac{2}{105}\langle r^4\rangle \tag{6.39}$$

It is important to note that the radial integral q must be treated as another constant since the radial wavefunctions for the transition metal ions in complexes are not known accurately, though very good SCF wavefunctions are now available for the first and second series of transition metal ions. There is enough experimental evidence by now to show that these SCF wavefunctions of free ions are not adequate for these ions when they occur in complexes. However, q always appears multiplied by D, and thus Dq is usually treated as a semiempirical parameter and is determined by experiment. Recently, however, Richardson and his co-workers[6] have tried to calculate Dq *ab initio* for the iron group of complexes, using Roothaan's open shell procedure.[7] The reader is referred to Roothaan's paper for further details.

Since D is different for different coordinations in a cube, the Dq's for different symmetries (O, O_h, T_d) bear the ratio[2]

$$(Dq)_{O_h} : (Dq)_{T_d} : (Dq)_O = 1 : \tfrac{4}{9} : \tfrac{8}{9} \tag{6.40}$$

Let us now concentrate in some detail on the ligand field of octahedral symmetry O_h, which arises from sixfold coordination. If we consider V_c, given by (6.38), as an operator operating on the (θ, ϕ) part of the d functions $|m\rangle$, where $m = -2, \ldots, +2$, the energy levels in the presence of V_c are given by the roots of the matrix $\langle m|V_c|m'\rangle$. If (001) is taken as the axis of quantization (z-axis), the matrix elements between the states of the d electron are given by[8]

$$\begin{aligned}
\langle \pm 2|V_c|\pm 2\rangle &= Dq\\
\langle \pm 2|V_c|\mp 2\rangle &= 5\,Dq\\
\langle \pm 1|V_c|\pm 1\rangle &= -4\,Dq\\
0|V_c|0\rangle &= 6\,Dq
\end{aligned} \tag{6.41}$$

If, instead, the axis of quantization is the threefold axis (1 1 1), then

$$\begin{aligned}
\langle \pm 2|V_c|\pm 2\rangle &= -\frac{2}{3}\,Dq\\
\langle \pm 1|V_c|\pm 1\rangle &= \frac{8}{3}\,Dq\\
\langle 0|V_c|0\rangle &= -4\,Dq\\
\langle \pm 2|V_c|\mp 1\rangle &= \pm\frac{10\sqrt{2}}{3}\,Dq
\end{aligned} \tag{6.42}$$

Table 6.4

Eigenvalues and Eigenfunctions of the terms $^2T_{2g}$ and ^2E_g

		Wavefunctions	
Term	Energy	Fourfold axis as the axis of quantization	Threefold axis as the axis of quantization
$^2T_{2g}$	$-4Dq$	$\xi=\frac{i}{\sqrt{2}}(\|1\rangle+\|-1\rangle)$	$x_+=-\sqrt{\tfrac{2}{3}}\left\|-2\right\rangle-\frac{1}{\sqrt{3}}\left\|1\right\rangle$
		$\eta=-\frac{1}{\sqrt{2}}(\|1\rangle-\|-1\rangle)$	$x_-=\sqrt{\tfrac{2}{3}}\left\|2\right\rangle-\frac{1}{\sqrt{3}}\left\|-1\right\rangle$
		$\zeta=-\frac{i}{\sqrt{2}}(\|2\rangle-\|-2\rangle)$	$x_0=\|0\rangle$
^2E_g	$6Dq$	$\theta=\|0\rangle$	$u_+=-\frac{1}{\sqrt{3}}\left\|-2\right\rangle+\sqrt{\tfrac{2}{3}}\left\|1\right\rangle$
		$\epsilon=\frac{1}{\sqrt{2}}(\|2\rangle+\|-2\rangle)$	$u_-=\frac{1}{\sqrt{3}}\left\|2\right\rangle+\sqrt{\tfrac{2}{3}}\left\|-1\right\rangle$

Diagonalizing these matrices given by (6.41) and (6.42), we obtain the required eigenvalues and eigenfunctions of the terms $^2T_{2g}$ and ^2E_g. These are given in Table 6.4. Thus, by the action of the cubic field, the fivefold degeneracy of the d electron is partially removed, and we get a twofold (^2E_g) and a threefold ($^2T_{2g}$) degenerate level. If we now introduce the spin-orbit interaction, the degeneracy of the $^2T_{2g}$ level is further removed (in which case we have to go to the crystal double group representation O_h'), whereas the ^2E_g state remains unsplit, as shown in Figure 6.3. The spin-orbit interaction will be discussed in detail in Chapter 10.

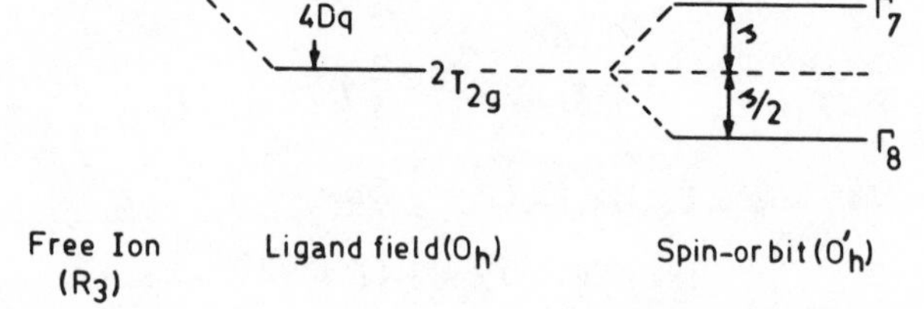

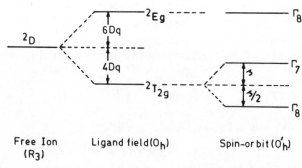

Fig. 6.3 Splitting of the ^2D state by ligand field of O_h symmetry and the spin-orbit interaction.

6.6 LIGAND FIELD OF D_{4h} SYMMETRY

A ligand field of tetragonal (D_{4h}) symmetry is produced when the octahedron is pulled along one of the fourfold axes (C_4), which is taken to be the z-axis. The result of such a distortion or deviation from O_h is usually treated as a perturbation over the octahedral field, so that we can write

$$V = V_c + V_T \tag{6.43}$$

where V_c is given by (6.38) and V_T represents the tetragonal field. In the point charge model, when the ligand positions in D_{4h} symmetry for sixfold coordination are $(a,\pi/2,0)$, $(a,\pi/2,3\pi/2)$, $(a,\pi/2,\pi)$, $(a,\pi/2,\pi/2)$, $(b,0,0)$ and $(b,\pi,0)$, the potential is derived in the same way[8] as for O_h and is given by

$$\begin{aligned} V(r) &= \sqrt{4\pi}\, Ze^2\left(\frac{4}{a}+\frac{2}{b}\right)Y_0^0+\sqrt{\frac{4\pi}{5}}\,2Ze^2r^2\left(\frac{1}{b^3}-\frac{1}{a^3}\right)Y_2^0 \\ &\quad +\sqrt{\frac{4\pi}{9}}\,\frac{Ze^2}{2}r^4\left(\frac{4}{b^5}+\frac{3}{a^5}\right)Y_4^0+\sqrt{\frac{4\pi}{9}}\,\frac{Ze^2}{a^5}\sqrt{\frac{35}{8}}\,r^4\left(Y_4^4+Y_4^{-4}\right) \\ &= V_c + A_2^0 r^2 Y_2^0 + B_4^0 r^4 Y_4^0 \\ &= V_c + V_T \end{aligned} \tag{6.44}$$

In the operator-equivalent form,[9] V_T is given by

$$V_T = Ds\left(l_z^2-2\right) - Dt\left(\tfrac{35}{12}l_z^4-\tfrac{155}{12}l_z^2+6\right) \tag{6.45}$$

where Ds, Dt are the tetragonal field parameters to be determined from experiment; $Ds \propto r^2$ and $Dt \propto r^4$, and their magnitudes can roughly be estimated from $\langle r^2\rangle$ and $\langle r^4\rangle$, in the point charge model, provided that the wavefunctions are known.

Now let us consider the problem of a single d electron placed in the tetragonal field. Since this field consists of the cubic field with some distortion V_T, the cubic field terms ($^2T_{2g}$, ^2E_g) undergo further splitting by V_T. The term $^2T_{2g}$ is split into a doublet, ^2E_g, and a singlet, $^2B_{2g}$, whereas ^2E_g is split into two singlets, $^2A_{1g}$ and $^2B_{1g}$, belonging to the irreducible representations of D_{4h}. Thus a d electron, placed in a ligand field of D_{4h} symmetry, splits into an orbital doublet (^2E_g) and three orbitally nondegenerate levels ($^2B_{2g}$, $^2A_{1g}$, $^2B_{1g}$). Making use of the basic matrix elements of the axial field, which can easily be calculated using (6.45), we have

$$\begin{aligned} \langle \pm 2|V_T|\pm 2\rangle &= 2Ds - Dt \\ \langle \pm 1|V_T|\pm 1\rangle &= -Ds + 4Dt \\ \langle 0|V_T|0\rangle &= -2Ds - 6Dt \end{aligned} \tag{6.46}$$

and the energies of the different levels become

$$\left.\begin{aligned} E(e_g) &= -4\,Dq - Ds + 4\,Dt \\ E(b_{2g}) &= -4\,Dq + 2\,Ds - Dt \\ E(a_{1g}) &= 6\,Dq - 2\,Ds - 6\,Dt \\ E(b_{1g}) &= 6\,Dq + 2\,Ds - Dt \end{aligned}\right\} \tag{6.47}$$

That the cubic (O_h) orbitals go over directly to the bases of irreducible representations of D_{4h} is evident from the transformation properties (see Appendix V). Therefore ξ and η correspond to the e_g orbitals, ζ to b_{2g}, θ to a_{1g}, and ϵ to b_{1g}. If the splittings of $^2T_{2g}$ and ^2E_g are denoted by δ and μ, respectively, then, from energy expressions (6.47),

$$\begin{aligned} \delta &= 3\,Ds - 5\,Dt \\ \mu &= 4\,Ds + 5\,Dt \end{aligned} \tag{6.48}$$

This is true for $Ds > 0$ and $Dt > 0$, that is, in the case of an elongated octahedron. To maintain the center of gravity rule, the energies of these orbitals,[2] in terms of δ and μ, are

$$\left.\begin{aligned} E(\xi,\eta) &= -\frac{\delta}{3} \\ E(\zeta) &= \frac{2\delta}{3} \\ E(\theta) &= -\frac{\mu}{2} \\ E(\epsilon) &= \frac{\mu}{2} \end{aligned}\right\} \tag{6.49}$$

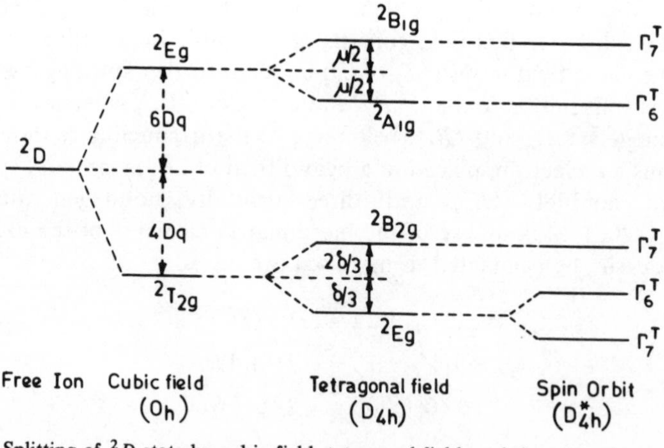

Fig. 6.4 Splitting of ^2D state by cubic field, tetragonal field, and the spin-orbit interaction.

These levels of D_{4h} symmetry are further split by the spin-orbit interaction, and as a result we get five Kramers' doublets, belonging to the Γ_6^T and Γ_7^T irreducible representations of the tetragonal double group D'_{4h}. The splitting of a d electronic state is shown schematically in Figure 6.4. Distortion from the O_h symmetry is also possible by compressing the out-of-plane ligands. In that case $Ds<0$, $Dt<0$, and the level structure of (6.49) becomes reversed.

6.7 LIGAND FIELD OF D_{3d} SYMMETRY

If the octahedron is distorted along the trigonal (1 1 1) axis, a splitting of the octahedral levels similar to that found in the tetragonal case takes place. The symmetry is lowered from octahedral (O_h) to trigonal (D_{3d} or C_{3v}). In this case it is convenient to take (1 1 1) as the z-axis, which is considered to be the axis of quantization. Here, as in Section 6.6, the ligand field is made up of a cubic and an axial field:

$$V = V_c + V_\tau \tag{6.50}$$

where V_τ is the trigonal field, and has the same operator-equivalent form as V_T, given in (6.45). The only difference is that the axis of quantization here is the (1 1 1) axis, and the parameters $D\sigma$ and $D\tau$ are used instead of Ds and Dt.

Because of the stretching of the sixfold coordinated cube along the (1 1 1) axis, the octahedral level ${}^2T_{2g}$ is split into an orbital doublet, ${}^2E_g$, and an orbital singlet, ${}^2A_{1g}$, belonging to the irreducible representations of D_{3d}. The octahedral level ${}^2E_g$, however, remains unsplit in D_{3d}. Thus a single d electron in a ligand field of D_{3d} symmetry splits into two orbital doublets ${}^2E_g$ and one singlet ${}^2A_{1g}$. Starting from the weak-field matrix elements:[9]

$$\begin{aligned}
\langle \pm 2|V_\tau|\pm 2\rangle &= 2D\sigma - D\tau \\
\langle \pm 1|V_\tau|\pm 1\rangle &= -D\sigma + 4D\tau \\
\langle 0|V_\tau|0\rangle &= -2D\sigma - 6D\tau
\end{aligned} \tag{6.51}$$

the matrix elements of V_τ between the trigonal terms are

$$\left.\begin{aligned}
\langle {}^2E_g(t_{2g})|V_\tau|{}^2E_g(t_{2g})\rangle &= -4Dq + D\sigma + \tfrac{2}{3}D\tau \\
\langle {}^2E_g(e_g)|V_\tau|{}^2E_g(e_g)\rangle &= 6Dq + \tfrac{7}{3}D\tau \\
\langle {}^2E_g(t_{2g})|V_\tau|{}^2E_g(e_g)\rangle &= \sqrt{2}\left(D\sigma - \tfrac{5}{3}D\tau\right) \\
&= \langle {}^2E_g(e_g)|V_\tau|{}^2E_g(t_{2g})\rangle \\
\langle {}^2A_{1g}(t_{2g})|V_\tau|{}^2A_{1g}(t_{2g})\rangle &= -4Dq - 2D\sigma - 6D\tau
\end{aligned}\right\} \tag{6.52}$$

Since the cubic bases go directly over to the lower symmetry representation (D_{3d}), as is evident from the transformation properties (Appendix V), the orbitals (x_+, x_-) and (u_+, u_-) belong to ^2E_g and x_0 to $^2A_{1g}$ irreducible representations. Again, there is a mixing between the two ^2E_g terms via the axial distortion, and the energies of these two terms are obtained by diagonalizing the matrix

	$^2T_{2g}$	^2E_g
$^2T_{2g}$	$-4Dq + D\sigma + \frac{2}{3}D\tau$	$\sqrt{2}\,(D\sigma - \frac{5}{3}D\tau)$
^2E_g		$6Dg + \frac{7}{3}D\tau$

When this matrix is diagonalized, the energies are given by

$$E(^2T_{2g}) = Dq + \frac{D\sigma + 3D\tau}{2} - \frac{1}{2}\Delta^{1/2}$$

$$E(^2Eg) = Dq + \frac{1}{2}(D\sigma + 3D\tau) + \frac{1}{2}\Delta^{1/2} \tag{6.54}$$

where

$$\Delta = 100\,Dq^2 - \frac{20}{3}Dq(3D\sigma - 5D\tau) + (3D\sigma - 5D\tau) \tag{6.55}$$

The levels in the D_{3d} ligand field are further split by the spin-orbit perturbation, and we get five Kramers' doublets belonging to the irreducible representations Γ_4^τ, Γ_5^τ, and Γ_6^τ of the trigonal double group D_{3d}'. The representation Γ_4^τ is doubly degenerate, whereas Γ_5^τ and Γ_6^τ are nondegenerate. But these nondegenerate representations are Kramers conjugate and constitute a Kramers' doublet. The splitting of the d level in the trigonal field is shown schematically in Figure 6.5.

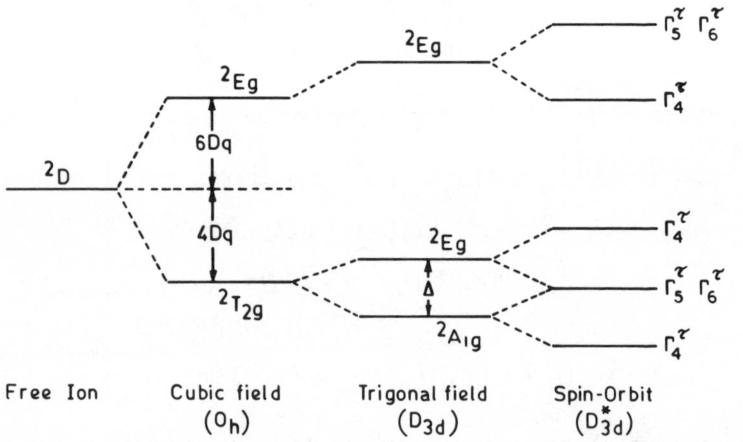

Fig. 6.5 Splitting of ^2D state by cubic field, trigonal field of D_{3d} symmetry, and the spin-orbit interaction.

The diagram of Figure 6.5 is also true when the octahedron is compressed along the trigonal axis. As in a tetragonal distortion if the octahedron is compressed along the threefold axis, the order of the two lowest terms becomes inverted.

6.8 LIGAND FIELD OF T_d SYMMETRY

As already mentioned, the regular tetrahedron (point group T_d) belongs to the cubic point group. We have already seen that the symmetry operations are E, $8C_3$, $3C_2$, $6S_4$, and $6\sigma_d$. As long as we stay inside the manifold spanned by the d orbitals, the crystal field potential $V(T_d)$ is identical with that of V_c, and the d orbitals split into a threefold degenerate t_2 level and a two fold degenerate e level, just as in the case of octahedral symmetry. This is evident from the transformation properties of these two point groups, given in Appendix V. Notice that we have to drop the subscript g in the orbital designations because of the lack of a center of symmetry in the molecule.

The magnitude of the splitting in a tetrahedral molecule, as compared to that in an octahedral molecule, was evident in (6.51) for a point charge model. It is also important to note that the level structure is inverted in going from an octahedron to a tetrahedron. This reversal was first pointed out by Gorter.[10]

Since the character table for T_d and O are identical, all the theory developed for the O_h point group is applicable to T_d molecules also. The only necessary change is to reverse the sign of Dq. This holds strictly as long as we stay inside the d orbitals, but a complication arises in that the tetrahedral molecule does not have a center of symmetry. It can be shown that the expansion of the tetrahedral potential contains an odd (third-order) potential in T_d, in contrast to what is found in O_h. This term cannot, of course, have matrix elements between two d orbitals, but it may very well mix the d and p states.

Having dealt with all the important point groups of interest to us, we now give, in Table 6.5, the splitting of the one-electron levels in various symmetries. The results we have obtained so far for a single electron in various types of orbitals also apply to the behavior of terms arising from groups of electrons, as we remarked earlier. Table 6.6 gives the ground terms of the free ions of the iron group of elements, together with the irreducible representations into which the terms decompose under a crystal field of O_h symmetry.

In Table 6.5 we have used small letters to represent the states of a single electron in environments of various symmetries. The small letters $s, p, d, f, \ldots$ represent the states in the free atom. Similarly, we shall use capital letters to represent the states into which the environment splits the terms of the free ion. Thus, for example, an F state of a free ion will be split into the states A_2, T_1, and T_2 when the ion is placed in the center of a tetrahedral environment.

The subscripts g and u in Table 6.5, from the German words *gerade* and *ungerade*, mean "even" and "odd" under reflection at the center of symmetry.

Table 6.5

Splitting of levels under crystal field of different symmetries

Type of level	Octahedral crystal field, O_h	Tetrahedral crystal field, T_d	Tetragonal crystal field, D_{4h}	Trigonal crystal field, D_3
s	a_{1g}	a_1	a_{1g}	a_1
p	t_{1u}	t_1	$a_{2u}+e_u$	a_2+e
d	e_g+t_{2g}	$e+t_2$	$a_{1g}+b_{1g}+b_{2g}+e_g$	a_1+2e
f	$a_{2u}+t_{1u}+t_{2u}$	$a_2+t_1+t_2$	$a_{2u}+b_{1u}+b_{2u}+2e_u$	a_1+2a_2+2e
g	$a_{1g}+e_g+t_{1g}+t_{2g}$	$a_1+e+t_1+t_2$	$2a_{1g}+a_{2g}+b_{1g}+b_{2g}+2e_g$	$2a_1+a_2+3e$

Table 6.6

Splitting of terms in octahedral crystal field

Free ion	Ground Term	Decomposition in O_h symmetry	Lowest lying representation
Ti^{3+}, V^{2+}	$3d^1, {}^2D$	${}^2E_g+{}^2T_{2g}$	${}^2T_{2g}$
Ti^{2+}, V^{3+}	$3d^2, {}^3F$	${}^3A_{2g}+{}^3T_{2g}+{}^3T_{1g}$	${}^3T_{1g}$
V^{2+}, Cr^{3+}	$3d^3, {}^4F$	${}^4A_{2g}+{}^4T_{2g}+{}^4A_{1g}$	${}^4A_{2g}$
Cr^{2+}, Mn^{3+}	$3d^4, {}^5D$	${}^5E_g+{}^5T_{2g}$	${}^5E_g$
Mn^{2+}, Fe^{3+}	$3d^5, {}^6S$	${}^6A_{1g}$	${}^6A_{1g}$
Fe^{2+}	$3d^6, {}^5D$	${}^5E_g+{}^5T_{2g}$	${}^5T_{2g}$
Co^{2+}	$3d^7, {}^4F$	${}^4A_{2g}+{}^4T_{2g}+{}^4T_{1g}$	${}^4T_{1g}$
Ni^{2+}	$3d^8, {}^3F$	${}^3A_{2g}+{}^3T_{2g}+{}^3T_{1g}$	${}^3A_{2g}$
Cu^{2+}	$3d^9, {}^2D$	${}^2E_g+{}^2T_{2g}$	${}^2E_g$

The use of g and u is governed by the following rules. If the point group of the environment possesses no center of symmetry, no subscripts are used since they are meaningless. When the environment possesses a center of symmetry, the subscript is determined by the type of orbital. All atomic orbitals for which the orbital quantum number l is even, for example, $s, d, g, \ldots$, being centrosymmetric, are of g character. All atomic orbitals for which l is odd, for example, $p, f, h, \ldots$, being antisymmetric to inversion, are of u character.

Which of the levels will lie lowest in each case is ultimately decided from study of the magnetic anisotropy as found from experiments or from paramagnetic resonance studies. The effect of the static crystal field on the ground terms of the $3d^n$ ions embedded in a crystal is to split the terms, and consequently the orbital magnetic moment becomes quenched (partially or totally), giving the "spin-only" value. To derive the expression for the paramagnetic susceptibility one has to apply the magnetic perturbation on the lowest lying representation given in Table 6.6. Fortunately this is not difficult to do, since the wavefunctions for the different levels belonging to particular irreducible representations are also known from the symmetry considerations, as discussed earlier in this chapter. In most cases it is possible to confine oneself to the lowest representation; the upper representations, being nearly 10^4 cm^{-1} away, will contribute very little to the magnetic susceptibility since they will not be appreciably populated even at room temperature. For the iron group of complexes having the $3d^n$ configuration, however, it is very important to realize that the spin-orbit coupling constant is comparable in magnitude to the crystal field of symmetry lower than O_h. Thus in the next stage one has to include the perturbation due to the tetragonal or trigonal part of the crystal field, along with the spin-orbit interaction. In the last stage one considers the Zeeman interaction on the lowest lying energy levels in order to arrive at the paramagnetic susceptibility.

We shall now deal with the situation where the lowest representation is orbitally degenerate. For the nondegenerate representation lying lowest, for example, in $Ni^{2+}(3d^8)$, the magnetic properties can be described adequately, to the first approximation, by a spin Hamiltonian that we shall describe very briefly in Chapter 8. For this aspect we refer the reader to the excellent book by Abragam and Bleaney.[11]

REFERENCES

1. H. Bethe, *Ann. Phys.* **3**, 133 (1929); English translation, Consultants Bureau, New York, 1958.
2. S. Sugano, Y. Tanabe, and H. Karimura, *Multiplets of Transition-Metal Ions in Crystal*, Academic Press, New York, 1970.
3. W. G. Penney and R. Schlapp., *Phys. Rev.* **41**, 194 (1932).
4. B. R. Judd, *Operator Techniques in Atomic Spectroscopy*, McGraw-Hill, New York, 1967.
5. H. L. Schlafer and G. Gliemann, *Basic Principles of Ligand Field Theory*, (translated by D. F. Ilten), Wiley-Interscience, London, 1959.
6. J. W. Richardson, T. F. Soules, D. M. Vaught, and R. R. Powell, *Phys. Rev.* **B4**, 1721 (1971).
7. C. C. J. Roothaan, *Rev. Mod. Phys.* **32**, 179 (1960).
8. C. J. Ballhausen, *Introduction to Ligand Field Theory*, McGraw-Hill, New York, 1962.
9. T. M. Dunn, *Trans. Faraday Soc.* **57**, 1441 (1961).
10. C. J. Gorter, *Phys. Rev.* **42**, 437 (1932).
11. A. Abragam and B. Bleaney, *Electron Paramagnetic Resonance of Transition Ions*, Clarendon Press, Oxford, England, 1970.
12. J. S. Griffith, *The Theory of Transition Metal Ions*, Cambridge University Press, London, 1961.

7

Magnetic Behavior of $3d^n$ Complexes

Before the magnetic behavior of the iron group of complexes having electronic configuration $3d^n$ outside the closed shell of electrons is investigated, following the method of Abragam and Pryce, [1] a few words regarding the inorganic complexes to be discussed in this chapter will be helpful.

Detailed experimental investigations of the magnetic anisotropy and susceptibility of the inorganic complexes of the transition metal ions during the last three decades have revealed that the crystalline electric fields in most of these complexes deviate, in general, from octahedral symmetry; in fact, it is almost impossible to find a perfectly octahedral system. Even when the six ligands are exactly the same, many factors may destroy their most regular arrangement about the cation. As a result, complexes show small or appreciable magnetic and optical anisotropies, depending on different cases. This observation necessitates a discussion of the effects of departures from octahedral symmetry. These departures are usually treated as perturbations upon the higher symmetry. The most common distortion of the crystalline field of octahedral symmetry is the presence of a tetragonal or trigonal field component; these distortions were treated in Section 6.6 and 6.7, respectively. In tetragonal distortion the complex may contain four identical ligands arranged in a square at whose center the cation is located (See Figure 5.4). Above and below this plane, two other ligands may be placed at equal distances from the cation. The symmetry is now that of a distorted octahedron, denoted by D_{4h}. Similarly, if the octahedron is distorted along the trigonal axis, the symmetry of the molecular complex is lowered from O_h to D_{3d} or C_{3v} (see Figure 5.1). In both of these cases, the threefold degenerate levels in O_h, T_{1g}, and T_{2g} split into one single and one twofold degenerate levels. The doubly degenerate level E_g, however, remains degenerate in D_{3d} or C_{3v} symmetry, in contrast to what takes place in going from O_h to D_{4h}. In Chapter 6 these situations were examined in detail.

The splitting of the triply degenerate levels T_{1g} and T_{2g} is denoted in this chapter by Δ. This Δ is *not* to be confused with the magnitude of the crystal field of O_h symmetry, which was conventionally written as $10Dq$ in Chapter 6. The method of Abragam and Pryce[1] will now be described to explain the magnetic susceptibilities and anisotropies of a few typical inorganic complexes whose ground state is either T_{1g} or T_{2g} in symmetry.

To see this method clearly let us start with the most general, but approximate, Hamiltonian, given by

$$\mathcal{H} = \mathcal{H}_F + V + \zeta \mathbf{L}\cdot\mathbf{S} + \mu_B \mathbf{H}(\mathbf{L} + 2\mathbf{S}) \tag{7.1}$$

where $\mathcal{H}_F$ represents the free-ion Hamiltonian; V, the ligand field that arises from the electrostatic field of the neighboring ligands, usually called the ligand field; the third term, the spin-orbit interaction; and the last term, the Zeeman interaction. We assume, of course, that the eigenvalues and the eigenfunctions of the free ions embedded in the ligand field of O_h symmetry are known. The second term of (7.1) represents the ligand field and can be written as

$$V = V_C + V_T \tag{7.2}$$

where V_C represents the octahedral, and V_T the tetragonal or trigonal ligand field. It will be shown that we now have nonvanishing matrix elements of the spin-orbit interaction in the ground manifold, and it is found that their general order of magnitude is the same as those of V_T. We must therefore treat $\mathbf{L}\cdot\mathbf{S}$ and V_T on the same footing.[2]

We have to solve first the eigenvalue problem $\mathcal{H}_F + V$, and this will give us a group of low-lying states separated in energy by V_T, which we shall take as defining our ground manifold. This manifold is not exactly the same as the ground manifold under the action of $\mathcal{H}_F + V_C$ because of the admixture of upper levels arising from the nondiagonal matrix elements of V_T. We are interested in the cases where the orbital degeneracy is always threefold (representations T_{1g} and T_{2g}). The E_g state is orbitaly nonmagnetic and hence will not be considered. Next we consider the effect of the spin-orbit coupling within this ground manifold but ignore the matrix elements to the upper multiplets. This gives us a set of eigenstates which are completely specified apart from the possible twofold Kramers degeneracy for the cases having an odd number of electrons. Definite magnetic moments are associated with these eigenstates and these determine the magnetic properties of the ion. Next we include the Zeeman iteraction. The magnetic behavior of the ion is now sensitive to ζ and the splitting arising from V_T, which we shall denote by Δ. The effective Hamiltonian, as it is called, is given by

$$\mathcal{H} = \Delta(1 - l_z'^2) - \alpha\zeta l_z' s_z - \alpha'\zeta(l_x' s_x + l_y' s_y) \tag{7.3}$$

where Δ is the strength of the ligand field having tetragonal or trigonal symmetry, l'_z is the z-component of the fictitious orbital angular momentum l' ($l'=1$) operating in the ground manifold, and α, α' are the effective orbital Lande' g-factors in the axial and perpendicular directions denoting the admixtures of the upper orbital levels. Also, ζ is the spin-orbit coupling constant, which is here assumed to be isotropic but can very well be anisotropic in certain cases where we can write $\zeta_{\parallel}$ and $\zeta_{\perp}$ for the axial and perpendicular directions. The advantage of defining the fictious angular momentum is that $l'_z + s_z$ is a constant of motion, and its eigenvalues m can be used to classify the various states.

The Abragam-Pryce procedure depends in essence on setting up "pseudo" atomic state functions and then taking over the appropriate features from the theory of atomic spectra. As an example, to show the p^n isomorphism, which means that the three d orbitals spanning the T_{2g} representations can be related to the three p orbitals, we shall treat an F state in a cubic ligand field.

Let us, first of all, construct the required wavefunctions of an orbital F state in O_h symmetry.

In the absence of spin-orbit coupling the ground state of the nd^3 configuration is 28-fold degenerate and involves the state vectors $|^4F; M_S, M_L\rangle$, the degeneracy being in M_S and M_L. In the presence of a crystal field of O_h symmetry, an additional (perturbing) term V_C is added to the Hamiltonian. The crystal field (which is electrostatic in nature) does not interact with the electron spins, and the degeneracy in M_S remains unaffected—only the degeneracy in M_L is affected. Then, for a given value of M_S, the perturbed state vectors will be (according to the first-order perturbation theory) linear combinations of $|^4F; M_S, M_L\rangle$ with different M_L values, the combining coefficients being independent of M_S. Then a general perturbed state vector is written, for a particular M_S, as

$$\psi_C(^4F; M_S) = \sum_{M_L=-3}^{3} \mathfrak{a}(M_L)|^4F; M_S, M_L\rangle \tag{7.4}$$

The coefficient $\mathfrak{a}(M_L)$ now satisfy, according to the first-order perturbation theory (in the degenerate case), the following matrix equation:

$$(A - IE_c)\mathfrak{a} = 0 \tag{7.5}$$

where A is the 7×7 matrix, $\{\langle^4F; M_S M'_L|V_C|^4F; M_S, M_L\rangle\}$
I is the 7×7 unit matrix
$\mathfrak{a}$ is the column vector:

$$\begin{bmatrix} \mathfrak{a}(+3) \\ \mathfrak{a}(+2) \\ \mathfrak{a}(+1) \\ \mathfrak{a}(0) \\ \mathfrak{a}(-1) \\ \mathfrak{a}(-2) \\ \mathfrak{a}(-3) \end{bmatrix}$$

E_c is the additional (perturbation) energy corresponding to the state vector $\psi(M_S)$.

Let us do all our calculations taking $\frac{3}{2}$ as the particular value of M_S. In Chapter 2 we constructed the states, which are given by

$$|^4F;\tfrac{3}{2},3\rangle=\left(\overset{+}{2}\ \overset{+}{1}\ \overset{+}{0}\right)$$
$$|^4F;\tfrac{3}{2},2\rangle=\left(\overset{+}{2}\ \overset{+}{1}\ \overset{+}{-1}\right)$$
$$|^4F;\tfrac{3}{2},1\rangle=\sqrt{\tfrac{3}{5}}\left(\overset{+}{2}\ \overset{+}{0}\ \overset{+}{-1}\right)-\sqrt{\tfrac{2}{5}}\left(\overset{+}{1}\ \overset{+}{2}\ \overset{+}{-2}\right)$$

By continuing the usual procedure of operating with $\mathbf{L}^-$, we easily construct the remaining state vectors:

$$|^4F;\tfrac{3}{2},0\rangle=\frac{2}{\sqrt{5}}\left(\overset{+}{2}\ \overset{+}{0}\ \overset{+}{-2}\right)+\frac{1}{\sqrt{5}}\left(\overset{+}{1}\ \overset{+}{0}\ \overset{+}{-1}\right)$$
$$|^4F;\tfrac{3}{2},-1\rangle=-\sqrt{\tfrac{3}{5}}\left(\overset{+}{-2}\ \overset{+}{0}\ \overset{+}{1}\right)+\sqrt{\tfrac{2}{5}}\left(\overset{+}{-1}\ \overset{+}{-2}\ \overset{+}{2}\right)$$
$$|^4F;\tfrac{3}{2},-2\rangle=-\left(\overset{+}{-2}\ \overset{+}{-1}\ \overset{+}{1}\right)$$
$$|^4F;\tfrac{3}{2},-3\rangle=-\left(\overset{+}{-2}\ \overset{+}{-1}\ \overset{+}{0}\right)$$

From Chapter 6 we have the following nonvanishing matrix elements of V_c between single-electron nd states (with the same m_s):

$$\langle\pm2|V_c|\pm2\rangle=Dq$$
$$\langle\pm2|V_c|\mp2\rangle=5\,Dq$$
$$\langle\pm1|V_c|\pm1\rangle=-4\,Dq$$
$$\langle0|V_c|0\rangle=6\,Dq$$

Using these results and the state vectors given above, we can easily evaluate the elements of the matrix A. Thus we have

$$\langle^4F;\tfrac{3}{2},3|V_c|^4F;\tfrac{3}{2},3\rangle$$
$$\equiv\langle\overset{+}{2}\ \overset{+}{1}\ \overset{+}{0}|V_c|\overset{+}{2}\ \overset{+}{1}\ \overset{+}{0}\rangle$$
$$=\langle2|V_c|2\rangle+\langle1|V_c|1\rangle+\langle0|V_c|0\rangle$$
$$=Dq-4\,Dq+6\,Dq=3\,Dq$$
$$\langle^4F;\tfrac{3}{2},3|V_c|^4F;\tfrac{3}{2},2\rangle$$
$$\equiv\langle\overset{+}{2}\ \overset{+}{1}\ \overset{+}{0}|V_c|\overset{+}{2}\ \overset{+}{1}\ \overset{+}{-1}\rangle$$
$$=\langle0|V_c|-1\rangle=0$$

$$\langle {}^4F;\tfrac{3}{2},3|V_c|{}^4F;\tfrac{3}{2},1\rangle$$
$$\equiv\langle \overset{+}{2}\ \overset{+}{1}\ \overset{+}{0}\,|V_c|\sqrt{\tfrac{3}{5}}\left(\overset{+}{2}\ \overset{+}{0}\ \overset{+}{-1}\right)-\sqrt{\tfrac{2}{5}}\left(\overset{+}{1}\ \overset{+}{2}\ \overset{+}{-2}\right)\rangle$$
$$=-\sqrt{\tfrac{3}{5}}\ \langle 1|V_c|-1\rangle+\sqrt{\tfrac{2}{5}}\ \langle 0|V_c|-2\rangle=0$$

$$\langle {}^4F;\tfrac{3}{2},3|V_c|{}^4F;\tfrac{3}{2},0\rangle$$
$$=-\frac{2}{\sqrt{5}}\langle 1|V_c|-2\rangle+\frac{1}{\sqrt{5}}\langle 2|V_c|-1\rangle=0$$

$$\langle {}^4F;\tfrac{3}{2},3|V_c|{}^4F;\tfrac{3}{2},-1\rangle$$
$$=\sqrt{\tfrac{3}{5}}\ \langle 2|V_c|-2\rangle=\sqrt{\tfrac{3}{5}}\cdot 5\,Dq=\sqrt{15}\ Dq$$

Proceeding in this way, we evaluate the other matrix elements and write the whole matrix A in the block form:

M_L' \ M_L	3	-1	-3	1	2	-2	0
3	$3\,Dq$	$\sqrt{15}\ Dq$	0	0	0	0	0
-1	$\sqrt{15}\ Dq$	Dq	0	0	0	0	0
-3	0	0	$3\,Dq$	$\sqrt{15}\ Dq$	0	0	0
1	0	0	$\sqrt{15}\ Dq$	Dq	0	0	0
2	0	0	0	0	$-7\,Dq$	$5\,Dq$	0
-2	0	0	0	0	$5\,Dq$	$-7\,Dq$	0
0	0	0	0	0	0	0	$6\,Dq$

(7.6)

From these results we note an important fact:

$$\langle {}^4F;\tfrac{3}{2},M_L'|V_c|{}^4F;\tfrac{3}{2},M_L\rangle=0$$

unless $M_L'-M_L=0$ or ±4. This fact actually follows directly from the Wigner-Eckart theorem, as shown below.

We have the form of V_C:

$$V_C=\text{Const.}\left[\ Y_4^0+\sqrt{\tfrac{5}{15}}\ \left(Y_4^4+Y_4^{-4}\right)\right]$$

Then, by using the Wigner-Eckart theorem, we have

$$\langle {}^4F;M_S,M_L'|V_C|{}^4F;M_S,M_L\rangle$$
$$=\text{Const.}\Big[\langle 3M_L'|40,3M_L\rangle+\sqrt{\tfrac{5}{15}}\ \{\langle 3M_L'|44,3M_L\rangle$$
$$+\langle 3M_L'|4-4,3M_L\rangle\}\Big]\langle {}^4F\|Y_4\|{}^4F\rangle$$

where all three Wigner coefficients will simultaneously vanish unless $M_L' - M_L = 0$ or ± 4.

The matrix equation (7.5) implies that the secular equation,

$$\det|A - IE_c| = 0 \tag{7.7}$$

must be satisfied. Using the matrix A, given in (7.6), we find the secular equation, after factorization, to be

$$(E_c - 6\,Dq)^3(E_c + 2\,Dq)^3(E_c + 12\,Dq) = 0$$

from which we obtain three distinct values of E_c:

$$E_c = 6\,Dq, \qquad -2\,Dq, \qquad -12\,Dq \tag{7.8}$$

each of the first two roots being threefold degenerate. We now consider these cases one by one.

For $E_c = 6\,Dq$, the matrix equation (7.5) gives the following four independent ordinary equations:

$$\begin{gathered} a(-1) = \sqrt{\tfrac{3}{5}}\ a(3) \\ a(2) = 0, \qquad a(-2) = 0 \\ a(-3) = \sqrt{\tfrac{5}{3}}\ a(1) \end{gathered}$$

Hence the three coefficients $a(3)$, $a(1)$, and $a(0)$ can take arbitrary values. Then we can construct three a-vectors orthogonal to each other as follows:

$$\begin{bmatrix} a(3) \\ 0 \\ 0 \\ 0 \\ \sqrt{\tfrac{3}{5}}\ a(3) \\ 0 \\ 0 \end{bmatrix}, \quad \begin{bmatrix} 0 \\ 0 \\ 0 \\ a(0) \\ 0 \\ 0 \\ 0 \end{bmatrix}, \quad \begin{bmatrix} 0 \\ 0 \\ a(1) \\ 0 \\ 0 \\ 0 \\ \sqrt{\tfrac{5}{3}}\ a(1) \end{bmatrix}$$

The corresponding normalized state vectors are then

$$\begin{aligned} \phi_x &= \sqrt{\tfrac{5}{8}}\ |{}^4F; \tfrac{3}{2}, 3\rangle + \sqrt{\tfrac{3}{8}}\ |{}^4F; \tfrac{3}{2}, -1\rangle \\ \phi_y &= |{}^4F; \tfrac{3}{2}, 0\rangle \\ \phi_z &= \sqrt{\tfrac{3}{8}}\ |{}^4F; \tfrac{3}{2}, 1\rangle + \sqrt{\tfrac{5}{8}}\ |{}^4F; \tfrac{3}{2}, -3\rangle \end{aligned} \tag{7.9}$$

Thus this orthonormal set of three state vectors corresponds to the threefold degenerate energy value, $E_c = 6\,Dq$.

Similarly, for the other threefold degenerate energy value, $E_c = -2Dq$, we construct the following orthonormal set of three state vectors:

$$\begin{aligned}
\psi_x &= \sqrt{\tfrac{3}{8}}\,|^4F;\tfrac{3}{2},3\rangle - \sqrt{\tfrac{5}{8}}\,|^4F;\tfrac{3}{2},-1\rangle \\
\psi_y &= \sqrt{\tfrac{5}{8}}\,|^4F;\tfrac{3}{2},1\rangle - \sqrt{\tfrac{3}{8}}\,|^4F;\tfrac{3}{2},-3\rangle \\
\psi_z &= \frac{1}{\sqrt{2}}|^4F;\tfrac{3}{2},2\rangle + \frac{1}{\sqrt{2}}|^4F;\tfrac{3}{2},-2\rangle
\end{aligned} \tag{7.10}$$

Finally, for $E_c = -12Dq$, we obtain from (7.5) the following equations:

$$a(3) = a(1) = a(0) = a(-1) = a(-3) = 0$$

$$a(2) = -a(-2)$$

Hence, in this case, we can construct only one normalized state vector:

$$\chi = \frac{1}{\sqrt{2}}|^4F;\tfrac{3}{2},2\rangle - \frac{1}{\sqrt{2}}|^4F;\tfrac{3}{2},-2\rangle \tag{7.11}$$

which is thus associated with the nondegenerate energy value, $E_c = -12Dq$.

In the nd^3 configuration each electron has $l=2$, and hence the parity of any of the wave functions constructed above is $(-1)^{2+2+2} = 1$, that is, any wavefunction will remain invariant under inversion. Then the sets of wavefunctions (7.9), (7.10), and (7.11) span only the g representations of the O_h group. [A g representation of dimension λ has the character $(2\lambda+1)$ for the inversion operation (i), whereas a u representation of the same dimension has the character $-(2\lambda+1)$ for i.] We have, for the O_h group, only two three-dimensional irreducible g representations, namely, T_{1g} and T_{2g}. Now, according to our construction, we have two three-dimensional sets, $\{\phi_x, \phi_y, \phi_z\}$ and $\{\psi_x, \psi_y, \psi_z\}$.Hence one of the following two cases will be true:

1. Both sets of functions span T_{1g} (or T_{2g}).
2. One set spans T_{1g}, and the other spans T_{2g}.

From the character table for O_h (Appendix V) we note that the characters of T_{1g} and T_{2g} for the C_4 operation are $+1$ and -1, respectively. According to the discussion in Chapter 5, the C_4 operation (fourfold rotation about the z-axis) can be represented in terms of the z-component of the orbital angular momentum operator (vector) $\mathbf{L}$:

$$C_4 = e^{-(i/\hbar)(\pi/2)\mathbf{L}_z}$$

Then we have the following transformations:

$$C_4\phi_x = e^{-(i/\hbar)(\pi/2)\mathbf{L}_z}\left\{\sqrt{\tfrac{5}{8}}\ |^4F;\tfrac{3}{2},3\rangle + \sqrt{\tfrac{3}{8}}\ |^4F;\tfrac{3}{2},-1\rangle\right\}$$
$$= e^{-i(\pi/2)3}\sqrt{\tfrac{5}{8}}\ |^4F;\tfrac{3}{2},3\rangle + e^{i(\pi/2)}\sqrt{\tfrac{3}{8}}\ |^4F;\tfrac{3}{2},-1\rangle$$
$$= e^{i(\pi/2)}\phi_x = i\phi_x$$
$$C_4\phi_y = e^{-(i/\hbar)(\pi/2)\mathbf{L}_z}|^4F;\tfrac{3}{2},0\rangle$$
$$= e^{-(i/\hbar)(\pi/2)0}|^4F;\tfrac{3}{2},0\rangle = \phi_y$$
$$C_4\phi_z = e^{-(i/\hbar)(\pi/2)\mathbf{L}_z}\left\{\sqrt{\tfrac{3}{8}}\ |^4F;\tfrac{3}{2},1\rangle + \sqrt{\tfrac{5}{8}}\ |^4F;\tfrac{3}{2},-3\rangle\right\}$$
$$= e^{-i(\pi/2)}\sqrt{\tfrac{3}{8}}\ |^4F;\tfrac{3}{2},1\rangle + e^{i(\pi/2)3}\sqrt{\tfrac{5}{8}}\ |^4F;\tfrac{3}{2},-3\rangle$$
$$= e^{-i(\pi/2)}\phi_z = -i\phi_z$$

or, in matrix notation,

$$C_4\begin{pmatrix}\phi_x\\ \phi_y\\ \phi_z\end{pmatrix} = \begin{pmatrix} i & 0 & 0\\ 0 & 1 & 0\\ 0 & 0 & -i\end{pmatrix}\begin{pmatrix}\phi_x\\ \phi_y\\ \phi_z\end{pmatrix}$$

Thus, for the C_4 operation, the trace of the representation spanned by (ϕ_x,ϕ_y,ϕ_z) is $(i+1-\mathrm{i})=1$, which is actually the trace of the representation T_{1g} for the C_4 operation. Hence we conclude that (ϕ_x,ϕ_y,ϕ_z) span the representation T_{1g} of O_h.

Next, we have, similarly,

$$C_4\psi_x = i\psi_x$$
$$C_4\psi_y = -i\psi_y$$

and

$$C_4\psi_z = e^{i\pi}\psi_z = -\psi_z$$

Hence

$$C_4\begin{pmatrix}\psi_x\\ \psi_y\\ \psi_z\end{pmatrix} = \begin{pmatrix} i & 0 & 0\\ 0 & -i & 0\\ 0 & 0 & -1\end{pmatrix}\begin{pmatrix}\psi_x\\ \psi_y\\ \psi_z\end{pmatrix}$$

Thus, for the C_4 operation, the trace of the representation spanned by (ψ_x,ψ_y,ψ_z) is $(i-i-1)=-1$, which is actually the trace of the representation

T_{2g} for the C_4 operation, and therefore (ψ_x, ψ_y, ψ_z) span the representation T_{2g} of O_h.

Finally, we see that

$$C_4\chi = -\chi$$

Hence the trace of the one-dimensional representation spanned by χ is -1 for the C_4 operation, and this representation is really A_{2g}.

Now, rewriting the required wavefunctions given by (7.9), (7.10), and (7.11) in terms of the Cartesian coordinates, we get the wavefunctions in the following form:

$$T_2\begin{cases} \psi_x = \dfrac{\sqrt{105}}{2}x(y^2-z^2) \\ \psi_y = \dfrac{\sqrt{105}}{2}y(z^2-x^2) \\ \psi_z = \dfrac{\sqrt{105}}{2}z(x^2-y^2) \end{cases}$$

$$T_1\begin{cases} \phi_x = \dfrac{\sqrt{7}}{2}x(2x^2-3y^2-3z^2) \\ \phi_y = \frac{1}{2}\sqrt{7}\, y(2y^2-3z^2-3x^2) \\ \phi_z = \frac{1}{2}\sqrt{7}\, z(2z^2-3x^2-3y^2) \end{cases} \tag{7.12}$$

and

$$A_2\chi = \sqrt{105}\, xyz$$

These wavefunctions are normalized and transform in the cubic point group as indicated above. We now form the new linear combinations:

$$T_2\begin{cases} \psi_1 = -\dfrac{1}{\sqrt{2}}(\psi_x + i\psi_y) \\ \psi_0 = \psi_z \\ \psi_{-1} = \dfrac{1}{\sqrt{2}}(\psi_x - i\psi_y) \end{cases} \tag{7.13}$$

$$T_1\begin{cases} \phi_1 = -\dfrac{1}{\sqrt{2}}(\phi_x + i\phi_y) \\ \phi_0 = \phi_z \\ \phi_{-1} = \dfrac{1}{\sqrt{2}}(\phi_x - i\phi_y) \end{cases} \tag{7.14}$$

If we now operate by $\mathbf{L}_z$ on (7.14), we have

$$L_z\phi_1 = \tfrac{3}{2}\phi_1 - \tfrac{1}{2}\sqrt{15}\,\psi_{-1}$$

$$L_z\phi_0 = -\frac{6i}{\sqrt{15}}\chi$$

$$L_z\phi_{-1} = \tfrac{3}{2}\phi_{-1} + \frac{\sqrt{15}}{2}\psi_1 \qquad (7.15)$$

By operating upon (7.14) similarly with $\mathbf{L}_x$ and $\mathbf{L}_y$ and without considering any admixture from other states, we have

$$\mathbf{L}_z\begin{bmatrix}\phi_1\\ \phi_0\\ \phi_{-1}\end{bmatrix} = \left(-\frac{3}{2}\right)\begin{bmatrix}1 & 0 & 0\\ 0 & 0 & 0\\ 0 & 0 & -1\end{bmatrix}\begin{bmatrix}\phi_1\\ \phi_0\\ \phi_{-1}\end{bmatrix}$$

$$\mathbf{L}_y\begin{bmatrix}\phi_1\\ \phi_0\\ \phi_{-1}\end{bmatrix} = \left(-\frac{3}{2}\right)\frac{i}{\sqrt{2}}\begin{bmatrix}0 & 1 & 0\\ -1 & 0 & 1\\ 0 & -1 & 0\end{bmatrix}\begin{bmatrix}\phi_1\\ \phi_0\\ \phi_{-1}\end{bmatrix} \qquad (7.16)$$

$$\mathbf{L}_x\begin{bmatrix}\phi_1\\ \phi_0\\ \phi_{-1}\end{bmatrix} = \left(-\frac{3}{2}\right)\frac{i}{\sqrt{2}}\begin{bmatrix}0 & 1 & 0\\ 1 & 0 & 1\\ 0 & 1 & 0\end{bmatrix}\begin{bmatrix}\phi_1\\ \phi_0\\ \phi_{-1}\end{bmatrix}$$

Now, by writing p_1, p_0, and p_{-1} for p electrons having $m_l = 1, 0$, and -1, and operating similarly with $\mathbf{L}_x, \mathbf{L}_y$, and $\mathbf{L}_z$, we can exactly reproduce the right-hand side of (7.16) except for the factor $-\frac{3}{2}$. Thus it follows that the set ϕ_1, ϕ_0, and ϕ_{-1} behaves as an atomic P state having $L=1$ but with an effective orbital Lande′ factor equal to $-\frac{3}{2}$.

Having dealt with the Abragam-Pryce method in some detail, let us use the above modified Hamiltonian to derive the algebraic expressions for the paramagnetic susceptibilities of the complexes of the $3d^1$, $3d^2$, $3d^6$, and $3d^7$ configurations in the next sections.

7.1 MAGNETIC PROPERTIES OF Ti^{3+} ($3d^1$) ALUM:

The isomorphous alums of the trivalent ions Ti^{3+}, V^{3+}, and Cr^{3+} have been the subject of extensive magnetic and other studies.[4-6] It has been concluded from these studies that the crystalline fields in all these alums, which arise mainly from a trigonally elongated octahedral cluster of six water molecules surrounding the trivalent ion, are more or less similar in nature. To explain the nearly "spin-only" value of the magnetic moment and its close Curie-law temperature dependence in both Ti^{3+} and V^{3+} alums, the trigonal field is assumed to include comparable proportions of contributions from the Jahn-Teller distortions of the water cluster and from distortions induced by atoms outside the cluster.[6]

In deriving the magnetic susceptibility we follow the work by Bose, Chakravarty, and Chatterjee.[7]

Under the cubic field alone, the $3d^1\ ^2D$ ground state of the Ti^{3+} ion is split into $^2T_{2g}$ lying about 20,000 cm^{-1} below the ^2E_g level. The trigonal component of the ligand field splits the triplet into a single Kramers' spin doublet, lying lowest, and two others of coincident energy, lying Δ cm^{-1} above the lowest one, with the assumption that the lowest singlet level is at $-\frac{2}{3}\Delta$ and the upper doublet at $\frac{1}{3}\Delta$. Under these circumstances the contribution from the upper ^2E_g orbital is negligible to a good degree of approximation. Also, no other multiplet states lie sufficiently near the ground state for consideration.

In setting up the secular equation for the spin-orbit interaction that transforms like T_1 in O_h and the trigonal distortion, we need to know the matrix elements of $\zeta\,\mathbf{L}\cdot\mathbf{S}$ (u_ξ, u_η, and u_ζ) where ζ is along the body diagonal, ξ is a direction perpendicular to this axis passing through the (octahedral)z-axis, and η is perpendicular to both, so that ξ, η, and ζ form the right-handed system. Following Pryce and Runciman,[8] we find that our appropriate trigonal orbital states are

$$|+\rangle = -\frac{1}{\sqrt{3}}(\omega|yz\rangle + \omega^2|zx\rangle + |xy\rangle)$$

$$|0\rangle = \frac{1}{\sqrt{3}}(|yz\rangle + |zx\rangle + |xy\rangle)$$

$$|-\rangle = \frac{1}{\sqrt{3}}(\omega^2|yz\rangle + \omega|zx\rangle + |xy) \qquad (7.17)$$

where $\omega = 2\pi i/3$ and $|yz\rangle$, and so on are given by[9]

$$|yz\rangle = N[\,|d,yz\rangle + \lambda\{|p,z,2\rangle - |p,z,5\rangle + |p,y,3\rangle - |p,y,6\rangle\}\,]$$

$$|zx\rangle = N[\,|d,zx\rangle + \lambda\{|p,z,1\rangle - |p,z,4\rangle + |p,x,3\rangle - |p,x,6\rangle\}\,]$$

$$|xy\rangle = N[\,|d,xy\rangle + \lambda\{|p,y,1\rangle - |p,y,4\rangle + |p,x,2\rangle - |p,x,5\rangle\}\,] \qquad (7.18)$$

in which N is the normalization factor, and λ is a measure of the amount of admixture of the p orbits of the ligand water oxygens with the d orbits of Ti^{3+}. Then, inclusive of covalency (we shall deal with covalency in detail in connection with the molecular orbital theory in Chapter 11), the nonzero matrix elements of u_ξ, u_η, and u_ζ are

$$\langle +|u_\zeta|+\rangle = -\langle -|u_\zeta|-\rangle = -\zeta_\parallel$$

$$\langle +|u_\xi|0\rangle = \langle 0|u_\xi|-\rangle = -\sqrt{2}\,\zeta_\perp \qquad (7.19)$$

$$\langle +|u_\eta|0\rangle = \langle 0|u_\eta|-\rangle = i\sqrt{2}\,\zeta_\perp$$

Since we assume a small departure of our actual orbits from pure d orbits, the spin-orbit coupling parameter ζ_i will be different from its free-ion value. We may also presume that the overlap of the p and d charge clouds is influenced by the symmetry of the structure. Thus both ζ_i and the effective orbital g-factor[9] k_i should partake of the trigonal symmetry of the ligand field and may have different values along ($i=\zeta$) and perpendicular ($i=\xi=\eta$) to the trigonal axis. Including the twofold spin degeneracy of each of the three orbital states, $|+\rangle, |0\rangle$, and $|-\rangle$, and solving the secular determinant, we finally get the three Kramers' doublets:

$$\left.\begin{aligned} \psi_+ &= a|+,-\tfrac{1}{2}\rangle - b|0,\tfrac{1}{2}\rangle \\ \psi_- &= a|-,\tfrac{1}{2}\rangle + b|0,-\tfrac{1}{2}\rangle \\ \psi'_+ &= |-,-\tfrac{1}{2}\rangle \\ \psi'_- &= |+,\tfrac{1}{2}\rangle \\ \psi''_+ &= b|+,\tfrac{1}{2}\rangle + a|0,\tfrac{1}{2}\rangle \\ \psi''_- &= b|-,\tfrac{1}{2}\rangle - a|0,-\tfrac{1}{2}\rangle \end{aligned}\right\} \tag{7.20}$$

with the following energies:

$$\begin{aligned} E_1 &= \tfrac{1}{2}\left[\left(\tfrac{1}{2}\zeta_{\|} - \tfrac{1}{3}\Delta\right) - D^{1/2}\right] \\ E_2 &= \tfrac{1}{2}\left[\left(\tfrac{1}{2}\zeta_{\|} - \tfrac{1}{3}\Delta\right) + D^{1/2}\right] \\ E_3 &= \tfrac{1}{3}\Delta - \tfrac{1}{2}\zeta_{\|} \end{aligned} \tag{7.21}$$

where

$$D = \left(\tfrac{1}{2}\zeta_{\|} - \tfrac{1}{3}\Delta\right)^2 + 4\left(\tfrac{2}{9}\Delta^2 + \tfrac{1}{3}\Delta\zeta_{\|} + \tfrac{1}{2}\zeta_{\perp}^2\right)$$

To calculate the g values for the lowest Kramers doublet, we apply the first-order magnetic perturbation:

$$\mathcal{H}_m = \mu_B \mathbf{H}(\mathbf{L} + 2\mathbf{S}) \tag{7.22}$$

and the g values are given by

$$\begin{aligned} g_{\|} &= 2|\langle\psi_+|L_z + 2S_z|\psi_+\rangle| \\ g_+ &= 2|\langle\psi_+|L_x + 2S_x|\psi_-\rangle| \end{aligned} \tag{7.23}$$

which, after calculation, take the forms

$$\begin{aligned} g_{\|} &= 2|b^2 - (1+k_{\|})a^2| \\ g_{\perp} &= 2|b^2 + \sqrt{2}\, k_{\perp} ab| \end{aligned} \tag{7.24}$$

where we assume, as before, that

$$\langle +|L_\zeta|+\rangle = -\langle -|L_\zeta|-\rangle = -k_\parallel$$

$$\langle +|L_\eta|0\rangle = \langle 0|L_\eta|-\rangle = i\sqrt{2}\,k_\perp$$

$$\langle +|L_\xi|0\rangle = \langle 0|L_\xi|-\rangle = -\sqrt{2}\,k_\perp$$

in which

$$k_\parallel = 1 - \frac{2\lambda_\parallel^2}{1+8\lambda_\parallel S_\parallel + 4\lambda_\parallel^2}$$

$$k_\perp = 1 - \frac{2\lambda_\perp^2}{1+8\lambda_\perp S_\perp + 4\lambda_\perp^2}$$

where $S_i(i=\parallel$ or $\perp)$ is the overlap integral of the d orbital with a p orbital[9] (See Chapter 11).

The experimental g values obtained by Bleaney and his co-workers[5] from paramagnetic resonance between 4.2 and 2.5°K are as follows:

$$g_\parallel = 1.25 \pm 0.02$$

$$g_\perp = 1.14 \pm 0.02 \tag{7.25}$$

The g values may be fitted exactly with any value of Δ in the range 30 to 200 cm^{-1}, with the corresponding values of the parameters ζ_i reduced by 5 to 10% of the free-ion value and the values of K_i between 0.85 and 0.6, consistent with the values of S_i, lying between 0.1 and 0.2.[9] For example, we obtain agreement with the g values where

$$\left.\begin{aligned} &\Delta = 170\ \text{cm}^{-1}\\ &\zeta_\parallel = 142.6\ \text{cm}^{-1}\\ &\zeta_\perp = 151.4\ \text{cm}^{-1}\\ &k_\parallel = 0.8(S_z = 0.110)\\ &k_\perp = 0.58(S_x = 0.224)\end{aligned}\right\} \tag{7.26}$$

On the other hand, Bleaney and his co-workers[5] have fitted the g values with $\Delta \simeq 50$ cm^{-1} and $k_i \simeq 0.7$. Thus the experimental g values could not be fitted by the Abragam-Pryce theory with any reasonable choice of Δ and ζ.

Calculating the first-order magnetic perturbations on the three Kramers' doublets, we get, for the mean susceptibility, the expression

$$K = \sum_j \tfrac{1}{3}(K_{1j} + K_{2j}) \tag{7.27}$$

where

$$K_{1j}=\frac{B\mu_B^2}{k}\left[G_{1j}+G'_{1j}e^{-(E_3-E_1)/kT}+G''_{1j}e^{-(E_2-E_1)/kT}\right]$$

$$K_{2j}=2B\mu_B^2\left[G_{2j}+G'_{2j}e^{-(E_3-E_1)/kT}+G''_{2j}e^{-(E_2-E_1)/kT}\right]$$

where

$$j=x,y,z \qquad \text{and} \qquad B=\frac{N}{\sum e^{-E_i/kT}}$$

$$G_{1x}=G_{1y}=2|\sqrt{2}\,k_{\perp}ab+b^2|^2$$

$$G'_{1x}=G'_{1y}=0$$

$$G''_{1x}=G''_{1y}=2|-\sqrt{2}\,k_{\perp}ab+b^2|^2$$

$$G_{1z}=2|b^2-(1+k_{\parallel})a^2|^2$$

$$G'_{1z}=2|k_{\parallel}-1|^2$$

$$G''_{1z}=2|a^2+(1+k_{\parallel})b^2|^2$$

$$G_{2z}=\frac{2}{E_2-E_1}|ab(2+k_{\parallel})|^2$$

$$G'_{2z}=0$$

$$G''_{2z}=\frac{2}{E_2-E_1}|ab(2+k_{\parallel})|^2$$

$$G_{2x}=G_{2y}=\frac{2}{E_3-E_1}\left|a+\frac{k_{\perp}}{\sqrt{2}}b\right|^2+\frac{2}{E_2-E_1}\left|-ab+\frac{k_{\perp}}{\sqrt{2}}(ab+b^2)\right|^2$$

$$G'_{2x}=G'_{2y}=-\frac{2}{E_3-E_1}\left|a+\frac{k_{\perp}}{\sqrt{2}}b\right|^2-\frac{2}{E_2-E_3}\left|b-\frac{k_{\perp}}{\sqrt{2}}a\right|^2$$

$$G''_{2x}=G''_{2y}=-\frac{2}{E_2-E_1}\left|ab+\frac{k_{\perp}}{\sqrt{2}}(ab+b^2)\right|^2-\frac{2}{E_2-E_3}\left|b-\frac{k_{\perp}}{\sqrt{2}}a\right|^2$$

It is important to note that the mean susceptibility obtained experimentally[10] between 100 and 300°K can be fitted only if we take $\Delta=800\ \text{cm}^{-1}$, keeping the values of the other parameters about the same as given in (7.26). The closeness of the calculated and the experimental values of the square of the effective moment, p_f^2, given by

$$p_f^2=\frac{3k}{N\mu_B^2}\cdot K_iT, \qquad i=\parallel \text{ or } \perp \tag{7.28}$$

at different temperatures between 100 and 300°K, is shown in Table 7.1. The magnitude of Δ to fit the g values in the helium range does not give any fit at all with our p_f^2. For $\Delta=800\ \text{cm}^{-1}$ and the other parameters the same as in (7.26), the g values are as follows: $g_{\parallel}=1.919$, $g_{\perp}=1.775$. Unfortunately, these

Table 7.1

Variation of p_f^2 with temperature for Ti^{3+} alum

T (°K)	p_f^2 (experimental)	p_f^2 (theoretical)
300	2.849	2.875
240	2.793	2.807
200	2.763	2.755
140	2.724	2.680
100	2.693	2.625

g values cannot be tested directly since no resonance value can be obtained for Ti^{3+} alum at or near room temperature.

The magnetic behavior of Ti^{3+} alum in the liquid air range is very different from that in the liquid helium range. The trigonal field separation Δ obviously changes from a value of about 800 cm^{-1} to about 170 cm^{-1} or less between 300 and 4.2°K.

To explain the very short spin-lattice relaxation time of about 10^{-7} sec in Ti^{3+} alum, Van Vleck[6] has shown that in the liquid air range [11] one has to assume that the "Raman process" is the main mechanism of the spin-lattice relaxation in Ti^{3+} alum, and Δ should be of the order of 1000 cm^{-1}. In the liquid helium range, to explain the relaxation time of about 10^{-3} sec,[12] Δ has to be about 100 cm^{-1}, and the Raman process is still predominant. It may be remarked that the spin-lattice relaxation time is dependent on the trigonal ligand field, which couples the spins of the ground state (though they are apparently quite free by virtue of the Kramers degeneracy) to the lattice through the spin-orbit interaction, and the decrease in the relaxation time from 10^{-3} to 10^{-7} sec with decreasing temperature clearly shows that the trigonal field has considerably decreased at low temperatures. The two values of Δ are thus in satisfactory agreement with those deduced from the paramagnetic relaxation data. Such a drastic change in Δ with temperature, first shown by Chakravarty from magnetic studies, may arise, however, from thermal expansion of the crystal and a small reorientation of the clusters in the unit cell with temperature.[12,13] We shall show subsequently for the complexes of Fe^{2+} and Co^{2+} that the lower symmetric part of the ligand field also changes with temperature.

In the absence of more detailed data on the entire range from 300 to 1°K it is not possible to say whether the changes in the ligand field, as observed, take place continuously as a function of thermal expansion of the lattice or are due to a sudden phase transition occurring at some intermediate temperatures.

7.2 MAGNETIC SUSCEPTIBILITY OF V^{3+} COMPLEXES

Following the procedure of Abragam and Pryce,[1] Chakravarty[14] has calculated the susceptibility of V^{3+} salts, and we shall follow his method. The main advantage of this method is that, in the susceptibility, we can also take into account the contribution of the upper orbital levels in a straightforward manner.

The ground state of the V^{3+} ion is $3d^2, {}^3F$. Another term, ${}^3P$, arises from the same configuration ($3d^2$) of the free ion but lies about 13,300 cm^{-1} higher.[15] Except for these, there is no term of sufficient importance lying in the neighborhood. X-ray analysis of Al^{3+} and Cr^{3+} alums shows that the field on the trivalent cation arises from a trigonally distorted octahedron of six water molecules surrounding the ion. The trigonal axis of each ion coincides with one of the four diagonals of the unit cube, which contains four such ions. Then the cubic ligand field splits the ${}^3F$ state into a triplet—${}^3T_{1g}$, lying lowest, and ${}^3T_{2g}$ and ${}^3A_{2g}$, lying above it. The comparatively small trigonal field now splits each of the two triplets into a singlet and a doublet; the singlet must be lower if the susceptibility is found to obey the Curie law with an approximate "spin-only" value of the moment.[6] Furthermore, the effect of the spin-orbit coupling on the orbital levels, each with threefold spin degeneracy, is to split them into three nondegenerate and three doubly degenerate spin levels and to cause an admixture between them. We now proceed to find the fine-structure levels of the Stark pattern under a ligand field of trigonal symmetry.

We assume the basic orbital states to be $|1\rangle$, $|0\rangle$, and $|-1\rangle$, and the appropriate effective Hamiltonian discussed earlier in this section to be

$$\mathcal{H}=\Delta(1-l_z'^2)-\alpha\zeta l_z's_z-\alpha'\zeta(l_x's_x+l_y's_y) \qquad (7.29)$$

where Δ stands for the energy difference between ϕ_z (singlet) and ϕ_{xy} (doublet) (see Figure 7.1). In (7.29), α and α' are the effective Landé factors, which are dependent on the admixtures of the excited orbitals in the axial and perpendicular directions, and ζ is the spin-orbit coupling constant. The fine-structure levels are classified by $m(\pm 2, \pm 1, 0)$, which are the eigenvalues of $l_z'+s_z$. The secular matrix for $\mathcal{H}$ breaks up into

$$\left.\begin{array}{ll}
\begin{bmatrix} \alpha\zeta & -\alpha'\zeta & 0 \\ -\alpha'\zeta & \Delta & -\alpha'\zeta \\ 0 & -\alpha'\zeta & \alpha\zeta \end{bmatrix}, & m=0 \\[2ex]
\begin{pmatrix} 0 & -\alpha'\zeta \\ -\alpha'\zeta & \Delta \end{pmatrix}, & m=\pm 1 \\[2ex]
\begin{pmatrix} \Delta & -\alpha'\zeta \\ -\alpha'\zeta & 0 \end{pmatrix}, & m=\pm 1 \\[2ex]
-\alpha'\zeta, & m=\pm 2
\end{array}\right\} \qquad (7.30)$$

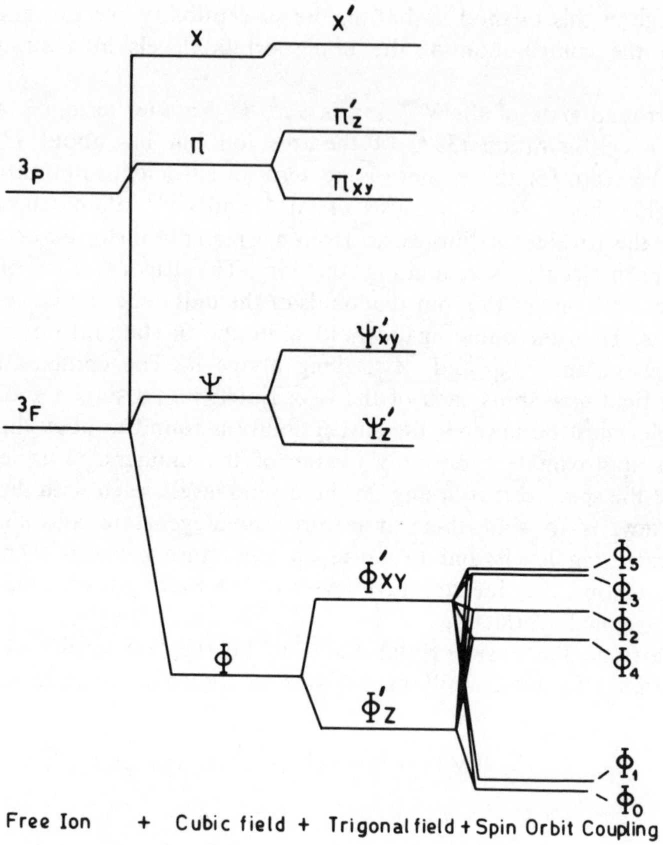

Fig. 7.1 The energy level diagram for V^{3+} ion.

The eigenvalues are therefore

$$\left.\begin{aligned} E_0 &= \tfrac{1}{2}(\Delta + \alpha\zeta - S_0) \\ E_1 &= \tfrac{1}{2}(\Delta - S_1) \\ E_2 &= \tfrac{1}{2}(\Delta + S_1) \\ E_3 &= \alpha\zeta \\ E_4 &= -\alpha'\zeta \\ E_5 &= \tfrac{1}{2}(\Delta + \alpha\zeta + S_0) \end{aligned}\right\} \tag{7.31}$$

where

$$S_0 = \left[(\Delta - \alpha\zeta)^2 + 8\alpha'^2\zeta^2\right]^{1/2}$$

$$S_1 = \left[\Delta + 4\alpha'^2\zeta^2\right]^{1/2}$$

The corresponding eigenstates are thus

$$\left.\begin{aligned}
\phi_0 &= a|1,-1\rangle + b|0,0\rangle + a|-1,1\rangle \\
\phi_1 &= c|1,0\rangle + d|0,1\rangle \\
\phi_1' &= c|-1,0\rangle + d|0,-1\rangle \\
\phi_2 &= d|1,0\rangle - c|0,1\rangle \\
\phi_2' &= d|-1,0\rangle - c|0,-1\rangle \\
\phi_3 &= \frac{1}{\sqrt{2}}\left|1,-1\right\rangle - \frac{1}{\sqrt{2}}\left|-1,1\right\rangle \\
\phi_4 &= |1,1\rangle \\
\phi_4' &= |-1,-1\rangle \\
\phi_5 &= \frac{1}{\sqrt{2}}b\left|1,-1\right\rangle - \sqrt{2}\,a|0,0\rangle + \frac{1}{\sqrt{2}}b\left|-1,1\right\rangle
\end{aligned}\right\} \tag{7.32}$$

The number of independent coefficients in the expressions given above is reduced to only four by virtue of the symmetry existing in the $m=0, \pm 1, \pm 2$ states and of the orthogonality and the normalization of the states.

It should be noted here that, only when Δ is negative, the lowest state is a singlet ($m=0$), and immediately above it, there is a doublet ($m=\pm 1$). The remaining levels are well separated above these two (Figure 7.1). We then proceed to calculate the magnetic susceptibility, considering only the lowest two levels where $m=0$ and $m=\pm 1$. The contributions from the other levels are negligible, since the next higher levels are approximately 1300 cm^{-1} above the lowest triplet, as will be seen later.

Using Van Vleck's well-known formula for susceptibility (discussed in details in Chapter 4), we obtain for the mean gram-ionic susceptibility K (which is the same as χ for the crystal per gram-ion),

$$\begin{aligned}
K &= \tfrac{1}{3}(K_1 + K_2 + K_3) \\
&= \frac{8N\mu_B^2}{3k}\,\frac{(A^2/4T) + (2B^2k/D)(e^{D/kT}-1)}{2+e^{D/kT}} + K_c
\end{aligned} \tag{7.33}$$

where

$$A^2 = (2d^2 - c^2)^2$$

$$B^2 = \left(ac + bd - \frac{\alpha'}{2}bc - \frac{\alpha'}{2}ad\right)^2$$

$$D = \tfrac{1}{2}(S_0 - S_1 - \alpha\zeta)$$

and K_c is the susceptibility coming from the upper excited levels. The actual calculation of K_c will be shown later. Figure 7.2 gives the variation of 10^2KT

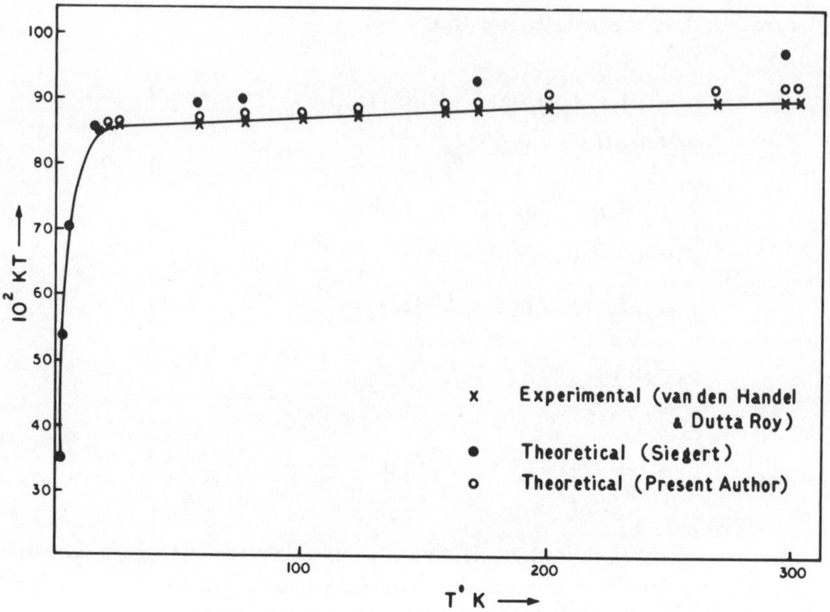

Fig. 7.2 The variation of $10^2 KT$ with temperature of V^{3+} alum.

with T obtained (theoretically) by Chakravarty, using the values

$$\left.\begin{aligned} A^2 &= 3.842 \\ B^2 &= 0.810 \\ D &= 4.8 \text{ cm}^{-1} \\ K_c &= 248.2 \times 10^{-6} \end{aligned}\right\} \tag{7.34}$$

so as to provide a fit with the experimental points of Van den Handel and Siegert[4] and of Dutta Roy, Chakravarty, and Bose.[10] Now, by trial, we choose values for α, α', Δ, and ζ that will yield for A^2, B^2, and D, the values given in (7.34). We find that approximately

$$\left.\begin{aligned} \Delta &= -1390 \text{ cm}^{-1} \\ \alpha &= 1.105 \\ \alpha' &= 1.350 \\ \zeta &= 64 \text{ cm}^{-1} \end{aligned}\right\} \tag{7.35}$$

It is very interesting that, to support Siegert's finding, the value of ζ required to give a fit is 40% less than the free-ion value of 104 cm^{-1}. But the important difference is that the Δ obtained here is more than twice the value given by Van den Handel and Siegart.

Figure 7.1 shows the schematic position of the energy levels for the values of the parameters given in (7.34) and (7.35) with the approximate energies (in

cm^{-1}) as follows:

$$\left.\begin{aligned}\phi_0 &= -1400.15\\ \phi_1 &= -1395.35\\ \phi_4 &= -70.72\\ \phi_2 &= 5.35\\ \phi_3 &= 70.72\\ \phi_5 &= 80.87\end{aligned}\right\} \tag{7.36}$$

We shall now obtain the theoretical expression for K_c, the high-frequency contribution to susceptibility, and the g values and evaluate them for the parameters chosen above.

In this calculation we consider the effect of the next four upper levels on the basic levels ϕ_0 and $\phi_{\pm 1}$. Proceeding in a manner, similar to that used for the susceptibility we get

$$K_c = \frac{N\mu_B^2 B}{3}\left[\left\{\frac{16[c-(\alpha'/2)d]^2}{E_4 - E_1} + \frac{4c^2d^2(2+\alpha)^2}{E_2 - E_1} + \frac{4\left(c+\frac{1}{2}d\right)^2}{E_3 - E_1}\right.\right.$$
$$\left. + \frac{(2bc-4ad+2ac-bd)^2}{2(E_5 - E_1)}\right\}$$
$$\left. + \left\{\frac{16[ad-bc-(\alpha'/2)bd+(\alpha'/2)ac]^2}{E_2 - E_0} + \frac{4(2+\alpha)^2a^2}{E_3 - E_0}\right\}e^{D/kT}\right] \tag{7.37}$$

where

$$B = (2 + e^{D/kT})^{-1}$$

When the values of the coefficients are substituted, K_c takes the form

$$K_c = \frac{N\mu_B^2}{3}\,\frac{(0.004631) + (0.005041)e^{6.9/T}}{2 + e^{6.9/T}}$$
$$= 276\cdot 8 \times 10^{-6} \qquad (\text{for } T = 300°K) \tag{7.38}$$

This result is in very good agreement with the value of 248.2×10^{-6} found from experiment, as against the obviously too high value of 453.1×10^{-6} calculated by Van den handel and Siegert.[4]

The expressions for the g values are given by

$$\begin{aligned} g_{\parallel} &= \langle \phi_1 | -\alpha l_z' + 2S_z | \phi_1 \rangle \\ g_{\perp} &= \langle \phi_1 | -\alpha' l_x' + 2S_x | \phi_0 \rangle \end{aligned} \tag{7.39}$$

Evaluating the matrix elements, we get

$$g_{\|}=(2d^2-\alpha c^2)$$
$$g_{\perp}=2\left(ac+db-\frac{\alpha'}{2}bc-\frac{\alpha'}{2}ad\right) \tag{7.40}$$

For the specified values of the parameters, we get $g_{\|}=1.96$, $g_{\perp}=1.80$. These results agree with the resonance values subsequently obtained by Pryce and Runciman.[8]

The fact that, to obtain agreement of the theory with experiment, one must use a value of ζ about 40% lower than the free-ion value suggests that it may not be correct to treat the complex $[V(H_20)_6]^{3+}$ according to a purely ionic model. Such discrepancies between experiment and theory have been observed by Van den Handel and Siegart[4] for V^{3+}, Abragam and Pryce[1] for Cu^{2+}, and Griffiths and Owen[16] for Ni^{2+}. Owen[15] has shown, however, that, by postulating a certain amount of overlap of the $3d$ wavefunctions of the paramagnetic ion with the σ and π wavefunctions of the ligand oxygens, this discrepancy can be accounted for. The magnitude of the "covalency factor" depends on the strength of bonding. For example, in Ni^{2+} salts, ζ in the crystal is found to be smaller by 10 to 20% than its free-ion value, as against 40% for V^{3+}, so that the apparent reduction of ζ in the crystal is greater in M^{3+} ions that in M^{2+} ions. We shall study the effect of covalency in these complexes when we discuss the molecular orbital theory in Chapter 11.

7.3 MAGNETIC SUSCEPTIBILITY OF $Fe^{2+}(3d^6)$ COMPLEXES

In dealing with this problem, we refer to the work of Konig and Chakravarty.[17]

In most high-spin compounds of Fe^{2+}, which were investigated by methods of structure analysis, and likewise in those of Co^{3+}, the microsymmetry about the central ion is essentially octahedral with some tetragonal or trigonal distortion. The ligand field potential may thus be written as

$$V=V_c+V(T) \tag{7.41}$$

The predominant cubic field splits the $^5D(3d^6)$ ground term into $^5T_2$ and the approximately 10,000 cm^{-1} higher lying ^5E. If the crystal field axes are chosen to pass through the ligand ions, $V(T)$ corresponds to a tetragonal symmetry. The $^5T_2$ term splits into ^5E and $^5B_2$, the separation being denoted by δ (see Figure 7.3). If the distortion is of trigonal symmetry, the z-axis is usually taken along the (1, 1, 1) direction of the octahedron, and then the $^5T_2$ term splits into ^5E and $^5A_1$. Since it can be shown that, with distortions of both tetragonal and trigonal symmetry, either one of the resulting levels may be lowest, in what follows only a distinction between $\delta>0$ and $\delta<0$ will be

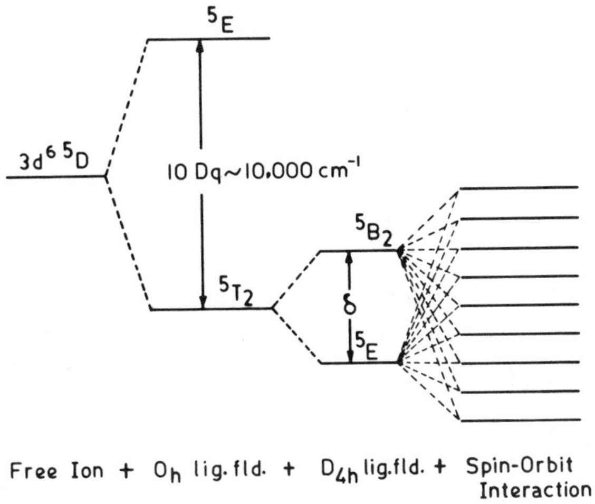

Fig. 7.3 Energy level diagram for a ^5D (d^6) term under the influence of a tetragonal field ($\delta > 0$) and of the spin-orbit interaction (not to scale).

made. The sign of δ will be chosen as positive whenever the orbital doublet ^5E is lowest. Also, the off-diagonal matrix elements of the trigonal potential between the low-lying and the higher ^5E terms will be disregarded, thus limiting our discussion to $\delta \ll 10Dq$.

An additional splitting is introduced by the spin-orbit coupling. In general, the spin-orbit coupling constant λ used here will be different from that of the free ion, since we assume a departure of our actual orbits from pure d orbits.[9] The only additional modification required by partly covalent bonding involves the operator of Zeeman energy. It has been shown by Tinkham[18] that the matrix elements of the operator **L** are related by

$$\langle \phi_n | \mathbf{L} | \phi_m \rangle = x_{nm} \langle d_n | \mathbf{L} | d_m \rangle \tag{7.42}$$

where ϕ_n designates the molecular orbitals, and x_{nm} the tensor components of the orbital reduction.[9] The magnetic operator thus obtained is of the form $\mu_B H(x\mathbf{L}+2\mathbf{S})$.

The energies within the $^5T_2$ term may be calculated on the basis of Abragam and Pryce's modification, using the structural isomorphism between $^5T_2$ and ^5P. The matrix elements of the Hamiltonian

$$\mathcal{H} = \delta(1 - l'^2_z) - \lambda_{\|} l'_z S_z - \lambda_{\perp}(l'_x s_x + l'_y S_y) \tag{7.43}$$

have to be evaluated within the 15 states $|M_L M_S\rangle$ of a ^5P term, classified according to their values of M_J. The matrices of the ligand field and of the spin-orbit coupling may be written as

$$\left.\begin{array}{ll}
\begin{array}{ccc} |-1,1\rangle & |0,0\rangle & |1,-1\rangle \end{array} & \\
\begin{bmatrix} \lambda_{\parallel} & -\sqrt{3}\,\lambda_{\perp} & 0 \\ -\sqrt{3}\,\lambda_{\perp} & \delta & -\sqrt{3}\,\lambda_{\perp} \\ 0 & -\sqrt{3}\,\lambda_{\perp} & \lambda_{\parallel} \end{bmatrix}, & M_J=0 \\
\begin{array}{ccc} |1,0\rangle & |0,1 & |-1,2\rangle \end{array} & \\
\begin{bmatrix} 0 & -\sqrt{3}\,\lambda_{\perp} & 0 \\ -\sqrt{3}\,\lambda_{\perp} & \delta & -\sqrt{2}\,\lambda_{\perp} \\ 0 & -\sqrt{2}\,\lambda_{\perp} & 2\lambda_{\parallel} \end{bmatrix}, & M_J=\pm 1 \\
\begin{array}{cc} |0,2\rangle & |1,1\rangle \end{array} & \\
\begin{bmatrix} \delta & -\sqrt{2}\,\lambda_{\perp} \\ -\sqrt{2}\,\lambda_{\perp} & -\lambda_{\parallel} \end{bmatrix}, & M_J=\pm 2 \\
\begin{array}{c} |1,2\rangle \\ -2\lambda_{\parallel}, \end{array} & M_J=\pm 3
\end{array}\right\} \tag{7.44}$$

The first matrix ($M_J=0$) may be factorized and gives the energies

$$\begin{aligned} E_0 &= \tfrac{1}{2}\left[(\lambda_{\parallel}+\delta)-A^{1/2}\right] \\ E_0' &= \tfrac{1}{2}\left[(\lambda_{\parallel}+\delta)+A^{1/2}\right] \\ E_0'' &= \lambda_{\parallel} \end{aligned} \tag{7.45}$$

where $A=[(\lambda_{\parallel}-\delta)^2+24\lambda_{\perp}^2]$.

The energies for $M_J=\pm 1$ are determined by the roots of the cubic equation

$$x^3-(2+\eta)x^2+(2\eta-5\epsilon^2)x+6\epsilon^2=0 \tag{7.46}$$

where $x=E_i/\lambda_{\parallel}$, E_i being E_1, E_1'', and E_1', respectively; $\eta=\delta/\lambda_{\parallel}$; and $\epsilon=\lambda_{\perp}/\lambda_{\parallel}$. The energies are labeled so that E_1 becomes the lowest one, followed by E_1'' and E_1' (see Figure 7.3):

$$E_1<E_1''<E_1' \tag{7.47}$$

The remaining energies are given by

$$\begin{aligned} E_2 &= \tfrac{1}{2}(\delta - \lambda_{\|} - B^{1/2}) \\ E_2' &= \tfrac{1}{2}(\delta - \lambda_{\|} + B^{1/2}) \\ E_3 &= -2\lambda_{\|} \end{aligned} \tag{7.48}$$

where $B = [(\delta + \lambda_{\|})^2 + 8\lambda_{\perp}^2]$

The corresponding wavefunctions may be written as

$$\left.\begin{aligned} \psi_1 &= a_1|1,0\rangle + b_1|0,1\rangle + c_1|-1,2\rangle \\ \psi_{-1} &= a_1|-1,0\rangle + b_1|0,-1\rangle + c_1|1,-2\rangle \\ \psi_0 &= a_0|1,-1\rangle + b_0|0,0\rangle + a_0|-1,1\rangle \\ \psi_0'' &= \frac{1}{\sqrt{2}}|1,-1\rangle - \frac{1}{\sqrt{2}}|-1,1\rangle \\ \psi_1'' &= a_1''|1,0\rangle + b_1''|0,1\rangle + c_1''|-1,2\rangle \\ \psi_{-1}'' &= a_1''|-1,0\rangle + b_1''|0,-1\rangle + c_1''|1,-2\rangle \\ \psi_2 &= a_2|0,2\rangle + b_2|1,1\rangle \\ \psi_{-2} &= a_2|0,-2\rangle + b^2|-1,-1\rangle \\ \psi_3 &= |1,2\rangle \\ \psi_{-3} &= |-1,-2\rangle \\ \psi_2' &= b_2|0,2\rangle - a_2|1,1\rangle \\ \psi_{-2}' &= b_2|0,-2\rangle - a_2|-1,-1\rangle \\ \psi_1' &= a_1'|1,0\rangle + b_1'|0,1\rangle + c_1'|-1,2\rangle \\ \psi_{-1}' &= a_1'|-1,0\rangle + b_1'|0,-1\rangle + c_1'|1,-2\rangle \\ \psi_0 &= \frac{1}{\sqrt{2}} b_0|1,-1\rangle - \sqrt{2}\, a_0|0,0\rangle + \frac{1}{\sqrt{2}} b_0|-1,1\rangle \end{aligned}\right\} \tag{7.49}$$

where the coefficients a_1, b_1, c_1, and so on are obtained from the set of linear equations corresponding to (7.44) and the normalization condition, and are given by

$$\begin{aligned} a_1 &= -\frac{\sqrt{3}\,(2\lambda_{\|} - E_1)}{\sqrt{2}\, E_1} c_1 \\ b_1 &= \frac{(2\lambda_{\|} - E_1)}{\sqrt{2}\,\lambda_{\perp}} c_1 \\ c_1 &= \left[\frac{3(2\lambda_{\|} - E_1)^2}{2E_1^2} + \frac{(2\lambda_{\|} - E_1)^2}{2\lambda_{\perp}^2} + 1\right]^{-1/2} \end{aligned} \tag{7.50}$$

The coefficients a', b', and c' are obtained from (7.50) by replacing E_1 by E_1'; similarly, a_1'', b_1'', and c_1'' are obtained by replacing E_1 by E_1''. Also,

$$a_0 = \frac{\sqrt{3}\,\lambda_\perp}{\lambda_\parallel - E_0} b_0$$

$$b_0 = \left[\frac{6\lambda_\perp^2}{(\lambda_\parallel - E_0)^2} + 1 \right]^{-1/2}$$

and (7.51)

$$a_2 = \frac{\sqrt{2}\,\lambda_\perp}{\delta - E_2} b_2$$

$$b_2 = \left[\frac{2\lambda_\perp^2}{(\delta - E_2)^2} + 1 \right]^{-1/2}$$

The combined effect of an axial field and of the spin-orbit interaction on a cubic $^5T_2$ term thus produces nine levels, namely, three singlets and six doublets. In the case of $\delta > 0$, these levels are drawn in Figure 7.3 in the order of increasing energy. If, however, an axial field having $\delta < 0$ is operative, the order of the levels will be changed. In particular, ϕ_1, which is lowest for $\delta > 0$, will be replaced by ϕ_0, and so on. Finally, the remaining degeneracy is lifted by the application of a magnetic field.

The magnetic susceptibility up to second order is determined from a well-known expression given by Van Vleck;[27]

$$K = N \frac{\sum\limits_{n,m} \left[\left(E_{nm}^{(1)}\right)^2 / kT - 2E_{nm}^{(2)} \right] e^{-E_n^o/kT}}{\sum\limits_n e^{-E_n^o/kT}} \tag{7.52}$$

where N is the Avogadro number, and k is the Boltzmann constant. To obtain the principal susceptibilities for systems with axial symmetry, (7.52) has to be calculated, separately, both in the direction parallel and in the direction perpendicular to the principal axis. The appropriate magnetic operators may be written as

$$\mu_\parallel = \mu_B(-x_\parallel l_z' + 2S_z)$$

and

$$\mu_\perp = \mu_B\left(-\tfrac{1}{2}x_\perp l_-' - \tfrac{1}{2}x_\perp l_+' + s_- + s_+\right) \tag{7.53}$$

respectively, where $l_\pm = l_x \pm il_y$, $s_\pm = s_x \pm is_y$.

Performing the calculation outlined above, one obtains the expressions for the principal molar susceptibilities for $\delta > 0$ and $\delta < 0$. Since these expressions are very lengthy and complicated, we do not present them here.[17]

The numerical results of our calculations are conveniently presented in terms of the principal magnetic moments $P_{\parallel}$ and $P_{\perp}$. The moments P_i ($i = \parallel$ or $\perp$) are related to $K_{\parallel}$ and $K_{\perp}$ by

$$P_i^2 = \frac{3k}{N\mu_B^2} K_i T \tag{7.54}$$

where the K's are given by (7.52) with (7.53). The moments are unique functions of the parameters kT/λ, δ/λ, and x, which have the following range of values:

$$\begin{array}{llll} \dfrac{kT}{\lambda}: & -0.01 \text{ to } -5.0 & & \\ \dfrac{\delta}{\lambda}: & 0 \text{ to } -30.0 & \text{for } \delta > 0 & \\ & 0 \text{ to } +30.0 & \text{for } \delta < 0 & \\ x_{\parallel} \text{ and } x_{\perp}: & 1.0 \text{ to } 0.5 & & \end{array} \tag{7.55}$$

On the basis of the theoretical calculations given above, a large number of Fe^{2+} complexes have been analyzed and successfully interpreted. For details the reader is referred to the papers of Konig and Chakravarty[17] and of Konig, Chakravarty, and Madeja.[19]

Let us now find the g values. When the magnetic field is applied along the z-direction (symmetry axis) and the x-direction of the Fe^{2+} ion, we have, respectively,

$$\begin{aligned} g_{\parallel} &= 2|\langle\psi_1| - xl'_z + 2S_z|\psi_1\rangle| \\ g_{\perp} &= 2|\langle\psi_1| - x'l'_x + 2S_x|\psi_{-1}\rangle| \end{aligned} \tag{7.56}$$

for the lowest $|\pm 1\rangle$ level. Performing the calculation, we get

$$\begin{aligned} g_{\parallel} &= 2(5c_1^2 + 2b_1^2 - a_1^2) \\ g_{\perp} &= 0 \end{aligned} \tag{7.57}$$

where the coefficients a_1, b_1, and c_1 are given in (7.50). For $\delta = 270\ \text{cm}^{-1}$, we get

$$\begin{aligned} g_{\parallel} &= 8.989 \\ g_{\perp} &= 0 \end{aligned}$$

in excellent agreement with the experimental g values at 20°K given by Tinkham.[18]

7.4 MAGNETIC PROPERTIES OF HYDRATED $Co^{2+}(3d^7)$ COMPLEXES

In dealing with this problem, we follow closely the work of Bose, Chakravarty, and Chatterjee.[20]

In a modification of Schlapp and Penny's theory[21] for Co^{2+} Tutton salts, Abragam and Pryce[1] assumed that the small anisotropic part of the ligand field, comparable in magnitude to the spin-orbit coupling, when superimposed upon the octahedral ligand field acting upon the Co^{2+} ion, has approximate tetragonal symmetry with the tetragonal axis elongated.[22] In the Abragam-Pryce theory, which considers the lowest orbital triplet state to be described by an effective orbital quantum number $l'=1$ and the Landé orbital-splitting factor to be similarly replaced by the anisotropic effective values α and α' (discussed earlier), the secular determinant has been diagonalized for the spin-orbital problem, yielding the fine-structure energy levels for the Co^{2+} ion (Figure 7.4). Thus the part of the orbital angular momentum operator **L** that is semidiagonal in the manifold of the lowest triplet has the following components:

$$L_x = -\alpha' l'_x, \qquad L_y = -\alpha' l'_y, \qquad L_z = -\alpha l'_z \tag{7.58}$$

where l'_x, l'_y, and l'_z are the components of the angular momentum operator with a quantum number equal to unity. The Co^{2+} ion possesses the electronic configuration $3d^7$, whose ground term is ^4F. There is also another term, ^4P, coming from the same configuration, which lies 14,000 cm^{-1} above the ground term. In hydrated Co^{2+} salts, the ^4F term splits into $^4A_2$, $^4T_2$, and $^4T_1$ terms, whereas the upper lying ^4P term remains unsplit and is denoted by $^4T_1$. The splittings of the ^4F term are denoted by χ, ψ, and ϕ, and the ^4P term is denoted by π, in Figure 7.4. Since there are nonvanishing matrix elements between ^4F and ^4P under the same symmetry, the lowest state of Co^{2+} in the hydrated salts contains an admixture of ^4P. Abragam and Pryce have derived expressions for the g values that are complicated functions of Δ, α and α', and ζ. To fit the g values determined by paramagnetic resonance measurements, they obtained $\Delta = 1000 \pm 300$ cm^{-1}, $\alpha = 1.10$, $\alpha' = 1.55$, and $\zeta = -180$ cm^{-1}. Following their work, Bose, Chakravarty, and Chatterjee[20] have been successful in interpreting the magnetic data for Co^{2+} ions in Tutton salts, which we describe now.

We consider first the effect of the spin-orbit coupling on the lowest orbital triplet, ϕ'. The effective orbital angular momentum l' operates on the ground manifold of ϕ' with the effective Landé factors $-\alpha$ and $-\alpha'$ in the axial and perpendicular directions, respectively. Since we assume an approximate tetragonal symmetry of the $(Co^{2+}, 6H_20)$ octahedron, our basic orbital states

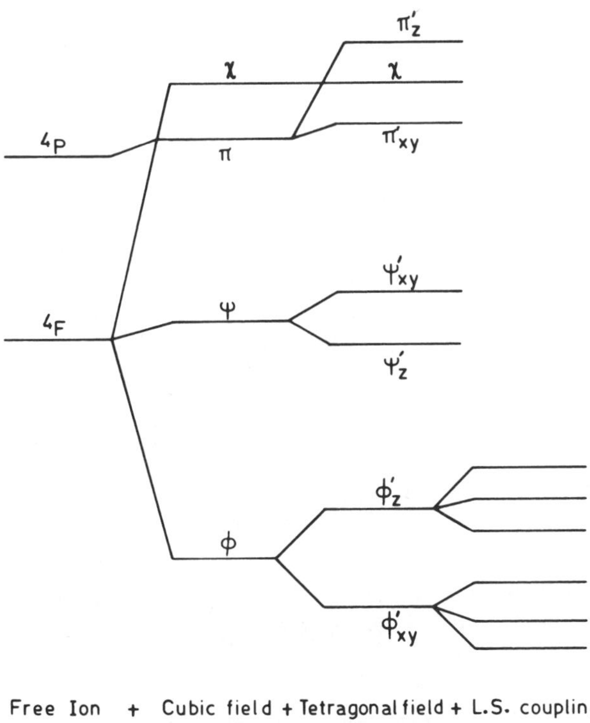

Fig. 7.4 Energy levels for Co^{2+} in a field of tetragonal symmetry. (Taken from Abragam and Pryce.[1])

are

$$|1\rangle = -\frac{1}{\sqrt{2}}(\phi'_x + i\phi'_y)$$

$$|0\rangle = \phi'_z \tag{7.59}$$

$$|-1\rangle = \frac{1}{\sqrt{2}}(\phi'_x - i\phi'_y)$$

and the appropriate effective Hamiltonian is

$$\mathcal{H} = \Delta(1 - l_z'^2) - \alpha\zeta l'_z s_z - \alpha'\zeta(l'_x s_x + l'_y s_y) \tag{7.60}$$

where Δ stands for the energy difference between ϕ'_{xy} (doublet) and ϕ'_z (singlet). Introducing the spin in the basic orbital states of (7.59) and operating the Hamiltonian on them, we find that the energy matrix for $\mathcal{H}$

breaks up into

$$\begin{bmatrix} \frac{3}{2}\alpha\zeta & -\sqrt{\frac{3}{2}}\ \alpha'\zeta & 0 \\ -\sqrt{\frac{3}{2}}\ \alpha'\zeta & \Delta & -\sqrt{2}\ \alpha'\zeta \\ 0 & -\sqrt{2} & \frac{1}{2}\alpha\zeta \end{bmatrix}, \qquad m=\pm\tfrac{1}{2}$$

$$\begin{bmatrix} \Delta & -\sqrt{\frac{3}{2}}\ \alpha'\zeta \\ -\sqrt{\frac{3}{2}}\ \alpha'\zeta & -\frac{1}{2}\alpha\zeta \end{bmatrix}, \qquad m=\pm\tfrac{3}{2} \tag{7.61}$$

$$-\tfrac{3}{2}\alpha\zeta, \qquad m=\pm\tfrac{5}{2}$$

The fine-structure levels are classified by $m(\pm\frac{1}{2}, \pm\frac{3}{2}, \pm\frac{5}{2})$, which are the eigenvalues of $l_z'+s_z$. The eigenvalues are

$$E_3=\tfrac{1}{2}\left(\Delta-\tfrac{1}{2}\alpha\zeta-A^{1/2}\right)$$

$$E_4=-\tfrac{3}{2}\alpha\zeta \tag{7.62}$$

$$E_5=\tfrac{1}{2}\left(\Delta-\tfrac{1}{2}\alpha\zeta+A^{1/2}\right)$$

where

$$A=\left(\Delta+\tfrac{1}{2}\alpha\zeta\right)^2+6\alpha'^2\zeta^2$$

and, the energies E_1, E_2, and E_6 for the $|\pm\frac{1}{2}\rangle$ state are given by the roots of the cubic equation

$$x^3-(2\alpha+\eta)x^2+\left(2\alpha\eta+\tfrac{3}{2}\alpha^2-\tfrac{7}{2}\alpha'^2\right)x-\tfrac{3}{4}\alpha^2\eta+\tfrac{15}{4}\alpha\alpha'^2=0$$

Here $\eta=\Delta/\zeta$ and $x=E_i/\zeta$, where $i=1$, 2, and 6, E_i being the energy in cm^{-1}.

The corresponding eigenstates are thus

$$
\left.
\begin{aligned}
\phi_{1/2} &= a|-1,\tfrac{3}{2}\rangle + b|0,\tfrac{1}{2}\rangle + c|1,-\tfrac{1}{2}\rangle \\
\phi_{-1/2} &= a|1,-\tfrac{3}{2}\rangle + b|0,-\tfrac{1}{2}\rangle + c|-1,\tfrac{1}{2}\rangle \\
\phi'_{1/2} &= a'|-1,\tfrac{3}{2}\rangle + b'|0,\tfrac{1}{2}\rangle + c'|1,-\tfrac{1}{2}\rangle \\
\phi'_{-1/2} &= a'|1,-\tfrac{3}{2}\rangle + b'|0,-\tfrac{1}{2}\rangle + c'|-1,\tfrac{1}{2}\rangle \\
\phi''_{3/2} &= a''|0,\tfrac{3}{2}\rangle + b''|1,\tfrac{1}{2}\rangle \\
\phi''_{-3/2} &= a''|0,-\tfrac{3}{2}\rangle + b''|-1,-\tfrac{1}{2}\rangle \\
\phi'''_{5/2} &= |1,\tfrac{3}{2}\rangle \\
\phi'''_{-5/2} &= |-1,-\tfrac{3}{2}\rangle \\
\phi^{\text{iv}}_{3/2} &= a^{\text{iv}}|0,\tfrac{3}{2}\rangle + b^{\text{iv}}|1,\tfrac{1}{2}\rangle \\
\phi^{\text{iv}}_{-3/2} &= a^{\text{iv}}|0,-\tfrac{3}{2}\rangle + b^{\text{iv}}|-1,-\tfrac{1}{2}\rangle \\
\phi^{\text{v}}_{1/2} &= a^{\text{v}}|-1,\tfrac{3}{2}\rangle + b^{\text{v}}|0,\tfrac{1}{2}\rangle + c^{\text{v}}|1,-\tfrac{1}{2}\rangle \\
\phi^{\text{v}}_{-1/2} &= a^{\text{v}}|1,-\tfrac{3}{2}\rangle + b^{\text{v}}|0,-\tfrac{1}{2}\rangle + c^{\text{v}}|-1,\tfrac{1}{2}\rangle
\end{aligned}
\right\} \quad (7.63)
$$

where the coefficients satisfy the unitary transformation properties to diagonalize the secular determinant and these are given by

$$
\left.
\begin{aligned}
a &= \frac{\sqrt{\tfrac{3}{2}}\,\alpha'\zeta}{\tfrac{3}{2}\alpha\zeta - E_1}\,b \\
c &= \frac{\sqrt{2}\,\alpha'\zeta}{\tfrac{1}{2}\alpha\zeta - E_1}\,b \\
b &= \left[\frac{\tfrac{3}{2}\alpha'^2\zeta^2}{\left(\tfrac{3}{2}\alpha\zeta - E_1\right)^2} + \frac{2\alpha'^2\zeta^2}{\left(\tfrac{1}{2}\alpha\zeta - E_1\right)^2} + 1\right]^{-1/2}
\end{aligned}
\right. \quad (7.64)
$$

Similarly, for a', b', and c' we have to replace E_1 by E_2, for a^{v}, b^{v}, and c^{v} we have to replace E_1 by E_6. Also

$$
\begin{aligned}
a'' &= \frac{\sqrt{\tfrac{3}{2}}\,\alpha'\zeta}{\Delta - E_3}\,b'' \\
b'' &= \left[\frac{\tfrac{3}{2}\alpha'^2\zeta^2}{(\Delta - E_3)^2} + 1\right]^{-1/2}
\end{aligned}
\quad (7.65)
$$

Here, again, to obtain a^{iv} and b^{iv} we have to replace E_3 by E_5.

The paramagnetic gram-ionic susceptibility is obtained in the usual way by applying the magnetic perturbation on the basic states of (7.63) and by using the Van Vleck formula. These calculations are quite lengthy and complicated, and therefore we do not present them here. The reader is refered to Bose, Chakravarty, and Chatterjee[20] for further details.

The crystals that have been studied are $Co(NH_4)_2(SO_4)_2 \cdot 6H_2O$, $Co(NH_4BeF_4)_2$, $6H_2O$, and $Co(KSO_4)_2$, $6H_2O$, and the results are presented in Table 7.2, where the experimental values are given in parentheses. It is very clear from Table 7.2 that, *unless* the anisotropic field parameter Δ is varied, neither

Table 7.2[a]

Variation of effective moment with temperature and with Δ for a number of Co complexes

Temperature (°K)	Δ (cm^{-1})	$P_{\parallel}^2$	$P_{\perp}^2$	$P_{\parallel}^2 - P_{\perp}^2$	$g_{\parallel}$	$g_{\perp}$
$Co(NH_4SO_4)_2 \cdot 6H_2O$ ($\alpha = 1.08$, $\alpha' = 1.53$, $\zeta = -180$ cm^{-1})						
290	1000	30.00 (29.19)	19.27 (18.75)	10.73 (10.44)	—	—
77.2	1200	35.59 (34.33)	11.85 (10.88)	23.74 (23.45)	—	—
20	1220	—	—	—	6.33 (6.45)	3.12 (3.06)
14.5	1230	31.48 (29.80)	7.93 (6.65)	23.55 (23.15)	—	—
$Co(NH_4BeF_4)_2 \cdot 6H_2O$ ($\alpha = 1.095$, $\alpha' = 1.460$, $\zeta = -180$ cm^{-1})						
296.7	850	29.94 (29.66)	19.85 (19.59)	10.09 (10.07)	—	—
185.8	700	31.46 (29.58)	19.28 (17.56)	12.18 (12.02)	—	—
86.8	680	29.49 (28.09)	15.30 (13.65)	14.19 (14.44)	—	—
$Co(KSO_4)_2 6H_2O$ ($\alpha = 1.245$, $\alpha' = 1.280$, $\zeta = -180$ cm^{-1})						
300	250	29.77 (29.49)	22.33 (23.08)	7.44 (7.41)	—	—
200	270	31.42 (30.14)	20.72 (19.42)	10.70 (10.72)	—	—
100	300	30.19 (29.23)	16.34 (15.19)	13.85 (14.04)	—	—
20	740	—	—	—	6.80 (6.60)	3.01 (2.71)

[a]The values in parentheses are the experimental values.

the absolute susceptibilities nor the anisotropies at all ranges of temperatures can be brought into agreement with the experimental values. However, α and α' are found to be the same over the entire temperature range, as is to be expected since Δ changes by only about 200 cm^{-1} in this range. It is interesting to note that a similar large change in Δ has also been observed in Ti^{3+} alum.[7]

When the magnetic field is applied along the z-direction, that is, along the symmetry axis, and along the x-direction of the Co^{2+} ion, we have, by definition, in the first order,

$$g_{\|}^{(1)}=2|\langle\phi_{1/2}|-\alpha l_z'+2S_z|\phi_{1/2}\rangle|$$

and

$$g_{\perp}^{(1)}=2|\langle\phi_{1/2}|-\alpha l_x'+2S_x|\phi_{-1/2}\rangle|$$

respectively, for the lowest Kramers' doublet. Performing the calculation, we get

$$\begin{aligned} g_{\|}^{(1)}&=2\left[(a^2-c^2)\alpha+3a^2+b^2-c^2\right]\\ g_{\perp}^{(1)}&=2\left[-\sqrt{2}\,bc\alpha'+2\sqrt{3}\,ac+2b^2\right]\end{aligned} \tag{7.66}$$

The effect of the spin-orbit coupling in admixing the upper orbital levels to ϕ' gives a second-order correction $g^{(2)}$ to the g values, given by[1]

$$\begin{aligned} g_{\|}^{(2)}&=(3a^2-c^2)\nu_1+b^2\nu_2+(\sqrt{6}\,ab-\sqrt{8}\,bc)\nu_3\\ g_{\perp}^{(2)}&=b^2\nu_4+c^2\nu_5+\sqrt{3}\,ac\nu_6+\sqrt{2}\,bc\nu_7\end{aligned} \tag{7.67}$$

Therefore

$$\begin{aligned} g_{\|}&=g_{\|}^{(1)}+g_{\|}^{(2)}\\ g_{\perp}&=g_{\perp}^{(1)}+g_{\perp}^{(2)}\end{aligned} \tag{7.68}$$

According to Abragam and Pryce,[1] this gives a correction of the order of 4% to the g values.

The paramagnetic resonance in Co^{2+} salts can be observed only near 20°K, owing to strong spin-lattice interaction at higher temperatures. The theoretical g values and also the available resonance g values (20°K) for the two sulfate salts (Table 7.2) are in reasonable agreement. In the potassium salt, though the value of $\Delta=740$ cm^{-1} at 20°K may not be very exact, there is a rather big jump from the value of 300 cm^{-1} at 100°K. The susceptibility and the g values in the intermediate range, when available, may be helpful in clarifying this point.

Our discussions in Sections 7.1 to 7.4 cover four typical paramagnetic ions having electronic configurations $3d^1$ (Ti^{3+}), $3d^2$ (V^{3+}), $3d^6$ (Fe^{2+}), and $3d^7$ (Co^{2+}), all of which are octahedrally coordinated with six water molecules in the crystal of the corresponding highly magnetically dilute salts. These have triply degenerate orbital states, either T_1 or T_2, in the crystalline Stark pattern. The main intrinsic differences between, and similarities of, the ions in the crystals are as follows:

1. The Ti^{3+} and V^{3+} ions are trivalent, whereas Fe^{2+} and Co^{2+} are divalent, so that the magnitudes of the cubic (O_h) ligand fields are rather different in the two sets.
2. The Ti^{3+} and Fe^{2+} ions are in D states so that, both of them have the T_{2g} triplet lowest, whereas V^{3+} and Co^{2+} are in F states, so that the T_{1g} triplet is lowest.
3. It is found that all the salts other than those of Co^{2+} show a fair amount of overlap of the $3d$ charge cloud with the p and s charge clouds of the surrounding O atoms. This overlap is so large in Ti^{3+} that, to explain the experimental results, the ligand field model of Van Vleck has to be replaced by the molecular orbital model of Stevens[9] and others. In V^{3+}, though the ligand field model may still be taken as a fair approximation, as much as 40% covalency reduction in spin-orbit coupling is needed to fit the experimental facts. In Fe^{2+} this reduction is about 20%, whereas in Co^{2+} it is inappreciable.
4. On the whole, our theoretical treatments in the different cases, following the methods of Abragam and Pryce, Stevens, and others, reveal the typical similarities and differences in the magnetic susceptibility, magnetic anisotropy, and paramagnetic resonance and in the optical absorption spectra. But a major departure from the usual theory occurs because of the effect of the anisotropic thermal expansion of the crystal lattice, which was pointed out earlier by Bose and his co-workers[12,13] and by Chakravarty.[14] According to these authors, such thermal expansion can linearly displace the ligand atoms surrounding the paramagnetic ions and also rotate the ligand clusters as a whole inside the crystals, thus changing not only the primary Jahn-Teller anisotropies (see Chapter 12) of the ligand cluster fields but also the induced anisotropies due to the closepacked atoms outside the ligand cluster. According to Van Vleck,[6] the Jahn-Teller distortion and the induced distortion of the ligand cluster should be comparable in Ti^{3+}, V^{3+}, Fe^{2+}, and Co^{2+}, and so, in studying the magnetic behaviors of the crystals at different temperatures, the thermal expansion of the lattice may change both types of distortion to an appreciable extent and cause departures from the usual theory. We find for V^{3+} that at high and low temperatures the effect of the thermal expansion in the anisotropic field is masked to a considerable extent by the usual departures from the Curie law predicted by the theory. This is also true for Ti^{3+} at high temperatures, but at low temperatures the anisotropic field

constant is found to change considerably. The same thing happens in Fe^{2+} and Co^{2+}. To correlate these changes, direct knowledge of the thermal changes of lattice parameters is necessary but is as yet unavailable. Our major important findings are listed in Table 7.3.

Before concluding, let us briefly discuss the contributions of W. M. Walsh, Jr., and his co-workers[24] and D. Walsh and his colleagues[25,26] which shed much light on future theoretical developments in regard to inorganic complexes of the first series of the transition metal ions, in which we are interested. From the investigations described above (see Table 7.3) it has become very clear that the static crystal field theory is inadequate to describe the situations in these complexes, even when we assume a certain reasonable amount of covalency.[9] In order to obtain agreement with the paramagnetic susceptibility from liquid helium up to room temperature and with the g values found by paramagnetic resonance experiments,[25] the lower symmetric crystal field parameter has to change continuously with temperature, to a lesser or greater extent, depending on a particular complex. Along with Van Vleck,[6] we think it likely that the thermal expansion also causes a change in the crystal field, as discussed earlier. Confirmation of this possibility is readily found in the important preliminary investigation of W. M. Walsh, Jr., and his co-workers described in Ref. 24. They have also found effects of the temperature variation of the crystalline field in several paramagnetic ions in simple crystals. The experimental data were analyzed for both implicit (thermal expansion) and explicit (lattice vibration) temperature dependences, and the following results were obtained. In MgO the g shifts of the two F-state ions, V^{2+} and Cr^{3+}, increase, and the cubic field splittings of the two S-state ions, Mn^{2+} and Fe^{3+}, decrease, with increasing temperature almost exactly as would be expected from the thermal expansion alone. The axial crystal field splitting of locally compensated Cr^{3+} ions also increases with temperatures at a rate attributable primarily to the thermal expansion. On the other hand, these authors did not find appreciable explicit (lattice vibration) temperature dependence of the crystal field parameters in MgO. This observation agrees, in general, with our earlier findings from studies of the magnetic susceptibilities and anisotropies in different systems (see Table 7.3 and the references cited).

In this connection let us also discuss the work of D. Walsh and his colleagues.[24,25] They have shown, for the complex $[Ti(H_2O)_6]^{3+}$, that by including the Ham effect (for details see Section 12.4), which includes the interaction between an E_g vibrational mode and the ground ${}^2T_{2g}$ electronic state, the g values (which are reduced from those of the free ion) obtained by resonance measurements can be successfully interpreted. The effect of the Jahn-Teller mechanism and the Ham effect are discussed in detail in Section 12.4. But let us see very briefly here what the Ham effect means. Ham has shown that, when the cluster approximation consisting of the transition metal ion and its nearest neighbors and the E_g normal mode of vibration are used,

Table 7.3

Pertinent parameters from Ligand Field Theory calculations for some of the free ions of the Fe group and for the same ions enbedded in crystals.

Configu-ration	Ion	Complex	ζ, free ion (cm^{-1})	ζ, crystal (cm^{-1})	Orbital reduction factor k in crystal	Dq (cm^{-1})	Variation of anisotropic crystal field with temperature (cm^{-1})	g, crystal	Covalency (%)	Reference
$3d^1$	Ti^{3+}	Alum		148.5	0.7	2030	170–800 (4.2–2.5°K)(100–300°K)	1.18*, 1.82†	30	*a*
$3d^2$	V^{3+}	Alum	104	64	0.62	1800	1400 (no variation)	1.85	40	*b*
$3d^3$	Cr^{3+}	MgO	91	63	0.7	1760	Negligible	1.97	30	*c*
		Alum		57	0.62				40	*c*
$3d^3$	V^{2+}	MgO	55	34	0.61	1180	—		40	*d*
		Tutton salts		444	0.79			1.99	20	*e*
$3d^6$	Fe^{2+}	Tutton salts	−103	−80	0.80	1000	270–650 (20°K)(300°K)	8.99 (20°K)	20	*f*
$3d^7$	Co^{2+}	$Co(NH_4SO_4)_2 \cdot 6H_2O$	−180	−180	Nil	1000	1265–1030 (14.5°K)(290°K)	4.03 (20°K)	—	*g*
$3d^7$	Co^{2+}	$Co(NH_4BeF_4)_2 \cdot 6H_2O$		−180	Nil	1000	720–864 (86.8°K)(296.7°K)	6.56	—	*g*
		$Co(KSO_4)_2 \cdot 6H_2O$		−180	Nil	1000	720–280 (20°K)(300°K)	—	—	*g*
$3d^8$	Ni^{2+}	MgO	−324	−250	0.77	850	—	2.25	23	*h*
		Tutton salts		−270	0.83	850	—	2.25	17	*i*
		Magnetic measurements		−250	0.77	850	—	2.25	23	*j*
$3d^9$	Cu^{2+}	Tutton salts	−829	−695	0.84	1120	—	2.23	16	*k*

*Helium Temp; †High Temp.

a. A. Bose, A. S. Chakravarty, and R. Chatterjee, *Proc. Roy. Soc.* (*London*) *A*, 255, **145** (1960).
b. A. S. Chakravarty, *Proc. Phys. Soc. A*, **74**, 711 (1959).
c. W. Low, *Phys. Rev.* **105**, 801 (1957)
d. W. Low, *Phys. Rev.* **101**, 1827 (1956)
e. B. Bleaney, D. J. E. Ingram, and H. E. D. Scovil, *Proc. Phys. Soc. A*, **64**, 601 (1951).
f. E. Konig and A. S. Chakravarty, *Theor. Chim. Acta* (*Berlin*) **9**, 151 (1967).
g. A. Bose, A. S. Chakravarty, and R. Chatterjee, *Proc. Roy. Soc.* (*London*) *A*, **261**, 43 (1961); K. D. Bowers and J. Owen, *Rep. Prog. Phys.* **18**, 304 (1955).
h. W. Low, *Phys. Rev.* **109**, 247 (1958).
i. J. H. E. Griffiths and J. Owen, *Proc. Roy. Soc.* (*London*) *A*, **213**, 459 (1952).
j. A. Bose, S. C. Mitra, and S. K. Dutta, *Proc. Roy. Soc.* (*London*) *A*, **239**, 165 (1957)

the vibronic (vibrational-electronic) spectrum remains the same as in the absence of the Jahn-Teller interaction except for the reduction by the Jahn-Teller energy, E_{JT}, common to all states, where $E_{JT}=A^2/2\mu\omega^2$ (Section 12.4). In this equation A is the Jahn-Teller coupling coefficient, μ is the effective mass of the cluster, and ω is the E_g normal-mode frequency. Because of the vibronic coupling, the vibronic eigenfunctions are products of the T_2 electronic wavefunctions and the simple harmonic oscillator functions for a displaced two-dimensional oscillator belonging to the representation E_g. The matrix elements of the various electronic operators between the vibronic states are then the products of the electronic matrix element and the oscillator-overlap integral. It can be shown that, within the vibronic ground state and to first order in perturbation theory, the orbital momentum operator, the spin-orbit coupling, and the lower symmetric crystal field (tetragonal or trigonal) are all diminished from their values in the absence of Jahn-Teller coupling by $\gamma=e^{-x/2}$, where $x=3E_{JT}/\hbar\omega$. We find out more about these details in Chapter 12. Here it is important to note that the g values of Ti^{3+} alum [as we saw in (7.25)] may be adequately explained by invoking the Jahn-Teller mechanism. As we saw in Section 7.1, however, g values can also be explained by including a small amount of covalency (10%), which is not at all unreasonable in such complexes.[9] D. Walsh and his co-workers, however, did not study the magnetization at different temperatures, and we do not know whether the Ham effect alone can adequately explain the temperature variation of the magnetization, besides in addition to explaining the g values at a particular temperature.

In our opinion it is impossible to obtain agreement with the temperature variation of magnetization found in such complexes *until* and *unless* the temperature variations of the crystal field parameters are taken into consideration. Neither the Ham effect nor the inclusion of covalency can help us in this matter. Only the thermal expansion of the crystal lattice can do it, as was rightly pointed out by Van Vleck.[6]

A complete analysis of the effect of the thermal motion of the surrounding ions on the crystal field parameters of a paramagnetic ion should be derived from the ion-crystal Hamiltonian, including a quantized phonon field. Such a fundamental calculation is hopelessly impracticable, however, because of the tremendous complexity. A simpler approach would be to determine, in principle, the changes in the observable (crystal field) energy level splittings produced by the small static displacements of the neighboring ions from their ideal equilibrium positions. As a result of these displacements, the point symmetry at the site of the paramagnetic ion would be reduced and the Hamiltonian appropriate to such a distorted situation would contain terms of the lower symmetric crystal field, as well as the modified values of the higher symmetry parameters appropriate to the static unstrained geometry. At the next stage one should consider the displacements from the equilibrium positions as time-dependent coupled variables that are specified by the amplitudes and polarizations of the normal vibrational modes of the crystal.

Here one should also include the local-mode modifications of the host vibrational spectrum because of the impurity nature of the guest paramagnetic species. We shall not discuss these ideas any further at this point; the reader is referred to Section 12.4 for further details.

REFERENCES

1. A. Abragam and M. H. L. Pryce, *Proc. Roy. Soc. (London) A*, **205**, 135 (1951).
2. M. H. L. Pryce, *Proc. Phys. Soc. (London) A* **63**, 25 (1950).
3. H. Bethe, *Ann. Phys.* **3**, 135 (1929); English Translation, Consultants Bureau, New York, 1958.
4. J. Van den Handel and A. Siegert, *Physica* **4**, 871 (1937).
5. B. Bleaney, G. S. Bogle, A. H. Cooke, R. J. Duffus, M. C. M. O'Brien, and K. W. H. Stevens, *Proc. Phys. Soc. (London) A* **68**, 57 (1955).
6. J. H. Van Vleck, *J. Chem. Phys.* **7**, 61 (1939); *Ibid.* **8**, 787 (1940).
7. A. Bose, A. S. Chakravarty, and R. Chatterjee, *Proc. Roy. Soc. (London) A*, **255**, 145 (1960).
8. M. H. L. Pryce and W. A. Runciman, *Disc. Faraday Soc.* **26**, 34 (1958).
9. K. W. H. Stevens, *Proc. Roy. Soc. (London) A*, **219**, 542 (1953).
10. S. K. Dutta Roy, A. S. Chakravarty, and A. Bose, *Indian J. Phys.* **33**, 483 (1959).
11. C. J. Gorter, *Phys. Rev.* **42**, 437 (1932).
12. A. Bose, S. C. Mitra, and S. K. Datta, *Proc. Roy. Soc. (London) A*, **239**, 165 (1957).
13. A. Bose, S. C. Mitra, and S. K. Datta, *Proc. Roy. Soc. (London) A*, **248**, 153 (1958).
14. A. S. Chakravarty, *Proc. Phys. Soc. (London)* **74**, 711 (1959).
15. J. Owen, *Proc. Roy. Soc. (London) A*, **227**, 183 (1955).
16. J. H. E. Griffiths and J. Owen, *Proc. Roy. Soc. (London) A*, **213**, 459 (1952).
17. E. Konig and A. S. Chakravarty, *Theor. Chim. Acta (Berlin)* **9**, 151 (1967).
18. M. Tinkham, *Proc. Roy. Soc. (London) A*, **236**, 535 (1956).
19. E. Konig, A. S. Chakravarty, and K. Madeja, *Theor. Chim. Acta (Berlin)* **9**, 171 (1967).
20. A. Bose, A. S. Chakravarty, and R. Chatterjee, *Proc. Roy. Soc. (London) A*, **261**, 43 (1961).
21. R. Schlapp and W. G. Penney, *Phys. Rev.* **42**, 666 (1932).
22. B. Bleaney and D. J. E. Ingram, *Proc. Roy. Soc. (London) A*, **208**, 143 (1951).
23. A. S. Chakravarty and R. Chatterjee, *Indian J. Phys.* **33**, 531 (1959).
24. W. M. Walsh, Jr., Jean Jeener, and N. Bloembergen, *Phys. Rev. A*, **139**, 1338 (1965).
25. N. Rumin, C. Vincent, and D. Walsh, *Phys. Rev. B*, **7**, 1811 (1973).
26. Y. H. Shing, C. Vincent, and D. Walsh, *Phys. Rev. B*, **9** 340 (1974).
27. J. H. Van Vleck, *The Theory of Electric and Magnetic Susceptibilities*, Oxford University Press, 1932.

8

The Spin Hamiltonian

In Chapter 7 we examined the magnetic properties of the $3d^n (n=1,2,6,7)$ systems. These have, as the ground state, the orbital triplets denoted by T_{1g} or T_{2g} in O_h symmetry in addition to the usual spin degeneracy. The ground states of the remaining systems, that is, $3d^3\ {}^4A_{2g}$ (V^{2+}, Cr^{3+}), $3d^4\ {}^5E_g$ (Cr^{2+}, Mn^{3+}), $3d^5\ {}^6A_{1g}$(Mn^{2+}, Fe^{3+}), and $3d^8\ {}^3A_{2g}$(Ni^{2+}), all have the orbitally nondegenerate state (A_2 or A_1) as the ground state with the exception of the $3d^4$ system, which has ${}^5E_g$ as the ground state. In these systems therefore the susceptibility should be isotropic and conform to the "spin-only" value since the upper excited states are too far from the ground state to make a significant contribution to the magnetic properties, and the only way these upper excited orbital states can contribute to the g value and susceptibility is in the second order via the spin-orbit interaction.

To describe adequately the magnetic properties of these orbitally nondegenerate ground states with spin degeneracy alone, we have to use the "spin Hamiltonian" concept, which has been developed by Abragam and Pryce and their collaborators.[1–4] We now give a simple derivation of the spin Hamiltonian before examining the specific systems discussed above.

In the case of a paramagnetic crystal, the unperturbed system is characterized by certain configurational variables and has energy levels that are nothing but the orbital levels. Owing to the spin degeneracy, each orbital level is $(2S+1)$-fold degenerate. It will be assumed that the lowest orbital level, in which we are interested at present, is nondegenerate (either A_{1g} or A_{2g}). Operators referring to spin S and nuclear variables I are treated as noncommuting algebraic quantities, that is, no representation is chosen for either S or I. The result is that, instead of obtaining the usual perturbed energy, an expression is obtained that involves components of S and I. This is called the spin Hamiltonian, and the actual energies are the eigenvalues of these new operators. We shall now consider the different perturbations on the nondegenerate orbital level.

The spin-orbit perturbation is represented, as usual, by

$$\mathcal{H}_{so} = \sum_i \zeta(r_i)\mathbf{l}_i\cdot\mathbf{s}_i = \zeta\mathbf{L}\cdot\mathbf{S} \tag{8.1}$$

where ζ is the spin-orbit coupling coefficient. The magnetic perturbation is represented by

$$\mathcal{H}_m = \sum_j \mu_B\mathbf{H}(\mathbf{l}_j + 2\mathbf{s}_j) = \mu_B\mathbf{H}(\mathbf{L}+2\mathbf{S}) \tag{8.2}$$

The spin-spin interaction is represented by

$$\mathcal{H}_{ss} = \sum_{j,k} 4\mu_B^2\left[\frac{\mathbf{S}\cdot\mathbf{S}'}{r_{jk}^3} - \frac{3(\mathbf{S}\cdot\mathbf{R})(\mathbf{S}'\cdot\mathbf{R})}{r_{jk}^5}\right]$$

which, when replaced by appropriate angular momentum operators, takes the form

$$\mathcal{H}_{ss} = -\rho\left[(\mathbf{L}\cdot\mathbf{S})^2 + \tfrac{1}{2}(\mathbf{S}\cdot\mathbf{L}) - \tfrac{1}{3}L(L+1)S(S+1)\right] \tag{8.3}$$

It is to be noted that the effect of this term is very small, and for the calculation of the susceptibilities of salts of the iron group of elements, we can safely neglect this effect. The coefficient ρ of the spin-spin interaction term has been calculated by Pryce[1] and also by Chakravarty[5] for the iron group of elements. Of course, if one is interested in the nuclear hyperfine structure, this term should definitely be included. The total Hamiltonian then becomes

$$\begin{aligned}\mathcal{H} = V_C + \left(\zeta - \tfrac{1}{2}\rho\right)(\mathbf{L}\cdot\mathbf{S}) - \rho(\mathbf{L}\cdot\mathbf{S})^2 \\ + \mu_B\mathbf{H}(\mathbf{L}+2\mathbf{S})\end{aligned} \tag{8.4}$$

where V_c represents the O_h ligand field perturbation.

If we wish to take account of the second-order perturbation effects arising from the upper excited states of $\mathcal{H}_F$ (the free-ion Hamiltonian), we must consider the matrix elements connecting the ground manifold to higher states. However, the separation between the energy levels of V_c is much greater than the magnitude of the other terms in $\mathcal{H}$ (8.4). We may therefore proceed one stage further in our perturbation calculations by treating these terms as a small perturbation on V_c. Further development of the theory now depends to a considerable amount on whether or not the lowest level of V_c is degenerate. As we have said, we shall consider the lowest level to be orbitally nondegenerate. The first-order contribution from the remaining terms to the energy is then given by replacing all orbital variables in (8.4) by their expectation values.

We take the ground orbital state as $|0\rangle$ and the excited states as $|n\rangle$ with energies $E(n)$, where $n=1,2,3,\ldots$. Now, applying the first-order perturbation theory, we have

$$\begin{aligned}\delta_1 E_0 &= \langle 0|\mathcal{H}|0\rangle \\ &= \langle 0|\zeta(\mathbf{L}\cdot\mathbf{S}) - \rho\left\{(\mathbf{L}\cdot\mathbf{S})^2 + \tfrac{1}{2}(\mathbf{L}\cdot\mathbf{S})\right\} \\ &\quad + \mu_B \mathbf{H}\cdot\mathbf{L} + 2\mu_B \mathbf{H}\cdot\mathbf{S}|0\rangle \end{aligned} \tag{8.5}$$

Now, since the ground state is an orbital singlet state with no orbital moment, the expectation value of **L** is zero, that is,

$$\langle 0|\mathbf{L}_x|0\rangle = \langle 0|\mathbf{L}_y|0\rangle = \langle 0|\mathbf{L}_z|0\rangle = 0 \tag{8.6}$$

Thus

$$\delta_1 E_0 = \langle 0| -\rho(\mathbf{L}\cdot\mathbf{S})^2|0\rangle + \langle 0|2\mu_B \mathbf{H}\cdot\mathbf{S}|0\rangle \tag{8.7}$$

Now the expectation value of **L** is zero, while that of the quadratic combinations $\mathbf{L}_i\mathbf{L}_j$, giving a symmetric tensor, is denoted by

$$\tfrac{1}{2}\langle 0|\mathbf{L}_i\mathbf{L}_j + \mathbf{L}_j\mathbf{L}_i|0\rangle \equiv \tfrac{1}{3}L(L+1)\delta_{ij} + l_{ij} \tag{8.8}$$

with $l_{ii}=0$. Hence we finally get

$$\delta_1 E_0 = -\rho l_{ij}\mathbf{S}_i\mathbf{S}_j + 2\mu_B \mathbf{H}\cdot\mathbf{S} \tag{8.9}$$

But the second-order contributions arising from the spin-orbit interaction and its joint effect with other terms cannot be neglected. We now proceed to calculate the second-order perturbation as follows:

$$\begin{aligned}\delta_1 E &= -\sum_{n\neq 0} \frac{\langle 0|\mathcal{H}|n\rangle\langle n|\mathcal{H}|0\rangle}{E(n)-E(0)} \\ &= -\sum_{n\neq 0} \frac{\langle 0|\zeta(\mathbf{L}\cdot\mathbf{S}) + \mu_B\mathbf{H}\cdot\mathbf{L}|n\rangle\langle n|\zeta(\mathbf{L}\cdot\mathbf{S}) + \mu_B\mathbf{H}\cdot\mathbf{L}|0\rangle}{E(n)-E(0)} \\ &= -\sum_{n\neq 0} \frac{\langle 0|\mathbf{L}_i|n\rangle\langle n|\mathbf{L}_j|0\rangle}{E(n)-E(0)}\left[(\mu_B\mathbf{H}_i + \zeta\mathbf{S}_i)(\mu_B\mathbf{H}_j + \zeta\mathbf{S}_j)\right] \\ &= -\Lambda_{ij}\left[\mu_B^2\mathbf{H}_i\cdot\mathbf{H}_j + 2\zeta\mu_B\mathbf{H}_i\cdot\mathbf{S}_j + \zeta^2\mathbf{S}_i\cdot\mathbf{S}_j\right]\end{aligned} \tag{8.10}$$

So the total Hamiltonian (neglecting the nuclear interaction terms) is given by

$$\mathcal{H} = D_{ij}\mathbf{S}_i\cdot\mathbf{S}_j + \mu_B g_{ij}\mathbf{H}_i\cdot\mathbf{S}_j - \mu_B^2\Lambda_{ij}\mathbf{H}_i\cdot\mathbf{H}_j \tag{8.11}$$

where

$$\Lambda_{ij} = \Lambda_{ji} = \sum_{n \neq 0} \frac{\langle 0|\mathbf{L}_i|n\rangle\langle n|\mathbf{L}_j|0\rangle}{E(n) - E(0)} \tag{8.12}$$

$$D_{ij} = -\rho l_{ij} - \zeta^2 \Lambda_{ij} \tag{8.13}$$

and

$$g_{ij} = 2(\delta_{ij} - \zeta \Lambda_{ij}) \tag{8.14}$$

When the ligand field possesses axial symmetry, for example, tetragonal or trigonal symmetry, the various tensors are characterized by two principal values, each parallel and perpendicular to the symmetry axis. Taking the latter to be the z-axis, we may write

$$\mathcal{H} = D\left[S_z^2 - \tfrac{1}{3}S(S+1)\right] + \mu_B g_{\parallel} H_z S_z + \mu_B g_{\perp}(H_x S_x + H_y S_y) \tag{8.15}$$

It is not difficult to show that in the orthorhombic case we obtain a more general expression

$$\mathcal{H} = D\left[S_z^2 - \tfrac{1}{3}S(S+1)\right] + \mu_B(g_x H_x S_x + g_y H_y S_y + g_z H_z S_z) + E(S_x^2 - S_y^2) \tag{8.16}$$

The parameters D and E for $3d^3$, $3d^5$, and $3d^8$ systems have been calculated by Chakravarty[5] and by Sharma, Das, and Orbach[6] and will be described later in this chapter.

Before proceeding further, let us calculate Λ_{ij} (8.12) for the ^2D state of Cu^{2+}, whose configuration is $3d^9$. In an octahedral field the ^2D state splits into $^2T_{2g}$ and ^2E_g, where ^2E_g lies lowest. The separation between $^2T_{2g}$ and ^2E_g is conventionally denoted by $10Dq$, which is approximately 11,200 cm^{-1} for typical Cu^{2+} salts, for example, $CuSO_4 \cdot 5H_2O$. The structural data indicate that each Cu^{2+} ion is surrounded by four water molecules arranged in an approximate square planar configuration, with two polar sulfate oxygens in *trans*position. Hence the ligand field is of D_{4h} symmetry, and the electronic levels are given schematically in Figure 8.1. The wavefunctions of the doublet and three singlets are as follows [see (6.7) and (6.8)]:

$$\left.\begin{aligned} {}^2B_{1g} &= \frac{\sqrt{15}}{2}(x^2 - y^2) \\ {}^2A_{1g} &= \frac{\sqrt{5}}{2}(3z^2 - r^2) \end{aligned}\right\} \quad E_g \text{ symmetry}$$

$$\left.\begin{aligned} {}^2B_{2g} &= \sqrt{15}\, xy \\ {}^2E_g &= \sqrt{15}\, yz \text{ or } \sqrt{15}\, xz \end{aligned}\right\} \quad T_{2g} \text{ symmetry}$$

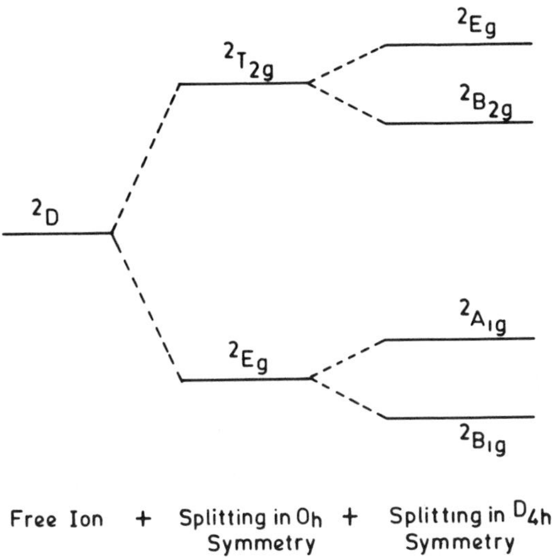

Fig. 8.1 Energy level diagram of sixfold coordinated Cu^{2+} ion (not to scale).

Note that, as we descend to D_{4h} symmetry from O_h symmetry, we have $E_g \rightarrow A_{1g} + B_{1g}$ and $T_{2g} \rightarrow B_{2g} + E_g$, that is, the doubly degenerate E_g state splits into two nondegenerate levels, and the triply degenerate T_{2g} splits into one nondegenerate and one twofold degenerate level. Using these wavefunctions, we see that

$$\mathbf{L}_z|^2B_{1g}\rangle = 2i\sqrt{15}\ xy = 2i|^2B_{2g}\rangle$$

Thus

$$\langle^2B_{2g}|\mathbf{L}_z|^2B_{1g}\rangle = 2i$$

and

$$\langle^2B_{2g}|\mathbf{L}_x|^2B_{1g}\rangle = \langle^2B_{2g}|\mathbf{L}_y|^2B_{1g}\rangle = 0$$

Hence

$$\Lambda_{zz} = \Lambda_{\|} = -\frac{\langle^2B_{2g}|\mathbf{L}_z|^2B_{1g}\rangle\langle^2B_{1g}|\mathbf{L}_z|^2B_{2g}\rangle}{E(^2B_{1g}) - E(^2B_{2g})}$$

$$= \frac{4}{\Delta E}$$

Similarly,

$$\langle^2E_g|\mathbf{L}_x|^2B_{1g}\rangle = \langle^2B_{1g}|\mathbf{L}_x|^2E_g\rangle = -i$$

$$\therefore \Lambda_{xx} = \Lambda_{yy} = \frac{1}{\Delta E}$$

Assuming that

$$E(^2B_{1g}) - E(^2B_{2g}) = E(^2B_{1g}) - E(^2E_g) \simeq \Delta E$$

we have

$$g_\parallel = 2(1 - \zeta\Lambda_\parallel) = 2\left(1 - \frac{4\zeta}{\Delta E}\right)$$

$$g_\perp = 2\left(1 - \frac{\zeta}{\Delta E}\right) \tag{8.17}$$

But if the ground state is $^2A_{1g}$ instead of $^2B_{1g}$ (Figure 8.1), then, since

$$\mathbf{L}_z|^2A_{1g}\rangle = 0$$

and

$$\langle^2E_g|\mathbf{L}_x|^2A_{1g}\rangle = -i\sqrt{3}$$

we have, as before,

$$g_\parallel = 2$$

$$g_\perp = 2\left(1 - \frac{3\zeta}{\Delta E}\right) \tag{8.18}$$

Now the experimental g values[7] of a typical Cu^{2+} salt, such as $Cu(NH_3)_4 \cdot 6H_2O$, are given by $g_\parallel = 2.22$ and $g_\perp = 2.05$, values which show definitely that the ground state is $^2B_{1g}$ and not $^2A_{1g}$. Hence the measurements of the g-factors can uniquely decide the ground state of a particular complex.

One way in which the g values of a particular ion can be measured is by the paramagnetic resonance method, which was discovered by Zavoisky.[20] Considering a set of paramagnetic ions that are noninteracting and each of which possesses a single unpaired electron, we obtain from the paramagnetic resonance experiment a "spin-only" magnetic dipole moment of $\frac{1}{2}g_s\mu_B$, where μ_B is the Bohr magneton and $g_s = 2.00229$. In a steady magnetic field $\mathbf{H}$ each dipole can orient itself parallel or antiparallel to $\mathbf{H}$ with energies $-\frac{1}{2}g_s\mu_B H$ and $+\frac{1}{2}g_s\mu_B H$, respectively. Now magnetic dipole transitions can be induced between these two energy levels by applying a high-frequency magnetic field, polarized perpendicular to $\mathbf{H}$ with frequency ν such that the quantum of high-frequency energy equals the separation between the levels, that is, $h\nu = g_s\mu_B H$. Then a resonance absorption corresponding to the dipoles being flipped from the parallel to the antiparallel states takes place, and there is an induced emission corresponding to the reverse process. If the system remains in thermal equilibrium, a lesser number of spins are in the parallel state because it is of lower energy, so that there is a net absorption of high-frequency power.

The paramagnetic resonance spectrum of a transition group ion in a crystal is generally more complex than that for the simple case discussed above. The lowest group of energy levels, that is, the ones between which microwave transitions are induced, depends in a complicated way on the particular ion, the symmetry and strength of the ligand field, and also the spin-orbit interaction and other effects such as the hyperfine interaction between the electrons and the nucleus.

The lowest group of electronic states is characterized by a fictitious quantum number S', even though each state is really a complicated mixture of the spin and the orbital wavefunctions of the free ion.[1] Called the effective or fictitious spin of the system, S' is defined by equating $(2S'+1)$ to the number of electronic levels in the lowest group. In general, $S=S'$. The values of S and S' for the various ions of the iron group of elements under the usual ligand fields in the hydrated salts are listed in Table 8.1. The use of an effective spin means that, for temperatures where only the $(2S'+1)$ lowest levels are appreciably populated, the magnetic ion is treated like a magnetic dipole, which has $(2S'+1)$ orientations in an applied magnetic field and each energy level is associated with one orientation. But the effective magnetic moment of this dipole does not correspond to a "spin-only" value; it differs by the factor g_i/g_s, where g_i is the spectroscopic splitting factor defined according to (8.14).

The energy levels, however, often have initial splittings in zero magnetic field [see (8.11)]. This means that the dipole cannot be treated as free, and that, in addition to the interaction with the applied magnetic field, energy terms must also be added to represent the electrostatic interaction between the magnetic electrons and the ligand field, which may give Stark splittings if $S>\frac{1}{2}$. The sum of all the terms, written as a sum of energy operators to be applied to the effective spin states, constitutes the spin Hamiltonian $\mathcal{H}$ for the system. The actual energies are then the eigenvalues E_0 which satisfy the operator equation

$$\mathcal{H}\Psi=E_0\Psi \tag{8.19}$$

where Ψ represents the wavefunctions of the effective spin states.

Table 8.1

Values of S and S' for the ground state of $M(H_2O)_6$ complexes where $M=3d^n (n=1 \text{ to } 9)$

Configuration:	d^1	d^2	d^3	d^4	d^5	d^6	d^7	d^8	d^9
Ions:	Ti^{3+}	Ti^{2+}	V^{2+}	Cr^{2+}	Mn^{2+}				
	V^{4+}	V^{3+}	Cr^{3+}	Mn^{3+}	Fe^{3+}	Fe^{2+}	Co^{2+}	Ni^{2+}	Cu^{2+}
Lowest orbital state in O_h symmetry:	T_{2g}	T_{1g}	A_{2g}	E_g	A_{1g}	T_{2g}	T_{1g}	A_{2g}	E_g
S	$\frac{1}{2}$	1	$\frac{3}{2}$	2	$\frac{5}{2}$	2	$\frac{3}{2}$	1	$\frac{1}{2}$
S'	$\frac{1}{2}$	1	$\frac{3}{2}$	2	$\frac{5}{2}$	$\frac{1}{2}$	$\frac{1}{2}$	1	$\frac{1}{2}$

Quite often there are initial splittings of the effective spin levels when $H=0$, although there is always at least a twofold degeneracy if the free-ion spin is half-integral (Kramers' theorem). The most important cause of these splittings is the Stark effect arising from the asymmetrical ligand fields. For example, a field of axial symmetry may give splitting along the symmetry axis, which is formally represented by the operator

$$D\left[S_z^2-\tfrac{1}{3}S(S+1)\right]$$

in the Hamiltonian (8.16). Perpendicular to the symmetry axis the splitting is represented by $E(S_x^2-S_y^2)$. The term $D[S_z^2-\frac{1}{3}S(S+1)]$ implies that, with $H=0$ and $S=1$, for example, the levels characterized by $M_S=\pm 1$ are separated from the level $M_S=0$ by D; similarly, with $H=0$ and $S=\frac{3}{2}$, the levels $M_S=\pm\frac{3}{2}$ are separated from the levels $M_S=\pm\frac{1}{2}$ by $D[(\frac{3}{2})^2-(\frac{1}{2})^2]=2D$. The Hamiltonian

$$\mathcal{H}=g_z\mu_B HS_z+D\left[S_z^2-\tfrac{1}{3}S(S+1)\right] \tag{8.20}$$

with $S=1$ and $H\|z$, thus gives energy $\pm g_z\mu_B H+\frac{1}{3}D$ for $M_S=\pm 1$ levels, and energy $-\frac{2}{3}D$ for $M_S=0$ levels. The allowed transitions, $\Delta M_S=\pm 1$, then occur in fields H, such that $g_z\mu_B H\pm D=h\nu$. Thus there is fine structure in the spectrum, consisting of two absorption lines separated in the magnetic field by $2D/g_z\mu_B$. Experimentally, the Stark splittings of the lowest levels of the transition group ions in crystals are found to have values ranging between zero and a few reciprocal centimeters.

With a strong magnetic field parallel to z and with z taken as the axis of quantization, the effective spin states can be characterized by the magnetic quantum number M_S. The eigenfunctions are written in the form $|M_S\rangle$, giving $(2S+1)$ states: $|S\rangle, |S-1\rangle, \ldots, |-S\rangle$. If there are zero-field splittings between the levels, then, with a weak magnetic field parallel to z, each state, in general, is of the form

$$|\psi\rangle=a|S\rangle+b|S-1\rangle+\cdots \tag{8.21}$$

with

$$a^2+b^2+\cdots=1$$

One way of formulating the problem of finding the eigenvalue for a state is to say that values of a, b, and so on must be found such that the operation $\mathcal{H}|\psi\rangle$ reproduces the same wavefunction multiplied by a number. This number is the required eigenvalue. However, it is not necessary to know the coefficients $a, b, \ldots$ in order to find the energy, as we shall see in the following.

The method for finding the eigenvalues can be shown by an example. Let us find the eigenvalues of

$$\mathcal{H}=\mu_B g_z H_z S_z + D\left[S_z^2-\tfrac{1}{3}S(S+1)\right]+E\left(S_x^2-S_y^2\right)$$

for the $S=1$ state and **H** parallel to z-direction. Finding the matrix elements in the state $|1\rangle$, $|0\rangle$, and $|-1\rangle$ in the usual way, we obtain the following energy matrix:

$$\begin{bmatrix} g_z\mu_B H+\frac{1}{3}D & 0 & E \\ 0 & -\frac{2}{3}D & 0 \\ E & 0 & -g_z\mu_B H+\frac{1}{3}D \end{bmatrix} \tag{8.22}$$

Solving (8.22), we get the energies

$$\begin{aligned} E_1&=\tfrac{1}{3}D+\left(g_z^2\mu_B^2H^2+E^2\right)^{1/2} \\ E_2&=\tfrac{1}{3}D-\left(g_z^2\mu_B^2H^2+E^2\right)^{1/2} \\ E_3&=-\tfrac{2}{3}D \end{aligned} \tag{8.23}$$

Similarly, we can find the energies when the magnetic field is parallel to the x- and y-directions, respectively.

Koster and Statz[8] have suggested an alternative to the spin Hamiltonian method for treating the behavior of a paramagnetic ion under the combined influence of the crystal field and an applied magnetic field. They approach this problem from a somewhat different point of view. Instead of trying to explain both the zero-field splitting and the behavior of the energy levels with the magnetic field in terms of perturbations of the free-ion levels, they start from the exact eigenstates of the paramagnetic ion in the crystal field and describe the behavior of the ion in a magnetic field in terms of these states, taking advantage of whatever symmetry they may have.

In a free atom, for sufficiently small magnetic fields (by "small" is meant that the splitting due to the magnetic field is small compared with the distance between the ground level and the next excited level), the Hamiltonian for the interaction with the magnetic field, $\mu_B\mathbf{H}(\mathbf{L}+2\mathbf{S})$, can be replaced by $g\mu_B\mathbf{H}\cdot\mathbf{J}$, where **J** is the total angular momentum. This replacement can be made because the matrix elements are taken between the states of a given angular momentum and because the operator $\mathbf{L}+2\mathbf{S}$ transforms like a vector. For the purpose of computing the matrix elements, this operator can be replaced by an appropriate constant times an operator with the same transformation properties (e.g., **J**). Once this replacement is made, the Hamiltonian matrix can be written at once because of the ease with which the matrix elements of **J** can be found. In other words, the matrix of the operator $\mathbf{L}+2\mathbf{S}$

in a manifold of states of a given total angular momentum $\mathbf{J}(-J \leqslant M_J \leqslant J)$ is just a constant times a known matrix. This is a special case of the Wigner-Eckart theorem, which was discussed earlier. This theorem states that the matrix of an operator that transforms irreducibly under the complete rotation group (i.e., like spherical harmonics), taken between states of a given **J**, is a constant times a known matrix. The matrix elements in the known matrix are the familiar Clebsch-Gordan (C-G) coefficients. Koster[9] has also shown that a somewhat similar situation exists for groups other than the full rotation group. This fact is useful to describe the behavior of the energy levels of a paramagnetic ion in a crystal field when a magnetic field is applied. To describe this behavior, however, more parameters are required than are present in a spin Hamiltonian. This is due, of course, to the less restrictive physical assumptions and should yield a more accurate description of the magnetic field splittings of the levels.

This method is not restricted in its applicability to assumptions about the strength of the crystal field. Since the considerations used are essentially symmetry arguments, no specific assumptions need be made about the detailed nature of the wavefunctions of the paramagnetic ions. All effects due to admixture of the higher states of the atom, as well as contributions due to binding with neighboring nonmagnetic ions, are automatically included. The method can also be applied to levels arising from the orbital S states of the free atom in the same way as it can be applied to all other states.

There is one disadvantage of the method: the difficulty of fitting the experimental data to the larger number of parameters. It should be kept in mind, however, that much of the fitting of the experimental paramagnetic resonance data has been carried out in a very narrow range of frequencies. As more extensive and accurate experimental data become available, it will very probably be necessary to use this approach to describe the experimental findings adequately. A possible example of the inadequacy of the conventional spin Hamiltonian is given by Muller[10], who has investigated the ^6S state of the Fe^{3+} ion in $SrTiO_3$ crystals. He has found discrepancies of 5 to 10% when measuring at only one wavelength of 3.2 cm. Since not much work has been done using the Koster-Statz approach during the last two decades, it will not be discussed in detail at the present moment. Needless to say, however, this approach is very important for an accurate description of experimental findings.

8.1 MAGNETIC PROPERTIES OF $3d^8$ COMPLEXES

As an example we now consider the case of $[Ni(H_2O)_6]$, SiF_6, which is a well-studied problem. The ground state of the $3d^8$ system in a cubic field is $^3A_{2g}$, that is, an orbitally nondegenerate state with threefold spin degeneracy. The next excited triplet state is $^3T_{2g}$. A tetragonal or trigonal field will split the orbitally degenerate states—in particular, the excited $^3T_{2g}$ state—into a twofold and a single level. The spin-orbit interaction perturbation then couples the ground state with the excited states. Hence the spin multiplet of the ground state will feel the splitting of the excited states indirectly. These spin levels will therefore undergo a small splitting known as the zero-field splitting. Without the influence of the excited states the spin triplets would be degenerate since the lowest orbital level is nondegenerate.

To begin, we shall consider the spin-orbit interaction, since the $^3A_{2g}$ state mixes with the $^3T_{2g}$ state via the spin-orbit coupling. Let us treat the $3d^8$ system as the hole equivalent of the $3d^2$ system, and consequently the orbital part of the ground state in the strong-field approximation (to be dealt with in considerable detail in Chapter 9) is given by

$$\psi(A_{2g}) = |(x^2-y^2)(z^2)| \tag{8.24}$$

Augmenting (8.24) with $S=1$ and remembering that such a state will produce a $T_1 \otimes A_2 = T_2'$ in the double group O', which equals a Γ_5 level, we find that the resulting three components are given (see Appendix VI) by

$^3A_{2g}(\Gamma_5)$:

$$\begin{aligned}
\theta_1 &= \left|\overset{+}{(x^2-y^2)} \quad \overset{+}{(z^2)}\right| \\
\theta_2 &= \frac{1}{\sqrt{2}}\left|\overset{+}{(x^2-y^2)} \quad \overset{-}{(z^2)}\right| + \frac{1}{\sqrt{2}}\left|\overset{-}{(x^2-y^2)} \quad \overset{+}{(z^2)}\right| \qquad (8.25) \\
\theta_3 &= \left|\overset{-}{(x^2-y^2)} \quad \overset{-}{(z^2)}\right|
\end{aligned}$$

The wavefunctions of the first excited triplet state, $^3T_{2g}$, can be found by descending in symmetry from O_h to D_{4h}, augmenting the spin $S=1$, as before, and picking out the three linear combinations transforming like Γ_5 in O'. With the help of the tables[15] in Appendix VIII, we can show that

$$\begin{aligned}
{}^3T_{2g}(\Gamma_5^a) &= \frac{1}{\sqrt{2}}[\Psi_2\psi_{-1} + i\Psi_3\psi_1] \\
(\Gamma_5^b) &= \frac{1}{\sqrt{2}}[\Psi_2\psi_0 + i\Psi_1\psi_1] \qquad (8.26) \\
(\Gamma_5^c) &= \frac{1}{\sqrt{2}}[\Psi_3\psi_0 - \Psi_1\psi_{-1}]
\end{aligned}$$

where

$$\Psi_1 = |(xy)(z^2)|$$

$$\Psi_2 = \frac{1}{\sqrt{2}}\left[-\tfrac{1}{2}|(xz)(z^2)| - \frac{\sqrt{3}}{2}|(xz)(x^2-y^2)|\right]$$

$$-\frac{i}{\sqrt{2}}\left[-\tfrac{1}{2}|(yz)(z^2)| + \frac{\sqrt{3}}{2}|(yz)(x^2-y^2)|\right]$$

$$\Psi_3 = \frac{i}{\sqrt{2}}\left[-\tfrac{1}{2}|(xz)(z^2) - \frac{\sqrt{3}}{2}|(xz)(x^2-y^2)|\right]$$

$$-\frac{1}{\sqrt{2}}\left[-\tfrac{1}{2}|(yz)(z^2) + \frac{\sqrt{3}}{2}|(yz)(x^2-y^2)|\right]$$

are the orbital parts, and

$$\psi_1 = \alpha\alpha$$

$$\psi_2 = \frac{1}{\sqrt{2}}(\alpha\beta + \beta\alpha) \tag{8.27}$$

$$\psi_3 = \beta\beta$$

are the three spin states corresponding to $S=1$.

Now it is easy to find[11] the matrix elements of $\lambda\mathbf{L}\cdot\mathbf{S}$ between the components of $^3A_{2g}$ and $^3T_{2g}$, which are given by

$$\langle\Gamma_5^n(^3A_{2g})|\lambda\mathbf{L}\cdot\mathbf{S}|\Gamma_5^n(^3T_{2g})\rangle = 2\sqrt{2}\,\lambda, \qquad n=a,b,c \tag{8.28}$$

If it is assumed that $E(^3T_{2g}) - E(^3A_{2g}) = 10Dq$, the first-order correction to the wavefunction of the ground state is given by

$$\Phi(\Gamma_5^n) = \Gamma_5^n(^3A_{2g}) - \frac{2\sqrt{2}\,\lambda}{10Dq}\Gamma_5^n(^3T_{2g}) \tag{8.29}$$

with $n=a$, b, and c.

Let us now calculate the g-factor for the ground state of cubic Ni^{2+} complexes. Using (8.26), we get

$$\langle\theta_1|\mathbf{L}_z + 2\mathbf{S}_z|\theta_1\rangle = 2$$

$$\langle\theta_2|\mathbf{L}_z + 2\mathbf{S}_z|\theta_2\rangle = 0$$

$$\langle\theta_3|\mathbf{L}_z + 2\mathbf{S}_z|\theta_3\rangle = -2$$

Thus the g-factor is 2 in the ground state of $^3A_{2g}$. Now, considering the admixture from the upper ${}^3T_{2g}$ state, we get, by using (8.29),

$$\langle \Phi(\Gamma_5^a)|\mathbf{L}_z + 2\mathbf{S}_z|\Phi(\Gamma_5^a)\rangle = 2 - \frac{8\lambda}{10Dq} \tag{8.30}$$

Hence we have, to the first order in λ,

$$g = 2 - \frac{8\lambda}{10Dq} \tag{8.31}$$

The experimental g value for $Ni(H_2O)_6^{2+}$ is 2.25, which means that to comply with experiment one must assume that the spin-orbit coupling constant λ is diminished from its free-ion value of $-324\ cm^{-1}$ to only about $-270\ cm^{-1}$, a reduction[12, 13] of about 17%. This reduction of λ from its free-ion value is due to covalency which we shall discuss in connection with the molecular orbital theory in Chapter 11.

To calculate the susceptibility we use (4.48), which gives us the low-frequency term. Thus, using (8.31) in (4.48), we have

$$\chi_{\text{low freq}} = \frac{8N\mu_{\mathrm{B}}^2}{3kT}\left(1 - \frac{4\lambda}{10Dq}\right)^2 \tag{8.32}$$

The high-frequency term can be obtained by finding the matrix element of $\mu_{\mathrm{B}}(\mathbf{L}+2\mathbf{S})$ between ${}^3A_{2g}$ and ${}^3T_{2g}$. Using the wavefunctions given earlier, we can easily show that

$$\chi_{\text{high freq}} = \frac{8N\mu_{\mathrm{B}}^2}{10Dq} \tag{8.33}$$

Adding (8.32) and (8.33), we have, for the susceptibility of an octahedral Ni^{2+} complex,

$$\chi = \frac{8N\mu_{\mathrm{B}}^2}{3kT}\left(1 - \frac{4\lambda}{10Dq}\right)^2 + \frac{8N\mu_{\mathrm{B}}^2}{10Dq} \tag{8.34}$$

Note that in (8.33) we have not included the effect of the mixing of ${}^3A_{2g}$ and ${}^3T_{2g}$ via the spin-orbit coupling. This is not important here because the ${}^3T_{2g}$ level is located $10Dq\ cm^{-1}$ above the ground state and hence can be neglected.

Equation (8.34), with the help of (8.31), can be conveniently written as

$$\mu_{\text{eff}}(Ni^{2+}) = 2g^2 + \frac{24kT}{E({}^3T_2) - E({}^3A_2)} \tag{8.35}$$

where $10Dq$ has been replaced simply by $E({}^3T_2) - E({}^3A_2)$. We anticipate and say that the g value is available from electron resonance experiments, and

$E(^3T_2)-E(^3A_2)$ from optical measurements.[21] The g value in Ni^{2+} is isotropic and is equal to approximately 2.25, and $E(^3T_2)-E(^3A_2)\simeq 8500\ \text{cm}^{-1}$ for a typical Ni^{2+} salt. Now, taking $kT=200\ \text{cm}^{-1}$, which is approximately the room temperature value, we obtain from (8.35) $\mu_{\text{eff}}=3.27$, compared with the experimental value[13] of 3.2. Similarly, for Cr^{3+} in $Cr(H_2O)_6^{3+}$, since $g\simeq 1.98$ and $E(^4T_{2g}-{}^4A_{2g})\simeq 17{,}400\ \text{cm}^{-1}$, we have $\mu_{\text{eff}}=3.87$, compared with the experimental value[22] of 3.88. In this case, of course, the numerical factor in the first term of (8.35) can be easily shown to be $\frac{15}{4}$ instead of 2.

If, in addition to the cubic field, the Ni^{2+} complex also possesses a lower symmetric field of tetragonal or trigonal symmetry superimposed, we can use the spin Hamiltonian developed earlier, where we assume the applied magnetic field to be parallel to the axis of quantization z. The spin Hamiltonian is given by (8.20), and the energies are given by

$$E_0=-\tfrac{2}{3}D \quad \text{for the} \quad M_S=0 \quad \text{level} \tag{8.36}$$

and

$$E_{\pm}=\tfrac{1}{3}D\pm g_z\mu_B H \qquad \text{for the} \quad M_S=\pm 1 \quad \text{levels}$$

Using, again, (4.48) with (8.36), we get

$$\chi_{\parallel}=\frac{2N\mu_B^2 g_{\parallel}^2}{kT}\cdot\frac{1}{2+e^{D/kT}} \tag{8.37}$$

Since $kT\gg D$, $e^{D/kT}\simeq 1+(D/kT)$, we get

$$\chi_{\parallel}=\frac{2N\mu_B^2 g_{\parallel}^2}{3kT}\left(1-\frac{D}{kT}\right) \tag{8.38}$$

where D is measured in degrees Kelvin. The sign of D can be decided uniquely by measuring the susceptibility over a suitable range of temperature. For Ni^{2+} salts, however, D is found to be negative, meaning that $M_S=\pm 1$ levels lie below the $M_S=0$ level. Apart from the crystal field, the spin-spin interaction also contributes to the zero-field splitting parameter D, not only in $3d^8$ systems but also in $3d^3$ and $3d^5$ systems.[1, 5, 6] We shall discuss these in Section 8.2.

We shall not discuss any other system where the orbitally nondegenerate level lies lowest. There has been exhaustive work on $3d^3$ (Cr^{3+}) systems, mainly by Tanabe, Sugano, and Kamimura;[15] they have also been discussed at length by Ballhausen[14] and by Griffith.[11]

Before the origin of the zero-field splittings in $3d^3$, $3d^5$, and $3d^8$ systems is discussed, a few remarks should be made regarding the paramagnetic susceptibilities of the $3d^n$ octahedral complexes.

Apart from the usual high-frequency contribution from the ground configuration, there may also be high-frequency contribution from the first excited configurations. This contribution to susceptibility has been estimated by Desai and Chakravarty.[16] The magnitude of this contribution, depending on the particular ion and the particular complex, varies roughly from 1 to 9% of the room temperature susceptibility value and is thus quite appreciable. This means that the magnitudes of the ligand field parameters, the spin-orbit coupling constant ζ in the crystal, and so on, usually obtained from the ground configuration alone, will be altered somewhat in the temperature range from 200 to 300°K when this additional contribution is taken into account. The reader is referred to Ref. 16 for further details.

Here it suffices to say that the magnetic properties of the $3d^n$ complexes are, by now, well understood, apart from a few points here and there, if, in addition to the ionic model we have been considering so long, one takes into account the molecular field-theoretic approach treated in Chapter 11. Even there, one has to work out many of the finer details in these systems.

8.2 ZERO-FIELD SPLITTING OF S-STATE IONS

Van Vleck and Penney[17] first discussed in a qualitative manner the origin of the zero-field splitting of S-state ions under the influence of a crystalline environment. By using the spin-orbit interaction and a crystal field of axial symmetry, they showed that it is necessary to advance to a higher order in the perturbation theory before any splitting of the level is obtained. Paramagnetic resonance has shown that small splittings do exist, and Pryce[1] showed in 1950 that one could obtain a finite contribution in relatively low order by the use of the spin-spin interaction and the admixtures of states outside the ground configuration. The Pryce mechanism for the $3d^5$ ions (Mn^{2+}, Fe^{3+}) involved the matrix element of the spin-spin interaction Hamiltonian $\mathcal{H}_{ss}$ between the $^6S(3d^5)$ and $^6D(3d^4\,4s)$ states, and the matrix element of the axial-field potential $\mathcal{H}_{ax}$ between the $^6D(3d^4\,4s)$ and $^6S(3d^5)$ states. Thus

$$D_P \propto \frac{\langle {}^6S(3d^5)|\mathcal{H}_{ax}|{}^6D(3d^4 4s)\rangle\langle {}^6D(3d^4 4s)|\mathcal{H}_{ss}|{}^6S(3d^5)\rangle}{E({}^6S)-E({}^6D)}$$

Watanabe[18] performed the first quantitative calculation of both spin-orbit and spin-spin contributions. We shall not discuss his approach, however because it has been found not to be very important in its contribution to the zero-field splitting.[5, 6]

Subsequently, Blume and Orbach[19] (BO) considered the axial splitting of S-state ions in a deformed cubic host. They proposed a mechanism involving the spin-orbit admixture of the excited $|^4P\rangle$ state into the ground $|^6S\rangle$ state, and the first-order matrix element of the axial and rhombic fields. Normally such terms would vanish, but in this case the term would be nonvanishing because of the mixing of the excited quartet states by the cubic crystal field. More specifically, the $^4\Gamma_4$ component of $|^4P\rangle$ is strongly admixed with the $^4\Gamma_4$ components of $|^4G\rangle$ and $|^4F\rangle$ states by the cubic crystal field. Nonvanishing first-order matrix elements of the axial and rhombic field potentials exist between the $|^4P\rangle$, $|^4F\rangle$, and $|^4G\rangle$ components of the admixed excited $^4\Gamma_4$ level. Thus

$$D_{\mathrm{BO}} \propto \frac{\langle^6S|\mathcal{H}_{so}|^4\Gamma_4\rangle\langle^4\Gamma_4|\mathcal{H}_{ax}|^4\Gamma_4\rangle\langle^4\Gamma_4|\mathcal{H}_{so}|^6S\rangle}{\left[E(^4\Gamma_4)-E(^6S)\right]^2}$$

In addition to these, there are other mechanisms that also contribute to the zero-field splittings, but since these are of minor importance, they are not included here.

The appropriate Hamiltonian for our purpose is

$$\mathcal{H} = \mathcal{H}_F + \mathcal{H}_c + \mathcal{H}_{so} + \mathcal{H}_{ss} + V_2^0 + V_2^2 + V_4^0 + V_4^2$$

where $\mathcal{H}_F$ is the free-ion Hamiltonian, $\mathcal{H}_c$ the crystal field of cubic symmetry, $\mathcal{H}_{so}$ the spin-orbit interaction, and $\mathcal{H}_{ss}$ the spin-spin interaction; the rest are the axial and rhombic terms. We now write the proper expressions for these terms:

$$\mathcal{H}_c = B_4^0 \sum_i r_i^4 \left\{ Y_4^0(i) + \sqrt{\tfrac{5}{14}}\left[Y_4^4(i) + Y_4^{-4}(i)\right]\right\}$$

$$\mathcal{H}_{so} = \sum_i \zeta(r_i)\mathbf{l}_i\cdot\mathbf{s}_i$$

$$\mathcal{H}_{ss} = \frac{g^2\mu_{\mathrm{B}}^2}{a_0^3} \sum_{i<j} \frac{\mathbf{S}_i\cdot\mathbf{S}_j}{r_{ij}^3} - \frac{3(\mathbf{S}_i\cdot\mathbf{r}_{ij})(\mathbf{S}_j\cdot\mathbf{r}_{ij})}{r_{ij}^5}$$

$$V_2^0 = -B_2^0\left(\frac{4\pi}{5}\right)^{1/2} \sum_i r_i^2 Y_2^0(i)$$

$$V_2^2 = -B_2^2\left(\frac{4\pi}{5}\right)^{1/2} \sum_i r_i^2\left[Y_2^2(i) + Y_2^{-2}(i)\right]$$

$$V_4^0 = -B_4^{0'}\left(\frac{4\pi}{9}\right)^{1/2} \sum_i r_i^4 Y_4^0(i)$$

$$V_4^2 = -B_4^2\left(\frac{4\pi}{9}\right)^{1/2} \sum_i r_i^4\left[Y_4^2(i) + Y_4^{-2}(i)\right]$$

Using these Hamiltonians in D_{BO}, we find that the zero-field splitting parameters D and E in the Blume-Orbach mechanism are given by

$$D_{\mathrm{BO}} = -(B_4^0)' \frac{\sqrt{5}}{36} \langle r^4 \rangle [\zeta^2 p_{\alpha\gamma}(2p_{\alpha\alpha} - 2p_{\alpha\beta})]$$

where

$$p_{\alpha\alpha} = \sum_{i=1}^{3} \frac{\alpha_i^2}{\Delta_i}$$

$$p_{\alpha\beta} = \sum_{i=1}^{3} \frac{\alpha_i \beta_i}{\Delta_i}$$

$$p_{\alpha\gamma} = \sum_{i=1}^{3} \frac{\alpha_i \gamma_i}{\Delta_i}$$

These p_{ij}'s are obtained by diagonalizing[6] the cubic field matrix of the excited quartets of character Γ_4 coming from the ^4P, ^4F, and ^4G states. Similarly,

$$E_{\mathrm{BO}} = -B_4^2 \frac{\sqrt{2}}{6} \langle r^4 \rangle [\zeta^2 p_{\alpha\gamma}(2p_{\alpha\alpha} - p_{\alpha\beta})]$$

It is interesting that

$$\frac{E_{\mathrm{BO}}}{D_{\mathrm{BO}}} = 6 \cdot \sqrt{\frac{2}{5}} \cdot \frac{B_4^2}{(B_4^0)'}$$

which is independent of the strength of the cubic admixtures of the excited quartet states.

Next we deal with the spin-spin mechanism. To obtain the spin-spin contribution to the axial and rhombic field splitting, $\mathcal{H}_{ss}$ is written as

$$\begin{aligned}\mathcal{H}_{ss} = -\tfrac{1}{2} \frac{g^2 \mu_B^2}{a_0^3} \sum_{i<j} r_{ij}^5 \big[&(3z_{ij}^2 - r_{ij}^2)(3S_i^z S_j^z - \mathbf{S}_i \cdot \mathbf{S}_j) \\ &+ 3(x_{ij}^2 - y_{ij}^2)(S_i^x S_j^x - S_j^y S_j^y) \\ &+ 6(S_i^x S_j^y + S_i^y S_j^x) x_{ij} y_{ij} \\ &+ 6(S_i^y S_j^z + S_i^z S_j^y) y_{ij} z_{ij} \\ &+ 6(S_i^z S_j^x + S_i^x S_j^z) z_{ij} x_{ij} \big]\end{aligned}$$

Constructing the ^6S ground determinant out of the perturbed one-electron orbitals and taking the matrix element of $\mathcal{H}_{ss}$ between M_S and M_S', it can be shown, after some manipulations, that

$$D_{ss} = -\frac{g^2\mu_B^2}{20a_0}\left\langle {}^6S \left| \sum_{i<j} \frac{3(z_{ij}^2 - r_{ij}^2)}{r_{ij}^5} \right| {}^6S \right\rangle'$$

and

$$E_{ss} = -\frac{g^2\mu_B^2}{20a_0}\left\langle {}^6S \left| \sum_{i<j} \frac{3(x_{ij}^2 - y_{ij}^2)}{r_{ij}^5} \right| {}^6S \right\rangle'$$

where the prime indicates the use of perturbed one-electron orbitals in constructing the ground Slater determinant.

Instead of presenting the mathematical details of the computations of D and E, which can be found in Sharma, Das, and Orbach,[6] it is sufficient to discuss the final result. Specific application of their calculations has been made to Mn^{2+} in ZnF_2 and MnF_2. They have found that the most important contribution in these cases comes from the Blume-Orbach mechanism and the next most significant contribution from the spin-spin interaction mechanism of Pryce. The other mechanisms do not contribute much to D and E (see Table VI of Ref. 6). In this connection the reader should also consult the work of Chakravarty.[5]

REFERENCES

1. M. H. L. Pryce, *Proc. Phys. Soc. A*, **63**, 25 (1950).
2. A. Abragam and M. H. L. Pryce, *Proc. Roy. Soc. (London) A*, **205**, 135 (1951).
3. B. Bleaney and K. W. H. Stevens, *Rep. Prog. Phys.* **16**, 108 (1953).
4. K. W. Bowers and J. Owen, *Rep. Prog. Phys.* **18**, 350 (1955).
5. A. S. Chakravarty, *J. Chem. Phys.* **39**, 1004 (1963).
6. R. R. Sharma, T. P. Das, and R. Orbach, *Phys. Rev* **149**, 257 (1966).
7. E. H. Carlson and R. D. Spence, *J. Chem. Phys.* **24**, 471 (1956).
8. G. F. Koster and H. Statz, *Phys. Rev.* **113**, 445 (1959).
9. G. F. Koster, *Phys. Rev.* **109**, 227 (1958).
10. K. A. Muller, *Helv. Phys. Acta* **31**, 173 (1958).
11. J. S. Griffith, *The Theory of Transition Metal Ions*, Cambridge University Press, London, England, 1961.
12. J. H. E. Griffiths and J. Owen, *Proc. Roy. Soc.* (London) *A*, **213**, 459 (1952).
13. A. Bose, S. C. Mitra, and S. K. Dutta, *Proc. Roy. Soc.* (London) *A*, **248**, 153 (1958).
14. C. J. Ballhausen, *Introduction to Ligand Field Theory*, McGraw-Hill, New York, 1962.
15. S. Sugano, Y. Tanabe, and H. Kamimura, *Multiplets of Transition Metal Ions in Crystals*, Academic Press, New York, 1970.
16. V. P. Desai and A. S. Chakravarty, *Indian J. Phys.* **42**, 506 (1968).
17. J. H. Van Vleck and W. G. Penney, *Phil. Mag.* **17**, 961 (1934).
18. H. Watanabe, *Prog. Theor. Phys.* **18**, 405 (1957).
19. M. Blume and R. Orbach, *Phys. Rev* **127**, 1587 (1961).
20. E. Zavoisky, *J. Phys. USSR* **9**, 211 (1945).
21. A. S. Chakravarty, *J. Phys. Chem.* **74**, 4347 (1970).
22. S. K. Dutta Roy, Thesis (unpublished), Calcutta University, 1959.

9

The Strong-Field Coupling Scheme

So far we have done all our calculations on the so-called weak-field formalism. This procedure is advantageous only when V_c is small, so that there are not many terms to take care of and practically no configuration mixing. By clinging to the atomic terms, we did not give due regard to the changes brought about by the formation of bonds between the central ion and its ligands. In other words, the itinerancy of electrons in the molecular framework could not be faithfully reflected in the atomic parameters. The chief advantage was that the mathematical apparatus to treat this formalism is readily available in the theory of angular momentum, the irreducible spherical tensors, and the theory of complex spectra.

Now, however, we shall follow a more general approach that is more advantageous for the situation when $V_c \geqslant e^2/r_{ij} > \zeta \mathbf{l}\cdot\mathbf{s}$ holds good to a great extent (see Section 6.1). This approach, known as the strong-field coupling scheme, was first proposed by Tanabe and Sugano.[1] Before discussing the intermediate coupling scheme, which adequately describes the situation in the complexes of the palladium and platinum groups, we must study the strong-field scheme because it is easy to proceed to the intermediate coupling scheme after starting from the strong-field formalism.

Let us begin by writing the Hamiltonian for an atom (ion) placed in a crystal field having octahedral (O_h) symmetry, which is given by

$$\mathcal{H} = \mathcal{H}_0 + V_c + \sum_{i<j} \frac{e^2}{r_{ij}} + \mathcal{H}_{so} \tag{9.1}$$

where $\mathcal{H}_0$ represents the free-ion Hamiltonian, V_c the cubic field, e^2/r_{ij} the interelectronic repulsion term (let us call it V_e, for the time being) and $\mathcal{H}_{so}$ the spin-orbit interaction term. Our problem is now to find the splitting of a g-fold degenerate unperturbed energy level E_0 (corresponding to $\mathcal{H}_0$) under the perturbation of $V_c + V_e + \mathcal{H}_{so}$. According to the first-order perturbation theory for the degenerate case, this is the problem of diagonalizing $V_c + V_e + \mathcal{H}_{so}$ with respect to a set of g basis functions associated with the unperturbed level.

In the weak-field coupling scheme (Chapter 6), we took these basis functions to be the wavefunctions $|E_0 i S L M_S M_L\rangle$ associated with the free-ion terms $(i S L)$ involved in E_0. Then V_e was fully diagonal with respect to these functions, the diagonal elements being the free-ion term energies $E_e(iSL)$ (referred to E_0) in the absence of $\mathcal{H}_{so}$. Hence the actual problem was to diagonalize $V_c + \mathcal{H}_{so}$ with respect to these free-ion wavefunctions. As discussed earlier, this scheme is advantageous only when $V_c \ll V_e$, because in that case we can neglect any possible configuration mixing involving two or more free-ion terms, and can calculate the splitting of each free-ion term individually under $V_c + \mathcal{H}_{so}$, or, at most, can consider configuration mixings in groups of closely spaced terms.

As discussed earlier, the chief disadvantage of the weak-field coupling scheme is that it cannot be generalized to give due regard to the changes brought about by the formation of chemical bonds between the central ion and its ligands, which constitute one of the main interests in the magnetically dilute system of inorganic complexes, especially of the transition metal ions of the palladium and platinum groups. Such generalization is, however, possible, as remarked in the beginning, by another approach, which is known as the strong-field coupling scheme. In the most general case, only the symmetry of the ligand field determines the number of independent parameters involved in the treatment of the problem according to this scheme. In particular, if, in this treatment, we employ the single-electron orbitals associated with the free-ion configuration (with the unperturbed energy E_0), these parameters will reduce to those of the weak-field scheme, and also, in that case, both schemes will yield the same results, provided that no approximation is made.

In the strong-field scheme V_c is fully diagonal with respect to the initial basis functions chosen, and hence the actual problem is to diagonalize $V_e + \mathcal{H}_{so}$ with respect to these functions.

9.1 ELECTRONIC CONFIGURATIONS AND THE BASIS FUNCTIONS

We shall describe the strong-field scheme with reference to the $(\bar{n}d)^N$ electronic configuration in a transition metal ion, which pertains to the subject of our main interest. We first develop the subject for a ligand field of octahedral (O_h) symmetry, and later use the same basis functions in problems involving fields of lower symmetry.

As discussed in Chapter 6, two irreducible representations, denoted to T_{2g} and E_g of the O_h group, appear when one reduces the representation $D^{(2)}$ of the full rotation group. This implies that a single-electron d level with fivefold orbital degeneracy (excluding spin, of course) will be split under a crystal field having O_h symmetry into a triply degenerate level t_{2g} and a doubly degenerate level e_g. We denote these two sets of orbitals [see (6.7) and (6.8)] by

$$|t_{2g}\xi\rangle = \frac{i}{\sqrt{2}}(d_1 + d_{-1})$$

$$|t_{2g}\eta\rangle = -\frac{1}{\sqrt{2}}(d_1 - d_{-1}) \qquad (9.2)$$

$$|t_{2g}\zeta\rangle = -\frac{i}{\sqrt{2}}(d_2 - d_{-2})$$

and

$$|e_g\theta\rangle = d_0 \qquad (9.3)$$

$$|e_g\epsilon\rangle = \frac{1}{\sqrt{2}}(d_2 + d_{-2})$$

The orbitals given in (9.2) and (9.3) are all real and hence advantageous in setting up the formulations. The energies of the single-electron t_{2g} and e_g levels are given by

$$\mathcal{E}(t_{2g}) = \mathcal{E}^0_{\bar{n}d} + \langle t_{2g}|V_c|t_{2g}\rangle \qquad (9.4a)$$

and

$$\mathcal{E}(e_g) = \mathcal{E}^0_{\bar{n}d} + \langle e_g|V_c|e_g\rangle \qquad (9.4b)$$

respectively. Here

$$\langle t_{2g}|V_c|t_{2g}\rangle = \langle\xi|V_c|\xi\rangle = \langle\eta|V_c|\eta\rangle = \langle\zeta|V_c|\zeta\rangle \qquad (9.5a)$$

and

$$\langle e_g|V_c|e_g\rangle = \langle\theta|V_c|\theta\rangle = \langle\epsilon|V_c|\epsilon\rangle \qquad (9.5b)$$

Here $\mathcal{E}^0_{\bar{n}d}$ is obviously the energy of the $\bar{n}d$ state of a hydrogen-like atom with a general central-force potential, $V^0(r)$, representing the average effect of the core inside the $\bar{n}d$-shell of the atom (ion) in question.

If $v_c(\mathbf{r})$ denotes the ligand field energy of a single electron at position $\mathbf{r}$, the total ligand field energy of the system is given by

$$V_c = \sum_{i=1}^{N} v_c(\mathbf{r}_i) \qquad (9.6)$$

where $\mathbf{r}_i$ is the position vector of the ith electron. Here we should remember that a spherically symmetric field can always accompany a pure O_h field. The only effect of this field, however, is to shift all the energy levels by the same amount, and this effect can thus be incorporated with $\mathcal{E}^0_{\bar{n}d}$, as a result of which v_c will be considered to be of pure octahedral symmetry.

The energy separation between the t_{2g} and e_g groups of levels is conventionally denoted by $10Dq$ (see Chapter 6), and thus we write

$$\begin{aligned}\mathcal{E}(e_g)-\mathcal{E}(t_{2g}) &\equiv \langle e_g|v_c|e_g\rangle - \langle t_{2g}|v_c|t_{2g}\rangle \\ &= 10Dq\end{aligned} \tag{9.7}$$

Using (6.41), we can easily show that

$$\langle e_g|v_c|e_g\rangle = 6Dq \tag{9.8}$$

and

$$\langle t_{2g}|v_c|t_{2g}\rangle = -4Dq$$

Hence, using (9.4) and (9.8), we have

$$\tfrac{1}{5}\left[2\mathcal{E}(e_g)+3\mathcal{E}(t_{2g})\right] = \mathcal{E}^0_{\bar{n}d} \tag{9.9}$$

This is the well-known "center-of-gravity rule," which states that the center of gravity of the split sublevels remains unchanged with respect to the original unperturbed level (see Appendix X).

So far we have been considering a single d electron in a cubic field. Now what happens when the number of electrons is increased? These electrons now have to be accommodated in the t_{2g} and e_g sublevels, and this can be done in various ways. Since, by including the electron spin, the d level is tenfold degenerate (fivefold due to orbit and twofold due to spin), the d level can accommodate, at most, 10 electrons. Since we are interested in the d^N ($N \geqslant 1$) configuration, the most general distribution is given by $t^n_{2g}e^m_g$ with $n+m=N$ and $n \leqslant 6, m \leqslant 4$. In the absence of electron-electron interactions, the $t^n_{2g}e^m_g$ configuration gives a $({}^6C_n \cdot {}^4C_m)$-fold degenerate level with energy

$$\begin{aligned}E_c(n,m) = N\mathcal{E}^0_{\bar{n}d} &+ n\langle t_{2g}|v_c|t_{2g}\rangle \\ &+ m\langle e_g|v_c|e_g\rangle\end{aligned} \tag{9.10}$$

Our next task is to construct the antisymmetric wavefunctions for each configuration described above. These wavefunctions will be constructed according to the $S\Gamma M\gamma$ scheme so that a wavefunction designated by $|\alpha S\Gamma M\gamma\rangle$ is a simultaneous eigenfunction of $\mathbf{S}^2$ and S_z with the eigenvalues $\hbar^2S(S+1)$ and $\hbar M$, respectively, and also belongs to the γth component of the irreducible representation Γ of the symmetry group of the crystal field.

9.1.1 Wavefunctions for the t_{2g}^n configuration

We first consider the simplest case of the t_{2g}^2 configuration. We have six single-electron spin-orbitals for the t_{2g} level, namely, $\xi^+, \xi^-, \eta^+, \eta^-, \zeta^+, \zeta^-$. Using these in pairs, we construct the following ${}^6C_2 = 15$ determinantal (antisymmetric) microstates for the t_{2g}^2 configuration:

$$|\xi^+\eta^+|, |\xi^-\eta^-|, |\eta^+\zeta^+|, |\eta^-\zeta^-|, |\zeta^+\xi^+|, |\zeta^-\xi^-| \tag{9.11a}$$

$$|\xi^+\eta^-|, |\xi^-\eta^+|, |\eta^+\zeta^-|, |\eta^-\zeta^+|, |\zeta^+\xi^-|, |\zeta^-\xi^+| \tag{9.11b}$$

$$|\xi^+\xi^-|, |\eta^+\eta^-|, |\zeta^+\zeta^-| \tag{9.11c}$$

These will be denoted, in general, by

$$|\phi(t_{2g}m_1\gamma_1)\phi(t_{2g}m_2\gamma_2)|$$

where $\phi(t_{2g}m\gamma)$ is a one-electron spin-orbital in the t_{2g} orbital ($m = \pm\frac{1}{2}$, $\gamma = \xi, \eta, \zeta$). Thus, for example, $\phi(t_{2g} - \frac{1}{2}\eta) = \eta^-$ in the notation given above.

Now we obtain the $(S\Gamma M\gamma)$ wavefunctions of the t_{2g}^2 configuration as the following linear combinations of the determinantal states:

$$|t_{2g}^2 S\Gamma M\gamma\rangle = \overline{N}\sum|\phi(t_{2g}m_1\gamma_1)\phi(t_{2g}m_2\gamma_2)| \cdot\langle\tfrac{1}{2}m_1\tfrac{1}{2}m_2|SM\rangle\cdot\langle T_2\gamma_1T_2\gamma_2|\Gamma\gamma\rangle \tag{9.12}$$

where $\overline{N}$ is the appropriate normalization constant. The direct product representation $T_{2g}\otimes T_{2g}$ is decomposed into the following irreducible representations (see Appendix V, Table 5):

$$T_{2g}\otimes T_{2g} = A_{1g} + E_g + T_{1g} + T_{2g} \tag{9.13}$$

Again, S can have only two possible values, 0 and 1. Then the possible sets of $S\Gamma$ that can occur in the $(S\Gamma M\gamma)$ states are all the possible combinations of $S=0$, 1 and $\Gamma = A_1, E, T_1, T_2$. However, some of these $S\Gamma$ sets will not be allowed, as can be seen from the following example. (Since we always deal with g representations of O_h, the subscript g will be dropped henceforth.) From (9.12) we get

$$|t_2^2\,{}^3T_2 1\xi\rangle$$

$$\equiv|t_2^2 S=1, \Gamma=T_2, M=1, \gamma=\xi\rangle$$

$$=\overline{N}\sum_{m_1,m_2,\gamma_1,\gamma_2}|\phi(t_2m_1\gamma_1)\phi(t_2m_2\gamma_2)|\langle\tfrac{1}{2}m_1\tfrac{1}{2}m_2|11\rangle\langle T_2\gamma_1T_2\gamma_2|T_2\xi\rangle$$

$$=\overline{N}\sum_{\gamma_1,\gamma_2}|\phi(t_2\tfrac{1}{2}\gamma_1)\phi(t_2\tfrac{1}{2}\gamma_2)|\langle T_2\gamma_1T_2\gamma_2|T_2\xi\rangle$$

since $\langle\frac{1}{2}m_1\frac{1}{2}m_2|11\rangle=1$. Now, using the $T_2\otimes T_2$ table of C-G coefficients given in Appendix VI, we obtain

$$\langle t_2^{2\,3}T_1 1\xi\rangle=\bar{N}\frac{1}{\sqrt{2}}\left[|\eta^+\zeta^+|+|\zeta^+\eta^+|\right]=0$$

Similarly, it can be shown that all the wavefunctions $|t_2^{2\,3}T_2M\gamma\rangle$ must vanish. Thus we conclude that the term ${}^3T_2$ is not allowed for the t_2^2 configuration. In this way we can show that the terms ${}^3A_1$, ${}^3E$, and ${}^1T_1$ are also not allowed. The allowed terms of the t_2^2 configuration are, therefore, ${}^1A_1$, ${}^1E$, ${}^3T_1$, and ${}^1T_2$. The total number of states in these allowed terms is $1+2+9+3=15$, a result that is quite expected since there are just ${}^6C_2=15$ states in the t_2^2 configuration. The wavefunctions for each allowed term are constructed by using (9.12). We now give two examples.

1. First we have

$$\begin{aligned}|t_2^{2\,3}T_1 1z\rangle&=\bar{N}\sum_{\gamma_1\gamma_2}|\phi(t_2\tfrac{1}{2}\gamma_1)\phi(t_2\tfrac{1}{2}\gamma_2)|\langle T_2\gamma_1T_2\gamma_2|T_1z\rangle\\&=\bar{N}\left[\frac{1}{\sqrt{2}}\left|\xi^+\eta^+\right|+\left(-\frac{1}{\sqrt{2}}\right)\left|\eta^+\xi^+\right|\right]\\&=|\xi^+\eta^+|\end{aligned}\qquad(9.14)$$

by using Appendix VI. From this we can derive $|t_2^{2\,3}T_1 0z\rangle$ and $|t_2^{2\,3}T_1-1z\rangle$ by successive operations of $\mathbf{S}^-$.

2. Next we consider

$$\begin{aligned}|t_2^{2\,1}E\theta\rangle&=\bar{N}\sum_{m_1,\gamma_1,m_2\gamma_2}|\phi(t_2m_1\gamma_1)\phi(t_2m_2\gamma_2)|\\&\quad\cdot\langle\tfrac{1}{2}m_1\tfrac{1}{2}m_2|00\rangle\langle T_2\gamma_1T_2\gamma_2|E\theta\rangle\\&=\bar{N}\frac{1}{\sqrt{2}}\sum_{\gamma_1\gamma_2}\langle T_2\gamma_1T_2\gamma_2|E\theta\rangle\cdot\\&\quad\cdot\left[|\phi(t_2\tfrac{1}{2}\gamma_1)\phi(t_2-\tfrac{1}{2}\gamma_2)|-|\phi(t_2-\tfrac{1}{2}\gamma_1)\phi(t_2\tfrac{1}{2}\gamma_2)|\right]\\&=\bar{N}\frac{1}{\sqrt{2}}\left[-\frac{1}{\sqrt{6}}\{|\xi^+\xi^-|-|\xi^-\xi^+|\}\right.\\&\quad\left.-\frac{1}{\sqrt{6}}\{|\eta^+\eta^-|-|\eta^-\eta^+|\}+\frac{2}{\sqrt{6}}\{|\zeta^+\zeta^-|-|\zeta^-\zeta^+|\}\right]\\&=\frac{1}{\sqrt{6}}\left[-|\xi^+\xi^-|-|\eta^+\eta^-|+2|\zeta^+\zeta^-|\right]\end{aligned}\qquad(9.15)$$

The other wavefunctions of the t_2^2 configuration can be similarly constructed. The results are given in Appendix VII.

Next we consider the t_2^3 configuration. We start from an imaginary configuration $t_2^2 t_2'$, where t_2' is a single-electron level with threefold orbital degeneracy having components ξ', η', ζ', which span T_{2g} in O_h, but this t_2' level is taken to be different from the original t_2 level. This $t_2^2 t_2'$ configuration then involves $15 \times 6 = 90$ wavefunctions, given by the products of the 15 wavefunctions of the original t_2^2 configuration and the 6 single-electron spin-orbitals, $\xi^{+\prime}, \xi^{-\prime}, \eta^{+\prime}, \eta^{-\prime}$, and $\zeta^{+\prime}, \zeta^{-\prime}$, associated with the t_2' level. Again we require suitable linear combinations of these products according to the usual $S\Gamma M\gamma$ scheme.

A $(S\Gamma M\gamma)$ wavefunction for the $t_2^2 t_2'$ configuration can be obtained from the wavefunctions of a given term $S_0\Gamma_0$ of the t_2^2 configuration, according to the following linear combination:

$$|t_2^2(S_0\Gamma_0)t_2', S\Gamma M\gamma\rangle = \sum_{M_0, \gamma_0, m_3, \gamma_3} |t_2^2 S_0\Gamma_0 M_0\gamma_0\rangle \phi(t_2' m_3 \gamma_3) \cdot \langle S_0 M_0 \tfrac{1}{2} m_3 | S M\rangle \langle \Gamma_0\gamma_0 T_2\gamma_3 | \Gamma\gamma\rangle \tag{9.16}$$

where $|t_2^2 S_0\Gamma_0 M_0\gamma_0\rangle$ is a wavefunction of t_2^2 involving electrons 1 and 2, and $\phi(t_2' m_3\gamma_3)$ is a spin-orbital of the added electron (electron 3) in the t_2' level.

Thus, from a term $S_0\Gamma_0$ of t_2^2, we obtain the terms $S\Gamma$ of $t_2^2 t_2'$, with $S = S_0 + \frac{1}{2}$ and $S = S_0 - \frac{1}{2}$ (for $S_0 \neq 0$) and Γ being an irreducible representation in $\Gamma_0 \otimes T_2$. Since there are 4 terms, ${}^1A_1$, ${}^1E$, ${}^1T_2$, and ${}^3T_1$ of t_2^2, we obtain 15 terms of $t_2^2 t_2'$, as shown in Table 9.1.

Expanding $|t_2^2\, S_0\,\Gamma_0\, M_0\,\gamma_0\rangle$ in terms of Slater determinants $|\phi(t_2 m_1\gamma_1)\phi(t_2 m_2\gamma_2)|_{(1,2)}$ by using (9.12), we can rewrite (9.16) as

$$|t_2^2(S_0\Gamma_0)t_2', S\Gamma M\gamma\rangle' = \sum_{\substack{m_1 m_2 m_3 \\ \gamma_1\gamma_2\gamma_3}} \mathcal{C}(m_1\gamma_1 m_2\gamma_2 m_3\gamma_3, S_0\Gamma_0, S\Gamma M\gamma) \cdot |\phi(t_2 m_1\gamma_1)\phi(t_2 m_2\gamma_2)|_{(1,2)} \cdot \phi_3(t_2' m_3\gamma_3) \tag{9.17}$$

where $\mathcal{C}$'s are the numerical coefficients. Here the subscript (1,2) of the Slater determinant indicates that it involves the coordinates of electrons 1 and 2

Table 9.1

Allowed terms of $\{t_2^2(S_0\Gamma_0)t_2', S\Gamma\}$

$S_0\Gamma_0$	$S\Gamma$
${}^1A_1$	${}^2T_2$
${}^1E$	${}^2T_1, {}^2T_2$
${}^1T_2$	${}^2A_1, {}^2E, {}^2T_1, {}^2T_2$
${}^3T_1$	${}^2A_2, {}^2E, {}^2T_1, {}^2T_2$ ${}^4A_2, {}^4E, {}^4T_1, {}^4T_2$

belonging to the t_2^2 configuration, and $\phi(t_2' m_3 \gamma_3)$ involves the coordinates of electron 3.

We note that (9.17) is antisymmetric with respect to the exchange of electrons 1 and 2, but not with respect to the exchange of electrons 1 and 3 or 2 and 3. To make (9.17) totally antisymmetric, it is sufficient to take the following linear combination:

$$\sum_{\substack{m_1 m_2 m_3 \\ \gamma_1 \gamma_2 \gamma_3}} \mathcal{C}(m_1\gamma_1 m_2\gamma_2 m_3\gamma_3, S_0\Gamma_0, S\Gamma M\gamma)$$

$$\cdot\big[|\phi(t_2 m_1 \gamma_1)\phi(t_2 m_2 \gamma_2)|_{(1,2)} \cdot \phi_3(t_2' m_3 \gamma_3)$$

$$- |\phi(t_2 m_1 \gamma_1)\phi(t_2 m_2 \gamma_2)|_{(1,3)} \cdot \phi_2(t_2' m_3 \gamma_3)$$

$$- |\phi(t_2 m_1 \gamma_1)\phi(t_2 m_2 \gamma_2)|_{(2,3)} \cdot \phi_1(t_2' m_3 \gamma_3)\big] \tag{9.18}$$

Combination (9.18) is also a $(S\Gamma M\gamma)$ wavefunction for the $t_2^2 t_2'$ configuration; moreover, it is totally antisymmetric. According to the property of determinants, it can be reexpressed as

$$|t_2^2(S_0\Gamma_0)t_2', S\Gamma M\gamma\rangle = \sum_{\substack{m_1 m_2 m_3 \\ \gamma_1 \gamma_2 \gamma_3}} \mathcal{C}(m_1\gamma_1 m_2\gamma_2 m_3\gamma_3, S_0\Gamma_0, S\Gamma M\gamma)$$

$$\cdot |\phi(t_2 m_1 \gamma_1)\phi(t_2 m_2 \gamma_2)\phi(t_2' m_3 \gamma_3)| \tag{9.19}$$

Now let us find the $(S\Gamma M\gamma)$ wavefunctions for the t_2^3 configuration, which can be obtained by replacing t_2' in (9.19) by t_2. When we do that, we know that some of the terms given in Table 9.1 will no longer be allowed. It can be easily shown that the allowed terms of t_2^3 will be greatly reduced from the 15 given in Table 9.1. The nonvanishing wavefunctions obtained by the replacement $t_2' \rightarrow t_2$ in (9.19) are usually not normalized, and therefore we have to normalize them to obtain the normalized wavefunctions $|t_2^3, S\Gamma M\gamma\rangle$. Let us now work out three typical examples.

1. First we have

$$|t_2^2(^3T_1)t_2', {}^4T_1 \tfrac{3}{2} x\rangle' = \sum_{\substack{M_0 m_3 \\ \gamma_0 \gamma_3}} |t_2^2\, {}^3T_1 M_0 \gamma_0\rangle \phi_3(t_2' m_3 \gamma_3) \langle 1 M_0 \tfrac{1}{2} m_3 | \tfrac{3}{2} \tfrac{3}{2}\rangle$$

$$\cdot \langle T_1 \gamma_0 T_2 \gamma_3 | T_1 x\rangle$$

$$= \sum_{\gamma_0 \gamma_3} |t_2^2\, {}^3T_1 1 \gamma_0\rangle \phi_3\big(t_2' \tfrac{1}{2} \gamma_3\big) \langle T_1 \gamma_0 T_2 \gamma_3 | T_1 x\rangle$$

$$= \frac{1}{\sqrt{2}} |t_2^2\, {}^3T_1 1 y\rangle \phi_3\big(t_2' \tfrac{1}{2} \zeta\big) + \frac{1}{\sqrt{2}} |t_2^2\, {}^3T_1 1 z\rangle \phi_3\big(t_2' \tfrac{1}{2} \eta\big)$$

$$= \frac{1}{\sqrt{2}} |\zeta^+ \xi^+|_{(1,2)} \cdot \zeta^{+\prime}(3) + \frac{1}{\sqrt{2}} |\xi^+ \eta^+|_{(1,2)} \cdot \eta^{+\prime}(3)$$

by using Appendix VII. Then the corresponding antisymmetrized function will be, by using (9.18),

$$|t_2^2(^3T_1)t_2',{}^4T_1\tfrac{3}{2}x\rangle=\frac{1}{\sqrt{2}}|\zeta^+\xi^+\zeta^{+\prime}|+\frac{1}{\sqrt{2}}|\xi^+\eta^+\eta^{+\prime}|$$

which vanishes by the replacement $\zeta'\to\zeta$, $\eta'\to\eta$. Thus the $^4T_1$ term of t_2^3 is forbidden.

2. Next let us consider

$$\begin{aligned}|t_2^2(^3T_1)t_2',{}^4A_2\tfrac{3}{2}a_2\rangle' &= \sum_{\substack{M_0m_3\\ \gamma_0\gamma_3}} |t_2^2\,{}^3T_1M_0\gamma_0\rangle\phi_3(t_2'm_3\gamma_3)\langle 1M_0\tfrac{1}{2}m_3|\tfrac{3}{2}\tfrac{3}{2}\rangle\\ &\quad\cdot\langle T_1\gamma_0T_2\gamma_3|A_2a_2\rangle\\ &= -\frac{1}{\sqrt{3}}|t_2^2\,{}^3T_1\,1x\rangle\phi_3\big(t_2'\tfrac{1}{2}\xi\big)-\frac{1}{\sqrt{3}}|t_2^2\,{}^3T_1\,1y\rangle\phi_3\big(t_2'\tfrac{1}{2}\eta\big)\\ &\quad-\frac{1}{\sqrt{3}}|t_2^2\,{}^3T_1\,1z\rangle\phi\big(t_2'\tfrac{1}{2}\zeta\big)\\ &= -\frac{1}{\sqrt{3}}\big[|\eta^+\zeta^+|_{(1,2)}\cdot\xi^{+\prime}(3)+|\zeta^+\xi^+|_{(1,2)}\cdot\eta^{+\prime}(3)\\ &\quad+|\xi^+\eta^+|_{(1,2)}\cdot\zeta^{+\prime}(3)\big]\end{aligned}$$

The corresponding antisymmetric function is

$$|t_2^2(^3T_1)t_2',{}^4A_2\tfrac{3}{2}a_2\rangle=-\frac{1}{\sqrt{3}}\big[|\eta^+\zeta^+\xi^{+\prime}|+|\zeta^+\xi^+\eta^{+\prime}|+|\xi^+\eta^+\zeta^{+\prime}|\big]$$

Then, by making the replacement $\xi'\to\xi$, $\eta'\to\eta$, $\zeta'\to\zeta$ and normalizing the result, we obtain the corresponding wavefunction of t_2^3:

$$|t_2^3\,{}^4A_2\tfrac{3}{2}a_2\rangle=-|\xi^+\eta^+\zeta^+| \tag{9.20}$$

3. As an example of obtaining the nonvanishing identical wavefunctions from different sets of $S_0\Gamma_0$, let us calculate $|t_2^3\,{}^2T_1\tfrac{1}{2}x\rangle$ from $S_0\Gamma_0={}^1E$, $^1T_2$,

and $^3T_1$.

$$
\begin{aligned}
&|t_2^2(^1E)t_2', {}^2T_1\tfrac{1}{2}x\rangle' \\
&\quad = \sum_{\gamma_0\gamma_3} |t_2^{2\,1}E\gamma_0\rangle \phi_3\big(t_2'\tfrac{1}{2}\gamma_3\big)\langle E\gamma_0 T_2\gamma_3|T_1x\rangle \\
&\quad = -\frac{\sqrt{3}}{2}|t_2^{2\,1}E\theta\rangle\, \phi_3\big(t_2'\tfrac{1}{2}\xi\big) - \tfrac{1}{2}|t_2^{2\,1}E\epsilon\rangle\phi_3\big(t_2'\tfrac{1}{2}\xi\big) \\
&\quad = -\frac{\sqrt{3}}{2}\cdot\frac{1}{\sqrt{6}}\big[-|\xi^+\xi^-|_{(1,2)} - |\eta^+\eta^-|_{(1,2)} + 2|\zeta^+\zeta^-|_{(1,2)}\big]\cdot\xi^{+\prime}(3) \\
&\qquad -\frac{1}{2}\cdot\frac{1}{\sqrt{2}}\big[|\xi^+\xi^-|_{(1,2)} - |\eta^+\eta^-|_{(1,2)}\big]\cdot\xi^{+\prime}(3)
\end{aligned}
$$

Then, antisymmetrizing, we get

$$
\begin{aligned}
|t_2^2(^1E)t_2', {}^2T_1\tfrac{1}{2}x\rangle' = \frac{1}{2\sqrt{2}}&\big[|\xi^+\xi^-\xi^{+\prime}| + |\eta^+\eta^-\xi^{+\prime}| - 2|\zeta^+\zeta^-\xi^{+\prime}|\big] \\
-\frac{1}{2\sqrt{2}}&\big[|\xi^+\xi^-\xi^{+\prime}| - |\eta^+\eta^-\xi^{+\prime}|\big]
\end{aligned}
$$

and the corresponding normalized wavefunction of t_2^3 is

$$
|t_2^{3\,2}T_1\tfrac{1}{2}x\rangle = \frac{1}{\sqrt{2}}\big[|\xi^+\eta^+\eta^-| - |\xi^+\zeta^+\zeta^-|\big] \tag{9.21}
$$

Proceeding similarly, it can be shown that, taking $S_0\Gamma_0 = {}^1T_2$ and $^3T_1$, we shall obtain exactly the same expression for $|t_2^{3\,2}T_1\tfrac{1}{2}x\rangle$ as is given by (9.21).

The other nonvanishing wavefunctions of the t_2^3 configuration can be obtained in the same manner, and we see that only four terms, $^4A_2$, ^2E, $^2T_1$, and $^2T_2$ survive. The wavefunctions for ^2E, $^2T_1$, or $^2T_2$ can be constructed from various sets of $S_0\Gamma_0$, and they will be identical to each other. The total number of states in these four allowed terms is $4+4+6+6=20={}^6C_3$, which is just the number of states in the t_2^3 configuration. The wavefunctions of t_2^3 are also given in Appendix VII for the highest values of M. Those for the lower values of M can be obtained as usual by successive operations with $\mathbf{S}^-$.

Finally we consider the t_2^n configuration with $n>3$. The method described above can easily be extended to obtain the wavefunctions of t_2^{n+1} from those of t_2^n ($n=3,4,5$). However, it can be shown that the allowed terms of t_2^{6-n} are just those of t_2^n, and there is a simple correlation between the wavefunctions for a term of t_2^{6-n} and those for the same term of t_2^n. We do not give the actual deductions here[2] but simply describe the procedure.

We have the linear combination

$$|t_2^n S\Gamma M\gamma\rangle = \sum_i \mathcal{C}_i(t_2^n S\Gamma M\gamma) D_i^n \tag{9.22}$$

where the D_i^n are the ${}^6C_n$ normalized determinantal microstates associated with the t_2^n configuration. Then the wavefunction $|t_2^{6-n} S\Gamma - M\gamma\rangle$ of the t_2^{6-n} configuration will be given[2] by the following linear combination:

$$|t_2^{6-n} S\Gamma - M\gamma\rangle = (-1)^{S-M} \sum_i \mathcal{C}_i(t_2^n S\Gamma M\gamma)^* D_i^{6-n} \tag{9.23}$$

where D_i^{6-n} is the normalized cofactor of D_i^n within the six-dimensional determinantal state $|\xi^+\xi^-\eta^+\eta^-\zeta^+\zeta^-|$, which is obviously the only state $|t_2^6\,{}^1A_1\rangle$ of the t_2^6 configuration. Then

$$|t_2^6\,{}^1A_1\rangle \equiv |\xi^+\xi^-\eta^+\eta^-\zeta^+\zeta^-| = q^{-1/2} \sum_{i=1}^{q} D_i^n D_i^{6-n} \tag{9.24}$$

with

$$q = {}^6C_n = \frac{6!}{n!(6-n)!}$$

The state $|t_2^{6-n} S\Gamma - M\gamma\rangle$ is said to be complementary to the state $|t_2^n S\Gamma M\gamma\rangle$, and vice versa. Let us illustrate the application of relation (9.23) with the help of some examples.

The terms of t_2^4 will be the same as those of t_2^2, namely, ${}^1A_1$, ${}^1E$, ${}^3T_1$, and ${}^1T_2$. From Appendix VII we have

$$|t_2^2\,{}^1A_1\rangle = \frac{1}{\sqrt{3}}\left[|\xi^+\xi^-| + |\eta^+\eta^-| + |\zeta^+\zeta^-|\right]$$

Now the cofactors of $|\xi^+\xi^-|$, $|\eta^+\eta^-|$, and $|\zeta^+\zeta^-|$ in $|\xi^+\xi^-\eta^+\eta^-\zeta^+\zeta^-|$ are $|\eta^+\eta^-\zeta^+\zeta^-|$, $|\xi^+\xi^-\zeta^+\zeta^-|$, and $|\xi^+\xi^-\eta^+\eta^-|$, respectively. Thus we have, according to (9.23),

$$|t_2^4\,{}^1A_1\rangle = \frac{1}{\sqrt{3}}\left[|\eta^+\eta^-\zeta^+\zeta^-| + |\zeta^+\zeta^-\xi^+\xi^-| + |\xi^+\xi^-\eta^+\eta^-|\right] \tag{9.25}$$

Again, we have

$$|t_2^2\,{}^3T_1 - 1z\rangle = |\xi^-\eta^-|$$

The corresponding complementary state is

$$|t_2^4\,{}^3T_1\,1z\rangle = -|\xi^+\eta^+\zeta^+\zeta^-| \tag{9.26}$$

since $(-1)^{S-M}=(-1)^{1-1}=1$. The other wavefunctions of the t_2^4 configuration can be similarly obtained from those of t_2^2. These are also listed in Appendix VII.

The t_2^5 configuration has only one term, $^2T_2$, which is the term of t_2^1. Using (9.23), we obtain the following wavefunctions for the t_2^5 configuration:

$$
\begin{aligned}
|t_2^{5\,2}T_2 \pm \tfrac{1}{2}\xi\rangle &= |\xi^{\pm}\eta^{+}\eta^{-}\zeta^{+}\zeta^{-}| \\
|t_2^{5\,2}T_2 \pm \tfrac{1}{2}\eta\rangle &= |\xi^{+}\xi^{-}\eta^{\pm}\zeta^{+}\zeta^{-}| \\
|t_2^{5\,2}T_2 \pm \tfrac{1}{2}\zeta\rangle &= |\xi^{+}\xi^{-}\eta^{+}\eta^{-}\zeta^{\pm}|
\end{aligned} \tag{9.27}
$$

9.1.2 Wavefunctions for the e_g^m Configuration

We have four single-electron spin-orbitals for the e_g level, namely, θ^+, θ^-, ϵ^+, and ϵ^-. Using these in pairs, we construct the following $^4C_2=6$ determinantal microstates for the e_g^2 configuration:

$$|\theta^+\epsilon^+|, |\theta^-\epsilon^-|, |\theta^+\epsilon^-|, |\theta^-\epsilon^+|, |\theta^+\theta^-|, |\epsilon^+\epsilon^-| \tag{9.28}$$

which will be denoted, in general, by

$$|\phi(e_g m_1 \gamma_1)\phi(e_g m_2 \gamma_2)|$$

Now, as in the case of the t_2^2 configuration, we obtain the $(S\Gamma M\gamma)$ wavefunctions of the e_g^2 configuration as follows:

$$
\begin{aligned}
|e^2 S\Gamma M\gamma\rangle = \bar{N} \sum_{\substack{m_1 m_2 \\ \gamma_1\gamma_2}} & |\phi(e m_1\gamma_1)\phi(e m_2\gamma_2)| \\
& \cdot\langle \tfrac{1}{2}m_1 \tfrac{1}{2}m_2|SM\rangle\langle e\gamma_1 e\gamma_2|\Gamma\gamma\rangle
\end{aligned} \tag{9.29}
$$

where, again, we omit the subscript g without causing any confusion. Now, since

$$E\otimes E = A_1 + A_2 + E \tag{9.30}$$

the possible terms of the e^2 configuration are $^1A_1$, $^3A_1$, $^1A_2$, $^3A_2$, ^1E, and ^3E. However, as in the case of t_2^2, here also some of these terms will yield vanishing wavefunctions and hence will be forbidden. We give an example. Let us take

$$
\begin{aligned}
|e^{2\,3}A_1 1 a_1\rangle &= \bar{N} \sum_{\substack{m_1 m_2 \\ \gamma_1\gamma_2}} |\phi(e m_1\gamma_1)\phi(e m_2\gamma_2)|\langle \tfrac{1}{2}m_1 \tfrac{1}{2}m_2|1\,1\rangle\langle e\gamma_1 e\gamma_2|A_1 a_1\rangle \\
&= \bar{N}\frac{1}{\sqrt{2}}\left[|\theta^+\theta^+| + |\epsilon^+\epsilon^+|\right] = 0
\end{aligned}
$$

This wavefunction vanishes since both determinants vanish. Thus the term $^3A_1$ is not allowed. Similarly, we can show that the terms $^1A_2$ and ^3E are also forbidden. The only allowed terms are $^1A_1$, $^3A_2$, and ^1E. The total number of states in these allowed terms is $1+3+2=6$, as expected. As usual, our next task is to construct the wavefunctions for these allowed terms by using (9.29). We give an example:

$$|e^{2\,1}E\theta\rangle = \bar{N}\sum_{\substack{m_1m_2\\ \gamma_1\gamma_2}} |\phi(em_1\gamma_1)\phi(em_2\gamma_2)|\langle\tfrac{1}{2}m_1\tfrac{1}{2}m_2|00\rangle\langle e\gamma_1 e\gamma_2|E\theta\rangle$$

$$= \bar{N}\sum_{\gamma_1\gamma_2}\langle e\gamma_1 e\gamma_2|E\theta\rangle\left[|\phi(e\tfrac{1}{2}\gamma_1)\phi(e-\tfrac{1}{2}\gamma_2)| - |\phi(e-\tfrac{1}{2}\gamma_1)\phi(e\tfrac{1}{2}\gamma_2)|\right]$$

$$= \frac{1}{\sqrt{2}}\left[-|\theta^+\theta^-| + |\epsilon^+\epsilon^-|\right] \tag{9.31}$$

by using Appendix VI. The other wavefunctions of the e^2 configuration and also those for e^3 and e^4 can be similarly obtained and are presented in Appendix VII.

9.1.3 Wavefunctions for the $t_{2g}^n e_g^m$ Configuration

The $(S\Gamma M\gamma)$ wavefunctions for the $t_{2g}^n e_g^m$ configuration can be constructed from those for t_2^n and e^m. Let $S_1\Gamma_1$ be a term of t_2^n, and $S_2\Gamma_2$ be a term of e^m. Corresponding to this pair of terms, we can have terms $S\Gamma$ of the $t_2^n e^m$ configuration, with

$$S = S_1+S_2, S_1+S_2-1, \ldots, |S_1-S_2|$$

and Γ being an irreducible representation in the product $\Gamma_1 \otimes \Gamma_2$. We consider the following linear combination of the products of the $(t_2^n S_1\Gamma_1)$ and $(e^m S_2\Gamma_2)$ wavefunctions. We can write

$$|t_2^n(S_1\Gamma_1)e^m(S_2\Gamma_2), S\Gamma M\gamma\rangle'$$

$$= \sum_{\substack{M_1M_2\\ \gamma_1\gamma_2}} |t_2^n S_1\Gamma_1 M_1\gamma_1\rangle|e^m S_2\Gamma_2 M_2\gamma_2\rangle\langle S_1M_1S_2M_2|SM\rangle\langle\Gamma_1\gamma_1\Gamma_2\gamma_2|\Gamma\gamma\rangle \tag{9.32}$$

This function belongs to the γth row of the irreducible representation Γ; also, it is a simultaneous eigenfunction of $\mathbf{S}^2$ and S_z with the eigenvalues $\hbar^2S(S+1)$ and $\hbar M$, respectively. But this function is not totally antisymmetric with respect to the exchange of electrons. Now, expanding $|t_2^n S_1\Gamma_1 M_1\gamma_1\rangle$ and

$|e^m S_2\Gamma_2 M_2\gamma_2\rangle$ in terms of Slater determinants, we rewrite (9.32) as

$$|t_2^n(S_1\Gamma_1)e^m(S_2\Gamma_2), S\Gamma M\gamma\rangle' = \sum_{\substack{m_1, m_2, \ldots, m_N \\ \gamma_1, \gamma_2, \ldots, \gamma_N}} \mathcal{C}(m_1\gamma_1 m_2\gamma_2 \cdots m_N\gamma_N, S_1\Gamma_1 S_2\Gamma_2, S\Gamma M\gamma) \cdot |\phi(t_2 M_1\gamma_1)\phi(t_2 m_2\gamma_2)\cdots\phi(t_2 m_n\gamma_n)|_{(1,2,\ldots,n)} \cdot |\phi(em_{n+1}\gamma_{n+1})\phi(em_{n+2}\gamma_{n+2})\cdots\phi(em_N\gamma_N)|_{(n+1,n+2,\ldots,N)} \tag{9.33}$$

which is analogous to (9.17). Here the subscripts $(1,2,\ldots,n)$, and $(n+1, n+2,\ldots,N)$ indicate that the first Slater determinant involves the coordinates of electrons $1,2,\ldots,n$ belonging to the t_2^n configuration, and the second determinant involves the coordinates of electrons $n+1, n+2,\ldots,N$ of the e^m configuration. Now, to make (9.33) totally antisymmetric, we take a linear combination analogous to (9.18), and finally obtain the required antisymmetric $(S\Gamma M\gamma)$ wavefunction for the $t_2^n e^m$ configuration as

$$|t_2^n(S_1\Gamma_1)e^m(S_2\Gamma_2), S\Gamma M\gamma\rangle = \bar{N} \sum_{\substack{m_1, m_2, \ldots, m_N \\ \gamma_1, \gamma_2, \ldots, \gamma_N}} \mathcal{C}(\cdots)|\phi(t_2 m_1\gamma_1)\cdots\phi(t_2 m_n\gamma_n) \cdot\phi(em_{n+1}\gamma_{n+1})\cdots\phi(em_N\gamma_N)| \tag{9.34}$$

where $\mathcal{C}(\cdots)$ is the same as in (9.33).

Let us first consider the t_2e configuration. We have, from Appendix VI, $T_2 \otimes E = T_1 + T_2$. Hence the possible terms of t_2e are ${}^1T_1$, ${}^3T_1$, ${}^1T_2$, and ${}^3T_2$. Using (9.32) to (9.34), we find that the wavefunctions for t_2e are given by

$$|t_2e\, S\Gamma M\gamma\rangle \equiv |t_2({}^2T_2)e({}^2E), S\Gamma M\gamma\rangle = \bar{N} \sum_{\substack{m_1m_2 \\ \gamma_1\gamma_2}} |\phi(t_2 m_1\gamma_1)\phi(e m_2\gamma_2)|\langle \tfrac{1}{2}m_1\tfrac{1}{2}m_2|SM\rangle \cdot\langle T_2\gamma_1 E\gamma_2|\Gamma\gamma\rangle \tag{9.35}$$

Thus, for example,

$$|t_2e\,{}^1T_1\gamma\rangle = \bar{N}\sum_{\gamma_1\gamma_2}\langle T_2\gamma_1 E\gamma_2|T_1\gamma\rangle \cdot\left[\frac{1}{\sqrt{2}}|\phi(t_2\tfrac{1}{2}\gamma_1)\phi(e-\tfrac{1}{2}\gamma_2)| - \frac{1}{\sqrt{2}}|\phi(t_2-\tfrac{1}{2}\gamma_1)\phi(e\tfrac{1}{2}\gamma_2)|\right] = \frac{1}{\sqrt{2}}\left[|\zeta^+\epsilon^-| - |\zeta^-\epsilon^+|\right] \tag{9.36}$$

Table 9.2

Wavefunctions $\vert t_2 e\, S\Gamma M\gamma\rangle$
$\vert t_2e\,{}^3T_1\,1\,z\rangle = \vert\zeta^+\epsilon^+\vert$
$\vert t_2e\,{}^3T_2\,1\,\zeta\rangle = \vert\zeta^+\theta^+\vert$
$\vert t_2e\,{}^1T_1\,z\rangle = \frac{1}{\sqrt{2}}[\vert\zeta^+\epsilon^-\vert - \vert\zeta^-\epsilon^+\vert]$
$\vert t_2e\,{}^1T_2\,\zeta\rangle = \frac{1}{\sqrt{2}}[\vert\zeta^+\theta^-\vert - \vert\zeta^-\theta^+\vert]$

All the $(S\Gamma M\gamma)$ wavefunctions of the t_2e configuration can be similarly constructed and are given in Table 9.2. In this table only one component of each type is listed. The other components can be easily obtained by the operation of $\mathbf{S}^-$ and by appropriate cyclic changes of (ξ,η,ζ), $(\theta_x,\theta_y,\theta_z)$, and $(\epsilon_x,\epsilon_y,\epsilon_z)$, where (see Table V.4)

$$\theta_x = -\tfrac{1}{2}\theta + \frac{\sqrt{3}}{2}\epsilon, \qquad \theta_y = -\tfrac{1}{2}\theta - \frac{\sqrt{3}}{2}\epsilon, \qquad \theta_z = \theta$$

and (9.37)

$$\epsilon_x = -\frac{\sqrt{3}}{2}\theta - \tfrac{1}{2}\epsilon, \qquad \epsilon_y = \frac{\sqrt{3}}{2}\theta - \tfrac{1}{2}\epsilon, \qquad \epsilon_z = \epsilon$$

Thus, for example, we have

$$|t_2e\,{}^3T_1\,1\,x\rangle = |\xi^+\epsilon_x^+| = -\frac{\sqrt{3}}{2}|\xi^+\theta^+| - \tfrac{1}{2}|\xi^+\epsilon^+|$$

$$|t_2e\,{}^3T_2\,1\,\xi\rangle = |\xi^+\theta_x^+| = -\tfrac{1}{2}|\xi^+\theta^+| + \frac{\sqrt{3}}{2}|\xi^+\epsilon^+|$$

$$\begin{aligned}|t_2e\,{}^1T_1\,x\rangle &= \frac{1}{\sqrt{2}}[|\xi^+\epsilon_x^-| - |\xi^-\epsilon_x^+|] \\ &= \frac{1}{\sqrt{2}}\left[-\frac{\sqrt{3}}{2}|\xi^+\theta^-| - \tfrac{1}{2}|\xi^-\epsilon^-| \right. \\ &\qquad \left. -\frac{\sqrt{3}}{2}|\xi^-\theta^+| - \tfrac{1}{2}|\xi^-\epsilon^+|\right], \text{ etc.}\end{aligned}$$

Let us now consider the t_2e configuration. As we saw earlier, the allowed terms of t_2^2 are ${}^1A_1$, ${}^1E$, ${}^1T_2$, and ${}^3T_1$. Then the allowed terms of t_2e are obtained according to the following decompositions:

$$\begin{aligned}
{}^1A_1 \otimes {}^2E &= {}^2E \\
{}^1E \otimes {}^2E &= {}^2A_1 + {}^2A_2 + {}^2E \\
{}^1T_2 \otimes {}^2E &= {}^2T_1 + {}^2T_2 \\
{}^3T_1 \otimes {}^2E &= {}^2T_1 + {}^2T_2 + {}^4T_1 + {}^4T_2
\end{aligned} \tag{9.38}$$

The wavefunctions for these terms can be obtained by using the formulas given earlier in this section. We give some examples:

$$\begin{aligned}
|t_2^2({}^1E)e, {}^2A_1 \tfrac{1}{2} a_1\rangle' &= \sum |t_2^2\,{}^1E\gamma_1\rangle \phi_3(e\tfrac{1}{2}\gamma_3)\langle E\gamma_1 E\gamma_3|A_1 a_1\rangle \\
&= \frac{1}{\sqrt{2}}|t_2^2\,{}^1E\theta\rangle\phi_3(e\tfrac{1}{2}\theta) + \frac{1}{\sqrt{2}}|t_2^2\,{}^1E\epsilon\rangle\phi_3(e\tfrac{1}{2}\epsilon) \\
&= \frac{1}{\sqrt{2}}\frac{1}{\sqrt{6}}\left[-|\xi^+\xi^-|_{(1,2)} - |\eta^+\eta^-|_{(1,2)} + 2|\zeta^+\zeta^-|_{(1,2)}\right]\cdot\theta^+(3) \\
&\quad + \frac{1}{\sqrt{2}}\frac{1}{\sqrt{2}}\left[|\xi^+\xi^-|_{(1,2)} - |\eta^+\eta^-|_{(1,2)}\right]\cdot\epsilon^+(3)
\end{aligned}$$

where we have used the wavefunctions of t_2^2 given in Appendix VII. Then the corresponding antisymmetric and normalized wavefunction of $t_2^2e({}^2A_1)$ will be

$$\begin{aligned}
|t_2^2({}^1E)e, {}^2A_1 \tfrac{1}{2} a_1\rangle = \frac{1}{2}\Bigg[&\frac{1}{\sqrt{3}}\{2|\zeta^+\zeta^-\theta^+| - |\xi^+\xi^-\theta^+| - |\eta^+\eta^-\theta^+|\} \\
&+ |\xi^+\xi^-\epsilon^+| - |\eta^+\eta^-\epsilon^+|\Bigg]
\end{aligned} \tag{9.39}$$

In this way the $(S\Gamma M\gamma)$ wavefunctions for all the configurations $t_2^m e^n$ can be similarly constructed. The details of this procedure can be obtained from Ref. 2. These wavefunctions are listed in Appendix VIII for future use.

We saw earlier that the terms of the t_2^{6-n} configuration $(0 \leqslant n \leqslant 6)$ are the same as those of the t_2^n configuration. A similar situation holds for the e^{4-m} and e^m configurations $(0 \leqslant m \leqslant 4)$. In other words, corresponding to a term $\{t_2^n(S_1\Gamma_1)e^m(S_2\Gamma_2), S\Gamma\}$ of $t_2^n e^m$, there is a term $\{t_2^{6-n}(S_1\Gamma_1)e^{4-m}(S_2\Gamma_2), S\Gamma\}$ of $t_2^{6-n}e^{4-m}$. The states

$$|t_2^n(S_1\Gamma_1)e^m(S_2\Gamma_2), S\Gamma M\gamma\rangle \qquad \text{and} \qquad |t_2^{6-n}(S_1\Gamma_1)e^{4-m}(S_2\Gamma_2), S\Gamma - M\gamma\rangle$$

are said to be complementary to each other with respect to the closed shell state:

$$|t_2^6 e\,{}^1A_1\rangle = |\xi^+\xi^-\eta^+\eta^-\zeta^+\zeta^-\theta^+\theta^-\epsilon^+\epsilon^-|$$

Let

$$|t_2^n(S_1\Gamma_1)e^m(S_2\Gamma_2), S\Gamma M\gamma\rangle = \sum_i \mathcal{C}_i[t_2^n(S_1\Gamma_1)e^m(S_2\Gamma_2), S\Gamma M\gamma]D_i^{n,m} \tag{9.40}$$

where $D_i^{n,m}$ are the ${}^6C_n \cdot {}^4C_m$ normalized microstates associated with the configuration $t_2^n e^m$. Then it can be shown[2] that

$$|t_2^{6-n}(S_1\Gamma_1)e^{4-m}(S_2\Gamma_2), S\Gamma - M\gamma\rangle = (-1)^{nm}\mu_1\mu_2(-1)^{S-M}$$
$$\cdot \sum_i \mathcal{C}_i^*[t_2^n(S_1\Gamma_1)e^m(S_2\Gamma_2), S\Gamma M\gamma]D_i^{6-n,4-m} \tag{9.41}$$

where $D_i^{6-n,4-m}$ is the normalized cofactor of $D_i^{n,m}$ within $|t_2^6e^4\,{}^1A_1\rangle$, and

$$\begin{aligned}
&\text{(i)}\ \mu_1 = -1 && \text{for } n=3 \text{ and } S_1\Gamma_1 = {}^4A_2,\ {}^2E,\ {}^2T_1 \\
&\qquad\ \ = \ \ 1 && \text{otherwise} \\
&\text{(ii)}\ \mu_2 = -1 && \text{for } m=2 \text{ and } S_2\Gamma_2 = {}^1E,\ {}^3A_2 \\
&\qquad\ \ = \ \ 1 && \text{otherwise}
\end{aligned} \tag{9.42}$$

9.2 MATRIX ELEMENTS, $S\Gamma$-BLOCKS, AND THE ENERGY LEVELS

In this section we shall derive the matrix elements of the crystal field, using the strong-field wavefunctions derived in Section 9.1. Since the crystal field of octahedral (O_h) symmetry belongs to the irreducible representation A_{1g}, we have the following relations:

$$\langle t_{2g}\gamma_1|V_c(\mathbf{r})|t_{2q}\gamma_2\rangle = \langle t_{2g}|V_c|t_{2g}\rangle\delta\gamma_1\gamma_2$$
$$\langle e_g\gamma_1|V_c(\mathbf{r})|e_g\gamma_2\rangle = \langle e_g|V_c|e_g\rangle\delta\gamma_1\gamma_2 \tag{9.43}$$

and

$$\langle t_{2q}\gamma_1|V_c(\mathbf{r})|e_g\gamma_2\rangle = 0$$

The matrix elements of $V_c(\mathbf{r})$ between two determinantal microstates will be evaluated in terms of (9.43). We now have the following cases:

I.

$$\langle\phi_{k_1}\phi_{k_2}\cdots\phi_{k_N}|V_c(\mathbf{r})|\phi_{k_1}\phi_{k_2}\cdots\phi_{k_N}\rangle = \sum_{i=1}^{N}\langle\phi_{k_i}|V_c(\mathbf{r})|\phi_{k_i}\rangle \tag{9.44}$$

II.

$$\langle \phi_{k_1}\phi_{k_2}\cdots\phi_{k_N}|V_c(\mathbf{r})|\phi_{k'_1}\phi_{k'_2}\cdots\phi_{k'_N}\rangle = \langle \phi_{k_1}|V_c(\mathbf{r})|\phi_{k'_1}\rangle \qquad \text{for } k'_1 \neq k_1 \quad (9.45)$$

III.

$$\langle \phi_{k_1}\phi_{k_2}\cdots\phi_{k_N}|V_c(\mathbf{r})|\phi_{k'_1}\phi_{k'_2}\cdots\phi_{k'_N}\rangle = 0 \quad (9.46)$$

where at least two of $k_1, k_2, \ldots, k_N$ are not involved in $\{k'_1, k'_2, \ldots, k'_N\}$ in Case III.

From the above relations we see at once that $V_c(\mathbf{r})$ is fully diagonal with respect to the wavefunctions $|t_2^n e^m i, S\Gamma M\gamma\rangle$, that is,

$$\langle t_2^n e^m i, S\Gamma M\gamma|V_c(\mathbf{r})|t_2^{n'} e^{m'} i', S'\Gamma' M'\gamma'\rangle = \mathcal{E}_C(n,m)\delta_{nn'}\delta_{mm'}\delta_{ii'}\delta_{SS'}\delta_{\Gamma\Gamma'}\delta_{MM'}\delta_{\gamma\gamma'} \quad (9.47)$$

where

$$\mathcal{E}_C(n,m) = n\langle t_{2g}|V_c(\mathbf{r})|t_{2g}\rangle + m\langle e_g|V_c(\mathbf{r})|e_g\rangle \quad (9.48)$$

and in the d-orbital approximation we have the further simplification (6.41). In (9.47) the index i is introduced in the wavefunction in order to distinguish the different wavefunctions with the same $(S\Gamma M\gamma)$ set for a given configuration. According to our preceding discussions, this situation can occur only if both n and m are nonzero and at least one of them is greater than unity. In that case i represents the pair of terms $S_1\Gamma_1$ and $S_2\Gamma_2$ in the wavefunction $|t_2^n(S_1\Gamma_1)e^m(S_2\Gamma_2), S\Gamma M\gamma\rangle$.

In the absence of the interelectronic repulsion term V_e, the energy levels of the system are given by

$$E_C(n,m) = E_0 + \mathcal{E}_C(n,m) \quad (9.49)$$

depending on n and m only; $E_0 = N\mathcal{E}_{nd}^0$ is the energy of the original unperturbed level corresponding to $\mathcal{H}_0$ in (9.1).

An (n,m) energy level (9.49) is $({}^6C_n \cdot {}^4C_m)$-fold degenerate and will be split in the presence of V_e. If $V_e \ll V_c$, the splitting of an (n,m) level will be small compared to the separation between the adjacent (n,m) levels. In that case the splitting of an (n,m) level can be calculated without reference to the other levels. This means that the energies of the sublevels in a split (n,m) level and the corresponding wavefunctions can be evaluated by diagonalizing V_e with respect to the set of functions $|t_2^n e^m i, S\Gamma M\gamma\rangle$ for the particular (n,m) pair considered.

However, this approximate treatment cannot be applied, in the general case, when $V_c \simeq V_e$. Then the exact treatment is to find the splitting of the original ${}^{10}C_N$-fold degenerate $(\bar{n}d)^N$ level, with energy $E_0 = N\mathcal{E}^0_{\bar{n}d}$, under $V_c + V_e$. We now have to diagonalize $(V_c + V_e)$ with respect to the full set of functions $|t_2^n e^m i, S\Gamma M\gamma\rangle$, with $n+m=N$. The matrix element of V_e will be of the following form:

$$\langle t_2^n e^m i, S\Gamma M\gamma | V_e | t_2^{n'} e^{m'} i', S'\Gamma' M'\gamma'\rangle = \mathcal{C}(nmi, n'm'i', S\Gamma)\delta_{\Gamma\Gamma'}\delta_{SS'}\delta_{MM'}\delta_{\gamma\gamma'} \tag{9.50}$$

From (9.47) and (9.50) we see that the matrix of $V_c + V_e$ is in the block form; $(2S+1)\cdot\lambda(\Gamma)$ *identical* blocks are associated with each $S\Gamma$-term, and these may be called $S\Gamma$-blocks. The rows (and columns) in an $S\Gamma$-block are labeled by *nmi*, and the $(nmi, n'm'i')$th element of the block is

$$(S\Gamma)_{nmi,n'm'i'} = \mathcal{E}_C(n,m)\delta_{nn'}\delta_{mm'}\delta_{ii'} + \mathcal{C}(nmi, n'm'i', S\Gamma) \tag{9.51}$$

By diagonalizing an $S\Gamma$-block, we obtain the characteristic roots $\mathcal{E}_{Ce}(nmi, S\Gamma)$; then the energies of the sublevels of the split $(\bar{n}d)^N$ level will be given by

$$E_{Ce}(nmi, S\Gamma) = E_0 + \mathcal{E}_{Ce}(nmi, S\Gamma) \tag{9.52}$$

and the corresponding wavefunctions will be

$$|nmi, S\Gamma M\gamma\rangle_{Ce} = \sum_{n'm'i'} A_{n'm'i'}(nmi, S\Gamma)|t_2^{n'} e^{m'} i', S\Gamma M\gamma\rangle \tag{9.53}$$

where

$$A_{nmi}(S\Gamma) \equiv \text{set}\{A_{n'm'i'}(nmi, S\Gamma)\}$$

are the orthonormal eigenvectors of the $S\Gamma$-block corresponding to the characteristic roots $\mathcal{E}_{Ce}(nmi, S\Gamma)$.

The sublevel $E_{Ce}(nmi, S\Gamma)$ is $(2S+1)\cdot\lambda(\Gamma)$-fold degenerate when the root $\mathcal{E}_{Ce}(nmi, S\Gamma)$ appears only once. But if this particular root appears r times, the (accidental) degeneracy of the level $E_{Ce}(nmi, S\Gamma)$ will be $r(2S+1)\cdot\lambda(\Gamma)$.

We note that the level *nmi* in $\mathcal{E}_{Ce}(nmi, S\Gamma)$ is not at all obvious. The level is actually chosen by looking at the results of the approximate treatment for the case $V_c \gg V_e$. According to the approximate treatment, the energy levels are given by

$$\begin{aligned} E_{C+e}(nmi, S\Gamma) &= E_0 + \mathcal{E}_C(n,m) + \mathcal{E}_e(nmi, S\Gamma) \\ &= E_C + \mathcal{E}_e(nmi, S\Gamma) \end{aligned} \tag{9.54}$$

Here $\mathcal{E}_e(nmi, S\Gamma)$ are the characteristic roots of the matrix

$$\langle t_2^n e^m i, S\Gamma M\gamma | V_e | t_2^n e^m i', S\Gamma M\gamma\rangle$$

where the rows (and columns) of the matrix are labeled by i. Now, in the limit $10Dq\to\infty$, the energy levels obtained by the exact treatment must coincide with those from the approximate treatment. Then we choose the label nmi in $\mathcal{E}_{Ce}(nmi, S\Gamma)$ so that, in the limit $10Dq\to\infty$, it coincides with $\mathcal{E}_C(n,m)+\mathcal{E}_e(nmi, S\Gamma)$.

The wavefunctions $|t_2^n e^m i, S\Gamma M\gamma\rangle$ are linear combinations of the N-dimensional determinantal microstates

$$|\phi_{k_1}\phi_{k_2}\cdots\phi_{k_N}|$$

where

$$\phi_{k_i}=\xi^+,\xi^-,\eta^+,\eta^-,\zeta^+,\zeta^-,\theta^+,\theta^-,\epsilon^+,\epsilon^-$$

Hence the matrix element (9.50) will be a linear combination of integrals of the form

$$\langle\phi_{k_1}\phi_{k_2}\cdots\phi_{k_N}|V_e|\phi_{k_1'}\phi_{k_2'}\cdots\phi_{k_N'}\rangle$$
$$=\sum_{i<j=1}^{N}\int\{\phi_{k_1}^*\phi_{k_2}^*\cdots\phi_{k_N}^*\}\left|\frac{e^2}{\mathbf{r}_{ij}}\right|\{\phi_{k_1'}\phi_{k_2'}\cdots\phi_{k_N'}\}\,d\tau \quad (9.55)$$

where the integration is carried over both the coordinate space and the spin space of the N electrons.

We saw in Chapter 2 that the problem of evaluating the matrix elements of (9.55) is actually the problem of calculating the various two-electron space integrals, $\langle\phi_i\phi_j||\phi_k\phi_l\rangle$,

$$\langle\phi_i\phi_j||\phi_k\phi_l\rangle=\delta(m_{s_i}m_{s_k})\,\delta(m_{s_j}m_{s_l})\langle\psi_i\psi_j||\psi_k\psi_l\rangle \quad (9.56)$$

where

$$\langle\psi_i\psi_j||\psi_k\psi_l\rangle=\int\psi_i^*(\mathbf{r}_1)\psi_j^*(\mathbf{r}_2)\frac{e^2}{\mathbf{r}_{12}}\psi_k(\mathbf{r}_1)\psi_l(\mathbf{r}_2)\,d\tau_1\,d\tau_2 \quad (9.57)$$

In particular,

$$C(\psi_i\psi_j)=\langle\psi_i\psi_j||\psi_i\psi_j\rangle \quad (9.58)$$

is called the Coulomb integral, and

$$J(\psi_i\psi_j)=\langle\psi_i\psi_j||\psi_j\psi_i\rangle \quad (9.59)$$

is called the exchange integral. It can be shown that

$$C(\psi_i\psi_j)\geqslant J(\psi_i\psi_j)\geqslant 0 \quad (9.60)$$

Here the functions ψ_i are the five functions ξ, η, ζ, θ, and ϵ, which are all real. Hence, in addition to the obvious equality

$$\langle \psi_I \psi_j || \psi_k \psi_l \rangle = | \psi_j \psi_i || \psi_l \psi_k \rangle \tag{9.61}$$

we have also the following equalities:

$$\langle \psi_i \psi_j || \psi_k \psi_l \rangle = \langle \psi_k \psi_j || \psi_i \psi_l \rangle = \langle \psi_i \psi_l || \psi_k \psi_j \rangle = \langle \psi_k \psi_l || \psi_i \psi_j \rangle$$

The four identical integrals in (9.62) can obviously be denoted by

$$\{\psi_i \psi_k, \psi_j \psi_l\} = \int \psi_i^*(\mathbf{r}_1)\psi_k(\mathbf{r}_1)\frac{e^2}{r_{12}}\psi_j^*(\mathbf{r}_2)\psi_l(\mathbf{r}_2)\,d\tau_1\,d\tau_2 \tag{9.63}$$

while (9.61) becomes

$$\{\psi_i \psi_k, \psi_j \psi_l\} = \{\psi_j \psi_l, \psi_i \psi_k\} \tag{9.64}$$

In terms of this modified (or contracted) notation, we have

$$C(\psi_i \psi_j) = \{\psi_i \psi_i, \psi_j \psi_j\} \tag{9.65}$$

and

$$J(\psi_i \psi_j) = \{\psi_i \psi_j, \psi_i \psi_j\} \tag{9.66}$$

The number of possible Coulomb integrals is $5 + {}^5C_2 = 15$, whereas the number of possible exchange integrals is only ${}^5C_2 = 10$; the number of possible integrals $\{\psi_i \Psi_k, \psi_j \psi_l\}$ with $\psi_i \psi_k \neq \psi_j \psi_l$ is

$$5 \times {}^5C_2 + \tfrac{1}{2}\{{}^5C_2 \times {}^5C_2 - {}^5C_2\} = 95$$

Many of these integrals will vanish, however, as we shall now see.

The products of the one-electron orbitals $\xi, \eta, \zeta, \theta, \epsilon$, taken in pairs, may be classified[3] by their behavior under the group D_4 with the fourfold axis along the z-axis (see Appendix V). We have the following results:

$$\left.\begin{aligned} \frac{1}{\sqrt{2}}(\xi^2+\eta^2), \zeta^2, \theta^2, \epsilon^2 &\subset A_1 \\ \zeta\epsilon &\subset A_2 \\ \frac{1}{\sqrt{2}}(\xi^2-\eta^2), \theta\epsilon &\subset B_1 \\ \xi\eta, \zeta\theta &\subset B_2 \\ \xi\theta, \eta\zeta, \xi\epsilon &\subset Ex \\ \eta\theta, \zeta\xi, \eta\epsilon &\subset Ey \end{aligned}\right\} \tag{9.67}$$

where $f \subset \Gamma\gamma$ means that the function f is $\pm$ the γ-component basis of the irreducible representation Γ of point group symmetry D_4. It is evident from (9.67) that we have five noninteracting sets: (A_1B_1), A_2, B_2, Ex and Ey, with ξ^2 and η^2 belonging to both A_1 and B_1. Now, obviously, $\{\psi_i\psi_k,\psi_j\psi_l\}$ is zero whenever $\psi_i\psi_k$ and $\psi_j\psi_l$ belong to two different sets. Thus $\{\zeta\epsilon,\psi_j\psi_l\}$ is always zero unless $\psi_j\psi_l=\zeta\epsilon$, since only $\zeta\epsilon$ belongs to A_2.

Next we investigate the number of independent parameters needed to specify the nonvanishing $\{\psi_i\psi_k,\psi_j\psi_l\}$ in the most general case. The function $\psi_i\psi_k$ (or $\psi_j\psi_l$) belongs to one of the three configurations e^2, t_2e, and t_2^2. Clearly, no operation of O_h can turn a function belonging to one particular configuration into a function of another. Hence there are six independent types of $\{\psi_i\psi_k,\psi_j\psi_l\}$, which may be written as e^2,e^2; e^2,t_2e; e^2,t_2^2; t_2e,t_2e; t_2e,t_2^2; and t_2^2,t_2^2. Thus, for example, in an integral $\{\psi_i\psi_k,\psi_j\psi_l\}$ of type e^2,t_2^2, function $\psi_i\psi_k$ belongs to e^2 and $\psi_j\psi_l$ belongs to t_2^2, or vice versa. According to classification (9.67), we see that all integrals of type e^2,t_2e are zero.

We can easily find the number of independent parameters necessary to specify the integrals of a given type. Thus, for example, the number of independent parameters required for integrals of type e^2,t_2^2 is equal to the number of times the unit representation A_1 occurs in the direct product of the symmetrized squares $[E^2]$ and $[T_2^2]$. Now $[E^2]=A_1+E$ and $[T_2^2]=A_1+E+T_2$; hence there are *two* independent parameters for integrals of type e^2,t_2^2. Similarly, we find the number of independent parameters to be 2, 2, 1, and 3 for types e^2,e^2; t_2e,t_2e; t_2e,t_2^2; and t_2^2,t_2^2, respectively. The relations involving integrals of a given type can be obtained by applying the generators of O_h. As an example, let us consider type e^2,t_2^2.

From classification (9.67) we find that

$$0=\left\{\frac{1}{\sqrt{2}}(\xi^2+\eta^2),\theta\epsilon\right\}=\frac{1}{\sqrt{2}}\{\xi^2,\theta\epsilon\}+\frac{1}{\sqrt{2}}\{\eta^2,\theta\epsilon\}$$

Thus

$$\{\xi^2,\theta\epsilon\}=-\{\eta^2,\theta\epsilon\}=a,\qquad \text{say} \tag{9.68}$$

Again,

$$0=\left\{\frac{1}{\sqrt{2}}(\xi^2-\eta^2),\theta^2\right\}=\left\{\frac{1}{\sqrt{2}}(\xi^2-\eta^2),\epsilon^2\right\}$$

which gives, respectively,

$$\{\xi^2,\theta^2\}=\{\eta^2,\theta^2\} \tag{9.69}$$

and

$$\{\xi^2,\epsilon^2\}=\{\eta^2,\epsilon^2\}=b,\qquad \text{say} \tag{9.70}$$

As we have already seen, only two independent parameters are required in this case, and we consider a and b to be these two parameters. Then all integrals of type e^2, t_2^2 can be expressed in terms of only a and b. Thus we have

$$\begin{aligned}\{\xi^2,\theta^2\} &\equiv C_4^x\{\xi^2,\theta^2\} = \{C_4^x\xi^2, C_4^x\theta^2\}\\ &= \left\{\xi^2,\left(-\tfrac{1}{2}\theta-\frac{\sqrt{3}}{2}\epsilon\right)^2\right\}\\ &= \tfrac{1}{4}\{\xi^2,\theta^2\}+\tfrac{3}{4}\{\xi^2,\epsilon^2\}+\frac{\sqrt{3}}{2}\{\xi^2,\theta\epsilon\} \qquad (9.71)\end{aligned}$$

Hence

$$\{\xi^2,\theta^2\} = \{\xi^2,\epsilon^2\}+\tfrac{2}{3}\{\xi^2,\theta\epsilon\} = b+\frac{2}{\sqrt{3}}a \qquad (9.72)$$

by using (9.68) and (9.70). In (9.71) we have used the fact that, when C_4^x operates on θ, we can show that

$$C_4^x\theta = -\tfrac{1}{2}\theta-\frac{\sqrt{3}}{2}\epsilon$$

by using the group theory (see Table V.4). The two other nonvanishing integrals of type e^2, t_2^2 are, similarly,

$$\begin{aligned}\{\zeta^2,\theta^2\} &\equiv C_4^x\{\zeta^2,\theta^2\} = \left\{\eta^2,\left(-\tfrac{1}{2}\theta-\frac{\sqrt{3}}{2}\epsilon\right)^2\right\}\\ &= \tfrac{1}{4}\{\eta^2,\theta^2\}+\tfrac{3}{4}\{\eta^2,\epsilon^2\}+\frac{\sqrt{3}}{2}\{\eta^2,\theta\epsilon\}\\ &= \tfrac{1}{4}\left(b+\frac{2}{\sqrt{3}}a\right)+\tfrac{3}{4}b+\frac{\sqrt{3}}{2}(-a)\\ &= b-\frac{1}{\sqrt{3}}a \qquad (9.73)\end{aligned}$$

and

$$\begin{aligned}\{\zeta^2,\epsilon^2\} &\equiv C_4^x\{\zeta^2,\epsilon^2\} = \left\{\eta^2,\left(-\frac{\sqrt{3}}{2}\theta+\tfrac{1}{2}\epsilon\right)^2\right\}\\ &= \tfrac{3}{4}\{\eta^2,\theta^2\}+\tfrac{1}{4}\{\eta^2,\epsilon^2\}-\frac{\sqrt{3}}{2}\{\eta^2,\theta\epsilon\}\\ &= \tfrac{3}{4}\left(b+\frac{2}{\sqrt{3}}a\right)+\tfrac{1}{4}b-\frac{\sqrt{3}}{2}(-a)\\ &= b+\sqrt{3}\,a \qquad (9.74)\end{aligned}$$

Table 9.3

Integrals $\{\psi_i\psi_k, \psi_j\psi_l\}$ of type e^2, t_2^2

$\psi_i\psi_k$ \ $\psi_j\psi_l$	ξ^2	η^2	ζ^2	$\xi\eta$	$\eta\zeta$	$\zeta\xi$
θ^2	$b+\frac{2}{\sqrt{3}}a$	$b+\frac{2}{\sqrt{3}}a$	$b-\frac{1}{\sqrt{3}}a$	0	0	0
ϵ^2	b	b	$b+\sqrt{3}\,a$	0	0	0
$\theta\epsilon$	a	$-a$	0	0	0	0

The integrals of the other types, e^2,e^2; t_2e,t_2e; t_2e,t_2^2; and t_2^2,t_2^2, can be similarly obtained by using classification (9.67) and the transformation properties under the generators of O_h listed in Appendix V.

The final result is that all the nonvanishing integrals $\{\psi_i\psi_k,\psi_j\psi_l\}$ can be expressed in terms of 10 independent parameters. It is important to remember that the parameters in the integrals of one type are not involved in the integrals of another type. The results are given in Tables 9.3 to 9.7. These results, obtained by group-theoretical reasoning alone, are the most general ones, and it is important to realize that they do not depend on the actual nature of the orbitals ξ, η, ζ, θ, and ϵ.

Thus, in the most general case, energy levels (9.54) are expressed in terms of 12 parameters: $a,b,\ldots,p$, $\langle t_{2g}|V_c|t_{2g}\rangle$, and $\langle e_g|V_c|e_g\rangle$. In the simplest case of d-orbital approximation, the 10 parameters $a,b,\ldots,p$, will reduce to 3 atomic parameters: the Slater-Condon parameters F_0, F_2, and F_4 or the Racah parameters A, B, and C, and the difference between $\langle t_{2g}|V_c|t_{2g}\rangle$ and $\langle e_g|V_c|e_g\rangle$ will be given by $10Dq$ [equation (9.7)]. Thus, in the simplest case, the energy levels will be expressed in terms of only *four* parameters.

We now derive the results for the case where the functions $\psi_i(\mathbf{r})$ are linear combinations of the d orbitals [equations (9.2) and (9.3)], so that the integrals (9.63) are actually linear combinations of integrals of the following type,

Table 9.4

Integrals $\{\psi_i\psi_k, \psi_j\psi_l\}$ of type e^2, e^2

$\psi_i\psi_k$ \ $\psi_j\psi_l$	θ^2	ϵ^2	$\theta\epsilon$
θ^2	e'	$e'-2f'$	0
ϵ^2	$e'-2f'$	e'	0
$\theta\epsilon$	0	0	f'

Table 9.5

Integrals $\{\psi_i\psi_k, \psi_j\psi_l\}$ of type t_2e, t_2e

$\psi_i\psi_k$ \ $\psi_j\psi_l$	$\xi\theta$	$\eta\theta$	$\zeta\theta$	$\xi\epsilon$	$\eta\epsilon$	$\zeta\epsilon$
$\xi\theta$	g	0	0	$-h$	0	0
$\eta\theta$	0	g	0	0	h	0
$\zeta\theta$	0	0	$g+\sqrt{3}\,h$	0	0	0
$\xi\epsilon$	$-h$	0	0	$g+\dfrac{2}{\sqrt{3}}h$	0	0
$\eta\epsilon$	0	h	0	0	$g+\dfrac{2}{\sqrt{3}}h$	0
$\zeta\epsilon$	0	0	0	0	0	$g-\dfrac{1}{\sqrt{3}}h$

involving only the d orbitals:

$$\langle m_1 m_2 \| m_1' m_2' \rangle = \int d\mathbf{r}_1\, d\mathbf{r}_2\, \phi^*_{dm_1}(\mathbf{r}_1)\phi^*_{dm_2}(\mathbf{r}_2)\frac{e^2}{r_{12}}\phi_{dm_1'}(\mathbf{r}_1)\phi_{dm_2'}(\mathbf{r}_2) \tag{9.75}$$

Now, using the usual expansion of $1/r_{12}$ in spherical harmonics [equation (3.21)] and

$$\phi_{dm}(\mathbf{r}) = R_{\bar{n}d}(\mathbf{r})\, Y_2^m(\theta,\phi)$$

one can easily evaluate the matrix element (9.75) in terms of the Racah parameters. The method of doing this was outlined in Chapter 3. All the nonvanishing integrals $\{\psi_i\psi_k, \psi_j\psi_l\}$ in Tables 9.3 to 9.7 can now be expressed in terms of the Racah parameters. We give a few examples.

Table 9.6

Integrals $\{\psi_i\psi_k, \psi_j\psi_l\}$ of type t_2e, t_2^2

$\psi_i\psi_k$ \ $\psi_j\psi_l$	ξ^2	η^2	ζ^2	$\xi\eta$	$\eta\zeta$	$\zeta\xi$
$\xi\theta$	0	0	0	0	i'	0
$\eta\theta$	0	0	0	0	0	i'
$\zeta\theta$	0	0	0	$-2i'$	0	0
$\xi\epsilon$	0	0	0	0	$-\sqrt{3}\,i'$	0
$\eta\epsilon$	0	0	0	0	0	$\sqrt{3}\,i$

Table 9.7

Integrals $\{\psi_i\psi_k, \psi_j\psi_l\}$ of type t_2^2, t_2^2

$\psi_i\psi_k$ \ $\psi_j\psi_l$	ξ^2	η^2	ζ^2	$\xi\eta$	$\eta\zeta$	$\zeta\xi$
ξ^2	j'	k'	k'	0	0	0
η^2	k'	j'	k'	0	0	0
ζ^2	k'	k'	j'	0	0	0
$\xi\eta$	0	0	0	p	0	0
$\eta\zeta$	0	0	0	0	p	0
$\zeta\xi$	0	0	0	0	0	p

1. First we have

$$C(\xi,\xi)=\langle\xi\xi||\xi\xi\rangle$$
$$=\int d\mathbf{r}_1\,d\mathbf{r}_2\left(-\frac{i}{\sqrt{2}}\right)\left[\phi_{d1}^*(\mathbf{r}_1)+\phi_{d-1}^*(\mathbf{r}_1)\right]\left(-\frac{i}{\sqrt{2}}\right)$$
$$\cdot\left[\phi_{d1}^*(\mathbf{r}_2)+\phi_{d-1}^*(\mathbf{r}_2)\right]\frac{e^2}{\mathbf{r}_{12}}\left(\frac{i}{\sqrt{2}}\right)\left[\phi_{d1}(\mathbf{r}_1)+\phi_{d-1}(\mathbf{r}_1)\right]$$
$$\cdot\left(\frac{i}{\sqrt{2}}\right)\left[\phi_{d1}(\mathbf{r}_2)+\phi_{d-1}(\mathbf{r}_2)\right]$$
$$=\tfrac{1}{4}\left[\langle 1\,1||1\,1\rangle+\langle -1\,-1||-1\,-1\rangle+\langle 1\,-1||1\,-1\rangle\right.$$
$$\left.+\langle -1\,1||-1\,1\rangle+\langle 1\,-1||-1\,1\rangle+\langle -1\,1||1\,-1\rangle\right] \tag{9.76}$$

Here we have used the fact that $\langle m_1m_2||m_1'm_2'\rangle$ is nonvanishing only when $m_1+m_2=m_2'+m_2'$. The integrals in (9.76) are not difficult to evaluate. With the help of Tables 2.1 to 2.3 we find that

$$\langle 1\,1||1\,1\rangle=\langle -1\,-1||-1\,-1\rangle=\langle 1\,-1||1\,-1\rangle=\langle -1\,1||-1\,1\rangle$$
$$=F^0+\tfrac{1}{49}F^2+\tfrac{16}{441}F^4 \tag{9.77}$$

and

$$\langle 1\,-1||-1\,1\rangle=\langle -1\,1||1\,-1\rangle=\tfrac{6}{49}F^2+\tfrac{40}{441}F^4$$

Thus

$$C(\xi\xi)=\{\xi\xi,\xi\xi\}=F_0+4F_2+36F_4 \tag{9.78}$$

Here, to simplify the results, we have used the new parameters

$$F_0=F^0,\qquad F_2=\tfrac{1}{49}F^2,\qquad F_4=\tfrac{1}{441}F^4 \tag{9.79}$$

2. Similarly, it can be shown that

$$\begin{aligned} C(\theta\theta) &= F_0 + 4F_2 + 36F_4 \\ C(\zeta\theta) &= F_0 - 4F_2 + 6F_4 \\ C(\xi\eta) &= F_0 - 2F_2 - 4F_4 \\ J(\xi\eta) &= \langle \xi\eta \| \eta\xi \rangle = 3F_2 + 20F_4 \\ \langle \xi\theta \| \xi\epsilon \rangle &= 2\sqrt{3}\, F_2 - 10\sqrt{3}\, F_4 \end{aligned} \tag{9.80}$$

It would be a good exercise for the reader to derive these matrix elements. Comparing them with those in Table 9.7, we find that

$$\begin{aligned} j' &= F_0 + 4F_2 + 36F_4 = A + 4B + 3C \\ k' &= F_0 - 2F_2 - 4F_4 \ = A - 2B + C \\ p &= 3F_2 + 20F_4 \qquad = 3B + C \end{aligned} \tag{9.81}$$

where the results are expressed in terms of the Racah parameters A, B, C, defined as

$$\begin{aligned} A &= F_0 - 49F_4 \\ B &= F_2 - 5F_4 \\ C &= 35F_4 \end{aligned} \tag{9.82}$$

By using the matrix elements of (9.80) and Tables 9.3 to 9.7, we obtain all 10 parameters $a, b, \ldots, p$, which are listed in Table 9.8. The resulting expressions of $\{\psi_i\psi_k, \psi_j\psi_l\}$ are given in Table 9.9. In these tables the Racah parameter C should not be confused with the Coulomb integral $C(ij)$.

Table 9.8

The Coulomb and the exchange integrals involving the strong-field d wavefunctions

$C(\xi\xi) = C(\eta\eta) = C(\zeta\zeta) = C(\theta\theta) = C(\epsilon\epsilon)$
$\quad = A + 4B + 3C$
$C(\xi\eta) = C(\eta\zeta) = C(\zeta\xi) = C(\xi\epsilon) = C(\eta\epsilon)$
$\quad = A - 2B + C$
$C(\xi\theta) = C(\eta\theta) = A + 2B + C$
$C(\zeta\theta) = C(\theta\epsilon) = A - 4B + C$
$C(\zeta\epsilon) = A + 4B + C$
$J(\xi\eta) = J(\eta\zeta) = J(\zeta\xi) = J(\xi\epsilon) = J(\eta\epsilon)$
$\quad = 3B + C$
$J(\zeta\theta) = J(\theta\epsilon) = 4B + C$
$J(\xi\theta) = J(\eta\theta) = B + C$
$J(\zeta\epsilon) = C$

Table 9.9

Matrix elements of $\{\psi_i\psi_k, \psi_j\psi_l\}$ for d electrons

$\psi_i\psi_k$	$\psi_j\psi_l$	$\{\psi_i\psi_k, \psi_j\psi_l\}$
$\xi\xi$	$\theta\epsilon$	$2\sqrt{3}\ F_2-10\sqrt{3}\ F_4=2\sqrt{3}\ B$
$\eta\eta$	$\theta\epsilon$	$-2\sqrt{3}\ F_2+10\sqrt{3}\ F_4=-2\sqrt{3}\ B$
$\xi\eta$	$\zeta\theta$	$2\sqrt{3}\ F_2-10\sqrt{3}\ F_4=2\sqrt{3}\ B$
$\zeta\xi$	$\eta\theta$	$\sqrt{3}\ F_2-5\sqrt{3}\ F_4=\sqrt{3}\ B$
$\zeta\xi$	$\eta\epsilon$	$3F_2-15F_4=3B$
$\eta\zeta$	$\xi\theta$	$\sqrt{3}\ F_2-5\sqrt{3}\ F_4=\sqrt{3}\ B$
$\eta\zeta$	$\xi\epsilon$	$-3F_2+15F_4=-3B$
$\eta\theta$	$\eta\epsilon$	$\sqrt{3}\ F_2-5\sqrt{3}\ F_4=\sqrt{3}\ B$
$\xi\theta$	$\xi\epsilon$	$-\sqrt{3}\ F_2+5\sqrt{3}\ F_4=-\sqrt{3}\ B$

9.3 ENERGY LEVELS OF THE d^N SYSTEM

We shall now find the energy levels of a d^N system, using the fundamentals given in Section 9.2. We shall begin with the d^2 system, for which we shall have three possible configurations: t_2^2, e^2, and t_2e. The corresponding energy levels, in the absence of V_e, are

$$\begin{aligned} E_c(2,0)&=E_0+\mathcal{E}_c(2,0)=E_0-8Dq\\ E_c(0,2)&=E_0+\mathcal{E}_c(0,2)=E_0+12Dq\\ E_c(1,1)&=E_0+\mathcal{E}_c(1,1)=E_0+2Dq \end{aligned} \tag{9.83}$$

respectively, in the d-orbital approximation. Equation (9.83) can be easily obtained from (9.49) and (9.48). First we consider the case where $V_e \ll V_c$ and then find the additional splitting of each of these levels separately under the perturbation of V_e.

9.3.1 Splitting of the $E_c(2,0)$ Level

From Appendix VII we find that the t_2^2 configuration gives the terms ${}^1A_1$, ${}^1E$, ${}^3T_1$, and ${}^1T_2$. Therefore this configuration will be split in the following way with energies [see (9.54)]

$$\begin{aligned} E_{c+e}(20,{}^1A_1)&=E_c(2,0)+\mathcal{E}_e(20,{}^1A_1)\\ E_{c+e}(20,{}^1E)&=E_c(2,0)+\mathcal{E}_e(20,{}^1E)\\ E_{c+e}(20,{}^3T_1)&=E_c(2,0)+\mathcal{E}_e(20,{}^3T_1)\\ E_{c+e}(20,{}^1T_2)&=E_c(2,0)+\mathcal{E}_e(20,{}^1T_2) \end{aligned} \tag{9.84}$$

The degeneracies of these levels are 1, 2, 9, and 3, respectively. We have, as usual,

$$\mathcal{E}_e(20,{}^1A_1)=\langle t_2^{2\,1}A_1|V_e|t_2^{2\,1}A_1\rangle$$

$$=\tfrac{1}{3}\langle\xi^+\xi^- +\eta^+\eta^- +\zeta^+\zeta^-|V_e|\xi^+\xi^- +\eta^+\eta^- +\zeta^+\zeta^-\rangle$$

$$=\frac{1}{3}\sum_{(\xi\eta\zeta)}\big[\langle\xi^+\xi^-|V_e|\xi^+\xi^-\rangle$$

$$+\langle\xi^+\xi^-|V_e|\eta^+\eta^-\rangle+\langle\xi^+\xi^-|V_e|\zeta^+\zeta^-\rangle\big]$$

where $\Sigma_{(\xi\eta\zeta)}$ means summation over the terms obtained by cyclic permutations of ξ, η, and ζ. Hence, by using (9.55) to (9.57), we obtain

$$\mathcal{E}_e(20,{}^1A_1)=\frac{1}{3}\sum_{(\xi\eta\zeta)}\big[\langle\xi\xi\|\xi\xi\rangle+\langle\xi\xi\|\eta\eta\rangle$$

$$+\langle\xi\xi\|\zeta\zeta\rangle\big]$$

$$=\frac{1}{3}\sum_{(\xi\eta\zeta)}\big[C(\xi\xi)+J(\xi\eta)+J(\zeta\xi)\big]$$

$$=C(\xi\xi)+2J(\xi\eta)=F_0+10F_2+76F_4 \tag{9.85}$$

by using (9.78) and 9.80). Similarly,

$$\mathcal{E}_e(20,{}^1E)=\tfrac{1}{2}\big[C(\xi\xi)+C(\eta\eta)-2J(\xi\eta)\big]$$

$$=F_0+F_2+16F_4 \tag{9.86}$$

$$\mathcal{E}_e(20,{}^3T_1)=C(\xi\eta)-J(\xi\eta)$$

$$=F_0-5F_2-24F_4 \tag{9.87}$$

and

$$\mathcal{E}_e(20,{}^1T_2)=C(\xi\eta)+J(\xi\eta)$$

$$=F_0+F_2+16F_4 \tag{9.88}$$

It would be a good exercise for the reader to derive (9.86) to (9.88).

Thus we have calculated the energies of the sublevels of the split $E_c(2,0)$ level. In other words, we have calculated the term energies of the t_2^2 configuration.

9.3.2 Splitting of the $E_c(0,2)$ Level

The e^2 configuration has the terms ${}^1A_1$, ${}^3A_2$, and ${}^1E$. Hence we have

$$
\begin{aligned}
E_{c+e}(02,{}^1A_1) &= E_c(0,2) + \mathcal{E}_e(02,{}^1A_1)\\
E_{c+e}(02,{}^3A_2) &= E_c(0,2) + \mathcal{E}_e(02,{}^3A_2)\\
E_{c+e}(02,{}^1E) &= E_c(0,2) + \mathcal{E}_e(02,{}^1E)
\end{aligned}
\tag{9.89}
$$

By using again the wavefunctions of the e^2 configuration, given in Appendix VII, we have

$$
\begin{aligned}
\mathcal{E}_e(02,{}^1A_1) &= \langle e^2\,{}^1A_1|V_e|e^2\,{}^1A_1\rangle\\
&= \tfrac{1}{2}\langle \theta^+\theta^- + \epsilon^+\epsilon^-|V_e|\theta^+\theta^- + \epsilon^+\epsilon^-\rangle\\
&= \tfrac{1}{2}\left[C(\theta\theta) + C(\epsilon\epsilon) + 2J(\theta\epsilon)\right]\\
&= F_0 + 8F_2 + 51F_4
\end{aligned}
\tag{9.90}
$$

$$
\begin{aligned}
\mathcal{E}_e(02,{}^3A_2) &= \langle e^{2\,3}A_2|V_e|e^{2\,3}A_2\rangle\\
&= \langle \theta^+\epsilon^+|V_e|\theta^+\epsilon^+\rangle\\
&= C(\theta\epsilon) - J(\theta\epsilon)\\
&= F_0 - 8F_2 - 9F_4
\end{aligned}
\tag{9.91}
$$

and, similarly,

$$
\begin{aligned}
\mathcal{E}_e(02,{}^1E) &= \tfrac{1}{2}\left[C(\theta\theta) + C(\epsilon\epsilon) - 2J(\theta\epsilon)\right]\\
&= F_0 + 21F_4
\end{aligned}
\tag{9.92}
$$

9.3.3 Splitting of the $E_c(1,1)$ Level

The t_2e configuration has the terms ${}^3T_1, {}^3T_2, {}^1T_1$, and ${}^1T_2$. Hence this level is split into the corresponding sublevels with energies

$$
\begin{aligned}
E_{c+e}(11,{}^3T_1) &= E_c(1,1) + \mathcal{E}_e(11,{}^3T_1)\\
E_{c+e}(11,{}^3T_2) &= E_c(1,1) + \mathcal{E}_e(11,{}^3T_2)\\
E_{c+e}(11,{}^1T_1) &= E_c(1,1) + \mathcal{E}_e(11,{}^1T_1)\\
E_{c+e}(11,{}^1T_2) &= E_c(1,1) + \mathcal{E}_e(11,{}^1T_2)
\end{aligned}
\tag{9.93}
$$

Using the proper wavefunctions (Appendix VII), we have

$$
\begin{aligned}
\mathcal{E}_e(11,{}^3T_1) &= \langle \zeta^+\epsilon^+|V_e|\zeta^+\epsilon^+\rangle\\
&= C(\zeta\epsilon) - J(\zeta\epsilon)\\
&= F_0 + 4F_2 - 69F_4
\end{aligned}
\tag{9.94}
$$

Similarly,

$$\mathcal{E}_e(11,{}^3T_2)=\langle\zeta^+\theta^+|V_e|\zeta^+\theta^+\rangle$$
$$=C(\zeta\theta)-J(\zeta\theta)$$
$$=F_0-8F_2-9F_4 \tag{9.95}$$

$$\mathcal{E}_e(11,{}^1T_1)=C(\zeta\epsilon)+J(\zeta\epsilon)$$
$$=F_0+4F_2+F_4 \tag{9.96}$$

and

$$\mathcal{E}_e(11,{}^1T_2)=C(\zeta\theta)+J(\zeta\theta)$$
$$=F_0+21F_4 \tag{9.97}$$

We now follow the exact treatment by considering the different terms one by one.

First, the term ${}^1A_1$ occurs in the configurations t_2^2 and e^2. Hence we shall have to diagonalize the corresponding two-dimensional matrix (the ${}^1A_1$-block) of (V_c+V_e) in order to obtain the exact energies of the two ${}^1A_1$ terms. The two diagonal elements of the matrix were obtained in (9.85) and (9.90). These are

$$E_c(2,0)+\mathcal{E}_e(20,{}^1A_1)=-8Dq+F_0+10F_2+76F_4$$

and

$$E_c(0,2)+\mathcal{E}_e(02,{}^1A_1)=12Dq+F_0+8F_2+51F_4$$

We now have to find the nondiagonal element between $t_2^2({}^1A_1)$ and $e^2({}^1A_1)$, which is, as before, given by

$$\langle t_2^2\,{}^1A_1|V_e|e^2\,{}^1A_1\rangle=\frac{1}{\sqrt{3}}\frac{1}{\sqrt{2}}\langle\xi^+\xi^-+\eta^+\eta^-+\zeta^+\zeta^-|V_e|\theta^+\theta^-\epsilon^+\epsilon^-\rangle$$
$$=\frac{1}{\sqrt{6}}\left[J(\xi\theta)+J(\eta\theta)+J(\zeta\theta)+J(\xi\epsilon)+J(\eta\epsilon)+J(\zeta\epsilon)\right]$$
$$=\sqrt{6}\,(2F_2+25F_4)$$

Then the required secular equation is

${}^1A_1$	$t_2^2({}^1A_1)$	$e^2({}^1A_1)$
	$E_c(2,0)+\mathcal{E}_e(20,{}^1A_1)$	$\sqrt{6}\,(2F_2+25F_4)$
	$\sqrt{6}\,(2F_2+25F_4)$	$E_c(0,2)+\mathcal{E}_e(02,{}^1A_1)$

(9.98)

The solutions of (9.98) are

$$\mathcal{E}_{ce}(20,{}^1A_1)=E_c(2,0)+\mathcal{E}_e(20,{}^1A_1)-\sqrt{6}\,(2F_2+25F_4)\cot\theta$$
$$\mathcal{E}_{ce}(02,{}^1A_1)=E_c(0,2)+\mathcal{E}_e(02,{}^1A_1)+\sqrt{6}\,(2F_2+25F_4)\cot\theta \qquad (9.99)$$

where

$$\tan 2\theta=\frac{2\sqrt{6}\,(2F_2+25F_4)}{E_c(2,0)-E_c(0,2)+\mathcal{E}_e(20,{}^1A_1)-\mathcal{E}_e(02,{}^1A_1)}$$
$$=\frac{2\sqrt{6}\,(2F_2+25F_4)}{-20Dq+(2F_2+25F_4)} \qquad (9.100)$$

The wavefunctions corresponding to these energy levels are given [see (9.47)] by

$$|20,{}^1A_1\rangle_{ce}=\sin\theta\,|t_2^2\,{}^1A_1\rangle-\cos\theta\,|e^2\,{}^1A_1\rangle$$
$$|02,{}^1A_1\rangle_{ce}=\cos\theta\,|t_2^2\,{}^1A_1\rangle+\sin\theta\,|e^2\,{}^1A_1\rangle \qquad (9.101)$$

These results follow from the fact that the two solutions of the matrix eigenequation

$$\begin{pmatrix} H_{11}-\mathcal{E} & H_{12} \\ H_{21} & H_{22}-\mathcal{E} \end{pmatrix}\begin{pmatrix} a_1 \\ a_2 \end{pmatrix}=0 \qquad (9.102)$$

are

$$\mathcal{E}=H_{11}-H_{12}\cot\theta, \qquad \begin{pmatrix} a_1 \\ a_2 \end{pmatrix}=\begin{pmatrix} \sin\theta \\ -\cos\theta \end{pmatrix} \qquad (9.103)$$

and

$$\mathcal{E}=H_{22}+H_{12}\cot\theta, \qquad \begin{pmatrix} a_1 \\ a_2 \end{pmatrix}=\begin{pmatrix} \cos\theta \\ \sin\theta \end{pmatrix} \qquad (9.104)$$

where

$$\tan 2\theta=\frac{2H_{12}}{H_{11}-H_{22}}$$

with $\pi\leqslant 2\theta\leqslant\pi/2$ for the negative values of $\tan 2\theta$, and $3\pi/2\leqslant 2\theta<\pi$ for the positive values.

From (9.100) we see that $2\theta\to\pi$ in the limit $Dq\to\infty$, and hence $\cot\theta$ vanishes in this limit. Then from (9.99) we see that

$$\mathcal{E}_{ce}(20,{}^1A_1)\to E_c(2,0)+\mathcal{E}_e(20,{}^1A_1)$$

and

$$\mathcal{E}_{ce}(02,{}^1A_1) \rightarrow E_c(0,2) + \mathcal{E}_e(02,{}^1A_1)$$

in the limit $Dq \rightarrow \infty$, thereby justifying these labels.

For large Dq we have

$$E_{c+e}(20,{}^1A_1)[\equiv E_0 + \mathcal{E}_c(2,0) + \mathcal{E}_e(20,{}^1A_1)]$$
$$< E_{c+e}(02,{}^1A_1)[\equiv E_0 + \mathcal{E}_c(0,2) + \mathcal{E}_e(02,{}^1A_1)]$$

so that $\tan 2\theta$ is negative and $\pi \leqslant 2\theta \leqslant \pi/2$. Then $\cot\theta$ is positive and hence, according to (9.91),

$$\mathcal{E}_{ce}(20,{}^1A_1) < \mathcal{E}_c(2,0) + \mathcal{E}_e(20,{}^1A_1)$$

and

$$\mathcal{E}_{ce}(02,{}^1A_1) > \mathcal{E}_c(0,2) + \mathcal{E}_e(02,{}^1A_1)$$

For small Dq, however,

$$E_{c+e}(20,{}^1A_1) > E_{c+e}(02,{}^1A_1)$$

so that $\tan 2\theta$ is positive ($3\pi/2 \leqslant 2\theta \leqslant \pi$) and $\cot\theta$ is negative. Thus, according to (9.99),

$$\mathcal{E}_{ce}(20,{}^1A_1) > \mathcal{E}_c(2,0) + \mathcal{E}_e(20,{}^1A_1)$$

and

$$\mathcal{E}_{ce}(02,{}^1A_1) < \mathcal{E}_c(0,2) + \mathcal{E}_e(02,{}^1A_1)$$

Thus we have the following general conclusion: If, according to the approximate treatment, we have two different terms with the same $S\Gamma$ (in two different configurations), the effect of the exact treatment is to depress the lower term and to raise the higher one, which amounts to saying that two different terms with the same $S\Gamma$ repel each other through the configuration interaction. A proof of this is included in Appendix X.

The second, or ${}^1E$, term occurs in t_2^2 and e^2. Hence we have a two-dimensional ${}^1E$-block with the diagonal elements

$$\mathcal{E}_c(2,0) + \mathcal{E}_e(20,{}^1E) = -8Dq + F_0 + F_2 + 16F_4$$
$$\mathcal{E}_c(0,2) + \mathcal{E}_e(02,{}^1E) = 12Dq + F_0 + 21F_4$$

The nondiagonal element between the t_2^2 and e^2 configurations is given by

$$\begin{aligned}\langle t_2^{2\,1}E\epsilon|V_e|e^{2\,1}E\epsilon\rangle &= \tfrac{1}{2}\langle \xi^+\xi^- - \eta^+\eta^-|V_e|\theta^+\epsilon^- + \theta^-\epsilon^+\rangle \\ &= \tfrac{1}{2}[\langle\xi\xi\|\theta\epsilon\rangle + \langle\xi\xi\|\epsilon\theta\rangle - \langle\eta\eta\|\theta\epsilon\rangle - \langle\eta\eta\|\epsilon\theta\rangle] \\ &= \{\xi\theta,\xi\epsilon\} - \{\eta\theta,\eta\epsilon\} \\ &= -2\sqrt{3}\,F_2 + 10\sqrt{3}\,F_4\end{aligned}$$

To obtain the energies and the eigenfunctions for the ^1E term, including the configuration interaction, we proceed in exactly the same way as we did for the $^1A_1$ term. Similarly, we shall proceed in the same way for the $^3T_1$ and $^1T_2$ terms because these terms occur in t_2^2 and t_2e. It would be a good exercise for the reader to derive the energies and the eigenfunctions for these cases.

Using all these results, we can now easily obtain the energies of the terms of the d^2 configuration, which separates into different $S\Gamma$-blocks as shown in this section. Exactly similar procedures can be adopted to derive the energies for other $d^n (n=3$, 4, and 5) configurations which are given in terms of the cubic field parameter, $10Dq$, and the Racah parameters, A, B, and C. We shall not give these details here; the reader is referred to the excellent book by Sugano, Tanabe, and Kamimura.[2]

9.4 SOME USEFUL PROPERTIES OF THE MATRIX ELEMENTS OF V_e

Because of (9.49) we have

$$\langle t_2^n(S_1\Gamma_1)e^m(S_2\Gamma_2), S\Gamma M\gamma|V_e|t_2^{n-k}(S_1'\Gamma_1')e^{m+k}(S_2'\Gamma_2'), S\Gamma M\gamma\rangle = 0 \qquad (9.105)$$

for $|k|>2$. Using this result, we can at once predict that many of the matrix elements encountered in the preceding sections will vanish. For example,

$$\left.\begin{aligned}&\langle t_2^{3\,2}ES\gamma|V_e|e^{3\,2}ES\gamma\rangle = 0\\ &\langle t_2^{4\,3}T_1S\gamma|V_e|t_2e^{3\,3}T_1S\gamma\rangle = 0\\ &\langle t_2^{4\,1}T_2\gamma|V_e|t_2e^{3\,1}T_2\gamma\rangle = 0\\ &\langle t_2^{4\,1}A_1|V_e|e^{4\,1}A_1\rangle = 0\\ &\langle t_2^3(^1E)e,{}^1A_1|V_e|e^{4\,1}A_1\rangle = 0\\ &\langle t_2^{5\,2}T_2S\gamma|V_e|t_2^2(^1T_2)e^{3\,2}T_2|S\gamma\rangle = 0\\ &\text{etc.}\end{aligned}\right\} \qquad (9.106)$$

According to (9.50), we have

$$\langle t_2^n(S_1\Gamma_1)e^m(S_2\Gamma_2), S\Gamma M\gamma|V_e|t_2^{n'}(S_1'\Gamma_1')e^{m'}(S_2'\Gamma_2'), S\Gamma M\gamma\rangle$$
$$= \mathcal{C}(nS_1\Gamma_1, mS_2\Gamma_2, n'S_1'\Gamma_1', m'S_2'\Gamma_2', S\Gamma) \tag{9.107}$$

Now let us write

$$\langle t_2^{6-n}(S_1\Gamma_1)e^{4-m}(S_2\Gamma_2), S\Gamma M\gamma|V_e|t_2^{6-n'}(S_1'\Gamma_1')e^{4-m'}(S_2'\Gamma_2'), S\Gamma M\gamma\rangle$$
$$= \mathcal{C}'(nS_1\Gamma_1, mS_2\Gamma_2, n'S_1'\Gamma_1', m'S_2'\Gamma_2', S\Gamma) \tag{9.108}$$

In the light of the discussion at the end of Section 9.1, it can be shown[2] that

$$\mathcal{C}'(nS_1\Gamma_1, mS_2\Gamma_2, n'S_1'\Gamma_1', m'S_2'\Gamma_2', S\Gamma)$$
$$= (G_0 - NG_1)\delta_{nn'}\delta_{mm'}\delta(S_1\Gamma_1, S_1'\Gamma_1')\delta(S_2\Gamma_2, S_2'\Gamma_2')$$
$$+ (-1)^{nm}(-1)^{n'm'}\mu_1\mu_2\mu_1'\mu_2'$$
$$\cdot \mathcal{C}(nS_1\Gamma_1, mS_2\Gamma_2, n'S_1'\Gamma_1', m'S_2'\Gamma_2', S\Gamma) \tag{9.109}$$

where

$$G_0 = \langle \xi^2\eta^2\zeta^2\theta^2\epsilon^2|V_e|\xi^2\eta^3\zeta^2\theta^2\epsilon^2\rangle \tag{9.110}$$

with

$$\xi^2 = \xi^+\xi^-, \text{ etc.}$$

and

$$G_1 = \sum_{j=1}^{10} \left[C(\phi_i\phi_j) - J(\phi_i\phi_j) \right] \tag{9.111}$$

with

$$\phi_i, \phi_j = \xi^+, \eta^+, \zeta^+, \theta^+, \epsilon^+, \xi^-, \eta^-, \zeta^-, \theta^-, \epsilon^-$$

It is easy to evaluate (9.110) in terms of the Racah parameters. We can show that

$$G_0 = 5(9A - 14B + 7C) \tag{9.112}$$

It can also be shown that G_1 is independent of the particular spin-orbital ϕ_i of (9.111), and we find that

$$G_1 = 9A - 14B + 7C = \tfrac{1}{5}G_0 \tag{9.113}$$

The constants μ_1, μ_1' and μ_2, μ_2' appearing in (9.109) were defined at the end of Section 9.1.

Relation (9.109) can be conveniently used for the d^5 system, in which the configurations $t^{n}_2e^m$ and $t_2^{6-n}e^{4-m}$ always occur in pairs $(n+m=5)$. Thus, if the matrix element (9.107) is calculated, the associated matrix element (9.108) will also be known because of (9.109). We note that, in this case, $G_0-NG_1=G_0-5G_1=0$, because of (9.113), and also $nm+n'm'=n(5-n)+n'(5-n')$, which is always even, so that $(-1)^{nm}(-1)^{n'm'}$ is always $+1$. Thus, for the d^5 system, relation (9.109) reduces to

$$\mathcal{C}'(nS_1\Gamma_1, mS_2\Gamma_2, n'S_1'\Gamma_1', m'S_2'\Gamma_2', S\Gamma) = \mu_1\mu_2\mu_1'\mu_2'\mathcal{C}(nS_1\Gamma_1, mS_2\Gamma_2, n'S_1'\Gamma_1', m'S_2'\Gamma_2', S\Gamma) \tag{9.114}$$

9.5 ENERGY LEVELS OF THE d^{10-N} SYSTEM $(N<5)$ AND THE CONCEPT OF HOLES

According to (9.40) and (9.41), the terms of the d^{10-N} system are actually the terms of the d^N system, that is, corresponding to a term $\{t_2^n(S_1\Gamma_1)e^m(S_2\Gamma_2), S\Gamma\}$ of the d^N system $(n+m=N)$, there is a term $\{t_2^{6-n}(S_1\Gamma_1)e^{4-m}(S_2\Gamma_2), S\Gamma\}$ of the d^{10-N} system. The matrix elements (9.108) for the d^{10-N} system are obtained from the corresponding elements (9.107) for the d^N system by using relation (9.109). In fact, in the present case $(N<5)$ the factor $(-1)^{nm}(-1)^{n'm'}\mu_1\mu_2\mu_1'\mu_2'$ will always turn out to be $+1$ for the nonvanishing elements. Thus we find that

$$\begin{aligned}\mathcal{C}'(nS_1\Gamma_1, mS_2\Gamma_2, n'S_1'\Gamma_1', m'S_2'\Gamma_2', S\Gamma) &= (G_0-NG_1)\delta_{nn'}\delta_{mm'}\delta(S_1\Gamma_1, S_1'\Gamma_1')\delta(S_2\Gamma_2, S_2'\Gamma_2') \\ &\quad + \mathcal{C}(nS_1\Gamma_1, mS_2\Gamma_2, n'S_1'\Gamma_1', m'S_2'\Gamma_2', S\Gamma)\end{aligned} \tag{9.115}$$

for $n+m=n'+m'=N(N<5)$.

Also, we have

$$\begin{aligned}\mathcal{E}_c(6-n, 4-m) &= -4Dq(6-n)+6Dq(4-m) \\ &= -(-4n+6m)Dq \\ &= -\mathcal{E}_c(n,m)\end{aligned} \tag{9.116}$$

Relations (9.115) and (9.116) imply that the $S\Gamma$-block for the d^{10-N} system $(N<5)$ will be obtained from the $S\Gamma$-block for the d^N system just by changing the sign of Dq and adding the constant term (G_0-NG_0) to the diagonal elements. This, in turn, implies that the term energies for d^{10-N} will be obtained from the corresponding term energies for d^N by changing the sign of Dq and adding the constant term (G_0-NG_1). Now, the change of sign of Dq is equivalent to the change of sign of the ligand field potential V_c.

Hence we can say that the d^{10-N} electron system behaves just like a d^N positron system, except for an energy shift as a whole. Such fictitious positrons with arbitrary mass are usually called holes.

9.6 VARIATION OF THE ENERGY LEVELS

According to the results derived in Sections 9.4 and 9.5, we see that for a given electron system d^N a constant additive term $G(N)$ appears in each diagonal element of every $S\Gamma$-block:

$$\left.\begin{aligned} G(2)&=A\\ G(3)&=3A\\ G(4)&=6A\\ G(5)&=10A\\ G(6)&=6A+G_0-4G_1\\ G(7)&=3A+G_0-3G_1\\ G(8)&=A+G_0-2G_1 \end{aligned}\right\} \tag{9.117}$$

We also observe that the parameter A appears only in this constant term $G(N)$ and nowhere else in a $S\Gamma$-block. Then the expression for a term energy of d^N will be of the following form:

$$\begin{aligned} E_{ce}(nmi, S\Gamma) = N\,\mathcal{E}^0_{\bar{n}d} + G(N) \\ + \bar{\mathcal{E}}_{ce}(nmi, S\Gamma; B, C, Dq) \end{aligned} \tag{9.118}$$

Now we cannot determine experimentally the absolute value of a term energy. We can find only the transition energy:

$$\begin{aligned} \Delta_{ce}(nmi, S\Gamma \to n'm'i', S'\Gamma') &= E_{ce}(n'm'i', S'\Gamma') - E_{ce}(nmi, S\Gamma) \\ &= \bar{\mathcal{E}}_{ce}(n'm'i', S'\Gamma'; B, C, Dq) \\ &\quad - \bar{\mathcal{E}}_{ce}(nmi, S\Gamma; B, C, Dq) \end{aligned} \tag{9.119}$$

which depends only on the three parameters B, C, and Dq. Then it is important to find theoretically the variation of a relative term energy $\bar{\mathcal{E}}_{ce}(nmi, S\Gamma; B, C, Dq)$ with these three parameters. Since the elements of the $S\Gamma$-block contain the parameters in linear combinations, $\bar{\mathcal{E}}_{ce}(nmi, S\Gamma)/B$ (the relative term energy in the unit of B) will depend on two parameters, $\gamma = C/B$ and Dq/B. From the spectroscopically determined values of B and C for the free iron-group ions ($\bar{n}=3$), we observe[2] that the values of γ fall in the relatively narrow range from 4 to 5, that is, γ is almost independent of both the atomic number and the number of electrons of the iron-group ions. This fact may also be theoretically explained[2] by assuming the single d-electron wavefunction in an iron-group ion to be a hydrogen-like wavefunction.

It should be emphasized that the radial functions $R_{\bar{n}d}(r)$ in crystals and complex ions are not necessarily equal to the radial functions of the $\bar{n}d$ orbitals in the free ions. However, it seems reasonable to assume that the deviation from the free-ion orbitals is not large, and that, in particular, the deviation of γ in crystals (or complexes) from that in free ions will be small. We can assume γ to remain constant at the free-ion value and find the variation of $\mathcal{E}_{ce}(nmi, S\Gamma)/B$ with Dq/B. From the original $S\Gamma$-blocks we obtain the corresponding reduced $S\Gamma$-blocks by subtracting the constant term $G(N)$ from the diagonal elements and dividing all the elements by B. The eigenvalues of a reduced $S\Gamma$-block are $\overline{\mathcal{E}}_{ce}(nmi, S\Gamma)/B$. Tanabe and Sugano computed these for each d^N system, and the results have been represented graphically, showing their continuous variation with Dq/B. In these calculations appropriate free-ion (empirical) values for γ were used for the different d^N systems. Thus, for example, in the calculations for the d^3 system the value of γ used was 4.50, which is spectroscopic free-ion value in Cr^{3+}. In this graphical representation of Tanabe and Sugano, the energies for a d^N system at a given value of Dq are always measured from the lowest energy level of the system at that value of Dq. We shall not deal with the octahedral field any further; the reader is referred to the excellent book by Sugano, Tanabe, and Kamimura[2] for details. Let us now proceed to the fields of lower symmetry.

9.7 FIELDS OF LOWER SYMMETRY

We shall now consider the case of a ligand field with a symmetry lower than that of an octahedral field having O_h symmetry. In reality, it is almost impossible to find a perfect octahedral system. There are many crystals and complexes in which, to a first approximation, the transition metal ions are surrounded octahedrally by ligands. Accurate measurements reveal, however, that the site symmetry of the metal ion is lower than cubic. Deviations from octahedral symmetry can occur in various ways, giving rise to lower symmetries such as tetragonal ($D_{4h}, D_4, C_4, C_{4h}, C_{4v}$), trigonal ($D_{3d}, D_3, C_{3v}$), and rhombic. From the perfect octahedral system we can obtain a D_{4h} system (with the fourfold axis along the z-axis) through one or both of the following two distortions:

1. Shifting the two ligands on the $\pm z$-axes by equal and opposite amounts along the z-axis.
2. Replacing the two ligands on the $\pm z$-axes by two other ligands (X), so that the formula for the complex becomes, for example, *trans*-ML_4X_2.

Again, from the perfect octahedral arrangement we can also obtain a D_{3d} system by displacing the three ligands on the $+x, +y, +z$-axes by δ along the (1 1 1) direction and displacing the other three ligands by the same amount along the $(\bar{1}\,\bar{1}\,\bar{1})$ direction. We have already given these details in Chapter 6, and there is no reason to repeat them here. Let us now study the splitting of the energy levels by the crystal field of tetragonal and trigonal symmetries in the strong-field coupling scheme.

9.7.1 The *SD*-Blocks in the Lower Symmetry Group

The ligand-field potential acting on the central metal atom (ion) can now be written as

$$V_G = V_{O_h} + V_G^0 \tag{9.120}$$

where V_{O_h} is the potential due to octahedral symmetry, and V_G^0 is a pure lower symmetry potential containing no octahedral part.

For a d^N system we have the usual octahedral terms, $\{t_2^n(S_1\Gamma_1)\, e^m(S_2\Gamma_2), S\Gamma\}$, which may be abbreviated as $\{nmi, S\Gamma\}, i$ representing the pair $S_1\Gamma_1, S_2\Gamma_2$. Now, the descent in symmetry $O_h \to G$ causes the reduction of an octahedral representation Γ into the representations $D^{(\alpha)}$ of the lower symmetry group G:

$$\Gamma \to \sum_{\alpha} a_{\alpha}^{\Gamma} D^{(\alpha)} \tag{9.121}$$

For the present situation no lower symmetry representation can occur more than once in such a reduction, that is, $a_{\alpha}^{\Gamma} \subset (0,1)$. For example, the reductions under descents in symmetry $O_h \to D_{4h}$ and $O_h \to D_{3d}$ are given in Table 9.10. Obviously, an even representation will always be reduced to even representations, and an odd representations to odd ones. Hence the suffixes g and u have been omitted in Table 9.10.

The energy levels of the d^N system under the lower symmetry potential V_G^0 will be classified into the lower symmetry terms, *SD*, as follows.

Table 9.10

Descents in symmetry from O_h

$O_h \to D_{4h}$ (four-fold axis along z)	$O_h \to D_{3d}$ (three-fold axis along (1 1 1))
$A_1 \to A_1$	$A_1 \to A_1$
$A_2 \to B_1$	$A_2 \to A_2$
$E \to A_1 + B_1$	$E \to E$
$T_1 \to A_2 + E$	$T_1 \to A_2 + E$
$T_2 \to B_2 + E$	$T_2 \to A_1 + E$

For a given S we collect all the octahedral terms $\{nmi, S\Gamma_D\}$ involving the octahedral representations Γ_D, each of which gives the particular lower symmetry representation D on reduction according to (9.121). Then, for each term $\{nmi, S\Gamma_D\}$, we can find a linear combination of the associated wave-functions:

$$\begin{array}{l} nmi, S\Gamma_D \rightarrow S\,D\,M\,\delta\rangle = \sum\limits_{\gamma} \alpha(\delta D, \Gamma_D \gamma)|nmi, S\Gamma_D M\gamma\rangle \\ \qquad O_h \rightarrow G \end{array} \tag{9.122}$$

which acts on the δ-basis functions of the representation D in G.

The matrix of V_G^0 with respect to functions (9.122) will be diagonal in S, M, D, δ:

$$\begin{array}{l} \langle nmi, S\Gamma_D \rightarrow S\,D\,M\,\delta | V_G^0 | n'm'i', S'\Gamma'_{D'} \rightarrow S'\,D'\,M'\,\delta' \rangle \\ \qquad O_h \rightarrow G \qquad\qquad\qquad\qquad O_h \rightarrow G \\ \qquad = B'(nmi\Gamma_D, n'm'i'\Gamma'_{D'}, SD)\,\delta_{SS'}\delta_{MM'}\delta_{DD'}\delta_{\delta\delta'} \end{array} \tag{9.123}$$

The matrix of V_{O_h} with respect to functions (9.122) will be completely diagonal, that is,

$$\begin{array}{l} \langle nmi, S\Gamma_D \rightarrow S\,D\,M\,\delta | V_{O_h} | n'm'i', S'\Gamma'_{D'} \rightarrow S'\,D'\,M'\,\delta' \rangle \\ \qquad O_h \rightarrow G \qquad\qquad\qquad\qquad O_h \rightarrow G \\ \qquad = \mathcal{E}_C(n,m)\,\delta_{nn'}\delta_{mm'}\delta_{ii}\delta_{ss''}\delta_{MM'}\delta_{DD'}\delta_{\delta\delta'}\delta_{\Gamma_D\Gamma'_{D'}} \end{array} \tag{9.124}$$

where $\mathcal{E}_C(n,m)$ is as given in (9.48).

Finally, the matrix of V_e with respect to functions (9.122) will be diagonal in $S, M, D, \delta, \Gamma_D$:

$$\begin{array}{l} \langle nmi, S\Gamma_D \rightarrow S\,D\,M\,\delta | V_e | n'm'i', S'\Gamma'_{D'} \rightarrow S'\,D'\,M'\,\delta' \rangle \\ \qquad O_h \rightarrow \Gamma \qquad\qquad\qquad\qquad O_h \rightarrow G \\ \qquad = \mathcal{C}(nmi, n'm'i', S\Gamma_D)\,\delta_{SS'}\delta_{MM'}\delta_{DD'}\delta_{\delta\delta'}\delta_{\Gamma_D\Gamma'_{D'}} \end{array} \tag{9.125}$$

where this $\mathcal{C}(\cdots)$ is the same $\mathcal{C}(\cdots)$ appearing in (9.50).

From relations (9.123) to (9.125) we see that the matrix of

$$V_G + V_e \equiv V_{O_h} + V_G^0 + V_e$$

with respect to functions (9.122) is in the block form, $(2S+1)\cdot\lambda(D)$, identical blocks being associated with a given lower symmetry term SD. These may be called the SD-blocks. The rows (and columns) in an SD-block are labeled by $\{nmi, \Gamma_D\}$. By diagonalizing a SD-block we obtain the actual energy levels, $E_{Ge}(nmi, S\Gamma_D \rightarrow SD)$, corresponding to the lower symmetry SD term, and these levels are $(2S+1)\cdot\lambda(D)$-fold degenerate. The label $\{nmi,\Gamma_D\}$ in $E_{Ge}(nmi,S\Gamma_D\rightarrow SD)$ is chosen in such a way that, at $V_G^0=0$, $E_{Ge}(nmi,S\Gamma_D\rightarrow SD)$ reduces to the octahedral energy level $E_{ce}(nmi,S\Gamma_D)$. Thus an octahedral level $E_{ce}(nmi,S\Gamma)$ will be split under V_G^0 into the levels $E_{Ge}(nmi,S\Gamma\rightarrow SD_\Gamma^{(\alpha)}), \alpha=1,2,\ldots,$ where we have the reduction $\Gamma=\Sigma_\alpha D_\Gamma^{(\alpha)}$, under the descent $O_h\rightarrow G$.

The procedure described above gives the exact energy levels. However, we can apply approximate treatments in the following special cases.

(1) $$V_{O_h}\gg V_G^0, V_e$$

In this case we can neglect both $B'(\cdots)$ and $\mathcal{C}(\cdots)$ for $n'\neq n$ and/or $m'\neq m$, since their absolute values will be much smaller than $|\mathcal{E}_C(n,m)-\mathcal{E}_C(n'm')|$. Then each SD-block of V_G+V_e is reduced into (nm,SD)-subblocks, and by diagonalizing them, we obtain the energy levels as

$$E_{C+G^0e}(nmi,S\Gamma_D\rightarrow SD)=E_c(n,m)+\mathcal{E}_{G^0e}(nmi,S\Gamma_D\rightarrow SD) \quad (9.126)$$

where $\mathcal{E}_{G^0e}(\cdots)$ are the characteristic roots of the (nm, SD)-subblock with the elements

$$\langle \underset{O_h\rightarrow G}{nmi,S\Gamma_D\rightarrow S\,D\,M\,\delta}|V_G^0+V_e|\underset{O_h\rightarrow G}{nmi',S\Gamma'_D\rightarrow S\,D\,M\,\delta}\rangle$$
$$=B'(nmi\Gamma_D,nmi',SD)+\mathcal{C}(nmi,nmi',S\Gamma_D)\,\delta_{\Gamma_D\Gamma'_D} \quad (9.127)$$

Physically we might say that the energy levels (9.126) are obtained by considering the splitting of each strong-field octahedral level $E_c(n,m)$ separately under the effect of $(V_G^0+V_e)$, this splitting being, for the present case, much smaller than the separation between the adjacent $E_c(n,m)$ levels. Again, the labels i and Γ_D in $\mathcal{E}_{G^0e}(nmi,S\Gamma_D\rightarrow SD)$ are chosen in such a way that, at $V_G^0=0, \mathcal{E}_{G^0e}(nmi,S\Gamma_D\rightarrow SD)$ coincides with $\mathcal{E}_e(nmi,S\Gamma_D)$ [recall (9.54) and the associated discussions].

(2) $$V_G^0\ll V_e\ll V_{O_h}$$

In this case we neglect $B'(\cdots)$ for $n'm'\neq nm$ and/or $\Gamma'_D\neq\Gamma_D$. Also we neglect $\mathcal{C}(\cdots)$ for $n'm'\neq nm$. Then each SD-block of $V_{O_h}+V_G^0+V_e$ is reduced into $(nm\Gamma_D,SD)$-subblocks. By diagonalizing them, we obtain the energies as

$$E_{C+e+G^0}(nmi,S\Gamma_D\rightarrow SD)=E_c(n,m)+\mathcal{E}_{e+G^0}(nmi,S\Gamma_D\rightarrow SD) \quad (9.128)$$

where $\mathcal{E}_{e+G^0}(nmi, S\Gamma_D \to SD)$ are the characteristic roots of the $(nm\Gamma_D, SD)$-subblock with the elements

$$\langle nmi, S\Gamma_D \to S D M \delta | V_G^0 + V_e | nmi', S\Gamma_D \to S D M \delta \rangle$$

$$O_h \to G \qquad\qquad\qquad\qquad O_h \to G$$

$$= B'(nmi\Gamma_D, nmi'\Gamma_D, SD) + \mathcal{C}(nmi, nmi', S\Gamma_D) \qquad (9.129)$$

The wavefunction corresponding to the root $\mathcal{E}_{e+G_0}(nmi, S\Gamma_D \to SD)$ will also be obtained as

$$|nm, S\Gamma_D \to S D M \delta \rangle_i = \sum_j a_j^i |nmj, S\Gamma_D \to S D M \delta \rangle$$

$$O_h \to G \qquad\qquad\qquad\qquad O_h \to G$$

$$= \sum_\gamma a(\delta D, \Gamma_D \gamma) \sum_j a_j^i |nmj, S\Gamma_D M \gamma \rangle \qquad (9.130)$$

by using (9.122).

Instead of diagonalizing matrix (9.129), we could diagonalize a bigger matrix:

$$\langle nmi, S\Gamma_D M \gamma | V_G^0 + V_e | nmi', S\Gamma_D M \gamma' \rangle \qquad (9.131)$$

and obtain the characteristic roots $\mathcal{E}_{e+G^0}(nmi, S\Gamma_D \to SD^{(\alpha)})$ for different i and different lower symmetry representations $D^{(\alpha)}, \alpha = 1, 2, \ldots$, that are contained in the O_h representation denoted by Γ_D. The roots will be $\lambda(D^{(\alpha)})$-fold degenerate, and the corresponding $\lambda(D^{(\alpha)})$ basis functions will be obtained, according to (9.130), as

$$|n, m, S\Gamma_D \to S D^{(\alpha)} M \delta \rangle_i = \sum_\gamma a(\delta D^{(\alpha)}, \Gamma_D \gamma) \sum_j a_j^i |nmj, S\Gamma_D M \gamma \rangle$$

$$(9.132)$$

Thus from (9.132) we can obtain the coefficients appearing in combinations (9.122). This is a useful method of constructing, from the basis functions of a given octahedral representation Γ, the linear combinations that act as the basis functions of the lower symmetry representation $D^{(\alpha)}$ contained in Γ_D. This method is particularly simple if, for a given Γ_D, we can find any system d^N that involves a congifuration $t_2^n e^m$ giving rise to a single $S\Gamma_D$ term with any suitable S. We shall find an application of this method later.

9.7.2 V_G^0 as a Linear Combination of the Basis Functions for the O_h Symmetric Group

In general, any lower symmetry potential, denoted by V_G^0, will be given by a linear combination of the basis functions for the O_h group, that is,

$$V_G^0 = \sum_{\bar{\Gamma}\bar{\gamma}} a_{\bar{\gamma}}^{\bar{\Gamma}} V_{\bar{\gamma}}^{\bar{\Gamma}} \tag{9.133}$$

where $V_{\bar{\gamma}}^{\bar{\Gamma}}$ is a $\bar{\gamma}$-basis function (in general, not normalized) for the irreducible representation $\bar{\Gamma}$ of O_h, and the $a_{\bar{\gamma}}^{\bar{\Gamma}}$ are just numbers. Here $\bar{\Gamma}$ is a representation that yields the unit representation of G on reduction through the descent $O_h \to G$. Of course, $\bar{\Gamma} \neq A_{1g}$, since V_G^0 does not contain an octahedral part.

From Table 9.10 we see that E_g in O_h yields the unit representation A_{1g} in D_{4h}. Hence $V_{D_{4h}}^0$ will be a suitable linear combination of two basis functions spanning E_g in O_h. In fact, the θ-basis function of E_g in O_h belongs to A_{1g} in D_{4h}. Thus, for example, $\theta = d_0 \sim d_{z^2}$, as given by (9.3), remains invariant under the operations of D_{4h} (i.e., tetragonal symmetry). Then we have

$$V_{D_{4h}}^0 = V_\theta^{E_g} \tag{9.134}$$

Similarly, from Table 9.10 we see that T_{2g} in O_h yields A_{1g} in D_{3d} (i.e., trigonal symmetry). Hence $V_{D_{3d}}^0$ will be a suitable linear combination of three basis functions spanning T_{2g} of O_h. Now, according to (9.2), we see that $\xi + \eta + \zeta \sim xy + yz + zx$ remains invariant under the operations of D_{3d}. Hence $V_{D_{3d}}^0$ will be given by

$$V_{D_{3d}}^0 = V_\xi^{T_{2g}} + V_\eta^{T_{2g}} + V_\zeta^{T_{2g}} \tag{9.135}$$

The D_3 group consists of all the operations of D_{3d} except the inversion, and hence the unit representation A_1 of D_3 will be obtained on the reduction of either T_{2g} or T_{2u} of O_h. Therefore $V_{D_3}^0$ will be given by

$$V_{D_3}^0 \left(V_\xi^{T_{2g}} + V_\eta^{T_{2g}} + V_\zeta^{T_{2g}} \right) + \left(V_\xi^{T_{2u}} + V_\eta^{T_{2u}} + V_\zeta^{T_{2u}} \right) \tag{9.136}$$

Note that the second part of (9.136) is of odd parity. For the same reason as given above, we have

$$V_{D_4}^0 = V_\theta^{E_g} + V_\theta^{E_u} \tag{9.137}$$

The cubic (point group symmetry O) and tetrahedral (point group symmetry T_d) crystal fields are given by

$$V_O^0 \qquad \text{or} \qquad V_{T_d}^0 = V^{A_{1g}} + V^{A_{1u}} \tag{9.138}$$

An orthorhombic (C_{2v}, D_2, or D_{2h}) system will be obtained from the octahedral system by making distortions along

(i) the [100] and [010] directions

or

(ii) the [110] and [1$\bar{1}$0] directions

The corresponding expressions for the rhombic potentials are, respectively,

$$V^0_{Rh(\text{i})} = V^{E_g}_{\epsilon} + V^{E_u}_{\epsilon}$$

and (9.139)

$$V^0_{Rh(\text{ii})} = V^{T_{2g}}_{\zeta} + V^{T_{2u}}_{\zeta}$$

The odd-parity part in (9.139) is zero or nonzero, according as the rhombic symmetry involved contains a center of inversion or does not. Rhombic symmetry is the lowest symmetry possible in an octahedral complex, and its effects are usually very small.

We note that the odd-parity part of a lower symmetry potential V^0_G [e.g., the potentials (9.136) to (9.139)] *does not* contribute to the matrix elements of V^0_G between octahedral wavefunctions that are of even parity. Hence only the even-parity part of V^0_G is responsible for the splittings and shifts of the octahedral terms of a d^N system. However, the odd-parity part of V^0_G is responsible[2] for slightly allowing the parity-forbidden electric-dipole transitions, with which we shall not deal (see Chapter 12).

9.7.3 Matrix Elements of V^0_G (Application of Wigner-Eckart Theorem)

Because of (9.133) the matrix element of V^0_G between two octahedral states will be of the following form, according to the Wigner-Eckart theorem, given in Chapter 5:

$$\begin{aligned}\langle t^n_{2g}(S_1\Gamma_1)e^m(S_2\Gamma_2), S\Gamma M\gamma | V^0_G | t^{n'}_2(S'_1\Gamma'_1)e^{m'}(S'_2\Gamma'_2), S'\Gamma' M'\gamma'\rangle \\ = \delta_{SS'}\delta_{MM'} \sum_{\bar{\Gamma}\bar{\gamma}} a^{\bar{\Gamma}}_{\bar{\gamma}} [\lambda(\Gamma)]^{-1/2} \langle \Gamma\gamma | \Gamma'\gamma'\bar{\Gamma}\bar{\gamma}\rangle \\ \cdot \langle t^n_2(S_1\Gamma_1)e^m(S_2\Gamma_2), S\Gamma \| V^{\bar{\Gamma}} \| t^{n'}_2(S'_1\Gamma'_1)e^{m'}(S'_2\Gamma'_2), S'\Gamma'\rangle\end{aligned} \quad (9.140)$$

The C-G coefficients have the properties

$$[\lambda(\Gamma)]^{-1/2}\langle\Gamma\gamma|\Gamma'\gamma'\bar{\Gamma}\bar{\gamma}\rangle = \epsilon(\Gamma\bar{\Gamma}\Gamma')[\lambda(\Gamma')]^{-1/2}\langle\Gamma'\gamma'|\Gamma\gamma|\bar{\Gamma}\bar{\gamma}\rangle \quad (9.141)$$

where

$$\epsilon(\Gamma A_1 \Gamma') = \epsilon(\Gamma T_2 \Gamma') = 1 \tag{9.142a}$$

$$\epsilon(\Gamma A_2 \Gamma') = \epsilon(\Gamma T_1 \Gamma') = -1 \tag{9.142b}$$

and

$$\epsilon(\Gamma E \Gamma') = 1 \tag{9.142c}$$

except for

$$\epsilon(T_1 E T_2) = \epsilon(T_2 E T_1) = -1 \tag{9.142d}$$

For real $V_{\bar{\gamma}}^{\bar{\Gamma}}$ we have

$$\langle \alpha \Gamma \gamma | V_{\bar{\gamma}}^{\bar{\Gamma}} | \alpha' \Gamma' \gamma' \rangle = \langle \alpha' \Gamma' \gamma' | V_{\bar{\gamma}}^{\bar{\Gamma}} | \alpha \Gamma \gamma \rangle$$

This gives, by using the Wigner-Eckart theorem and (9.141), the following relation:

$$\langle \alpha \Gamma \| V^{\bar{\Gamma}} \| \alpha' \Gamma' \rangle = \epsilon(\Gamma \bar{\Gamma} \Gamma') \langle \alpha' \Gamma' \| V^{\bar{\Gamma}} \| \alpha \Gamma \rangle \tag{9.143}$$

Equations (9.142b) and (9.143) show that

$$\langle \alpha \Gamma \| V^{\bar{\Gamma}} \| \alpha \Gamma \rangle = 0 \qquad \text{for} \quad \bar{\Gamma} = A_2 \text{ and } T_1 \tag{9.144}$$

In (9.133) each potential $V_{\bar{\gamma}}^{\bar{\Gamma}}$ for a many-electron system is given as the sum of the corresponding potentials for individual electrons i:

$$V_{\bar{\gamma}}^{\bar{\Gamma}} = \sum_i v_{\bar{\gamma}}^{\bar{\Gamma}}(\mathbf{r}_i) \tag{9.145}$$

Then, since the many-electron states, $|t_2^n(S_1\Gamma_1)e^m(S_2\Gamma_2), S\Gamma M\gamma\rangle$, are given by linear combination of Slater determinants, the matrix elements in (9.140) can always be expressed in terms of the single-electron reduced matrices $\langle t_2 \| v^{\bar{\Gamma}} \| t_2 \rangle$, $\langle e \| v^{\bar{\Gamma}} \| e \rangle$, and $\langle t_2 \| v^{\bar{\Gamma}} \| e \rangle$. Thus we have the following general results:

$$\langle t_2^n(S_1\Gamma_1)e^m(S_2\Gamma_2), S\Gamma \| V^{\bar{\Gamma}} \| t_2^{n-k}(S_1'\Gamma_1')e^{m+k}(S_2'\Gamma_2'), S\Gamma' \rangle = 0 \qquad \text{for } |k| \geqslant 2 \tag{9.146}$$

$$\langle t_2^n(S_1\Gamma_1)e^m(S_2\Gamma_2), S\Gamma \| V^{\bar{\Gamma}} \| t_2^{n-1}(S_1'\Gamma_1')e^m(S_2'\Gamma_2'), S\Gamma' \rangle = C_0 \langle t_2 \| v^{\bar{\Gamma}} \| e \rangle \tag{9.147}$$

$$\langle t_2^n(S_1\Gamma_1)e^m(S_2\Gamma_2), S\Gamma \| V^{\bar{\Gamma}} \| t_2^n(S_1'\Gamma_1')e^m(S_2'\Gamma_2'), S\Gamma' \rangle = C_1 \langle t_2 \| v^{\bar{\Gamma}} \| t_2 \rangle + C_2 \langle e \| v^{\bar{\Gamma}} \| e \rangle \tag{9.148}$$

where C_0, C_1, and C_2 are numerical coefficients depending on the states of interest and $\bar{\Gamma}$. From (9.148) and (9.144) we see that

$$\langle t_2^n(S_1\Gamma_1)e^m(S_2\Gamma_2), S\Gamma \| V^{\bar{\Gamma}} \| t_2^n(S_1'\Gamma_1')e^m(S_2'\Gamma_2'), S\Gamma' \rangle = 0$$

for (9.149).

$$\bar{\Gamma} = A_2 \text{ and } T_1$$

We now give two examples illustrating relations (9.147) and (9.148). First, we have, according to (9.140),

$$\begin{aligned}\langle t_2^{3\,2}E\tfrac{1}{2}\theta | V_\zeta^{T_2} | t_2^2({}^1T_2)e, {}^2T_2\tfrac{1}{2}\zeta \rangle \\ &= \frac{1}{\sqrt{2}} \langle E\theta | T_2\zeta\, T_2\zeta \rangle \langle t_2^{3\,2}E \| V^{T_2} \| t_2^2({}^1T_2)e, {}^2T_2 \rangle \\ &= \frac{1}{\sqrt{2}} \frac{2}{\sqrt{6}} \langle t_2^{3\,2}E \| V^{T_2} \| t_2^2({}^1T_2)e, {}^2T_2 \rangle \quad (9.150)\end{aligned}$$

Again, by using the expressions for these two wavefunctions from Appendix VIII, we get

$$\begin{aligned}\langle t_2^{3\,2}E\tfrac{1}{2}\theta | V_\zeta^{T_2} | t_2^2({}^1T_2)e, {}^2T_2\tfrac{1}{2}\zeta \rangle \\ &= \tfrac{1}{2} \langle \xi^+\eta^-\zeta^+ - \xi^-\eta^+\zeta^+ | V_\zeta^{T_2} | \xi^+\eta^-\theta^+ - \xi^-\eta^+\theta^+ \rangle \\ &= \langle \zeta | v_\zeta^{T_2} | \theta \rangle \\ &= \frac{1}{\sqrt{3}} \langle t_2 \| v^{T_2} \| e \rangle \langle T_2\zeta | E\theta\, T_2\zeta \rangle \\ &= \frac{1}{\sqrt{3}} \langle t_2 \| v^{T_2} \| e \rangle \quad (9.151)\end{aligned}$$

From (9.150) and (9.151), we get therefore

$$\langle t_2^{3\,2}E \| V^{T_2} \| t_2^2({}^1T_2)e, {}^2T_2 \rangle = \langle t_2 \| v^{T_2} \| e \rangle \quad (9.152)$$

Next, we consider

$$\begin{aligned}\langle t_2^2({}^3T_1)e, {}^4T_2\tfrac{3}{2}\zeta | V_\theta^E | t_2^2({}^3T_1)e, {}^4T_2\tfrac{3}{2}\zeta \rangle \\ &= \frac{1}{\sqrt{3}} \langle t_2^2({}^3T_1)e, {}^4T_2 \| V^E \| t_2^2({}^3T_1)e, {}^4T_2 \rangle\end{aligned}$$

(9.153)

Again, by using the explicit expressions for the wavefunctions (Appendix VIII), we get

$$\langle t_2^2(^3T_1)e, {}^4T_2\tfrac{3}{2}\zeta|V_\theta^E|t_2^2(^3T_1)e, {}^4T_2\tfrac{3}{2}\zeta\rangle$$

$$=\langle \xi^+\eta^+\epsilon^+|V_\theta^E|\xi^+\eta^+\epsilon^+\rangle$$

$$=\langle \xi|v_\theta^E|\xi\rangle+\langle \eta|v_\theta^E|\eta\rangle+\langle \epsilon|v_\theta^E|\epsilon\rangle$$

$$=\frac{1}{\sqrt{3}}\langle t_2\|v^E\|t_2\rangle[\langle t_2\xi|T_2\xi E\theta\rangle+\langle t_2\eta|T_2\eta E\theta\rangle]$$

$$+\frac{1}{\sqrt{2}}\langle e\|v^E\|e\rangle\langle E\epsilon|E\epsilon E\theta\rangle$$

$$=-\frac{1}{\sqrt{3}}\langle t_2\|v^E\|t_2\rangle+\tfrac{1}{2}\langle e\|v^E\|e\rangle \tag{9.154}$$

Thus, from (9.153) and (9.154), we get

$$\langle t_2^2(^3T_1)e, {}^4T_2\|V^E\|t_2^2(^3T_1)e, {}^4T_2\rangle=-\langle t_2\|v^E\|t_2\rangle+\frac{\sqrt{3}}{2}\langle e\|v^E\|e\rangle \tag{9.155}$$

Since the matrix of the ligand field potential is Hermitian and real, we can show, because of (9.40) and (9.41), that the matrix elements of $V_{\bar{\gamma}}^{\bar{\Gamma}}$ in the complementary states are related to each other as follows:

$$\langle t_2^{6-n}(S_1\Gamma_1)e^{4-m}(S_2\Gamma_2), S\Gamma M\gamma|V_{\bar{\gamma}}^{\bar{\Gamma}}|t_2^{6-n'}(S_1'\Gamma_1')e_e^{4-m'}(S_2'\Gamma_2'), S\Gamma' M\gamma'\rangle$$

$$=F_0'\,\delta_{nn'}\delta_{mm'}\delta_{S_1\Gamma_1,S_1'\Gamma_1'}\delta_{S_2\Gamma_2,S_2'\Gamma_2'}-(-1)^{nm}(-1)^{n'm'}$$

$$\cdot\mu_1\mu_2\mu_1'\mu_2'\langle t_2^n(S_1\Gamma_1)e^m(S_2\Gamma_2), S\Gamma M\gamma|V_{\bar{\gamma}}^{\bar{\Gamma}}|t_2^{n'}(S_1'\Gamma_1')$$

$$\cdot e^{m'}(S_2'\Gamma_2'), S\Gamma' M\gamma'\rangle \tag{9.156}$$

where $n+m=n'+m'=N$ and

$$F_0'=2\sum_{\alpha=\xi,\eta,\zeta,\theta,\epsilon}\langle\alpha|v_{\bar{\gamma}}^{\bar{\Gamma}}|\alpha\rangle \tag{9.157}$$

which is a constant appearing in the diagonal elements of matrix (9.156) and has the effect of shifting all the energy levels of the d^{10-N} system by the same amount. Hence we can omit the constant, since, as we mentioned earlier, we are interested only in the relative positions of the energy levels in a given system. Then, from (9.156), we get (by omitting F_0')

$$\langle t_2^{6-n}(S_1\Gamma_1)e^{4-m}(S_2\Gamma_2),S\Gamma\| V^{\bar{\Gamma}}\| t_2^{6-n'}(S_1'\Gamma_1')e^{4-m'}(S_2'\Gamma_2'),S\Gamma'\rangle$$
$$= -(-1)^{nm}(-1)^{n'm'}\mu_1\mu_2\mu_1'\mu_2'\langle t_2^n(S_1\Gamma_1)e^m(S_2\Gamma_2)\|$$
$$\cdot V^{\bar{\Gamma}}\| t_2^{n'}(S_1'\Gamma_1')e^{m'}(S_2'\Gamma_2'),S\Gamma'\rangle \qquad (9.158)$$

For $N\neq 5$ we shall always have

$$(-1)^{nm}(-1)^{n'm'}\mu_1\mu_2\mu_1'\mu_2' = +1$$

and in particular, for $n=n'=3, m'=m=0$, we get from (9.158)

$$\langle t_2^3e^4,S\Gamma\| V^{\bar{\Gamma}}\| t_2^3e^4,S\Gamma'\rangle = -\langle t_2^3 S\Gamma\| V^{\bar{\Gamma}}\| t_2^3 S\Gamma'\rangle$$

From this equation, because of (9.40) and (9.41), we obtain

$$\langle t_2^3 S\Gamma\| V^{\bar{\Gamma}}\| t_2^3 S\Gamma'\rangle = -\mu_1\mu_1'\langle t_2^3 S\Gamma\| V^{\bar{\Gamma}}\| t_2^3 S\Gamma'\rangle$$

From the preceding equation we can show that

$$\langle t_2^3 S\Gamma\| V^{\bar{\Gamma}}\| t_2^3 S\Gamma\rangle = 0$$

and (9.159)

$$\langle t_2^3 S\Gamma\| V^{\bar{\Gamma}}\| t_2^3 S\Gamma'\rangle \neq 0$$

for $S\Gamma={}^2E$, ${}^2T_1$, and $S\Gamma'={}^2T_2$, and vice versa. Equation (9.159) follows since

$$\begin{aligned} &\mu_1 \text{ (or } \mu_1') = -1 &&\text{for}\\ &n \text{ (or } n') = 3 &&\text{and}\\ &S_1\Gamma_1 \text{ (or } S_1'\Gamma_1') = {}^4A_2, {}^2E, {}^2T_1 &&\\ &\mu_1 \text{ (or } \mu_1') = 1 &&\text{otherwise}\end{aligned}$$

Similarly, for $n=n'=0$, $m=m'=2$, we can show that

$$\langle e^2 S\Gamma\| V^{\bar{\Gamma}}\| e^2 S\Gamma\rangle = 0$$

and (9.160)

$$\langle e^2 S\Gamma\| V^{\bar{\Gamma}}\| e^2 S\Gamma'\rangle \neq 0$$

only for the combinations $S\Gamma={}^1E$ and $S\Gamma'={}^1A_1$, and vice versa.

Finally, for $N=5$ we always have $(-1)^{nm}(-1)^{n'm'}=1$, and for $n\neq 3$, $n'\neq 3$ we have $\mu_1=\mu_2=\mu_1'=\mu_2'=1$, so that

$$\langle t_2^{6-n}(S_1\Gamma_1)e^{4-m}(S_2\Gamma_2),S\Gamma\| V^{\bar{\Gamma}}\| t_2^{6-n'}(S_1'\Gamma_1')e^{4-m'}(S_2'\Gamma_2'),S\Gamma'\rangle$$
$$= -\langle t_2^{n}(S_1\Gamma_1)e^{m}(S_2\Gamma_2),S\Gamma\| V^{\bar{\Gamma}}\| t_2^{n'}(S_1'\Gamma_1')e^{m'}(S_2'\Gamma_2'),S\Gamma'\rangle \tag{9.161}$$

For $N=5$, with $n=3, n'=3$, we have

$$\langle t_2^{3}(S_1\Gamma_1)e^{2}(S_2\Gamma_2),S\Gamma\| V^{\bar{\Gamma}}\| t_2^{3}(S_1'\Gamma_1')e^{2}(S_2'\Gamma_2'),S\Gamma'\rangle$$
$$= -\mu_1\mu_2\mu_1'\mu_2'\langle t_2^{3}(S_1\Gamma_1)e^{2}(S_2\Gamma_2),S\Gamma\| V^{\bar{\Gamma}}\| t_2^{3}(S_1'\Gamma_1') \cdot e^{2}(S_2'\Gamma_2'),S\Gamma'\rangle \tag{9.162}$$

In particular, it can be shown that

$$\langle t_2^{3}(S_1\Gamma_1)e^{2}(S_2\Gamma_2),S\Gamma\| V^{\bar{\Gamma}}\| t_2^{3}(S_1\Gamma_1)e^{2}(S_2\Gamma_2),S\Gamma'\rangle=0 \tag{9.163}$$

It can be clearly seen from the equations given above that the diagonal matrix elements of any low-symmetry potential in the states of the half-filled shell configurations, t_2^3, e^2, and $t_2^3e^2$, vanish. This has the important consequence that the spectral lines due to the transitions between the terms of the same half-filled configuration are not broadened by the vibrational fluctuation of the low-symmetry fields.

9.7.4 Some Examples

We shall now work out a few important examples to clarify the ideas contained in this chapter. Let us find the energy levels of a single d-electron under a D_{4h} field.

The two octahedral levels of the d^1 system are ${}^2T_2$ and ${}^2E$. From Table 9.10 we find the reductions, $T_2\rightarrow B_2+E$ and $E\rightarrow A_1+B_1$, under the descent $O_h\rightarrow D_{4h}$. Hence the six-fold degenerate octahedral level ${}^2T_2$ is split into two levels, ${}^2T_2\rightarrow {}^2B_2, {}^2E$, with degeneracies 2 and 4, respectively. The other octahedral level, ${}^2E$, with four-fold degeneracy, is split into ${}^2E\rightarrow {}^2A_1+{}^2B_1$. Since there cannot be any configuration mixing, the two octahedral levels are split individually. We first calculate the splitting of the ${}^2T_2$ level.

The matrix elements of v_{4h}^0 are given, as usual, by

$$\langle t_2 M\gamma|v_{4h}^0|t_2 M\gamma'\rangle=\frac{1}{\sqrt{3}}\langle t_2\|v^E\|t_2\rangle\langle T_2\gamma|T_2\gamma' E\theta\rangle \tag{9.164}$$

Using the $T_2\otimes E$ table of Appendix VI, we obtain the nonvanishing matrix elements in (9.164):

$$\begin{aligned}\langle t_2 M\xi|v^0_{4h}|t_2 M\xi\rangle &= \langle t_2 M\eta|v^0_{4h}|t_2 M\eta\rangle \\ &= -\tfrac{1}{2}\langle t_2 M\zeta|v^0_{4h}|t_2 M\zeta\rangle \\ &= -\frac{1}{2\sqrt{3}}\langle t_2\|v^E\|t_2\rangle \end{aligned} \tag{9.165}$$

Thus matrix (9.164) is diagonal, giving a nondegenerate root and a doubly degenerate root. Their energies are

$$E_{D_{4h}}(^2T_2 \to {}^2B_2) = \mathcal{E}(t_2) + \frac{1}{\sqrt{3}}\langle t_2\|v^E\|t_2\rangle$$

and (9.166)

$$E_{D_{4h}}(^2T_2 \to {}^2E) = \mathcal{E}(t_2) - \frac{1}{2\sqrt{3}}\langle t_2\|v^E\|t_2\rangle$$

Proceeding in an exactly similar way for the octahedral ^2E level, we can show that

$$\begin{aligned} E_{D_{4h}}(^2E \to {}^2A_1) &= \mathcal{E}(e) - \tfrac{1}{2}\langle e\|v^E\|e\rangle \\ E_{D_{4h}}(^2E \to {}^2B_1) &= \mathcal{E}(e) + \tfrac{1}{2}\langle e\|v^E\|e\rangle \end{aligned} \tag{9.167}$$

Let us now find the splittings of the octahedral levels $^2T_2$ and ^2E of the d^1 system under a D_{3d} field. Since, from Table 9.10, we have $T_2 \to A_1 + E$ and $E \to E$, there will be configuration mixing between the split *two* ^2E levels. Let us first consider the case $v^0_{D_{3d}} \ll v_{O_h}$, where we can neglect the configuration mixing. In this case we have

$$\begin{aligned}\langle t_2 M\gamma|v^0_{D_{3d}}|t_2 M\gamma'\rangle = \frac{1}{\sqrt{3}}\langle t_2\|v^{T_2}\|t_2\rangle \\ \sum_{\delta=\xi,\eta,\zeta}\langle T_2\gamma|T_2\gamma' T_2\delta\rangle\end{aligned} \tag{9.168}$$

Using again the $T_2 \otimes T_2$ table of C-G coefficients in Appendix VI, we have

$$\begin{aligned}\langle t_2 M\xi|v^0_{D_{3d}}|t_2 M\eta\rangle &= \langle t_2 M\eta|v^0_{D_{3d}}|t_2 M\zeta\rangle \\ &= \langle t_2 M\zeta|v^0_{D_{3d}}|t_2 M\xi\rangle \\ &= \frac{1}{\sqrt{6}}\langle t_2\|v^{T_2}\|t_2\rangle \\ &= a, \qquad \text{say}\end{aligned} \tag{9.169}$$

Then matrix (9.168) is written as

$$\begin{array}{c|ccc} & \xi & \eta & \zeta \\ \hline \xi & 0 & a & a \\ \eta & a & 0 & a \\ \zeta & a & a & 0 \end{array} \tag{9.170}$$

On diagonalizing this matrix, we obtain a nondegenerate root and a doubly degenerate root:

$$\mathcal{E}_{D^0_{3d}}(^2T_2 \rightarrow {}^2A_1) = 2a \qquad \text{(nondegenerate)}$$

and (9.171)

$$\mathcal{E}_{D^0_{3d}}(^2T_2 \rightarrow {}^2E) = -a \qquad \text{(doubly degenerate)}$$

Since the first level in (9.171) does not undergo any configuration mixing, it can be written as

$$E_{D_{3d}}(^2T_2 \rightarrow {}^2A_1) = \mathcal{E}(t_{2g}) + 2a \tag{9.172}$$

The two trigonal ^2E levels are evaluated by diagonalizing a two-dimensional ^2E-block, which can be written as

$$\begin{array}{|cc} \begin{matrix} |^2T_2 \rightarrow 2EM\epsilon\rangle \\ O_h \rightarrow D_{3d} \end{matrix} & |eM\epsilon\rangle \\ \hline \mathcal{E}(t_2) - a & b \\ b & \mathcal{E}(e) \end{array} \tag{9.173}$$

where

$$a = \frac{1}{\sqrt{6}} \langle t_2 \| v^{T_2} \| t_2 \rangle$$
$$b = \frac{1}{\sqrt{2}} \langle t_2 \| v^{T_2} \| e \rangle \tag{9.174}$$

On diagonalizing the matrix (9.173) we obtain the required energies as

$$E_{D_{3d}}(^2T_2 \rightarrow {}^2E) = \tfrac{1}{2}[\mathcal{E}(e) + \mathcal{E}(t_2) - a - p]$$
$$E_{D_{3d}}(^2E \rightarrow {}^2E) = \tfrac{1}{2}[\mathcal{E}(e) + \mathcal{E}(t_2) - a + p] \tag{9.175}$$

where

$$p^2 = \left[\{\mathcal{E}(e) - \mathcal{E}(t_2) + a\}^2 + 4b^2\right]$$

For $|\mathcal{E}(e)-\mathcal{E}(t_2)|\gg b$ these two energies reduce, respectively, to

$$E_{C+D_{3d}^0}(^2T_2\rightarrow{}^2E)=\mathcal{E}(t_2)-a$$

$$E_{C+D_{3d}^0}(^2E\rightarrow{}^2E)=\mathcal{E}(e) \tag{9.176}$$

which are the values obtained by neglecting the configuration mixing.

We must stop at this point but should remind ourselves that many more examples of this type can be worked out easily in the strong-field coupling scheme. For details we refer the reader to the excellent book by Sugano, Tanabe, and Kamimura.[2] We now proceed to see how the spin-orbit interaction is treated in the strong-field coupling scheme by using the Wigner-Eckart theorem.

9.8 SPIN-ORBIT INTERACTION IN THE STRONG-FIELD COUPLING SCHEME (APPLICATION OF THE WIGNER-ECKART THEOREM)

In this section we deal briefly with the spin-orbit interaction in the strong field coupling scheme, where we use the Wigner-Eckart theorem. This theorem was treated in detail in Chapter 5, and therefore a full treatment is not given here.

The spin-orbit Hamiltonian can be written, in general, as

$$\begin{aligned}\mathcal{H}_{so}&=\sum_i \xi(\mathbf{r}_i)\mathbf{l}_i\cdot\mathbf{s}_i\\&=\sum_i h_{so,i}\end{aligned} \tag{9.177}$$

where the summation is over all the electrons i of the system. Here, of course, we have neglected the residual-spin-other-orbit interaction,[4] which has no additive terms whose matrix elements are proportional, throughout a given configuration, to $\Sigma_i\mathbf{l}_i\cdot\mathbf{s}_i$. Now, what happens if the system is placed in a ligand field with point group symmetry G, say? Then, in the most general case, we can replace the spherically symmetric function $\xi(r)$ in (9.177) by a function $\xi(\mathbf{r})$ with the same symmetry. This generalization is in agreement with the condition that $\mathcal{H}_{so}$ should remain invariant under the operations of G. Let us consider the ligand field with octahedral symmetry.

We have already constructed the basis functions, $|\alpha S\Gamma M\gamma\rangle$, according to the representations Γ of the O_h group. With respect to these basis functions, V_c and V_e are diagonal in $S\Gamma M\gamma$. But this no longer holds for $\mathcal{H}_{so}$, simply because $\mathcal{H}_{so}$ does not belong to the unit representation A_{1g} of O_h. In fact, it belongs to the unit representation of the double group (the spinor group) O_h', associated with O_h. The idea of the double group was introduced in Chapter 5. As discussed in Chapter 5, the identity element E of a double group is a rotation by 4π, and, the rotation by 2π is denoted by R. Then the number of elements in a double group G' is *twice* that in the corresponding simple group G. Although the number of elements is doubled, the number of classes is *not* necessarily so. There are eight classes in the cubic double group O', and the character table for O' is given in Appendix V. The complete multiplication table for the O' group is also given in Appendix V.

With this introduction let us consider an octahedral term, $\{\alpha S\Gamma\}$, of the d^N system in the absence of $\mathcal{H}_{so}$. We have the reduction

$$D^{(S)}=\sum_i \Gamma_i' \tag{9.178}$$

under the descent $R_3'\to O_h'$. Then

$$D^{(S)}\otimes\Gamma=\sum_i \Gamma_i'\otimes\Gamma=\sum_j \Gamma_j' \tag{9.179}$$

According to this reduction into the representations Γ_j' of O_h', the octahedral term $\{\alpha S\Gamma\}$ will be split into the corresponding terms, $\{\alpha\Gamma_j'\}$, with respect to the double group O_h'. Some of these terms, of course, may be accidentally degenerate. For example, a 12-fold degenerate octahedral level, $\{\alpha^4T_1\}$, is split in the following way:

$$D^{(3/2)}\otimes T_1=\Gamma_8\otimes T_1=\Gamma_6+\Gamma_7+2\Gamma_8 \tag{9.180}$$

Similarly, $^6A_1$ is reduced to

$$D^{(5/2)}\otimes A_1=(\Gamma_7+\Gamma_8)\otimes A_1=\Gamma_7+\Gamma_8 \tag{9.181}$$

We also note that a ^2E term will not be split since

$$D^{(1/2)}\otimes E=\Gamma_6\otimes E=\Gamma_8 \tag{9.182}$$

Now, to apply the Wigner-Eckart theorem, let us rewrite (9.177) as

$$h_{so,j}=\xi(\mathbf{r}_j)\left[\mathbf{l}_{jx}\mathbf{s}_{jx}+\mathbf{l}_{jy}\mathbf{s}_{jy}+\mathbf{l}_{jz}\mathbf{s}_{jz}\right] \tag{9.183}$$

$$\begin{aligned}&=\frac{1}{\sqrt{2}}\left[-v_{+1x}^{1T_1}(j)+iv_{+1y}^{1T_1}(j)\right]\\&+\frac{1}{\sqrt{2}}\left[v_{-1x}^{1T_1}(j)+iv_{-1y}^{1T_1}(j)\right]\\&+v_{0z}^{1T_1}(j)\end{aligned} \tag{9.184}$$

where

$$\begin{aligned}v_{\pm 1x}^{1T_1}(j)&=-\mathbf{s}_{j\pm1}\mathbf{t}_{jx}\\v_{\pm 1y}^{1T_1}(j)&=\mathbf{s}_{j\pm1}\mathbf{t}_{jy}\\v_{0z}^{1T_1}(j)&=\mathbf{s}_{j0}\mathbf{t}_{jz}\end{aligned} \tag{9.185}$$

with

$$\mathbf{s}_{j\pm1} = \mp\frac{1}{\sqrt{2}}(\mathbf{s}_{jx} \pm i\mathbf{s}_{jy})$$

$$\mathbf{s}_{j0} = \mathbf{s}_{jz} \tag{9.186}$$

and

$$\mathbf{t}_{jx} = \xi(\mathbf{r}_j)\mathbf{l}_{jx}, \qquad \mathbf{t}_{jy} = \xi(\mathbf{r}_j)\mathbf{l}_{jy}, \qquad \mathbf{t}_{jz} = \xi(\mathbf{r}_j)\mathbf{l}_{jz} \tag{9.187}$$

Then we have

$$\mathcal{H}_{so} = \sum_j h_{so,j} = \frac{1}{\sqrt{2}}\left[-\mathrm{v}^{1T_1}_{+1x} + i\mathrm{v}^{1T_1}_{+1y}\right] + \frac{1}{\sqrt{2}}\left[\mathrm{v}^{1T_1}_{-1x} + i\mathrm{v}^{1T_1}_{-1y}\right] + \mathrm{v}^{1T_1}_{0z} \tag{9.188}$$

where

$$\mathrm{v}^{1T_1}_{q\bar{\gamma}} = \sum_j v^{1T_1}_{q\bar{\gamma}}(j), \qquad (q=1,0,-1; \quad \bar{\gamma}=x,y,z) \tag{9.189}$$

Note that, since the spin operator $\mathbf{s}_j$ transforms like a vector, $\mathbf{s}_{jq}$ $(q=1,0,-1)$ transforms like Y_1^q in the *spin* space. Again, $\mathbf{t}_{j\bar{\gamma}}(\bar{\gamma}=x,y,z)$ acts as the $\bar{\gamma}$ basis of the irreducible representation T_1 of the cubic group in the position coordinate space. Thus $v^{1T_1}_{q\bar{\gamma}}(j)$ and hence $\mathrm{v}^{1T_1}_{q\bar{\gamma}}$ transform like $|\alpha\, {}^3T_1 q\bar{\gamma}\rangle$. Then, using the Wigner-Eckart theorem (see Chapter 5), we find that

$$\langle \alpha S\Gamma M\gamma|\mathrm{v}^{1T_1}_{q\bar{\gamma}}|\alpha' S'\Gamma' M'\gamma'\rangle = \left[(2S+1)\lambda(\Gamma)\right]^{-1/2}\langle\alpha S\Gamma\|\mathrm{v}^{1T_1}\|\alpha' S'\Gamma'\rangle \cdot\langle SM|S'M'1q\rangle\langle\Gamma\gamma|\Gamma'\gamma' T_1\bar{\gamma}\rangle \tag{9.190}$$

where $\langle SM|S'M'1q\rangle$ are the Wigner coefficients and $\langle\Gamma\gamma|\Gamma'\gamma' T_1\bar{\gamma}\rangle$ are the C-G coefficients.

Because of (9.189) the reduced matrices in (9.190) can be expressed in terms of the single-electron reduced matrices $\langle t_2\|v^{1T_1}\|t_2\rangle$ and $\langle t_2\|v^{1T_1}\|e\rangle$.

In the d-orbital approximation these reduced matrices can be expressed in terms of ζ and ζ', respectively, as can be seen in the following way. Since

$$\langle t_2\tfrac{1}{2}\eta|v^{1T_1}_{0z}|t_2\tfrac{1}{2}\xi\rangle = (2.3)^{1/2}\langle\tfrac{1}{2}\tfrac{1}{2}|\tfrac{1}{2}\tfrac{1}{2}10\rangle\langle T_2\eta|T_2\xi T_1 z\rangle\langle t_2\|v^{1T_1}\|t_2\rangle = -\tfrac{1}{6}\langle t_2\|v^{1T_1}\|t_2\rangle \tag{9.191}$$

and also

$$\langle t_2\tfrac{1}{2}\eta|v^{1T_1}_{0Z}|t_2\tfrac{1}{2}\xi\rangle = \langle t_2\tfrac{1}{2}\eta|\xi(\mathbf{r})\mathbf{l}_z\mathbf{s}|t_2\tfrac{1}{2}\xi\rangle = -\frac{i}{2}\zeta \tag{9.192}$$

we have

$$\langle t_2 \| v^{1T_1} \| t_2 \rangle = 3i\zeta \tag{9.193}$$

And since

$$\langle t_2 - \tfrac{1}{2}\xi | v^{1T_1}_{-1x} | e\tfrac{1}{2}\epsilon \rangle = \tfrac{1}{6}\langle t_2 \| v^{1T_1} \| e \rangle \tag{9.194}$$

and also

$$\langle t_2 - \tfrac{1}{2}\xi | v^{1T_1}_{-1x} | e\tfrac{1}{2}\epsilon \rangle = \langle t_2 - \tfrac{1}{2}\xi | \frac{1}{\sqrt{2}} (\mathbf{s}_x - i\mathbf{s}_y)\xi(\mathbf{r})\mathbf{l}_x | e\tfrac{1}{2}\epsilon \rangle$$

$$= \frac{1}{\sqrt{2}} \zeta' \left[\tfrac{1}{2}(-i) - \tfrac{1}{2}i(i)(-i) \right] = -\frac{i}{\sqrt{2}} \zeta' \tag{9.195}$$

we also have

$$\langle t_2 \| v^{1T_1} \| e \rangle = -3\sqrt{2}\,\zeta' \tag{9.196}$$

From (1.193) and (1.196) it is evident that the reduced matrices of the spin-orbit interaction are purely imaginary. Therefore

$$\langle \alpha S \Gamma M \gamma | \mathrm{v}^{1T_1}_{q\bar{\gamma}} | \alpha' S' \Gamma' M' \gamma' \rangle^* = -\langle \alpha S \Gamma M \gamma | \mathrm{v}^{1T_1}_{q\bar{\gamma}} | \alpha' S' \Gamma' M' \gamma' \rangle \tag{9.197}$$

Now, from (9.185) to (9.189), we at once see that

$$\left(\mathrm{v}^{1T_1}_{q\bar{\gamma}} \right)^\dagger = (-1)^q \mathrm{v}^{1T_1}_{-q\bar{\gamma}} \tag{9.198}$$

Hence

$$\langle \alpha S \Gamma M \gamma | \mathrm{v}^{1T_1}_{q\bar{\gamma}} | \alpha' S' \Gamma' M' \gamma' \rangle = -(-1)^q \langle \alpha' S' \Gamma' M' \gamma' | \mathrm{v}^{1T_1}_{-q\bar{\gamma}} | \alpha S \Gamma M \gamma \rangle \tag{9.199}$$

Now, applying the Wigner-Eckart theorem to (9.199), and using the following relations:

$$[\lambda(\Gamma)]^{-1/2} \langle \Gamma \gamma | \Gamma' \gamma' T_1 \bar{\gamma} \rangle = -[\lambda(\Gamma')]^{-1/2} \langle \Gamma' \gamma' | \Gamma \gamma T_1 \bar{\gamma} \rangle$$

and

$$(2S+1)^{-1/2} \langle SM | S'M' 1q \rangle = (-1)^q (-1)^{S-S'} (2S+1)^{-1/2} \cdot \langle S'M' | SM 1 -q \rangle$$

we obtain the useful relation

$$\langle \alpha S \Gamma \| \mathrm{v}^{1T_1} \| \alpha' S' \Gamma' \rangle = (-1)^{S-S'} \langle \alpha' S' \Gamma' \| \mathrm{v}^{1T_1} \| \alpha S \Gamma \rangle \tag{9.200}$$

All the nonvanishing reduced matrices $\langle \alpha S\Gamma \| \mathrm{v}^{1T_1} \| \alpha' S'\Gamma' \rangle$ occurring in N-electron systems ($N=1,2,\ldots,5$) have been calculated by Sugano, Tanabe, and Kamimura[2] in terms of $\langle t_2 \| v^{1T_1} \| t_2 \rangle$ and $\langle t_2 \| v^{1T_1} \| e \rangle$. As discussed earlier, the reduced matrices in systems with ($N>5$) are related to those of $d^N(N<5)$ systems and therefore need not be mentioned here. This completes the application of the Wigner-Eckart theorem to the spin-orbit interaction of the d^N system of the transition metal ions.

REFERENCES

1. Y. Tanabe and S. Sugano, *J. Phys. Soc. (Japan)* **9**, 753, 766 (1954).
2. S. Sugano, Y. Tanabe, and H. Kamimura, *Multiplets of the Transition Metal Ions in Crystals*, (Academic Press, New York, 1970).
3. J. S. Griffith, *Theory of Transition Metal Ions*, Cambridge University Press, London, (1961); *The Irreducible Tensor Methods of Molecular Symmetry Groups*, Prentice-Hall, Englewood Cliffs, N.J., (1962).
4. M. Blume and R. E. Watson, *Proc. Roy. Soc. (London) A*, **270**, 127 (1962); *Ibid, A*, **271**, 565 (1963).

10

Compounds of the Palladium and Platinum Groups

While discussing the weak-field and strong-field schemes, we pointed out that the relative strengths of the cubic field, the interelectronic repulsion, and the spin-orbit interaction can be roughly divided into three catagories [the inequalities given by (6.3)]. We also described the procedure for calculating the many-electron states for Cases I and II. Most of the transition metal ions having a $3d$ outer shell fall under Cases I and II. Case III corresponds to a relatively strong cubic field and spin-orbit interaction but a weaker interelectronic repulsion. In going through the periodic table, we notice that the interelectronic repulsion among the outer shell electrons gets weaker while the spin-orbit interaction becomes stronger. In atomic spectroscopy this situation is known as the breakdown of the Russell-Saunders scheme. The appropriate coupling scheme to use in such cases is the *jj*-coupling scheme, also known as the intermediate coupling scheme. Exactly this situation is encountered in compounds of the palladium ($4d^n$) and platinum ($5d^n$) groups of compounds. We can take the Racah parameter B, the spin-orbit interaction constant ζ, and the cubic ligand field parameter Dq as the indicators of the respective interactions. The values of these parameters for the triply ionized ions of the three series of transition metal ions are given in Table 10.1, where for Mo^{3+} and W^{3+} we have extrapolated the values given in Griffith.[1] These values show that, while the applicability of the weak-field scheme for the ions of the third series is of doubtful validity, any calculation based on the strong-field scheme should include the configuration mixing due to the strong spin-orbit interaction. Alternatively, one may inquire whether it is possible to develop a scheme analogous to the *jj*-coupling scheme in atomic spectroscopy. Such a procedure can be developed and is outlined below. Before doing this, however, it will be helpful to review the earlier attempts at incorporating the spin-orbit interaction in the calculation.

Table 10.1

Order of magnitudes of the Racah parameter B, spin-orbit coupling constant ζ, and the cubic field parameter Dq for the triply ionized ions of the transition metals

Configuration	Ion	B (cm^{-1})	ζ (cm^{-1})	Dq (cm^{-1})
$3d^3$	Cr^{3+}	1030	273	1800
$4d^3$	Mo^{3+}	500	820	2000
$5d^3$	W^{3+}	800	3000	2800

The elements belonging to the second and third transition series are listed in Table 10.2, along with their electronic configurations. In the table [Kr] and [Xe] denote the closed shell configurations of krypton and xenon, respectively.

That the spin-orbit interaction can change the magnetic behavior of the $4d^n$ and $5d^n$ compounds was pointed out by Van Vleck[2] as far back as 1932. While studying the anomalous magnetic properties of the palladium and platinum series of ions, he showed by qualitative arguments that the large spin-orbit interaction is responsible for the complicated behavior of these ions. He also showed that, just as the electrostatic field quenches the orbital angular momentum of the ion, the combined action of the electrostatic and the spin-orbit interaction can quench the total angular momentum. Later Liehr[3] and Moffit and his co-workers[4, 5] studied the cases where the spin-orbit interaction is strong. Liehr considered an ion with a single electron; he treated the electrostatic (the cubic field and the lower symmetric crystal fields) and the spin-orbit interactions on the same footing and stressed the

Table 10.2

Elements belonging to the second and third transition series

Second transition series				Third transition series			
		Configurations				Configurations	
Elements		$4d$	$5s$	Elements		$5d$	$6s$
Y	[Kr]	1	2	La	[Xe]	1	2
Zr		2	2	Lu [Xe]	$(4f)^{14}$	1	2
Nb		4	1	Hf		2	2
Mo		5	1	Ta		3	2
Tc		5	2	W		4	2
Ru		7	1	Re		5	2
Rh		8	1	Os		6	2
Pd		10	0	Ir		7	2
Ag		10	1	Au		10	1

importance of the role of the spin-orbit interaction in the properties of the $4d^n$ and $5d^n$ ions. Moffitt and his co-workers, from a similar point of view, analyzed the absorption spectra of the hexafluorides of the $5d^n$ ($n=2$, 3, and 4) ions. These studies may be said to have used the intermediate coupling scheme in the true sense of the word, where a number of authors still use the strong-field coupling scheme but take into account the configuration interaction due to the spin-orbit interaction. Because of the availability of modern high-speed digital computers with large memories, it is possible to diagonalize the complete spin-orbit interaction matrices constructed from the strong-field bases. Contrary to the claims that are made, such calculations cannot be considered as truly based on the intermediate coupling scheme for the following reasons.

Since the wavefunctions belonging to different coupling schemes are related by a unitary transformation, diagonalization of the complete matrices using any of the schemes will give identical results. However, the appropriateness of a particular scheme lies in its ability to give "good" results when only a few terms or configurations are included in the calculation. In this sense the basic orbitals used to build the many-electron functions are of importance.

The three schemes use different starting orbitals. In the weak-field scheme, the starting orbital is d and the zeroth-order wave functions are $|d,m_l=2\rangle$, $|d,1\rangle$, $|d,0\rangle$, $|d,-1\rangle$, and $|d,-2\rangle$. The zeroth-order Hamiltonian consists of the free-ion Hamiltonian, given by

$$\mathcal{H}_w = \mathcal{H}_{\text{core}} - \sum_i \frac{Ze^2}{r_i} + \sum_{i \neq j} \frac{e^2}{r_{ij}} \tag{10.1}$$

where the summation is over all the d electrons. Here $\mathcal{H}_{\text{core}}$ involves the electrons in the closed shells. In the strong-field scheme the starting orbitals are t_2 and e, described earlier. The zeroth-order functions are $|t_2\xi\rangle$, $|t_2\eta\rangle$, $|t_2\zeta\rangle$, $|e\theta\rangle$, and $|e\epsilon\rangle$. The zeroth-order Hamiltonian is given by

$$\mathcal{H}_s = \mathcal{H}_{\text{core}} - \sum_i \frac{Ze^2}{r_i} + \sum_i V_c(i) \tag{10.2}$$

The symmetry group of $\mathcal{H}_s$ is O_h, whereas in the weak-field scheme it is the full rotation group R_3. In the intermediate coupling scheme the starting orbitals are Γ_8^u, Γ_7, and Γ_8^l, which are explained in the next section. The zeroth-order Hamiltonian now includes the spin-orbit interaction:

$$\mathcal{H}_I = \mathcal{H}_{\text{core}} - \sum_i \frac{Ze^2}{r_i} + \sum_i V_c(i) + \zeta \sum_i \mathbf{l}_i \cdot \mathbf{s}_i \tag{10.3}$$

Because of the spin operators in the Hamiltonian given by (10.3), the symmetry group of $\mathcal{H}_I$ is now the octahedral double group, denoted by O_h'.

It follows from what has just been said that, although the strong-field calculations, including the configuration mixing via the spin-orbit interaction, can be applied to the $4d^n$ and $5d^n$ ions, the strong-field scheme is not identical with the intermediate coupling scheme. From the computational point of view, however, the strong-field approach has certain advantages, which will be evident from the following discussions.

10.1. STRONG LIGAND FIELD OCTAHEDRAL COMPLEXES

We list in Table 10.3 the important ions of the palladium and platinum groups, together with ions of the $3d$ group that are known to occur in strong ligand field octahedral coordination.

The wavefunctions for the $4d$ and $5d$ electrons have the same angular dependences as those for the $3d$ electrons, but their radial wavefunctions differ greatly since the $4d$ functions have one node and the $5d$ have two nodes, while there is no node for the $3d$ electrons. Near the nucleus the radial wavefunctions have an amplitude which increases rapidly with the atomic number Z, so that larger values of the hyperfine interactions are expected and are actually found in the $4d$ group and particularly in the $5d$ group. Values of the spin-orbit coupling parameter ζ, which are similarly dependent on the amplitude of the wavefunctions near the nucleus, are also noticeably larger for the $4d$ ions, the values given in Table IX.1 being due to Blume, Freeman, and Watson[6] and to Dunn.[7] No detailed calculations have been made for the $5d$ ions, but values of ζ in the range of 2000 to 3000 cm^{-1} have been estimated.[1]

We shall discuss the theory and the results for the octahedral complexes. In 1935, Van Vleck[8] discussed the problem of the strong ligand field complexes

Table 10.3

Ions of the platinum and palladium groups, together with those of the iron group, known to occur in strong ligand field octahedral coordination

Configuration:	d^1	d^2	d^3	d^4	d^5	d^6	d^7	d^8	d^9
$3d^n$			Mn^{4+}						
			Cr^{3+}	Mn^{3+}	Fe^{3+}	Co^{3+}	Ni^{3+}	Cu^{3+}	
			V^{2+}		Mn^{2+}	Fe^{2+}			
$4d^n$	Mo^{5+}								
	Nb^{4+}	Mo^{4+}	Tc^{4+}	Ru^{4+}	Ru^{4+}				
	Zr^{3+}	Nb^{3+}	No^{3+}	Tc^{3+}	Ru^{3+}	Rh^{3+}	Pd^{3+}	Ag^{3+}	
	Y^{2+}	Zr^{2+}	Nb^{2+}	Mo^{2+}	Tc^{2+}	Ru^{2+}	Rh^{2+}	Pd^{2+}	Ag^{2+}
					Mo^{+}			Rh^{+}	
$5d^n$			Re^{4+}	Os^{4+}		Pt^{4+}			
	La^{2+}					Ir^{3+}	Pt^{3+}		

in connection with the explanation of the magnetic properties of the iron-group cyanides. The contributions from the orbital magnetic moments obtained by incorporating the effects of spin-orbit coupling were calculated by Kotani[9] and especially by Kamimura and his colleagues[10] for the $4d^n$ and $5d^n$ compounds.

Calculations of $\Delta(=10Dq)$ are obviously much more difficult for the $4d$ and $5d$ groups than for the $3d$ group. Such a calculation of Δ has been attempted by Richardson and his co-workers[11] for the $3d$ systems, using Roothaan's SCFMO procedure for open shells. We shall discuss this work in Chapter 11 in more detail. The values of $\langle r^4 \rangle$, as well as B and C/B for the $4d$ and $5d$ groups, calculated by De, Desai, and Chakravarty,[12] are given in Tables IX.2 and IX.3. These authors have found that values of $\langle r^4 \rangle$ for the $4d$ group are very much larger than those for the analogous ions of the $3d$ group (the relevent wavefunctions are given by Burns,[13] Freeman and Watson,[14] and Tucker[15]) so that a larger ionic contribution to Δ would be expected. For the $5d$ group, since the SCF wave functions are not available, nothing much can be said about the $\langle r^4 \rangle$ values, but certainly the B and C/B values appear to be very reasonable in spite of the fact that they have used Burn's wavefunctions, as can be seen by comparing Tables IX.2 and IX.3. Whatever be the origin of the large ligand field splitting Δ, its magnitude, relative to the Racah parameter B, determines the order in which the d orbitals are filled. It can easily be seen that the configurations are the same for d^1, d^2, d^3, d^8, and d^9, irrespective of whether Δ is big or small, but quite significant differences are expected for the d^4 to d^7 ions. In a strong ligand field it is energetically preferable for the electrons first to occupy the t_2 oribital so that filling of this shell continues, subject to the restraints imposed by the Pauli principle. This gives a closed $(t_2)^6$ subshell at d^6, and such ions have singlet ground states for both orbit and spin, giving only a small temperature-independent susceptibility. The d^5 ion resembles t_2^1 except that it has one "hole" instead of one electron in the t_2 shell. Similarly, d^4 resembles t^2, with two holes instead of two electrons. The d^3 ion has an orbital singlet with $S=\frac{3}{2}$, which is the maximum spin multiplicity allowed in the t_2 shell, and its magnetic properties correspond to the "spin-only" value, similar to the situation for the half-filled d^5 shell in a weak ligand field complex. We thus have the situation shown in Table 10.4 for the strong-field $4d$ and $5d$ compounds. After d^6, additional electrons must go into the e_g shell, which is rather high in energy so that compounds of this kind tend to be chemically unstable and therefore occur rather rarely.

The four ions d^1, d^2, d^4, and d^5 all have triplet orbital ground states[16] which can be regarded as manifolds with $l'=1$. The orbital reduction factor k due to the covalency effect would play a greater role here than in the weak ligand field $3d$ group in order to allow for the bonding effects. This situation has been discussed by Owen and Thornley[17] and also will be examined by us in connection with the molecular orbital theory in Chapter 11.

Table 10.4

Ground states for ions of the 4*d* and 5*d* groups in a strong ligand octahedral field

Configuration	Ground state
d^1	$^2T_2$
d^2	$^3T_1$
d^3	$^4A_2$
d^4	$^3T_1$
d^5	$^2T_2$
d^6	$^1A_1$
d^7	^2E
d^8	$^3A_2$
d^9	^2E

The spin-orbit coupling is reversed in sign for the "hole" configurations d^4 and d^5 compared to the coupling for d^2 and d^1. In an octahedral field the states J into which the (LS) manifold splits are shown in Figure 10.1. Again, the value of the spin-orbit coupling is likely to be lower than that for the free ion because of the bonding effects.

In this connection we should mention the work done by Kamimura and his colleagues[10] on complexes of the palladium and platinum groups. They used the strong-field formalism and diagonalized the cubic field and the spin-orbit matrices simultaneously using only the ground t_2 orbital. They did not, however, take into the account the effect of the lower symmetric crystal field or the effect of covalency in their calculations, whereas it is well known that the effect of bonding is very pronounced in these compounds. Under the circumstances not much importance can be attached to such a calculation in view of the discussion at the beginning of this chapter.

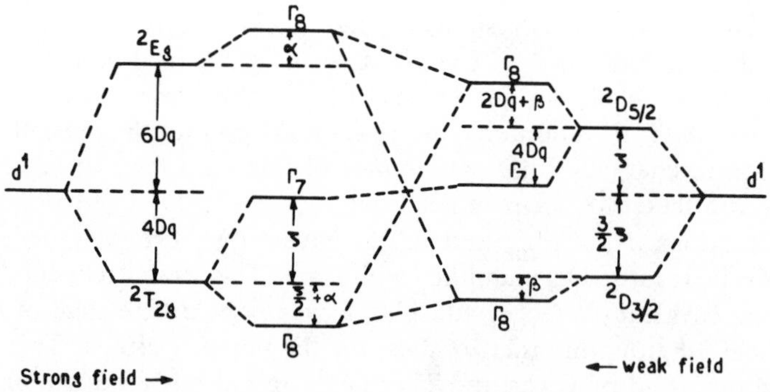

Fig. 10.1 Energy level diagram of the term splitting of an octahedrally coordinated d^1 ion with spin-orbit coupling; α, β denote the corrections after diagonalization.

10.2 THE WEAK-FIELD APPROACH

Let us start from the weak-field limit and consider an ion with a single d electron that has state ^2D. If we include the spin-orbit interaction, the ^2D state is tenfold degenerate since the number of the degeneracy is given by $(2l+1)$ $(2s+1)$. The resulting states can be classified by the quantum numbers l, s, j, and m_j. Also, j will have two values, $l+s=\frac{5}{2}$ and $l-s=\frac{3}{2}$, with m_j ranging from $-\frac{5}{2}$ to $+\frac{5}{2}$ and $-\frac{3}{2}$ to $+\frac{3}{2}$, respectively. Now if the ligand field of O_h symmetry is included, the tenfold degeneracy of the ^2D state will be partially lifted, and we have a sixfold degenerate $^2D_{5/2}$ state and a fourfold state given by $^2D_{3/2}$. The characters for these two representations ($j=\frac{3}{2}$ and $\frac{5}{2}$) in the octahedral double group are given below in tabular form.

	E	R	$4C_3$ $4C_3^2R$	$4C_3R$ $4C_3^2$	$4C_2$ $4C_2R$	$3C_4$ $3C_4^3R$	$2C_4R$ $3C_4^3$	$6C_2$ $6C_2R$
$\chi_{j=3/2}$	4	-4	-1	1	0	0	0	0
$\chi_{j=5/2}$	6	-6	0	0	0	$-\sqrt{2}$	$\sqrt{2}$	0

Comparison of these characters with those of O_h', given in Appendix V.3 shows that the representation for $j=\frac{3}{2}$ is Γ_8 (fourfold), and that the representation for $j=\frac{5}{2}$ is a reducible one. It decomposes into Γ_7 (twofold) and Γ_8 (fourfold). Thus we have

$$\begin{aligned}\Gamma\left(j=\tfrac{3}{2}\right)&=\Gamma_8\\ \Gamma\left(j=\tfrac{5}{2}\right)&=\Gamma_7+\Gamma_8\end{aligned} \tag{10.4}$$

We thus expect that, under the influence of an octahedral field, the term $^2D_{3/2}$ remains unsplit but $^2D_{5/2}$ splits into Γ_7 and Γ_8, as is shown schematically in Figure 10.2. If we now include the spin-orbit interaction:

$$\mathcal{H}_{so}=\zeta\mathbf{l}\cdot\mathbf{s}=\zeta\left[l_zs_z+\tfrac{1}{2}(l_-s_++l_+s_-)\right] \tag{10.5}$$

the nonzero matrix elements of the spin-orbit interaction are as follows:

$$\begin{aligned}\zeta&=\langle\overset{+}{2}|\mathcal{H}_{so}|\overset{+}{2}\rangle=\langle-\overset{-}{2}|\mathcal{H}_{so}|-\overset{-}{2}\rangle=\langle-\overset{+}{2}|\mathcal{H}_{so}|-\overset{+}{2}\rangle\\ &=-\langle\overset{-}{2}|\mathcal{H}_{so}|\overset{-}{2}\rangle=2\langle\overset{+}{1}|\mathcal{H}_{so}|\overset{+}{1}\rangle=2\langle-\overset{-}{1}|\mathcal{H}_{so}|-\overset{-}{1}\rangle\\ &=-2\langle-\overset{+}{1}|\mathcal{H}_{so}|-\overset{+}{1}\rangle=-\langle\overset{-}{1}|\mathcal{H}_{so}|\overset{-}{1}\rangle=\langle\overset{+}{1}|\mathcal{H}_{so}|\overset{-}{2}\rangle\\ &=\langle-\overset{+}{2}|\mathcal{H}_{so}|-\overset{+}{1}\rangle\end{aligned} \tag{10.6}$$

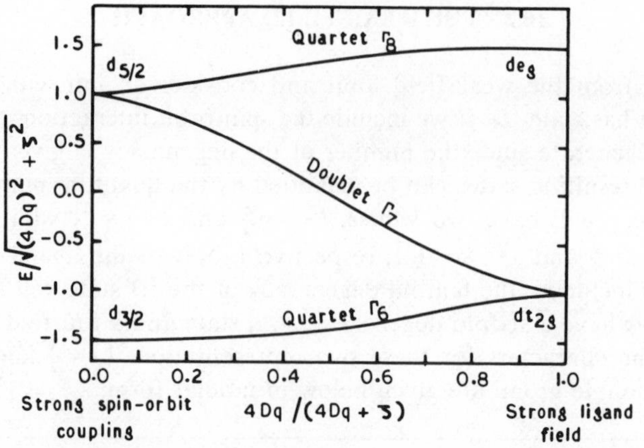

Fig. 10.2 The coupling scheme for a single d electron in an octahedral field.

and

$$\langle \overset{+}{0}\,|\mathcal{H}_{so}|\,\overset{-}{1}\,\rangle=\sqrt{\tfrac{3}{2}}\;\zeta=\langle -\overset{+}{1}\,|\mathcal{H}_{so}|\,\overset{-}{0}\,\rangle$$

The spin-orbit matrix, when diagonalized, gives the energies ζ and $-\frac{3}{2}\zeta$. The corresponding wavefunctions are given in Table IX.4. Perturbing these wavefunctions with the cubic field, we find that ${}^2D_{3/2}$ (Γ_8) is unsplit, but ${}^2D_{5/2}$ splits into Γ_7 and Γ_8 of O_h' with energies $-4Dq$ and $2Dq$, respectively. The wavefunctions and the energies of Γ_7 and Γ_8 representations are listed in Table IX.5; the Γ_8 wavefunctions coming from ${}^2D_{3/2}$ are the same as those in Table IX.4. It is important to note that the cubic field causes mixing of the *two* Γ_8 wavefunctions coming from ${}^2D_{3/2}$ and ${}^2D_{5/2}$ and the matrix element is $-2\sqrt{6}\ Dq$, as can be readily verified. Thus, as a result of spin-orbit interaction and the ligand field of cubic symmetry, the energy of the Γ_7 term is $\zeta-4Dq$, and the wavefunctions are given in Table IX.5. The energies of the Γ_8 terms, however, are obtained by diagonalizing the matrix[3] given in Table IX.6. After diagonalization the energies of $E(\Gamma_8^u)$ and $E(\Gamma_8^l)$, where u and l stand for upper and lower, respectively, are

$$\begin{aligned} E(\Gamma_8^u)&=-\tfrac{1}{4}\zeta+Dq-\tfrac{1}{4}\Delta_1\\ E(\Gamma_7)&=\zeta-4Dq\\ E(\Gamma_8^l)&=-\tfrac{1}{4}\zeta+Dq+\tfrac{1}{4}\Delta_1 \end{aligned}\tag{10.7}$$

where

$$\Delta_1^2=25\zeta^2+40\zeta\cdot Dq+400Dq^2$$

The corresponding wavefunctions are

$$\psi_1 = a_1\Gamma_8(d_{5/2}) + b_1\Gamma_8(d_{3/2})$$
$$\psi = a_2\Gamma_8(d_{5/2}) + b_2\Gamma_8(d_{3/2}) \qquad (10.8)$$

where the coefficients a_1, b_1, etc., are

$$a_i = \frac{2\sqrt{6}\,Dq}{\zeta + 2Dq - E_i}\cdot b_i$$

$$b_i = \left[1 + \frac{384}{\{5\alpha + 4 + (25\alpha^2 + 40\alpha + 400)^{1/2}\}^2}\right]^{-1/2}, \qquad i = 1,2;\ \alpha = \frac{\zeta}{Dq}$$

When the lower symmetric field of tetragonal or trigonal symmetry is included Γ_8 is found to be further split into two doublets, and the energies of the five Kramers' doublets are obtained by diagonalizing the matrices listed in Table IX.6. The resulting wavefunctions corresponding to these doublets are the eigenfunctions of the problem.

10.3 THE INTERMEDIATE COUPLING SCHEME

We now start from the strong-field limit. The first step gives the states of the ion in the ligand field without the spin-orbit coupling; in the next stage the spin-orbit coupling is taken into account. Under the influence of the octahedral ligand field, the tenfold degenerate ^2D term of a d^1 ion splits into ^2E_g (fourfold) and $^2T_{2g}$ (sixfold) states. The wavefunctions of these are products of the space part and the spin part. We now have to resort to the double group representation, O_h'. In this representation we have, including the spin $S=\frac{1}{2}$, which is Γ_6 in O_h',

$$\Gamma(e_g)\otimes\Gamma\left(s=\tfrac{1}{2}\right) = \Gamma_3\otimes\Gamma_6 = \Gamma_8$$

and

$$\Gamma(t_{2g})\otimes\Gamma\left(s=\tfrac{1}{2}\right) = \Gamma_5\otimes\Gamma_6 = \Gamma_7 + \Gamma_8 \qquad (10.9)$$

These results are shown schematically on the left-hand side of Figure 10.2.

Starting from the strong-field limit and considering the spin-orbit interaction as a perturbation over the cubic field, we have given the terms, their energies, and the wavefunctions[3, 19, 20] in Table IX.7. It can be shown that the *two* Γ_8's, that is, Γ_8^l and Γ_8^u, become mixed up by the spin-orbit interaction, and the matrix element is $\sqrt{\frac{3}{2}}\,\zeta$. The energies of these two Γ_8's are obtained by diagonalizing the matrix given in the table cited above. Diagonalizing this matrix yields, of course, the same energy values as those obtained in the weak-field approach. But the wavefunctions will now be different, and they are given by

$$|\Gamma_8^l a\rangle = \cos\theta|\Gamma_8(t_{2g})a\rangle + \sin\theta|\Gamma_8(e_g)a\rangle$$
$$|\Gamma_8^u a\rangle = -\sin\theta|\Gamma_8(t_{2g})a\rangle + \cos\theta|\Gamma_8(e_g)a\rangle \qquad (10.10)$$
$$|\Gamma_7 b\rangle = |\Gamma_7(t_{2g})b\rangle$$

where

$$a = \tfrac{3}{2}, \tfrac{1}{2}, -\tfrac{1}{2}, -\tfrac{3}{2}; \qquad b = \tfrac{1}{2}, -\tfrac{1}{2}$$

and

$$\tan 2\theta = \frac{-\sqrt{6}\,\zeta}{10Dq + \frac{1}{2}\zeta}$$

As before, when the lower symmetric field of tetragonal or trigonal type is included, the term Γ_8 splits further into two Kramers' doublets, as discussed earlier, and the energies are obtained by diagonalizing the matrices given in Table IX.8. Exactly as in the weak-field approach, the wavefunctions corresponding to the doublets are now the eigenfunctions for calculating the magnetic properties of the d^1 system.

The strong-field and weak-field approaches are mainly determined by the splitting. In the weak-field limit it is $\frac{5}{2}\zeta$, whereas in the strong-field limit it is $10Dq$. If $10Dq \gg \frac{5}{2}\zeta$, the strong-field approach must be used; otherwise the weak-field approach is more appropriate. If, however, $10Dq \simeq \frac{5}{2}\zeta$, the intermediate coupling scheme is more appropriate. Hence, by varying $4Dq/(4Dq + \zeta)$, we can plot the variation of Γ_8^l, Γ_8^u, and Γ_7; such a plot is given in Figure 10.2.

We now use the orbitals given by (10.10) to construct the many-electron configurations and states. We will use small γ instead of capital Γ. Two-electron configurations are $(\gamma_8^l)^2$, $\gamma_8^l\gamma_7$, $\gamma_8^l\gamma_8^u$, γ_7^2, $\gamma_7\gamma_8^u$, and $(\gamma_8^u)^2$. The allowed terms are easier to obtain. Since the exclusion principle operates only between equivalent electrons, it is only necessary to know the allowed terms of γ_7^2, γ_8^2, γ_8^3, and γ_8^4. When the direct product $\Gamma_7 \otimes \Gamma_7$ is decomposed, there is only one symmetric representation, and this is labeled by A_1. Hence the allowed term of γ_7^2 is A_1. Similarly, the allowed terms of γ_8^2 are A_1, E, and T_2. Since γ_8 is fourfold degenerate, it can accommodate at most four electrons. Thus γ_8^3 can be considered as representing one hole in the γ_8^4 configuration. Therefore the allowed term of γ_8^3 is Γ_8. The allowed term of γ_8^4 is again A_1. A general

n-electron configuration in the intermediate coupling scheme is

$$(\gamma_8^l)^{m_1}(\gamma_7)^{m_2}(\gamma_8^u)^{m_3} \qquad \text{with} \quad m_1+m_2+m_3=n \tag{10.11}$$

The allowed terms of such a configuration can be found by taking the product of the allowed terms in the individual $(\gamma_8^l)^{m_1}$, $(\gamma_7)^{m_2}$, and $(\gamma_8)^{m_3}$ configurations.

The allowed terms of the two-electron configurations are as follows:

$$\left.\begin{array}{l} (\gamma_8^l)^2 : A_1, E, T_2 \\ (\gamma_8^l\gamma_7) : E, T_1, T_2 \\ (\gamma_7)^2 : A_1 \\ (\gamma_8^u\gamma_7) : E, T_1, T_2 \\ (\gamma_8^u)^2 : A_1, E, T_2 \\ (\gamma_8^l\gamma_8^u) : A_1, A_2, E, 2T_1, 2T_2 \end{array}\right\} \tag{10.12}$$

The matrix elements of the one-electron and the two-electron operators follow the same pattern as that for the strong-field scheme. The Wigner-Eckart theorem applies here also, but now one encounters some complications. For example, consider the matrix elements of an operator transforming as a T_1 representation in O_h', in the subspace of functions transforming as Γ_8 in O_h'. Since, in the product $T_1 \otimes \Gamma_8, \Gamma_8$ is repeated (Appendix V.5), we have

$$\begin{aligned} \langle \Gamma_8^a | O_\alpha^{T_1} | \Gamma_8^b \rangle = {} & \langle \Gamma_8 \| O^{T_1} \| \Gamma_8 \rangle^a \lambda(\Gamma_8)^{-1/2} \langle \Gamma_8 a; \Gamma_8 b\, T_1 \alpha \rangle^a \\ & + \Gamma_8 \| O^{T_1} \| \Gamma_8 \rangle^b \lambda(\Gamma_8)^{-1/2} \langle \Gamma_8 a; \Gamma_8 b\, T_1 \alpha \rangle^b \end{aligned} \tag{10.13}$$

Thus we get two reduced matrix elements and two vector-coupling coefficients. Apart from this difference, the procedures adopted to calculate the matrix elements are identical in the two coupling schemes.

10.4 THE PSEUDO-SPHERICAL APPROXIMATION[4, 10]

Let us now use the pseudo-spherical approximation in the intermediate coupling scheme. In this approximation one neglects the γ_8^u orbital in constructing the configurations. Some authors also neglect the e_g state, which does not appreciably alter the energy eigenvalues; its admixture with t_{2g} $(=\sin\theta)$ is, however, responsible for the magnetic properties. This can easily be seen from the fact that a Γ_8 state purely from t_{2g} has no magnetic moment.[4] The admixture $(\simeq\sin\theta)$ is approximately equal to

$$\sin\theta \sim \frac{\left(\frac{3}{2}\right)^{1/2}\zeta}{10Dq} \simeq 0\cdot 1 \tag{10.14}$$

for $Dq \simeq 3000\ \text{cm}^{-1}$ and $\zeta = 2000\ \text{cm}^{-1}$.

Table 10.5

Terms of the two-electron system in *jj* coupling scheme

Configuration	Terms: Spherical symmetry	O_h symmetry
$(t_{1/2})^2$	$J=0$	A_1
$(t_{3/2}t_{1/2})$	$J=2$	E, T_2
	$J=1$	T_1
$(t_{3/2})^2$	$J=0$	A_1
	$J=2$	E, T_2

If we neglect the γ_8^u orbital, we need consider only the γ_8^l and γ_7 orbitals. We note that the $j=\frac{3}{2}$ state of spherical symmetry (e.g., the $^2P_{3/2}$ or the $^2D_{3/2}$ state) transforms as the Γ_8 representation in O_h'. Also, $j=1/2$ transforms as Γ_6 in O_h'. If we multiply a function transforming as A_2 in O_h' with $j=\frac{1}{2}$ functions, it transforms as Γ_7 in O_h' (e.g., $|A_2a_2\rangle\cdot|j=\frac{1}{2}, m_j=+\frac{1}{2}\rangle$ and $|A_2a_2\rangle\cdot|j=\frac{1}{2}, m_j=-\frac{1}{2}\rangle$). Denoting the γ_8 and γ_7 orbitals by $t_{3/2}$ and $t_{1/2}$, respectively, we can reduce the intermediate coupling scheme for the d^n systems to that for the *jj*-coupling scheme in the atomic p^n configuration. This is similar to the t_2-p isomorphism, discussed first by Abragam and Pryce[16] and later more rigorously by Griffith.[21] We call this procedure the pseudo-spherical approximation. In this approximation the energies of the terms can be expressed in terms of a single parameter $(3B+C)$ instead of two Racah parameters in the d^n case. By using *jj*-coupling scheme, we obtain the terms of the two-electron system, which are given in Table 10.5.

When complete calculations are made, including the γ_8^u orbital, the ordering given above is roughly confirmed. This subject will be discussed further in Chapter 12.

10.5 WAVEFUNCTIONS OF MANY-ELECTRON CONFIGURATIONS

In a preceding section we mentioned how one can determine the many-electron configurations and the corresponding terms. In the following, we outline the procedure for constructing the wavefunctions and tabulate them for the two-electron system.

Basically, the procedure is the same as that used for strong-field wavefunctions, that is, we use the relevant coupling coefficients and then antisymmetrize the total wavefunction. On the other hand, since we have written γ_8^l and γ_8^u in terms of the $\gamma_8(t_{2g})$ and $\gamma_8(e_g)$ orbitals [(equation 10.10)], we write the wavefunctions of the states $|(\gamma_8^l)^2\Gamma\gamma\rangle$ in terms of wavefunctions:

$$|\gamma_8(t_2)^2\Gamma\gamma\rangle, \qquad |\gamma_8(t_2)\gamma_8(e)\Gamma\gamma\rangle, \qquad \text{and} \qquad |\gamma_8(e)^2\Gamma\gamma\rangle$$

This is done by writing the determinantal function for $|(\gamma_8^l)^2\Gamma\gamma\rangle$ and substituting (10.10) for the γ_8^l and γ_8^u orbitals. Although the procedure is straightforward, it is tedious. These wavefunctions have been evaluated for the two-electron system[22] and are as follows:

$$|(\gamma_8^l)^2A_1\rangle=\cos^2\theta|\gamma_8(t_2)^2A_1\rangle+\sqrt{2}\sin\theta\cos\theta|\gamma_8(t_2)\gamma_8(e)A_1\rangle$$
$$+\sin^2\theta|\gamma_8(e)^2A_1\rangle$$
$$|(\gamma_8^u)^2A_1\rangle=\sin^2\theta|\gamma_8(t_2)^2A_1\rangle-\sqrt{2}\sin\theta\cos\theta|\Gamma_8(t_2)\gamma_8(e)A_1\rangle$$
$$+\cos^2\theta|\gamma_8(e)^2A_1\rangle$$
$$|\gamma_8^u\gamma_8^lA_1\rangle=-\sqrt{2}\sin\theta\cos\theta|\gamma_8(t_2)^2A_1\rangle+(\cos^2\theta-\sin^2\theta)|\gamma_8(t_2)\gamma_8(e)A_1\rangle$$
$$+\sqrt{2}\sin\theta\cos\theta|\gamma_8(e)^2A_1\rangle$$
$$|(\gamma_8^l)^2E\rangle=\cos^2\theta|\gamma_8(t_2)^2E\rangle+\sqrt{2}\sin\theta\cos\theta|\gamma_8(t_2)\gamma_8(e)E\rangle$$
$$+\sin^2\theta|\gamma_8(e)^2E\rangle$$
$$|(\gamma_8^u)^2E\rangle=\sin^2\theta|\gamma_8(t_2)^2E\rangle-\sqrt{2}\sin\theta\cos\theta|\gamma_8(t_2)\gamma_8(e)E\rangle$$
$$+\cos^2\theta|\gamma_8(e)^2E\rangle$$
$$|\gamma_8^l\gamma_8^uE\rangle=-\sqrt{2}\sin\theta\cos\theta|\gamma_8(t_2)^2E\rangle+(\cos^2\theta-\sin^2\theta)|\gamma_8(t_2)\gamma_8(e)E\rangle$$
$$+\sqrt{2}\sin\theta\cos\theta|\gamma_8(e)^2E\rangle$$
$$|\gamma_8^l\gamma_7E\rangle=\cos\theta|\gamma_8(t_2)\gamma_7(t_2)E\rangle+\sin\theta|\gamma_8(e)\gamma_7(t_2)E\rangle$$
$$|\gamma_8^u\gamma_7E\rangle=-\sin\theta|\gamma_8(t_2)\gamma_7(t_2)E\rangle+\cos\theta|\gamma_8(e)\gamma_7(t_2)E\rangle$$
$$|\gamma_8^l\gamma_8^uT_1\rangle^{a,b}=|\gamma_8(t_2)\gamma_8(e)T_1\rangle^{a,b}$$
$$|\gamma_8^l\gamma_7T_1\rangle=\cos\theta|A\rangle+\sin\theta|B\rangle$$
$$|\gamma_8^u\gamma_7T_2\rangle=-\sin\theta|A\rangle+\cos\theta|B\rangle$$

where

$$A=|\gamma_8(t_2)\gamma_7(t_1)T_1\rangle$$
$$B=|\gamma_8(e)\gamma_7(t_2)T_1\rangle$$

$$|(\gamma_8^l)^2T_2\rangle=\cos^2\theta|P\rangle+\sqrt{2}\sin\theta\cos\theta|Q\rangle+\sin^2\theta|R\rangle$$
$$|(\gamma_8^u)^2T_2\rangle=\sin^2\theta|P\rangle-\sqrt{2}\sin\theta\cos\theta|Q\rangle+\sin^2\theta|R\rangle$$
$$|\gamma_8^u\gamma_8^lT_2\rangle=-\sqrt{2}\sin\theta\cos\theta|P\rangle+(\cos^2\theta-\sin^2\theta)|Q\rangle$$
$$+\sin^2\theta|R\rangle$$

where

$$|P\rangle=|\gamma_8(t_2)^2T_2\rangle$$
$$|Q\rangle=|\gamma_8(t_2)\gamma_8(e)T_2\rangle$$
$$|R\rangle=|\gamma_8(e)^2T_2\rangle$$
$$|\gamma_8^u\gamma_8^lT_2\rangle'=|\gamma_8(t_2)\gamma_8(e)T_2\rangle'$$
$$|\gamma_8^l\gamma_7T_2\rangle=\cos\theta|P'\rangle+\sin\theta|Q'\rangle$$
$$|\gamma_8^u\gamma_7T_2\rangle=-\sin\theta|P'\rangle+\cos\theta|Q'\rangle$$

where

$$|P'\rangle=|\gamma_8(t_2)\gamma_7(t_2)T_2\rangle$$
$$|Q'\rangle=|\gamma_8(e)\gamma_7(t_2)T_2\rangle$$

and

$$|\gamma_8^l\gamma_8^u|A_2\rangle=|\gamma_8(t_2)\gamma_8(e)A_2\rangle$$

10.6 RELATION BETWEEN THE STRONG-FIELD AND INTERMEDIATE COUPLING SCHEMES

In Section 10.5 we wrote the wavefunctions of the terms from configurations $(\gamma_8^l)^2$, $(\gamma_8^l\gamma_7)$, and so on, in terms of $\gamma_8(t_2^2)$, $\gamma_8(e)^2$, and so on. The purpose was to establish the relation between the strong-field states and the intermediate coupling scheme states. This approach also enables us to evaluate the electrostatic matrices easily. Since the orbitals in the intermediate coupling scheme are diagonal in the cubic field and the spin-orbit interaction, the electrostatic interaction (e^2/r_{ij}) matrices are necessary to calculate the term energies. In the most general form of the strong-field scheme, we have already seen that there are 10 independent electrostatic integrals (Table 9.1). Whereas in the strong-field scheme we had two independent orbitals, t_2 and e, in the intermediate coupling scheme we have three: γ_8^l, γ_7, and γ_8^u. Consequently, in this case we have more independent electrostatic integrals. Unfortunately, the experimental data from absorption spectroscopy are insufficient to give these many integrals. The practice has been to represent the electrostatic interaction between the electrons by two integrals, F_2 and F_4, or the Racah parameters, B and C. This is equivalent to assuming that the orbitals (t_2 and e or γ_8^l, γ_7, and γ_8^u) have the same radial functions. Then, instead of independently constructing the electrostatic matrices for the intermediate coupling scheme, we can use those for the strong-field scheme, provided that the relationship between the two schemes is known. As we have already derived both the strong-field orbitals (t_2 and e) and the intermediate coupling scheme orbitals $(\gamma_8^l, \gamma_7, \gamma_8^u)$ from the same d orbitals, they must be related by a unitary matrix. The elements of this matrix, however, depend on the strength of the cubic field and the spin-orbit interaction. From this, it follows that the unitary matrix which transforms the strong-field states to those of the intermediate coupling scheme will also involve the cubic field parameter Dq and the spin-orbit interaction constant ζ. On the other hand, the transformation matrix connecting the t_2 and e oribitals and $\gamma_8(t_2)$, $\gamma_7(t_2)$, and $\gamma_8(e)$ does *not* involve these parameters. The same is true of the many electron states constructed by using the $\gamma_8(t_2)$, $\gamma_7(t_2)$, and $\gamma_8(e)$ orbitals and the strong-field states. This approach involves higher order coupling coefficients. Evaluation of these is a fairly tedious procedure,but they need be calculated only once. Using these transformation matrices, we can write the wavefunctions of the type $|\gamma_8(t_2)^2\Gamma\rangle$, $|\gamma_8(t_2)\gamma_7\Gamma\rangle$, and so on, in terms of the strong-field states, for example,

$$|\gamma_8(t_2)^2A_1\rangle=\sqrt{\tfrac{2}{3}}\;|t_2^2(^1A_1)A_1\rangle-\frac{1}{\sqrt{3}}|t_2^2(^3T_1)A_1\rangle \qquad (10.15)$$

Complete matrices for the configurations of the two-electron terms are given in Table IX.9. Note that the transformation coefficients depend on the symmetry of the term Γ but not on the component γ. The methods outlined in this and the preceding section enable us to calculate the energy levels in the intermediate coupling scheme.

10.7 MAGNETIC MOMENTS

Since the wavefunctions of the γ_8^l and γ_7 states are now known in both the strong-field and the intermediate coupling schemes, it is a simple and straightforward matter to calculate the magnetic moment, and hence the susceptibility, for compounds of the palladium and platinum groups. To calculate the magnetic moment, we use the operator

$$\boldsymbol{\mu}=\mu_B(k\mathbf{L}+2\mathbf{S}) \qquad (10.16)$$

where k is the orbital reduction factor, which measures in a semiempirical way the extent of covalence due to the overlap of the charge clouds of the central metal ion with those of the ligands. We shall discuss this factor in detail in Chapter 11 on molecular orbital theory.

These matrix elements have been calculated and tabulated by De.[23] From these, the paramagnetic susceptibility and hence the effective moments are calculated in the usual way. The variation of effective moment with temperature and also the variation of the strength of the ligand field with the spin-orbit interaction have been studied by varying the parameters kT/ζ and $Dq/(Dq+\zeta)$. These variations, starting from the strong-field limit, are shown in Figure 10.3. They have been calculated in the weak-field scheme also. It is

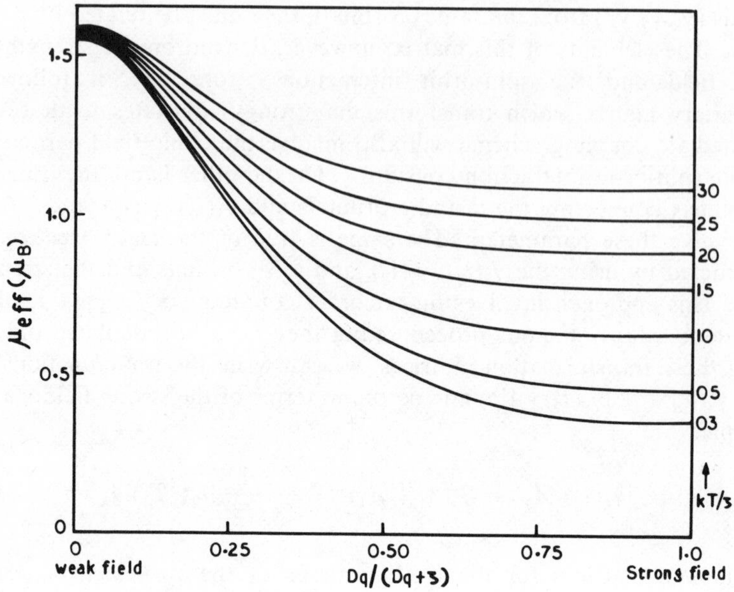

Fig. 10.3 Variation of μ_{eff} from strong field limit.

seen that the two approaches give identical results when configuration interaction is taken fully into account, a finding that is expected and was discussed in the introduction to this chapter. This is true in the case of energy calculations. For the calculation of the magnetic moment in the second order, where the two terms, γ_8 and γ_7, are connected, however, the above statement is not true because, although the strong-field and the weak-field γ_8 orbitals are related by a unitary transformation, the γ_7 orbitals are not. The effective magnetic moments in the strong-field and weak-field limits, however, vary with temperature in a similar fashion, as shown by Kotani.[24] Kotani's formula for the weak-field case is

$$n^2 = \frac{3}{5(3+2^{-5x/2})}\left[\left(63+\frac{8}{5x}\right)+\left(8-\frac{8}{5x}\right)e^{-5x/2}\right] \qquad (10.17)$$

where $x = \zeta/kT$ and $n = \mu_{\text{eff}}/\mu_{\text{B}}$. Equation (10.17) shows that n^2 in the high-temperature and the low-temperature limits is 3.0 and 3.55, respectively. Similarly, the the strong-field case, Kotani's formula for the effective number of Bohr magnetons is

$$n^2 = \frac{8+(3x-8)e^{-3x/2}}{x(2+e^{-3x/2})} \qquad (10.18)$$

From (10.18) we find that n is zero when $kT/\zeta=0$. This is so because, when ζ/kT is large at very low temperatures, the contribution from $\gamma_7(t_{2g})$ is insignificant and the magnetic moment is given by the moment of γ_8^I, which is zero. This is not true, however, in the intermediate field, where we consider the spin-orbit admixture of the $\gamma_8(e_g)$ state which has a nonzero magnetic moment. This is also evident from calculations done by De, Desai, and Chakravarty.[25] When there is a considerable admixture of the excited $\gamma_8(e_g)$ with the ground term $\gamma_8(t_{2g})$, the magnetic moment differs from zero by a considerable amount. We shall now use the intermediate coupling scheme to interpret the magnetic and optical properties of $4d^1$ and $5d^2$ systems.

10.8 MAGNETIC PROPERTIES OF THE $4d^1$ CONFIGURATION

Let us, following De, Desai, and Chakravarty,[25] examine the magnetic properties of the $4d^1$ configuration. The tenfold degenerate ^2D ground state of the configuration [Kr] $4d^1$, under the combined effect of the cubic field and the spin-orbit interaction, splits into one Γ_7 and two Γ_8 irreducible representations of the octahedral double group O_h', as discussed earlier. In octahedral coordination the lowest and the highest of these states have symmetry Γ_8, and the state in between has symmetry Γ_7. The symmetry-adapted wavefunctions for these states were given in Section 10.2. Although no X-ray data exist for some of the complexes under investigation, the possibility of a tetragonal distortion cannot be ruled out since state Γ_8 is Jahn-Teller sensitive [26, 28] (Chapter 12). The tetragonal part of the ligand field has been treated as a perturbation in the usual way. As a result, the states belonging to Γ_8 symmetry split into two levels of Γ_6 and Γ_7 symmetry , while those of Γ_7 remain unsplit.

The paramagnetic susceptibility has been calculated by assuming a Hamiltonian of the form

$$\mathcal{H}=\mathcal{H}_1+\mu_B H(k\mathbf{L}+2\mathbf{S}) \tag{10.19}$$

where

$$\mathcal{H}_1=V_c+V_T+\zeta\mathbf{l}\cdot\mathbf{s}$$

and k is the orbital reduction factor. First $\mathcal{H}_1$ is diagonalized, and then the second term is treated as a perturbation. To set up the energy matrix of H_1, the strong-field bases have been used, and the symmetry-adapted wavefunctions have already been given in Section 10.2. The energy matrices are as follows:

$$
\begin{array}{c|cc}
\Gamma_6 & \Gamma_8^{\pm 1/2}(t_{2g}) & \Gamma_8^{\pm 1/2}(e_g) \\
\hline
 & -4Dq-\frac{1}{2}\zeta-Ds+4Dt & \sqrt{\frac{3}{2}}\,\zeta \\
 & & 6Dq-2Ds-6Dt
\end{array}
$$

$$
\begin{array}{c|ccc}
\Gamma_7 & \Gamma_8^{\pm 3/2}(t_{2g}) & \Gamma_7^{\pm 1/2}(t_{2g}) & \Gamma_8^{\pm 3/2}(e_g) \\
\hline
 & -4Dq-\frac{1}{2}\zeta+Ds+\frac{3}{2}Dt & \mp\frac{\sqrt{2}}{3}(3Ds-5Dt) & \sqrt{\frac{3}{2}}\,\zeta \\
 & & -4Dq+\zeta+\frac{7}{3}Dt & 0 \\
 & & & 6Dq+2Ds-Dt
\end{array}
$$

As can be seen from the above, the matrix elements are in terms of Dq (cubic) and Ds and Dt (tetragonal) parameters. The 10×10 matrix, on account of symmetry, breaks up into two 3×3 blocks and two 2×2 blocks. These are diagonalized to obtain the energies and the wavefunctions in the absence of the external field H. From these wavefunctions the susceptibility has been calculated by using the well-known Van Vleck formula discussed earlier.

The theory given above has been applied to estimate the range of the crystal field parameters that give a good fit with the experimental curves for compounds having Mo^{5+} in a ligand field of D_{4h} symmetry. This theory has also been applied to explain the g values of Mo^{5+} in $CaWO_4$, which has the structure of a distorted tetrahedron.[29] For all the compounds under investigation, $Dq \simeq 2500\ cm^{-1}$ and $\zeta = 800\ cm^{-1}$ (reduced by about 20% from the free-ion value of $1030\ cm^{-1}$ estimated by Dunn[7]). Fixing these parameters and assuming $k \sim 0.85$, one can vary Ds and Dt with temperature to obtain a good fit with experiment in the range of 100 to 300°K. Since the ligand field

Table 10.6

Variations of Ds, $Dt(cm^{-1})$, assuming that $Dq = 2500\ cm^{-1}$, $\zeta = 800\ cm^{-1}$, and $k = 0.85$.
The experimental values are given in parentheses.

Compounds	Ds (cm^{-1})	Dt (cm^{-1})	Splitting of the lowest Γ_8 (cm^{-1})
{ $CsMoF_6$, $NaMoF_6$ }	230–430	300–560	620–1240
$RbMoF_6$	340–800	450–1040	970–2460
$KMoF_6$	430–600	560–780	1240–1800
Mo^{5+} in $CaWO_4$	1140	1490	$g_{\parallel} = 1.998$ (1.987) $g_{\perp} = 1.884$ (1.887)

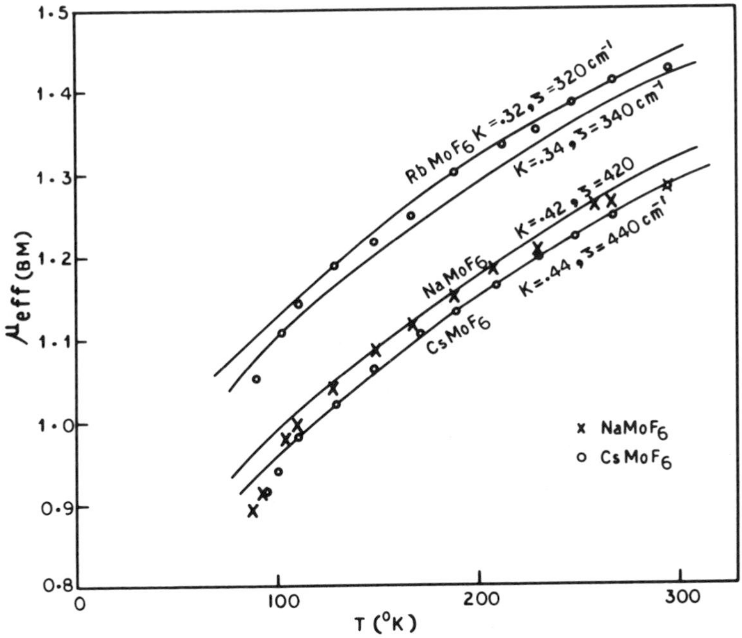

Fig. 10.4 Curve of μ_{eff} versus T. $Dq = 2400\ cm^{-1}$.

of cubic symmetry does not explain the experimental results until and unless a very low value of ζ (300 to 500 cm^{-1}) and a high value of k (50 to 70%) are assumed, De, Desai, and Chakravarty have tried to fit the experiments by assuming an axial distortion for the complexes in question. Such departure from cubic symmetry is also found in the case of a hexachlororhenium compounds. The results are summarized in Table 10.6.

The variations of μ_{eff} with temperature for the compounds mentioned above are given in Figure 10.4. It is evident from Table 10.6 that Ds and Dt show a marked variation with temperature, which was observed for the first time for $3d^1$ complexes by Bose, Chakravarty, and Chatterjee[30] and for $4d^1$ and $5d^1$ complexes by Hare, Bernal and Gray[31] and by Brisden and his co-workers.[32]

10.9 OPTICAL AND MAGNETIC PROPERTIES OF OsF_6

In this section we apply the methods developed in Section 10.3 on the intermediate coupling scheme to study the optical and magnetic properties of d^2 system. We consider OsF_6, which has been studied by Moffit and his co-workers.[4,5] They have studied the optical absorption spectra and the infrared spectra of the OsF_6 molecule, along with other hexafluoride molecules. It has been established by Raman and infrared vibrational spectra that the symmetry group of the molecule is O_h. We take Os^{6+} to be at the center of an octahedron with F^- at the corners. The ground state has symmetry E, hence is Jahn-Teller sensitive. No static distortion is indicated in the infrared spectra of the molecule, but there is evidence of the dynamic Jahn-Teller effect.[5] For the purpose of our analysis we consider the system as a $5d^2$ ion in an octahedral field. Magnetic susceptibility data were obtained by Hargreaves and Peacock[33] and Lewis,[33] who were interested in the oxidation state of the ion. Moffitt and his co-workers have interpreted the electronic transitions in the pseudo-spherical approximation (see Section 10.4) of the intermediate coupling scheme. Eisenstein[34] generalized the earlier work of Liehr and Ballhausen[35] on the d^2 system and interpreted both the spectroscopic and the magnetic data. He used additional Racah parameters (different A,B,C parameters for different configurations) and one more spin-orbit interaction constant. In all, he used 11 parameters (1 cubic field, 2 spin-orbit, 8 Racah parameters) to interpret the absorption spectrum and 2 more covalency parameters to calculate the magnetic susceptibility.

Desai and Chakravarty[36] undertook the study of this system to illustrate the use of the intermediate coupling scheme. They used a simple picture and only four parameters (Dq,ζ,B,C) in their calculations and obtained good agreement with experimental results. The usefulness of the intermediate coupling scheme as a tool to study such systems thus becomes evident.

The configurations and allowed terms of a d^2 system were given in (10.12). The energy levels were calculated by the procedure described in Section 10.3. To display the interplay of the ζ and Dq parameters, we plot the energy levels in Figure 10.5 as a function of the ratio $x=4Dq/(Dq+\zeta)$. The region around $x=0$ corresponds to the weak-field, and the region around $x=1$ to the strong-field, picture. The B and C values are indicated on the diagram. As there is a considerable near-degeneracy among the levels (particularly E and T_2) the levels have been displayed in two parts. This near-degeneracy also shows that the pseudo-spherical approximation of the intermediate coupling scheme as used by Moffitt and his co-workers[4,5] is a good first-order approximation. In this approximation the lowest levels E and T_2 are degenerate, as they come from the $J=2$ state (see Section 10.2). However, the calculations of Moffitt *et al.*[4] show that the separation varies with the parameters, but the E level always lies lower than the T_2 level. This separation is important in regard to the magnetic properties, which will be discussed later in this section.

To interpret the absorption spectra, Desai and Chakravarty estimated the approximate magnitudes of the ζ, B, C, and Dq parameters from the peaks at ~4000, ~8000, ~35,000 and ~40,000 cm^{-1}, which are due to transitions from the ground configuration, $(\gamma_8^l)^2$, to the excited configurations, $\gamma_8^l\gamma_7$, γ_7^2, $\gamma_8^l\gamma_8^u$, $\gamma_7\gamma_8^u$, $(\gamma_8^u)^2$. Their values are as follows: $\zeta\sim3000$, $Dq\sim3000$, $B\sim400$, $C\sim1400$ (all in cm^{-1}). These authors also vary these parameters, and their final calculated levels and assignments are given in Table 10.7.

From the spectra it is evident that the electronic transitions are associated with vibrational progressions. The peak at 3936 cm^{-1} is attributed to a "hot"

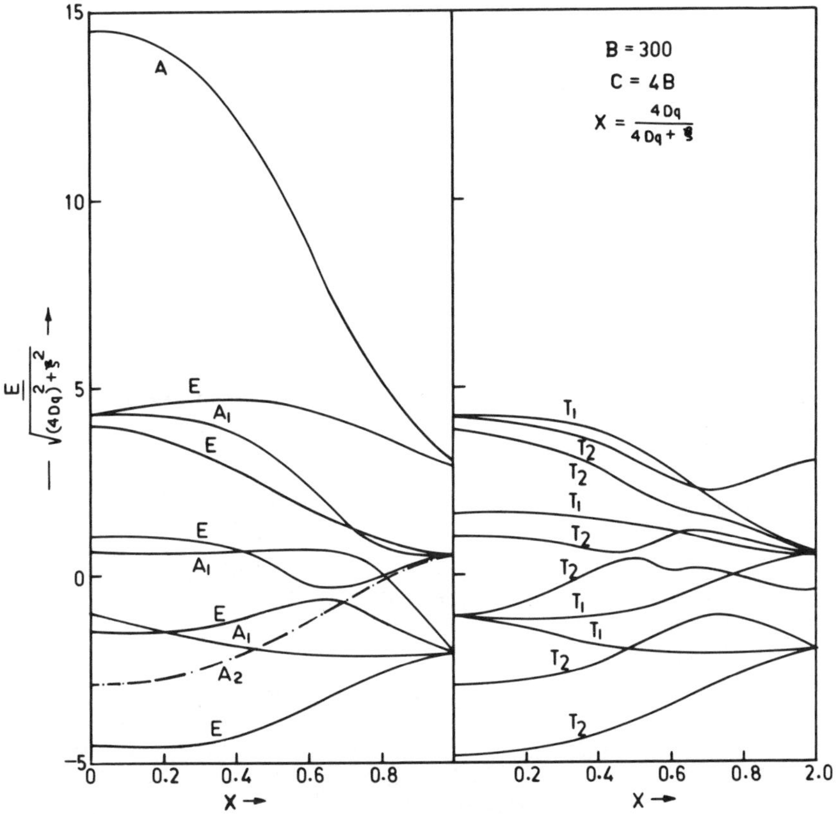

Fig. 10.5 E versus X.

peak. As Desai and Chakravarty would like to keep the intermediate coupling scheme at the simplest level, they have not attempted a detailed fitting. With only four parameters, a close fit is not possible for all the levels. Improvement is possible if one takes more electron repulsion parameters (B, C), but this is tantamount to attributing different electron repulsion integrals to different intermediate coupling scheme configurations.

The calculation of the energy levels shows that the ground state has an E symmetry. Hence it is nonmagnetic, and all of the magnetic moment comes from the excited T_2 state and the high-frequency term between the E and the T_2 states. The other states are separated by more than 3900 cm^{-1}. Hence they are neglected in our present consideration.

The ground state E comes mainly from the ground configuration, $(\gamma_8^l)^2$. Its admixture from the other E states from configurations $\gamma_8^l\gamma_7$, $\gamma_8^l\gamma_8^u$, $\gamma_7\gamma_8^u$, and $(\gamma_8^u)^2$ can be neglected. It may be remembered that in the intermediate coupling scheme the configuration interaction is solely due to the Coulomb interaction between the electrons. Similarly, one should consider only the T_2

Table 10.7

Observed and calculated energy levels of OsF_6
$B=300\ cm^{-1}$, $C=1500\ cm^{-1}$, $\zeta=3000\ cm^{-1}$, $Dq=3000\ cm^{-1}$

Energy levels of OsF_6 (experimental)	Calculated values (cm^{-1})	Values from Ref. 5 (pseudo-spherical approximation) (cm^{-1})
0	$0(E)$	
	$184(T_2)$	$0(J=2)$
3,936		
4,316		
4,371		
4,587	$4{,}301(A_1)$	$3{,}742(J=0)$
4,815	$4{,}895(T_1)$	$4{,}193(J=1)$
5,013		
5,092		
5,750		
8,163		
8,482	$8{,}045(T_2)$	$8{,}086(J=2)$
9,090	$8{,}243(E)$	
9,488		
9,996		
16,892		
17,301	$15{,}706(A_1)$	$16{,}644(J=0)$
17,778		
35,700		
	$35{,}112(A_2)$	
	$35{,}658(T_2)$	
	$36{,}600(T_1)$	
40,800	$39{,}368(T_2)$	
	$42{,}789(E)$	

state from the ground configuration, $(\gamma_8^l)^2$. Taking the axis of quantization along the direction of the field, one can write the perturbation as

$$\mathcal{H}_m = \mu_B(\mathbf{L}_z + 2\mathbf{S}_z)\cdot\mathbf{H} \tag{10.20}$$

Van Vleck's formula is used to obtain the expression for the susceptibility. The first-order and the second-order Zeeman energies are calculated from the roots of the perturbation matrix, which is as follows:

	$\|E\theta\rangle$	$\|E\epsilon\rangle$	$\|T_2\xi\rangle$	$\|T_2\eta\rangle$	$\|T_2\zeta\rangle$
$\langle E\theta\|$	0	0	0	0	0
$\langle E\epsilon\|$	0	0	0	0	C_1H
$\langle T_2\xi\|$	0	0	Δ	C_2H	0
$\langle T_2\eta\|$	0	0	C_2^*H	Δ	0
$\langle T_2\zeta\|$	0	C_1^*H	0	0	Δ

where

$$C_1 = \langle E\epsilon | L_z + 2S_z | T_2\zeta \rangle$$
$$C_2 = \langle T_2\xi | L_z + 2S_z | T_2\eta \rangle \qquad (10.21)$$

and Δ is the separation between the E and the T_2 levels. Since we have used a real basis, C_1 and C_2 are purely imaginary and are as follows:

$$C = i\left(\tfrac{4}{3}\cos^4\theta + \frac{4\sqrt{2}}{\sqrt{3}}\sin^3\theta\cos\theta\right)$$

$$C_2 = \sqrt{2}\, i\left(\tfrac{4}{3}\cos^2\theta\sin^2\theta + 2\sin^4\theta - 2\cos^2\theta\sin^2\theta + \frac{4\sqrt{2}}{\sqrt{3}}\sin^3\theta\cos\theta\right) \qquad (10.22)$$

where

$$\tan\theta = \frac{\sqrt{6}\,\zeta}{10Dq + \frac{1}{2}\zeta}$$

These equations are obtained as follows. First the wavefunctions given in Section 10.3 are used. Then, using Table IX.7, we write these wavefunctions in terms of the strong-field states and evaluate the matrix elements of the operator $(\mathbf{L}_z + 2\mathbf{S}_z)$ between these states. This procedure enables us to write the analytic expressions for C_1 and C_2. This is one of the advantages of the intermediate coupling scheme. In this connection the susceptibility expression

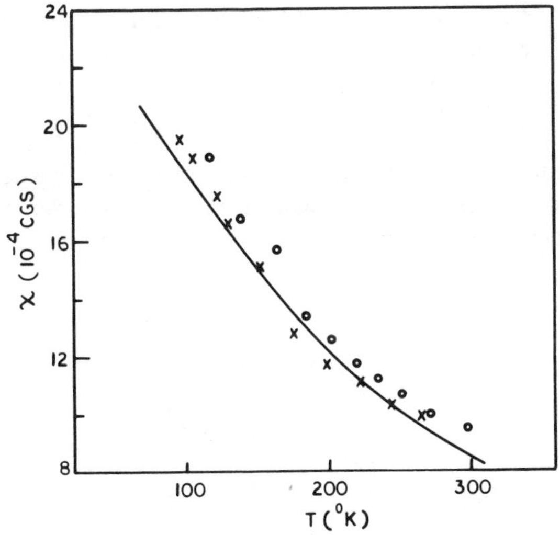

Fig. 10.6 Variation of χ with T for OsF_6.

Table 10.8

Experimental and calculated values of the susceptibility, $10^6\chi$ (in cgs units), of OsF_6

The calculated values are obtained by using the following values of the parameters (cm^{-1}): $B=300$, $C=1500$, $Dq=3000$, $\zeta=3000$.

T (°K)	$10^6\chi$ (calculated)	$10^6\chi$ (experimental)[a]
297.0	870	943
273.0	942	996
267.8	959	(995.4)
252.0	1015	1062
246.0	1038	(1024)
235.0	1082	1118
222.7	1136	(1107)
220.0	1148	1172
202.0	1238	1255
198.6	1256	(1174)
184.0	1339	1332
175.5	1395	(1279)
164.0	1468	1564
152.0	1554	(1504)
138.0	1664	1675
129.6	1734	(1652)
120.8	1810	(1759)
117.0	1844	(1886)
105.9	1944	(1890)
81.5	2160	(2333)

[a] G. B. Hargreaves and R.D. Peacock, *Proc. Chem. Soc. (London)* **49**, 85 (1959). The values in parentheses were obtained by W. B. Lewis, quoted in this reference.

[see (10.24)] may be compared with the numerical expression given by Eisenstein on p. 316 of Ref. 34. It can be shown that Δ depends on Dq, ζ, B, and C. The eigenvalues of the perturbation matrix are given by

$$\left.\begin{aligned} E_1 &= 0 \\ E_2 &= \Delta + \frac{1}{\Delta} C_1^2 H^2 \\ E_3 &= \Delta - \frac{1}{\Delta} C_1^2 H^2 \\ E_4 &= \Delta + |C_2| H \\ E_5 &= \Delta - |C_2| H \end{aligned}\right\} \qquad (10.23)$$

where $C_1^2 = C_1^* C_1$ and $|C_2| = (C_2^* C_2)^{1/2}$. The separation Δ was assumed to be large compared to the Zeeman splitting. Therefore the susceptibility can be written as

$$\chi = \frac{N\mu_B^2}{2+3e^{-\Delta/kT}}\left(\frac{C_2^2}{kT}e^{-\Delta/kT} + \frac{2C_1^2}{\Delta} - \frac{2C_1^2}{\Delta}e^{-\Delta/kT}\right) \tag{10.24}$$

For different sets of values of B, C, ζ, and Dq, one can obtain $\sin\theta$ and $\cos\theta$, using (10.10), and from the diagonalization of the energy matrices, Δ was obtained. These values were used to obtain χ at various temperatures. A plot of χ versus T is given in Figure 10.6, where we have indicated the experimental values of Hargreaves and Peacock[33] by small circles and those of Lewis[33] by $\times$'s. In this calculation the following values of the parameters were used: $B = 320$ cm^{-1}, $C = 5B$, $Dq = 3400$ cm^{-1}, $\zeta = 3413$ cm^{-}1. The separation Δ between the E and T_2 states is found to be 208 cm^{-1}. The susceptibility values for a different set of these parameters (those used in Table 10.7) are presented in Table 10.8, and it is evident that, as long as $\Delta \sim 200$, the susceptibility values do not change much.

In conclusion we may say that the intermediate coupling scheme is, indeed, a better approach, even when used in its simplest form. Its usefulness becomes more convincing, however, when the analysis is extended to $4d^n$ and $5d^n (n \geqslant 3)$ ions.

10.10 EXTENSION OF THE INTERMEDIATE COUPLING SCHEME FOR MORE THAN TWO ELECTRONS IN THE *d* SHELL

In the earlier sections of this chapter we considered the intermediate coupling scheme for the d^2 system. The transformation matrices (similar to those given in Table IX.9) become very tedious to evaluate. Alternatively, the electrostatic matrices for an n-electron system can be calculated from those for an $(n-1)$-electron system. The necessary relations for the atomic case were derived by Racah.[37] Similar relations for the strong-field configurations were derived by Tanabe and Sugano.[38] The proof of these relations is given by Griffith in his book.[39] It is straightforward, though somewhat tedious, to extend these relations to the intermediate coupling scheme, which we give in this section.

We write the general term in the intermediate coupling scheme as

$$|\gamma_1^{n_1}(\delta_1)\gamma_2^{n_2}(\delta_2)\gamma_3^{n_3}(\delta_3)\delta\rangle$$

where $n_1 + n_2 + n_3 = n$, the total number of electrons, and γ_1, γ_2, and γ_3 are, respectively, the γ_8^u, γ_7, and γ_8^l orbitals, given earlier. Thus the overall term δ is due to n_1 electrons in the γ_1 orbital, n_2 electrons in γ_2, and n_3 electrons in γ_3. The allowed terms in $\gamma_1^{n_1}$, $\gamma_2^{n_2}$, and $\gamma_3^{n_3}$ are δ_1, δ_2, and δ_3, respectively. This completely specifies the most general term. Let

$$G_n = \sum_{i<j}^{n} \frac{e^2}{r_{ij}}$$

and

$$G_{n-1} = \sum_{i<j}^{n-1} \frac{e^2}{r_{ij}}$$

Then the matrix elements of an n-electron system and an $(n-1)$-electron system are related by the following expression:

$$\langle \gamma_1^{n_1}(\delta_1)\gamma_2^{n_2}(\delta_2)\gamma_3^{n_3}(\delta_3)\delta|G_n|\gamma_1^{n_1'}(\delta_4)\gamma_2^{n_2'}(\delta_5)\gamma_3^{n_3'}(\delta_6)\delta\rangle$$

$$= \frac{\sqrt{n_1 n_1'}}{n-2} \sum_{\delta',\delta'',\bar{\delta}} \langle \gamma_1,\delta'\delta_{23}(\bar{\delta})\delta|\gamma_1\delta'(\delta_1)\delta_{23}\delta\rangle^*$$

$$\cdot\langle \gamma_1\gamma_1^{n-1}(\delta')|\}\gamma_1^{n_1}\delta_1\rangle\langle \gamma_1,\delta''\delta_{56}(\bar{\delta})\delta|\gamma_1\delta''(\delta_4)\delta_{56}\delta\rangle$$

$$\cdot\langle \gamma_1\gamma_1^{n_1'-1}(\delta'')|\}\gamma_1^{n_1'}\delta_4\rangle\langle \gamma_1^{n_1-1}(\delta')\gamma_2^{n_2}(\delta_2)\gamma_3^{n_3}(\delta_3)\bar{\delta}|$$

$$\cdot G_{n-1}|\gamma_1^{n_1'-1}(\delta'')\gamma_2^{n_2'}(\delta_5)\gamma_3^{n_3'}(\delta_6)\bar{\delta}\rangle$$

$$+(-1)^{n_1+n_1'}\frac{\sqrt{n_2 n_2'}}{n-2} \sum_{\delta',\delta'',\bar{\delta}} \langle \gamma_2,\delta'\delta_{13}(\bar{\delta})\delta|\gamma_2\delta'(\delta_2)\delta_{13}\delta\rangle^*$$

$$\cdot\langle \gamma_2\gamma_2^{n_2-1}(\delta')|\}\gamma_2^{n_2}\delta_2\rangle^*\langle \gamma_2,\delta''\delta_{46}(\bar{\delta})\delta|\gamma_2\delta''(\delta_5)\delta_{46}\delta\rangle$$

$$\cdot\langle \gamma_2\gamma_2^{n_2'-1}(\delta'')|\}\gamma_2^{n_2'}(\delta_5)\rangle\langle \gamma_1^{n_1}(\delta_1)\gamma_2^{n_2-1}(\delta')\gamma_3^{n_3}(\delta_3)$$

$$\bar{\delta}|G_{n-1}|\ \gamma_1^{n_1'}(\delta_4)\gamma_2^{n_2'-1}(\delta'')\gamma_3^{n_3'}(\delta_6)\bar{\delta}\rangle$$

$$+\frac{\sqrt{n_3 n_3'}}{n-2} \sum_{\delta',\delta'',\bar{\delta}} \langle \delta_{12},\delta'\gamma_3(\delta_3)\delta|\delta_{12}\delta'(\bar{\delta})\gamma_3\delta\rangle$$

$$\cdot\langle \gamma_3^{n_3-1}(\delta')\gamma_3|\}\gamma_3^{n_3}\delta_3\rangle^*\langle \delta_{45},\delta''\gamma_3(\delta_6)\delta|\delta_{45}\delta''(\bar{\delta})\gamma_3\delta\rangle^*$$

$$\cdot\langle \gamma_3^{n_3'-1}(\delta'')\gamma_3|\}\gamma_3^{n_3'}(\delta_6)\rangle$$

$$\cdot\langle \gamma_1^{n_1}(\delta_1)\gamma_2^{n_2}(\delta_2)\gamma_3^{n_3-1}(\delta')\bar{\delta}|G_{n-1}|\gamma_1^{n_1'}(\delta_4)\gamma_2^{n_2'}(\delta_5)\gamma_3^{n_3'-1}$$

$$(\delta'')\bar{\delta}\rangle$$

where δ_{ij} is any one of the representations in the product $\delta_i \times \delta_j$. Also, $n_1+n_2+n_3=n_1'+n_2'+n_3'=n$. The brackets

$$\langle \Gamma_1,\Gamma_2\Gamma_3(\Gamma_{23})\Gamma|\Gamma_1\Gamma_2(\Gamma_{12})\Gamma_3\Gamma\rangle$$

are the coefficients in the octahedral group, analogous to the Racah coefficients. The brackets $\langle \gamma_1^{n_1-1}(\delta')\gamma_1| \} \gamma_1^{n_1}\delta \rangle$ are the fractional parentage coefficients. From the vector coupling coefficients, one can calculate the Racah coefficients. As far as the fractional parentage coefficients are concerned, the only nontrivial ones occur for the γ_8^3 configuration. The only allowed term for this configuration is Γ_8. We can write

$$|\gamma_8^3\Gamma\rangle = \sum_{\Gamma'} \langle \gamma_8^2(\Gamma')\gamma_8| \} \gamma_8^3\Gamma_8 \rangle |\gamma_8^2(\Gamma')\gamma_8\rangle$$

The fractional parentage coefficients satisfy the relation

$$\sum_{\Gamma} \langle \gamma_8^2(\Gamma)\gamma_8| \} \gamma_8^3\Gamma_8 \rangle \langle \gamma_8\gamma_8(\Gamma)\gamma_8\Gamma_8|\gamma_8, \gamma_8\gamma_8(\Gamma')\Gamma_8\rangle = 0$$

for all Γ' that are forbidden in γ_8^2. By solving the set of equations given above, one obtains the following fractional parentage coefficients:

$$\langle \gamma_8^2(A_1)\gamma_8| \} \gamma_8^3\Gamma_8 \rangle = \frac{1}{\sqrt{6}}$$

$$\langle \gamma_8^2(E)\gamma_8| \} \gamma_8^3\Gamma_8 \rangle = \frac{1}{\sqrt{3}}$$

$$\langle \gamma_8^2(T_2)\gamma_8| \} \gamma_8^3\Gamma_8 \rangle^a = \frac{1}{\sqrt{10}}$$

$$\langle \gamma_8^2(T_2)\gamma_8| \} \gamma_8^3\Gamma_8 \rangle^b = -\sqrt{\frac{2}{5}}$$

The last two coefficients are superscripted since $T_2 \times \Gamma_8$ give *two* Γ_8 representations and hence *two* coefficients. The results already derived and the particle-hole complementarity should enable one to calculate the electrostatic matrices in the intermediate coupling scheme. These are all that are necessary since the spin-orbit and the crystal field energies will be diagonal and therefore trivial to evaluate. The most tedious task is to evaluate the Racah coefficients. It is worth investigating whether symmetry relations of the type derived by Griffith[39] are obtainable when the double group representations appear in the coefficients.

REFERENCES

1. J. S. Griffith, *The Theory of Transition Metal Ions*, Cambridge University Press, London, England, 1961.
2. J. H. Van Vleck, *The Theory of Electric and Magnetic Susceptibilities*, Oxford University Press, Oxford, England 1932.

3. A. D. Liehr, *J. Phys. Chem.* **64**, 43 (1960).
4. W. E. Moffitt, G. L. Goodman, M. Fred, and B. Weinstock, *Mol. Phys.* **2**, 109 (1959).
5. B. Weinstock and G. L. Goodman, *Adv. Chem. Phys.* **9**, 169 (1965).
6. M. Blume, A. J. Freeman, and R. E. Watson, *Phys. Rev. A*, **134**, **320** (1964).
7. T. M. Dunn, *Trans. Faraday Soc.* **57**, 1441 (1961).
8. J. H. Van Vleck, *J. Chem. Phys.* **3**, 807 (1935).
9. M. Kotani, *Prog. Theor. Phys. (Osaka)*, Suppl. 14, p. 1 (1960).
10. H. Kamimura, S. Koide, H. Sekiyama, and S. Sugano, *J. Phys. Soc. Japan* **15**, 1264 (1960).
11. J. W. Richardson, T. F. Soules, D. M. Vaught, and R. R. Powell, *Phys. Rev. B* **4**, 1721 (1971). See also other references in this work.
12. I. De, V. P. Desai, and A. S. Chakravarty, *Indian J. Phys.* **48**, 1133 (1974).
13. G. Burns, *Phys. Rev.* **128**, 2121 (1962); *J. Chem Phys.* **41**, 1521 (1964).
14. A. J. Freeman and R. E. Watson, *Magnetism*, Vol. IIA, G. T. Rado and H. Suhl (Eds.), Academic Press, New York, 1965, p. 167.
15. E. B. Tucker, *Phys. Rev.* **143**, 264 (1966).
16. A. Abragam and M. H. L. Pryce, *Proc. Roy. Soc. (London) A*, **205**, 135 (1951).
17. J. Owen and J. H. M. Thornley, *Rep. Prog. Phys.* **29**, Part II, 675 (1966).
18. A. Abragam and B. Bleaney, *Electron Paramagnetic Resonance of Transitions Ions*, Oxford University Press, Oxford, England, 1970.
19. C. J. Ballhausen, *Introduction to Ligand Field Theory*, McGraw-Hill, New York, 1962.
20. H. L. Schlafer and G. Gliemann, *Basic Principles of Ligand Field Theory* (translated by D. F. Ilten), Wiley-Interscience, London, 1969.
21. J. S. Griffith, *Trans. Faraday Soc.* **54**, 1109 (1958); *Disc. Faraday Soc.* **26**, 81 (1958).
22. V. P. Desai, Ph.D. Thesis, University of Calcutta, 1976.
23. I. De, Ph.D. Thesis, University of Calcutta, 1975.
24. M. Kotani, *J. Phys. Soc. Japan* **4**, 293 (1949).
25. I. De, V. P. Desai, and A. S. Chakravarty, *Phys. Rev. B* **8**, 3769, (1973).
26. G. C. Allen and K. D. Warren, *Struct. Bonding* **9**, 49 (1971).
27. H. H. Classen, G. L. Goodman, J. H. Holloway, and H. Selig, *J. Chem. Phys.* **53**, 341 (1970).
28. A. A. Kiseljov, *Proc. Phys. Soc. (London)* **2**, 270 (1969).
29. R. W. G. Wyckoff, *Crystal Structures*, Vol. 3, Interscience, New York, 1965.
30. A. Bose, A. S. Chakravarty, and R. Chatterjee, *Proc. Roy. Soc. (London) A* **255**, 145 (1960).
31. C. H. Hare, I. Bernal, and H. B. Gray, *Inorg. Chem.* **1**, 828 (1962).
32. B. J. Brisden, D. A. Edwards, D. J. Machin, K. S. Murray, and R. A. Walton, *J. Chem. Soc. A*, p. 1825 (1967).
33. G. B. Hargreaves and R. D. Peacock, *Proc. Chem. Soc. (London)*, **44**, 85 (1959).
34. J. C. Eisenstein, *J. Chem. Phys.* **34**, 310 (1961).
35. A. D. Liehr and C. J. Ballhausen, *Ann. Phys.* **6**, 134 (1959).
36. V. P. Desai and A. S. Chakravarty, *Proceedings of Nuclear Phys. and S. S. Phys. Symp.*, BARC (Bombay) **3**, 563 (1972); The complete investigation is in course of publication in *Phys. Rev. B.* (1980).
37. G. Racah, *Phys. Rev.* **63**, 367 (1943).
38. Y. Tanabe and S. Sugano, *J. Phys. Soc. Japan* **9**, 753 (1954).
39. J. S. Griffith, *Irreducible Tensor Method for Molecular Symmetry Groups*, Prentice-Hall, Englewood Cliffs, N.J., 1962.

11

Molecular Orbital Theory

In Chapter 10 we mentioned the inadequacy of the point charge model for the complexes of the transition metal ions, especially compounds of the $4d$ and $5d$ groups. This model has given way to the modern ligand field theory, in which the key parameters such as Dq, ζ, B, and C can, in principle, be rigorously calculated using the molecular orbital theory, although very often a semiempirical approach is adopted to make the calculations simpler and easily tractable. In this chapter we briefly describe first how the above parameters are interpreted in the framework of the molecular orbital theory, postponing an *ab initio* calculation done by Richardson and his co-workers, which we shall briefly describe in Section 11.8. We shall also see how the orbital reduction factor can account for the covalence, which is very important in the $4d$ and $5d$ groups of complexes. Finally, we give one variant of the molecular orbital (MO) theory that is increasingly used to calculate the energy levels of a molecular complex containing a transition metal ion.

There has been considerable progress in our understanding of the nature of chemical bonds in the complexes of the transition metal ions. There are two methods for treating the bonding between atoms: the valence bond method of Pauling, and the molecular orbital method of Mulliken, Slater, and Van Vleck. If no approximations are made, both approaches lead to the same result, as has been shown by Van Vleck and Sherman.[1] However, some approximation is always necessary at one stage or another. If the atoms in a complex are not far removed from their equilibrium positions, the MO method leads to correct physical properties of the complexes, especially if the configuration interaction is taken into account.[2] This is the reason why we deal with the MO theory to study the molecular complexes.

11.1 GENERAL CONSIDERATIONS

The first step consists of choosing suitable molecular orbitals. These are written as a linear combination of atomic orbitals (LCAO), given by

$$\Psi_{MO} = \sum_i c_i \chi_i \tag{11.1}$$

where c_i are the coefficients to be determined, and χ_i are the atomic orbitals (AOs) centered at the atoms constituting the molecule. The number of AOs and the type of AOs to be included in (11.1) depend, of course, on the nature of the molecule and the problem. To make full use of the symmetry of the molecule and anticipating that we will be interested in a complex containing a metal ion situated at the center with the neighboring ligands in a symmetrical position, we can rewrite (11.1) as

$$\Psi_\gamma^\Gamma = \alpha \phi_\gamma^\Gamma + \beta \chi_\gamma^\Gamma \tag{11.2}$$

The molecular orbital Ψ_γ^Γ, the metal orbital ϕ_γ^Γ, and a linear combination of ligand atomic orbitals χ_γ^Γ all transform as the Γth irreducible representation and γth row of the symmetry group to which the molecule belongs. The terms α and β are the coefficients to be determined. The coefficients of the linear combination χ_γ^Γ are given solely by the geometry of the molecule and its symmetry type under the transformations of the group ($\Gamma\gamma$). Unlike α and β, these are not subject to variation. They can be obtained by the standard procedures of group theory,[3] and we shall not elucidate them here.

While constructing the MOs, the orbitals of the inner core of electrons of the central metal atom, for example, $1s^2 2s^2 2p^6 3s^2 3p^6$ of a $3d$ metal and $1s^2$ of ligands, are omitted from consideration; neglecting them does not affect the results as these inner electrons do not take part in bonding. Among the metal orbitals, the nd, $(n+1)s$ and $(n+1)p$ orbitals are taken into consideration. As we have already seen, the d orbitals span the T_{2g} and E_g representations of O_h, the symmetry group of the molecule. Each of the ligand contributes $2s$ and $2p$ orbitals for the bonding. The bonding may be such that the charge density is along the bond axis; in other words, the nodal surface of the orbital is normal to the bond axis. Such bonding is known as σ bonding and the corresponding MO as the σ orbital by analogy with the nomenclature used for diatomic molecules.

The s orbitals always form σ bonds. The other type of orbitals, the π orbitals, have maximum charge density along an axis perpendicular to the bond axis; the nodal surface contains the bond axis. Metal e_g orbitals form σ bonds, and t_{2g} orbitals form π bonds.

There are two methods by which the coefficients α and β in (11.2) can be obtained. One method is to extend the Hartree-Fock self-consistent field (HFSCF) calculation to molecular clusters. This was first formulated by

Roothaan[4] for molecules and later applied to ML_6 clusters of transition metal ions by a number of authors.[5–8] The other method is a semiempirical one.[3]

In the SCF method the coefficients in the linear combination (11.2) can be interpreted in terms of "covalency." They can be calculated separately for σ and π bonding and (it is hoped) compared to experimental results. On the other hand, the semiempirical method also gives the coefficients, but usually the results are interpreted in terms of charge densities. From the computational point of view, however, this method is comparatively simple, and the energy levels of the complex can be easily determined.

First let us briefly outline the SCF method. If we denote by Ψ_i an orbital of the type given by (11.2), a given state (normally we are interested in the ground and a few excited states) of n electrons can be represented by a linear combination of Slater determinants constructed from the orbitals Ψ_i and the spin functions α_s and β_s, say. We denote this spin-orbital state by Ψ. The coefficients of the linear combination are found by the requirements that Ψ be an eigenfunction of the operator $\mathbf{S}^2$ (S being the total spin) that and it transforms according to an irreducible representation of the symmetry group of the Hamiltonian ($\Gamma\gamma$). The latter requirement corresponds to the demand, in atomic spectroscopy, that the total wavefunction be an eigenfunction of the operator $\mathbf{L}^2$ (L being the total orbital angular momentum).

By using the condition for stationary states:

$$\delta\langle\Psi|\mathcal{H}|\Psi\rangle=0 \tag{11.3}$$

with the subsidiary condition

$$\langle\psi_i|\psi_j\rangle=\delta_{ij} \tag{11.4}$$

one obtains a set of integrodifferential equations for the orbitals. The Hamiltonian in (11.3) can be written as the sum of one-particle and two-particle operators:

$$\mathcal{H}=\sum_{i=1}^{n}\mathbf{f}_i+\sum_{i<j}^{n}\mathbf{g}_{ij} \tag{11.5}$$

Thus (11.3) and (11.4) lead to the set of equations

$$\mathbf{h}_i\psi_i=\varepsilon_i\psi_i+\lambda_{ij}\psi_j \tag{11.6}$$

where $\mathbf{h}_i$ is a single-electron Hamiltonian and can be written in terms of the matrix elements of the operators $\mathbf{f}$ and $\mathbf{g}$. We shall learn more about the SCF MOs in Section 11.8.

Now let us, for the sake of simplicity, assume that there are a single metal orbital ϕ and a single ligand orbital χ. This leads to two molecular orbitals, one of which predominantly contains the metal orbital ϕ and the other χ:

$$\psi_1=(N_1)^{-1/2}(\phi-\lambda\chi) \tag{11.7}$$

and

$$\psi_2=(N_2)^{-1/2}(\chi+\gamma\phi) \tag{11.8}$$

where N_1 and N_2 are the normalization constants. The coefficients λ and γ, related to each other by the orthogonality condition (11.4), are given by

$$\lambda=\frac{\gamma+S}{1+\lambda S}$$

and (11.9)

$$N_1=1-2S\lambda+\lambda^2, \qquad N_2=1+2S\gamma+\gamma^2$$

where S is the overlap integral $\langle\chi|\phi\rangle$. Here λ is known as the covalency parameter since it gives the charge delocalization from the metal ion to the ligand. The orbital given by (11.7) is known as the antibonding orbital, and (11.8) as the bonding orbital. The ligand $2s$ and $2p$ orbitals have lower energy than the metal d, s, and p orbitals. Hence the bonding orbitals are stabilized, whereas the antibonding orbitals are destabilized.

We shall now confine our discussion to the case where the central metal atom attaches n atoms arranged in some symmetrical fashion (octahedron, tetrahedron, square, etc.) characteristic of a crystallographic point group. Let $\Psi(\Gamma)$ be a wavefunction of the central metal atom which has the proper symmetry, that is, whose transformation scheme under the covering operations of the group is characteristic of some irreducible representation Γ. Let ψ_i be a wavefunction of the attached atom i. We shall assume that only one orbital state need be considered for each attached atom and that this state either is an s state or else is symmetric (as in a $2p_\sigma$ bond) about the line joining the attached atom to the central atom. The method of MOs, in its simplest form, seeks to construct solutions of the form

$$\Psi=\psi(\Gamma)+\sum_i a_i\psi_i \tag{11.10}$$

The coefficients a_i must be chosen so that $\sum_i a_i\psi_i$ transforms according to the irreducible representation Γ. Now the important point is that bases for only certain irreducible representations can be constructed from the linear combinations of the ψ_i. To determine these, one ascertains the group characters associated with the transformation scheme, usually reducible, of the original attached wavefunctions before the linear combinations are taken. It is easy to do this, since the character χ_D for a covering operation D is simply equal to q, say, where q is the number of atoms left invariant by D. Since D leaves q atoms undisturbed and completely rearranges the others, the diagonal sum involved in the character will contain unity q times and will have zeros for the other entries. We saw in Chapter 5 how the group characters under a prescribed symmetry can be obtained for important point group symmetries.

We now consider a complex containing six atoms octahedrally arranged, that is, located at the centers of the six cube-faces. We shall follow the excellent article by Van Vleck[9] to describe the molecular orbitals. Let us first consider the so-called σ bonds in an octahedral complex (Figure 11.1). These orbitals are symmetric about the line joining the ligand and the central atom and can consequently be s, p_z, d_{z^2} orbitals. There are six such bonds, and therefore we can construct six linearly independent orbitals using these wavefunctions. We shall now see how to construct such combinations.

Since the symmetry group is O_h, the characters associated with the arrangement of attached atoms can be shown to be as follows:

	E	$8C_3$	$3C_2$	$6C_4$	$6C_2'$	i	$8iC_3$	$3iC_2$	$6iC_4$	$6iC_2'$
χ	6	0	2	2	0	0	0	4	0	2

Here $\chi(C_4)$, for instance, means the character for the symmetry operation consisting of rotation about one of the fourfold or principal cubic axes by $2\pi/4$. Any rotation about such an axis leaves two atoms invariant, and hence $\chi(C_2)=\chi(C_4)=2$. On the other hand, $\chi(C_2')=\chi(C_3)=0$, since no atoms are

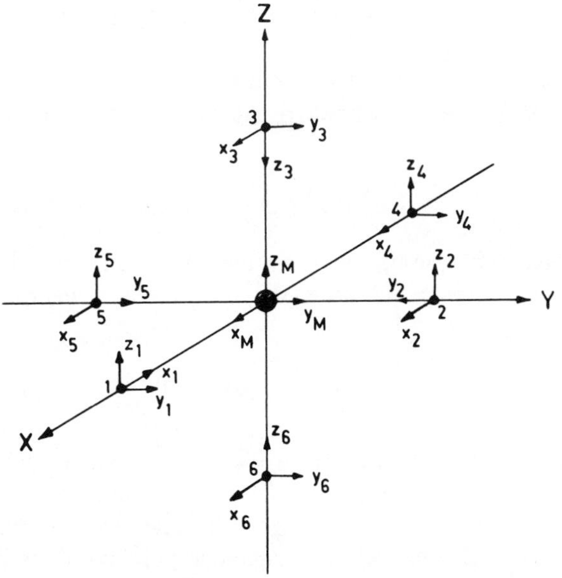

Fig. 11.1 The σ and π bonds on O_h complex. (Taken from Richardson et al.[58])

left invariant under rotations about the twofold or secondary cubic axes (surface diagonals) or about the threefold axes (body diagonals). Inversion in the center of symmetry is denoted by i. By using tables of characters for the group O_h (given in Appendix III), one finds that the irreducible representations contained in the character scheme given above, are a_{1g}, e_g, and t_{1u}. Also, the central-ion s orbitals transform as a_{1g} and p orbitals as t_{1u}, and d orbitals as e_g and t_{2g} (Chapter 6). Thus we see that the bonding scheme is a d^2sp^3 hybridization. The two d orbitals to be used for σ bonds are therefore the e_g and not the t_{2g} orbitals.

The exact form of the ligand orbitals can be found by using group theory. We shall not go into the details here. Having found the various linear combinations of the ligand orbitals transforming like irreducible representations of the point group of the molecule we must next find the irreducible representations spanned by the various types of central orbitals. The irreducible representations corresponding to the various kinds of central orbitals are as follows:

Orbitals:	s	p	d	f
Representations:	a_{1g}	t_{1u}	e_g, t_{2g}	a_{2u}, t_{1u}, t_{2u}

In Table 11.1 we have collected the various possibilities for the molecular orbital formation, using the s, p, and d orbitals of the metal atom. Now, from Table 11.1, let us make a linear combination of the metal and the ligand orbitals to form a MO, say for one of the components having e_g symmetry; it is

$$\Psi_{x^2-y^2} = \alpha d_{x^2-y^2} + \tfrac{1}{2}\beta(-x_1 + y_2 + x_4 - y_5) \tag{11.11}$$

In (11.11) α and β are mixing coefficients. If the metal and the combination of the ligand orbitals are separately normalized, the coefficients α and β must obey the relation

$$\alpha^2 + \beta^2 + 2\alpha\beta S = 1 \tag{11.12}$$

where S denotes the overlap between the metal and the ligand orbitals, $\langle\Psi_M|\Psi_L\rangle$. If they overlap but little, the value of S is nearly zero, so that approximately

$$\beta^2 = 1 - \alpha^2 \tag{11.13}$$

We thus see that the transition from the crystal field theory to the ligand field theory depends on the value of α. If $\alpha \cong 1$, we have the crystal field case. If, on the other hand, $\alpha < 1$, the complex deviates from strict crystal field theory, whereas $\alpha^2 = 0.5$ for what is called a pure covalent complex—the ferricyanides, for example.

Table 11.1

Molecular orbitals in O_h symmetry
The numbering of the ligand orbitals is shown in Figure 11.1.

Representation, Γ	Components, γ	Metal-centered orbitals, ϕ	Ligand-centered orbitals,[a,b] χ
a_{1g}		ns	$6^{-1/2}N(\sigma_{a_{1g}})[\sigma_1+\sigma_2+\sigma_3+\sigma_4+\sigma_5+\sigma_6]$
t_{1u}	x	np_x	$2^{-1/2}N(\sigma_{t_{1u}})[\sigma_1-\sigma_4]$
			$\frac{1}{2}N(\pi_{t_{1u}})[x_2+x_3+x_5+x_6]$
	y	np_y	$2^{-1/2}N(\sigma_{t_{1u}})[\sigma_2-\sigma_5]$
			$\frac{1}{2}N(\pi_{t_{1u}})[y_1+y_3+y_4+y_6]$
	z	np_z	$2^{-1/2}N(\sigma_{t_{1u}})[\sigma_3-\sigma_6]$
			$\frac{1}{2}N(\pi_{t_{1u}})[z_1+z_2+z_4+z_5]$
e_g	θ	nd_{z^2}	$12^{-1/2}N(\sigma_{eg})[-\sigma_1-\sigma_2+2\sigma_3-\sigma_4-\sigma_5+2\sigma_6]$
	ε	$nd_{x^2-y^2}$	$\frac{1}{2}N(\sigma_{eg})[\sigma_1-\sigma_2+\sigma_4-\sigma_5]$
t_{2g}	ξ	nd_{yz}	$\frac{1}{2}N(\pi_{t_{2g}})[z_2+y_3-z_5-y_6]$
	η	nd_{xz}	$\frac{1}{2}N(\pi_{t_{2g}})[z_1+x_3-z_4-x_6]$
	ζ	nd_{xy}	$\frac{1}{2}N(\pi_{t_{2g}})[y_1+x_2-y_4-x_5]$
t_{1g}	x		$\frac{1}{2}N(\pi_{t_{1g}})[z_2-y_3-z_5+y_6]$
	y		$\frac{1}{2}N(\pi_{t_{1g}})[-z_1+x_3+z_4-x_6]$
	z		$\frac{1}{2}N(\pi_{t_{1g}})[y_1-x_2-y_4+x_5]$
t_{2u}	ξ		$\frac{1}{2}N(\pi_{t_{2u}})[x_2-x_3+x_5-x_6]$
	η		$\frac{1}{2}N(\pi_{t_{2u}})[-y_1+y_3-y_4+y_6]$
	ζ		$\frac{1}{2}N(\pi_{t_{2u}})[z_1-z_2+z_4-z_5]$

[a]Here σ_i is any AO that is rotationally symmetric about the M-L_i bond axis; x_i, y_i, z_i refer to the Cartesian components of degenerate π-type AOs referred to the M-L_i bond axis. See Figure 11.1.

[b]$N(\sigma_{a_{1g}})=[1+4\langle\sigma_1|\sigma_2\rangle+\langle\sigma_1|\sigma_4\rangle]^{-1/2}$
$N(\sigma_{t_{1u}})=[1-\langle\sigma_1|\sigma_4\rangle]^{-1/2}$
$N(\sigma_{eg})=[1-2\langle\sigma_1|\sigma_2\rangle+\langle\sigma_1|\sigma_4\rangle]^{-1/2}$
$N(\pi_{t_{1u}})=[1+2\langle z_1|z_2\rangle+\langle z_1|z_4\rangle]^{-1/2}$
$N(\pi_{t_{2g}})=[1+2\langle y_1|x_2\rangle-\langle y_1|y_4\rangle]^{-1/2}$
$N(\pi_{t_{1g}})=[1-2\langle y_1|x_2\rangle-\langle y_1|y_4\rangle]^{-1/2}$
$N(\pi_{t_{2u}})\ \ [1-2\langle z_1|z_2\rangle+\langle z_1|z_4\rangle]^{-1/2}$

Now, to determine the mixing coefficients α and β, for example, for the e_g orbital, we have to solve a quadratic secular equation. The lower root represents a lower energy than is found in the individual systems. Hence we call it a bonding orbital. The upper root, on the other hand, yields a higher energy than is found for the free atoms. Hence we call it an antibonding orbital. The antibonding orbital is usually distinguished from the bonding orbital by means of an asterisk. By convention, the signs of α and β are both taken to be positive for the bonding orbital, whereas they differ for the antibonding orbital. In O_h there is only one case (the t_{1u} orbital) where we obtain a 3×3 secular equation, the solution of which gives us three roots—one bonding, one nearly nonbonding, and one antibonding orbital. One more point to note is that, since the π orbitals do not point toward the central atom, they usually give much weaker bonding than do the σ orbitals. In many cases we may therefore neglect the π bonds in the bonding scheme.

Let us now go through the quantum-mechanical techniques of solving the molecular problem presented by the transition metal complexes within the framework of MO theory. The atomic and molecular problems proceed along very similar lines. The Hamiltonian operators are identical, except for the inclusion of nuclear repulsion terms and attraction terms between the electrons and the additional nuclei. The n-electron operator is recast into a sum of one-electron operators in an analogous fashion. There is thus a one-to-one correspondence between the AOs of the metal atom and the MOs of the molecule. The one-electron Schrödinger equations are then solved for the MOs (using the variation principle where necessary) and for their energies.

The electrons in the complex ion are next assigned to the derived MOs (Figure 11.2). Such an assignment defines the molecular electronic configura-

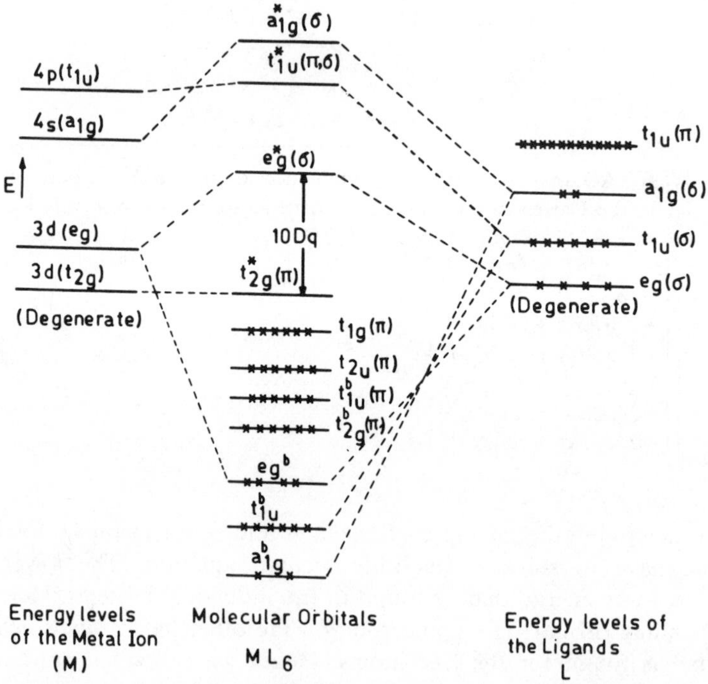

Fig. 11.2 Molecular orbital energy level scheme for an O_h complex.

tion. For the cluster $M^{z+}L_6$, with A being the atomic number of M, belonging to the transition series, and L being a first-row ion, there are $A-Z+6\times10$ electrons to be accommodated in MOs. One may assume that the 11 lowest energy MOs will be essentially undisturbed free-ion inner shell AOs of M and L. The valence MOs, which may be approximated by the linear combinations of the $3d$, $4s$, and $4p$ and the ligand-centered orbitals of Table 11.1, are given appropriate group-theoretic symbols and assigned a principal quantum number. Thus the $(60+A-Z)$ electron configurations of principal interest are as follows:

$$\{(1s_M)^2(2s_M)^2(2p_M)^6(1s_{L_1})^2(1s_{L_2})^2(1s_{L_3})^2(1s_{L_4})^2(1s_{L_5})^2(1s_{L_6})^2$$
$$(3s_M)^2(3p_M)^6\}\{(1a_{1g})^2(1e_g)^4(1t_{1u})^6\}$$
$$\{(2a_{1g})^2(2t_{1u})^6(2_{eg})^4\}$$
$$\{(3t_{1u})^6(1t_{2g})^6\}$$
$$\{(1t_{1g})^6(1t_{2u})^6\}\{(2t_{2g})^x(3e_g)^y\}$$

where brackets have been introduced to indicate groups of MOs having similar bonding character. The first group contains localized innershell or core AOs; the second consists of MOs largely $2s_L$ in character; the third group is largely $2p_{\sigma_L}$ in character; the fourth is $M-L$ π bonding; the fifth is nonbonding ligand $2p_\pi$; and the sixth contains $x+y=A+Z$ electrons in antibonding MOs that are largely $3d$ in character and gives rise to most of the chemical and physical properties of interest.

From Table 11.1 it is evident that a general MO has the form

$$\psi_i = a_i\phi_i + b_i\chi(\sigma_i) + c_i\chi(\pi_i)$$

where the ϕ_i, $\chi(\sigma_i)$, and $\chi(\pi_i)$ are already defined in the table. An inspection of Table 11.1 reveals that the t_{1g} and t_{2u} MOs are identically equal to the ligand symmetry orbitals (SOs) since there are no metal orbitals with which to combine. These two $\pi-$type orbitals therefore are nonbonding and are completely specified by the symmetry of the molecule. Upon determining the coefficients by the variational method, one energy level is obtained per coefficient, so to speak. We now discuss in a qualitative fashion the nature of these energy levels and the way in which they are influenced by changes in the type of ligand and metal atom.[11]

First let us examine a hypothetical complex ion where there is no possibility of π bonding. The MOs therefore reduce to the form

$$\psi_i = a_i\phi_i + b_i\chi(\sigma_i)$$

Now consider a hypothetical complex in which the bonds are roughly nonpolar, that is, in a bonding MO an electron has equal probability of being at the metal and being at some part of the ligand. In other words, under these circumstances the metal and the ligands are approximately equal in electronegativity. Then it follows that $a_i \approx b_i$. The bonding MOs (corresponding to the lower roots of the secular equations) will then be considerably lower in energy than the AOs of either the metal or the ligand from which the MO was formed. The antibonding MOs (corresponding to the higher roots of the secular equations), on the other hand, will be of considerably higher energy.

We now consider the effect of causing the ligands to be more electronegative, that is, the energy of the ligand AOs to become more negative with respect to the metal orbitals. It is intuitively obvious and is borne out also by calculation that the bonding electrons shift out more "into" the region of ligands. Thus the bonds become more polar. Conversely, the antibonding electrons concentrate more on the metal atom. The electronic structure is described by the same sort of functions but with altered coefficients; here $b_i \gg a_i$ in the bonding MOs.

The electrons of the complex are assigned to the derived MOs by the familiar "Aufbau" principle of atomic structure. In a typical complex all nonbonding MOs, as well as the six bonding MOs (the d^2sp^3 combination), are occupied by electron pairs. The remaining electrons (always equal to the number of valence shell electrons originally present in the transition metal ion) are distributed as follows. The first three go unpaired into the triply degenerate t_{2g} level (Hund's rule of maximum multiplicity on the molecular scale). If there are more electrons to be accommodated, they either pair up with those already in the t_{2g} MOs or go unpaired into the antibonding e_g MOs, maintaining maximum multiplicity. The choice (and this is the crux of the whole problem) depends on the energy separation between the t_{2g} and the e_g levels in comparison to the energy of inverting the spin of an electron in the field of the others. For the transition elements, the latter energy is of the order of 2 to 3 eV.[12]

When the possibility of π bonding is also considered, the system of energy levels becomes somewhat more complicated. From Table 11.1 it is seen that one new t_{1u} level is formed; furthermore the t_{2g} level, formerly nonbonding, interacts with π-type orbitals of the ligands, giving rise to two triply degenerate levels, one bonding and the other antibonding. These are also shown in Figure 11.2. It is seen that the inclusion of the π bonding admits the possibility of additional stabilization of the molecule and, at the same time, tends to decrease the separation between the antibonding e_g and t_{2g} levels.

Most six-coordinated complexes of the transition metals are closely octahedral insofar as the σ-bonding system is concerned. On the other hand, the nature of the π-type interactions varies. In the simplest of complexes, such as $M(F_6)^{3-}$, the π interactions, as described in the preceding paragraph, arise from the lone-pair electrons in the pure p_π orbitals of the ligands. In another case, $M(NH_3)_6^{3+}$, for example, the p_π orbitals of each N atom are already involved in bonding to the H atoms.

In the hydrated iron group of salts $[M(H_2O)_6]$, where $M=3d^n$, MO wavefunctions identical to those of Van Vleck, as discussed above, can be used, but it should be remembered that the bonding of the e_g central orbits with $\psi(\sigma)$ orbits on the ligands is usually fairly weak in contrast to the bonding in the $4d^n$ and $5d^n$ complexes. For this case, $\psi(\sigma)$ includes $2p_\sigma$ and $2s$ orbits on O atoms, and choosing the signs so that the $\psi(\sigma)$ wavefunction is positive in the region of overlap with e_g wavefunction, and neglecting terms containing overlap as discussed earlier, we can write the antibonding and bonding wavefunctions as[13]

Antibonding:

$$\psi_{3z^2-r^2}=\alpha d_{3z^2-r^2}-\frac{(1-\alpha^2)^{1/2}}{2\sqrt{3}}(-2z_3+2z_6+x_1+y_2-x_4-y_5)$$

$$\psi_{x^2-y^2}=\alpha d_{x^2-y^2}-\frac{(1-\alpha^2)^{1/2}}{2}(-x_1+y_2+x_4-y_5) \qquad (11.14a)$$

Bonding:

$$\psi^*_{3z^2-r^2}=(1-\alpha^2)^{1/2}d_{3z^2-r^2}+\frac{\alpha}{2\sqrt{3}}(-2z_3+2z_6+x_1+y_2-x_4-y_5)$$

$$\psi^*_{x^2-y^2}=(1-\alpha^2)^{1/2}d_{x^2-y^2}+\frac{\alpha}{2}(-x_1+y_2+x_4-y_5) \qquad (11.14b)$$

The oxygen-ligand orbits are presumed to be lower in energy than the central e_g (with components $d_{3z^2-r^2}$ and $d_{x^2-y^2}$) orbits; and when there is some admixing (i.e., when $\alpha<1$), the ligand $\psi(\sigma)$ and the metal e_g energy levels can be thought to repel each other so that the energy of e_g is raised and that of $\psi(\sigma)$ is lowered. Equations (11.14a) are thus taken to represent the antibonding combinations of the orbits, and (11.14b) the bonding combinations. There will be four more bonding and four antibonding combinations of the orbits with the central $4s$ and $4p_x$, $4p_y$, and $4p_z$ orbits. These are all shown in Figure 11.2.

In an analogous way the central metal t_{2g} (with components d_{xy}, d_{yz}, and d_{zx}) can combine with ligand $\psi(\pi)$ orbits, π denoting unit angular momentum about the bond axis. In this case there will be $2p_\pi$ orbits on the O atoms. There are now three bonding and three antibonding combinations; one such antibonding combination is, for example,

$$\psi_{xy}=\beta d_{xy}-(1-\beta^2)^{1/2}\tfrac{1}{2}(x_2+y_1-x_5-y_4) \qquad (11.15)$$

where, as before, terms containing overlap are neglected. Such bonding is also shown to be of importance in octahedral complexes, especially in complexes of the $4d$ and $5d$ groups, where the σ bonding is very strong.[13–15]

Thus the presence of bonding has two main effects. First, it increases the splitting between the e_g and t_{2g} orbits, and in this respect it is similar in effect to an increase in the crystal field strength.[15] Second, the e_g orbits are modified so that the electrons in these orbits are partially transferred to the attached ligand atoms of the complex. The analogous modification of the t_{2g} orbits is assumed to be much smaller. If the first effect is large, the e_g orbits will always be unoccupied because of their higher energy, which implies that the ground state for $n>3$ (n is the number of d electrons) may not be the state of maximum spin, as expected from Hund's rule. For example, for $n=5$ the configuration t_{2g}^5 with $S=\frac{1}{2}$ (low spin) may be lower in energy than $t_{2g}^3e_g^2$ with $S=\frac{5}{2}$ (high spin), which is normally the ground state, because the high energy of the two e_g electrons more than compensates for the lowering of energy achieved by keeping the electron spins parallel. This is exactly what happens in the ferricyanides treated by Van Vleck[9] and in many complexes of the $4d$ and $5d$ groups studied by Griffiths, Owen, and Ward,[16] but in only one common hydrated iron group of complex $[Co(H_2O)_6]^{3+}$, t_{2g}^6, $S=0$. For the hydrated complexes of the transition metal ions it can be shown from the magnetic data[13] that the energy separation between t_{2g} and e_g single-electron states is much bigger for M^{3+} ions than for M^{2+} ions. This suggests that the σ bonding is stronger for M^{3+} ions, a view that is also confirmed by the magnetic data. This stronger bonding presumably arises because there is a greater tendency for electrons on the ligand atoms to move into the central M ion in bonding molecular orbits in order to even out the charge distribution. This effect might reasonably be expected to reduce the Coulomb interactions e^2/r_{ij} between the d electrons because the charge cloud is spread out to a greater degree over the entire complex, as a result of which the Racah parameters become reduced from the free-ion values, especially in complexes where the bonding is strong (e.g., in compounds of the $4d$ and $5d$ groups).

A detailed treatment of configurations d^1, d^2, d^6, and d^7, from the standpoint of paramagnetic susceptibility, was given in Chapter 7. Study of these configurations reveals that there are systematic discrepancies in the crystal field theory and that the orbital magnetic moment in many cases is smaller than would be expected on a purely ionic model. The discrepancies can be explained partly by introducing weak σ bonds into the $[M(H_2O)_6]$ complex, so that there is some charge transfer between the paramagnetic ion and the surrounding ligands, and partly by assuming that the magnitude of the anisotropic ligand field changes with temperature because of anisotropic thermal expansion of the crystal lattice.[17] From the MO standpoint we shall study the orbital reduction factor in Section 11.3. Let us discuss now the calculation of the $10Dq$ parameter, using the MO theory.

11.2 CALCULATION OF 10*Dq*

From Figure 11.2 it is seen that above the t_{2g} orbital is the e_g orbital. In the usual crystal field theory the energy difference between the e_g and the t_{2g}

orbitals is written as $10Dq$, but in the MO approach $10Dq$ must be interpreted as the energy difference between the antibonding $e_g^*(\sigma)$ and $t_{2g}^*(\pi)$ levels. However, the magnitude of $10Dq$ in the ligand field theory depends entirely on the strength of the bonding. Thus formally we may write

$$10Dq \equiv \langle e_g^a|\mathbf{h}|e_g^a\rangle - \langle t_{2g}^a|\mathbf{h}|t_{2g}^a\rangle \tag{11.16}$$

where $\mathbf{h}$ is a single-electron Hamiltonian whose exact form depends on the particular case. Some authors, instead of using single-electron orbitals in (11.16), define $10Dq$ as the difference in energy between many-electron states. With the explicit form of $\mathbf{h}$, it is possible to identify the various contributions to $10Dq$, such as the point charge and the correction due to the spread of ligand wavefunctions. Kleiner[20] found that the point charge contribution to Dq for chrome alum is $-550\ \text{cm}^{-1}$, whereas the experimental value is 1750 cm^{-1}. Tanabe and Sugano[21] obtained $Dq = 4750\ \text{cm}^{-1}$ for the same compound. Both these results are thus in serious disagreement with experiment. The reason for Kleiner's disagreement is his neglect of the exchange forces between the metal and the ligand orbitals. The inclusion of this effect by Tanabe and Sugano corrects the sign of Dq, but the magnitude is almost three times that of the experimental value. Generally it is found that the point charge (ionic) contribution is only part of the total splitting.

There have been several theoretical estimates of Dq (ionic) for the complex $Ni^{2+}F_6^-$ in $KNiF_3$. The results are in the range[22–24]

$$10Dq\ (\text{ionic}) = 2300 \pm 300\ \text{cm}^{-1} \tag{11.17}$$

whereas the experimental value is $7250\ \text{cm}^{-1}$. Thus the ionic model accounts for only about one third of the observed splitting, whereas the remaining two thirds ($\sim 5000\ \text{cm}^{-1}$) is generally attributed to the effect of covalent bonding.

In the MO framework the effect of the covalent bonding making a contribution to the crystal field splitting corresponds to the difference between the antibonding energy shifts for σ and π orbits, which we shall discuss in considerable detail at the end of this section. Using in (11.16), the admixture coefficients for an octahedral complex given by (11.21), we can show approximately that the covalent contribution to $10Dq$ is given by

$$\begin{aligned} 10Dq_{\text{cov}} &= \Delta E_\sigma - \Delta E_\pi \\ &\simeq (\gamma_\sigma^2 - \gamma_\pi^2)(E_d - E_p) + \gamma_s^2(E_d - E_s) \end{aligned} \tag{11.18}$$

Although (11.18) is very approximate, it suggests that $10Dq$ may depend sensitively on the relative strengths, $\gamma_\sigma^2 - \gamma_\pi^2$, of the p_σ and p_π bonding. For the complex $Ni^{2+}F_6^-$, to be described later, we have approximately $\gamma_\sigma^2 - \gamma_\pi^2 \simeq 6\%$ and[71] $E_d - E_p \sim 10$ eV. Hence the term in (11.18) contributes about $5000\ \text{cm}^{-1}$ to $10Dq_{\text{cov}}$. It is interesting to note that this is about what is required to explain the experimental results. This estimate is, of course, very crude and therefore should not be taken too seriously.

Thus the failure of the purely electrostatic crystal field theory to predict the magnitude of $10Dq$ is now recognized as being due to the neglect of covalence. It is clear from the discussion above that the relative extent of involvement of the metal t_{2g} and e_g orbitals in bond formation will profoundly influence the extent of their energetic separation since the e_g orbitals interact in a σ-sense with the ligand s and p orbitals, whereas the t_{2g} orbitals interact only in a π-sense with the ligand p_π orbitals. Such a covalent bonding process would be expected to bring about a reduction in the effective central field of the metal and to allow a spreading out of the metal radial function.

Table 11.2 gives some Dq values for the $3d$, $4d$, and $5d$ series of complexes. It can easily be seen from this table that the values of Dq determined experimentally for the divalent ions of the first transition series are about 1000 cm^{-1}, and for the trivalent ions are about 1700 cm^{-1}. They are larger in the second and third series. The increase of Dq for the trivalent ions is qualitatively consistent with our findings from the magnetic susceptibility (Chapter 7) since the trivalent ions are found to be more covalent than the divalent ions. The variation of Dq from one ion to another in the same valence state, however, is not yet completely understood.

Next we consider what happens when the metal ion is fixed and the ligands are varied. Some specific examples are given in Table 11.2, where it is seen that Dq increases steadily from left to right for the ligands. With a few exceptions it is always possible to arrange the ligands in the following series:

$$I^- < Br^- < Cl^- < F^- < H_2O < NH_3 < CN^- \tag{11.19}$$

such that Δ $(=10Dq)$ increases from left to right for a given metal ion. This is known as the spectrochemical series, first observed by Jorgensen.[25] This sequence cannot be explained by the simple assumption that Dq increases as the covalency increases because one would expect Br^- and Cl^- to be more covalent than F^-.

To explain the experimental value of $10Dq$ for the complex $(NiF_6)^{4-}$ in $KNiF_6$, Rimmer[33] in 1965 did a calculation by applying the idea of configuration interaction and indeed obtained a nice agreement with experiment. The configuration interaction approach to the theory of covalency in magnetic salts can be regarded as an extension of the MO theory. There is an important advantage of this method, which lies in the fact that it allows for various kinds of numerically important electron-correlation effects. Let us consider the example of $Ni^{2+}F^-$. After the transfer of an electron from the F^- site to Ni^{2+}, the system becomes Ni^+F according to the configuration interaction (CI) method. The unpaired spin on the ligand is then in a $2p$ orbit on a neutral F atom. In the MO method the wavefunction is built up from a $3d$ orbit on Ni^{2+} and a $2p$ orbit on F^-. The ligand unpaired spin is then usually interpreted as being in a $2p$ orbit on a F^- ion. In explaining the experimental ligand hyperfine structure, however, one must decide whether to assume orbits on F or F^-. Since the CI method gives better results than the

Table 11.2

Dq values of kd^n ($k=3,4,5$) complexes[a]

Configuration	Ion	Ligand					
		Br^-	Cl^-	F^-	H_2O	NH_3	CN^-
$3d^1$	Ti^{3+}				2030		
$3d^2$	V^{3+}				1840		
$3d^3$	V^{2+}				1240		
	Cr^{3+}		1380	1520	1740	2160	2670
	Mn^{4+}			2180			
$3d^5$	Mn^{2+}		750	840	850		
	Fe^{3+}			1400	1430		
$3d^6$	Fe^{2+}				1040		3140
	Co^{3+}				1820	2290	3350
$3d^7$	Co^{2+}				930	1010	
$3d^8$	Ni^{2+}	700	720	730	850	1080	
$3d^9$	Cu^{2+}		650		1260	1510	
$4d^3$	Mo^{3+}		1920	2350			
	Tc^{4+}			2840			
$4d^5$	Ru^{3+}		2160				
$4d^6$	Rh^{3+}	1900		2230	2200	3410	
	Pd^{4+}			2590			
$4d^8$	Ag^{3+}			1840			
$5d^3$	Re^{4+}		2750	3200			
	Ir^{3+}	2310		2500			
	Pt^{4+}			3300			

[a]From C. K. Jorgensen, *Absorption Spectra and Chemical Bonding in Complexes*, Pergamon Press, Oxford, England, 1962; D. S. McClure, *Electronic Spectra of Molecules and Ions in Crystals*, Academic Press, New York, 1959; G. L. Allen and K. D. Warren, *Struct. Bonding* **9**, 49 (1971).

MO method, it seems likely that the assumption of neutral F is the better one. However, until more calculations are performed on different types of complexes, it is difficult to say for sure whether the configuration interaction, along with the ionic contribution, can explain the observed crystal field splitting, as well as the unpaired spin densities at the ligand sites. We shall not describe the CI method in any detail; the reader is referred to the paper by Hubbard, Rimmer, and Hopgood.[34] Instead we shall study in Section 11.6 another semiempirical method, first proposed by Wolfsberg and Helmholz, which offers an approach to a quantitative evaluation of the nature of the electronic interactions in transition metal compounds.

Before leaving this section, let us, for example, see how Sugano and Shulman[5] estimated the covalence and the cubic field splitting parameter $10Dq$ in ionic crystals within the framework of the self-consistent field molecular orbital (SCFMO) theory. They did the calculation, of course, with particular reference to $KNiF_3$, as discussed earlier in this section. It should be pointed out here that, although they were successful in obtaining a reasonable agreement with experiment, no theoretically rigorous derivation of the Hartree-Fock Hamiltonian was attempted at that time. Later Sugano and Tanabe[7] remedied this deficiency and studied from a theoretical standpoint the questions of what types of expression should be used to estimate the covalency and $10Dq$ and what kind of Hamiltonian should be used in such a formalism. This study was based on the SCFMO theory for open shells proposed by Roothaan, which we described in some detail in Chapter 3. Sugano and Tanabe, in this investigation, showed that the effect of the rearrangement among the virtually transferred electron and the *d*-electrons is essential for an understanding of covalence. One can take this effect of rearrangement into account in a simple fashion in the Heitler-London framework, in which one considers an admixed excited configuration where the AOs of the metal and the ligand ions are different from those in the ground configuration. This formalism gives a reasonable value of the covalence and also of the cubic field splitting $10Dq$. With these preliminary remarks let us now discuss these investigations in some details.

It is well known that, to get the orbital energies E and the MO wavefunctions Ψ, one has to solve the Hartree-Fock equation, given by

$$\mathbf{h}\Psi = E\Psi \tag{11.20}$$

where $\mathbf{h}$ is the Hartree-Fock Hamiltonian. We shall see in the following discussion that this Hamiltonian will involve the coefficients of the molecular orbitals γ, given below. The standard procedure for solving (11.20) is to fix γ in the Hamiltonian and then vary γ in the wavefunctions so as to minimize the energy. We have the final solution when, after a number of successive iterations, γ in the Hamiltonian becomes identically equal to γ in the wavefunctions.

As discussed earlier, for d electrons in a crystal field of O_h symmetry, the molecular orbitals of interest are two antibonding MOs, given by

$$\begin{aligned}\Psi_e^a &= (N_e)^{-1/2}(\phi_e - \lambda_s \chi_s - \lambda_\sigma \chi_\sigma) \\ \Psi_t^a &= (N_t)^{-1/2}(\phi_t - \lambda_\pi \chi_\pi)\end{aligned} \tag{11.21}$$

where the ϕ's are the metal orbitals and the χ's are appropriate linear combinations of the ligand atomic orbitals. These were listed in Table 11.1. The subscripts t and e are abbreviations for t_{2g} and e_g, which are, as already seen, Mulliken's notations for the irreducible representations of the cubic (O_h) group, and σ and π denote the p_σ and p_π orbitals of the ligands, respectively. Both the ϕ's and the χ's are normalized, so that we have

$$\begin{aligned}N_e &= 1 - 2\lambda_s S_s + 2\lambda_\sigma S_\sigma + \lambda_s^2 + \lambda_\sigma^2 \\ N_t &= 1 - 2\lambda_\pi S_\pi + \lambda_\pi^2\end{aligned} \tag{11.22}$$

where S denotes the overlap integral:

$$S = \langle \phi | \chi \rangle \tag{11.23}$$

Similarly, the bonding orbitals which are orthogonal to the antibonding MOs in (11.21) are given by

$$\begin{aligned}\Psi_{es}^b &= (N_e')^{-1/2}(\chi_s + \gamma_s \phi_e + \gamma_{s\sigma} \chi_\sigma) \\ \Psi_{e\sigma}^b &= (N_e'')^{-1/2}(\chi_\sigma + \gamma_\sigma \phi_e + \gamma_{\sigma s} \chi_s) \\ \Psi_t^b &= (N_t')^{-1/2}(\chi_\pi + \gamma_\pi \phi_t)\end{aligned} \tag{11.24}$$

It is important to note that in the so-called ionic crystals the orbitals of (11.24) have predominantly the nature of ligand orbitals. We now assume that λ, γ, and S are small quantities of the order of ϵ ($\epsilon \ll 1$), and neglect, therefore, small quantities of higher order, and obtain

$$\begin{aligned}\lambda_s &= \gamma_s + S_s \\ \lambda_\sigma &= \gamma_\sigma + S_\sigma \\ \lambda_\pi &= \gamma_\pi + S_\pi\end{aligned} \tag{11.25}$$

These can be easily derived from the orthogonalities

$$\langle \Psi_e^a | \Psi_{es}^b \rangle = \langle \Psi_e^a | \Psi_{e\sigma}^b \rangle = \langle \Psi_t^a | \Psi_t^b \rangle = 0 \tag{11.26}$$

To obtain the crystal field splitting parameter $10Dq$, let us assume, along with Sugano and Shulman, that the antibonding and bonding wavefunctions given by

$$\Psi^a = (N_1)^{-1/2}(\phi - \lambda\chi) \tag{11.27}$$

$$\Psi^b = (N_2)^{-1/2}(\chi + \gamma\phi) \tag{11.28}$$

which are the exact eigenfunctions of (11.20). Putting (11.27) in (11.20) and integrating, after multiplication by ϕ from the left, we get

$$\mathbf{h}\Psi^a = E^a\Psi^a$$

or

$$\mathbf{h}N_1^{-1/2}(\phi - \lambda\chi) = E^a N_1^{-1/2}(\phi - \lambda\chi) \tag{11.29}$$

or

$$\langle\phi|\mathbf{h}|\phi\rangle - \lambda\langle\phi|\mathbf{h}|\chi\rangle = E^a\{\langle\phi|\phi\rangle - \lambda\langle\phi|\chi\rangle\}$$

or

$$\langle\phi|\mathbf{h}|\phi\rangle - \lambda\langle\phi|\mathbf{h}|\chi\rangle = E^a\{1 - \lambda S\}$$

or

$$E^a = \frac{1}{1-\lambda S}\left[\langle\Phi|\mathbf{h}|\phi\rangle - \lambda\langle\phi|\mathbf{h}|\chi\rangle\right] \tag{11.30}$$

Similarly, multiplying by χ from the left and integrating, we get

$$\langle\chi|\mathbf{h}|\phi\rangle - \lambda\langle\chi|\mathbf{h}|\chi\rangle = E^a\{\langle\chi|\phi\rangle - \lambda\langle\chi|\chi\rangle\}$$

or

$$E^a = \frac{\langle\chi|\mathbf{h}|\phi\rangle - \lambda\langle\chi|\mathbf{h}|\chi\rangle}{S - \lambda} \tag{11.31}$$

From (11.30) and (11.31) we easily obtain

$$E^a = \frac{1}{1-\lambda^2}\left[\langle\phi|\mathbf{h}|\phi\rangle - \lambda^2\langle\chi|\mathbf{h}|\chi\rangle\right] \tag{11.32}$$

In a similar fashion, inserting (11.28) in (11.20), we get

$$E^b = \frac{1}{1+\gamma S}\left[\langle\chi|\mathbf{h}|\chi\rangle + \gamma\langle\phi|\mathbf{h}|\chi\rangle\right] \tag{11.33}$$

$$= \frac{1}{1-\gamma^2}\left[\langle\chi|\mathbf{h}|\chi\rangle - \gamma^2\langle\phi|\mathbf{h}|\phi\rangle\right] \tag{11.34}$$

In our problem the Ψ_t's have the forms given in (11.27) and (11.28), but the Ψ_e's have the more complicated forms given in Table 11.1. But, as shown by Sugano and Shulman, even for the Ψ_e's we can use (11.30) and (11.31), provided that we replace χ, S, and λ by χ_e, S_e, and λ_e, respectively, which are now defined as follows:

$$\begin{aligned} \chi_e &= \mu_s \chi_s + \mu_\sigma \chi_\sigma \\ S_e &= \mu_e S_s + \mu_\sigma S_\sigma \\ \lambda_e &= \frac{\lambda_s}{\mu_s} = \frac{\lambda_\sigma}{\mu_\sigma} \end{aligned} \tag{11.35}$$

where

$$\mu_s^2 + \mu_\sigma^2 = 1$$

Now the cubic crystal field splitting parameter, $\Delta = 10Dq$, is defined as

$$\Delta = E_e^a - E_t^a \tag{11.36}$$

We may now write

$$E_e^a = \frac{1}{1 - \lambda_\sigma S_e}\left[\langle\phi_e|\mathbf{h}|\phi_e\rangle - \lambda_e\langle\phi_e|\mathbf{h}|\chi_e\rangle\right]$$

$$E_t^a = \frac{1}{1 - \lambda_t S_t}\left[\langle\phi_t|\mathbf{h}|\phi_t\rangle - \lambda_t\langle\phi_t|\mathbf{h}|\chi_t\rangle\right]$$

Therefore

$$\begin{aligned} E_e^a - E_t^a &= (1 + \lambda_e S_e)\left[\langle\phi_e|\mathbf{h}|\phi_e\rangle - \lambda_e\langle\phi_e|\mathbf{h}|\chi_e\rangle\right] \\ &\quad - (1 + \lambda_t S_t)\left[\langle\phi_t|\mathbf{h}|\phi_t\rangle - \lambda_t\langle\phi_t|\mathbf{h}|\chi_t\rangle\right] \\ &= \left[\langle\phi_e|\mathbf{h}|\phi_e\rangle - \langle\phi_t|\mathbf{h}|\phi_t\rangle\right] \\ &\quad - \left[\lambda_e\langle\phi_e|\mathbf{h}|\chi_e\rangle - \lambda_t\langle\phi_t|\mathbf{h}|\chi_t\rangle\right] \\ &\quad + \left[(\lambda_e S_e - \lambda_t S_t)\langle\phi_e|\mathbf{h}|\phi_e\rangle\right] \end{aligned} \tag{11.37}$$

where we have neglected terms containing λ^2 and assumed that $\langle\phi_t|\mathbf{h}|\phi_t\rangle \simeq \langle\phi_e|\mathbf{h}|\phi_e\rangle$. Here, of course, we have used the fact that the difference $[\langle\phi_e|\mathbf{h}|\phi_e\rangle - \langle\phi_t|\mathbf{h}|\phi_t\rangle]$ is of the same order as $\lambda\langle\phi|\mathbf{h}|\chi\rangle$, which has the order ϵ^2 when the order of $\langle\chi|\mathbf{h}|\chi\rangle$ is unity. By using relations (11.25), we can rewrite (11.37) as follows:

$$\begin{aligned} \Delta &= \left[\langle\phi_e|\mathbf{h}|\phi_e\rangle - \langle\phi_t|\mathbf{h}|\phi_t\rangle\right] \\ &\quad + \left[(\lambda_s S_s + \lambda_\sigma S_\sigma - \lambda_\pi S_\pi)\langle\phi_e|\mathbf{h}|\phi_e\rangle\right] \\ &\quad - \left[S_s\langle\phi_e|\mathbf{h}|\chi_s\rangle + S_\sigma\langle\phi_e|\mathbf{h}|\chi_\sigma\rangle - S_\pi\langle\phi_t|\mathbf{h}|\chi_\pi\rangle\right] \\ &\quad - \left[\gamma_s\langle\phi_e|\mathbf{h}|\chi_s\rangle + \gamma_\sigma\langle\phi_e|\mathbf{h}|\chi_\sigma\rangle - \gamma_\pi\langle\phi_t|\mathbf{h}|\chi_\pi\rangle\right] \end{aligned} \tag{11.38}$$

The first bracket of (11.38) represents the point charge contribution, Kleiner's[20] correction, and the exchange interaction between the metal and the ligand electrons. The second bracket is derived from the renormalization. The third bracket comes from the nonorthogonality between the metal and ligand orbitals, and the last bracket from the covalence, which is measured by γ. Phillips[60] argued that the terms in the third bracket cancel Kleiner's correction and the exchange terms in the first bracket. Shulman and Sugano[22] have shown numerically that this argument is approximately correct, although the main contribution to Δ comes from the fourth bracket representing the covalence term, and not from the point charge term.

A number of very useful relations also derived by Shulman and Sugano can easily be derived from the relations given above. For example, for t_{2g} orbitals, λ_π is determined by (11.31) and (11.32) and is given by

$$\lambda_\pi = \frac{[-\langle\phi_t|\mathbf{h}|\chi_\pi\rangle + S_\pi\langle\phi_t|\mathbf{h}|\phi_t\rangle]}{[\langle\phi_t|\mathbf{h}|\phi_t\rangle - \langle\chi_\pi|\mathbf{h}|\chi_\pi\rangle]} \tag{11.39}$$

Similarly, by equating (11.33) and (11.34), we have

$$\gamma_\pi = \frac{[-\langle\phi_t|\mathbf{h}|\chi_\pi\rangle - S_\pi\langle\chi_\pi|\mathbf{h}|\chi_\pi\rangle]}{[\langle\phi_t|\mathbf{h}|\phi_t\rangle - \langle\chi_\pi|\mathbf{h}|\chi_\pi\rangle]} \tag{11.40}$$

Also we can show that

$$\gamma_s = \frac{[-\langle\phi_e|\mathbf{h}|\chi_s\rangle + S_s\langle\chi_s|\mathbf{h}|\chi_s\rangle]}{[\langle\phi_e|\mathbf{h}|\phi_e\rangle - \langle\chi_s|\mathbf{h}|\chi_s\rangle]} \tag{11.41}$$

$$\gamma_\sigma = \frac{[-\langle\Phi_e|\mathbf{h}|\chi_\sigma\rangle + S_\sigma\langle\chi_\sigma|\mathbf{h}|\chi_\sigma\rangle]}{[\langle\phi_e|\mathbf{h}|\phi_e\rangle - \langle\chi_\sigma|\mathbf{h}|\chi_\sigma\rangle]} \tag{11.42}$$

and

$$\gamma_{s\sigma} = \frac{\langle\chi_s|\mathbf{h}|\chi_\sigma\rangle}{[\langle\chi_s|\mathbf{h}|\chi_s\rangle - \langle\chi_\sigma|\mathbf{h}|\chi_\sigma\rangle]} \tag{11.43}$$

We are now in a position to apply the MO formalism presented above in the case of $KNiF_3$, as was done for the first time by Shulman and Sugano.

The Hamiltonian $\mathbf{h}$ is given by

$$\mathbf{h} = \mathbf{h}_0 + \mathbf{h}' + V_{\text{crys}} \tag{11.44}$$

where $\mathbf{h}_0$ represents the Hamiltonian of a purely ionic $Ni^{2+}F_6^-$ complex in which the overlap between the metal atom and the ligands are neglected ($S=0$), and $\mathbf{h}'$ is the correction due to deviation from this ionic model.

Following Tanabe and Sugano,[21] we have

$$\mathbf{h}_0 = -\tfrac{1}{2}\Delta_{kE} + V_M + V_L \qquad (11.45)$$

in which $-\frac{1}{2}\Delta_{kE}$ is the kinetic energy operator, V_M is the Coulomb and exchange interaction operator of all the Ni^{2+} electrons and the Ni^{2+} nucleus and is of the form

$$V_M = V_{\text{core}} + V_d^{\text{Coul}} + V_d^{\text{ex}} \qquad (11.46)$$

where V_{core} comes from the nucleus and the core electrons up to and including the $3p$ shell and has been assumed to be

$$V_{\text{core}} = \frac{-10 + 31.0e^{-4.73r}}{r} \qquad (11.47)$$

This is obtained by an analytical fit to Watson's[61] Hartree-Fock core potential of Ni^{2+}. Also, V_d^{Coul} and V_d^{ex} are given by

$$V_d^{\text{Coul}} = \sum_{\gamma=\xi\xi\eta\eta\zeta\theta\epsilon} \int \frac{|\phi_\gamma(2)|^2}{r_{12}}\, d\tau_2 \qquad (11.48)$$

and

$$V_d^{\text{ex}} = \sum_{\gamma=\xi\eta} \int \frac{\phi_\gamma^*(2)\phi_\gamma(1)P_{12}}{r_{12}}\, d\tau_2 \qquad (11.49)$$

where P_{12} is the permutation operator for electrons 1 and 2, and (ξ,η,ζ), (θ,ϵ) are the strong-field wavefunctions for t_{2g} and e_g, given in Chapter 10.

From (11.36) we have, for Ni^{2+}, the cubic field splitting parameter

$$\Delta\,(\equiv 10Dq) = E\left(t_{2g}^5 e_g^3, {}^3T_2\right) - E\left(t_{2g}^6 e_g^2, {}^3A_2\right)$$

$$= \langle\epsilon|\mathbf{h}|\epsilon\rangle - \langle\zeta|\mathbf{h}|\zeta\rangle \qquad (11.50)$$

The last line of relation (11.50) automatically follows from the detailed electronic configurations of the ζ component of the ${}^3T_2$ state and of the ${}^3A_2$ state, which can be written as

$${}^3A_2 = \xi^+\xi^-\eta^+\eta^-\zeta^+\zeta^-\theta^+\epsilon^+ \qquad (11.51)$$

and

$${}^3T_2 = \xi^+\xi^-\eta^+\eta^-\zeta^+\theta^+\epsilon^+\epsilon^-$$

In (11.45) V_L is the contribution from the six F ions and can be decomposed as follows:

$$V_L = V_L^p + V_L^K + V_L^{ex} \tag{11.52}$$

where V_L^p, the point charge contribution from the six ligands, can be written as

$$V_L^p = \sum_{i=1}^{6} \frac{1}{|\mathbf{r}-\mathbf{R}_i|} \tag{11.53}$$

This comes from a single negative point charge, assumed to be situated at $\mathbf{R}_i$ of the six F nuclei. As already discussed, V_L^K in (11.52) is Kleiner's additional potential due to imperfect screening of the ligand nuclear charge by the ligand electrons; it is expressed as

$$V_L^K = \sum_{i=1}^{6} - \frac{8}{|\mathbf{r}-\mathbf{R}_i|} + 2\int \frac{d\tau_2}{r_{12}} \sum_{k=2s,2p(x,y,z)} |\phi_{ik}(2)|^2 \tag{11.54}$$

In (11.52) V_L^{ex} is the exchange interaction operator, introduced by Tanabe and Sugano in the form

$$V_L^{ex} = -\sum_{i=1}^{6} \int \frac{d\tau_2}{r_{12}} \sum_{k=2s,2p(x,y,z)} \phi_{ik}^*(2)\phi_{ik}(1)P_{12} \tag{11.55}$$

It is important to note that V_L^{ex} makes an important contribution to Δ.

The Hamiltonian $\mathbf{h}'$ in (11.44) makes only a small contribution and does so only when the overlap integral S and the covalency γ are small; therefore it has been neglected by Sugano and Shulman. This is a reasonable assumption for ionic crystals of the first transition series. For the second and third transition series, however, this assumption is not valid since in these complexes both S and γ are appreciable.

In (11.44) V_{crys} represents the crystal field effects arising because the molecule is embedded in a crystal. Assuming that the electron of interest is localized in the molecule for which $\mathbf{h}$ is given, we are primarily interested in V_{crys} in the region of the molecule. As discussed earlier, in the crystal field theory we are allowed to neglect the periodicity of the crystalline lattice since experimentally it is well known that a cubic crystal field parameter is insensitive to the surroundings beyond the nearest-neighbor ligands. For example, the Δ's in $Ni(NH_3)_6^{2+}$ and $Ni(en)_6^{2+}$ are almost the same;[21] 10,300 and 10,800 cm^{-1}, respectively. From these considerations we can assume that V_{crys} is a constant in the region of the $[NiF_6]^{4-}$ molecule in which we are presently interested. Therefore, V_{crys} has been neglected here since it merely shifts the origin in the energy scale.

As long as V_{crys} in (11.44) is assumed to be constant, the eigenfunction satisfying (11.20) is a molecular orbital of the $[NiF_6]^{4-}$ molecule. Also, in ionic crystals, such as $KNiF_3$, it is a good starting approximation to assume that the molecular orbital Ψ is a linear combination of atomic orbitals. These were listed in Table 11.1. The AOs for the present problem are, of course, the Hartree-Fock solutions of Ni^{2+} and F^- ions.

The SCF wavefunctions for Ni^{2+} and F^- can be obtained from various sources.[61] The 3*d* function of Ni^{2+}, taken from Watson,[61] is

$$R_{3d}(r) = r^2(a_1e^{-\alpha_1 r} + a_2e^{-\alpha_2 r} + a_3e^{-\alpha_3 r} + a_4e^{-\alpha_4 r}) \tag{11.56a}$$

and the ones for F^- are

$$R_{2p}(r) = r(b_1e^{-\beta_1 r} + b_2e^{-\beta_2 r})$$

and (11.56b)

$$R_{2s}(r) = (c_1e^{-\gamma_1 r} + c_2e^{-\gamma_2 r})$$

where

$$\begin{aligned} a_1 &= 3.4096 & \alpha_1 &= 2.315 \\ a_2 &= 45.261 & \alpha_2 &= 4.523 \\ a_3 &= 129.48 & \alpha_3 &= 8.502 \\ a_4 &= 24.071 & \alpha_4 &= 15.01 \\ b_1 &= 15.671 & \beta_1 &= 3.7374 \\ b_2 &= 1.5742 & \beta_2 &= 1.3584 \\ c_1 &= -11.156 & \gamma_1 &= 8.70 \\ c_2 &= 10.805 & \gamma_2 &= 2.425 \end{aligned}$$

The next stage in the calculation is the evaluation of the relevant matrix elements, which we shall not consider. The reader can easily obtain them from Ref. 5.

We now present, in Table 11.3, different contributions to $10Dq$. A glance at the table will show that the point charge contribution to $10Dq$ is only 1390 cm^{-1}, while Kleiner's correction makes $10Dq$ negative. Thus the ionic model ($S=0$, $\gamma=0$), which includes only the contribution from the Ni diagonal term, predicts a negative value of $10Dq$, which disagrees with experiment. After the contributions from the Ni diagonal term we have listed the contributions from nonorthogonality. Here, of course, our assumption is that $\gamma=0$ but $S\neq 0$, meaning that the metal ion orbitals are orthogonalized to the ligand orbitals. It is also clear from the table that covalence makes the major contribution to $10Dq$ (5290 cm^{-1}), even in the so-called ionic complexes. Thus the triumph of calculating the $10Dq$ parameter, in the case of $KNiF_3$, by using the molecular orbitals in the Hartree-Fock scheme, for the first time,

Table 11.3

Various contributions to $10Dq$ parameter of $KNiF_3$[a]

Origin	Term	Contribution to $10Dq$ (cm^{-1})
Diagonal	V_L^P	+1390
	V_L^K	−2080
	I_L^{ex}	−2880
	Renormalization $(\lambda_e S_e - \lambda_t S_t)\langle\phi_e\|\mathbf{h}\|\phi_e\rangle$	+ 900
Nonorthogonality	$-S_s\mu_s\langle\phi_e\|\mathbf{h}_0\|\chi_s\rangle$	+2060
	$-S_\sigma\mu_\sigma\langle\phi_e\|\mathbf{h}_0\|\chi_\sigma\rangle$	+2390
	$+S_\pi\langle\phi_t\|\mathbf{h}_0\|\chi_\pi\rangle$	− 720
Covalency	$-\gamma_s\mu_s\langle\phi_e\|\mathbf{h}_0\|\chi_s\rangle$	+ 790
	$-\gamma_\sigma\mu_\sigma\langle\phi_e\|\mathbf{h}_0\|\chi_\sigma\rangle$	+6170
	$+\gamma_\pi\langle\phi_t\|\mathbf{h}_0\|\chi_\pi\rangle$	−1670
Totals		
Theoretical		6350
Experimental		7250

[a]From S. Sugano and R. G. Shulman, *Phys. Rev.* **130**, 517 (1963).

goes to Sugano and Shulman. The reason why this model calculation has been incorporated is to show that one can, in principle, calculate the crystal field parameters starting from the theoretical standpoint. We shall see more of such calculations at the end of this chapter, though very briefly.

11.3 ORBITAL REDUCTION FACTOR

The fact that the introduction of covalency into an ionic complex leads to a reduction in the orbital contribution to the magnetic moment was first recognized by Owen and Stevens[62] in the strongly bound complex $Ir(Cl^-)_6$, $5d^5(t_{2g}^5)$, where the experimental g value was found to be 10% smaller than predicted by the ionic model. Stevens[14] gave the basic theory, which was later extended by a number of authors. It is of interest that F. S. Ham has also shown that in some complexes the dynamic Jahn-Teller effect can give rise to an equally important reduction in the angular momentum. We shall deal with this aspect in the next chapter.

Let us see, in a simple way, how covalency reduces the orbital moment. In the ionic picture the orbital moment of an unpaired electron is that associated with the central d orbit. When covalent bonding is taken into consideration, a fraction f of the unpaired spin is transferred to the ligands and the amount of unpaired electron in the central d orbit is proportional to $(1-f)$. The orbital moment then is reduced by a factor $k \sim (1-f)$. This is the qualitative result obtained by a detailed calculation which we shall see below. Typically, $k \sim 0.8$ to 0.9 for the complexes of the $3d$ transition metal ions, and $k \sim 0.5$ for the complexes of the second and third transition metal ions.

In a formal way, we can define the orbital reduction factor by

$$k_{ij} = \frac{\langle \Psi_i | \mathbf{l} | \Psi_j \rangle}{\langle \phi_i | \mathbf{l} | \phi_j \rangle} \tag{11.57}$$

where $\mathbf{l}$ is the orbital angular momentum operator, the Ψ's are the molecular orbitals given by (11.21) and (11.24), and the ϕ's are the purely ionic d wavefunctions of the metal atom. It is important to realize that there can be only two reduction factors; these are $k_{\pi\pi}$ (within the t_{2g} orbitals) and $k_{\sigma\pi}$ (between the t_{2g} and e_g orbitals). The third possibility, $k_{\sigma\sigma}$, does not arise since the orbital angular momentum associated with σ-type orbitals is zero, that is, $\langle e_g^i | \mathbf{l} | e_g^j \rangle = 0$.

To calculate $k_{\pi\pi}$ and $k_{\sigma\pi}$, we consider the matrix element of the angular momentum operator $\mathbf{l}_z$. Using the ionic wavefunctions (9.2), we can show that

$$\langle t_{2g}\xi | \mathbf{l}_z | t_{2g}\eta \rangle = i$$

and (11.58)

$$\langle t_{2g}\zeta | \mathbf{l} | e_g \epsilon \rangle = 2i$$

The numerator of (11.57) is now calculated using the MOs given by (11.21) and (11.24).

Let us remind ourselves that to set up the appropriate MOs the usual procedure is to take the linear combinations of the s and p (σ and π) valence orbitals belonging to the six ligands (see Figure 11.1), which transform as various irreducible representations of the octahedral group. These can then be appropriately combined with the central-ion orbitals, which transform as the same irreducible representation. We need in our calculation only the antibonding orbitals that describe the distribution of the unpaired electron spins, and they are written explicitly as follows (see Table 11.1):

$$\left.\begin{aligned}
\Psi_\theta &= N_\sigma\big\{ e_g\theta - (12)^{-1/2}\lambda_\sigma(-2z_3+2z_6+x_1-x_4+y_2-y_5) \\
&\qquad - (12)^{-1/2}\lambda_s(2s_3+2s_6-s_1-s_2-s_4-s_5)\big\} \\
\Psi_\epsilon &= N_\sigma\big\{ e_g\epsilon - \tfrac{1}{2}\lambda_\sigma(-x_1+x_4+y_2-y_5)\big\} \\
\Psi_\xi &= N_\pi\big\{ t_{2g}\xi - \tfrac{1}{2}\lambda_\pi(z_2-z_5+y_3-y_6)\big\} \\
\Psi_\eta &= N_\pi\big\{ t_{2g}\eta - \tfrac{1}{2}\lambda_\pi(x_3-x_6+z_1-z_4)\big\} \\
\Psi_\zeta &= N_\pi\big\{ t_{2g}\zeta - \tfrac{1}{2}\lambda_\pi(y_1-y_4+x_2-x_5)\big\}
\end{aligned}\right\} \tag{11.59}$$

where the numbers 1,2,3,4,5,6 refer to ligands on the $x,y,z,-x,-y,-z$ axes of the octahedron, respectively (see Figure 11.1); x,y,z represent the p-type orbits of the outer shell of the ligands; and the s's represent the s orbits of the ligands. In (11.59), for example, x_1 has a lobe pointing toward the central ion, has zero angular momentum about the bond axis, and is a p_σ orbit. Similarly, y_1 and z_1 are p_π orbits. The normalization constants are easily calculated to be

$$N_\sigma^{-2}=1-4\lambda_\sigma S_\sigma-4\lambda_s S_s+\lambda_\sigma^2+\lambda_s^2$$
$$N_\pi^{-2}=1-4\lambda_\pi S_\pi+\lambda_\pi^2 \qquad (11.60)$$

where the S's are the overlap integrals, given by

$$S_\sigma=-\langle e_g\epsilon|x_1\rangle$$
$$S_s=-\langle e_g\epsilon|s_1\rangle \qquad (11.61)$$
$$S_\pi=\langle t_{2g}\zeta|y_1\rangle$$

Now, in order to obtain $k_{\pi\pi}$, we must evaluate the following matrix element:

$$N_\pi^2\langle t_{2g}\xi-\tfrac{1}{2}\lambda_\pi(z_2-z_5+y_3-y_6)|\mathbf{l}_z|t_{2g}\eta-\tfrac{1}{2}\lambda_\pi(x_3-x_6+z_1-z_4)\rangle \qquad (11.62)$$

In (11.62) there are some elements that can be evaluated most easily by noticing that $\langle t_{2g}\xi|\mathbf{l}_z|z_1\rangle$ must be purely imaginary since the wavefunctions are real and the orbital angular momentum operator $\mathbf{l}_z$ is imaginary. Hence, for example,

$$\langle t_{2g}\xi|\mathbf{l}_z|z_1\rangle=-\langle z_1|\mathbf{l}_z|t_{2g}\xi\rangle$$
$$=i\langle z_1|t_{2g}\eta\rangle=iS_\pi$$

Proceeding in a similar way, we can show that

$$-\langle t_{2g}\zeta|\mathbf{l}_z|x_1\rangle=\langle t_{2g}\zeta|\mathbf{l}_z|x_4\rangle=\langle t_{2g}\zeta|\mathbf{l}_z|y_2\rangle$$
$$=-\langle t_{2g}\zeta|\mathbf{l}_z|y_5\rangle=2iS_\sigma$$

and (11.63)

$$\langle t_{2g}\xi|\mathbf{l}_z|z_1\rangle=-\langle t_{2g}\xi|\mathbf{l}_z|z_4\rangle=-\langle t_{2g}\eta|\mathbf{l}_z|z_2\rangle$$
$$=\langle t_{2g}\eta|\mathbf{l}_z|z_5\rangle=iS_\pi$$

There is another type of elements in (11.62) which are also not difficult to evaluate since it is easy to operate with $\mathbf{l}_z$ on wavefunctions that lie on the z-axis, such as x_3 or y_6. On the other hand, elements of type $\langle y_2|\mathbf{l}_z|x_2\rangle$ are rather difficult to evaluate since this can be done only by transferring the

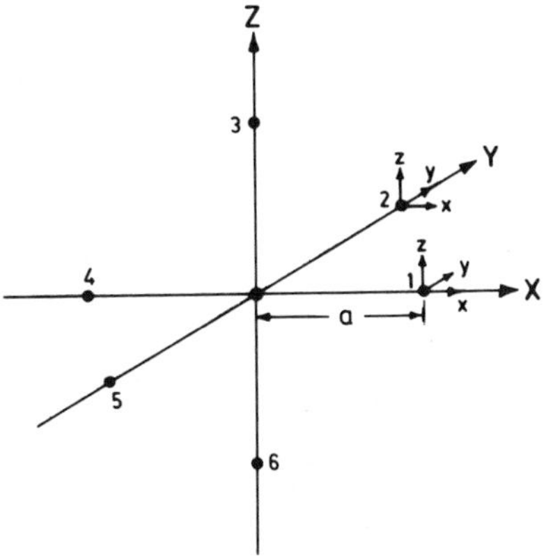

Fig. 11.3 Shifting of origin to ligand 1 (see text).

origin to ligand 2. With the help of (Figure 11.3), we can easily visualize what happens when the origin is transferred to ligand 1. In this case the coordinates at ligand 1 are related to those at the metal ion site at the center in the following way:

Ligand 1	Metal ion at the center
x	$=X-a$
y	$=Y$
z	$=Z$

Thus

$$\begin{aligned}\mathbf{l}_z &= -i\left(X\frac{\partial}{\partial Y} - Y\frac{\partial}{\partial X}\right)\\ &= -i\left[(a+x)\frac{\partial}{\partial y} - y\frac{\partial}{\partial x}\right]\\ &= -i\left(x\frac{\partial}{\partial y} - y\frac{\partial}{\partial x} + a\frac{\partial}{\partial y}\right)\\ &= \mathbf{l}_z(1) + a\mathbf{p}_y(1) \qquad (11.64)\end{aligned}$$

Similarly, by transferring the origin to ligand 2, we have

$$\mathbf{l}_z = \mathbf{l}_z(2) - a\mathbf{p}_x(2) \qquad (11.65)$$

In (11.64), $\mathbf{l}_z(1)$ is an angular momentum operator centered at position 1, $\mathbf{p}_y(1)$ is the y-component of the linear momentum at position 1, and the distance between the ligand and the center of the metal ion is denoted by a.

Let us now evaluate the element of type $\langle y_1|\mathbf{l}_z|x_1\rangle$ with the help of (11.64). We can easily show that

$$\langle y_1|\mathbf{l}_z|x_1\rangle = \langle y_1|\mathbf{l}_z(1) + ap_y(1)|x_1\rangle$$

$$= \langle y_1|\mathbf{l}_z(1)|x_1\rangle = +i \tag{11.66}$$

Proceeding similarly, we have

$$\langle x_2|\mathbf{l}_z|y_2\rangle = \langle x_5|\mathbf{l}_z|y_5\rangle = \langle x_3|\mathbf{l}_z|y_3\rangle = \langle x_6|\mathbf{l}_z|y_6\rangle$$

$$= -i$$

$$\langle y_2|\mathbf{l}_z|x_2\rangle = \langle y_5|\mathbf{l}_z|x_5\rangle = \langle y_3|\mathbf{l}_z|x_3\rangle = \langle y_6|\mathbf{l}_z|x_6\rangle$$

$$= +i \tag{11.67}$$

$$\langle y_1|\mathbf{l}_z|x_1\rangle = \langle y_4|\mathbf{l}_z|x_4\rangle = -\langle x_1|\mathbf{l}_z|y_1\rangle = -\langle x_4|\mathbf{l}_z|y_4\rangle$$

$$= +i$$

Now, using (11.63) and (11.67), we find that (11.62) is given by

$$iN_\pi^2\left(1 - 4\lambda_\pi S_\pi + \tfrac{1}{2}\lambda_\pi^2\right)$$

and finally, using (11.57) and (11.58), we have

$$K_{\pi\pi} = N_\pi^2\left(1 - 4\lambda_\pi S_\pi + \tfrac{1}{2}\lambda_\pi^2\right) \tag{11.68}$$

To obtain an analogous expression for $K_{\sigma\pi}$, we have to evaluate the following matrix element:

$$N_\sigma N_\pi \langle t_{2g}\zeta - \tfrac{1}{2}\lambda_\pi(y_1 - y_4 + x_2 - x_5)|\mathbf{l}_z|e_g\epsilon - \tfrac{1}{2}\lambda_\sigma(-x_1 + x_4 + y_2 - y_5)$$

$$- \tfrac{1}{2}\lambda_s(s_1 + s_4 - s_2 - s_5)\rangle$$

Proceeding in a way similar to that for $K_{\pi\pi}$ and using (11.57) and (11.58), we can easily show that

$$K_{\sigma\pi} = N_\sigma N_\pi\left(1 - 2\lambda_\pi S_\pi - 2\lambda_\sigma S_\sigma - \tfrac{1}{2}\lambda_\pi\lambda_\sigma\right) \tag{11.69}$$

Next we consider the contribution of the ligand s orbits, which we have so far neglected. For this, we can show that

$$\langle t_{2g}\zeta|\mathbf{l}_z|s_1 + s_4 - s_2 - s_5\rangle = 8iS_s \tag{11.70}$$

Here we must also evaluate the matrix elements like $\langle y_1|\mathbf{l}_z|s_1\rangle$. Using (11.64), we can show that

$$\begin{aligned}\langle y_1|\mathbf{l}_z|s_1\rangle &= \langle y_1|ap_y(1)|s_1\rangle \\ &= -ia\left\langle y_1\left|\frac{\partial}{\partial y}\right|s_1\right\rangle\end{aligned} \tag{11.71}$$

since $\langle y_1|\mathbf{l}_z(1)|s_1\rangle = 0$.

Now, using (11.70) and (11.71), we finally replace (11.69) by

$$K_{\sigma\pi} = N_\sigma N_\pi\Bigg(1 - 2\lambda_\pi S_\pi - 2\lambda_\sigma S_\sigma - 2\lambda_s S_s - \tfrac{1}{2}\lambda_\pi\lambda_\sigma - \tfrac{1}{2}\lambda_\pi\lambda_s a\left\langle y\left|\frac{\partial}{\partial y}\right|s\right\rangle\Bigg) \tag{11.72}$$

Expressions (11.68) and (11.72) can be further simplified if we neglect the difference between N_π^2, N_σ^2, and $N_\sigma N_\pi$. This is a reasonable approximation since all these quantities are close to unity. Therefore, making the approximation that the admixtures are small and using (11.60), we can rewrite (11.68) and (11.72) as

$$K_{\pi\pi} = 1 - \tfrac{1}{2}N_\pi^2\lambda_\pi^2 \tag{11.73}$$

and

$$K_{\sigma\pi} = 1 - \tfrac{1}{2}N_\sigma N_\pi\left(\lambda_\pi^2 + \lambda_\sigma^2 + \lambda_s^2 + \lambda_\pi\lambda_\sigma + \lambda_\pi\lambda_s a\left\langle y\left|\frac{\partial}{\partial y}\right|s\right\rangle\right) \tag{11.74}$$

Equations (11.73) and (11.74) were first derived by Tinkham.[63] But, in order to use (11.74), one needs to know the magnitude of the last term, about which various opinions exist in the literature. Tinkham assumed that it roughly canceled the last but one term, $\lambda_\sigma\lambda_\pi$, and this approximation was also used by Sugano and Shulman.[5] To show this cancellation, one has to use the Slater orbitals for the $2s$ and $2p$ wavefunction of the ligands—F^-, for example, for the complex $Ni^{2+}F_6^-$. If, however, the more realistic $2s$ and $2p$ Hartree-Fock wavefunctions, for example, the analytical wavefunctions of Sugano and Shulman,[5] are used, one can easily find that $a\langle y|\partial/\partial y|s\rangle = -1{\cdot}6$, where it is assumed that $a = 2.1$ Å, a reasonable separation between the metal and the F^- ligand.

We have thus found the expressions for the orbital reduction factors $K_{\pi\pi}$ and $K_{\sigma\pi}$. Let us now return to the example of $Ir^{4+}Cl_6^-$, $5d^5$ (t_{2g}^5) in order to obtain these factors from experiments. For the complex in question the g value computed[14] from the theory is given by

$$g = g_{\text{spin}} + g_{\text{orbit}} = \tfrac{2}{3} + \tfrac{4}{3}K_{\pi\pi} \tag{11.75}$$

Griffiths and Owen[64] found from the observed hyperfine structure of the chlorine ligands in this complex that $g=1.79$. Inserting this value in (11.75), one finds that $K_{\pi\pi}=0.84$. A similar analysis for the orbital reduction factor has been made by Thornley, Windsor, and Owen[65] for the complex $Co^{2+}F_6^-$ ($3d^7$), where the ground state is a mixture of both t_{2g} and e_g orbitals, so that both $K_{\pi\pi}$ and $K_{\sigma\pi}$ are involved. For complexes where the ground state is an orbital singlet, for example $Cr^{3+}F_6^-$ ($3d^3$) and $Ni^{2+}F_6^-$ ($3d^8$), the reduction of the orbital moment takes place via the spin-orbit coupling. In such cases the expressions for the g values are fairly complicated because of the effect of covalency in the spin-orbit coupling. We shall discuss this topic in the next section.

11.4 REDUCTION OF THE SPIN-ORBIT COUPLING

We saw in Chapter 7 that the spin-orbit coupling constant ζ is reduced from its free-ion value because of the effect of covalent bonding. The covalent bonding modifies the spin-orbit coupling in a way that is related to the reduction of the orbital momentum through the coefficients $K_{\pi\pi}$ and $K_{\sigma\pi}$, given by (11.73) and (11.74).

We saw in Chapter 4 that the spin-orbit interaction for one electron is of the form

$$\mathcal{H}_{so}=\zeta\mathbf{l}\cdot\mathbf{s} \tag{11.76}$$

where ζ is the spin-orbit coupling parameter. In the central field approximation, ζ depends on Z/r^3 where Z is the effective nuclear charge and r the distance of the electron from the nucleus. To introduce the effects of covalency, one might simply treat ζ as a constant in (11.76). In this case the covalency will reduce the orbital angular momentum operator $\mathbf{l}$, as we saw in Section 11.3, and there will be effective spin-orbit coupling parameters given by

$$\zeta_{\pi\pi}=K_{\pi\pi}\zeta, \qquad \zeta_{\sigma\pi}=K_{\sigma\pi}\zeta \tag{11.77}$$

Now the question arises, is ζ itself affected by covalency? Since $\zeta\sim 1/r^3$, the predominant contribution really comes from the parts of the d wavefunction that are close to the metal ion nucleus and therefore not much affected by covalency. If this is true, it is permissible to write

$$\zeta_{\text{crys}}\simeq\zeta_{\text{free ion}} \tag{11.78}$$

for ζ in (11.77). For the complex $Ni^{2+}F_6^-$, Sugano and Shulman[5] indeed found that their results could be fitted nicely by (11.77) with assumption (11.78).

However, there is another approach to this problem, proposed by Misetich and Buch,[19] who argue that the effective operator in (11.76) is not really **l** but looks more like $(Z/r^3)\mathbf{l}$, and therefore terms corresponding to regions near the ligand nuclei should also be included. This means that the spin-orbit coupling of the ligands will also contribute to the total interaction. One can therefore write the one-electron spin-orbit Hamiltonian for a complex as

$$\mathcal{H}_{so} = \zeta_d \mathbf{l}\cdot\mathbf{s} + \zeta_p \sum_i \mathbf{l}_i \cdot \mathbf{s}_i \tag{11.79}$$

where ζ_d is the spin-orbit coupling constant for the central d orbit and ζ_p that for the ligand p orbit; $\mathbf{l}$ and $\mathbf{l}_i$ are, respectively, the angular momentum operators centered about the central nucleus and about a ligand nucleus denoted by i; and the summation runs over all the ligands. We now have to calculate the matrix elements of $\zeta\mathbf{l}$ in order to find the effective spin-orbit coupling. This is exactly similar to the calculation we did in Section 11.3 to obtain the effective orbital angular momentum, except that all the cross terms between the central metal d orbits and the ligand $2s$ and $2p$ orbits are eliminated because of the localization of the **l**'s in (11.79). Also, the term $\langle y|\partial/\partial y|s\rangle$ in (11.72) does not arise since $\langle y|\mathcal{H}_{so}|s\rangle = 0$. By omitting these terms from (11.68) and (11.72), we have

$$\zeta_{\pi\pi} = N_\pi^2\left(\zeta_d + \tfrac{1}{2}\lambda_\pi^2 \zeta_p\right) \tag{11.80}$$

$$\zeta_{\sigma\pi} = N_\sigma N_\pi\left(\zeta_d - \tfrac{1}{2}\lambda_\sigma\lambda_\pi \zeta_p\right) \tag{11.81}$$

The reduction of the spin-orbit coupling can also be treated from the standpoint of the Wigner-Eckart theorem. In Chapter 9, while discussing the spin-orbit interaction, we showed that, if the Wigner-Eckart theorem is applied, *two* spin-orbit coupling constants are necessary to describe adequately the spin-orbit interaction for a d electron in a cubic field. These are defined by the following reduced matrix elements:

$$\langle t_{2g}\|\mathbf{l}\cdot\mathbf{s}\|t_{2g}\rangle = 3i\zeta$$

and (11.82)

$$\langle t_{2g}\|\mathbf{l}\cdot\mathbf{s}\|e_g\rangle = -3\sqrt{2}\, i\zeta'$$

In the case of pure d orbitals, of course, we have $\zeta = \zeta'$. But for a polycentric system the spin-orbit Hamiltonian is given by (11.79). Using (11.79) and (11.59), we find that the reduced matrix elements, after some lengthy manipulation, take the following forms:

$$\langle t_{2g}\|\mathbf{l}\cdot\mathbf{s}\|t_{2g}\rangle = \frac{3i\left(\zeta_d + \frac{1}{2}\lambda_\pi \zeta_p\right)}{N_\pi} \tag{11.83}$$

and

$$\langle t_{2g}\|\mathbf{l}\cdot\mathbf{s}\|e_g\rangle = -\frac{3\sqrt{2}\,i(\zeta_d - \frac{1}{2}\lambda_\sigma\lambda_\pi\zeta_p)}{(N_\sigma N_\pi)^{1/2}}$$

Using the expressions for the normalization constants given by (11.60), retaining terms up to the second order, and comparing these with (11.82), we finally get

$$\zeta \simeq [1 + (S_\pi^2 - \gamma_\pi^2)]\zeta_d + \tfrac{1}{2}\lambda_\pi^2\zeta_p$$

and (11.84)

$$\zeta' \simeq [1 + \tfrac{1}{2}(S_s^2 - \gamma_s^2) + \tfrac{1}{2}(S_\sigma^2 - \gamma_\sigma^2) + \tfrac{1}{2}(S_\pi^2 - \gamma_\pi^2)]\zeta_d$$

It would be a good exercise for the reader to derive relations (11.83) and (11.84).

So far we have not explicitly touched upon the normalization factors $N_\pi^{-1/2}$ and $N_\sigma^{-1/2}$. It is important to note that these factors are greater than unity when $\gamma = 0$ but less than unity when $\gamma > S$. They are, however, just unity when $\gamma = S$. Assuming that γ and S are small, we have from (11.60)

$$\begin{aligned} N_\sigma^{-1/2} &= 1 - \tfrac{1}{2}(\gamma_s^2 + \gamma_\sigma^2 - S_s^2 - S_\sigma^2) \\ N_\pi^{-1/2} &= 1 - \tfrac{1}{2}(\gamma_\pi^2 - S_\pi^2) \end{aligned} \qquad (11.85)$$

Inserting the calculated[5] values of γ and S for the $KNiF_3$ complex, we have, for example,

$$N_\sigma^{-1/2} = 0.968, \qquad N_\pi^{-1/2} = 0.988 \qquad (11.86)$$

We can now find the magnitude for $\zeta_{\sigma\pi}$, given by (11.81) for the $KNiF_3$ complex, provided that we know ζ_d for Ni^{2+} and ζ_p for F^-. For Ni^{2+}, $\zeta_d = 630\ \text{cm}^{-1}$, as found by Dunn,[66] and for F^-, Misetich and Buch[19] give $\zeta_p = 220\ \text{cm}^{-1}$. Using these values and (11.86), we find that $\zeta_{\sigma\pi} \simeq 0.94\zeta_d$, which means that the spin-orbit coupling due to the ligands has a negligible effect in this case and most of the reduction comes from the normalization factors. It is only in complexes with *heavier* ligands that this effect would be expected to be much greater. For example, $\zeta_p \simeq 2200\text{cm}^{-1}$ for Br^-, so in this case the effect would be 10 times greater.

From the measurements of the g values we can obtain clear evidence of the reduced spin-orbit coupling. Let us take the example of $Ni^{2+}F_6^-$. Misetich and Buch find that the expression for g in this complex can be written as

$$g = 2.0023 + \frac{4K_{\sigma\pi}|\zeta_{\sigma\pi}|}{10Dq} \qquad (11.87)$$

where $K_{\sigma\pi}$ is given by (11.74) and $\zeta_{\sigma\pi}$ by (11.81), and 2.0023 is the contribution from spin. Also, $10Dq$ is the energy difference

$$E[3d^8(t_{2g}^6e_g^2)] - E[3d^8(t_{2g}^5e_g^3)]$$

between the ground state and the first excited state. Assuming that $10Dq = 7380\ \text{cm}^{-1}$, a reasonable value for dilute nickel salts, $\zeta_d = 630\ \text{cm}^{-1}$, and $\zeta_p = 220\ \text{cm}^{-1}$, we obtain $g = 2.28$, which agrees excellently with the experimental value.

This example shows that the MO theory can indeed provide an explanation of the reduction of the spin-orbit coupling constants, although too much significance should not be attached to the exact numerical agreement because of the uncertainties involved in ζ_d, λ_π, λ_σ, and such quantities. It is still unknown whether, for an ion in a crystal, a correction to the metal ion should be made for the change in the effective nuclear charge due to the ligands. This sort of covalent electron transfer is an effect somewhat similar to the screening effect due to finite overlap.[67] We nevertheless conclude from our discussion that this type of g-value analysis lends really good general support to the MO theory for the transition metal complexes.

11.5 REDUCTION OF THE RACAH PARAMETERS

The Racah parameters B and C are also found to be smaller for ions in complexes than for free ions. This can be seen very easily from Table 11.4, where we list these Racah parameters for some of the free ions of the first transition series, as compared to those found in octahedral complexes. In principle, it is possible to calculate these integrals by using the MOs given in (11.59). We have already seen that these Racah parameters are related to the Slater integrals F_2 and F_4, given by

$$F^\kappa(dd) = 2N^4 \int_0^\infty dr_1 r_1^{-(\kappa-1)} R_d^2(r_1) \int_0^{r_1} dr_2 r_2^{\kappa+2} R_d^2(r_2) \qquad (11.88)$$

where $R_d(r) = Nr^2e^{-\kappa r}$, N being the normalization factor.

When molecular orbitals (11.59) are used, F_2 and F_4 become multicenter integrals and therefore are very difficult to evaluate. In fact, no reliable calculation has been done so far. The only plausible way of explaining the reduction of the Slater integrals is to assume that any term involving the ligand orbitals is zero. Then the reduction factor is simply given by the normalization factors of the MOs. Since the integrals contain four MOs, the reduction factor is given by $(N_\sigma^a)^{-n/2} \cdot (N_\pi^a)^{-2+n/2}$, where n is the number of electrons contained in the integral. As seen from relations (11.85), the reduction will take place only if the covalency parameters are larger than the corresponding overlap integrals.

Table 11.4

Racah parameters B and C for the free ions of the first transition series and the reduced values B' and C' for the same ions in octahedral complexes[a]

Ions:[b]	V^{2+}	Cr^{3+}	Mn^{2+}	Co^{2+}	Ni^{2+}
d^n:	d^3	d^3	d^5	d^7	d^8
B (cm^{-1}):	766	830	960	1115	1084
C (cm^{-1}):	2855	3430	3325	4366	4831
Host crystal:	MgO	MgO	MgO	MgO	$KMgF_3$
B' (cm^{-1}):	550[c]	688[d]	786[e]	845[f]	955[g]
C' (cm^{-1}):	2475[c]	3095[d]	3210[e]	3803[f]	4234[g]
B'/B:	0.72	0.83	0.82	0.76	0.88
C'/C:	0.87	0.90	0.97	0.87	0.88

[a]This table is based on Table 5 of J. Owen and J. H. M. Thornley, *Rep. Prog. Phys.* **29**, 675, 1966.
[b]J. S. Griffiths, *Theory of Transition Metal Ions*, Cambridge University Press, London, England, 1961.
[c]M. D. Sturge, *Phys. Rev.* **130**, 639 (1963).
[d]W. A. Runciman and K. A. Schroeder, *Proc. Roy. Soc. (London) A*, **265**, 489 (1962).
[e]G. W. Pratt and R. Coelho, *Phys. Rev.* **116**, 281 (1959); also see S. Koide and M. H. L. Pryce, *Phil. Mag.* **3**, 607 (1958).
[f]R. Pappalardo, D. L. Wood, and R. C. Linares, *J. Chem. Phys.* **35**, 2041 (1961).
[g]K. Knox, R. G. Shulman, and S. Sugano, *Phys. Rev.* **130**, 512 (1963).

This type of argument was first advanced by Koide and Pryce.[68] To analyze the optical spectra they introduced the covalency effect through the parameter ϵ', which is given by

$$1-\epsilon'=\frac{N_\pi^a}{N_\sigma^a}$$

This leads to

$$\epsilon'=(\gamma_\sigma^2+\gamma_s^2-\gamma_\pi^2)-(S_\sigma^2+S_s^2-S_\pi^2)$$

Experimentally, ϵ' for the Mn^{2+} ion is found to be in the range 0.03 to 0.05.

From the foregoing arguments we can write in a phenomenological way that

$$B'_{\text{comp}}=KB_{\text{free ion}}$$
$$C'_{\text{comp}}=KC_{\text{free ion}} \tag{11.89}$$

with

$$K=(N_\sigma^a)^{-n/2}\cdot(N_\pi^a)^{(n-4)/2}$$

where n is the number of e_g electrons in the integrals. Calculation of K then involves calculation of the overlap integrals and the covalency parameters. Whether or not K exceeds unity depends on the magnitudes of the parameters mentioned above. It is worth remembering that no firm conclusions should be drawn from the data (Table 11.4) available at present, although a semi-quantitative discussion of Ni^{2+} in $KNiF_3$, as presented earlier, shows that the reductions observed in the Racah parameters are in reasonable agreement with the prediction of an N^4 proportionality.

Using the basis that a covalent bonding process is expected to bring about a reduction in the effective central field of the metal and allowing for the spreading out of the metal radial function, one can rationalize the reduction of the Racah parameter B in metal complexes. Jorgensen[25] therefore introduced a parameter $\beta = B_{\text{comp}}/B_{\text{free ion}}$, known as the nephelauxetic (cloud-expanding) ratio, which affords a measure of the extent of the expansion of the radial function of the metal and also gives some indication of the tendencies toward covalency in a given complex. In general, in octahedral complexes one normally expects σ-type interactions to be more important than π-bonding contributions except for ligands such as CN^-, which have their own π systems. Consequently the e_g orbitals should be more involved in covalency than the t_{2g} orbitals, with the result that the e_g orbitals should undergo greater expansion of the radial function and should show greater reduction in the value of the parameter B. In O_h symmetry the representations e_g and t_{2g} are frequently abbreviated as γ_3 and γ_5, respectively, and interactions between two e_g electrons would thus involve a parameter β_{33}; interactions between two t_{2g} electrons, a parameter β_{55}; and interactions between an e_g and a t_{2g} electron, a parameter β_{35}. It follows from the above arguments that for the corresponding nephelauxetic ratios we should expect

$$\beta_{55} > \beta_{35} > \beta_{33}$$

We have so far considered some features of the information about β for a complex. To obtain β, $B_{\text{free ion}}$ has, of course, to be determined. Tanabe and Sugano[29] utilized a number of extrapolation techniques with spectroscopic data to obtain the B and C parameters for the $3d^n$ series. Since then, these values have been used for finding β, although a number of other compilations of the Slater integrals F_k (and hence B and C) have subsequently appeared.[26,27]

It is, of course, also possible to calculate the F_k integrals theoretically from HFSCF radial functions, but these values exceed those obtained empirically by a factor of roughly $(Z+3/Z+2)$, where Z is the ionic change of the cation.[28] Nevertheless, for fairly high ionic charges there exists a strong proportionality between the spectroscopic B values and the calculated average reciprocal radii $\langle r^{-1} \rangle$ of the Hartree-Fock function, as well as the dependence on Z, and Jorgensen,[28] modifying a suggestion of Racah, has shown that for the $3d$ group the variation of B (in cm^{-1}) with Z and with q

(the occupation number of the d shell) is well expressed by the relationship

$$B = 384 + 58q + 124(Z+1) - \frac{540}{Z+1} \tag{11.90a}$$

The results thus derived for integer values of Z and q agree closely with those of Tanabe and Sugano[29] and also allow the determination, by interpolation, of the effective cationic charges for various complexes.

For the 4d series the variation of B with Z and q is given by Jorgensen as

$$B = 472 + 28q + 50(Z+1) - \frac{500}{Z+1} \tag{11.90b}$$

which, for M^{n+} ($n=1$ to 4) ions, gives values of B closely similar to those obtained by using the assumption

$$B,\ M^{n+}\ (4d) = 0.66B,\ M^{n+}\ (3d) \tag{11.90c}$$

This is also in reasonable agreement with the values derived from the parameters.[30] For the 5d series the values of $B_{\text{free ion}}$ were obtained by assuming that

$$B,\ M^{n+}\ (5d) = 0.60B,\ M^{n+}\ (3d) \tag{11.90d}$$

More theoretical and experimental investigations are required, however, to understand adequately the apparent reduction of the Racah parameters in complexes of the transition metal ions.

11.6 WOLFSBERG–HELMHOLZ METHOD

In 1952 Wolfsberg and Helmholz[35] gave a semiempirical approach to calculate the energies of the molecular orbitals for the ground state and the first few excited states of permanganate, chromate, and the perchlorate ions. Since then, there has been a spurt of activity in applying this method to the octahedral complexes. Ballhausen and Gray[3] and their co-workers refined this method and utilized it to interpret the spectral data of complexes of the transition metal ions. Since the Wolfsberg-Helmholz (W-H) method incorporates many of the general concepts of chemical bonding such as overlap of the bonding orbitals and electronegativity, and has opened up a line of attack for spectral analysis, it is worthwhile to deal with it in some detail before proceeding to the more rigorous HFSCF MO method, developed mainly by Fenske, Richardson, and their co-workers.

Let $\mathcal{H}$ be the Hamiltonian for a single electron in the molecular skeleton. The MOs can be written in the fashion of (11.2):

$$\Psi_i^{\Gamma\gamma} = \alpha_i^{\Gamma\gamma}\Phi_\gamma^\Gamma + \sum_k \beta_i^{\Gamma\gamma,k}\chi^{\Gamma\gamma,k} \tag{11.91}$$

The suffix i denotes MOs of the same symmetry. The summation over k is necessary when more than one linear combination of ligand functions transforms as $|\Gamma\gamma\rangle$. Then the energy is given by

$$E=\langle \Psi_i^{\Gamma\gamma}|\mathcal{H}|\Psi_i^{\Gamma\gamma}\rangle \tag{11.92}$$

Now, the variational principle is applied to this energy with the subsidiary condition that the $\Psi_i^{\Gamma\gamma}$ are normalized. In (11.91) α's and β's are the variational parameters. To obtain the eigenvalues and the eigenvectors, it is necessary to solve secular equations of the form

$$|H_{ij}-G_{ij}E|=0 \tag{11.93}$$

where for simplicity we have condensed the three indices Γ,γ,i into one i, and the G_{ij} are the group overlap integrals. (Note that the $\chi^{\Gamma\gamma}$ are linear combinations of the ligand atomic orbitals.) Since the Hamiltonian is invariant under the symmetry group of the molecule, matrix (11.93) is in block form. In the case of a degenerate level, since the degenerate partners span the same irreducible representation, it is sufficient to solve for each type of orbital. Thus we have a_{1g}, e_g, t_{2g}, and t_{1u} secular equations in O_h symmetry. The reduction to the block form reduces the labor considerably.

At this stage the immediate problem is to estimate the Coulomb (H_{ii}) and the exchange (H_{ij}) integrals. The $H(i,i)$ terms correspond roughly to the Coulomb energy of an electron on the atom or a linear combination of atoms. Thus $H(4s,4s)$ is approximated by the ionization potential (IP) of the $4s$ electron from the metal atom, and $H(\sigma_1,\sigma_1)$, say, by the IP of an electron from a σ orbital of the ligand. These IPs are adjusted in a crude manner so as to correspond to the IPs of atoms (or ions) of a charge equal to that assigned to the atom in the molecule. This approximation consists of using the valence state ionization energies (VSIEs) of the ith orbital for the integral $H(i,i)$. The $H(i,j)$ terms are obtained from the formula

$$H(i,j)=\tfrac{1}{2}G_{ij}F^x\{H(i,i)+H(j,j)\} \tag{11.94}$$

where F^x is an empirical constant to be determined so as to give results in good agreement with experiment. For σ-type bonding $F^x=F^\sigma$, and for π-type bonding $F^x=F^\pi$. Wolfsberg and Helmholz used a value of F between 1 and 2. In place of (11.94), Ballhausen and Gray[3] have suggested

$$H(i,j)=-2G_{ij}(H_{ii}\cdot H_{jj})^{1/2} \tag{11.95}$$

From the geometry of the molecule, the overlap integrals G_{ij} can be evaluated and hence the eigenvalues of (11.93) can be determined. The refinement in the method consists of doing a self-consistent calculation for the electronic charge distribution among the orbitals. Suppose that initially there is a net charge q on the ion and charges d,s,p in the respective metal orbitals. These charges are not restricted to integers. To this point the VSIEs are known only for configurations where the occupancy of orbitals is integral, but it is necessary to know the VSIEs for fractionally occupied configurations as well. These have been derived by Ballhausen and Gray[3] [Equations (8.60), (8.61), and (8.62) of Ref. 3].

It is assumed that a metal VSIE for a particular configuration (d^n) can be represented by

$$\text{VSIE} = Aq^2 + Bq + C \tag{11.96}$$

where q is the charge on the metal. Such functions are calculated from Moore's tables (see Ref. 3). The ligand VSIEs are given fixed (reasonable) values close to the ionization potentials of the neutral atoms.

Let us now derive the expressions for the uncorrected (neglecting the ligand-ligand overlap) H_{ii}''s, which are essential in order to solve the secular equations (11.93). We use configurations of the type $d^d s^s p^p$, where d electrons are in the d orbital, s electrons are in the s orbital, and p electrons are in the p orbital. Usually these numbers are not whole numbers but are fractions. In such calculations one uses the data obtained from Moore's tables for configurations $d^{d_1}s^{s_1}p^{p_1}$, where d_1, s_1, and p_1 are nonnegative integers. To use data of this type, we have to write the configuration $d^d s^s p^p$ as a weighted combination of configurations of type $d^{d_k}s^{s_k}p^{p_k}$, where d_k, s_k, p_k are nonnegative integers and the subscript k denotes a particular configuration.

Our object now is to find the numbers a_k such that

$$(dsp) = \sum_k a_k (d_k s_k p_k) \tag{11.97}$$

Obviously, the numbers a_k depend on d, s, and p. Let us now take the total energy of configurations on both sides of (11.97). Thus the left-hand side of (11.97) is

$$E(dsp) = dE_d + sE_s + pE_p \tag{11.98}$$

where E_d, E_s, and E_p are the energies of the d, s, and p orbitals, respectively. Since there are d electrons in d orbitals, and so on, the total energy on the left-hand side of (11.97) is truly given by (11.98). Similarly, the right-hand side of (11.97) is given by

$$E(d_k s_k p_k) = d_k E_d + s_k E_s + p_k E_p \tag{11.99}$$

Inserting (11.98) and (11.99) in (11.97), we have

$$dE_d + sE_s + pE_p = \sum_k (a_k d_k) E_d + \sum_k (a_k s_k) E_s + \sum_k (a_k p_k) E_p \tag{11.100}$$

Now, equating coefficients of E_d, E_s, and E_p from both sides of (11.100), we have

$$d=\sum_k a_k d_k, \qquad s=\sum_k a_k s_k, \qquad p=\sum_k a_k p_k \tag{11.101}$$

To find a_k, we now have to solve the set of linear equations (11.101). From the theory of linear equations, we know that we can find a_k unambiguously *only if* there are three configurations. Obviously, if there are more than three configurations, there will be arbitrariness in the solution. For three configurations, however, (11.101) turns out to be

$$\begin{aligned} a_1d_1+a_2d_2+a_3d_3&=d\\ a_1s_1+a_2s_2+a_3s_3&=s\\ a_1p_1+a_2p_2+a_3p_3&=p \end{aligned} \tag{11.102}$$

Solving for a_1, a_2, and a_3, we finally get

$$\Delta a_1=\begin{vmatrix} d & d_2 & d_3\\ s & s_2 & s_3\\ p & p_2 & p_3 \end{vmatrix}$$

$$\Delta a_2=\begin{vmatrix} d_1 & d & d_3\\ s_1 & s & s_3\\ p_1 & p & p_3 \end{vmatrix} \tag{11.103}$$

$$\Delta a_3=\begin{vmatrix} d_1 & d_2 & d\\ s_1 & s_2 & s\\ p_1 & p_2 & p \end{vmatrix}$$

and

$$\Delta=\begin{vmatrix} d_1 & d_2 & d_3\\ s_1 & s_2 & s_3\\ p_1 & p_2 & p_3 \end{vmatrix}$$

Let us now find the expression for H'_{dd}. In this case we have

$$d^{n-s-p-q}s^s p^p=a_1(d^{n-q})+a_2(d^{n-1-q}s)+a_3(d^{n-1-q}p) \tag{11.104}$$

where $d+s+p=n-q$. Then we have

$d_1=n-q$	$d_2=n-1-q$	$d_3=n-1-q$
$s_1=0$	$s_2=1$	$s_3=0$
$p_1=0$	$p_2=0$	$p_3=1$

and therefore

$$\Delta a_1 = \begin{vmatrix} n-s-p-q & n-1-q & n-1-q \\ s & 1 & 0 \\ p & 0 & 1 \end{vmatrix}$$
$$= (1-s-p)(n-q)$$

Also,

$$\Delta = \begin{vmatrix} n-q & n-1-q & n-1-q \\ 0 & 1 & 0 \\ 0 & 0 & 1 \end{vmatrix} = n-q$$

Therefore

$$a_1 = 1-s-p \tag{11.105}$$

and, similarly,

$$a_2 = s, \qquad a_3 = p$$

Hence

$$\begin{aligned} -H'_{dd} = (d\text{VSIE}) = (1-s-p)(d\text{VSIE}: d^{n-q}) \\ + s(d\text{VSIE}: d^{n-1-q}s) \\ + p(d\text{VSIE}: d^{n-1-q}p) \end{aligned} \tag{11.106}$$

Similarly, for H'_{ss} we use

$$d^{n-s-p-q}s^s p^p = a_1(d^{n-1-q}s) + a_2(d^{n-2-q}s^2) + a_3(d^{n-2-q}sp)$$

Thus

$d_1 = n-1-q$	$d_2 = n-2-q$	$d_3 = n-2-q$
$s_1 = 1$	$s_2 = 2$	$s_3 = 1$
$p_1 = 0$	$p_2 = 0$	$p_3 = 1$

and

$$\Delta = \begin{Bmatrix} n-1-q & n-2-q & n-2-q \\ 1 & 2 & 1 \\ 0 & 0 & 1 \end{Bmatrix} = n-q$$

Proceeding exactly as before, we have

$$a_1 = 2-s-p, \qquad a_2 = s-1, \qquad a_3 = p \tag{11.107}$$

Thus

$$-H'_{ss}=(s\text{VSIE})=(2-s-p)(s\text{VSIE}: d^{n-1-q}s)$$
$$+(s-1)(s\text{VSIE}: d^{n-2-q}s^2)$$
$$+p(s\text{VSIE}: d^{n-2-q}sp) \qquad (11.108)$$

Similarly,

$$-H'_{pp}=(p\text{VSIE})=(2-s-p)(p\text{VSIE}: d^{n-1-q}p)$$
$$+(p-1)(p\text{VSIE}: d^{n-2-q}p^2)$$
$$+s(p\text{VSIE}: d^{n-2-q}sp) \qquad (11.109)$$

We thus arrive at the H_{ii}''s given by (11.106), (11.108), and (11.109).

Calculating the group overlap integrals G_{ij} by using the best available SCF wavefunctions and the known metal-ligand distance, one can then solve the secular equations (11.93). For each MO calculated, a Mulliken population analysis[36] is performed in which each overlap population is divided equally between the metal and the combined ligand orbitals. In subsequent cycles the input configuration is altered until a self-consistent charge distribution is obtained. This modified W-H method has been used by Ballhausen and Gray[3] to interpret the optical spectra and hence the Dq values of $3d^n$ complexes. This, in short, is the full story.

Now let us calculate the VSIEs. These are obtained by appropriately combining the values of E_{av} for the two configurations under consideration, together with the appropriate ionization potential; E_{av} is the weighted mean of the energies of the terms arising from a configuration relative to the

Table 11.5

$4d$ VSIEs (in units of 10^3 cm^{-1})

		$0\to+1$			$+1\to+2$		
Atom	n	$4d^n \to 4d^{n-1}$	$4d^{n-1}5s \to 4d^{n-2}5s$	$4d^{n-1}5p \to 4d^{n-2}5p$	$4d^{n-1} \to 4d^{n-2}$	$4d^{n-2}5s \to 4d^{n-3}5s$	$4d^{n-2}5p \to 4d^{n-3}5p$
Y	3	31.3	42.7	53.9	95.7	111.9	120.1
Zr	4	37.2	50.9	60.1	105.7	125.0	132.3
Nb	5	43.0	58.8	66.5	115.7	137.6	144.2
Mo	6	48.5	66.3	73.3	125.6	149.6	156.0
Tc	7	(54.0)	73.5	80.4	(135.5)	(161.1)	(167.5)
Ru	8	59.2	80.3	87.7	145.1	171.9	178.6
Rh	9	64.2	86.7	95.4	154.7	182.1	189.6
Pd	10	69.1	92.7	103.3	164.2	191.8	200.3
		$A=-0.08$	$A=-0.18$	$A=0.15$	$A=-0.05$	$A=-0.29$	$A=-0.12$
		$B=6.0$	$B=8.4$	$B=6.0$	$B=10.1$	$B=12.9$	$B=12.1$
		$C=31.3$	$C=42.7$	$C=53.9$	$C=105.8$	$C=125.1$	$C=132.3$
		SDV = 0.86	SDV = 2.79	SDV = 3.26	SDV = 5.42	SDV = 4.67	SDV = 4.42
		= 0.9	= 2.8	= 3.3	= 5.4	= 4.7	= 4.4

ground state of the atom or ion in question. The weighting factor is equal to the total degeneracy (spin × orbital) of the term. The VSIEs for the $3d^n$ atoms have been calculated by Ballhausen and Gray,[3] and there is no need to reproduce them here. Since no such table for the VSIEs is available for the $4d^n$ transition metal atoms, De, Desai, and Chakravarty,[37] following the method outlined by Ballhausen and Gray, have calculated the $4d$, $5s$, and $5p$ VSIEs, using data from Moore's tables for $q=0\rightarrow1$ and $1\rightarrow2$. These are listed in Tables 11.5, 11.6 and 11.7 for future use.

As is well known, the quantity known as the overlap integral is of considerable importance in the theory of molecular structure. The overlap integral S for a pair of overlapping AOs χ_a and χ_b of a pair of atoms a and b is defined for any value of the internuclear distance R by

$$S(\chi_a,\chi_b;R)=\int \chi_a^*\chi_b\,d\tau \tag{11.110}$$

Evaluation of the two-atom overlap integrals has been discussed, notably by Mulliken and his co-workers.[38] Tables of overlap integrals are also given by Jaffe and his co-workers,[39] Leifer, Cotton, and Leto,[40] and Craig and his co-workers.[41] Lofthus[42] has provided some additional master formulas. All these authors, however, have invariably used the orthogonalized Slater AOs without seeking more appropriate SCF AOs, although there are good reasons to believe that good SCF AOs may give quite different S values than Slater AOs, especially at large R values. This raises serious questions whether we should use these overlap integrals computed from Slater AOs. It might seem that we ought to proceed to compute SCF S values for suitable parameter ranges for the more important AO pairs. But again we should remember that the best SCF AOs for atoms in molecules may be considerably different from

Table 11.6

$5s$ VSIEs (in units of 10^3 cm^{-1})

		$0\rightarrow+1$			$+1\rightarrow+2$		
Atom	n	$4d^{n-1}5s \rightarrow 4d^{n-1}$	$4d^{n-2}5s^2 \rightarrow 4d^{n-2}5s$	$4d^{n-2}5s5p \rightarrow 4d^{n-2}5p$	$4d^{n-2}5s \rightarrow 4d^{n-2}$	$4d^{n-3}5s^2 \rightarrow 4d^{n-3}5s$	$4d^{n-3}5s5p \rightarrow 4d^{n-3}5p$
Y	3	50.2	55.0	64.8	101.5	107.4	120.2
Zr	4	53.0	59.8	69.0	106.0	123.6	127.0
Nb	5	55.4	63.9	72.8	110.9	(135.1)	133.0
Mo	6	57.4	67.2	76.0	116.1	(141.7)	(138.3)
Tc	7	58.9	(69.8)	78.8	(121.6)	(143.5)	(142.6)
Ru	8	59.9	71.7	81.0	127.4	140.4	146.1
Rh	9	60.6	72.8	82.8	133.5	(132.6)	(148.9)
Pd	10	60.8	73.1	84.0	139.9	(120.0)	(150.8)
		$A=-0.22$	$A=-0.37$	$A=-0.25$	$A=0.15$	$A=-2.40$	$A=-0.41$
		$B=3.10$	$B=5.2$	$B=4.5$	$B=4.7$	$B=13.8$	$B=6.4$
		$C=50.2$	$C=55.0$	$C=64.8$	$C=106.1$	$C=123.7$	$C=127.1$
		SDV$=0.53$ $=0.5$	SDV$=1.8$	SDV$=0.78$ $=0.8$	SDV$=7.7$	SDV$=0$	SDV$=3.3$

Table 11.7

5*p* VSIEs (in units of 10 cm^{-1})

		0→+1		+1→+2	
Atom	n	$4d^{n-1}5p \rightarrow 4d^{n-1}$	$4d^{n-2}5s5p \rightarrow 4d^{n-2}5s$	$4d^{n-2}5p \rightarrow 4d^{n-2}$	$4d^{n-3}5s5p \rightarrow 4d^{n-3}5s$
Y	3	35.3	38.6	75.1	85.3
Zr	4	35.5	39.6	78.8	89.5
Nb	5	35.5	40.5	82.0	93.3
Mo	6	35.3	41.3	84.7	(96.9)
Tc	7	34.8	42.1	(87.0)	(100.1)
Ru	8	34.2	42.7	88.8	102.9
Rh	9	33.4	43.3	90.2	(105.5)
Pd	10	32.4	43.8	91.1	(107.7)
		$A=-0.10$	$A=-0.04$	$A=-0.24$	$A=-0.16$
		$B=0.3$	$B=1.0$	$B=3.5$	$B=4.0$
		$C=35.3$	$C=38.6$	$C=78.8$	$C=89.5$
		SDV=2.8	SDV=0.8	SDV=2.6	SDV=3.3

those for free atoms, even though we have a very poor knowledge of the exact nature of these differences and of their variations in different molecular states. Even then, for molecular states with stable binding, we should probably expect the free-atom AOs to be modified in such a way as to correspond to increased, rather than decreased, overlaps, and this increase should be relative to SCF AOs. Hence the matter boils down to the fact that one should compute the overlap integrals using the SCF AOs for known internuclear separation. This matter, of course, deserves further serious investigation.

To calculate the overlap integrals, let us consider the atoms situated at A and B as pictured in Figure 11.4. The polar coordinates of an electron at P is an atomic orbital centered at A or B are as indicated in the figure, and the aximuthal angles $\phi_A = \phi_B = \phi$. If we change these polar coordinates to spheroidal coordinates α, β, ϕ, which are suitable for our purpose, we have

$$\alpha = \frac{r_A + r_B}{R'}, \qquad \beta = \frac{r_A - r_B}{R'}, \qquad \phi = \phi_A = \phi_B \tag{11.111}$$

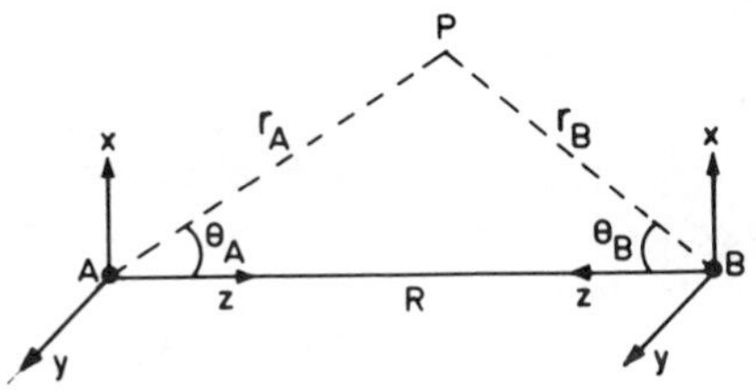

Fig. 11.4 Definition of elliptical coordinates α, β, and φ (see text).

The coordinate α ranges from 1 to ∞; β, from -1 to $+1$; and ϕ from 0 to 2π. Therefore

$$\cos\theta_A = \frac{1+\alpha\beta}{\alpha+\beta} \qquad \text{and} \qquad \cos\theta_B = \frac{1-\alpha\beta}{\alpha-\beta} \tag{11.112}$$

and the volume element

$$d\tau = \frac{R'^3}{8}(\alpha^2-\beta^2)\,d\alpha\,d\beta\,d\phi \tag{11.113}$$

The AOs are written in the form

$$\psi_A = R_{nl}(r_A)\,Y_l^m(\theta_A,\phi_A) \tag{11.114}$$

where the normalized angular parts of the wavefunctions are as follows:

For $l=1$,

$$Y_1^0 = \sqrt{\frac{3}{4\pi}}\,\cos\theta$$

$$Y_1^{\pm1} = \sqrt{\frac{3}{4\pi}}\,\sin\theta(\cos\phi \text{ or } \sin\phi) \tag{11.115}$$

For $l=2$,

$$Y_2^0 = \sqrt{\frac{5}{4\pi}}\cdot\tfrac{1}{2}(3\cos^2\theta-1)$$

$$Y_2^{\pm2} = \sqrt{\frac{5}{4\pi}}\sqrt{\frac{3}{4}}\cdot\sin^2\theta(\cos 2\phi \text{ or } \sin 2\phi)$$

$$Y_2^{\pm1} = \sqrt{\frac{5}{4\pi}}\,\sqrt{3}\cdot\sin\theta\cos\theta(\cos\phi \text{ or } \sin\phi)$$

and the radial parts are of the form of Slater orbitals:

$$R_{nl}(r) = Nr^{m+1}e^{-\mu r/a_0}$$

Let us now calculate the overlap integrals of σ and π character with s-, p-, and d-type orbitals. For any given AO pair, we want to calculate S as a function of R'. To accomplish this, the best procedure is first to set up for each AO pair a single master formula expressed in terms of suitable parameters, depending on the μ's of the two AOs and on R'. For this purpose Mulliken and his co-workers introduced two parameters, p and t, defined as follows:[38]

$$
\begin{aligned}
p &= \frac{\frac{1}{2}(\mu_A+\mu_B)R'}{a_0}\\
t &= \frac{\mu_A-\mu_B}{\mu_A+\mu_B}
\end{aligned}
\tag{11.116}
$$

The group overlap integrals in terms of α, β, p, and t are given in Refs. 38–43, and we need not list them here. For an octahedral complex, in which we are interested, we have the following group overlaps:[3]

$$
\left.
\begin{aligned}
G[a_{1g}(\sigma)] &= \int (n+1)S\frac{1}{\sqrt{6}}\sum_{i=1}^{6}\sigma_i\,d\tau = \sqrt{6}\,[\sigma,(n+1)S]\\
G[e_g(\sigma)] &= \int nd_{z^2}\frac{1}{2\sqrt{3}}(-2\sigma_3+2\sigma_6+\sigma_1+\sigma_2-\sigma_4-\sigma_5)\,d\tau\\
&= \sqrt{3}\;S(\sigma, nd_\sigma)\\
&= \int nd_{x^2-y^2}\tfrac{1}{2}(\sigma_1-\sigma_2+\sigma_4-\sigma_5)\,d\tau\\
G[t_{1u}(\sigma)] &= \int (n+1)p_x\frac{1}{\sqrt{2}}(\sigma_1-\sigma_4)\,d\tau\\
&= \sqrt{2}\;S[\sigma,(n+1)p_\sigma]\\
&= \int (n+1)p_y\frac{1}{\sqrt{2}}(\sigma_2-\sigma_5)\,d\tau\\
&= \int (n+1)p_z\frac{1}{\sqrt{2}}(\sigma_3-\sigma_6)\,d\tau\\
G[t_{2g}(\pi)] &= \int nd_{xz}\tfrac{1}{2}(y_1+x_3+x_4+y_6)\,d\tau\\
&= 2S(p_\pi, nd_\pi)\\
&= \int nd_{yz}\tfrac{1}{2}(x_2+y_3+y_5+x_6)\,d\tau\\
&= \int nd_{xy}\tfrac{1}{2}(x_1+y_2+y_4+x_5)\,d\tau\\
G[t_{1u}(\pi)] &= \int (n+1)p_x\tfrac{1}{2}(y_2+x_3-x_5-y_6)\,d\tau\\
&= 2S[p_\pi,(n+1)p_\pi]\\
&= \int (n+1)p_y\tfrac{1}{2}(x_1+y_3-y_4-x_6)\,d\tau\\
&= \int (n+1)p_z\tfrac{1}{2}(y_1+x_2-x_4-y_5)\,d\tau
\end{aligned}
\right\}
\tag{11.117}
$$

So far we have not considered the ligand-ligand overlap. If we include it, the correction factor, in the case of octahedral symmetry, can be obtained in the following way. Let us assume that the ligand functions are of the form[3]

$$\psi_i = \sum_\alpha a_{i\alpha}\phi_{i\alpha} \tag{11.118}$$

and therefore

$$\int \psi_i^2\, d\tau = \frac{1}{N_i^2} = \sum_\alpha a_{i\alpha}^2 + \sum_{\alpha\neq\beta} a_{i\alpha}a_{i\beta}S(i\alpha, i\beta)$$

For two different AOs, for example two orbitals on different atoms or different orbitals on the same atom, denoted by ϕ_k and ϕ_n, we have

$$H_{kn} = -2\sqrt{H_{kk}\cdot H_{nn}}\; S(k,n) \tag{11.119}$$

Using this assumption, we have

$$H_{ii} = \int (N_i\psi_i)H(N_i\psi_i)\,d\tau$$

$$= N_i^2 H'_{ii}\left[\sum_\alpha a_{i\alpha}^2 + 2\sum_{\alpha\neq\beta}\sum_\beta a_{i\alpha}a_{i\beta}S(i\alpha, i\beta)\right] \tag{11.120}$$

where $H'_{ii} = H_{i\alpha}H_{i\alpha}$ is the diagonal element uncorrected for ligand-ligand overlap. If ψ_i is already normalized, then, neglecting the ligand-ligand overlap, we can easily see that N_i gives the correction factor for normalization, which we tabulate for various representations in Table 11.8. In this table $S(\sigma_L, \sigma_L : 2R')$ denotes the overlap integral between two ligand σ functions at a distance $2R'$, where R' is the metal-ligand internuclear separation.

In (11.120) $\Sigma_\alpha a_{i\alpha}^2 = 1$. Writing

$$X_i = \sum_\alpha \sum_{\beta\neq\alpha} a_{i\alpha}a_{i\beta}S(i\alpha, i\beta) \tag{11.121}$$

we have

$$H_{ii} = H'_{ii}\left(\frac{1+2X_i}{1+X_i}\right)$$

$$= aH'_{ii}, \qquad \text{say} \tag{11.122}$$

If we neglect the ligand orbitals, then, of course, $H_{ii} = H'_{ii}$. We have to compute the off-diagonal elements H_{ij} from the uncorrected H'_{ii}'s. This can be

Table 11.8

The correction factor N_i for ligand-ligand overlap in O_h symmetry[a]

Representation	Ligand function	N_i
A_{1g}	$z_1+z_2+z_3+z_4+z_5+z_6$	$[1+S(\sigma_L,\sigma_L;2R')+2S(\sigma_L,\sigma_L;\sqrt{2}\,R')+2S(\pi_L,\pi_L;\sqrt{2}\,R')]^{-1/2}$
E_g	$z_1-z_2+z_4-z_5$	$[1+S(\sigma_L,\sigma_L;2R')-S(\sigma_L,\sigma_L;\sqrt{2}\,R')-S(\pi_L,\pi_L;\sqrt{2}\,R')]^{-1/2}$
T_{1g}	$x_2-y_3+y_5-x_6$	$[1-S(\sigma_L,\sigma_L;2R')-S(\sigma_L,\sigma_L;\sqrt{2}\,R')-S(\pi_L,\pi_L;\sqrt{2}\,R')]^{-1/2}$
T_{2g}	$y_1+x_3+x_4+y_6$	$[1-S(\pi_L,\pi_L;2R')+S(\sigma_L,\sigma_L;\sqrt{2}\,R')+S(\pi_L,\pi_L;\sqrt{2}\,R')]^{-1/2}$
T_{1u}	z_1-z_4	$[1-S(\sigma_L,\sigma_L;2R')]^{-1/2}$
	$y_2+x_3-x_5-y_6$	$[1+S(\pi_L,\pi_L;2R')+S(\pi_L,\pi_L;\sqrt{2}\,R')]^{-1/2}$
T_{2u}	$y_2-x_3-x_5+y_6$	$[1+S(\pi_L,\pi_L;2R')-2S(\pi_L,\pi_L;\sqrt{2}\,R')]^{-1/2}$

[a]From C. J. Ballhausen and H. B. Gray, *Molecular Orbital Theory*, Benjamin, New York, 1965.

shown easily since

$$G_{ij}=\int (N_i\psi_i)(N_j\psi_j)\,d\tau=N_iN_j\sum_{\alpha,\mu} a_{i\alpha}a_{j\mu}S(i\alpha,j\mu)$$

We have, from (11.119),

$$\begin{aligned}H_{ij}&=N_iN_j\sum_{\alpha,\mu} a_{i\alpha}a_{j\mu}\Big[-2(H_{i\alpha i\alpha}H_{j\mu j\mu})^{1/2}S(i\alpha,j\mu)\Big]\\&=-2(H'_{ii}\cdot H'_{jj})^{1/2}G_{ij}\end{aligned} \tag{11.123}$$

We now present the MO treatment of the complex $[TiF_6]^{3-}$ to obtain the electronic energy levels by assuming ideal octahedral coordination and taking into account the π bonding but neglecting the ligand-ligand interactions. This classic example which is due to Bedon, Horner, and Tyree,[14] shows beautifully how the W-H approximation can be applied in real systems with very reasonable success. Since we have already presented the theoretical details of this approximation, we need not elaborate on them here.

The MO energies may be obtained, as seen earlier, by solving the secular determinants

$$|H_{ij}-G_{ij}E|=0 \tag{11.124}$$

To solve for E, the energies, we must evaluate three quantities, namely, H_{ii}, H_{jj}, and G_{ij}. As discussed earlier, the G_{ij} terms represent the group overlap integrals, that is, the overlap of the ligand molecular orbitals with the Ti^{3+} atomic orbitals of appropriate symmetry. We have seen from (11.117) that the G_{ij}'s are functions of the atomic overlap integrals S_{ij}. Consider, for example, the calculation of G_{e_g} for the d_{z^2} orbital interacting with the appropriate set of F orbitals. We now rotate the T_i coordinates, in turn, to each of the six F coordinates. Since the axis of each F lies along the σ-bond direction (see Figure 11.1), we have

$$\begin{array}{lll} \sigma_1\colon x^2 \to z^2, & y^2 \to y^2, & z^2 \to x^2 \\ \sigma_2\colon x^2 \to x^2, & y^2 \to z^2, & z^2 \to y^2 \\ \sigma_3\colon x^2 \to y^2, & y^2 \to x^2, & z^2 \to z^2 \\ \sigma_4\colon x^2 \to z^2, & y^2 \to y^2, & z^2 \to x^2 \\ \sigma_5\colon x^2 \to y^2, & y^2 \to x^2, & z^2 \to z^2 \\ \sigma_6\colon x^2 \to y^2, & y^2 \to x^2, & z^2 \to z^2 \end{array}$$

Therefore, using Table 11.1,

$$\begin{aligned} G_{e_g} &= \int \left[\frac{1}{\sqrt{3}}(2z^2 - r^2) \right] \left[\frac{1}{\sqrt{3}}(-\sigma_3 + \sigma_6) - \frac{1}{2\sqrt{3}}(-\sigma_1 - \sigma_2 + \sigma_4 + \sigma_5) \right] d\tau \\ &= -\frac{1}{3}\int (3z^3 - y^2 - x^2 - z^2)\sigma_3\, d\tau + \frac{1}{3}\int (3z^2 - y^2 - x^2 - z^2)\sigma_6\, d\tau \\ &\quad + \frac{1}{6}\int (3z^2 - y^2 - x^2 - z^2)\sigma_1\, d\tau + \frac{1}{6}\int (3z^2 - y^2 - x^2 - z^2)\sigma_2\, d\tau \\ &\quad - \frac{1}{6}\int (3z^2 - y^2 - x^2 - z^2)\sigma_4\, d\tau - \frac{1}{6}\int (3z^2 - y^2 - x^2 - z^2)\sigma_5\, d\tau \end{aligned}$$

After such conversion all σ's are alike, and therefore the subscripts may be dropped. Then

$$\begin{aligned} G_{e_g} &= \int (2z^2 - y^2 - x^2)\sigma\, d\tau \\ &= \sqrt{3}\int \psi_{d_{z^2}}\psi_{F_\sigma}\, d\tau \\ &= \sqrt{3}\, S_{3d_\sigma, \mathrm{hyb}\,\sigma} \end{aligned} \tag{11.125}$$

In a similar manner all G_{ij}'s are determined as functions of S_{ij}. These have already been given in (11.117).

The S_{ij} values are obtained using the expression

$$S_{ij} = \int \psi_i \psi_j\, d\tau \tag{11.126}$$

The wavefunctions to be used should preferably be in the form of combinations of normalized Slater orbitals, so that the overlap integral tables[38–43] can be used to evaluate the individual terms. In some cases, however, these tables are insufficient, and it will be necessary to calculate the values using the master equations given by Kotani and his co-workers.[69] Using the *Ti* wavefunctions published by Richardson and his co-workers, Watson's SCF functions, and the fluorine wavefunctions of Allen and Lohr, all given in Ref. 44, we can obtain the atomic overlap integrals. They are, for this case,

$$\left.\begin{aligned} S_{3d_\sigma,2p_\sigma} &= 0.130 \\ S_{3d_\sigma,2s_\sigma} &= 0.161 \\ S_{4s_\sigma,2p_\sigma} &= 0.103 \\ S_{4s_\sigma,2s_\sigma} &= 0.219 \\ S_{4p_\sigma,2p_\sigma} &= 0.0765 \\ S_{4p_\sigma,2s_\sigma} &= 0.350 \\ S_{3d_\pi,2p_\pi} &= 0.101 \\ S_{4p_\pi,2p_\pi} &= 0.135 \end{aligned}\right\} \tag{11.127}$$

The overlap of Ti^{3+} atomic orbitals with fluorine σ-hybrid orbitals is obtained by using the relation

$$S_{3d_\sigma,\mathrm{hyb}\,\sigma} = \sqrt{0.13}\; S_{3d_\sigma,2s_\sigma} + \sqrt{0.87}\; S_{3d_\sigma,2p_\sigma} \tag{11.128}$$

from which we get

$$\begin{aligned} S_{3d_\sigma,\mathrm{hyb}\,\sigma} &= 0.180 \\ S_{4s_\sigma,\mathrm{hyb}\,\sigma} &= 0.175 \\ S_{4p_\sigma,\mathrm{hyb}\,\sigma} &= 0.174 \end{aligned} \tag{11.129}$$

The H_{ii} terms are estimated from the VSIE for the Ti ion of appropriate charge. Curves of ionization energy as a function of the charge on titanium for various electronic configurations are interpolated for fractional charge on titanium. The ionization potentials for various valence states of integral charge are obtained from Moore's tables[70] of atomic spectra. Since the F atoms will have some net negative charge, satisfactory VSIEs cannot be obtained from Moore's tables. Therefore the ionization energy for fluorine in TiF_6^{3-} has been estimated from the experimental ionization potentials for HF. Crude estimates of the ionization potentials used by Bedon, Horner, and Tyree[44] are $-127{,}200$, $-136{,}000$, $-272{,}200$, and $-154{,}500$ cm^{-1} for $2p_\pi$, $2p_\sigma$, $2s$ and hybσ, respectively.

The off-diagonal terms are estimated from (11.123). We now calculate the energy levels. The values of H_{ii}, H_{ij}, and G_{ij} are then substituted into (11.124), and the secular determinants are solved for E. The H_{ii} values used for titanium depend on both the charge and the electronic distribution selected.

The secular equations to be solved are

$$\begin{aligned} c_1(H_{11}-E)+c_2(H_{12}-G_{12}E)&=0 \\ c_1(H_{12}-G_{12}E)+c_2(H_{22}-E)&=0 \end{aligned} \tag{11.130}$$

along with the normalization conditions. Using the normalization condition

$$\sum c_i^2+2\sum c_ic_jG_{ij}=1 \tag{11.131}$$

for one-electron MOs and Mulliken's suggestion of dividing the overlap population equally between titanium and fluorine, we calculate the electronic population in each Ti atomic level. The net charge on titanium is calculated from the population. The calculated charge and population do not generally agree with the initially assumed ones, so a new charge and population have to be assumed. New H_{ii} values for Ti orbitals are then calculated, followed by new molecular orbitals, the c values, and, finally, a new charge and population. This iteration is repeated until the calculated charge and population agree with the assumed ones.

The final electron population on titanium found by Bedon, Horner, and Tyree[44] is given by $3d^{2.75}4s^{0.37}4p^{0.34}$. This yields an effective charge of $+0.54$ on Ti^{3+} and -0.59 on F^-. Table 11.9 gives the final values of the Coulomb energies used for Ti and F orbitals. Table 11.10 lists the energy levels of the molecular orbitals finally obtained by Bedon, Horner, and Tyree. They also found that the qualitative picture of the location of bonding and antibonding electrons is substantiated by the population distribution. In the t_{2g}^b level,

Table 11.9

H_{11} and H_{22} corresponding to $Ti^{+0.54}$ for the configuration $3d^{2.75}4s^{0.37}4p^{0.34}$[a]

Ion	Orbital	Energy (cm^{-1})
Ti	$3d$	−95,540
	$4s$	−83,340
	$4p$	−51,450
F	$2p_\pi$	−127,200
	$2p$	−136,900
	$2s$	−272,200
	hyb σ	−154,500

[a] From H. D. Bedon, S. M. Horner, and S. Y. Tyree, Jr., *Inorg. Chem.* **3**, 647 (1964).

Table 11.10

Molecular orbital energies for TiF_6^{3-} [a]

	Energy (cm^{-1})	
Orbital	Bonding	Antibonding
t_{2g}	−134,770	−78,790
e_g	−163,490	−61,280
a_{1g}	−163,670	−25,440
t_{1u}	−154,910	−23,810

[a] From H. D. Bedon, S. M. Horner, and S. Y. Tyree, Jr., *Inorg. Chem.* **3**, 647 (1964).

approximately 80% of the electrons are on the ligands, while 80% of the electrons in the t_{2g} antibonding level are located on the central metal ion.

The positions of the absorption bands in the electronic spectra of the three compounds studied by Bedon, Horner, and Tyree are shown in Table 11.11. It should be noted that the bands are separated by 3900 cm^{-1} in the first case and by only 2800 cm^{-1} in the second and third cases. This splitting may be due primarily to the Jahn-Teller distortion of the excited ^2E_g state (see Chapter 12 for a description of the Jahn-Teller effect). If the two bands are primarily due to the Jahn-Teller effect, the mean of the two transitions should be $10Dq$ for a pure octahedral field. It is evident from Table 11.11 that these mean values compare remarkably well with the calculated 17,510 cm^{-1} separation of the t_{2g} and e_g antibonding levels in the MO scheme. Although agreement with experiment has been good, there has been some criticism concerning the evaluation of the diagonal and off-diagonal elements involved in such a calculation, which we shall discuss in the next section.

Without going into further details of the extended W-H method, which has been described above, we should mention that similar calculations have been performed by many authors during the last decade, notably by Ballhausen and Gray.[3] They have used this method on two systems: MnO_4^- and CrF_6^{3-}.

Table 11.11

Position of the absorption bands of Ti^{3+} in various complexes

Compound	Position of absorption bands (cm^{-1})		$10Dq$ (cm^{-1})
1. $(NH_4)_3TiF_6$	19,010	15,110	17,060
2. Na_2KTiF_6	18,900	16,000	17,450
3. NaK_2TiF_6	18,900	16,100	17,500
Octahedral TiF_6^{3-}	17,510 (calculated)		17,510

Although they have also obtained qualitatively good agreement between theory and experiment in these cases, closer examination reveals certain facts that are a bit discordant. For example, for MnO_4^- a double σ (ligand) basis set $(2s, 2p)$ was used, whereas for CrF_6^{3-} only a single $2p$ function was used. If a double σ basis set is also used for CrF_6^{3-}, one gets a Dq value that is too large, while for MnO_4^- a single σ valence function gives Dq with a wrong sign. This proves that the W-H method in its present form is incapable of reproducing the $10Dq$ value. Ballhausen and Gray have discovered the (interesting) fact that for octahedral geometry it is sufficient to use a single σ function $(2p)$, whereas for tetrahedral geometry a double σ set is needed. From the preceding discussion it is clear that one must look for the deficiencies of the extended W-H method and then try to correct them.

Before considering the improvement of the W-H method, we should mention that a great many other metal complexes have been treated in the above-mentioned way, seemingly with reasonable results.[43–52]

11.7 IMPROVEMENT OF THE WOLFSBERG–HELMHOLZ METHOD

Although the modified W-H approach to MO theory for the transition metal complexes has had apparent success in recent applications (see the references cited at the end of Section 11.6), certain assumptions concerning the quantitative estimation of the diagonal and off-diagonal matrix elements employed in this method cannot be justified on the basis of more rigorous considerations. These approximations, as already shown,[52,53] overestimate the amount of covalency in the resulting molecular orbitals. For example, one such calculation using the modified W-H method on MnO_4^- indicates that the e antibonding orbital is approximately 60% ligand in character, whereas electron spin resonance results indicate that this orbital should involve not more than 10% ligand participation.

Let us, with Fenske,[53] probe a little deeper into the problem. For the ligand diagonal terms H_{jj}, two methods are usually employed. In the "free-ion" method the term is estimated from the VSIE in a manner analogous to that for the central ion and hence is charge dependent. In the "hybrid-ion" method the value for H_{jj} is estimated from the experimental ionization energy of an electron in a hybrid of the ligand. For example, for fluoride ligand the lowest ionization energy of an HF electron is used, but for oxides the ionization energy of an H_2O electron is employed. In this approach the H_{jj} is presumed to be independent of the charge on the ligand.

However, the use of hybrid ligand orbitals for the fluoride complex does not seem to be justified. It has been shown by Ballhausen and Gray[55] that for a fluoride complex the σ-bonding wavefunction is

$$\Psi_\sigma = \sqrt{0.13}\ \phi(2s) + \sqrt{0.87}\ \phi(2p) \tag{11.132}$$

and the overlap of this function with the $3d_\sigma$ orbital on the central metal is 0.180. However, there exists a function orthogonal to this one, which is presumed to be the nonbonding orbital for the lone-pair (LP) electrons on the fluorine, of the form

$$\Psi_{LP} = \sqrt{0.87}\,\phi(2s) - \sqrt{0.13}\,\phi(2p) \tag{11.133}$$

which has an overlap with the metal $3d_\sigma$ orbital of 0.103. But this is obviously inconsistent with the assumption of a nonbonding overlap. To be consistent in such a calculation, therefore, one must not ignore this interaction.

On the other hand, for the oxide ligands Viste and Gray[56] have shown that, when the individual $2s$ and $2p$ oxygen levels are considered, instead of hybrids, better agreement with the transition energies is obtained. Let us examine similarly the effect of the individual $2s$ and $2p$ orbitals on the energy levels of TiF_6^{3-}. In order that suitable comparisons could be made, the same set of values employed by Bedon, Horner, and Tyree[44] in their original calculation was used by Fenske. The data are presented in Table 11.12. It is apparent from this table that the excellent agreement between theory and experiment for the $10Dq$ value claimed by Beadon and his co-workers breaks down when individual $2s$ and $2p$ ligand orbital interactions are considered. Indeed, the theoretical value of $10Dq$ is almost twice the experimental result. Fenske also showed that the reason for such an increase in Dq may be that, when the absolute value of one diagonal term is much greater than the other,

Table 11.12

Results for TiF_6^{3-} by various methods[a]

Parameters	Hybrid orbital	Without ligand-ligand overlap	With ligand-ligand overlap: Hybrid method	With ligand-ligand overlap: Free-ion method
$10Dq$ (cm^{-1})	17,500[b]	28,300[c]	30,600	25,700
Self-consistent charge on metal	+0.54	+0.54	+0.59	+0.51
Electron configuration on metal	$e^{2.75}$ $s^{0.37}$ $p^{0.34}$	$d^{2.77}$ $s^{0.33}$ $p^{0.36}$	$d^{2.84}$ $s^{0.31}$ $p^{0.26}$	$d^{2.94}$ $s^{0.32}$ $p^{0.23}$

[a]R. F. Fenske, *Inorg. Chem.* **4**, 33 (1965).
[b]Results from H. D. Beadon, S. M. Horner, and S. Y. Tyree, Jr., *Inorg. Chem.* **3**, 647 (1964).
[c]Hybrid method: Fenske, *op. cit.*

for example, $|H_{jj}| \gg |H_{ii}|$, and the off-diagonal term is estimated by

$$H_{ij} = -2G_{ij}(H_{ii} \cdot H_{jj})^{1/2} \tag{11.134}$$

the bonding level is only slightly different from H_{jj} but the antibonding level is appreciably different from H_{ii}. This means that the inclusion of low-lying energy levels drastically affects the antibonding levels because of the inordinate importance given to the off-diagonal term by relation (11.134).

It is clear from the preceding discussion that the extended W-H method for determining the electronic energy levels in the transition metal complexes should be accepted with considerable reservation.

To improve on the W-H method, Fenske and Radtke[53,54] have suggested a procedure that is believed to be a substantial improvement. It is in better accord with theoretical considerations and therefore holds some promise for much better agreement with the experimental information. Let us deal with Refs. 53 and 54 in some detail in the following section.

11.8 LCAO-SCF MO CALCULATIONS

Let us begin by considering the one-electron energies of a closed shell system. The one-electron MOs, Ψ_i, as seen earlier, approximated by a linear combination of symmetry-adapted AOs, are given by

$$\Psi_i = a_i \chi_i + b_i \phi_i \tag{11.135}$$

where a_i and b_i are the coefficients, χ_i is a normalized metal function, and ϕ_i is the normalized symmetry-adapted combination of ligand wavefunctions associated with n ligands (Table 11.2):

$$\phi_i = \sum_{j=1}^{n} c_{ij} \rho_{ij} \tag{11.136}$$

where ρ_{ij} is an AO on the jth ligand. Assuming ρ_{ij} to be orthonormal, that is,

$$\int \rho_{ij} \rho_{ik} \, d\tau = \delta_{jk}$$

we have

$$\sum_j c_{ij}^2 = 1$$

Since the metal wavefunction χ_i is not orthogonal to ϕ_i, we have the group overlap integrals defined by

$$G(\chi_i, \phi_i) = \sum_j c_{ij} S(\chi_i, \rho_{ij}) \tag{11.137}$$

The charge distribution Ψ_i^2 is therefore,

$$\Psi_i^2 = a_i^2 \chi_i^2 + 2a_i b_i \sum c_{ij} \chi_i \rho_{ij} + b_i^2 \sum_j \sum_k c_{ij} c_{ik} \rho_{ij} \rho_{ik}$$

which, on integration over all space, gives

$$a_i^2 + 2a_i b_i \sum_j c_{ij} S(\chi_i, \rho_{ij}) + b_i^2 \sum c_{ij}^2 = 1 \tag{11.138}$$

Now, dividing the charge density, represented by the middle term, equally between the metal and the ligand centers, and using Mulliken's approximation,[36] we have

$$\begin{aligned} \bar{a}_i &= a_i^2 + a_i b_i \sum_j c_{ij} S(\chi_i, \rho_{ij}) \\ \bar{b}_i &= \sum_j b_i^2 c_{ij}^2 + a_i b_i c_{ij} S(\chi_i, \rho_{ij}) \end{aligned} \tag{11.139}$$

where $\bar{a}_i$ represents the fractional charge on the metal, and $\bar{b}_i$ the fractional charge on the ligands.

We now use another Mulliken approximation[38] for the estimation of the two-electron electrostatic interaction integrals (this is done in order to avoid the multicenter integrals, which are extremely difficult to evaluate accurately):

$$(\rho_A \rho_B | \rho_C \rho_C) \approx \tfrac{1}{2}\left[(\rho_A \rho_A | \rho_C \rho_C) + (\rho_B \rho_B | \rho_C \rho_C) \right] S(\rho_A, \rho_B) \tag{11.140}$$

where

$$(\rho_A \rho_B | \rho_C \rho_C) = \int \rho_A^*(1) \rho_B^*(1) \frac{1}{r_{12}} \rho_C^*(2) \rho_C(2)\, d\tau$$

and the ρ's are one-electron wavefunctions on atoms A, B, and C. Let us consider an integral of the type

$$\begin{aligned} (\Psi_i \Psi_j | \rho_{m1} \rho_{m1}) &= a_i^2 (\chi_i \chi_i | \rho_{m1} \rho_{m1}) + 2a_i b_i \sum_j c_{ij} (\chi_i \rho_{ij} | \rho_{m1} \rho_{m1}) \\ &\quad + b_i^2 \sum_j \sum_k c_{ij} c_{ik} (\rho_{ij} \rho_{ik} | \rho_{m1} \rho_{m1}) \end{aligned} \tag{11.141}$$

Using the Mulliken approximation, the last term simplifies to

$$b_i^2 \sum_j c_{ij}^2 (\rho_{ij}\rho_{ij}|\rho_{m1}\rho_{m1})$$

since $S(\rho_{ij}\rho_{ik}) = \delta_{jk}$, by earlier assumption. Similarly, the middle term is

$$2a_i b_i \sum_j c_{ij}(\chi_i \rho_{ij}|\rho_{m1}\rho_{m1}) = a_i b_i \sum_j c_{ij} S(\chi_i \rho_{ij})\left[(\chi_i\chi_i|\rho_{m1}\rho_{m1}) + (\rho_{ij}\rho_{ij}|\rho_{m1}\rho_{m1})\right]$$

Therefore, finally,

$$(\Psi_i \Psi_i|\rho_{m1}\rho_{m1}) = \left[a_i^2 + a_i b_i \sum_j c_{ij} S(\chi_i \rho_{ij})\right](\chi_i\chi_i|\rho_{m1}\rho_{m1})$$

$$+ \sum_j \left[b_i^2 c_{ij}^2 + a_i b_i c_{ij} S(\chi_i \rho_{ij})\right](\rho_{ij}\rho_{ij}|\rho_{m1}\rho_{m1})$$

$$= \bar{a}_i(\chi_i\chi_i|\rho_{m1}\rho_{m1}) + \sum_j \bar{b}_{ij}(\rho_{ij}\rho_{ij}|\rho_{m1}\rho_{m1}) \qquad (11.142)$$

From the above, it is very important to realize that the Coulomb integrals involving the molecular charge distribution, Ψ_i^2, can be greatly simplified by using the Mulliken approximations.

To evaluate the necessary Coulomb and the exchange integrals, we now write the Hamiltonian, which is represented conveniently in terms of the Fock operator:[4,11]

$$\mathcal{H} = -\tfrac{1}{2}\Delta_{\mathrm{KE}} + V_{\mathrm{M}} + \sum_j \cdot V_j \qquad (11.143)$$

where $-\frac{1}{2}\Delta_{\mathrm{KE}}$ represents the kinetic energy operator, and V_{M} is given by

$$V_{\mathrm{M}} = \sum_i \bar{a}_i\{2(\bar{\chi}_i\chi_i| - (\chi_i \cdot|\bar{\chi}_i\} - Z_{\mathrm{M}}\left(\frac{1}{r_{\mathrm{M}}}\right| \qquad (11.144)$$

in which $(\bar{\chi}_i\chi_i|$, $(\chi_i \cdot|\bar{\chi}_i$, and $Z_{\mathrm{M}}(1/r_{\mathrm{M}}|$ indicate the Coulomb, exchange, and nuclear attraction operators, respectively. In (11.143) V_j is the corresponding operator for the electrons and nucleus of the jth ligand and is given by

$$V_j = \sum_i \bar{b}_{ij}\{2(\rho_{ij}\rho_{ij}| - (\rho_{ij} \cdot|\rho_{ij}\} - Z_j\left(\frac{1}{r_j}\right| \qquad (11.145)$$

Note the position of the bar in the exchange term! The Coulomb operators $(\bar{\lambda}_j\lambda_j|$ and the exchange operators $(\lambda_j \cdot|\bar{\lambda}_j$ are defined by their operational effect on some molecular spin orbitals (MSOs) by

$$\{(\bar{\lambda}_j(\nu)\lambda_j(\nu)|\}\lambda_i(\mu)=\left\{\int\bar{\lambda}_j(\nu)\lambda_j(\nu)\frac{1}{r_{\mu\nu}}\,d\tau(\nu)\right\}\lambda_i(\mu)$$

or

$$\{\bar{\lambda}_j\lambda_j|\}\lambda_i=(\bar{\lambda}_j\lambda_j|\lambda_i$$

and

$$\{(\lambda_j(\nu)\cdot|\bar{\lambda}_j(\nu)\}\lambda_i(\mu)=\left\{\int\lambda_i(\nu)\bar{\lambda}_j(\nu)\frac{1}{r_{\mu\nu}}\,d\tau(\nu)\right\}\lambda_j(\mu)$$

or

$$\{(\lambda_j\cdot|\bar{\lambda}_j\}\lambda_i \begin{array}{ll} =(\lambda_i\bar{\lambda}_j|\lambda_j & \text{for spin } i=\text{spin } j \\ =0 & \text{for spin } i\neq j \end{array}$$

As before, we now have to find the diagonal and the off-diagonal elements of the secular equation (11.93). The diagonal terms involving the metal wavefunctions χ_i are

$$\begin{aligned} H_{ii}&=\langle\chi_i|\mathcal{H}|\chi_i\rangle=\langle\chi_i|-\tfrac{1}{2}\Delta_{\mathrm{KE}}+V_{\mathrm{M}}|\chi_i\rangle+\langle\chi_i|V_j|\chi_i\rangle \\ &=\epsilon_{\chi}(q_{\mathrm{M}})-\sum_{j=1}^{n}q_j\left(\frac{1}{r_j}\middle|\chi_i\chi_i\right) \end{aligned} \quad (11.146)$$

where the first term is the orbital energy of the metal electron in the free ion of charge q_{M}, and the second term is the crystal field potential due to the point charges q_j placed on the ligands. Both q_{M} and q_j are evaluated by Mulliken electron population analysis, and it is required that self-consistency be established between the initial and final calculated values of q_{M} and q_j.

The ligand diagonal terms, exclusive of ligand-ligand interaction, have the form

$$\langle\phi_i|\mathcal{H}|\phi_i\rangle=\mathrm{E}\rho_{i1}(q_1)-q_{\mathrm{M}}\left(\frac{1}{\mathbf{r}_{\mathrm{M}}}\middle|\rho_{i1}\rho_{i1}\right)+\sum_{j=2}^{n}q_j\left(\frac{1}{\mathbf{r}_j}\middle|\rho_{i1}\rho_{i1}\right) \quad (11.147)$$

where $\epsilon\rho_{i1}$ is the orbital energy of the electron in the ith orbital of ligand 1 and of charge q_1, and the second and third terms constitute the crystal field potential due to the metal ion of charge q_{M} and the other ligands with charges q_j, respectively.

Table 11.13

$10Dq$, Racah parameter B, and $\beta = B_{complex}/B_{free\ ion}$ for a number of octahedral and tetrahedral complexes of the $3d^n$ ions[a]

The numbers in parentheses are the experimental values.

	Complex										
	$TiCl_6^{3-}$	VCl_6^{3-}	$CrCl_6^{3-}$	$FeCl_6^{3-}$	$TiCl_4$	VCl_4	$FeCl_4^-$	$MnCl_4^{2-}$	$FeCl_4^{2-}$	$CoCl_4^{2-}$	$NiCl_4^{2-}$
M-L distance (Å)	2.47	2.45	2.34	2.38	2.18	2.03	2.19	2.33	2.27	2.25	2.27
Self-consistent charge on metal	0.60	0.49	0.44	0.14	0.33	0.23	0.34	0.20	0.24	0.09	0.08
$10Dq$ (eV)	1.66	1.49	1.47	1.25	1.02	0.96	0.63	0.45	0.52	0.48	0.44
	(1.71)	(1.54)	(1.57)	(1.14)	(0.93)	(1.12)	(0.62)	(0.45)	(0.50)	(0.41)	(0.44)
B (cm^{-1})	619	649	640				625	830		880	875
	(563)	(561)	(655)				(590)	(770)		(730)	(765)
β	0.70	0.68	0.59				0.58	0.95		0.88	0.82
	(0.65)	(0.62)	(0.65)				(0.58)	(0.86)		(0.75)	(0.72)

[a] R. F. Fenske and D. D. Radtke, *Inorg. Chem.* **7**, 479 (1968).

In terms of the single-ligand wavefunctions the off-diagonal matrix element is given by

$$\langle \rho_{i1}|\mathcal{H}|\chi_i\rangle = \epsilon_{\chi_i}(q_M)S(\rho_{i1}\chi_i) - \sum_{j=2} q_j\left(\frac{1}{r_j}\middle|\rho_{i1}\chi_i\right) + \sum_k \bar{b}_k\{2(\rho_{k1}\rho_{k1}|\rho_{i1}\chi_i) - (\rho_{k1}\rho_{i1}|\rho_{k1}\chi_i)\} - Z_1\left(\frac{1}{r_1}\middle|\rho_{i1}\chi_i\right) \tag{11.148}$$

In his earlier work,[53] Fenske used a reduction factor R_a in the off-diagonal matrix element, which was eliminated in the later work of Fenske and Radtke;[54] thus making their calculations free of all parameters. The concepts that led to the final forms of the diagonal and off-diagonal matrix elements are outlined in the references cited above.

Without going into the details of their calculations, we now discuss the final results that they obtained by applying their method to four octahedral and seven tetrahedral chloro complexes. The results are shown in Table 11.13, where we have presented only the values for $10Dq$, B, and β. From the table it is clear that the agreements with experimental values are very satisfactory. In another interesting paper by Fenske and his co-workers,[57] they also calculated the spin densities and B and β values for the MF_6^{3-} complexes where M = Ti, V, Cr, Fe, and Co and obtained good agreement with experiment. In the absence of more sophisticated *ab initio* calculations using the Roothaan-Hartree-Fock-SCF MO methods (which can, in principle, be applied to open shells as well and have, in fact, been made for octahedral complexes by Richardson and his co-workers[58,59] with some success), the semiempirical extended W-H methods of Fenske and his co-workers seem to provide a good, workable procedure for the transition metal complexes.

Since very few actual molecular calculations have been done using the Roothaan-Hartree-Fock open shell equations (one such example, mentioned above, is the calculations of Richardson and his co-workers), we prefer not to deal with this procedure in detail. A general outline of the procedure was given in Section 3.2.

REFERENCES

1. J. H. Van Vleck and A. Sherman, *Rev. Mod. Phys.* **7**, 167 (1935).
2. C. A. Coulson and I. Fisher, *Phil. Mag.* **40**, 386 (1949).
3. C. J. Ballhausen and H. B. Gray, *Molecular Orbital Theory*, Benjamin, New York, 1965.
4. C. C. J. Roothaan, *Rev. Mod. Phys.* **32**, 179 (1960).
5. S. Sugano and R. G. Shulman, *Phys. Rev.* **130**, 517 (1963).
6. R. E. Watson and A. J. Freeman, *Phys. Rev. A*, **134**, 152 (1964).
7. S. Sugano and Y. Tanabe, *J. Phys. Soc. Japan*, **20**, 1155 (1965).

8. T. F. Soules, J. W. Richardson, and D. M. Vaught, *Phys. Rev. B*, **3**, 2186 (1971).
9. J. H. Van Vleck, *J. Chem. Phys.* **3**, 803 (1935).
10. C. J. Ballhausen, *Introduction to Ligand Field Theory*, McGraw-Hill, New York, 1962.
11. J. W. Richardson and R. E. Rundle, "A Theoretical Study of the Electronic Structure of Transition Metal Complexes," Ames Laboratory, Iowa State College, ISC-830, U.S. Atomic Energy Commission, Technical Information Service Extension, Oak Ridge, Tenn., 1956.
12. L. E. Orgel, *Quart. Rev. (London)* **8**, 422 (1954).
13. J. Owen, *Proc. Roy. Soc. (London) A*, **227**, 183 (1953).
14. K. W. H. Stevens, *Proc. Roy. Soc. (London) A*, **219**, 542 (1953).
15. A. S. Chakravarty, *Proc. Phys. Soc.* **74**, 711 (1959).
16. J. H. E. Griffiths, J. Owen, and I. M. Ward, *Proc. Roy. Soc. (London) A*, **219**, 526 (1953).
17. A. Bose, A. S. Chakravarty, and R. Chatterjee, *Proc. Roy. Soc. (London) A*, **255**, 145 (1960).
18. J. Owen and J. H. M. Thornley, *Rep. Prog. Phys.* **29**, 675 (1966).
19. A. A. Misetich and T. Buch, *J. Chem. Phys.* **41**, 2524 (1964).
20. W. H. Kleiner, *J. Chem. Phys.* **20**, 1784 (1952).
21. Y. Tanabe and S. Sugano, *J. Phys. Soc. Japan* **11**, 864 (1956).
22. R. G. Shulmen and S. Sugano, *Phys. Rev.* **130**, 506 (1963).
23. R. E. Watson and A. J. Freeman, *Phys. Rev. A*, **134**, 152 (1964).
24. S. Sugano and Y. Tanabe, *J. Phys. Soc. Japan* **20**, 1155 (1965).
25. C. K. Jorgensen, *Absorption Spectra and Chemical Bonding in Complexes*, Pergamon Press, Oxford, England, 1962.
26. J. Hinze and H. H. Jaffe, *J. Chem. Phys.* **38**, 1834 (1963).
27. E. Tondello, G. De Michelis, L. Oleari, and L. Di Sipio, *Coord. Chem. Rev.* **2**, 65 (1967).
28. C. K. Jorgensen, *Helv. Chim. Acta, Fasciculus Extraordinarius Alfred Werner*, **131** (1967).
29. Y. Tanabe and S. Sugano, *J. Phys. Soc. Japan* **9**, 753, 766 (1954).
30. L. Di Sipio, E. Tondello, G. De Michelis, and L. Oleari, *Inorg. Chem.* **9**, 927 (1970).
31. D. S. McClure, *Electronic Spectra of Molecules and Ions in Crystals*, Academic Press, New York, 1959.
32. G. L. Allen and K. D. Warren, *Struct. Bonding* **9**, 49 (1971).
33. D. E. Rimmer, *Proceedings of the International Conference on Magnetism, Nottingham, 1964*, Institute of Physics and Physical Society, London, p. 337, 1965.
34. J. Hubbard, D. E. Rimmer, and F. R. A. Hopgood, *Proc. Phys. Soc.* **88**, 13 (1966).
35. M. Wolfsberg and L. Helmholz, *J. Chem. Phys.* **20**, 837 (1952).
36. R. S. Mulliken, *J. Chem. Phys.* **23**, 1833 (1955).
37. I. De, V. P. Desai, and A. S. Chakravarty, article to be published (1980).
38. R. S. Mulliken, C. A. Rieke, D. Orloff, and H. Orloff, *J. Chem. Phys.* **17**, 1248 (1949).
39. H. H. Jaffe, *J. Chem. Phys.* **21**, 258 (1953); H. H. Jaffe and G. O. Doak, *J. Chem. Phys.* **21**, 196 (1953); J. L. Roberts and H. H. Jaffe, *J. Chem. Phys.* **27**, 883 (1957).
40. L. Leifer, F. A. Cotton, and J. R. Leto, *J. Chem. Phys.* **28**, 364, 1253 (1958).
41. D. P. Craig, A. Maccoll, R. S. Nyholm, L. E. Orgel, and L. E. Sutton, *J. Chem. Soc.*, p. 354, 1954.
42. A. Lofthus, *Mol. Phys.* **5**, 105 (1962).
43. L. L. Lohr, Jr., and W. W. Lipscomb, *J. Chem. Phys.* **38**, 1607 (1963).
44. H. D. Bedon, S. M. Horner, and S. Y. Tyree, Jr., *Inorg. Chem.* **3**, 647 (1964).
45. H. D. Bedon, W. E. Hatfield, S. M. Horner, and S. Y. Tyree, Jr., *Inorg. chem.* **4**, 743 (1965).
46. W. E. Hatfield, H. D. Bedon, and S. M. Horner, *Inorg. Chem.* **4**, 1181 (1965).

47. T. H. Wirth, *Acta Chem. Scand.* **19**, 2261 (1965).
48. H. Basch, A. Viste, and H. B. Grey, *J. Chem. Phys.* **44**, 10 (1966).
49. D. H. Johansen and C. J. Ballhausen, *Mol. Phys.* **10**, 175 (1966).
50. L. L. Ingraham, *Acta Chem. Scand.* **20**, 283 (1966).
51. J. W. Moore, *Acta Chem. Scand.* **20**, 1154 (1966).
52. V. Valenti and J. P. Dahl, *Acta Chem. Scand.* **20**, 2387 (1966).
53. R. F. Fenske, *Inorg. Chem.* **4**, 33 (1965).
54. R. F. Fenske and D. D. Radtke, *Inorg. Chem.* **7**, 479 (1968).
55. C. J. Ballhausen and H. B. Gray, *Inorg. Chem.* **1**, 111 (1962).
56. H. Viste and H. B. Grey, *Inorg. Chem.* **3**, 113 (1964).
57. R. F. Fenske, K. G. Coulton, D. D. Radtke, and C. C. Sweeney, *Inorg. Chem.* **5**, 951, 960 (1966).
58. J. W. Richardson, T. F. Soules, D. M. Vaught, and R. R. Powell, *Phys. Rev. B*, **4**, 1721 (1971). See also other references quoted in this paper.
59. T. F. Soules, J. W. Richardson, and D. M. Vaught, *Phys. Rev. B*, **3**, 2186 (1971).
60. J. C. Phillips, *J. Phys. Chem. Solids* **11**, 226 (1959).
61. R. E. Watson, *Phys. Rev.* **118**, 1036 (1960).
62. J. Owen and K. W. H. Stevens, *Nature (London)* **171**, 836 (1953).
63. M. Tinkham, *Proc. Roy. Soc. (London) A*, **236**, 535 (1956).
64. J. H. E. Griffiths and J. Owen, *Proc. Roy. Soc. (London) A*, **226**, 96 (1954).
65. J. H. M. Thornley, C. G. Windsor, and J. Owen, *Proc. Roy. Soc. (London) A*, **284**, 252 (1965).
66. T. M. Dunn, *Trans. Faraday Soc.* **57**, 1441 (1961).
67. W. Marshall and R. Stuart, *Phys. Rev.* **123**, 2048 (1961).
68. S. Koide and M. H. L. Pryce, *Phil. Mag.* **3**, 607 (1958).
69. M. Kotani, A. Amemiya, E. Ishiguro, and T. Kimura, *Tables of Molecular Integrals*, Maruzen Co., Tokyo, 1963.
70. C. E. Moore, "Atomic Energy Levels," U.S. National Bureau of Standards, Circular 467, 1949 and 1952.
71. A. S. Chakravarty, *J. Phys. Chem.* **74**, 4347 (1970).

12

Optical Properties of ML_6 Complexes

Optical transitions are radiative transitions involving quanta having energies of the order of volts, which distinguish them from other electronic transition processes such as those occurring in the microwave or X-ray region. The optical spectra of molecules and ions in solids are normally observed in the range of a few thousand to about 100,000 wave numbers. The purpose of this chapter is to show the extent to which optical spectra have been, and can be, interpreted and the useful information they contain.

The electronic states of solids may be treated theoretically starting with either of the two zeroth-order approximations: the Bloch model and the Heitler-London model, which represent, respectively, completely delocalized and completely localized electronic states. Real systems fall somewhere between these extremes, and it is possible to make an approximate classification of the electronic states on the basis of which of the two models provides the better description. Molecular and ionic crystals usually have low-lying excited states, which are described best in terms of the states of the free molecule or ion. Metals and semiconductors, however, are best described starting with the free-electron wavefunctions. We shall be interested here in the electronic spectra of the transition metal ions in crystals.

We shall limit ourselves to the theory, for the interpretation of the spectra and for the derivation of useful empirical constants from the spectra. The ligand field theory has been developed in sufficient detail in the preceding chapters, and we shall show in this chapter how to find the crystal field constants from the spectra.

The transition metal complexes are characterized by their brilliant and varied colors and are aesthetically pleasing to the observer; for the spectroscopist the color variation presents an immediate challenge. Chemists, however, consider the remarkable color change of the compounds from the structural point of view, which suggests that this color change is accompanied by a change in the ligand arrangement, that is, coordination, in a compound. After the introduction of coordination theory by Werner,[1] investigations on the optical properties of compounds began.

The transition metal ions are characterized by incomplete *d*-shell electrons, which are responsible for determining the color of the complexes. These electrons, in the presence of the ligand field, undergo transitions from the ground state to the excited states of the ligand field terms by absorbing light and thus give rise to the ligand field bands. The complexes of the transition metal ions absorb light in the wave number range of 8000 to 40,000 cm^{-1}, and the absorption spectra are characterized by the presence of several bands in the near-infrared, visible, and ultraviolet regions. Some spectra show sharp lines besides the broad bands.

The optical absorption of both crystals and solutions containing these complex ions have been studied since the days of Werner by many physicists and chemists. It is known from their studies that the absorption spectra of solutions are quite similar to those of crystals containing the same complex ions as the solutions, except for an intense edge absorption ($\log \epsilon_{max} \sim 4$) in the ultraviolet region. This does not appear in the spectra of crystals and hence is considered to be due mainly to the absorption of the complex ion.

The observed absorption spectra of the transition metal complexes or compounds are explained in the frame work of the ligand field theory. The free-ion levels in the ligand field undergo Stark splitting, as discussed in Chapter 6, and the absorption bands are due to the transitions between the Stark components of the ground multiplet of the metal ion. Although the broad bands were explained reasonably in this way by Hartmann and his co-workers[2, 2a] they made no mention of the line spectra. Later Tanabe and Sugano[3] qualitatively explained the line widths of the observed spectra. According to them, the broad bands appear when there is a transition between the terms of different electron configurations, and sharp lines are due to a transition between the terms of the same electronic configuration.

To assign the spectra, we need to calculate the energy levels of the different strong-field terms. The electrostatic and the cubic field matrices have been constructed for d^n strong-field configurations by Tanabe and Sugano[4, 13] and also by Griffith.[5] The sort of matrix elements occurring in the presence of lower symmetric fields of the D_{4h} and D_{3d} types have been calculated by Perumareddi.[6, 7] He found these matrix elements to be correct by comparing them to the results in the weak-field limit. The band positions are then found by diagonalizing the matrices with a suitable choice of the Racah parameters and the ligand field parameters.

So far we have not considered the spin-orbit interaction, which has very little effect on the energy levels for $3d^n$ ions. For $4d^n$ and $5d^n$ ions, however, the spin-orbit interaction being large, there is considerable mixing among the strong-field terms via the spin-orbit interaction, and therefore the energies corresponding to the band positions are obtained by diagonalizing the Hamiltonian, which includes the spin-orbit interaction in addition to the ligand field and the electrostatic interactions. Moreover, in order to explain the spin-forbidden transitions, which are allowed by the spin-orbit interaction, one cannot neglect this interaction even for $3d^n$ ions. Spin-orbit matrices for the d^n strong-field terms have been calculated by Liehr and Ballhausen,[8] Eisenstein,[9] Runciman and Schroeder,[10] and Schroeder.[11, 12] To use these matrices along with the ligand field and the electrostatic matrices, one should be very careful about the phases of the wavefunctions, so that these phases are consistent throughout.

The transition metal ions in the octahedral complexes, having small Dq values, usually form high-spin complexes; those with large Dq form low-spin complexes. When Dq is large, instead of d orbitals, t_{2g} and e_g orbitals are considered to be filled by the electrons. The order of the energy levels, depending on the strength of the cubic ligand field, can be found in the energy level diagrams[3, 5] given by Tanabe and Sugano. From these diagrams one can easily find the ground states for the electronic configurations of the high-spin and the low-spin complexes. Once the ground state and the excited states of a particular complex are known, it is possible to explain the observed band positions.

12.1 NATURE OF TRANSITIONS

Transitions between the levels of d^n configurations are governed by the selection rules. The transitions are forbidden if any of these rules is violated. In Russell-Saunders coupling the free-ion LS terms are given by the allowed values of L and S obtained by coupling the electrons in the d^n configurations. In view of the spin multiplicity of the terms, the transitions are broadly classified into two classes. The transitions between terms having the same spin multiplicity (spin-allowed) are called the intrasystem transitions, and those between terms of different spin multiplicities (spin-forbidden) are called the intersystem transitions. The selection rule for L is $\Delta L = 0, \pm 1$, where ΔL is the change in the orbital angular momenta of the initial and the final states of transitions. The spin selection rule is $\Delta S = 0$, where ΔS is the change in the spin angular momenta of the initial and final states. This selection rule can be relaxed by considering the spin-orbit interaction. If we consider the transition between the levels designated by the total angular momentum J, then, in addition to the selection rules given above, there is another selection rule, $\Delta J = 0, \pm 1$, which of course excludes the $0 \to 0$ transition. For the spin-allowed transitions occurring within a given term when $L = 0$, the matrix elements of the dipole moment operator vanishes because the dipole moment operator is of odd parity and hence the direct product of the representation of this operator and that of the state concerned cannot contain any representation of the parity to which the state belongs. Thus all the electric-dipole transitions within a term are forbidden by the parity selection rule. This is Laporte's rule. On the other hand, the magnetic moment operator $\mathbf{M} = \mu_B(\mathbf{L} + 2\mathbf{S})$ and the quadrupole moment tensor $\mathbf{Q}_{ij} = -e(\mathbf{r}_i\mathbf{r}_j - \frac{1}{3}r^2\delta_{ij})$ belong to the even-parity representation; therefore the magnetic-dipole transitions, as well as the electric-quadrupole transitions, within a term are not forbidden by parity. Since the parity remains invariant under the ligand field splittings, the parity selection rule is valid even when the ligand field has inversion symmetry. Here we shall discuss the transitions between the multiplet terms of a ligand field of octahedral symmetry.

Let us first consider the transitions between terms of the same spin multiplicities, which are called the intrasystem transitions. Among the spin-allowed transitions there are three types having nonvanishing intensities:

1. The electric-dipole transition.
2. The magnetic-dipole transition.
3. The electric-quadrupole transition.

Since all the terms under consideration belongs to even parity, the electric-dipole transitions which are proportional to the absolute square of the matrix elements of the electric-dipole moment operator $\mathbf{P} = -e\Sigma_i \mathbf{r}_i$ are forbidden, as the parity of $\mathbf{P}$ is odd. This selection rule is called the parity selection rule. The parity selection rule is slightly relaxed if the octahedral symmetry either is slightly distorted by the presence of a weak low-symmetry field of odd parity or is instantaneously distorted by the presence of the lattice vibrations of certain modes. In the latter case the instantaneous distortion also brings in a weak low-symmetry field of odd parity. The odd-parity field admixes the even-parity states with the odd-parity states, resulting in nonvanishing matrix elements of the electric-dipole moment. Let us denote by $\langle V_{\text{odd}}\rangle$ the matrix element of the static odd-parity field V_{odd} between the even- and odd-parity states. The degree of admixture of an even-parity state with odd-parity states is given by $\langle V_{\text{odd}}\rangle/\Delta E_{eo}$, where ΔE_{eo} is the energy separation between the even-parity and the odd-parity states. Therefore the oscillator strength f_{el}^{forb} of the parity-forbidden transition is given by

$$f_{el}^{\text{forb}} \sim f_{el}^{\text{allow}} \cdot \left(\frac{\langle V_{\text{odd}} \rangle}{\Delta E_{eo}} \right)^2 \tag{12.1}$$

Here f_{el}^{allow} is the oscillator strength of the parity-allowed transition,

$$f_{el}^{\text{allow}} = \frac{8\pi^2 m}{3e^2 h} \nu\left(|p_x|^2 + |p_y|^2 + |p_z|^2\right) \tag{12.2}$$

where p_i $(i = x, y, z)$ is the matrix element of the ith component of the electric-dipole moment $\mathbf{P}$ between the initial and final states of transition. In many cases $f_{el}^{\text{allow}} \sim 1$, and to obtain the magnitude of f_{el}^{forb} it is necessary to estimate the magnitude of V_{odd}.

For the intrasystem transitions the magnetic-dipole transitions are generally allowed, since the magnetic-dipole moment

$$\mathbf{M} = -\mu_B \sum_i (\mathbf{l}_i + 2\mathbf{s}_i) \tag{12.3}$$

where μ_B is the Bohr magneton, is of even parity. The oscillator strength of the magnetic-dipole transition f_{m1} is given by

$$f_{m1} = \frac{8\pi^2 m}{3e^2 h} \nu\left(|m_x|^2 + |m_y|^2 + |m_z|^2\right) \tag{12.4}$$

where m_i $(i = x, y, z)$ is the matrix element of the ith component of the magnetic-dipole moment $\mathbf{M}$ between the initial and final states of the transition.[17]

In addition to the magnetic-dipole transitions, electric-quadrupole transitions are also allowed between states of the same parity. The oscillator strength of the electric-quadrupole transition is given by

$$f_{e2} = \frac{4\pi^4 m}{5c^2 e^2 h} \nu^3 \sum_{i,j = x,y,z} |q_{ij}|^2 \tag{12.5}$$

where q_{ij} is the matrix element of the electric-quadrupole moment tensor,

$$Q_{ij} = -e \sum_n \left(\mathbf{r}_{in}\mathbf{r}_{jn} - \tfrac{1}{3} r_n^2 \delta_{ij}\right) \tag{12.6}$$

between the initial and final states of the transition.[18]

Let us now consider the transitions between terms of different spin multiplicities, which are called the intersystem transitions. Since none of the electric-dipole, magnetic-dipole, and electric-quadrupole moments has nonvanishing matrix elements between the states of different spin multiplicities, the intersystem transitions are forbidden. This selection rule is known as the spin selection rule. The spin selection rule is slightly relaxed if the spin-orbit interaction is taken into account. The spin-orbit interaction connects terms with the resultant spins S and S', where $|S - S'| = 0, 1$. Therefore the term with S may have small components of the terms with $(S \pm 1)$ if the spin-orbit interaction is taken into account, and the presence of small spin-orbit admixture makes the transitions allowed. Thus the spin-forbidden transitions $S \rightleftharpoons S \pm 1$ are slightly allowed via the spin-orbit interaction. The degree of spin-orbit admixture of the S term with $(S \pm 1)$ terms is approximately given by $\langle V_{so} \rangle / \Delta E$, where $\langle V_{so} \rangle$ is the matrix element of the spin-orbit interaction between the S and $(S \pm 1)$ terms and ΔE is the energy separation between them. Once the spin forbiddenness is removed, as discussed here, all the transitions are allowed.

For the comparison of theory with experiment it is sometimes helpful to consider the transitions allowed by the configuration of the electrons. In the presence of a cubic ligand field, if the terms are well specified by the electron configuration $t_{2g}^m e_g^n$, the optical transitions are forbidden between the terms of $t_{2g}^m e_g^n$ and $t_{2g}^{m-k'} e_g^{n+k'}$ when $k' \geqslant 2$. This selection rule is known as the configuration selection rule[13] and is expressed by saying that electron jumps from one cubic orbital (t_{2g}) to the other (e_g) by more than one electron are forbidden. The explanation of this selection rule is as follows. Since the transition moments are one-electron operators, their matrix elements between the terms of configurations $t_{2g}^m e_g^n$ and $t_{2g}^{m-k'} e_g^{n+k'}$ are zero when $k' \geqslant 2$. Even in the case of electric-dipole transitions, when the one-electron operators, the electric-dipole moment, and the odd-parity perturbation V_{odd} are incorporated, this selection rule is valid. Of course, this rule cannot be applied when there is appreciable configuration mixing in the ligand field terms.[13]

The electric-dipole transitions between cubic field terms of even parity are allowed by odd-parity perturbation. To calculate the intensity of this transition one has to consider the admixture of even-parity with odd-parity states as a result of this perturbation.[19] If we take odd-parity perturbation as a lower symmetry field arising from the lattice vibration, this lower symmetric field is denoted by V_{odd}. To obtain V_{odd}, the ligand field $V(x,Q)$ is expanded in power series of Q, where x denotes the electron coordinate and Q is the normal coordinate of the vibrational displacements. Therefore

$$V(x,Q)=V_0(x)+\sum V_q^{(\kappa)}Q_q^{(\kappa)}+\frac{1}{2}\sum V_{qq'}^{(\kappa\kappa')}Q_q^{(\kappa)}Q_{q'}^{(\kappa)} \tag{12.7}$$

where $V_0(x)$ is the ligand field at equilibrium positions of the nuclei. Since

$$V(x,Q)=V(Rx,RQ) \tag{12.8}$$

where R denotes the symmetry operator, $V_q^{(\kappa)}$ and $Q_q^{(\kappa)}$ or $V_{qq'}^{(\kappa\kappa')}$ and $Q_q^{(\kappa)}Q_{q'}^{(\kappa')}$ have the same transformation properties. Now V_{odd} originates from the vibrational displacement with odd inversion symmetry T_{1u} and T_{2u}, and the main contribution to V_{odd} comes from $V_q(T_{1u})Q_q(T_{1u})$ and $V_q(T_{2u})Q_q(T_{2u})$. The magnitude of this contribution is of the order $(Q/r)V_0$, so that V_{odd} can be estimated from

$$V_{\text{odd}} \sim \frac{\delta R}{\bar{r}} V_0 \tag{12.9}$$

where δR is the zero-point amplitude of the vibration and $\bar{r}$ is the average of r. The zero-point amplitude of vibration is estimated from

$$2\pi^2 m\nu_0^2(\delta R)^2=\tfrac{1}{2}h\nu_0 \tag{12.10}$$

where m is the mass of the ligand ion, h is Planck's constant, and ν_0 is the frequency of the zero-point vibration. From the infrared absorption of some complexes, $\nu_0 \sim 10^{13}$ sec^{-1}, and assuming that $m \sim 10^{-23}$ g, $\bar{r} \sim 10^{-8}$ cm, and

$V_0 \sim 10^4\ \text{cm}^{-1}$, we have

$$V_{\text{odd}} \sim 10^3\ \text{cm}^{-1}$$

In view of the experimental data on the absorption spectra in the ultraviolet region, it is reasonable to assume that $\Delta E_{\text{eo}} \sim 10^5\ \text{cm}^{-1}$. Therefore

$$f_{e1}^{\text{forb}} \sim f_{e1}^{\text{allow}} \times 10^{-4} \sim 10^{-4}$$

since $f_{e1}^{\text{allow}} \sim 1$, as estimated from $f_{e1}^{\text{allow}} \sim (8\pi^2 m/3h)\nu \bar{r}^2$, where ν is assumed to be $6 \times 10^{14}\ \text{sec}^{-1}$ and m is the mass of the electron.

The weak low-symmetry ligand fields of odd parity also come from nuclear displacements, and in this case the displacements are static and are associated with the geometrical structure of the crystal or molecule. In many cases the static displacements are of the same order of magnitude as are the amplitudes of zero-point vibration. Therefore a static low-symmetry field of odd parity $\sim 10^4\ \text{cm}^{-1}$ leads to the same result for f_{e1}^{forb} as was given above. For the magnetic-dipole transitions, since the matrix element $m_i \sim \mu_B$, $f_{m1}^{\text{allow}} \sim 10^{-6}$, where ν is assumed to be $6 \times 10^{14}\ \text{sec}^{-1}$, as we are concerned with the transitions in the visible region. For the electric-quadrupole transitions the magnitude of $|q_{ij}|$ may be estimated from the approximate relation $q \sim e\langle r \rangle^2$; then, assuming $\langle r \rangle \sim 10^{-8}$ cm, we get $f_{e2}^{\text{allow}} \sim 10^{-7}$. In the cases of intersystem (spin-forbidden) transitions, $\langle V_{so} \rangle \sim 100\ \text{cm}^{-1}$, $\Delta E \sim 1000\ \text{cm}^{-1}$, and the oscillator strengths for the parity-forbidden electric-dipole and the electric-quadrupole transitions are $f_{e1}^{\text{forb}} \sim 10^{-7}$, $f_{m1}^{\text{forb}} \sim 10^{-9}$, and $f_{e2}^{\text{forb}} \sim 10^{-10}$, respectively. From the experimental data for the oscillator strengths it can be found that the electric-dipole transitions are more important than the other types, though these are parity forbidden and are allowed only by the odd ligand field.

Since there are two reasons for the occurrence of the parity-forbidden electric-dipole transitions, for broad bands it is necessary to know whether or not the transitions are accompanied by the excitation of the odd vibration. The difference between the transitions caused by the odd ligand field and the odd vibration lies in the temperature dependence of the absorption strength. The temperature dependence of the intensity of vibration-assisted transitions is given by[14]

$$f \sim \coth\left(\frac{h\nu}{2kT}\right) \qquad (12.11)$$

For the transitions assisted by the lower symmetric field of odd parity, the temperature dependence will be similar to that for the allowed transitions if the temperature variation of the odd ligand field is ignored. This difference is most definitely revealed in the polarization and the Zeeman pattern of the spectrum if they are observable. The electric- and magnetic-dipole transitions in a crystal may be distinguished experimentally by comparison of the axial and the transverse spectra.

In the study of absorption spectra or any other spectra, line width is very important. From the calculations of the energy levels it is obvious that the separations of these levels of a strong-field configuration are almost independent of the cubic splitting parameter in the range of large Dq/B, while separations of the levels belonging to different configurations are proportional to $10Dq$. This fact suggests that the transition energies between terms belonging to different configurations have a certain spread corresponding to the fluctuation of the cubic field that is caused by the nuclear vibration, whereas the transiton energies between terms of the same configuration are independent of the fluctuation of the cubic field. The magnitude of this fluctuation may be considered to be of the same order as $V_{\text{odd}} \sim 1000 \text{ cm}^{-1}$ at $T = 0°\text{K}$, if one can, of course, apply the same method of estimation as is used for V_{odd}. Therefore the widths of the bands coming from the transitions between the terms of different configurations are expected to be of the order $\sim 1000 \text{ cm}^{-1}$ even at $T = 0°\text{K}$. On the other hand, the spectral lines due to the transitions between terms of the same configuration are expected to be sharp.

In the present argument we have considered only a single mode of nuclear vibration, which preserves the cubic symmetry of the system. This mode is sometimes called a "breathing mode" of vibration. However, if the effects of the other vibrational modes are taken into account, the spectral lines arising from the same configuration may be broadened because of vibrational fluctuations of the low-symmetry fields. From study of the spectral width, it has been observed that the transitions within a given strong-field configuration give sharp spectra, whereas those between different strong-field configurations are broad. Since most of the intersystem transitions originate from the same configuration, and the intrasystem transitions from different configurations, it is reasonable to expect the intrasystem transitions to be broad and the intersystem transitions to be sharp.

12.2 OSCILLATOR STRENGTHS AND ABSORPTION COEFFICIENTS

If I is the intensity of a beam of monochromatic light passing through an absorption cell containing colored solution and provided with parallel windows spaced l cm apart, and I_0 is the intensity of light traversing a comparison cell with water or solvent without colored substance under similar circumstances, then, with the aid of the Lambert-Beer law, the optical density or absorbancy is defined as[15]

$$D = \log_{10}\left(\frac{I_0}{I}\right) = \epsilon c_0 l \tag{12.12}$$

where ϵ is the molar decadic extinction coefficient, and c_0 the concentration of absorption centers in moles per liter. The absorption coefficient is defined as

$$\alpha = \log\left(\frac{I_0}{I}\right)$$
$$= 2.303 \log_{10}\left(\frac{I_0}{I}\right)$$
$$= 2.303 \epsilon c_0 \quad (12.13)$$

as it corresponds to the reduction in intensity for a path length of unity. The relation between the oscillator strength and the absorption coefficient is given by

$$f = \frac{mc^2}{N\pi e^2} \int \alpha(\nu)\, d\nu \quad (12.14)$$

where N is the number of absorption centers per cubic centimeter, m the mass of the electron, c the velocity of light, and ν the frequency (cm^{-1}) absorbed. For solutions the oscillator strenght reduces to

$$f = \frac{(2\cdot 3\times 10^3)mc^2}{N_{Av}\pi e^2} \int \epsilon(\nu)\, d\nu \quad (12.15)$$

where N_{Av} is the Avogadro number and $c_0 = 1$ when N is replaced by N_{Av}.

The oscillator strength can roughly be estimated by assuming a Gaussian shape of the absorption curve. For the calculation of the oscillator strength, one has to measure the absorption coefficient or the molar extinction coefficient at the absorption peak and the half-width of the band. The half-width δ is the width of the absorption band (cm^{-1}) when ϵ, the molar decadic extinction coefficient, falls to half of its value at the band maximum ϵ_{max}. For symmetrical bands the following expression holds:

$$\nu_{max} - \bar{\nu} = \pm\frac{\delta}{2} \qquad \text{for} \quad \epsilon = \frac{\epsilon_{max}}{2} \quad (12.16)$$

where $\bar{\nu}_{max}$ is the wave number at the band maximum, and $\bar{\nu}$ is the wave number at $\epsilon_{max}/2$. Then the oscillator strengths are measured from the formula

$$f = 4.32\times 10^{-9} \int \epsilon(\bar{\nu})\, d\bar{\nu} \quad (12.17)$$

This shows that, by measuring the area under the $(\epsilon - \nu)$ curve, the oscillator strength can be determined.

The broad absorption bands of complexes generally have the shape of Gaussian error curves, and often a region of the spectrum can be successfully represented as a linear combination of such Gaussian curves. Kuhn and Braun proposed a Gaussian curve[15] in wave number $\bar{\nu}$ as

$$\epsilon = \epsilon_{\max} 2^{-4(\bar{\nu}-\bar{\nu}_{\max})^2/\delta^2} \qquad (12.18)$$

where δ is the half-width, and $\epsilon = \epsilon_{\max}/2$ when $\bar{\nu} - \bar{\nu}_{\max} = \pm(\delta/2)$. The area of a band in this representation is given by

$$\int_{-\infty}^{\infty} \epsilon(\bar{\nu})\, d\bar{\nu} = \epsilon_{\max} \int_{-\infty}^{\infty} e^{-4(\bar{\nu}-\bar{\nu}_{\max})^2 \ln 2/\delta^2}\, d(\bar{\nu} - \bar{\nu}_{\max})$$

$$= \frac{1}{2}\sqrt{\frac{\pi}{\ln 2}}\; \epsilon_{\max}\delta$$

$$= 1.0644\epsilon_{\max}\delta \qquad (12.19)$$

Hence, if a Gaussian shape of the absorption band is assumed, the oscillator strength is given by[16]

$$f = 4.6 \times 10^{-9} \epsilon_{\max}\delta \qquad (12.20)$$

Thus the oscillator strength can be roughly calculated from the product of the maximum extinction coefficient and the half-width δ.

12.3 THE VIBRONIC INTERACTION

The general principles of the symmetry and the mechanics that govern the appearance of vibronically allowed bands were laid down by Herzberg and Teller[20] in 1933. However, not until recent years has this theory been put on a concrete basis and applied in a semiquantitative fashion to problems of inorganic complexes. In the present calculations we have followed the scheme outlined by Liehr and Ballhausen,[21] but for the sake of completeness we shall attempt to give a general outline of the theory involved in such a calculation. The wavefunctions requisite for our purpose have been determined by standard techniques.

We now assume that the ligands can be approximated by point charges of a certain effective charge q_j and mass m associated with the jth ligand. The electrons at radius vector $\mathbf{r}_i$ will therefore be perturbed by the charge on the ligands (Figure 12.1).

From Figure 12.1 we get for the ith electron

$$\frac{1}{R} = |\mathbf{r}_i + \mathbf{s}_M - \mathbf{s}_j - \mathbf{a}_j|^{-1} \qquad (12.21)$$

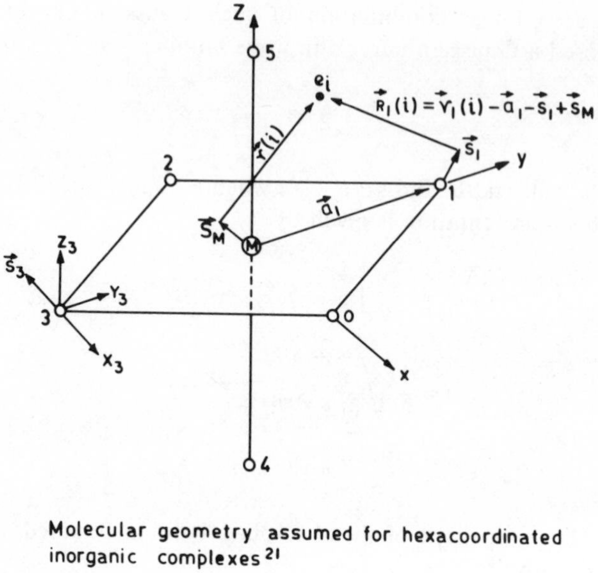

Fig. 12.1 Molecular geometry assumed for hexacoordinates in organic complexes. (Taken from Liehr and Ballhausen[21])

Expanding this expression, and keeping only the linear terms in the nuclear displacements $\mathbf{s}_M$ and $\mathbf{s}_j$, we obtain

$$\frac{1}{R}=\frac{1}{|\mathbf{r}_i-\mathbf{a}_j|}\left[1-\frac{(\mathbf{r}_j-\mathbf{a}_j)(\mathbf{s}_M-\mathbf{s}_j)}{|\mathbf{r}_i-\mathbf{a}_j|^2}\right] \tag{12.22}$$

The first-order perturbation term $\mathcal{H}^{(1)}$, which is a function of the normal coordinates, is given by

$$\mathcal{H}^{(1)}(Q_\kappa)=q_je^2\sum_{i,j}(\mathbf{s}_j-\mathbf{s}_M)\cdot\mathbf{\Delta}_{a_j}\left(\frac{1}{|\mathbf{r}_i-\mathbf{a}_j|}\right) \tag{12.23}$$

where $\mathbf{r}_i$ is the distance of the ith electron of charge $-e$, and a_j is the position vector of the jth ligand carrying a charge q_je.

By the first-order perturbation theory we now have, for the mixing coefficients,

$$\Gamma^\kappa_{d,p}=-\frac{\langle\psi^0_u(p)|\mathcal{H}^{(1)}(Q_\kappa)|\Psi^0_g(d)\rangle}{E_p-E_d} \tag{12.24}$$

with the corresponding first-order wavefunctions:

$$\Psi(d)=\Psi^0_g(d)+\sum_{\kappa,p}Q_\kappa\Gamma^\kappa_{d,p}\Psi^0_u(p) \tag{12.25}$$

Now, since the ligands are taken to be octahedrally situated point masses and point charges (Figure 12.1), we have the electronic potential energy V as

$$V(\mathbf{r}_i,\mathbf{a}_j) = -\sum_{i,j} \frac{q_j e^2}{|\mathbf{R}_j(i)|} = -\sum \frac{q_j e^2}{|\mathbf{r}_i - \mathbf{a}_j|} \tag{12.26}$$

In this approximation the mixing coefficient $\Gamma^{\kappa}_{d,p}$ is defined by the following relation:

$$\sum_{\kappa} Q_{\kappa}\Gamma^{\kappa}_{d,p} = (E_p - E_d)^{-1} \sum_{i} \sum_{j=0}^{5} (\mathbf{s}_j - \mathbf{s}_M)\nabla_{a_j} \int \Psi^0_u(p) \frac{q_j e^2}{|\mathbf{r}_i - \mathbf{a}_j|} \Psi^0_g(d)\,dr \tag{12.27}$$

the summations being performed over the i different orbitals, the several point charges, and the κ different normal oscillations. Since the potential of the ligand oscillations constituting the normal oscillations of the entire system will be taken to be harmonic, the intensity in terms of the oscillator strength f is given by[22]

$$f(d^n, \nu_\kappa \rightarrow d^{n'}, \nu_\kappa \pm 1) = 1.085 \times 10^{11} \left(\frac{N_{\nu_\kappa}}{N}\right) \Delta E_{d^n} \cdot |\langle \Psi(d), \nu_\kappa | \mathbf{r} | \Psi(d'), \nu_\kappa \pm 1 \rangle|^2 \tag{12.28}$$

where N_{ν_κ} is the molar population in the ground electronic state $\Psi(d)$ of the $(\nu_\kappa + 1)$th harmonic of the normal mode Q_κ (frequency ν_κ), and N is the Avogadro number. We therefore have[14]

$$f(d^n \rightarrow d^{n'}) = \sum_{\kappa} \sum_{\nu_\kappa} 1.085 \times 10^{11} \coth\left(\frac{h\nu_\kappa}{2kT}\right) \Delta E_{d^n} \cdot \left|\sum_{p} \langle 0|Q_\kappa|1\rangle \Gamma^{\kappa}_{d,p} \langle d'|\mathbf{r}|p\rangle + \Gamma^{\kappa}_{d',p} \langle d|\mathbf{r}|p\rangle\right|^2 \tag{12.29}$$

First, we evaluate the electronic integrals, using the antisymmetrized products contained in the $\Psi^0_g(d)$ and $\Psi^0_u(p)$ wavefunctions. Since $\mathcal{H}^{(1)}$ is a sum of one-electron perturbations, we need consider only the matrix elements between the individual d and p wavefunctions. By summing over the six ligands, we obtain

$$\mathcal{H}^{(1)} = q_j e^2 \sum_{j=0}^{5} (\mathbf{s}_j - \mathbf{s}_M) \nabla_{a_j} \sum_{l=0}^{\infty} \sum_{m=-l}^{+l} \frac{4\pi}{2l+1} \frac{r^l_<}{r^{l+1}_>} Y_l^{m*}(\theta_j \phi_j) Y_l^m(\theta, \phi) \tag{12.30}$$

We now find the matrix elements

$$\Gamma_{ij} = \langle (k+1)p_j | \mathcal{H}^{(1)}(Q_\kappa) | kd_i \rangle \tag{12.31}$$

where $i = xy, yz, xz, x^2-y^2, z^2$; $j = x, y, z$; $k = 3, 4, 5$; and $\kappa = 3a, 3b, 3c$; $4a, 4b, 4c$; $6a, 6b, 6c$. The general form of the matrix elements given in (12.31) can be determined by investigating the symmetry properties under reflections in the six $x=0$, $y=0$, $z=0$, $x=z$, $y=z$, and $x=y$ planes.[3] We shall not enter into a discussion of this here, since these matrix elements are given by Liehr in detail.[23] We quote his final expressions here:

$$\left.\begin{aligned}
\Gamma_{xy,x} &= \gamma_{x,3b}^{xy} S_{3b} + \gamma_{x,4b}^{xy} S_{4b} + \gamma_{x,6b}^{xy} S_{6b} \\
\Gamma_{xy,y} &= \gamma_{y,3c}^{xy} S_{3c} + \gamma_{y,3c}^{xy} S_{3c} + \gamma_{y,6c}^{xy} S_{6c} \\
\Gamma_{xy,z} &= 0 \\
\Gamma_{x^2-y^2,x} &= \gamma_{x,3c}^{x^2-y^2} S_{3c} + \gamma_{x,4c}^{x^2-y^2} S_{4c} + \gamma_{x,6c}^{x^2-y^2} S_{6c} \\
\Gamma_{x^2-y^2,y} &= \gamma_{y,3b}^{x^2-y^2} S_{3b} + \gamma_{y,4b}^{x^2-y^2} S_{4b} + \gamma_{y,6b}^{x^2-y^2} S_{6b} \\
\Gamma_{x^2-y^2,z} &= \gamma_{z,6a}^{x^2-y^2} S_{6a}
\end{aligned}\right\} \tag{12.32}$$

The remaining vibronic perturbation terms are readily obtained from (12.32) by applying a counterclockwise rotation of 120° about the (1 1 1) direction.[23] To obtain expressions (12.32) from (12.31), we expressed the nuclear displacements in terms of the symmetry coordinates after evaluating the electronic integrals. In O_h symmetry, τ_{1u} and τ_{2u} yield nonzero mixing coefficients.[21, 24] The algebraic forms of the vibronic coupling constants $\gamma_{x,3b}^{xy}$, and so on, appearing in (12.32), computed on the basis of a point charge model,[25] are given by

$$\left.\begin{aligned}
\gamma_{x,3b}^{xy} &= \gamma_{x,3c}^{xy} = -\frac{1}{\sqrt{5}}\frac{qe^2}{r_0}\left(\tfrac{9}{7}G_k^3 + 2G_k^1\right) \\
\gamma_{x,4b}^{xy} &= -\sqrt{\tfrac{2}{5}}\, qe^2\left(-\tfrac{3}{7}B_k^3 + B_k^1\right) \\
&= \gamma_{y,4c}^{xy} \\
\gamma_{x,6b}^{xy} &= \gamma_{y,6c}^{xy} = -\frac{3\sqrt{5}}{7}\frac{qe^2}{r_0}G_k^3 \\
\gamma_{x,3c}^{x^2-y^2} &= -\gamma_{y,3b}^{x^2-y^2} = -\frac{1}{\sqrt{5}}\frac{qe^2}{r_0}\left(-\tfrac{27}{14}G_k^3 + 2G_k^1\right) \\
\gamma_{x,4c}^{x^2-y^2} &= -\gamma_{y,4b}^{x^2-y^2} = -\sqrt{\tfrac{2}{5}}\, qe^2\left(\tfrac{9}{14}B_k^3 + B_k^1\right) \\
\gamma_{x,6c}^{x^2-y^2} &= -\gamma_{y,6b}^{x^2-y^2} = \frac{3\sqrt{5}}{14}\frac{qe^2}{r_0}G_k^3
\end{aligned}\right\} \tag{12.33}$$

where the integrals G_k^l and B_k^l ($l = 1, 3$: $k = 3, 4, 5$) are defined by[23]

$$G_k^l = \int_0^\infty R_{kd}(r)\frac{r_<^l}{r_>^{l+1}} R_{(k+1)p}(r) r^2\, dr \tag{12.34}$$

and

$$B_k^l = \frac{d}{dr_0} G_k^l$$

where r_0 is the metal-ligand bond distance.

In the electronic integrals given by (12.32), we now replace the symmetry coordinates S_κ with suitable linear combinations of the normal coordinates Q_κ. It is important to note that the τ_{1u} symmetry coordinates S_{jt} ($j=3,4$; $t=a,b,c$) are, however, not normal coordinates.[24] The normal coordinates of symmetry τ_{1u} can be determined only after assumptions have been made as to the potential energy of the system. If we assume Wilson's valence force potential, we obtain the following relations connecting the S_{jt} symmetry coordinates with Q_{jt} normal coordinates:[21, 24]

$$\begin{aligned} S_{3t} &= \frac{1}{\sqrt{2}}(L_{33}Q_{3t} + L_{34}Q_{4t}) \\ S_{4t} &= L_{43}Q_{3t} + L_{44}Q_{4t}, \qquad t=a,b,c \end{aligned} \tag{12.35}$$

where, taking the even vibrational frequencies ν_1, ν_2, and ν_5 equal to 300, 275, and 150 cm^{-1}, respectively, we obtain the odd frequencies,[21]* $\nu_3=137$, $\nu_4=356$, and $\nu_6=106$ cm^{-1}, and hence it can be shown that[26]

$$\begin{aligned} L_{33} &= 0.295816\times\sqrt{2}\,, \qquad & L_{34} &= 0.186319\sqrt{2} \\ L_{43} &= -0.027619, & L_{44} &= 0.296860 \end{aligned} \tag{12.36}$$

Substituting these in (12.29), we derive for kd^n octahedral complexes the final expressions for the oscillator strengths. We do not go into the details of the calculation, since it is lengthy though quite straightforward,[19] and instead give the final results.

The intensities of the hydrated complexes of the transition metal ions have been estimated[19] using the SCF wavefunctions of Richardson and his co-workers[27, 28] to evaluate the integrals, and we present these oscillator strengths in Table 12.1.

From the table we see that the calculated and the observed oscillator strengths are in very reasonable agreement. Our main aim was to see whether the point charge model and the vibronic mechanism are capable of explaining, at least qualitatively, the observed oscillator strengths of the crystal field bands of the $3d^n$ octahedral complexes. We can say that, within the limits of our approximations, we are successful in obtaining a semiquantitative understanding of the observed crystal field spectra. One source of error, and

*In the intensity calculations Chakravarty has used these values of the odd frequencies throughout. It is true that these values will change appreciably, depending on the ion and the associated ligands, but since these frequencies fall in the far-infrared region, experimental data are very scarce. Thus it can be said that much experimental work is needed to obtain a reliable force potential, from which will follow the normal vibrational frequencies for these complexes.

Table 12.1

Comparison of theoretical and experimental oscillator strengths of the $3d^n$ transition metal hydrated complexes

Configuration	Ion	Transition ($g \rightarrow g$)	Theoretical[a] $f \times 10^4$	Experimental[b] $f \times 10^4$
$3d^1$	Ti^{3+}	$^2T_2 \rightarrow {}^2E$	3.7577 (1.5688)	1.0
$3d^2$	V^{3+}	$^3T_1 \rightarrow {}^3T_1$	5.2929 (2.0457)	1.8
		$\rightarrow {}^3T_2$	3.0283 (1.3882)	1.93
$3d^3$	Cr^{3+}	$^4A_2 \rightarrow {}^4T_1$	8.0218 (2.8780)	1.98
		$\rightarrow {}^4T_2$	3.3329 (1.7174)	3.02
$3d^6$	Fe^{2+}	$^5T_2 \rightarrow {}^5E$	1.5794 (0.6519)	0.25
$3d^7$	Co^{2+}	$4T_1 \rightarrow {}^4T_2$	0.9107 (0.4112)	0.3
		$\rightarrow {}^4T_1$	3.0710 (1.1674)	0.75
$3d^8$	Ni^{2+}	$^3A_2 \rightarrow {}^3T_1$	3.7343 (1.3740)	0.35
		$\rightarrow {}^3T_2$	0.9700 (0.5020)	0.22
$3d^9$	Cu^{2+}	$^2E \rightarrow {}^2T_2$	5.6433 (2.2825)	2.9

[a]The values in parentheses are at 0 °K.
[b]The experimental f's are at about 20 °K for most cases.

probably the main source, is the odd vibrational frequencies, which vary widely for different complexes.* Moreover, the vibrational frequencies quoted are for the ML_6 complex and not for the crystal. Hence the contributions from the optical and acoustical modes in the vibrational frequencies, and hence in the oscillator strengths, cannot be ascertained until and unless experiments are done with crystalline complexes. Therefore appreciable experimental work which has not yet been done is highly desirable in this area.

Considering the approximations mentioned above, we regard the agreement with experiment as quite good. The vibronic theory outlined in this paper[19] for the kd^n electronic configurations of the transition metal complexes applies equally well to the f^n configurations of the rare earth complexes. A large number of experimental data on the absorption spectra of the rare earth complexes are now available, so that this type of calculation is very desirable.

*See footnote on p. 345

It is well known that the absorption spectra of the rare earths are very sharp compared to those of the transition metal ions, and therefore the oscillator strengths can be found very precisely from experiments, unlike the situation for the transition metal spectra. The dipolar and covalent model should be a much more realistic model in calculating the necessary integrals involved in such calculations, and future investigation with this model is highly desirable.

12.4 THE JAHN-TELLER EFFECT

By virtue of its importance in solid state physics the Jahn-Teller effect would merit a separate chapter. Because vibronic interaction is discussed in this chapter, however, it has been thought proper to include this effect here. In view of the fact that two excellent reviews[29, 30] on the Jahn-Teller effect have been published, we shall deal with it rather briefly, emphasizing only the important aspects without going into much detail.

Although the Jahn-Teller effect was predicted[31] by Jahn and Teller in 1937, unambiguous experimental evidence for its existence[32] was not obtained before 1950. In those early years of its verified existence around midcentury, the Jahn-Teller effect was found only in a very limited class of systems and proved to be elusive when searched for elsewhere, so much so that at one time Van Vleck[33] remarked that "one great merit of the Jahn-Teller effect is that it disappears when not needed." However, the situation is not as bad as this remark implies, and from 1952 onward experimental evidence for the existence of the Jahn-Teller effect has been forthcoming and will be described in this section.

As a general rule, it is found that transition metal ions whose ground states in O_h symmetry are orbitally degenerate form complexes whose site symmetry is lower, for example, tetragonal or rhombic. This lowering of symmetry may be due to many causes, such as crystal packing considerations for the complex within a lattice or repulsions between neighboring ligands. According to the Jahn-Teller effect, it is also to be expected on very general grounds. The theorem of Jahn and Teller states that, when the orbital state of an ion is degenerate for symmetry reasons, the ligands will experience forces distorting the nuclear framework until the ion assumes a configuration both of lower symmetry and of lower energy, thereby resolving the degeneracy; in other words, the Jahn-Teller effect is the intrinsic instability of an electronically degenerate complex against distortions that remove the degeneracy.

The Jahn-Teller theorem states that any electronically degenerate complex is unstable against a distortion that removes that degeneracy in the first order. If no such distortion is possible, only then the electronically degenerate level will be stable. This is the case of Kramers' degeneracy, which can be removed only by the application of a magnetic field. For linear molecules also, there are certain levels whose degeneracy cannot be removed by any idstortion in the first order. This case, of course, is of no interest in solids. Apart from the two cases just mentioned, we can always find a distortion that removes the degeneracy. Now the question is, Why does the instability arise? It arises simply because a linear splitting of a level necessarily leads to a state with lower energy than that of the unsplit level. Finally, the distortion is limited to a finite value by the quasi-elastic forces, covalent and electrostatic, that resist it. But what is more important to realize is that, since the original symmetry state would have been in a position of equilibrium in the absence of electronic degeneracy, the quasi-elastic contribution to the energy contains *no terms linear* in the distortion. As a result a new position of equilibrium is reached in which the local (site) symmetry is lower than the point group symmetry of the crystal, and, in general, this new symmetry will be low enough to remove all electronic degeneracy except the Kramers degeneracy. Hence there will be more than one position of equilibrium with equal energy, since a distortion that removes the electronic degeneracy must itself be degenerate. Thus in this process the original purely electronic degeneracy is replaced by a more complicated vibronic (vibrational-electronic) degeneracy.

Let us now put the foregoing qualitative argument on a mathematical basis. The instantaneous potential, seen by an electron in some nuclear framework (crystalline or molecular), can be divided into a static part $V_0(q)$, which is a function only of the mean nuclear positions, and a dynamic part $V(q,Q)$, say, which depends on the displacements Q of the nuclei from their mean positions. Expanding $V(q,Q)$ in powers of Q, we have

$$V(q,Q)=V_0(q)+\sum_{i,\alpha}\left(\frac{\partial V}{\partial Q_{i\alpha}}\right)Q_{i\alpha}+\frac{1}{2}\sum_{i,j,\alpha,\beta}\left(\frac{\partial^2 V}{\partial Q_{i\alpha}\partial Q_{j\beta}}\right)Q_{i\alpha}Q_{j\beta}+\cdots \tag{12.37}$$

where i,j label the nuclei, and α,β are their Cartesian displacements. Since the $Q_{i\alpha}$'s do not transform in a simple way under the operations of the symmetry group of the molecule, we choose linear combinations of the $Q_{i\alpha}$'s that do transform according to the irreducible representations of the group. These are known as the collective coordinates Q_κ. Hence (12.37) can be rewritten in terms of Q_κ as

$$V(Q)=V_0+\sum_{\kappa}\left(\frac{\partial V}{\partial Q_\kappa}\right)Q_\kappa+\frac{1}{2}\sum\left(\frac{\partial^2 V}{\partial Q_\kappa\partial Q_l}\right)Q_\kappa Q_l+\cdots \tag{12.38}$$

where the dependence of Q is understood. Since V is invariant under the operations of the symmetry group of the molecule, $\partial V/\partial Q_\kappa$ must transform in the same way as Q_κ. Methods for finding Q_κ in terms of $Q_{i\alpha}$ are given in Wilson, Decius, and Cross,[34] and we need not elaborate on them here.

We are now primarily interested in two types of transition metal complexes: the octahedral ML_6 complexes (O_h symmetry) and the tetrahedral ML_4 complexes (T_d symmetry). These two types of complexes are shown in Figure 12.2, and their collective coordinates Q_κ are listed in Tables 12.2 and 12.3. The nuclear motions associated with those Q_κ that are important for the Jahn-Teller effect are pictured in Figure 12.3 (for O_h symmetry only). In Tables 12.2 and 12.3, Q_2 represents tetragonal (even-parity) distortion of the octahedron or tetrahedron, lowering the symmetry from O_h to D_{4h} or from T_d to D_{2d}. (For group-theoretical aspects see Chapter 5.) Similarly, the combination $(Q_4+Q_5+Q_6)$ or $(Q_7+Q_8+Q_9)$ represents trigonal distortion lowering the symmetry to D_{3d} or C_{3v}.

So far, we have treated the electronic and the nuclear motions separately, but when we consider the Jahn-Teller effect we shall see that they are really inseparable. The eigenfunctions of the Jahn-Teller Hamiltonian, which is a

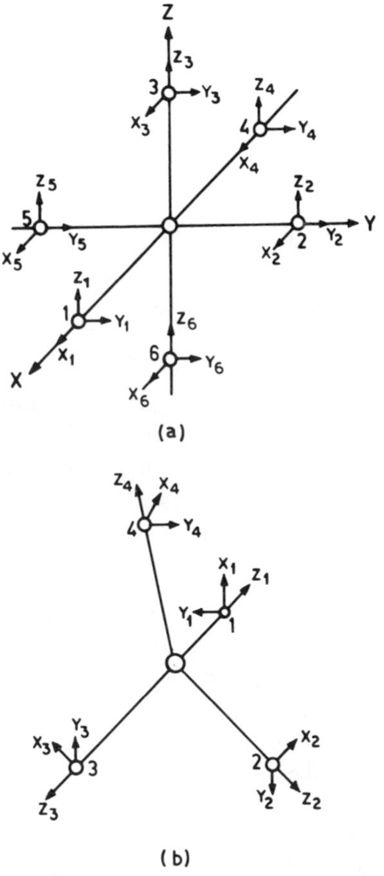

Fig. 12.2 (*a*) The octahedral ML_6 complex. (*b*) The tetrahedral ML_4 complex. Collective coordinates $Q\kappa$ are linear combinations of the Cartesian coordinates shown in the figures. (See also Tables 12.2 and 12.3. These figures are taken from Liehr.[35])

Table 12.2

Q_κ for the ML_6 complex (O_h symmetry)

Q_κ	Symmetry of normal coordinates	Linear combinations of Cartesian coordinates giving the Q_κ (see Figure 12.2*a*)
Q_1	a_{1g}	$\frac{1}{\sqrt{6}}(x_1-x_4+y_2-y_5+z_3-z_6)$
Q_2	$e_g(\theta)$	$\frac{1}{\sqrt{12}}(2z_3-2z_6-x_1+x_4-y_2+y_5)$
Q_3	$e_g(\epsilon)$	$\frac{1}{2}(x_1-x_4-y_2+y_5)$
Q_4	$t_{2g}(\xi)$	$\frac{1}{2}(z_2-z_5+y_3-y_6)$
Q_5	$t_{2g}(\eta)$	$\frac{1}{2}(x_3-x_6+z_1-z_4)$
Q_6	$t_{2g}(\zeta)$	$\frac{1}{2}(y_1-y_4+x_2-x_5)$
Q_7	$t_{1u}^a(x)$	$\frac{1}{2}(x_2+x_3+x_5+x_6)$
Q_8	$t_{1u}^a(y)$	$\frac{1}{2}(y_1+y_3+y_4+y_6)$
Q_9	$t_{1u}^a(z)$	$\frac{1}{2}(z_1+z_2+z_4+z_5)$
Q_{10}	$t_{1u}^b(x)$	$\frac{1}{\sqrt{2}}(x_1+x_4)$
Q_{11}	$t_{1u}^b(y)$	$\frac{1}{\sqrt{2}}(y_2+y_5)$
Q_{12}	$t_{1u}^b(z)$	$\frac{1}{\sqrt{2}}(z_3+z_6)$
Q_{13}	$t_{2u}(\xi)$	$\frac{1}{2}(x_2+x_5-x_3-x_6)$
Q_{14}	$t_{2u}(\eta)$	$\frac{1}{2}(y_3+y_6-y_1-y_4)$
Q_{15}	$t_{2u}(\zeta)$	$\frac{1}{2}(z_1+z_4-z_2-z_5)$

combined electron-nuclear Hamiltonian, are called the vibronic (vibrational-electronic) wavefunctions, denoted by Ψ. In the absence of the Jahn-Teller effect, the Born-Oppenheimer approximation[36] for Ψ is valid, that is, Ψ can be written as a product:

$$\Psi=\psi_n(q,Q)\phi(n,Q) \tag{12.39}$$

where q and Q are the electron and the nuclear coordinates, respectively. In (12.39), ψ_n is a solution of the Schrödinger equation for electrons with the nuclei fixed at their positions Q, whereas $\phi(n,Q)$ is a solution of the nuclear Schrödinger equation in which the electronic energy $E_n(Q)$ is added to the ordinary internuclear potential $V(Q)$. The wavefunction $\psi\phi$ is called a Born-Oppenheimer product. Although (12.39) is not strictly valid while considering the Jahn-Teller effect, we shall still be able, as a first approximation, to write our vibronic wavefunctions as linear combinations of a number (usually small) of Born-Oppenheimer products, and in most cases still be able to define an energy surface $E_n(Q)+V(Q)$ on which the nuclei move.

Table 12.3

Q_κ for the ML_4 complex (T_d symmetry)

Q_κ	Symmetry of normal coordinates	Linear combinations of Cartesian coordinates giving the Q_κ (see Figure 12.2*b*)
Q_1	a_1	$\frac{1}{2}(z_1+z_2+z_3+z_4)$
Q_2	$e(\theta)$	$\frac{1}{2}(x_1-x_2-x_3+x_4)$
Q_3	$e(\epsilon)$	$\frac{1}{2}(y_1-y_2-y_3+y_4)$
Q_4	$t_2^a(\xi)$	$\frac{1}{2}(z_1-z_2+z_3-z_4)$
Q_5	$t_2^a(\eta)$	$\frac{1}{2}(z_1+z_2-z_3-z_4)$
Q_6	$t_2^a(\zeta)$	$\frac{1}{2}(z_1-z_2-z_3+z_4)$
Q_7	$t_2^b(\xi)$	$\frac{1}{4}(x_2-x_1-x_3+x_4)+\frac{\sqrt{3}}{4}(y_2-y_1-y_3+y_4)$
Q_8	$t_2^b(\eta)$	$\frac{1}{4}(-x_2-x_1+x_3+x_4)+\frac{\sqrt{3}}{4}(y_2+y_1-y_3-y_4)$
Q_9	$t_2^b(\zeta)$	$\frac{1}{2}(x_1+x_2+x_3+x_4)$

Let us now, along with Sturge, prove the Jahn-Teller theorem by using the Hellmann-Feynman theorem,[37–40] which has an intuitive appeal. This can be related to the perturbation approach, which is usually adopted for practical applications of the theorem.

According to the Hellmann-Feynman theorem, one can write the total potential energy of a system of electrons and nuclei as a function of $V(q,Q)$ [equation (12.38)], and the first derivative $\partial V/\partial Q_k$ is nonvanishing. In other words, if F_κ is the generalized force acting on the nuclei, then, according to the Hellmann-Feynman theorem,

$$F_\kappa = -\left\langle i \left| \frac{\partial V}{\partial Q_\kappa} \right| i \right\rangle \tag{12.40}$$

where $|i\rangle$ is the electronic state.

If $F_\kappa \neq 0$ for a particular nuclear configuration, this configuration is unstable and will spontaneously distort until $F_\kappa = 0$.[40] The Jahn-Teller theorem states that, if $|i\rangle$ is a degenerate state when $Q_\kappa = 0$, the matrix element in (12.40) exists for some Q_κ, $\kappa \neq 1$, that is, for some nontotally symmetric coordinate. This can be proved easily by the elementary group theory, and the proof is left as an exercise for the reader.

It should be noted that the Hellmann-Feynman theorem, although physically appealing, is not very useful in dealing with a practical problem, simply because (12.40) is valid only for the exact electronic wavefunctions,[41] which are impossible to obtain. Also, to define the force F_κ in a degenerate electronic state, the wavefunctions $|i\rangle$ should be chosen in such a way that

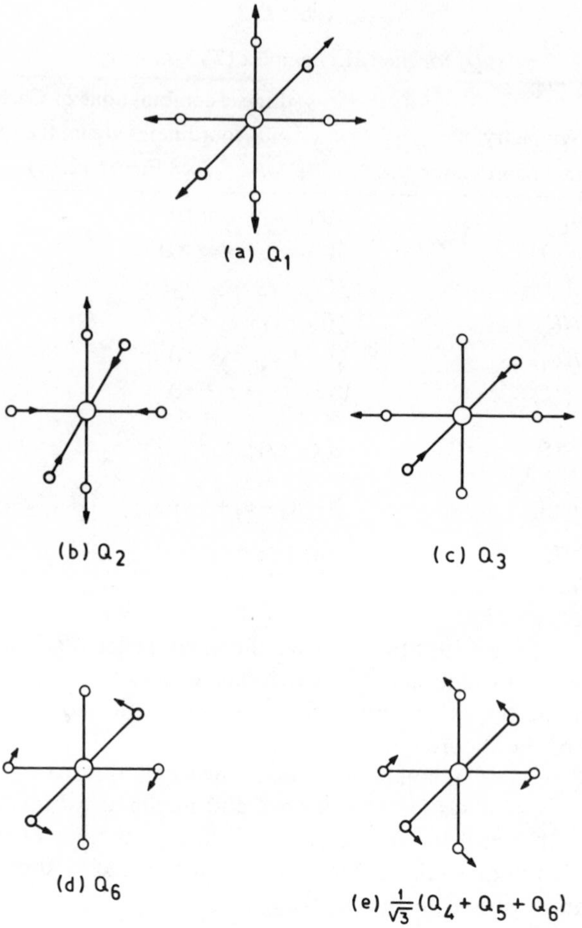

Fig. 12.3 Collective coordinates Q_κ for O_h symmetry. (Taken from Liehr[35])

the off-diagonal matrix elements

$$i\left\langle\left|\frac{\partial V}{\partial Q_\kappa}\right|j\right\rangle=0$$

for $i \neq j$. For a given Q_κ such combinations can always be found, but they will not, in general, be the correct combinations for any other Q_κ. To eliminate this difficulty, the perturbation theory, in the first order, is generally used. The outermost d electrons of the transition metal ions move in a potential $V(q,Q)$ that is a function of electronic coordinates q and nuclear coordinates Q. Expanding $V(q,Q)$ in a Taylor series about $Q_\kappa=0$, we can write the

Hamiltonian for the d electrons as

$$\mathcal{H} = \mathcal{H}_0 + \sum_{\kappa} \left(\frac{\partial V}{\partial Q_\kappa} \right) Q_\kappa + \cdots \tag{12.41}$$

where $\mathcal{H}_0$ represents the Hamiltonian when $Q_\kappa = 0$. Assuming that $\mathcal{H}_0$ has an m-fold degenerate eigenvalue $\bar{E}$ with the eigenfunctions transforming as Γ, which we shall use to label a level, the matrix elements of (12.41), to the first order in Q_κ, are given by[29]

$$\mathcal{H}_{ij}^{(1)} = \bar{E}\delta_{ij} + \sum_{\kappa} h_{ij}(\kappa) Q_\kappa \tag{12.42}$$

where $h_{ij}(\kappa) = \langle i | \partial V / \partial Q_\kappa | j \rangle$, and $|i\rangle$ and $|j\rangle$ are orthogonal substates of Γ. If Q_κ and $\partial V / \partial Q_\kappa$ transform as the lth component of $\bar{\Gamma}$, and $\bar{\Gamma}$ is contained in $[\Gamma^2]$, we have from the Wigner-Eckart theorem [see (5.30)]

$$h_{ij}(\kappa) = A_{\bar{\Gamma}} \begin{pmatrix} \Gamma & \Gamma & \bar{\Gamma} \\ i & j & l \end{pmatrix} \tag{12.43}$$

where

$$A_{\bar{\Gamma}} = \left\langle \Gamma \left\| \frac{\partial V}{\partial Q_\kappa} \right\| \Gamma \right\rangle \tag{12.44}$$

is the electron-lattice interaction constant. If $\bar{\Gamma} = \Gamma_1$, $h_{ij}(1) = A_1 \delta_{ij}$ and we get a uniform shift but no splitting. If $\Gamma \neq \Gamma_1$, $\mathrm{Tr}\, h_{ij}(\kappa) = 0$.[42]

Also, the nuclear Hamiltonian in the harmonic approximation is given by

$$\mathcal{H}_{\text{nuc}} = \frac{1}{2} \sum_{\kappa} \left[\frac{P_\kappa^2}{\mu_\kappa} + \mu_\kappa \omega_\kappa^2 Q_\kappa^2 \right] \tag{12.45}$$

where P_κ is the momentum operator conjugate to Q_κ, μ_κ is the effective mass, and ω_κ is the frequency of the κth normal mode. Now, adding (12.44) and (12.45), we have the first-order Hamiltonian for the Jahn-Teller effect, given by

$$\mathcal{H}_{ij}^{(1)} = \bar{E}\delta_{ij} + \sum_{\kappa} \left[\frac{1}{2} \left(\frac{P_\kappa^2}{\mu_\kappa} + \mu_\kappa \omega_\kappa^2 Q_\kappa^2 \right) \delta_{ij} + h_{ij}(\kappa) Q_\kappa \right] \tag{12.46}$$

where the last two terms define a potential surface in the Q_κ-space. It is also to be noted that the last term, which is linear in Q_κ, ensures that the minima of this potential are not at the origin. This is another way of stating the Jahn-Teller theorem. The last term of (12.46) is, in general, nondiagonal and causes the nuclear motion to mix the electronic states.

It has been shown by Sturge that (12.46) has solutions of the form

$$\Psi_m = \sum_n a_{mn}\psi_n(q,Q)\phi(n,Q) \tag{12.47}$$

where the coefficients a_{mn} form a square matrix, and ψ and ϕ were defined in (12.39). It is very important to realize that there is an infinity of possible ϕ's for a given ψ. Equations (12.47) can be shown to be accurate to order m/M, that is, the ratio of the electron to the nuclear mass. Also, the dimension of Ψ_m is the same as that of ψ_n. This means that the degeneracy of a level, given by the multiplicity of an irreducible representation to which the level belongs, cannot be reduced by the Jahn-Teller effect because the total Hamiltonian preserves its symmetry. Because of the Jahn-Teller effect, therefore, the electronic degeneracy is replaced by the vibronic degeneracy.

We have thus, very briefly, seen what the Jahn-Teller effect is, both qualitatively and quantitatively, and have also seen that any complex (other than a linear molecule) in a state with electronic (non-Kramers') degeneracy is, in principle, unstable against a distortion that can remove that degeneracy. A first-order perturbation treatment leads to the basic Hamiltonian (12.46), which consists of the electronic, the lattice, and the electron-lattice interaction terms. The basis states for (12.46) are the Born-Oppenheimer products, given by (12.39). The first two terms of (12.46) are diagonal, but the third term represents the electron-lattice interaction and can have off-diagonal elements, leading to a breakdown of the Born-Oppenheimer approximation.

Having dealt with the theoretical aspects of the Jahn-Teller effect, we now consider the question of when and how this effect is likely to be observed in reality. It is well known that an orbitally nondegenerate ground state with only spin degeneracy well below the upper excited states (e.g., the ^6S ground term of Mn^{2+}) is extremely insensitive to the distortion of its surroundings, and therefore the Jahn-Teller effect is unlikely to be observed in such situations. Hence it is reasonable to look for this effect in the orbitally degenerate ground states of the transition metal ions and in point defects. Even in such situations, where there is very weak coupling of the levels to the lattice, the Jahn-Teller effect may be too insignificant to be observed. An example of this kind of situation is given by the $^2E(t_{2g}^3)$ term of the octahedral Cr^{3+} complex.

As Sturge states, there are three ways, in general, by which the Jahn-Teller effect can have observable consequences. First, there may be a lowering of the site symmetry as a result of this effect, which may show up directly in the crystal structure, in spin resonance, in nuclear magnetic resonance (NMR), in the Mössbauer effect, and also in the magnetic susceptibility. To be detectable in this way, the mean distortion must be finite when averaged over a certain period of time. This permanent lowering of the site symmetry is known as the static Jahn-Teller effect. When the distortion averages out to zero in the characteristic time of the experiment, we have a dynamic Jahn-Teller effect.

Second, another effect that may be seen is a difference between the vibronic energy levels and the vibrational levels that the system would have if there were no Jahn-Teller effect. In the case of molecules, this aspect has attracted considerable attention.[42a] For an impurity or a defect in a solid, we have a very rough idea of the vibrational levels and can expect only to be able to identify such effects on the vibronic spectra if they are gross. For example, splittings of the broad optical absorption bands, in which the detailed vibrational structure is already washed out, can be characteristic of the Jahn-Teller effect.

Third, we might expect to see a reduction of the electronic operators when the matrix elements of the operators are calculated using the vibronic wavefunctions Ψ rather than the electronic wavefunctions ψ. This aspect we shall encounter a little later in this section.

The most extensive investigation of the structural consequences of the static Jahn-Teller effect is the work of Liehr,[35] where many types of molecular structures that are subject to a Jahn-Teller instability are treated in great detail. This structural aspect of the Jahn-Teller theorem, in particular the appearance at sufficiently low temperatures of stable distorted configurations having lower symmetry than that of the host crystal (the static Jahn-Teller effect), has provided guidance for experimental investigations by electron paramagnetic resonance (EPR) of systems that might possibly exhibit static Jahn-Teller effects.

Another way of looking at the effect was initiated by the late W. Moffitt and his students.[43, 44] They showed that, as a consequence of the coupling between the electronic energy levels and the nuclear displacements, which provides the driving force of the Jahn-Teller instability, a coupled motion of the electrons and the lattice vibrational modes occurs. For sufficiently strong coupling the complex may exhibit a static distortion as a limiting case; but if the coupling is weak, there is still a profound modification of the energy levels of the coupled vibronic system that is very different from what is obtained with zero coupling. In this situation the system is said to exhibit a dynamic Jahn-Teller effect.

It was shown by Ham[45a] in 1965 that the dynamic Jahn-Teller effect may contribute significantly to the interpretation of the EPR and the optical spectra of the paramagnetic complexes of the transition metal ions. He realized that the dynamic Jahn-Teller effect can explain successfully the anomalous EPR spectra obtained by Ludwig and Woodbury[46] for a number of transition metal ions in silicon having orbital triplet states in the interstitial sites of tetrahedral symmetry. These spectra showed a large quenching of the orbital contribution to the g value, which could not be explained on the basis of covalent bonding. The spectra did not show any indication that the paramagnetic ion was subjected to any departure from cubic symmetry. Ham showed that a dynamic Jahn-Teller effect can cause large changes in the matrix elements of the orbital angular momentum operator, the spin-orbit interaction, and so on, as a result of mixing of the vibrational and the electronic wavefunctions, which occurs even in the vibronic ground state.

This is now known popularly as the "Ham effect." Only in the limiting case of strong Jahn-Teller coupling, in which some of these matrix elements are quenched effectively to zero, does the system show a static Jahn-Teller effect. In this connection the reader should study a number of earlier papers[46-50] by other authors, who also noted such effects of dynamic quenching. Sturge[29, 51] has also observed in solids dynamical Jahn-Teller quenching of the electronic operators in the optical spectra of paramagnetic ions.

Since the dynamical Jahn-Teller quenching has important effects on the optical as well as the magnetic properties of the transition metal solids, we should like to devote some space to it. Reduction of the orbital angular momentum, as well as the spin-orbit coupling constant and other electronic operators, automatically follows if one uses the vibronic interaction mechanism, that is, if one calculates the matrix elements of the perturbing operators using the vibronic wavefunction Ψ rather than the electronic wavefunctions ψ [see (12.47)]. Ham[45b] has shown that this can drastically affect the values of the matrix elements, and in many cases this method proves to be the most powerful one for detecting the Jahn-Teller effect in solids.

From this point onward our discussion of the Jahn-Teller effect will be very brief and qualitative. For details the reader should consult Refs. 29 and 30, cited earlier.

Let us briefly examine the static Jahn-Teller effect in the doubly degenerate electronic states in an octahedral complex. In Cu^{2+} and Ni^{+} (configuration $3d^9$), Mn^{3+} and Cr^{2+} [weak-field $3d^4$ (t_2^3e) configuration], and Ni^{3+} and Pt^{3+} [strong-field $3d^7$ (t_2^6e) configuration], we have the E (doubly degenerate) ground term. Since e orbitals can form σ bonds (see Chapter 11) in octahedral coordination, they are very sensitive to nuclear displacements, and therefore strong Jahn-Teller effects may be expected and are also observed by experiment. Since the oribtal momentum is quenched in the E state, the complication due to the spin-orbit coupling does not arise. As discussed earlier, of 15 normal modes describing the distortion of an octahedron (Table 12.2) only the e_g mode splits the E term in the first order. Since the nuclei move in a potential $V(Q_2, Q_3)$, it can be shown that the first-order vibronic Hamiltonian is given by [from (12.46)]

$$\mathcal{H}^{\mathrm{I}} = -A\begin{bmatrix} -Q_3 & Q_2 \\ Q_2 & Q_3 \end{bmatrix} + \tfrac{1}{2}\mu\omega_e^2(Q_2^2+Q_3^2) + \frac{1}{2\mu}(P_2^2+P_3^2) \qquad (12.48)$$

where P_κ is the momentum conjugate to Q_κ, and the last term represents the nuclear kinetic energy. In (12.48) we have, of course, assumed a harmonic quasi-elastic potential in which the e_g vibrations have frequency ω_e and an effective mass μ, which is the mass of a ligand for an isolated octahedron. The eigenvalues of (12.48) can be easily shown to be

$$E = V + \frac{1}{2\mu}(P_2^2+P_3^2) \qquad (12.49)$$

where

$$V = \pm A\rho + \tfrac{1}{2}\mu\omega_e^2\rho^2 \tag{12.50}$$

In (12.50) we have written

$$Q_2 = \rho\cos\theta, \qquad Q_3 = \rho\sin\theta \tag{12.51}$$

From (12.50) it is apparent that $V(\rho, \theta)$ is a double-valued potential surface on which the nuclei move.

The Jahn-Teller energy at a radius

$$\rho_0 = \frac{|A|}{\mu\omega_e^2} \tag{12.52}$$

has been shown by Van Vleck[33] to be

$$E_{JT} = -\frac{A^2}{2\mu\omega_e^2} \tag{12.53}$$

The surface generated by the rotating parabola [equation 12.50] about the $\rho = 0$ axis is pictured in Figure 12.4. The eigenstates corresponding to (12.49) are given by

$$\begin{aligned} \psi^+ &= \psi_\theta \sin\left(\frac{\theta}{2}\right) + \psi_\epsilon \cos\left(\frac{\theta}{2}\right) \\ \psi^- &= \psi_\theta \cos\left(\frac{\theta}{2}\right) - \psi_\epsilon \sin\left(\frac{\theta}{2}\right) \end{aligned} \tag{12.54}$$

The fact that the energies given by (12.49) are independent of θ, so that the energy surfaces in Figure 12.4 have rotational symmetry about the origin,

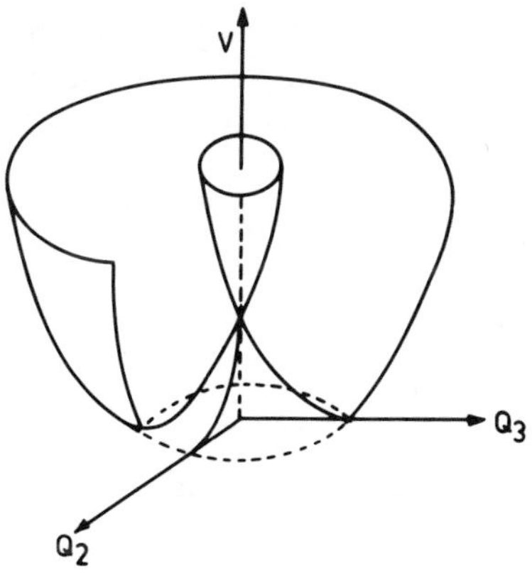

Fig. 12.4 The potential surface $V(Q_2, Q_3)$ for E state in O_h symmetry [equation (12.50)].

means that there are an infinite number of energetically equivalent configurations of minimum energy corresponding to all the points at the bottom of the trough, as shown in Figure 12.4. Thus, in this case, we do not obtain a discrete set of configurations of stable equilibrium as long as we use the Hamiltonian (12.48). When we include the kinetic energy of the lattice ions, we would expect to find that the system retains a dynamic behavior corresponding to some sort of motion around the trough, shown in Figure 12.4, even for strong Jahn-Teller coupling, provided that we confine ourselves to linearity, as in (12.48). Such behavior affects many important properties of the system, such as the Zeeman interaction, the hyperfine interaction, the strain, and the zero-point motion. It has been found that the dynamic nature of the Jahn-Teller effect can cause qualitative changes in the EPR spectrum even in the limiting case of a static Jahn-Teller effect.

To see more clearly the dynamic Jahn-Teller effects due to the zero-point ionic motion for a doublet state in cubic symmetry, let us begin from the full vibronic Hamiltonian for the doublet E electronic state, given by (12.48);

$$\mathcal{H}_{\text{vib}} = \mathcal{H}_0 \mathcal{J} + A(Q_\theta U_\theta + Q_\epsilon U_\epsilon) \tag{12.55}$$

where

$$\mathcal{H}_0 = E_0 + \frac{1}{2\mu}\left[P_\theta^2 + P_\epsilon^2 + \mu^2\omega^2(Q_\theta^2 + Q_\epsilon^2)\right] \tag{12.56}$$

Equation (12.56) has been written simply by replacing Q_2, Q_3 and P_2, P_3 in (12.48) by Q_θ, Q_ϵ, and so on. In (12.56), ω is the angular frequency approprate to the Q_θ, Q_ϵ mode of vibration, $\mathcal{H}_0$ is the unperturbed vibrational Hamiltonian in the absence of Jahn-Teller coupling, and μ is the effective mass appropriate to this mode. It has been shown by Ham[45] that for the cluster model, for example, the ML_6 complexes, $\mu = M$, where M is the mass of one of the nearest-neighbor ions. We assume here, of course, that this mode is a normal mode of vibration of the lattice; this assumption is a gross oversimplification of the actual situation in a real crystal. In (12.56) the force constant is given by $\mu\omega_e^2$, so that the Jahn-Teller energy may now be expressed in the form $E_{\text{JT}} = A^2/2\mu\omega_e^2$, as shown earlier in (12.53). In (12.55), $\mathcal{J}$, U_θ, U_ϵ, and $\mathcal{A}_2$ (not appearing here) are the Hermitian operators, which are defined to have the following matrix forms:

$$\mathcal{J} = \begin{pmatrix} +1 & 0 \\ 0 & +1 \end{pmatrix}, \qquad U_\theta = \begin{pmatrix} -1 & 0 \\ 0 & 1 \end{pmatrix}, \qquad U_\epsilon = \begin{pmatrix} 0 & +1 \\ +1 & 0 \end{pmatrix}$$

and

$$\mathcal{A}_2 = \begin{pmatrix} 0 & -i \\ +i & 0 \end{pmatrix}$$

in the basis ψ_θ and ψ_ϵ.

Our next problem is that we have to solve (12.53) in order to obtain the exact eigenstates and eigenvalues, but unfortunately this equation cannot be solved analytically. Therefore it has been solved for two extreme cases. For weak Jahn-Teller coupling, where $E_{JT} \ll \hbar\omega$, Moffitt and Thorson[44] have solved (12.53) by using the second-order perturbation theory. For strong Jahn-Teller coupling, where $E_{JT} \gg \hbar\omega$, the equation has been solved numerically by Longuet-Higgins and his co-workers[52] and also by Moffitt and Thorson.[53] The eigenstates Ψ_n may be expressed in terms of ψ^+ and ψ^-, given by (12.54), as

$$\Psi_n = \chi_{n1}(\rho,\theta)\psi^-(\rho,\theta) + \chi_{n2}(\rho,\theta)\psi^+(\rho,\theta) \qquad (12.57)$$

where ρ and θ were defined in (12.51). As remarked earlier, it is not possible to simplify relation (12.57) in general, into a simple Born-Oppenheimer product. For all values of A, the ground state is the doublet E, while at an energy Δ above the ground state there is another doublet consisting of A_1 and A_2. For $E_{JT} \ll \hbar\omega$ it can be shown that

$$\Delta = \hbar\omega\left(1 - \frac{2E_{JT}}{\hbar\omega}\right) \qquad (12.58)$$

and for $E_{JT} \gg \hbar\omega$

$$\Delta = \hbar\omega\left(\frac{\hbar\omega}{2E_{JT}}\right) \qquad (12.59)$$

Hence, in the limit of $E_{JT}/\hbar\omega$ being large, this excited state approaches the ground state asymptotically. In Figure 12.5 we present the variation of Δ with E_{JT} for linear Jahn-Teller coupling of intermediate strength.

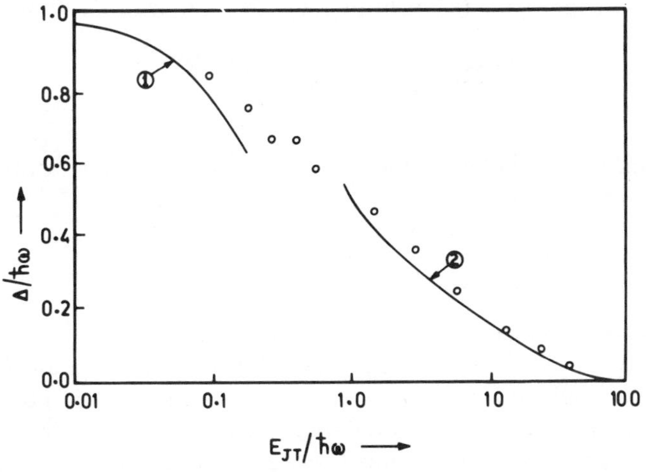

Fig. 12.5 Variation of excitation energy Δ with E_{JT} in the limit of linear Jahn-Teller coupling. Curve 1 represents (12.58), and curve 2 represents (12.59) of the text. The hollow circles represent the exact values from Longuet-Higgins and his coworkers.[52] (Taken from Ham[30])

Without going into any further detail, we shall see, along with Ham,[30] that the Jahn-Teller coupling will modify the values of the parameters that determine the splitting of the ground state under a given perturbation. Defining two real parameters p and q (the reduction parameters) in terms of the matrix elements of $\mathscr{C}_2$ and U_θ, U_ϵ within the ground state, Ham[45b] has shown, by using the perturbation theory, that, for $E_{JT} \ll \hbar\omega$,

$$p = 1 - \frac{4E_{JT}}{\hbar\omega}$$

$$q = 1 - \frac{2E_{JT}}{\hbar\omega} \tag{12.60}$$

while for $E_{JT} \gg \hbar\omega$ one obtains the asymptotic result[54]

$$p \simeq \left(\frac{\hbar\omega}{4E_{JT}}\right)^2$$

$$q \simeq \tfrac{1}{2}\left[1 + \left(\frac{\hbar\omega}{4E_{JT}}\right)^2\right] \tag{12.61}$$

In the limit $E_{JT} \gg \hbar\omega$, therefore, we have $p \simeq 0$, $q \simeq \frac{1}{2}$. Ham has shown that, as long as one restricts the case to linear Jahn-Teller coupling, we have

$$q = \tfrac{1}{2}(1 + p) \tag{12.62}$$

Figure 12.6 shows the exact values of p and q obtained for intermediate coupling strengths from the calculations by Child and Longuet-Higgins.[47] These are well approximated over the range $0 \cdot 1 \leqslant E_{JT}/\hbar\omega \leqslant 3 \cdot 0$ by

$$p = \exp\left[-1 \cdot 974\left(\frac{E_{JT}}{\hbar\omega}\right)^{0 \cdot 761}\right]. \tag{12.63}$$

Also shown in Figure 12.6 are the curves obtained by simple extrapolation of (12.60), using

$$p \simeq \exp\left(-\frac{4E_{JT}}{\hbar\omega}\right) \tag{12.64}$$

which is a good approximation for $E_{JT}/\hbar\omega \leqslant 0 \cdot 1$.

When we depart from the linear Jahn-Teller coupling, these results will obviously be modified considerably in the higher order. We do not propose to go into this aspect; the reader is referred to the work of Ham.[30]

Quite similarly, Ham has studied in detail the Jahn-Teller effects for a triplet state T_1 or T_2 coupled to the e and the t_2 vibrational modes, but we shall not go into these effects here.

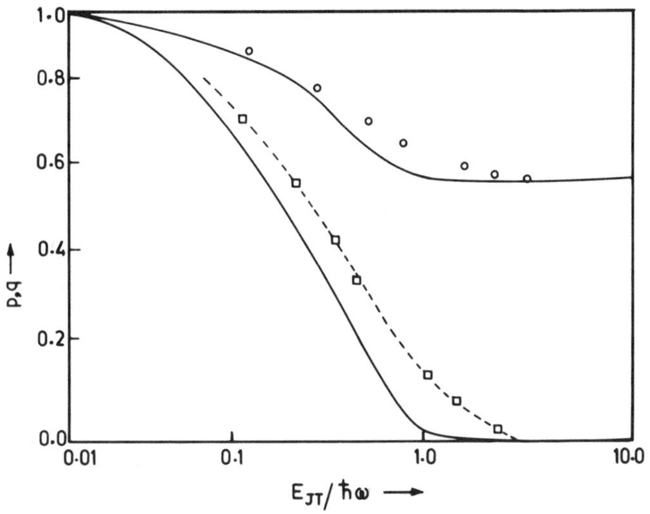

Fig. 12.6 Reduction factors p and q for the vibronic ground state E with linear Jahn-Teller coupling, as a function of E_{JT}. The solid curves show the variations of D and q given by (12.64) and (12.62), respectively. The squares and circles are the exact values from the calculations of Child and Longuet-Higgins,[47], while the dashed curve is given by (12.63). (Taken from Ham[30])

Let us now discuss the reduction of the orbital angular momentum and the spin-orbit coupling in a very qualitative way. Let us consider the angular momentum in a T term that is undergoing the Jahn-Teller effect. The nonvanishing matrix elements of $\mathbf{L}$ within a T term, having components ξ, η, and ζ, are given by

$$\begin{aligned}\langle\xi|\mathbf{L}_y|\zeta\rangle &= \langle\zeta|\mathbf{L}_x|\eta\rangle \\ &= \langle\eta|\mathbf{L}_z|\xi\rangle = i\hbar l \end{aligned} \tag{12.65}$$

where l is a number of order 1, depending on a particular term. Since $\mathbf{L}$ is a purely electronic operator, its matrix elements between the Born-Oppenheimer product states Ψ_i have the same form, that is,

$$\begin{aligned}\langle\Psi_1|\mathbf{L}_y|\Psi_3\rangle &= \langle\Psi_3|\mathbf{L}_x|\Psi_2\rangle \\ &= \langle\Psi_2|\mathbf{L}_z|\Psi_1\rangle = i\hbar l\gamma \end{aligned} \tag{12.66}$$

where

$$\gamma = \langle\phi_1|\phi_2\rangle$$

since the vibrational overlap integral γ is the same for any pair of unequal indices. Thus the oribtal momentum is reduced by the factor γ, which, in the harmonic approximation, is given by

$$\gamma = \langle \phi_1 | \phi_2 \rangle$$
$$= \exp\left(\frac{-3E_{JT}}{2\hbar\omega} \right) \tag{12.67}$$

Since **L** is involved in the spin-orbit interaction given by $\lambda \cdot \mathbf{L} \cdot \mathbf{S}$, exactly the same considerations apply here also, and therefore, the spin-orbit matrix elements are reduced in the same proportion as those of **L**. The same is true of any operator whose orbital part is exclusively off-diagonal in the real representation which is forced on the electronic term by the Jahn-Teller effect. By the "real representation" we mean the representation in which the Jahn-Teller interaction is diagonal. For example, the potential due to a trigonal distortion has T_2 symmetry (see Chapter 6), and all its matrix elements are off-diagonal in the (ξ, η, ζ) basis. Their values are therefore reduced by the same factor, γ. We have already seen, in Chapter 5 on the crystal field theory, that the T state splits, in first order, into E and A_1 states of trigonal symmetry (C_{3v}, D_{3d}, etc), and that the direction of quantization is the trigonal axis, which is different from the tetragonal axis of the Jahn-Teller distortion, but the splitting is reduced by the factor γ. Although this general reduction in off-diagonal elements was also observed by others (Refs. 46–51) much earlier, it was Ham who put it on a more rigorous mathematical basis. Hence it is now known as the Ham effect.

It is important to realize that in the limit when $\gamma \to 0$, the matrix elements of the off-diagonal operators go to zero and the situation then returns to the static Jahn-Teller effect. This means that the complex becomes frozen into a potential minimum and quenches the orbital angular momentum. A trigonal field then cannot distinguish between the three equivalent distortions, and therefore there is no trigonal splitting. Diagonal operators, on the other hand, such as a tetragonal crystal field, do not suffer this quenching; hence a stress parallel to a cubic axis can distinguish between the distortions parallel and perpendicular to that axis and therefore produces a splitting.

In 1960 Van Vleck[55] considered the opposite limit, that is, the spin-orbit coupling strong compared to the Jahn-Teller effect. Opik and Pryce[56] have also discussed the conditions under which sufficiently strong spin-orbit coupling will stabilize a complex against the Jahn-Teller distortion. This limit is likely to apply in the T terms of $4d$ and $5d$ transition metal complexes and other heavy ions where the spin-orbit splittings are large. In the rare earths this limit certainly holds, but the Jahn-Teller coupling is too weak to matter in the $4f^n$ configuration.

For the second and the third transition metal ions, the spin-orbit coupling is so large that even quite a strong Jahn-Teller effect may not mix the spin-orbit levels. In these complexes, therefore, each degenerate level (excluding Kramers' doublets, of course) may show its own Jahn-Teller effect, a Γ_3 level behaving like a doubly degenerate E term, and so on. The Jahn-Teller effect in a Γ_8 level has been discussed by Moffitt and Thorson[57] and also by Child.[58] Its behavior under an e_g distortion is essentially the same as that of a Γ_3 level. In the case of very strong spin-orbit coupling, not only the orbital angular momentum but also the total momentum of a T_1 or T_2 level is quenched by the Ham effect. The Ham effect also has remarkable consequences in the optical spectra, but we shall not go further into these details. The main consequence of the Ham effect is that, even when the spin-orbit coupling is not strong enough to prevent the Jahn-Teller distortion from occurring, the qualitative appearance of the energy levels is the same as in the absence of the Jahn-Teller effect, but the first-order splittings are reduced. Second-order splittings due to the Jahn-Teller effect are far more complicated, and we shall not go into them.

Before closing this section, we should like to discuss qualitatively (and very briefly) the cooperative Jahn-Teller effect, which is rapidly gaining importance in the solid state theory. By the "cooperative Jahn-Teller effect" we mean the effect on magnetically concentrated systems in which the individual Jahn-Teller ions cannot be regarded as independent. This phenomenon is exactly opposite to that in magnetically dilute systems, where the Jahn-Teller effect has been considered in isolated octahedral (ML_6) or tetrahedral (ML_4) complexes having one magnetic ion situated at the center of the molecular complex.

Just like a single ion, a pure crystal can be in an electronically degenerate state. The electronic state, in this case, transforms as an irreducible representation Γ of the *space* group (not the point group) of the crystal, and we may now have a Jahn-Teller distortion of the whole crystal, which should be analogous to that for an isolated complex. It is important to realize that the symmetric square $[\Gamma^2]$ must contain the irreducible representation of some possible distortion. In the absence of perturbations, such as impurities or excitons, this limits the possible distortions to $\kappa=0$ optical modes.[59] The irreducible representation of the distortions then is simply an irreducible representation of the parent point group. Birman[62] has considered the possibility of zone boundary distortions coupled to indirect excitons, for example, in diamond type crystals. It has been pointed out that in many structures there is no distortion satisfying this requirement. For example, a structure in which every ion is at a center of symmetry (e.g., the NaCl or the cubic perovskite structure) has no even-parity $\kappa=0$ modes and therefore cannot have a first-order Jahn-Teller coupling.

The proper treatment of Jahn-Teller effects when the coupling is with a continuous spectrum of lattice phonons rather than with a discrete set of vibrational modes is a theoretical problem still awaiting solution. It is important to realize that for a complex in a crystal one must, in principle, have coupling with such a phonon continuum even if there are also local modes of vibration. When the coupling is with a continuum of phonon states, we may still classify the vibronic eigenstates of the coupled system according to the irreducible representations of the group corresponding to the original undistorted symmetry of the site of the Jahn-Teller ion, since the full Hamiltonian describing this coupling has this symmetry. For very strong coupling between our Jahn-Teller ion and the lattice phonons, on the other hand, it is quite reasonable to suppose that the system has states in which the vibrational parts of the wavefunction have appreciable amplitude only for lattice ions close to the Jahn-Teller ion. It is very reasonable to think, in the case of the orbital doublet, that these localized states may be represented approximately in terms of the coupling of the degenerate electronic states with a single pair of localized modes of vibration belonging to E, primarily involving the nearest-neighbor lattice ions and having some effective frequency ω. These modes would thus be linear combinations of the normal modes of vibration of the unperturbed crystal. In terms of such modes, we surmise that a static Jahn-Teller effect occurs, as already described, and that the residual coupling with other lattice modes is relatively weak and occurs via the strain field that they produce. Slonczewski[60] has considered the formation of such a localized vibrational mode as a result of the Jahn-Teller coupling to the phonons, but in the absence of further calculations we do not know what effective frequency should be associated with this mode in various cases. Bates and his co-workers[61] have shown that the "cluster model" is justified as a good approximation to the real problem of a Jahn-Teller ion coupled to the entire lattice when the Jahn-Teller coupling is sufficiently strong. Such operator methods as those described by Bates and his co-workers should provide useful insights when there is Jahn-Teller coupling with the continuous spectrum of lattice phonons.

REFERENCES

1. A. Werner, *Z. Anorg. Chem.* **3**, 267 (1893).
2. E. E. Ilse and H. Hartmann, *Z. Phys. Chem.* **197**, 239 (1951); H. Hartmann and H. L. Schlafer, *ibid.* **197**, 115 (1951).

2a. H. Hartmann and H. L. Schlafer, *Naturforschung* **6a**, 751, 754, 760 (1951).

3. Y. Tanabe and S. Sugano, *J. Phys. Soc. Japan* **9**, 766 (1954).
4. Y. Tanabe and S. Sugano, *J. Phys. Soc. Japan* **9**, 753 (1954).
5. J. S. Griffith, *Theory of Transition Metal Ions*, Cambridge University Press, London, England, 1961.
6. J. R. Perumareddi, *J. Phys. Chem.* **71**, 3144 (1967).
7. J. R. Perumareddi, *J. Chem. Phys.* **71**, 3155 (1967).
8. A. D. Liehr and C. J. Ballhausen, *Ann. Phys.* **6**, 134 (1959).
9. J. C. Eisenstein, *J. Chem. Phys.* **34**, 310, 1628 (1961).
10. W. A. Runciman and K. A. Schroeder, *Proc. Roy. Soc. (London)* A, **265**, 489 (1962).
11. K. A. Schroeder, *J. Chem. Phys.* **37**, 2553 (1962).
12. K. A. Schroeder, *J. Chem. Phys.* **37**, 1587 (1962).
13. S. Sugano, Y. Tanabe, and H. Kamimura, *Multiplets of Transition-Metal Ions in Crystals*, Academic Press, New York, 1970.
14. C. J. Ballhausen, *Introduction to Ligand Field Theory*, McGraw-Hill, New York, 1962.

14a. T. S. Piper and R. L. Carlin, *J. Chem. Phys.* **33**, 608 (1960).

15. C. K. Jorgensen, *Absorption Spectra and Chemical Bonding in Complexes*, Pergamon Press, Oxford, 1962.
16. H. L. Schlafer and Gunter Gliemann, *Basic Principles of Ligand Field Theory* (translated by David F. Ilten), Interscience, New York, 1969.
17. V. P. Desai and A. S. Chakravarty, *Indian J. Phys.* **42**, 552 (1968).
18. A. S. Chakravarty, *Indian J. Phys.* **41**, 602 (1967).
19. A. S. Chakravarty, *J. Phys. Chem.* **74** 4347 (1970).
20. G. Herzberg and E. Teller, *Z. Phys. Chem. (Leipzig)* **21**, 410 (1933).
21. A. D. Liehr and C. J. Ballhausen, *Phys. Rev.* **106**, 1161 (1957).
22. R. S. Mulliken and C. A. Ricke, *Rep. Prog. Phys.* **8**, 231 (1941).
23. A. D. Liehr, *Adv. Chem. Phys.* **5**, 241 (1963).
24. A. D. Liehr and C. J. Ballhausen, *Ann. Phys. (New York)* **3**, 304 (1958).
25. C. J. Ballhausen and A. D. Liehr, *Mol. Phys.* **2**, 123 (1959).
26. C. J. Ballhausen, "Electron States in Complexes of the First Transition Group", Ph.D. Thesis, Copenhagen, 1958, p. 52.
27. J. W. Richardson, W. C. Nieuwpoort, R. R. Powell, and W. F. Edgell, *J. Chem. Phys.* **36**, 1057 (1962); J. W. Richardson, R. R. Powell, and W. C. Nieuwpoort, *J. Chem. Phys.*, **38**, 796 (1963).
28. J. W. Richardson, private communications (1964).
29. M. D. Struge, *Solid State Phys.* **20**, 91 (1967).
30. F. S. Ham, *Electron Paramagnetic Resonance*, S. Geschwrind (Ed.), Plenum Press, New York, 1972, p. 1.
31. H. A. Jahn and E. Teller, *Proc. Roy. Soc. (London) A*, *161*, 220 (1937).
32. B. Bleaney and K. D. Bowers, *Proc. Phys. Soc. (London) A*, **65**, 667 (1952).
33. J. H. Van Vleck, *J. Chem. Phys.* **7**, 61, 72 (1939).
34. E. B. Wilson, J. C. Decius, and P. C. Cross, *Molecular Vibrations*, McGraw-Hill, New York, 1955.
35. A. D. Liehr, *J. Phys. Chem.* **67**, 389 (1963).
36. M. Born and J. R. Oppenheimer, *Ann. Phys.* **84**, 457 (1927).
37. W. L. Clinton and B. Rice, *J. Chem. Phys.* **30**, 542 (1959).
38. H. Hellmann, *Quantenchemie*, Deuticke, Leipzig, 1937, p. 285.
39. R. P. Feynman, *Phys. Rev.* **56**, 340 (1939).
40. A. D. Liehr and H. L. Frisch, *J. Chem. Phys.* **28**, 1116 (1958).
41. M. L. Benston and B. Kirtman, *J. Chem. Phys.* **44**, 119, 126 (1966).
42. J. S. Griffith, *The Irreducible Tensor Method for Molecular Symmetry Groups*, Prentice-Hall, Englewood Cliffs, N.J., 1962.
42a. G. Herzberg, *Disc. Faraday Soc.* **35**, 7 (1963); *Electronic Spectra and Electronic Structure of Polyatomic Molecules*, Van Nostrand, Princeton, N.J., 1966.
43. W. Moffitt and A. D. Liehr, *Phys. Rev.* **106**, 1195 (1957).
44. W. Moffitt and W. Thorson, *Phys. Rev.* **108**, 1251 (1957).
45a. F. S. Ham, *Phys. Rev. A*, **138**, 1727 (1965).
45b. F. S. Ham, *Phys. Rev.* **166**, 307 (1968).
46. H. M. McConnell and A. D. McLachlan, *J. Chem. Phys.* **34**, 1 (1961).
47. M. S. Child and H. C. Longuet-Higgins, *Phil. Trans. Roy. Soc. (London) A*, **254**, 259 (1962).
48. I. B. Bersuker, *Sov. Phys. JETP* **16**, 933 (1963).
49. I. B. Bersuker and B. G. Vekhter, *Sov. Phys. Solid State* **5**, 1772 (1964).
50. C. J. Ballhausen, *Theor. Chim. Acta (Berlin)* **3**, 368 (1965).

51. W. C. Scott and M. D. Struge, *Phys. Rev.* **146**, 262 (1966).
52. H. C. Longuet-Higgins, U. Opik, M. H. L. Pryce, and R. A. Sack, *Proc. Roy. Soc. (London) A*, **244**, 1 (1958).
53. W. Moffitt and W. Thorson, *Calcul des Fonctions d'Onde Moleculaire*, R. Daudel, (Ed.), Rec. Mem. Centre Nat. Rech. Sci., Paris, 1958, p. 141.
54. F. I. B. Williams, D. C. Krupka, and D. P. Breen, *Phys. Rev.* **179**, 255 (1969).
55. J. H. Van Vleck, *Physica* **26**, 544 (1960).
56. U. Opik and M. H. L. Pryce, *Proc. Roy. Soc. (London) A*, **238**, 425 (1957).
57. W. Moffitt and W. Thorson, *Phys. Rev.* **108**, 1251 (1957).
58. M. S. Child, *Phil. Trans. Roy. Soc. (London) A,* **255**, 31 (1962).
59. N. N. Kristofel, *Sov. Phys. Solid State* (Eng. trans.) **6**, 2613 (1965).
60. J. C. Slonczewski, *Phys. Rev.* **131**, 1596 (1963).
61. C. A. Bates, J. M. Dixon, J. R. Fletcher, and K. W. H. Stevens, *J. Phys. C*, **1**, 859 (1968).
62. J. L. Birman, *Phys. Rev.* **125**, 1959 (1962).

13

Molecular Fields in Ordered Magnetic Solids

So far we have neglected the magnetic or other interactions between the magnetic ions. In other words, we have considered the metal ions as isolated. This treatment is, of course, justified for very dilute magnetic compounds, for example, the hydrated salts in a diamagnetic medium containing magnetic ions in small concentrations. However, the assumption that there is no coupling between the spins of different paramagnetic ions is never strictly correct. In the following chapters we shall see what happens when this coupling is taken into consideration.

It is well known that the classical theory of magnetism can envisage only three types of magnetic materials: diamagnetic, paramagnetic, and ferromagnetic. In addition, there is another type of magnetic material which is in such an ordered state that it is neither a paramagnet nor a ferromagnet. This type is known as antiferromagnetic. In this chapter we shall deal with ferromagnetism in solids, leaving antiferromagnetism to be discussed in Chapter 14.

Let us see what ferromagnetism is. The intense response to an applied magnetic field shown by a sample of pure or alloyed Fe, Ni, or Co is known as ferromagnetism. The magnetization induced by an applied field in a ferromagnetic sample is generally related in a complex way to the shape and the magnetic history of the sample, as well as to the strength of the magnetic field. The most important characteristic point is that the sample remains magnetized to some extent even when the field is withdrawn. This permanent magnetization is the most markedly characteristic property of ferromagnetic materials. Another characteristic property of a ferromagnetic sample is that, above a certain critical temperature known as the Curie temperature, the spontaneous magnetization vanishes and the sample becomes paramagnetic. This spontaneous magnetization indicates a long-range ordering of the directions of the magnetic moments of some of the electrons in the sample. The magnitude of the spontaneous magnetization is clearly revealed by experimental observations on single crystals. With a pure single-crystal sample it is possible to find a crystallographic direction (easy direction) in which a field of a few oersteds is sufficient to magnetize the crystal to within a few percent of the saturation intensity. In other crystallographic directions magnetization is much less easily induced. It is also observed that a ferromagnetic crystal is spontaneously magnetized to a high degree at every point but that different regions of the crystal are magnetized in different but crystallographically equivalent directions.

Let us take an example. An Fe crystal is cubic, and therefore there are six equivalent preferred directions parallel to the edges of the cubic elementary cell. A demagnetized Fe crystal is therefore subdivided into regions or "domains," as they are called, each spontaneously magnetized in one or another of the six equivalent preferred directions. If now a small field is applied in one of these six directions, the domains that are magnetized in this direction expand at the expense of the others, so that, in the ideal case, the crystal would become a single domain. This picturization, due originally to Weiss[6], accounts beautifully for the ease with which single crystals can be magnetized to saturation in their preferred directions. The existence of domains has been confirmed by direct observations, but we shall not go into these aspects any further. A good description of different aspects of magnetism in ordered magnetic systems can be found in Refs. 1–5.

It has already been remarked that samples of Fe, Ni, and Co and also simple compounds of these elements are paramagnetic above their Curie temperatures; this means that the magnetization induced by an applied field is in the direction of the field but is generally many orders of magnitude less than would be induced in a sample below the Curie temperature. As seen earlier, paramagnetism is measured by the volume susceptibility χ, which is the ratio of the induced magnetization to the inducing field. This is found to be markedly temperature dependent, decreasing with increasing temperature. At temperatures well above the Curie point, the susceptibility roughly follows a Curie-Weiss relation:

$$\chi = \frac{C}{T-\theta} \tag{13.1}$$

where C and θ are constants for a given material. If θ is zero, (13.1) goes over to the well-known Curie law. When T is comparable to or less than θ, relation (13.1) will not hold good, in which case, the dependence of χ on T is more complicated because the interactions may lead to some kind of ordering of the magnetic ions. The law (13.1) can be derived in a simple way if it is assumed that the influence of the interionic forces can be represented by an internal field $\mathbf{H}_{\text{int}}$ proportional to the prevailing magnetization $\mathbf{M}$ or, writing explicitly, $\mathbf{H}_{\text{int}} = \gamma \mathbf{M}$, where γ is the molecular field constant. The interionic forces are not necessarily magnetic; in fact, they are *electrostatic in nature*, giving rise to "exchange" interactions, which are generally responsible for the development of the internal field. The total effective field, therefore, is the sum of the applied magnetic field and the internal field $\mathbf{H}_{\text{int}}$. Thus

$$\mathbf{H} = \mathbf{H}_{\text{app}} + \mathbf{H}_{\text{int}} = \mathbf{H}_{\text{app}} + \gamma\mathbf{M} \tag{13.2}$$

where $\mathbf{H}_{\text{app}}$ is the applied field. We now substitute this expression for $\mathbf{H}$ in (4.33), which represents the magnetization of a paramagnetic gas, thus getting

$$\mathbf{M} = \frac{N\mu_B^2}{3kT} g_J^2 J(J+1)(\mathbf{H}_{\text{app}} + \gamma\mathbf{M}) = \frac{C}{T}(\mathbf{H}_{\text{app}} + \gamma\mathbf{M}) \tag{13.3}$$

whence

$$\chi = \frac{M}{H_{\text{app}}} = \frac{C}{T - C\gamma} = \frac{C}{T - \theta} \tag{13.4}$$

This is known as the Curie-Weiss law with the Weiss constant $\theta = C\gamma$. Thus θ is proportional to the molecular field constant γ; θ may be positive or negative, indicating that the exchange forces may, in some substances, favor antiparallel or parallel alignments, respectively, of the neighboring magnetic ions. However, an apparent Curie-Weiss behavior of material should not be taken as conclusive evidence of an exchange interaction between the magnetic ions because lower symmetric ligand fields can also simulate a nonzero θ, as will be discussed later in this and the next chapter.

It will be quite apparent later that θ is not an atomic property, but is due primarily to distortions by the interatomic forces (the ligand fields). This is shown, for example, by the fact that θ varies markedly with the compound in which a paramagnetic ion of a given valence occurs. From experiments it is found that, as a rule, the values of θ are lower in compounds of high magnetic dilution. Heisenberg's theory, which we shall discuss in this chapter, shows that the exchange interactions between the paramagnetic atoms or ions can also contribute to θ. We have already mentioned, however, that the exchange interaction plays a very insignificant role in the dilute magnetic systems that we have examined in great detail in earlier chapters. It is important to note here that, as we pass from magnetically dilute to ordered magnetic systems, the contribution of the crystalline electric fields of lower symmetry (lower than O_h symmetry, for instance) to θ decreases significantly, while the contribution due to the exchange interaction increases very markedly.

We shall see shortly that all the theories attempting to provide a microscopic explanation of ferromagnetism start with the Heisenberg exchange Hamiltonian:

$$\mathcal{H} = -\sum_{i \neq j} J_{ij} \mathbf{S}_i \cdot \mathbf{S}_j \tag{13.5}$$

where the summation is over all the magnetic ions in the lattice, and J_{ij} is the exchange integral between the ith and the jth ion. A simple derivation of (13.5), when the ions have no orbital degeneracy, will be given in the next section, but the general form of this expression, assuming the presence of both spin and orbital degeneracy, is not yet known. However, once the Hamiltonian is taken for granted, the problem is attacked at two levels. One approach investigates the quantum-mechanical and thermodynamical consequences of the Hamiltonian given by (13.5). These deal with the existence of phase transitions and the nature of such transitions. At another level, one may suitably parametrize the Hamiltonian so that it can be applied to real substances. For either approach there exists an enormous literature, and it is not the purpose of this book to give even a cursory account of the general problems involved.

When the Hamiltonian (13.5) is used as a model Hamiltonian, it is enormously successful, particularly for insulators: the oxides and the halides of the transition metal ions. The usual procedure is to assume J_{ij} to be nonvanishing only for the nearest neighbors (n.n). If the experimental data warrant, the next-nearest neighbors (n.n.n) are also considered. In the simplest models, J_{ij} is taken to be isotropic. Then the partition function for an assembly of ions is calculated in the usual way, and the magnetic properties are derived from it.

It is easy to see that, when J_{ij} is positive, the lowest energy state corresponds to the situation where the spins of all the ions are parallel. This is the case of ferromagnetism. But when J_{ij} is negative, the problem of calculating the ground state of the Hamiltonian given by (13.5) is complicated. However, heuristically one observes that the neighboring spins tend to be antiparallel, resulting in an antiferromagnetic substance.

In principle, the thermodynamic properties of an assembly of interacting ions having the Hamiltonian given by (13.5) can be derived, once the eigenvalues of the Hamiltonian are known. When we are interested in the magnetic properties, we require the response of the system to the external magnetic field. The Hamiltonian including the magnetic perturbation is given by

$$\mathcal{H} = -\sum_{i \neq j} J_{ij}\mathbf{S}_i \cdot \mathbf{S}_j - g\mu_B \sum_i (\mathbf{L}_i + 2\mathbf{S}_i) \cdot \mathbf{H} \tag{13.6}$$

where g is the Landé g-factor for the ion, and $\mathbf{H}$ the applied magnetic field.

The various methods of solving for (13.6) are (1) the molecular field (MF) method, (2) the method of spin waves, and (3) the series expansion method. We begin with the MF theory, originally due to Weiss.[6]

13.1 MOLECULAR FIELD THEORY OF ORDERED MAGNETIC SOLIDS

The main features of the ferromagnetic, the ferrimagnetic, and the antiferromagnetic materials are well known. The substances belonging to these classes have in common the fact that the orientations of their elementary magnetic dipoles are ordered spontaneously by some kind of mutual interaction. In ferromagnetic substances the dipoles are oriented in a single direction (the direction of the magnetic field), and all in the same sense. The ordering in ferrimagnetic and antiferromagnetic substances is more complex, however, the elementary magnetic dipoles being arranged in two or more interpenetrating sublattices that are spontaneously magnetized in different directions. The existence of permanent magnetism, the characteristic discontinuities in the variations of specific heat with temperature, and the observed patterns formed by the diffraction of slow neutrons from single crystals—all point to the existence of spontaneous ordering of different kinds in these substances. We shall presently see in this section that the MF theory can successfully explain, at least in a qualitative way, the peculiar properties mentioned above.

According to Weiss, the interaction responsible for the spontaneous ordering can be represented by an internal molecular field, which is assumed to be proportional to the degree of magnetization. To discuss these interactions and their representations in the MF theory we shall deal, first of all, with the case of ferromagnetism. Metallic Fe, Co, and Ni and also the rare earths Gd, Dy, Tb, Ho, and Er at low temperatures, as well as a number of alloys containing these elements, are all ferromagnetic. Ferromagnetic samples are subdivided into domains, each of which is spontaneously magnetized to a very high degree. It is an observed fact that the direction of a domain's spontaneous magnetization can be easily changed by the application of a magnetic field, but the magnitude of the spontaneous magnetization is only very weakly dependent on the strength of the magnetic field. Only by the application of heat can the spontaneous magnetization within a domain be decreased significantly, and it vanishes altogether at a critical temperature known as the Curie point.

The existence of permanent magnets is due to the spontaneous magnetization developed in the ferromagnetic material by the application of a magnetic field of very low intensity. For example, a soft magnetic material can be easily magnetized to saturation by an applied magnetic field of less than 1 oersted. On the other hand, in a paramagnetic sample a field of about several million oersteds is necessary to produce saturation. The Curie point up to which the spontaneous magnetization persists is fairly high for materials like cobalt, for which the Curie temperature T_C is observed to be 1393°K. Clearly, therefore, the interatomic forces responsible for the spontaneous magnetization inside a ferromagnetic sample are effectively equivalent to a field of several million oersteds, and it can be easily shown that the magnetic interactions between the elementary magnetic dipoles can produce magnetic forces only of the order of 2×10^3 oersteds and are thus much too small to account for the spontaneous magnetization.

It was shown by Heisenberg in 1926 that the strong interaction responsible for the spontaneous magnetization is due to the exchange interaction between the electrons, which tends to order the spin moments of two neighboring electrons either in parallel or in antiparallel alignment, depending on the details of their states of motion. The exchange forces are basically electrostatic in origin and essentially quantum mechanical in nature, having no classical analogue. These exchange forces, as shown by Heisenberg, may have an intensity many orders of magnitude greater than the ordinary magnetic forces between the spin moments. We shall deal with this topic adequately in a subsequent section. Now let us look into the details of the MF hypothesis advanced by Weiss in 1907.

As discussed earlier, Weiss assumed that each atom of a ferromagnetic material is a magnetic dipole and is acted upon by an internal magnetic field proportional to and parallel to the magnetization in the region surrounding it. This hypothesis accounted very successfully for the spontaneous magnetization and its observed variation with temperature. Although Weiss realized that his molecular field had to be more intense than the ordinary magnetic forces, as discussed earlier, he was unable to trace its origin. After Heisenberg, it is now known to be a representation of the net effects of the exchange interactions.

The above discussion applies also, with some modifications, to ferrimagnetic or antiferromagnetic materials. Spontaneous ordering occurs in such materials and persists in many cases up to temperatures well above 100°K, so that exchange forces can again be held responsible. The MF concept is easily extended to cover the interactions in these cases toward the antiparallel alignment of neighboring electrons or ions, as we shall see later.

13.2 MOLECULAR FIELD THEORY FOR FERROMAGNETISM

Ferromagnetic substances are usually metallic. This fact gives rise to the immediate problem of how to describe the states of motions of the electrons inside the metal. When atoms combine to form a metal, their outermost electrons become influenced by the neighboring atoms, whereas the inner electrons are hardly affected and remain localized on separate nuclei. The $4f$ magnetic electrons in the ferromagnetic rare earth metals, being shielded by the $5s$ and $5p$ electrons, probably remain tightly bound and therefore are localized on the parent nuclei. In Fe, Ni, and Co and in their alloys the $3d$ electrons responsible for ferromagnetism, being the outermost electrons, are in itinerant states or at least are intermediate between being localized on parent nuclei and being freely itinerant. These itinerant electrons move through the lattice, but their motions are *highly correlated* as a result of Coulomb repulsions between them. Such a situation is extremely difficult to analyze theoretically, as will be seen later. Molecular field theories have been developed for two limiting cases. The original Weiss model is applicable to electrons that are localized within the positive ions forming the lattice, while the collective electron model, due mainly to Stoner, deals with almost free itinerant electrons.

13.3 THE WEISS MODEL

We now describe the Weiss model for localized electrons. This model is based on the hypothesis that, when Fe, Ni, or Co atoms or the rare earth atoms condense to form metallic solids, the magnetic electrons retain their localized character, so that ions at the crystal lattice sites constitute the elementary dipoles which are ordered by the exchange forces between them.

Within the framework of the MF theory, we can derive the expression for the spontaneous magnetization by using the statistical theory of paramagnetism, which we develop now. If we assume that the magnetic dipoles are free to orient themselves, two influences are to be considered on them; one is the ordering influence of the molecular field, and the other is the disordering effect of thermal agitation. Spontaneous magnetization arises when there is a balance between these two influences. In Chapter 4 we derived a relation between the intensity of magnetization **M** of a paramagnetic sample and the applied magnetic field **H** that is given by

$$\mathbf{M}=M_m B_J(\mu_m \mathbf{H}/kT) \tag{13.7}$$

where M_m is the saturation magnetization that would be attained in an infinite field with all the dipoles of moment μ_m aligned, and $B_J(x)$ is the Brillouin function, given by

$$B_J(x)=\frac{2J+1}{2J}\coth\frac{2J+1}{2J}x-\frac{1}{2J}\coth\frac{x}{2J} \tag{13.8}$$

where J is the total angular momentum quantum number of the ions, in terms of which $\mu_m = g\mu_B J$. (This J is not to be confused with the exchange integral, which is also represented by the same symbol). Assuming only the spin of the electron contributing to the dipole moment, we have $J=\frac{1}{2}$, and in this case $B_J(x)$ reduces to

$$B_{1/2}(x)=\tanh x \tag{13.9}$$

The total magnetic field, which consists of the applied field $\mathbf{H}_{\text{app}}$ and the internal molecular field $\gamma\mathbf{M}$, is given by

$$\mathbf{H}=\mathbf{H}_{\text{app}}+\gamma\mathbf{M} \tag{13.10}$$

where γ is the molecular field constant. Since $\gamma\mathbf{M} \gg \mathbf{H}_{\text{app}}$, we have from (13.7)

$$\frac{\mathbf{M}_s}{M_m}=B_J\left(\frac{\mu_m \gamma M_s}{kT}\right) \tag{13.11}$$

where $\mathbf{M}_s$ represents the spontaneous magnetization due to the molecular field. Equation (13.11) represents the variation of $\mathbf{M}_s$ with T, but since $\mathbf{M}_s$ appears on both sides of the equation, we graphically plot both sides against

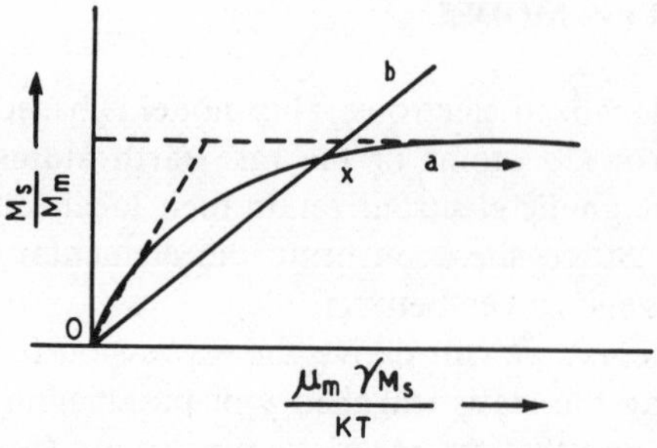

Fig. 13.1 Graphical determination of the spontaneous magnetization M_s.

$\mu_m \gamma M_s / kT$ and find the points at which the curves intersect (Figure 13.1). Thus, in Figure 13.1, b represents the left-hand side of (14.11), and a represents the right-hand side, for $J = \frac{1}{2}$. The simultaneous solutions are given by the points of coincidence of the two curves at the origin and at $\times$. The value of $\mathbf{M}_s$ at $\times$ represents the stable solution. It is easy to find, from the figure, that the critical temperature at which the spontaneous magnetization vanishes is the Curie temperature T_{C}, and its value in terms of the molecular field parameters is determined from the slope of the Brillouin function at the origin, which is $(J+1)/3J$. Equating this to the slope of the line, given by $kT/\mu_m \gamma M_m$, we have

$$\frac{kT_{\mathrm{C}}}{\mu_m \gamma M_m} = \frac{J+1}{3J}$$

or

$$T_{\mathrm{C}} = \frac{J+1}{3J} \frac{\gamma}{k} \mu_m M_m$$
$$= \frac{J(J+1)}{3k} g^2 \mu_{\mathrm{B}}^2 N \gamma \qquad (13.12)$$

since $\mu_m = g\mu_{\mathrm{B}} J$ and $M_m = Ng\mu_{\mathrm{B}} J$, where N is the number of dipoles per unit volume. Thus

$$T_{\mathrm{C}} = \gamma C \qquad (13.13)$$

where, C, the Curie constant, is given by

$$C = Ng^2 \mu_{\mathrm{B}}^2 \frac{J(J+1)}{3k} \qquad (13.14)$$

Now, rewriting (13.11), we have

$$\frac{M_s}{M_m} = B_J\left(\frac{3J}{J+1} \cdot \frac{T_{\mathrm{C}}}{T} \frac{M_s}{M_m} \right) \qquad (13.15)$$

From (13.15) we find that, for a particular J, the variation of M_s/M_m with T/T_C depends only on the form of the Brillouin function, and not on any other parameter, which would vary from one material to another. According to Weiss, this relation should be obeyed by all ferromagnetic elements. Experimental measurements[5] of M_s/M_m with T/T_C for several elements of the transition metals, notably Fe, Co, and Ni, fall between the Weiss theoretical curves for $J=1$ and for $J=\frac{1}{2}$. There are some deviations, of course.

Above the Curie temperature T_C, the ferromagnetic material goes over to the paramagnetic phase, and we can still use the MF approach to derive the Curie-Weiss law, provided that, at $T>T_C$, we retain $\mathbf{H}_{app}$ along with $\gamma\mathbf{M}$, since they are now of the same order of magnitude. The Curie-Weiss law is given, as before, by

$$\chi=\frac{M}{H_{app}}=\frac{C}{T-\theta} \tag{13.16}$$

where

$$C=\frac{Ng^2\mu_B^2J(J+1)}{3k}$$

and

$$\begin{aligned}\theta=\gamma C &= \text{positive for antiparallel spins}\\ &= \text{negative for parallel spins}\end{aligned}$$

Equation (13.16) indicates that a graph of $1/\chi$ with temperature should give a straight line with an intercept at T_C^*, the Curie temperature. It is sufficient to mention that the observed behavior of Fe, Ni, and Co do correspond to the Curie-Weiss variation except at temperatures near the Curie point.

Let us now discuss the limitations of the Weiss model. For ferromagnetic metals of the iron group, the value of the molecular field constant γ, deduced from the observed Curie temperature using (13.12), does not agree with that obtained from the paramagnetic Curie constant C using (13.16), if it is assumed that the ionic moment remains the same above and below T_C. For example, for nickel $\gamma \simeq$ 18,500 from (13.12), whereas it is $\simeq$13,400 from (13.16). On the other hand, if γ is assumed to be constant, the measured properties would suggest that the ionic moment is variable. The situation is similar for iron and cobalt. For iron, the ionic moment deduced from M_s at $T=0$ is $2.2\mu_B$, and that deduced from the paramagnetic Curie constant is $3.13\mu_B$. Another serious defect is that the ionic moment is *not* an integral number of Bohr magnetons for any of the metals of the transition elements. Each atom contributes, on the average, $2.22\mu_B$ in iron, $0.6\mu_B$ in nickel, and $1.72\ \mu_B$ in cobalt, whereas we should expect integral values for the moments, assuming spin contributions only (orbital contributions being quenched by the neighboring ligand field).

The deficiencies of the Weiss model described above can be compensated for, to a great extent, by three very important approaches: (1) the collective electron ferromagnetism of Stoner[7] and Wohlfarth,[8] (2) the generalized Heisenberg model, advocated by Van Vleck,[9] wherein there are a nonintegral number of spins per atom and the spins are continually being redistributed among the different lattice sites, and (3) the band-theoretic approach of Slater,[10] which shows that, using a single determinantal wavefunction composed of energy band orbitals, we have the familiar energy band theory given by approach 1, which may be fairly accurate at small internuclear distances. For large internuclear distances, however, it leads to quite wrong limiting behavior, since it gives a wavefunction involving large contributions by the ionic states. To eliminate these, Slater says, at large internuclear distances we must make linear combinations of different determinantal functions corresponding to different assignments of electrons to the orbitals. We shall describe these approaches one after the other, but it should be borne in mind that each has its own merits and shortcomings. With these remarks we now proceed to the exchange interaction mechanism, first proposed by Heisenberg in 1926.

13.4 THE EXCHANGE INTERACTION

It was remarked earlier in this chapter that, although Weiss realized that the molecular field had to be much more intense than could be accounted for by the ordinary magnetic forces, he was unable to trace its origin. It is now known to be a representation of the net effects of exchange interactions.

It was Heisenberg[11] who first recognized, in 1926, that the quantum theory would provide a physical explanation for the large magnitude of the Weiss field. Because of the requirement imposed by the Pauli exclusion principle that the wavefunction describing a system of electrons be antisymmetric, an extra electrostatic energy term appears in the Hamiltonian for the system. This term, depending on the overlap of the wavefunctions of the adjacent particles, is called the electrostatic exchange interaction. It also depends on the relative orientation of the spins of the adjacent particles and therefore contributes to the ferromagnetic effect. In a crystal in which each ion has a nonzero spin, such an effect is large enough to account for the magnitude of the Weiss field. It should be noted that the exchange interaction has no classical counterpart, and the success of the Heisenberg model represents one further striking triumph of the quantum theory.

Let us now outline the mathematical development of the Heisenberg theory. First we consider the Schrödinger wave equation for a system of two noninteracting atoms. The stationary states of each atom, designated by 1 and 2, are determined by identical wave equations:

$$\mathcal{H}_0^{(1)}\psi_n(\mathbf{r}_1)=E_n\psi_n(\mathbf{r}_1)$$
$$\mathcal{H}_0^{(2)}\psi_m(\mathbf{r}_2)=E_n\psi_m(\mathbf{r}_2) \tag{13.17}$$

where the quantum states identified by n and m are two of a complete set of eigenfunctions of $\mathcal{H}_0$, the unperturbed Hamiltonian, of the respective atoms. Combining the two equations (13.17), we get the wave equation for the system of two atoms, given by

$$\left[\mathcal{H}_0^{(1)}+\mathcal{H}_0^{(2)}\right]\psi_n(\mathbf{r}_1)\psi_m(\mathbf{r}_2)=(E_n+E_m)\psi_n(\mathbf{r}_1)\psi_m(\mathbf{r}_2) \tag{13.18}$$

where $\mathcal{H}_0^{(1)}$ operates only on $\psi(\mathbf{r}_1)$, and $\mathcal{H}_0^{(2)}$ only on $\psi(\mathbf{r}_2)$. Since n and m are any two of a complete set of states, (13.18), with n and m interchanged, is also a solution. Because of the fact that $\psi_n(\mathbf{r}_1)\psi_m(\mathbf{r}_2)$ and $\psi_m(\mathbf{r}_1)\psi_n(\mathbf{r}_2)$ are two distinct orthogonal solutions of the same differential equation, the energy eigenvalue (E_n+E_m) has a twofold degeneracy with respect to the interchange of n and m between the wavefunctions for the neighboring atoms. This degeneracy is removed by the Coulomb interaction $e^2/\mathbf{r}_{12}$, acting between the two atoms.

The effect of the perturbation $e^2/\mathbf{r}_{12}$ on the energy levels of the system may then be obtained by means of the standard technique of the degenerate perturbation theory. The twofold degeneracy is removed by diagonalizing the secular determinant:

$$\begin{vmatrix} \langle\mathcal{H}\rangle_{11}-E & \langle\mathcal{H}'\rangle_{12} \\ \langle\mathcal{H}'\rangle_{21} & \langle\mathcal{H}\rangle_{22}-E \end{vmatrix}=0 \tag{13.19}$$

where $\mathcal{H}'=e^2/\mathbf{r}_{12}$, $\mathcal{H}=\mathcal{H}_0+\mathcal{H}'$, and

$$\langle\mathcal{H}\rangle_{ij}=\int\psi_i^*\mathcal{H}\psi_j\,d\tau$$

Also,

$$\psi_1=\psi_n(\mathbf{r}_1)\psi_m(\mathbf{r}_2)$$
$$\psi_2=\psi_m(\mathbf{r}_1)\psi_n(\mathbf{r}_2) \tag{13.20}$$

are the zero-order eigenfunctions. Since

$$J_{mn}=\langle\mathcal{H}'\rangle_{12}=\langle\mathcal{H}'\rangle_{21}$$

$$=\int\psi_m^*(\mathbf{r}_1)\psi_n^*(\mathbf{r}_2)\frac{e^2}{\mathbf{r}_{12}}\psi_n(\mathbf{r}_1)\psi_m(\mathbf{r}_2)\,d\tau_1\,d\tau_2 \tag{13.21}$$

and

$$C_{mn} = \langle \mathcal{H}' \rangle_{11} = \langle \mathcal{H}' \rangle_{22}$$
$$= \int \psi_m^*(\mathbf{r}_1)\psi_n^*(\mathbf{r}_2)\frac{e^2}{\mathbf{r}_{12}}\psi_m(\mathbf{r}_1)\psi_n(\mathbf{r}_2)\,d\tau_1\,d\tau_2 \tag{13.22}$$

Secular equation (13.19) gives the two eigenvalues

$$E^{\pm} = E_{mn} + C_{mn} \pm J_{mn} \tag{13.23}$$

where

$$E_{mn} = E_m + E_n$$

The corresponding eigenfunctions of the total Hamiltonian:

$$\mathcal{H} = \mathcal{H}_0^{(1)} + \mathcal{H}_0^{(2)} + \mathcal{H}' \tag{13.24}$$

which diagonalize the determinant are the symmetric and antisymmetric functions

$$\psi_{\text{sym}} = \frac{1}{\sqrt{2}}(\psi_1 + \psi_2)$$
$$\psi_{\text{antisym}} = \frac{1}{\sqrt{2}}(\psi_1 + \psi_2) \tag{13.25}$$

The treatment presented thus far, neglects the fact that each of the interacting atoms has spin as well as space coordinates. Since the spin-orbit interaction has been neglected, there is no coupling between the particle spins and their orbital motions, and therefore the total wavefunction describing the two-atom system must be a product of the space part and the spin part.

For simplicity, let us assume that each atom has a single electron in the $3d$ shell. The spin of each electron is $\frac{1}{2}$ (in units of $\hbar$); and according to the way in which the angular momenta combine, the total spin of the system is characterized either by a triplet state ($S=1$) or by a singlet state ($S=0$). The possible values for the z-component of the total spin of two atoms are $1, 0, -1$. The spin functions are then described as follows:

$$|1,1\rangle = |\alpha_1\alpha_2\rangle$$
$$|1,0\rangle = \frac{1}{\sqrt{2}}[|\alpha_1\beta_2\rangle + |\beta_1\alpha_2\rangle]$$
$$|1,-1\rangle = |\beta_1\beta_2\rangle \tag{13.26}$$
$$|0,0\rangle = \frac{1}{\sqrt{2}}[|\alpha_1\beta_2\rangle - |\beta_1\alpha_2\rangle]$$

where the kets on the right-hand side are in the scheme $|s_1 m_{s_1}, s_2 m_{s_2}\rangle$, α_1 denotes $|s_1 m_{s_1}\rangle = |\frac{1}{2}\ \frac{1}{2}\rangle$, and β_1 denotes $|s_1 m_{s_1}\rangle = |\frac{1}{2}\ -\frac{1}{2}\rangle$. The first three spin functions in (13.26) are clearly symmetric with respect to an interchange of atoms 1 and 2 and hence correspond to $S=1$, while the fourth one describes the singlet state $S=0$.

The total wavefunction for the two-atom system is then given by the products of the space and the spin functions. Now, since the Pauli principle demands that the total wavefunction be antisymmetric, the symmetric spin function will be coupled only to the antisymmetric space function, and vice versa. Thus

$$\begin{aligned} {}^3\psi &= \frac{1}{\sqrt{2}}(\psi_1 - \psi_2)\chi(S=1) \\ {}^1\psi &= \frac{1}{\sqrt{2}}(\psi_1 + \psi_2)\chi(S=0) \end{aligned} \tag{13.27}$$

correspond, respectively, to

$$\begin{aligned} {}^3E &= E + C - J \\ {}^1E &= E + C + J \end{aligned} \tag{13.28}$$

and since

$$\mathbf{S}_1 \cdot \mathbf{S}_2 = \tfrac{1}{2}\left[(\mathbf{s}_1 + \mathbf{s}_2)^2 - \mathbf{s}_1^2 - \mathbf{s}_2^2\right]$$

$$\langle(\mathbf{s}_1 + \mathbf{s}_2)^2\rangle = S(S+1) = \begin{cases} 0 & \text{for singlet state} \\ 2 & \text{for triplet state} \end{cases}$$

and

$$\langle s_i^2 \rangle = \tfrac{1}{2}\left(\tfrac{1}{2} + 1\right) = \tfrac{3}{4}$$

we have

$$\langle \mathbf{S}_1 \cdot \mathbf{S}_2 \rangle = \begin{cases} -\frac{3}{4} & \text{for singlet state} \\ +\frac{1}{4} & \text{for triplet state} \end{cases}$$

Therefore (13.28) may be combined in the form

$${}^{1,3}E = \left(E + C - \tfrac{1}{2}J\right) - 2J\,\mathbf{S}_1 \cdot \mathbf{S}_2 \tag{13.29}$$

Thus the energy eigenvalues of the two-atom system can be expressed in terms of the spin coupling term depending on $\mathbf{S}_1 \cdot \mathbf{S}_2$. It is this term, according to Heisenberg, that gives rise to the Weiss field. In practice, it is very difficult, however, to evaluate the exchange integral J or even to determine its sign, since such a calculation requires the solution of the many-electron-atom problem. If the model is correct, the sign of J must be positive for the transition metals to produce ferromagnetism.

Heisenberg's calculation assumes that the atoms are in S states, but this is not really necessary since the orbital angular momentum is quenched by the asymmetrical crystal field in the solid. This assumption is usually valid in real magnetic systems, however, and we shall always make it. Heisenberg also assumes that a given atom has an appreciable exchange interaction only with the adjacent atoms and possesses z such neighbors equidistant from it. Thus $z=2$ for a linear chain, 4 for a quadratic surface grating, 6 for a simple cubic grating, 8 for a body-centered cubic grating, and so on. Let us further suppose that the valence electrons or electrons not in closed shells are in similar states. The part of the Hamiltonian involving the spin-spin coupling is then given by

$$\mathcal{H}_{\text{ex}} = -2J \sum_{z} \mathbf{S}_i \cdot \mathbf{S}_j \tag{13.30}$$

where J is the exchange integral between two valence electrons of adjacent atoms, and the summation extends over all neighboring pairs of atoms. Equation (13.30) follows since $\Sigma \mathbf{S}_k \cdot \mathbf{S}_l = \Sigma \mathbf{s}_k \cdot \Sigma \mathbf{s}_l = \mathbf{S}_i \cdot \mathbf{S}_j$, if k and l refer to different atoms i and j and if we sum over the valence electrons of both atoms.

The basic problem of Heisenberg's theory of ferromagnetism is to calculate the characteristic values of (13.30) and hence the energy states belonging to various resultant spins of the crystal.

The real problem in understanding ferromagnetism is to discover the nature of the interactions between ions that make their magnetic moments line up with one another. It has been seen that the mechanism is not the direct effect of one magnetic dipole on its neighbor, because the energy involved in this interaction is so small that the alignment could be destroyed by thermal activation at very low temperatures, whereas we know that ferromagnetism can persist up to very high temperatures. We have just seen that, in the case of two neighboring ions contributing one electron each, the resultant triplet and singlet states are given by (13.27). If the triplet state has the lower energy, that is, is the ground state, this automatically determines that the state has neighboring spins parallel, leading to a ferromagnetic effect. On the other hand, if the singlet state is the ground state, the neighboring spins are antiparallel, leading to an antiferromagnetic effect. We shall deal with antiferromagnetism in the next chapter.

So far we have been concerned with one electron per atom, so that $S=\frac{1}{2}$. Anderson,[14] however, gave treatments where each ion had $S>\frac{1}{2}$. For the whole system of ions we may write the exchange Hamiltonian $\mathcal{H}_{\text{ex}}$ more generally as

$$\mathcal{H}_{\text{ex}} = -\sum_{i \neq j} J_{ij} \mathbf{S}_i \cdot \mathbf{S}_j \tag{13.31}$$

To make the solution of (13.31) quantitative, we must be able to calculate the integral J. A usual simplification is to consider significant overlap only between the nearest neighbors. Most of the calculations have been done for the iron group of metals to try to demonstrate that the Heisenberg model can account for ferromagnetism, but unfortunately none of these efforts has been successful. It will be recalled that, for ferromagnetism to occur, J must be positive, but among the many calculations of J for Fe, Co, and Ni there is no agreement even as to its sign. In general, even in calculations in which J has turned out to be positive, its numerical value is far too small to account for the Curie temperatures that are observed. One of the calculations by Stuart and Marshall,[12] for example, gives a positive J, but its magnitude is 70 times too small. It seems clear that, in the iron group of elements at least, direct exchange between nearest neighbors in a localized model is not responsible for ferromagnetism.

Even in nonmetals, where antiferromagnetic interactions appear to be dominant, calculations suggest that direct exchange cannot always account for the coupling between the ions, because in many cases the distance between the nearest neighbors is too great for there to be appreciable overlap of the wavefunctions. The general ideas of exchange interaction have therefore been extended by the concept of indirect exchange, which was first suggested by Kramers.[13] When applied to nonmetals this concept is sometimes called superexchange. The general principle involved in this mechanism is that the exchange between two adjacent magnetic ions does not occur directly but takes place through the intermediacy of a diamagnetic ion. In the simplest terms, the exchange alignment takes place first between an electron on the magnetic ion and one on the diamagnetic ion, which then interacts with the next magnetic ion, producing a further alignment. The net effect is therefore an exchange coupling between the two magnetic ions. The idea of superexchange has been developed by Anderson[14] in a more general way, which we shall deal with in Chapter 14.3.

The general principle of superexchange has been extended to conductors, following a suggestion by Zener.[46] It is proposed that the interaction between the localised $3d$ or $4f$ electrons on different ions takes place by means of a coupling through the conduction electrons. We shall also discuss this idea later.

13.5 EXCHANGE ENERGY AND MOLECULAR FIELD

In this section we are concerned with the relationship between the exchange energy and the molecular fields. When two electrons are separated by a distance sufficiently large so that their wavefunctions do not overlap, the electrostatic energy is given by the well-known Coulomb formula. If the electrons are now brought closer, so that there is overlapping of the two charge clouds, an extra electrostatic energy appears, which is a consequence of the Pauli principle requirement that the wavefunction representing the two electrons be antisymmetric in coordinates of both spins and orbits. This extra electrostatic energy is the exchange energy, which we discussed earlier. It was also shown that the exchange energy of a pair of electrons depends strongly on the relative orientations of their spin moments. For given orbits, the spin dependence is determined by the expectation value of the scaler product of the spins of the two electrons, that is,

$$E_{\text{ex}} = -2J\langle \mathbf{s}_i \cdot \mathbf{s}_j \rangle \tag{13.32}$$

where $\langle\rangle$ indicates an average over the motion of the electrons, and $\mathbf{s}_i$ and $\mathbf{s}_j$ are the spin moments of the electrons. For two isolated electrons, $\langle \mathbf{s}_i \cdot \mathbf{s}_j \rangle$ can only be equal to $+\frac{1}{4}$ or $-\frac{3}{4}$ for parallel or antiparallel relative orientations, respectively. The magnitude of the exchange integral constant J is determined, in principle, by the nature and degree of the overlap of the wavefunctions of the electrons. It is important to note that the magnitude of J decreases rapidly with decreasing overlap and can, in principle, be positive or negative, as discussed earlier. A form exactly similar to that of (13.32) is obtained when we consider two atoms or ions instead of two electrons, and the resultant exchange energy is now determined by the resultant spin moments of the ions, $\mathbf{S}_i$ and $\mathbf{S}_j$, that is,

$$E_{\text{ex}} = -2\sum_{i \neq j} J_{ij}\langle \mathbf{S}_i \cdot \mathbf{S}_j \rangle \tag{13.33}$$

where J_{ij} is an average exchange integral for the overlapping ions.

In the nearest neighbor (n.n.) approximation, if the ith magnetic ion interacts equally with each of z nearest neighbors, situated at the jth site with spin moments $\mathbf{S}_j$, the total exchange energy is given by

$$E_{\text{ex}}^{z} = -2J\sum_{j}^{z} \langle \mathbf{S}_i \cdot \mathbf{S}_j \rangle \tag{13.34}$$

where the summation is over the z neighbors. If $\mathbf{p}$ is the dipole moment of the z neighbors, then, assuming that only the spin contributes to the magnetic moment, we have

$$\mathbf{p} = g\mu_{\text{B}}\sum_{j}^{z} \mathbf{S}_j \tag{13.35}$$

where g is the Landé g-factor, which is assumed to be the same for all z neighbors. Substituting (13.35) in (13.34), we have

$$E_{\text{ex}}^{z} = -\frac{2J}{g^2\mu_{\text{B}}^2}\langle \boldsymbol{\mu}_i \cdot \mathbf{p} \rangle \tag{13.36}$$

where $\boldsymbol{\mu}_i = g\mu_B \mathbf{S}_i$ is the magnetic moment of the ith ion. The intensity of magnetization $\mathbf{M}$ of a sample containing N magnetic ions per unit volume is given by $\mathbf{M} = n\bar{p}/z$, where $\bar{p}$ is the average of $\mathbf{p}$ over all groups of z ions in unit volume. Assuming no fluctuation of $\mathbf{p}$ from point to point, we may replace $\bar{p}$ by $\mathbf{p}$ and thus have

$$E_{\mathrm{ex}}^{z} = -\frac{2J}{g^2\mu_B^2}\frac{z}{N}\langle \boldsymbol{\mu}_i \cdot \mathbf{M} \rangle \tag{13.37}$$

Comparing this expression with the energy of a dipole of moment $\boldsymbol{\mu}_i$ in a hypothetical molecular field equal to $\gamma\mathbf{M}$, we have ultimately

$$\gamma = \frac{2J}{g^2\mu_B^2} \cdot \frac{z}{N} \tag{13.38}$$

This expresses the relationship of the exchange integral with the molecular field, due to Stoner,[15] who also gives the relations of the Curie temperature T_C and the Curie constant C as follows:

$$\begin{aligned} T_C &= \frac{2JzS(S+1)}{3k} \\ C &= \frac{4N\mu_B^2 S(S+1)}{3k} \end{aligned} \tag{13.39}$$

Stoner's substitution of the average for the instantenous values, which is equivalent to the neglect of any fluctuation among groups of z neighboring ions, from point to point in a sample at any instant, or from time to time at a particular point, obviously represents only an approximation. The exact nature of this approximation is yet unknown, so that, until more accurate methods of calculation are found, the above theory can scarcely be said to possess a rigorous or logical basis. As has been stressed by Neel,[16] the instantaneous molecular field acting on a given atom is by no means the same as the average field, and so the Weiss theory is equivalent, in a certain sense, to supposing that all fluctuations in the molecular field can be neglected. It has been shown by Neel that these fluctuations are responsible for the difference between the paramagnetic and ferromagnetic Curie points, which was discussed earlier. Such marked fluctuations in ferromagnetic as well as antiferromagnetic materials do exist near the Curie and Neel temperatures and, in fact, have been observed by the critical scattering of neutrons. The theory of the statistical behaviors of the assemblies of dipoles interacting through the exchange forces introduces a number of modifications of the MF hypothesis and is also supported by experimental observations. We shall not probe into this aspect, however, simply because no clear picture has yet emerged that can be included in a textbook like the present one.

13.6 THE COLLECTIVE ELECTRON THEORY

This model was devised by Stoner,[7] who first proposed that the 3*d* electrons of the transition metals, instead of being localized on particular nuclei, move relatively freely from ion to ion through the crystal lattice. In this model the freedom of movement of the conduction electrons are in fact exaggerated because the correlations of the motions of the itinerant electrons due to their Coulomb repulsions are neglected; in the Weiss model, on the other hand, the influence of correlation is overemphasized.

When atoms combine to form a solid, the outermost electrons are subjected to forces arising from the neighboring atoms and the result is the formation of chemical bonds. In the metallic type of bond each outermost electron loses its affinity for the particular nucleus it came from and becomes more or less free to move through the crystal lattice of the metal. The free-electron theory really imagines the electrons to move in a field of uniform potential with positive ions distributed uniformly throughout the lattice; the electrostatic attraction between these electrons and the positive ions situated on the lattice holds the metal lattice together. To find the magnetic properties of the electron gas we must first determine the permissible states of motion that such an electron may follow and then distribute the electrons over these states at thermal equilibrium and in the presence of a magnetic field. Each electron moves in the electrostatic field of the positive ions. The potential well in which each electron moves is shown, in one dimension, in Figure 13.2. The free electrons in the metal are the most energetic in the solid, and therefore their kinetic energy may exceed the potential barriers due to the positive ions. Hence, such electrons may be treated as though they are moving in a flat-bottomed potential well bounded by the reflecting walls, which coincide with the surfaces of the sample.

Just as the possible states of an electron moving in the field of a free atomic nucleus are characterized by discrete energies, so are the free electron states, but there is an important difference. In view of the fact that the potential barrier of a metal is very large compared to the atomic dimensions, the energy levels are very dense, so that the distance between them tends to zero as the sample volume tends to infinity. The spectrum is thus practically continuous, and consequently the perturbation energy is always larger than the distance between the levels. In this case one must, strictly speaking, use the perturbation theory for the continuous spectrum.

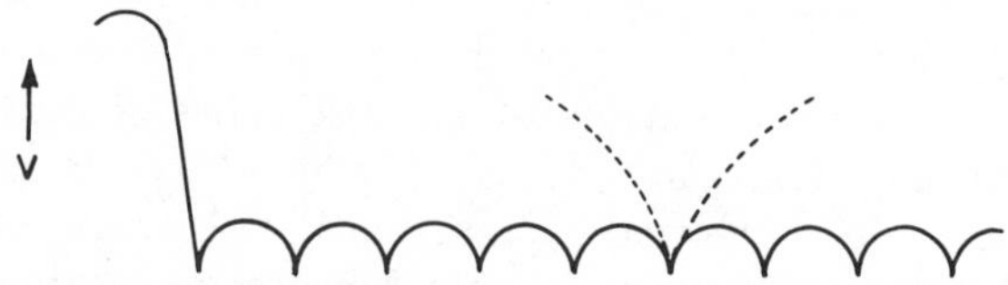

Fig. 13.2 Variation of electric potential inside a crystal. The dotted line represents the potential well due to a single ion.

Before solving the quantum-mechanical equation of motion to find the number of possible states for an electron, let us try to understand the problem of how electrons actually behave in the crystal of a metal. We imagine the electrons to be added to the crystal. For the moment we merely assume that the crystal has some definite structure and do not consider its details. The electrons will then find themselves in a periodic field which is characteristic of the structure concerned. If, for example, we consider a line passing through the centers of the atoms, the potential will vary as shown in Figure 13.2 with a singularity at each atomic nucleus. The potential energy of an electron will therefore diminish as the electron approaches a nucleus. We now have to consider how electrons will behave in a periodic field of this kind.

Let us begin by saying that, in the case of a free electron, we know that the momentum **p**, the velocity **v**, the mass m, and the wavelength λ associated with the electron are connected by the relation

$$\mathbf{p} = m\mathbf{v} = \frac{h}{\lambda} \tag{13.40}$$

where h is Planck's constant. We see, therefore, that, since the momentum is proportional to $1/\lambda$, that is, proportional to the wave number, we could equally well have described the situation in terms of a wave-number diagram instead of a momentum diagram. Actually the mathematician finds it more convenient to describe things in terms, not of the wave number $1/\lambda$, but of $2\pi/\lambda$, and this quantity is called the wave number **K**. Hence we can draw a diagram such as that in Figure 13.3, in which the three axes K_x, K_y, and K_z represent the components of the wave number **K** in the directions of the x-, y-, and z-axes. The state of the electron will be represented by a small cell in this **K**-space diagram. However, when we deal with electrons in the periodic field of a crystal, the wave number **K** is still defined by $\mathbf{K} = 2\pi/\lambda$, but **K** is no longer necessarily proportional to $m\mathbf{v}$—and this is where the theory becomes complicated.

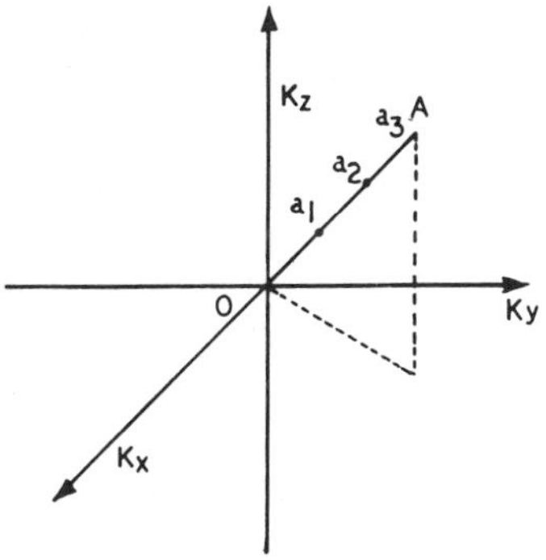

Fig. 13.3 The three-dimensional **K**-space inside a crystal. The three axes K_x, K_y and K_z of **K** in the x-, y-, and z-directions.

Now what happens if we introduce electrons into the crystal? They will naturally fill the lowest states around the origin of Figure 13.3, and the circle that bounds the occupied states will cut OA of the figure at gradually increasing distances from the origin. As we introduce more and more electrons, the region of occupied states will become larger and larger, and the wave numbers **K** of the highest occupied states will become greater and greater and the corresponding wavelengths smaller and smaller. Now let us return to the direction OA in the actual crystal (Figure 13.3), and let us suppose that there is a set of atomic planes in the crystal. Then, in the general case, an electron whose state is described by a wave number on OA (see Figure 13.3) will pass through these planes with only slight scattering. But if the electronic wavelength satisfies the Bragg equation:

$$n\lambda = 2d\sin\theta \tag{13.41}$$

where θ is the angle between the electron beam and the reflecting planes of spacing d, and n is an integer, a strong reflection takes place and the electron cannot travel through the lattice. For each direction OA inside a crystal, there will be a series of critical points in the wave-number diagram at which the wavelength satisfies the condition for a Bragg reflection of the electrons by one of the sets of planes in the lattice, and it can be shown that at each of these critical points there is an abrupt increase in the energy on passing to the next higher wave number. If, for example, a_1 is the first point on OA (Figure 13.3) at which the Bragg condition is satisfied, then, on passing from O to a_1, the energy increases continuously with the wave number (as said earlier, the successive energy states of a metal are so close together that we may regard them as forming a continuum), but at a_1 there is an abrupt increase in the energy. On passing from a_1 to a_2, the energy again increases continuously with the wave number, but at a_2 there is a second sudden increase. Then, if, as in Figure 13.3, a_1 represents a wave number that satisfies the Bragg equation for a reflection from a particular set of planes, it would seem that for different directions of OA the points a_1 must lie at different distances from the origin, because, if one varies θ in (13.41), one has to vary λ and hence **K**.

In the two-dimensional wave-number diagram, for example, the critical wave numbers at which the conditions for a Bragg reflection are satisfied lie on straight lines that form the boundaries of a polygon. For electron states inside this polygon the energy varies continuously with the wave number, but on passing through the boundary of the polygon there is an abrupt increase in the energy. A polygon of this kind is called a two-dimensional Brillouin zone, the shape of which must depend on the crystal structure. In other words, each type of crystal structure gives Brillouin zones of a characteristic type.

When we deal with a proper three-dimensional crystal, the general principle outlined above is exactly the same. For a three-dimensional crystal we construct a three-dimensional wave-number diagram like that of Figure 13.3,

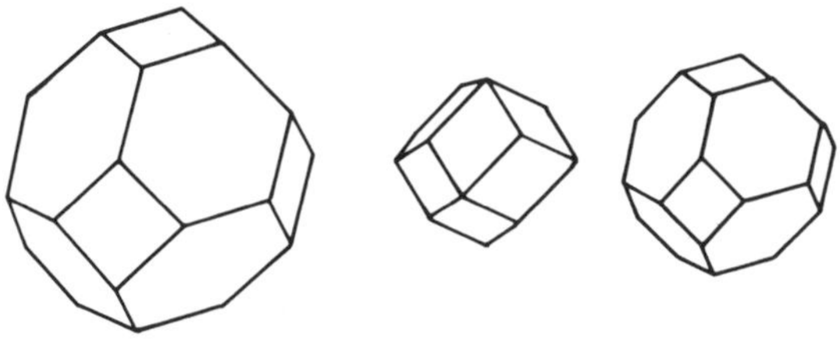

(a) (b) (c)

Fig. 13.4 (*a*) The first Brillouin Zone of a f.c.c. structure. (*b*) The first Brillouin zone of a b.c.c. structure. (*c*) The second Brillouin zone of a b.c.c. structure.

where K_x, K_y, and K_z represent the components of the wave numbers in the directions x, y, and z in the crystal in general, the axes being chosen to be parallel to the crystal axes. Then, for each electron in the three-dimensional **K**-space, on starting from the origin, the energy will increase continuously with the wave number until the critical point is reached at which the wavelength satisfies the condition for a Bragg reflection. These critical points for the different directions then lie on plane surfaces that bound three-dimensional Brillouin zones. Figure 13.4, for example, shows the first Brillouin zone for the face-centered cubic (f.c.c.) structure. The meaning of this diagram is that, for electron states with wave numbers lying within this zone, the energy increases continuously with the wave number, but at the surface of the zone there is an abrupt increase in energy. It must be emphasized that Figure 13.4 is a diagram in **K**-space.

The importance of this idea is that is shows that there is a distinct difference between the energy relations of electrons in the free-electron theory and those in the actual periodic field of a crystal. In the free-electron theory the whole process is continuous. As more and more electrons are introduced, they continually occupy states of increasing energies and the overall process is smooth and regular. In the actual crystal, where there is a periodic field, as increasing numbers of electrons are added, they fill the energy states in the wave-number space, two electrons per state, one with up spin and the other with down spin, but the sudden increase in energy that occurs on crossing the zone boundaries means that the process is no longer smooth and regular as regards the energy.

In filling the state, Pauli's principle still applies, and therefore only two electrons can enter a given state. At absolute zero an assembly of N electrons will then occupy the $N/2$ lowest electron states, each occupied state containing two electrons—one of each spin. It is therefore important to know how the electron states are distributed over the various energy ranges. For this purpose let us treat the free-electron gas in one dimension by using the Pauli principle. Consider an electron of mass m confined to a line of length L by infinite barriers at both ends of the line. To find the energy of the electron we have to solve the Schrödinger equation:

$$\mathcal{H}\psi_n = E_n\psi_n \tag{13.42}$$

where ψ_n is the wavefunction of the electron in state n, and $\mathcal{H}$ is the Hamiltonian. Since the electron is free, we can neglect the potential energy, and hence $\mathcal{H} = \mathbf{p}^2/2m$, where $\mathbf{p}$ is the momentum. Then we have

$$-\frac{\hbar^2}{2m}\frac{d^2\psi_n}{dx^2} = E_n\psi_n \tag{13.43}$$

Since the electron is confined to a line of length L, we can assume an infinite potential energy barrier at either end of the line, and therefore the boundary conditions are

$$\psi_n(0) = 0$$

and

$$\psi_n(L) = 0$$

It is easy to show that these conditions are satisfied if the wavefunction is sine-like, that is,

$$\psi_n = c\sin\left(\frac{n\pi}{L}x\right) \tag{13.44}$$

where c is a constant. Using this wavefunction in Schrödinger equation (13.42), we can easily show that

$$E_n = \frac{\hbar^2}{2m}\left(\frac{n}{2L}\right)^2 \tag{13.45}$$

for one dimension. The constant c must now be chosen in such a way that the electron must be somewhere on the line, meaning that we must have

$$\int_0^L \psi_n^*(x)\psi_n(x) = 1 \tag{13.46}$$

Evaluating this integral, we can easily show that $c = (2/L)^{1/2}$. Therefore the normalized wavefunction of the free electron (in one dimension) is given by

$$\psi_n(x) = \left(\frac{2}{L}\right)^{1/2}\sin\left(\frac{n\pi x}{L}\right) \tag{13.47}$$

What happens if we increase the number of electrons on the line? Before obtaining the answer, we must realize that we now have to use the Pauli principle, which states that no two electrons can have all their quantum numbers identical, that is, each quantum state can be occupied by at most one electron. This principle is applicable to solids as well. In a solid the quantum numbers of a free electron are n and m_s. For a given value of the integer n, m_s can have two values, $\pm\frac{1}{2}$ (in units of $\hbar$), corresponding to two spin orientations. Therefore each energy level, labeled by the quantum number n, can accommodate two electrons, one with up spin and the other with down spin. For example, suppose that there are six electrons; then in the ground state of the system the levels with $n=1,2,3$ are filled and the levels with $n\rangle 3$ are empty.

If we start filling the levels from the bottom ($n=1$) and continue filling the higher levels until all N (suppose that N is an even number) electrons are accommodated, we have the condition

$$2n_F=N \tag{13.48}$$

where n_F denotes the topmost filled energy level. The Fermi energy E_{F} is defined as the energy of the topmost filled level. Putting $n=n_F$ in (13.45), we have

$$E_{\mathrm{F}}=\frac{\hbar^2}{2m}\left(\frac{N}{4L}\right)^2 \tag{13.49}$$

The total energy E_0 of N electrons in the lowest energy state of the system can be found by summing E_n between $n=1$ and $n_F=\frac{1}{2}N$:

$$E_0=2\sum_{n=1}^{n=n/2} E_n=2\frac{\hbar^2}{2m}\left(\frac{1}{2L}\right)^2\sum_{n=1}^{n=N/2} n^2 \tag{13.50}$$

The factor 2 appears because there are two electrons in each level. The summation (13.50) can be shown to be $\simeq\frac{1}{3}(N/2)^3$, provided that $N\gg 1$. Using (13.49), we have therefore

$$E_0\simeq\tfrac{1}{3}NE_{\mathrm{F}} \tag{13.51}$$

We thus conclude that the average kinetic energy of free electrons in the ground state is one third of the Fermi energy—in one dimension, of course.

The density of states $N(E)$ is defined as the number of electronic states per unit energy range. Taking differentials in (13.45), we thus have

$$dE=\frac{\hbar^2}{m}\left(\frac{n}{2L}\right)\frac{dn}{dL} \tag{13.52}$$

Since there are two quantum states for each energy level, one with up spin and one with down spin, for a free-electron gas in one dimension, we have

$$N(E)=2\frac{dn}{dE}=4L\left(\frac{m}{2E}\right)^{1/2}\frac{1}{\hbar} \tag{13.53}$$

where we have used (13.45) to express n in terms of E.

We can easily extend the above procedure for a free-electron gas in three dimensions. In that case our wavefunctions must satisfy the periodic boundary conditions, that is, we now require the wavefunctions to be periodic in x,y,z with periodicity L. Thus

$$\psi(x+L,y,z)=\psi(x,y,z) \tag{13.54}$$

and similarly for the y and z coordinates. It can be easily seen that the wavefunctions satisfying the free-particle Schrödinger equation:

$$-\frac{\hbar^2}{2m}\left(\frac{\partial^2}{\partial x^2}+\frac{\partial^2}{\partial y^2}+\frac{\partial^2}{\partial z^2}\right)\psi_{\mathbf{K}}(\mathbf{r})=E_{\mathbf{K}}\psi_{\mathbf{K}}(\mathbf{r}), \tag{13.55}$$

the normalization condition over the volume $V=L^3$, and the periodicity condition given by (13.54) are of the form

$$\psi_{\mathbf{K}}(\mathbf{r})=\left(\frac{1}{V}\right)^{1/2}e^{i\mathbf{K}\cdot\mathbf{r}} \tag{13.56}$$

provided that the components of the wave vector **K** satisfy

$$K_i=0,\ \pm\frac{2\pi}{L},\ \pm\frac{4\pi}{L},\ \pm\cdots \tag{13.57}$$

where $i=x,y,z$. From (13.57) it is clear that the wave vector **K** is of the form $2n\pi/L$, where n is an integer, positive or negative. Obviously, the components of **K** are the quantum numbers of the problem, along with the quantum number m_s.

On substituting (13.56) in (13.55), we get

$$E_{\mathbf{K}}=\frac{\hbar^2}{2m}\mathbf{K}^2=\frac{\hbar^2}{2m}(K_x^2+K_y^2+K_z^2) \tag{13.58}$$

for the energy eigenvalue $E_{\mathbf{K}}$ of the state with wave vector **K**. The magnitude of the wave vector is related to the wavelength λ by

$$K=\frac{2\pi}{\lambda} \tag{13.59}$$

Since the linear momentum $\mathbf{p}$ may be represented by $\mathbf{p} = -i\hbar\nabla$, we get, for state (13.56),

$$\mathbf{p}\psi_{\mathbf{K}}(\mathbf{r}) = \hbar\mathbf{K}\psi_{\mathbf{K}}(\mathbf{r}) \tag{13.60}$$

so that the plane wave $\psi_{\mathbf{K}}$ is an eigenfunction of the linear momentum with the eigenvalue $\hbar\mathbf{K}$. The particle velocity in state $\mathbf{K}$ is given by

$$m\mathbf{v} = \hbar\mathbf{K} \tag{13.61}$$

In the ground state of a system of free electrons the occupied states may be represented as points inside a sphere in $\mathbf{K}$-space. The energy at the surface of the sphere is the Fermi energy, and the wave vectors at the Fermi surface have a magnitude K_{F} such that

$$E_{\mathrm{F}} = \frac{\hbar}{2m} K_{\mathrm{F}}^2 \tag{13.62}$$

It is clear from (13.57) that there is one allowed wave vector, that is, one distinct triplet of quantum numbers K_x, K_y, K_z for the volume element $(2\pi/L)^3$ of $\mathbf{K}$-space. Thus, in the sphere of volume $4\pi K_{\mathrm{F}}^3/3$, the total allowed number of states is

$$2 \cdot \frac{4\pi K_{\mathrm{F}}^3/3}{(2\pi/L)^3} = \frac{V}{3\pi^2} K_{\mathrm{F}}^3 = N \tag{13.63}$$

where the factor 2 on the left is due to two allowed values of m_s for each allowed value of $\mathbf{K}$. In (13.63) we have set the number of states equal to N, the number of electrons. Thus

$$K_{\mathrm{F}} = \left(\frac{3\pi^2 N}{V}\right)^{1/3} \tag{13.64}$$

It is important to note here that K_{F} depends only on the electron density and not on the mass.

From (13.62) and (13.64) we then have

$$E_{\mathrm{F}} = \frac{\hbar^2}{2m}\left(\frac{3\pi^2 N}{V}\right)^{2/3} \tag{13.65}$$

Equation (13.65) relates the Fermi energy to the electron density N/V and the mass m.

Now $N(E)$, the density of states in three dimensions, can be written as

$$N(E) = \frac{2V}{(2\pi)^3} \int \frac{ds_E}{|\mathrm{grad}_{\mathbf{K}} E|} \tag{13.66}$$

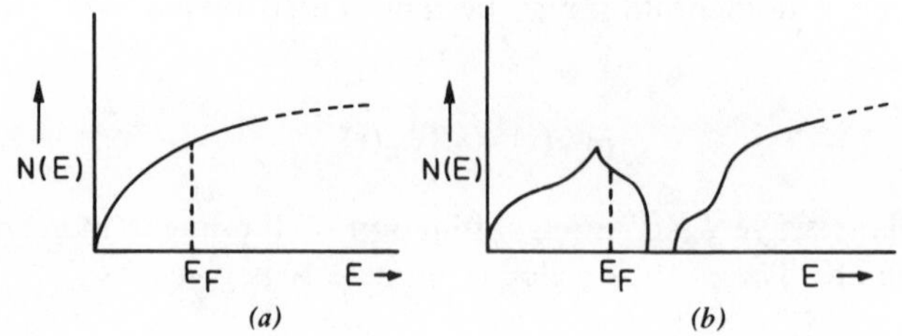

Fig. 13.5 (*a*) Variation of the density of states $N(E)$ with E, for free electrons. (*b*) The same as (*a*) but for electrons in a periodic field.

where V is the volume of the specimen, and ds_E is the element of area in **K**-space of the surface of constant energy E. For free electrons it follows from (13.58) that

$$|\text{grad}_{\mathbf{K}} E| = \frac{\hbar^2 K}{m}$$

and the surface of constant energy E is spherical with area $S_E = 4\pi K^2$. Hence it follows from (13.66) that

$$N(E) = \frac{2V}{(2\pi)^3} \cdot \frac{4\pi K^2}{\hbar^2 K/m} = \frac{V}{\pi^2} \frac{mK}{\hbar^2}$$

$$= \frac{V}{2\pi^2} \left(\frac{2m}{\hbar^2} \right)^{3/2} \cdot E^{1/2} \tag{13.67}$$

where we have used (13.58). Thus, by solving the quantum-mechanical equation of motion for free electrons, we have shown that the number of possible states for an electron having a total energy lying between E and $(E + dE)$ is given by (13.67). Figure 13.5 shows the variation of $N(E)$ with E for free electrons, which is obtained by plotting (13.67).

When we consider the electrons in a periodic field, the $N(E)$ curve will no longer be a simple parabola as it is in the free-electron theory [see (13.67)]. In filling the states, Pauli's principle still applies and only two electrons can enter a given state. Now, as the limit of the occupied states approaches the boundaries of the Brillouin zone, they will become distorted and there will be energy discontinuities at the zone boundary, so that the process is more complicated than in the free-electron theory. For example, the $N(E)$ curve no longer has the simple form of Figure 13.5, which represents $N(E) \propto E^{1/2}$ (a parabolic form), where E_F represents the Fermi energy up to which each state is filled with two electrons. Now what happens when the surface of the occupied states touches the zone boundary? If there is an abrupt increase in energy on crossing the zone boundary, then, when the occupied states reach the surface of the zone, the next electron added will go on filling the states in

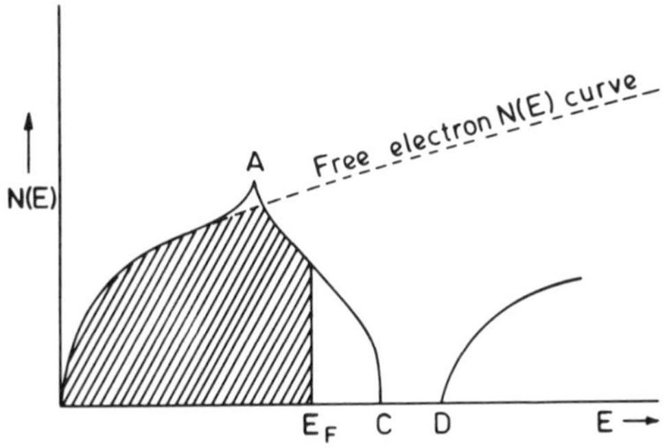

Fig. 13.6 Variation of the density of states with energy for electrons in a periodic field (schematic). All states up to the Fermi level, denoted by E_F, are occupied at low temperatures, as indicated by shading.

the first zone, but the electrons will not begin to enter the states in the second zone until the energy is sufficiently great to exceed the energy gap at the zone boundary. The occupied states for this part of the $N(E)$ curve are shown in Figure 13.6. The abrupt fall in the $N(E)$ curve at A is the result of there being no states, in this range of energy, for certain directions in **K**-space. Figure 13.6 is in a form suitable for nonferromagnetic material for electrons in a periodic field.

To see the effect of the exchange interaction in ferromagnetic metals, we must remember that an assembly of N electrons occupies the $N/2$ lowest energy levels, each occupied level containing two electrons, one of each spin. We might represent the state of affairs by Figure 13.7 *a*, in which the levels occupied by each spin (up and down) are shown separately. Each band is therefore treated as two half-bands, one comprising the states with up spins and the other the states with down spins. In the unmagnetic state (Figure 13.7 *a*), the half-bands will be similar because the same spatial states of motion are involved in both. Now suppose that we place the assembly in a magnetic field of strength H. The energy of each electron whose spin is parallel to the field will then be diminished by an amount $H\mu_B$, where μ_B is the Bohr magneton. Similarly, the energy of each electron with antiparallel spin will be increased by $H\mu_B$, and the state of affairs may be represented by Figure 13.7 *b*. It is clear that in this case the energies of the highest occupied antiparallel levels are greater than those of the highest occupied parallel spin levels, and a lower energy can be produced if some of the antiparallel electrons change their spin. It must be remembered, however, that an occupied level can contain only one electron of a given spin, and so electrons that change their spins will have to go into states above those previously occupied. It is clear, therefore, that a stage will be reached in which the decrease in magnetic energy on changing from antiparallel to parallel spin is balanced by the increase in kinetic energy, and a situation such as that shown in Figure 13.7 *c* will result. In this case there is clearly a preponderance of electrons with parallel spins

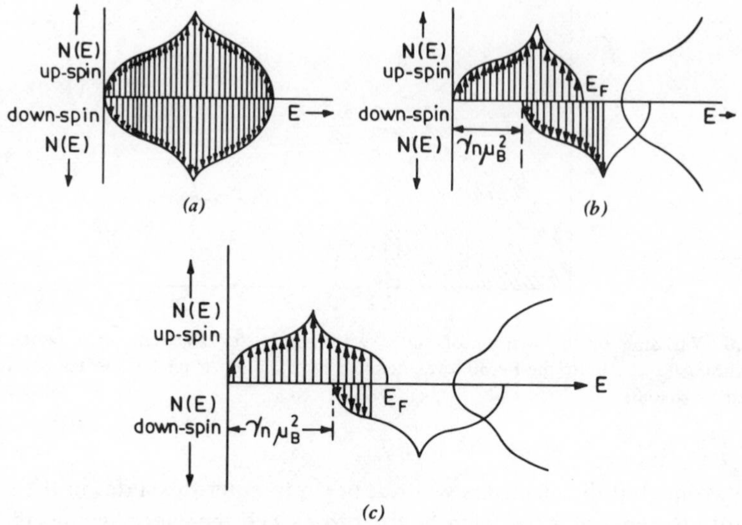

Fig. 13.7 Up-spin (↑) and down-spin (↓) states in a band. (*a*) Nonmagnetic state where the number of up-spin (↑) and down-spin (↓) states are equal. (*b*) Up-spin (↑) and down-spin (↓) states are relatively shifted in energy, and the Fermi energy E_F falls outside one of the half-bands. (*c*) Up-spin (↑) and down-spin (↓) states are relatively shifted, but E_F spans both the half-bands.

and hence a ferromagnetic effect. The relative displacement of the two half-bands, is caused by the exchange interaction, as we shall see below. Figure 13.7 *b* and 13.7 *c* are suitable for a discussion of ferromagnetism.

In terms of the molecular field framework we now recast the concepts stated above into formal equations. If n is the number of excess electrons in the upward band, the molecular field represented by $\gamma \mathbf{M}$ is given by $\gamma n \mu_B$, where γ is the molecular field constant. In solids we always assume, of course, that the orbital momentum of the electrons is quenched by the crystal fields. Now, what happens when electrons are transferred, one by one, from the downward to the upward band in a ferromagnetic sample initially containing equal numbers of electrons of both spin orientations? If we assume that $n/2$ electrons are transferred to the upward band, as a result of which there now are n electrons with parallel spins in the upward band, the exchange energy can be easily shown to be given by

$$E_{ex} = -\tfrac{1}{2}\gamma n \mu_B^2 \tag{13.68}$$

This exchange energy can be included in the density-of-states curves by relatively displacing the two half-bands by (13.68) on the energy axis.

Since, in the density-of-states curves, the levels are filled up to E_F, the Fermi energy, the density of states in one half-band is $\tfrac{1}{2}N(E_F)$. Now the transfer of one electron from the downward to the upward band reduces the exchange energy, using (13.68), by

$$\Delta E_{ex} = -2\gamma\mu_B^2 \tag{13.69}$$

However, by this process the kinetic energy of the electrons in the upward band will be increasing, and the increase can be shown to be given by

$$\Delta E_{KE} = \frac{\Delta E_K}{\Delta N}\Delta N = \frac{2V}{N(E_F)} \tag{13.70}$$

since the transfer will involve depopulation of $\Delta N \equiv V$ states, where V is the volume of the sample. Now, if because of such transfer of electrons from the downward to the upward band there is a decrease in energy, this process will continue, and as a result there will be an excess population in one half-band. If now the situation is such that

$$\Delta E_{ex} > \Delta E_{KE} \tag{13.71}$$

we have a sufficient condition for the spontaneous magnetization that is a direct outcome of the Pauli principle and the MF hypothesis. If somehow condition (13.71) is not fulfilled, there will be no spontaneous magnetization at all. For example, in a broad band with a consequent low density of states, the right-hand side of (13.71) may be greater than the left-hand side, in which case it is impossible to have spontaneous magnetization.

If, however, condition (13.71) is satisfied, there will be a spontaneous transfer of electrons from the downward to the upward band until the minimum energy is attained. Figures 13.7 *b* and 13.7 *c* illustrate the two possible situations. For *strong* exchange forces and *narrow* half-bands, the two half-bands may be shifted relatively to the extent that the upward band becomes completely filled, with the remaining electrons partly filling the downward band (Figure 13.7 *b*). If there is a total of $A/2$ states per unit volume in each half-band, we can show that

$$M_s = (A - N)\mu_B \tag{13.72}$$

at the absolute zero of temperature. Since $(A - N)$ is the number of unoccupied states or "holes" left in the downward band, we see that ferromagnetism arises because of the holes in the $3d$ bands of Fe, Ni, and Co. This was also shown by Stoner in his collective electron theory, which we shall discuss in the next section.

For *intermediate* exchange forces we may have the situation illustrated in Figure 13.7 *c*. In this case the transfer of electrons reaches a limit when the Fermi level in the upward band is at a point of sufficiently low density of states, whereupon further transfer of electrons to the upward band becomes energetically unfavorable. In this situation we have, at $T=0$,

$$M_s = n\mu_B = \frac{V}{2\gamma\mu_B}\left[\frac{1}{N(E_\uparrow)} - \frac{1}{N(E_\downarrow)}\right] \tag{13.73}$$

where $N(E_\uparrow)$ and $N(E_\downarrow)$ are the densities of states at the highest filled level in the upward and downward bands, respectively. Equation (13.73) shows that the equilibrium value of the spontaneous magnetization can be calculated if $N(E_\uparrow)$ and $N(E_\downarrow)$ of a solid are known.

It is now sufficiently clear that the free-electron model, unlike the simple Weiss model discussed earlier, can account for the observed nonintegral moments per atom. This is clearly realized from the fact that although each atom contributes the same number of electrons to the band, the fraction of electrons having upward spin is determined by the shape of the band, which is, in fact, a complicated function of the electronic energy.

To determine the magnetic properties we now have to determine the self-consistent distribution of electrons over the available states. The distribution is governed by Fermi-Dirac statistics, since this statistics takes full account of the Pauli exclusion requirement that no more than one electron may occupy any distinct quantum state. At absolute zero of temperature the Fermi-Dirac distribution takes the simple form that every state below a limiting energy E_F (the Fermi level) is occupied by one electron and every state above E_F is unoccupied; E_F is determined by the density-of-states curve and the total number of electrons present. Now, what happens when the temperature is increased? This is a standard problem in elementary statistical mechanics, and the solution is given by the Fermi-Dirac distribution function. As the temperature increases, more and more electrons become excited into states of higher energy. Within a band of energy of width of the order of kT, on each side of E_F, the distribution of electrons over the states becomes altered from that existing at absolute zero. As a result some energy levels immediately above E_F that were vacant at $T=0$ are occupied, and some levels immediately below E_F that were occupied at $T=0$ fall vacant. According to Fermi-Dirac statistics, the relative probability of a given state i being occupied at any instant of time is

$$F_i = \left[e^{(E_i-\mu)/kT}+1\right]^{-1} \tag{13.74}$$

This is the Fermi-Dirac distribution function.

The quantity μ is a function of temperature and should be chosen for a particular problem in such a way that the total number of particles in the system comes out correctly, that is, equal to N. If the energy levels are E_i, we must have $\Sigma_i F_i = N$ at all temperatures. In the integral form we have,

$$\int_0^\infty F_i N(E)\,dE = N \tag{13.75}$$

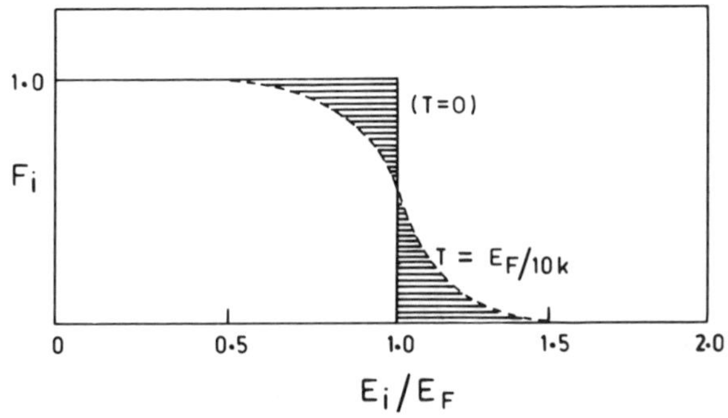

Fig. 13.8 Plot of the Fermi-Dirac distribution function F_i versus E_i/E_F for $T=0$ and for $kT=E_m/10$. The value of F_i gives the fraction of levels at a given energy that are occupied when the system is in thermal equilibrium. When the system is raised in temperature from $T=0$, electrons are transferred from the shaded region at $E_i/E_F<1$ to the shaded region at $E_i/E_F>1$.

at $T=0$, $\mu=E_F$ because, in the limit $T\to 0$, F_i changes discontinuously from the value 1 (filled) to the value 0 (empty) at $E_i=E_F=\mu$. The quantity μ is called the chemical potential, and we see that, at absolute zero, the chemical potential is equal to the Fermi energy. The situation is illustrated in Figure 13.8. At low temperatures μ is close to E_F in value. The high-energy tail of the distribution is the part for which $E_i-\mu \gg kT$, and we have, in this case $F_i \simeq e^{(\mu-E_i)/kT}$, which is the Boltzmann distribution.

Now, if n is the number of excess electrons in the upward band, then, using (13.74), we obtain

$$n=\int_{\text{up}} F_{\uparrow} N(E_{\uparrow})\,dE-\int_{\text{down}} F_{\downarrow} N(E_{\downarrow})\,dE \tag{13.76}$$

where $N(E_{\uparrow})$ and $N(E_{\downarrow})$ are the densities of states in the upward and downward bands, respectively. If the band shapes are known, n can be calculated, and then $M_S(=n\mu_B)$ can be estimated as a function of temperature.

13.7 COLLECTIVE ELECTRON FERROMAGNETISM OF STONER

Stoner's[7] collective electron ferromagnetism is based on a well-defined, clear-cut model that assumes a "standard" energy distribution of the states in a band. It is assumed that the electronic energy levels belong to the entire solid rather than the individual atoms and therefore can conveniently be handled by Fermi-Dirac statistics. In this model one can calculate the magnetization as a function of temperature. Since the forms of the electronic energy bands of different metals are very different, it is impossible to claim any generality of the results applicable to any real ferromagnetic material. In order for the results to be applicable with fair approximation, however, Stoner assumes that the band forms of all the metals approximate to a *parabolic* type over part of its range up to and somewhat beyond the top of the Fermi distribution. In other words, the density of states $N(E)$ is assumed to be of the form $N(E)=aE^{1/2}$, which is supported by the form of the band in nickel, computed by the Wigner-Seitz method.[17]

In discussing Stoner's theory it is better to retain his nomenclature in view of the extensions of his model by Wahlfarth and also by Lidiard, which we shall discuss later in this chapter. The magnetic moment can be calculated from the formula

$$M=-\left(\frac{\partial F}{\partial H}\right)_{T,V} \tag{13.77}$$

where F is the free energy. Using Fermi-Dirac statistics, we can write the free energy as

$$F=N\alpha+\Omega$$

where

$$\begin{aligned}\Omega=&-kT\int_0^\infty N(E)\log\left[1+\frac{\exp\{\alpha-(E-k\theta'\zeta-\mu_B H)\}}{kT}\right]dE\\&-kT\int_0^\infty N(E)\log\left[1+\frac{\exp\{\alpha-(E+k\theta'\zeta+\mu_B H)\}}{kT}\right]dE\end{aligned} \tag{13.78}$$

where $N(E)$ is the density of states in the energy intervals E and $(E+dE)$, and α is the chemical potential discussed earlier. The two integrals in (13.78) correspond to electrons with spins parallel and antiparallel, respectively, to the field **H**. Also, θ' denotes a sort of characteristic temperature that is a measure of the exchange interaction energy, and ζ is the relative magnetization, that is, the ratio of the magnetization to the value that would result if all the spins were perfectly parallel (i.e., $\zeta=M/N\mu_B$). To simplyfy the characteristic function (13.78), we use the following abbreviations:

$$x=\frac{E}{kT},\qquad \mu=\frac{k\theta'\zeta}{kT}=\frac{\theta'\zeta}{T},\qquad \mu'=\frac{\mu_B H}{kT},\qquad \eta=\frac{\alpha}{kT}$$

For bands of the parabolic form we have

$$N(E)=aE^{1/2} \tag{13.79}$$

and it is easy to show that the number of electrons in the band, denoted by N, is given by

$$N=2\int_0^{E_0} aE^{1/2}\,dE=\tfrac{4}{3}aE_0^{3/2} \tag{13.80}$$

where E_0 is the maximum electron energy when the electrons are in balanced pairs in the lowest energy states. The quantity E_0 has the physical significance of being the maximum particle energy at absolute zero. Thus we have

$$N(E)=\frac{3}{4}\left(\frac{N}{E_0^{3/2}}\right)E^{1/2} \tag{13.81}$$

Substituting this expression for $N(E)$ and using the abbreviations given above, we have

$$\begin{aligned}\Omega=-\tfrac{3}{4}NkT\left(\frac{kT}{E_0}\right)^{3/2}\int_0^{\infty}x^{1/2}\big[\log\{1+\exp(\eta-x+\mu+\mu')\}\\ \cdot\{1+\exp(\eta-x-\mu-\mu')\}\,dx\big]\end{aligned} \tag{13.82}$$

This formula is simplified further by using the following relation:

$$\begin{aligned}\int_0^{\infty}x^{1/2}\log(1+e^{\eta-x})\,dx&=\frac{2}{3}\int_0^{\infty}\frac{x^{3/2}\,dx}{e^{x-\eta}+1}\\ &=\tfrac{2}{3}F_{3/2}(\eta)\end{aligned}$$

Thus

$$\Omega=-\tfrac{1}{2}NkT\left(\frac{kT}{E_0}\right)^{3/2}\left[F_{3/2}(\eta+\mu+\mu')+F_{3/2}(\eta-\mu-\mu')\right] \tag{13.83}$$

Making use of the relation

$$F'_{3/2}(\eta)=\frac{d}{d\eta}F_{3/2}(\eta)=\frac{3}{2}\int_0^{\infty}\frac{x^{1/2}\,dx}{e^{x-\eta}+1}$$

we get the expressions for M' and N', given by

$$M'=\tfrac{3}{4}N\mu_B\left(\frac{kT}{E_0}\right)^{3/2}\left[F_{1/2}(\eta+\mu+\mu')-F_{1/2}(\eta-\mu-\mu')\right] \tag{13.84}$$

and

$$\begin{aligned}N'&=-\left(\frac{1}{kT}\right)\left(\frac{\partial\Omega}{\partial\eta}\right)\\ &=\tfrac{3}{4}N\left(\frac{kT}{E_0}\right)^{3/2}\left[F_{1/2}(\eta+\mu+\mu')+F_{1/2}(\eta-\mu-\mu')\right]\end{aligned}$$

These are the two fundamental equations of Stoner's paper, from which the appropriate forms for the various special cases will be derived, as will be seen below. One great advantage in writing the equations in the form of (13.84) is that the function $F_{1/2}(\eta)$ has been tabulated for a wide range of values of the argument by McDougall and Stoner.[18] Since only the function $F_{1/2}(\eta)$ has been used, we may drop the subscript, as well as the argument, and therefore write

$$F_{1/2}(\eta)=F(\eta)=F.$$

In the next stage Stoner deals with the situation above the Curie temperature T_C, where the material becomes paramagnetic. For $T>T_C$ we therefore have the molecular field $\mu=0$, and the relative magnetization is then given by

$$\zeta=\frac{M}{N\mu_B}=\frac{F(\eta+\mu')-F(\eta-\mu')}{F(\eta+\mu')+F(\eta-\mu')} \tag{13.85}$$

where $\mu'=\mu_B H/kT$.

When $(E_0/kT)\rightarrow 0, F(\eta)\rightarrow(\sqrt{\pi}\,/2)e^{\eta}$, and we get the classical result

$$\zeta=\tanh\frac{\mu_B H}{kT} \tag{13.86}$$

When μ is small, we retain only up to the first power of H (in 13.85), and thus have

$$\zeta=\mu'\left(\frac{F'}{F}\right) \tag{13.87}$$

where $F=F_{1/2}(\eta)$ and $F'=(d/d\eta)F_{1/2}(\eta)$, giving the Curie formula (for which $F'/F\rightarrow 1$)

$$\chi=\frac{N\mu_B^2}{kT} \tag{13.88}$$

In the paramagnetic phase, $T<T_C$, it is usual to consider $1/\chi$ as a function of temperature, but in presenting the theoretical results with Fermi-Dirac statistics it is more convenient to consider kT/E_0 as the variable, rather than T, since F and F' (the derivative of F with respect to η) are functions of kT/E_0. We may thus reexpress relation (13.87) in the following form:

$$\frac{1}{\zeta}=\frac{kT}{\mu_B H}\frac{F}{F'}$$

or (13.89)

$$\frac{1}{\zeta}\frac{\mu_B H}{E_0}=\frac{kT}{E_0}\frac{F}{F'}$$

where the quantity on the left-hand side is proportional to $1/\chi$. The theoretical curves corresponding to this relation have been computed by Stoner[19,20] and may be compared with the experimental findings.

Now, in the next stage, what happens when the molecular field is nonzero but small so that the magnetization is proportional to the external field? In this case both μ and μ' may be treated as small, and therefore we have, in place of (13.87),

$$\zeta=(\mu+\mu')\left(\frac{F'}{F}\right) \tag{13.90}$$

where $\mu=\theta'\zeta/T$ and $\mu'=\mu_B H/kT$. Substituting these in (13.90), we have,

$$\zeta=\frac{(\mu_B H/kT)(F'/F)}{1-(\theta' F'/TF)} \tag{13.91}$$

which, in terms of kT/E_0, can be reexpressed as

$$\frac{1}{\zeta}\frac{\mu_B H}{E_0}=\frac{kT}{E_0}\frac{F}{F'}-\frac{k\theta'}{E_0} \tag{13.92}$$

This relation expresses the important fact that the effect of the molecular field is to decrease $1/\chi$ $[=(1/\zeta)(\mu_B H/E_0)]$ by a constant amount $(=k\theta'/E_0)$, which is calculated in the absence of the molecular field. This important relation makes it possible to calculate χ for any value of $k\theta'/E_0$ above the Curie temperature T_C from χ in the absence of the molecular field. It is important to note that in the classical limit (13.92) can be shown to go over to the well-known Curie-Weiss law. At $T=T_C$ the left-hand side of (13.92) becomes zero, so that we have

$$\frac{k\theta'}{kT_C}=\frac{F}{F'} \tag{13.93}$$

We shall now deal with the spontaneous magnetization, for which we may consider $\mu'=0$. For this purpose the two fundamental equations given earlier can be conveniently written as

$$F(\eta+\mu)+F(\eta-\mu)=\frac{4}{3}\left(\frac{E_0}{kT}\right)^{3/2} \tag{13.94}$$

and

$$F(\eta+\mu)-F(\eta-\mu)=\frac{4}{3}\left(\frac{E_0}{kT}\right)^{3/2}\zeta \tag{13.95}$$

Also we write the equation defining μ:

$$\mu=\frac{\theta'}{T}\zeta \tag{13.96}$$

Our ultimate aim with these equations is to obtain ζ as a function of kT/E_0, for a series of given values of $k\theta'/E_0$. From (13.94) and (13.95) it is easy to show that

$$F(\eta+\mu)=\frac{2}{3}\left(\frac{E_0}{kT}\right)^{3/2}(1+\zeta) \tag{13.97}$$

and

$$F(\eta-\mu)=\frac{2}{3}\left(\frac{E_0}{kT}\right)^{3/2}(1-\zeta) \tag{13.98}$$

From these equations $(\eta+\mu)$ and $(\eta-\mu)$ may be easily found by inverse interpolation from the $F(\eta)$ tables given by McDougall and Stoner.[18] If the values of $(\eta+\mu)$ and $(\eta-\mu)$ fall outside the range of the tables, then, for given values of kT/E_0 and ζ, one can use the series expansions given in Ref. 18. Thus, knowing η and μ, one can obtain θ'/T, from which the characteristic temperature $k\theta'/E_0$ can be obtained. The relations outlined above are of great value because they show the dependence of magnetization on temperature resulting from the application of Fermi-Dirac statistics.

The Curie temperature T_C is the temperature at which the spontaneous magnetization vanishes, that is, at $T\to T_C$, $\zeta\to 0$, and hence $\mu\to 0$ also. In this situation (13.94) and (13.95) may be expanded, and we have, from (13.95),

$$\zeta=\frac{\mu F'+(\mu^3/6)F'''+\cdots}{F+(\mu^2/2)F''+\cdots} \tag{13.99}$$

and, in the limit $\zeta\to 0$, $\mu\to 0$,

$$\zeta=\mu\left(\frac{F'}{F}\right)=\left(\frac{\theta'}{T}\right)\left(\frac{F'}{F}\right)\zeta \tag{13.100}$$

Substituting T_C for T in (13.100), we get

$$\frac{\theta'}{T_C}=\frac{F}{F'} \tag{13.101}$$

This relation gives us the ferromagnetic Curie point.

The series forms of (13.101) for high and low temperatures can be obtained using the series expansions given in Stoner's paper.[7] For example, for $kT_C/E_0 > 1$, we have

$$\frac{\theta'}{T_C}=1+\sum_{r=1}^{\infty} b_r'' y^r \tag{13.102}$$

where

$$y=\frac{4}{3\sqrt{\pi}}\left(\frac{E_0}{kT_C}\right)^{3/2}$$

and

$$b_1''=\frac{1}{2\sqrt{2}},\qquad b_2''=-\frac{2}{3\sqrt{3}}+\frac{3}{8},\qquad b_3''=\frac{3}{8}+\frac{5\sqrt{2}}{16}-\frac{\sqrt{2}}{\sqrt{3}}$$

Expression (13.102) shows that, for $E_0/kT_C \to 0$, $\theta' \to T_C$, and it follows that the ratio θ'/T_C increases as kT_C/E_0 decreases.

For $kT_C/E_0 \ll 1$ it has been shown by Stoner that

$$\frac{k\theta'}{E_0}=\tfrac{2}{3}\left[1+\frac{\pi^2}{12}\left(\frac{kT_C}{E_0}\right)^2+\frac{3\pi^4}{80}\left(\frac{kT_C}{E_0}\right)^4+\cdots\right] \tag{13.103}$$

This is an important relation in the sense that it shows that there is a lower limit to $k\theta'/E_0$ for the spontaneous magnetization to occur at any temperature. We therefore have the necessary condition for the occurrence of ferromagnetism (for $kT_C/E_0>0$), which is

$$\frac{k\theta'}{E_0}>\frac{2}{3}$$

Similarly the upper limit for the occurrence of ferromagnetism has been shown to be $2^{-1/3}$. Thus we have the following condition for ferromagnetism:

$$\frac{2}{3}<\frac{k\theta'}{E_0}<\frac{1}{2^{1/3}} \tag{13.104}$$

In addition to these, there are many other useful relations in Stoner's paper, and we consider that it is better to refer the reader to Ref. 7 for the details of the calculations and here to go straight to a discussion of the results.

Figure 13.9 gives the important findings of Stoner, taken from his paper. The abscissa in this figure is the reduced temperature; the left-hand ordinate is the saturation magnetization measured relative to that at $T=0$, and the right-hand ordinate, applicable above the Curie point, is proportional to the reciprocal of the susceptibility. In connection with this figure it is important to note that the magnetization at $T=0$ is not necessarily the same as that corresponding to complete parallel alignment of the spins. Since $\zeta=M/N\mu_B$, complete parallelism would demand that $\zeta_0=1$. In 1939, Bloch[21] pointed out that ferromagnetism is not possible if the band spread is too large compared to the exchange integral. From Stoner's theory the critical value of $k\theta'/E_0$ is $\frac{2}{3}$. If $k\theta'/E_0$ is sufficiently large, $\zeta_0=1$, but in the interval $\frac{2}{3}<k\theta'/E_0<2^{-1/3}$

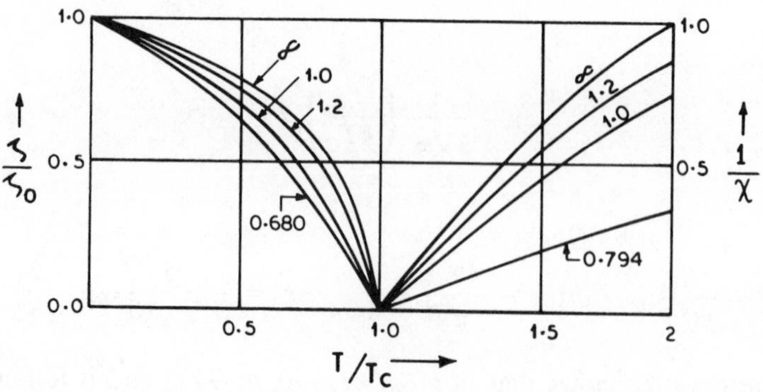

Fig. 13.9 Calculated ζ/ζ_0 below T_C and $1/\chi$ above T_C in units of $(\zeta_0\mu_B/kT_C)$. The various curves are for different values of $k\theta'/E_0$ which are proportional to the ratio of the exchange integral to the Fermi energy. The limiting case, $k\theta'/E_0=\infty$, corresponds to Weiss theory for $S=\frac{1}{2}$. (Taken from Stoner.[7])

the saturation ζ_0 at $T=0$ is between 0 and 1. It follows from Stoner's theory that the saturation magnetization, at absolute zero, need not be an integral number of Bohr magnetons in agreement with experimental observations.

So far, we have developed two important molecular field models—the Weiss and the Stoner. These two models really represent two extremes, while the reality is in between. The Stoner model allows full itinerancy of the magnetic electrons while ignoring the fact that electrons in metals, though moving from atom to atom through the entire crystal, are, in fact, correlated in their motions to a great extent as a result of the Coulomb repulsions between them. The Weiss model, on the other hand, is a localized model in which correlations are overemphasized. Both models have their merits and deficiencies. Some of the properties of ferromagnetics can be adequately understood by the Weiss theory, and others by the Stoner theory. In assessing the relative successes of the two models, it should be borne in mind that the molecular field assumption is itself an approximation to a more complicated statistical approach, quite apart from the question as to whether the magnetic electrons are localized or itinerant.

Let us examine the achievements of these two models. First, we see that both models reproduce the main qualitative features of ferromagnetism. As discussed earlier, the Weiss model, unfortunately, predicts integer values for the moments per atom at absolute zero, whereas we find from experiments $2.2\mu_B$ for iron, $0.6\mu_B$ for nickel, and $1.7\mu_B$ for cobalt. Also, there is no consistency in the apparent values of the moments per ion above and below the Curie temperature. Van Vleck[9] suggested that the introduction of some itinerancy into the Weiss model would make things all right, but no quantitative theory has been proposed so far. We shall discuss Van Vleck's ideas when we deal with the generalized Heisenberg method.

The itinerant electron model, on the other hand, can successfully account for the nonintegral moments per atom and reproduce the simple magnetic properties above and below T_C. Indeed, Wohlfarth[8] has shown that a single choice of values for the molecular field parameters in an itinerant model can nicely account for many of the thermal and magnetic properties of nickel and cobalt and a number of their alloys. Before discussing Wohlfarth's work, it is important to realize that the unresolved problems regarding the magnetic properties of solids arise from the more general difficulty of formulating a consistent theory of correlated itinerant motions.

13.8 EXCHANGE INTERACTION AND COLLECTIVE ELECTRON THEORY

In this section we shall see how the exchange interaction can be incorporated into the collective electron model of Stoner and thereby applied to actual metals. We shall closely follow the treatment of Wohlfarth[22] in this connection. The most suitable approach to the metallic solid, within the framework of the collective electron formalism, is through the treatment of many-electron atoms. To achieve this purpose, Brillouin[23] and others have made use of Hartree's self-consistent formalism, where the initial wavefunction for the metal is taken to be a determinant, the elements of which are the one-electron wavefunctions of the Bloch[24] type. Since the wavefunction is written in a determinantal form, the Pauli principle is automatically satisfied and may be written in the form

$$\Psi=(n!)^{-1/2}\begin{vmatrix}\phi_1(\mathbf{r},\mathbf{s}_1)\cdots\phi_1(\mathbf{r}_N,\mathbf{s}_N)\\ \vdots\\ \phi_n(\mathbf{r}_1,\mathbf{s}_1)\cdots\phi_n(\mathbf{r}_N,\mathbf{s}_N)\end{vmatrix}\tag{13.105}$$

where $\phi_n(\mathbf{r}_i,\mathbf{s}_i)$ is the wavefunction for an electron in the nth state, depending on the spatial coordinate $\mathbf{r}_i$ measured from a fixed origin, assumed to be same for all electrons, and the spin coordinate $\mathbf{s}_i$. Neglecting the spin-orbit coupling, ϕ can be separated into the space part and the spin part, given by

$$\phi_n(\mathbf{r}_i,\mathbf{s}_i)=\psi_n(\mathbf{r}_i)\sigma_n(\mathbf{s}_i)\tag{13.106}$$

Now, we write the space functions in the Bloch form

$$\psi_n(\mathbf{r}_i)=\exp(i\mathbf{K}_n\cdot\mathbf{r}_i)\,U(\mathbf{K}_n,\mathbf{r}_i)\tag{13.107}$$

where U has the periodicity of the lattice, and $\mathbf{K}_n$ is the electronic momentum.

The Hamiltonian for the system is given, as usual, by

$$\mathcal{H}=\sum_n\left\{\left(-\frac{\hbar^2}{2m}\right)\nabla_n^2+V_n(\mathbf{r}_i)\right\}+\frac{1}{2}\sum_{i<j}\frac{e^2}{r_{ij}}\tag{13.108}$$

where $V_n(\mathbf{r}_i)$ is the potential at $\mathbf{r}_i$ of the nuclei, and r_{ij} is the interelectronic distance $|\mathbf{r}_i - \mathbf{r}_j|$. The total energy is then given by

$$E = \int \cdots \int \Psi^* \mathcal{H} \Psi \, d\tau(\cdots \mathbf{r}_i \cdots \mathbf{s}_i \cdots) \tag{13.109}$$

Substituting for the Ψ's in (13.109), we have

$$\begin{aligned} E = & \sum_n \left(-\frac{\hbar^2}{2m}\right) \int \phi_n^*(\mathbf{r}_i, \mathbf{s}_i) \nabla_n^2 \phi_n(\mathbf{r}_i, \mathbf{s}_i) \, d\tau(\mathbf{r}_i, \mathbf{s}_i) \\ & + \sum_n \int \phi_n^*(\mathbf{r}_i, \mathbf{s}_i) V_n(\mathbf{r}_i) \phi_n(\mathbf{r}_i, \mathbf{s}_i) \, d\tau(\mathbf{r}_i, \mathbf{s}_i) \\ & + \frac{1}{2} \sum_{m,n} \int\int \frac{e^2}{r_{ij}} |\phi_n(\mathbf{r}_i, \mathbf{s}_i)|^2 |\phi_m(\mathbf{r}_j, \mathbf{s}_j)|^2 \, d\tau(\mathbf{r}_i, \mathbf{s}_i, \mathbf{r}_j, \mathbf{s}_j) \\ & - \frac{1}{2} \sum_{m,n} \int\int \frac{e^2}{r_{ij}} \phi_n^*(\mathbf{r}_i, \mathbf{s}_i) \phi_m^*(\mathbf{r}_j, \mathbf{s}_j) \phi_m(\mathbf{r}_i, \mathbf{s}_i) \phi_n(\mathbf{r}_j, \mathbf{s}_j) \, d\tau(\mathbf{r}_i, \mathbf{s}_i; \mathbf{r}_j, \mathbf{s}_j) \end{aligned} \tag{13.110}$$

The first term in (13.110) represents the kinetic energy E_1 of the electrons. Using the orthogonality conditions for the spin functions:

$$\int \sigma_n(\mathbf{s}_i) \sigma_n(\mathbf{s}_j) = \delta_{ij} \tag{13.111}$$

we have

$$E_1 = \sum_n \left(-\frac{\hbar^2}{2m}\right) \int \psi_n^*(\mathbf{r}_i) \nabla_n^2 \psi_n(\mathbf{r}_i) \, d\tau \tag{13.112}$$

The second and third terms in (13.110) represent the Coulomb energy due to the nuclei and the electronic charge cloud, respectively. These terms, though important in calculating the cohesive energies, do not contribute to magnetization and hence can be conveniently omitted.

The last term is the most important one for our purpose since it represents the exchange interaction integral J. Its appearance in E_1 is a direct consequence of the use of the determinantal wavefunction (13.105). Using (13.111), we can simplify the expression for J and write it as

$$\begin{aligned} J &= \sum_{m,n} J_{mn} \\ &= -\frac{1}{2} \sum_{m,n} \int\int \frac{e^2}{r_{ij}} \psi_n^*(\mathbf{r}_i) \psi_m^*(\mathbf{r}_j) \psi_m(\mathbf{r}_i) \psi_n(\mathbf{r}_j) \, d\tau \end{aligned} \tag{13.113}$$

This is nonvanishing only when the spins on i and j are parallel. For antiparallel spins J vanishes. Thus the exchange interaction is effective only between electrons of parallel spins. If there are N_1 electrons with up spin and N_2 electrons with down spin in a band, then

$$J=\sum_{m=1}^{N_1}\sum_{n=1}^{N_1} J_{mn}+\sum_{m=1}^{N_2}\sum_{n=1}^{N_2} J_{mn} \tag{13.114}$$

with J_{mn} given by (13.113).

The exchange interaction forces can usually be termed as the correlation forces since they can correlate the positions of the individual electrons[25, 26] that have parallel spins. It is also possible to have some correlation for the antiparallel spins, as shown by Wigner,[27] but this approach presents serious difficulties and is open to criticism. We shall therefore not deal with it.

The values of the relevant energies, given by (13.106), are easy to calculate for perfectly free electrons, for which $U(i\mathbf{K}_n\cdot\mathbf{r}_i)$ in (13.107) is constant, and the normalized one-electron wavefunctions take the form

$$\psi_n(\mathbf{r}_i)=V^{-1/2}\exp(i\mathbf{K}_n\cdot\mathbf{r}_i) \tag{13.115}$$

where V is the volume of the metal. The kinetic energy is given by (13.112), leading to

$$E_1=\sum_n E_n=\sum_n \frac{\hbar^2}{2m}\mathbf{K}^2 \tag{13.116}$$

At absolute zero of temperature, the Fermi surface in the momentum space is spherical[2] with radius $K_0=(3\pi^2N/V)^{1/3}$ in the absence of spontaneous magnetization (when there are as many up-spin electrons as there are down-spin electrons, so that $N_1=N_2$). If there is complete saturation, that is, all the spins are pointing in the same direction at $T=0$, then $K_0=(6\pi^2N/V)^{1/3}$. But when there are more electrons in the up-spin states than in the down-spin states, that is, $N_1>N_2$, there will be a net spontaneous magnetization, given by $N\mu_B\zeta$, where

$$\zeta=\frac{N_1-N_2}{N}=\frac{N_1-N_2}{N_1+N_2}$$

At $T=0$ it can be shown for perfectly free electrons that the kinetic energy (13.116) is given by

$$E_1(\zeta)=\tfrac{3}{10}NE_FF_{5/3}(\zeta) \tag{13.117}$$

where N is the total number of electrons in volume V of the sample and E_F is the zero-point energy (the Fermi energy), in the absence of spontaneous magnetization, given by[2]

$$E_{\mathrm{F}}=\frac{\hbar^2}{2m}\left(\frac{3\pi^2 N}{V}\right)^{2/3}$$

and

$$F_\nu(x)=(1+x)^\nu+(1-x)^\nu-2$$

Here the energies are measured[7] relative to the unmagnetized state, that is, $E_1(0)=0$.

The exchange interaction term, given by (13.113), for free electrons can be written as

$$J_{mn}=-\frac{1}{2}\left(\frac{e}{V}\right)^2\int\int_V (r_{ij})^{-1}\exp\{i(\mathbf{K}_m-\mathbf{K}_n)\cdot(\mathbf{r}_i-\mathbf{r}_j)\}\,d\tau_i\,d\tau_j$$
$$\cdot\rho_{\mathbf{K}_i}\rho_{\mathbf{K}_j}\,d\mathbf{K}_i\,d\mathbf{K}_j$$

where $\rho_{\mathbf{K}}\,d\mathbf{K}$ (equal to $4\pi V\mathbf{K}^2\,d\mathbf{K}$) is the number of states in the interval $d\mathbf{K}$. This integral, first evaluated by Bloch,[24] can be evaluated in a straightforward manner by considering two sets of spins (up and down) and summing over all the occupied states and may conveniently be put in the form

$$J(\zeta)=-\tfrac{3}{8}NE_jF_{4/3}(\zeta) \tag{13.118}$$

where

$$E_j=e^2\left(\frac{3N}{\pi V}\right)^{1/3}$$

Adding (13.117) and (13.118), we get the total energy for the free electrons at $T=0$, given by

$$\frac{E(\zeta)}{NE_0}=\tfrac{3}{10}\left[F_{5/3}(\zeta)-\frac{5E_j}{4E_0}F_{4/3}(\zeta)\right] \tag{13.119}$$

Differentiating (13.119) with respect to ζ, we can show that $E(\zeta)$ has a minimum for $\zeta=0$, so that, for $E_j/E_0<2$, there is no spontaneous magnetization but, for $E_j/E_0\geqslant 2$, the spontaneous magnetization has a maximum at $\zeta=0$ and attains its lowest value at $\zeta=1$. For $E_j/E_0\geqslant 2$, therefore, all the spins are parallel, giving complete saturation at $T=0$.

In the intermediate case we consider N_1 electrons having up spins and N_2 electrons having down spins, so that $N_1>N_2$. In this case the total exchange interaction is strictly proportional to the square of the magnetization (Stoner's assumption). The total energy, at absolute zero, can then be shown to be

$$\frac{E(\zeta)}{NE_0}=\tfrac{3}{10}F_{5/3}(\zeta)-\frac{1}{2}\left(\frac{k\theta'}{E_0}\right)\zeta^2 \tag{13.120}$$

where the second term represents the exchange interaction in the form $-\frac{1}{2}Nk\theta'\zeta^2$, due to Stoner.[7]

Equation (13.120), also due to Stoner, has been applied with considerable success to the interpretation of several magnetic and thermal properties of particular metals and alloys by Wohlfarth.[22, 28]

We now discuss the limitations of the collective electron model and its successes and failures. The formulas given above may be applied approximately to metals like sodium and copper where the outermost s electrons are very nearly free; they also apply to the s electrons of transition metals like nickel and palladium. It can be shown from (13.119) that the condition for ferromagnetism is given by $E_j/E_0>2$, which, after inserting (13.117) and (13.118) and the numerical values for the constants, gives

$$\frac{E_j}{E_0}=0.4604\left(\frac{A}{q\rho}\right)^{1/3}>2 \tag{13.121}$$

where A is the atomic weight, ρ the density, and q the number of electrons per atom. For $4s$ electrons in nickel, $q\simeq 0.6$ and hence $E_j/E_0\simeq 1.03$, showing clearly that the ferromagnetism of nickel cannot be due to the s electrons, which are the conduction electrons.

It has been seen that the ferromagnetic properties of metals and alloys may be ascribed to the "holes" in the $3d$ bands, and since these are, to a large extent, equivalent to electrons in otherwise empty bands, it may be permissible to consider the effects of the exchange interaction between holes in the same way as for the electrons. The main difference between the d and s bands of the transition metals, as shown by Slater[29] and also by Mott,[30] arises from the fact that the d band has a very much higher density of states than the s band. For nickel it has been found that, although the number of holes in the d band and the number of electrons in the s band are equal, the width of the unoccupied d band is 30 times less than that of the occupied s band. This has been inferred from low-temperature specific heat measurements. Since in nickel, the energy distribution of states near the Fermi limit in the d band may be approximately taken to be of the same form as that for the free electrons, we may consider the holes as quasi-free with an effective mass $m^*\simeq 30m$. Using this effective mass, we may write the unoccupied band width in the form

$$E_0=\left(\frac{\pi^2\hbar^2}{2m^*}\right)\left(\frac{3N}{\pi V}\right)^{2/3} \tag{13.122}$$

Now the question arises, How does the effective mass of the d electrons affect the exchange energy? It is reasonable to think that it would have an effect; otherwise the ratio E_j/E_0 would become impossibly large, of the order of 30 for nickel. Wohlfarth therefore suggests considering the wavefunctions of the holes in the d band to be of the form

$$\psi_n(\mathbf{r}_i) = V^{-1/2}\exp\left\{ i\left(\frac{m^*}{m}\right)^{1/2}\mathbf{K}_n\cdot\mathbf{r}_i \right\} \tag{13.123}$$

simply because this leads to the correct expression for E_0 as given in (13.122). With this wavefunction, going through the calculational steps given earlier yields the same exchange energy as in (13.118) but with

$$E_j = \left(\frac{me^2}{m^*}\right)\left(\frac{3N}{\pi V}\right)^{1/3} \tag{13.124}$$

Unfortunately, we again arrive at the same value for E_j/E_0 as is given in (13.121), which is the free-electron value, in spite of this effective mass approximation. But there has been some gain: θ', the characteristic temperature, is now reduced in the ratio m^*/m, giving, for nickel, $\theta' \simeq 700$ °K, which is, at least, of the correct order of magnitude as found from experimental observations and the Curie temperature.

It is clear from the discussion above that the exchange energy for free electrons is too small to produce ferromagnetism. The "quasi-free holes" approximation just described leads to no improvement either. To achieve a considerable amount of exchange energy for a particular metal, it is therefore necessary to consider in detail the actual wavefunctions of the metal, which should be able to distinguish effectively between the d and s electrons, leading in certain cases to ferromagnetism for the d but not for the s electrons. Not only do the proper wavefunctions have to be obtained, but also they should be sensitive enough to be able to distinguish between metals like nickel and palladium whose electronic band structures, as well as exchange energies, are strikingly similar. (From the experimentally observed magnetic properties of Ni-Pd alloys we find that $\theta' \simeq 1.2\times10^3$ °K for nickel and 1.0×10^3 °K for palladium—a very slight difference indeed.) With a view to finding the accurate wavefunctions of a metal we briefly discuss, along with Wohlfarth, some of the difficulties involved in the well-known approximate methods used to describe the behavior of electrons in metals in the collective electron formalism.

The first approximate method is due to Peierls[31] and Sommerfeld and Bethe[32] and is known as the nearly-free-electron approximation. This method, though not suitable for the transition metals, does demonstrate some of the difficulties, particularly those arising from the Brillouin zone structure, which may be of importance for the ferromagnetic alloys.

The one-electron Bloch wavefunctions (13.111) may be written in the form of a Fourier series:

$$\psi_n(\mathbf{r}_i)=\exp(i\mathbf{K}_n\cdot\mathbf{r}_i)\sum_m c_m(\mathbf{K}_n)\exp\left(\frac{2\pi i\mathbf{m}\cdot\mathbf{r}_i}{a}\right) \tag{13.125}$$

and the potential due to the lattice is given by

$$V=\sum_m \mathrm{v}_m \exp\left(\frac{2\pi i\mathbf{m}\cdot\mathbf{r}_i}{a}\right) \tag{13.126}$$

In this approximation V is, of course, very small. Calculation of the energies, using (13.101), by the well-known perturbation methods shows that all the c_m's are small compared with c_0, unless the Bragg reflection condition

$$\mathbf{m}\cdot\mathbf{K}_n=\frac{\pi}{a}|m|^2 \tag{13.127}$$

which indicates the presence of the Brillouin zone boundary, is satisfied. Near the zone boundary, (13.125) may be written as

$$\psi_n(\mathbf{r}_i)=\exp(i\mathbf{K}_n\cdot\mathbf{r}_i)\left\{c_0+c_m\exp\left(\frac{2\pi i\mathbf{m}\cdot\mathbf{r}_i}{a}\right)\right\} \tag{13.128}$$

with the normalization condition

$$c_0^2+c_m^2=\frac{1}{V}$$

It is not easy to determine the coefficients c_0 and c_m, since they are complicated functions of v_m and $\mathbf{K}_n$. However, it is easy to understand that $|c_m/c_0|$ is small if the Bragg condition is not satisfied and $|c_m/c_0|\to 1$ in the neighborhood of the zone boundary.

Considering, for example, two electrons with wavefunctions $\psi_n(\mathbf{r}_i)$ and $\psi_n'(\mathbf{r}_i')$ and using (13.125), we can show in a straightforward manner that the exchange integral can be expressed as

$$J=\sum_{n,n'}J_{nn'}=\left(\frac{V}{8\pi^3}\right)^2\int\int J(\mathbf{K}_n\cdot\mathbf{K}_{n'})\,d\tau_n\,d\tau_{n'} \tag{13.129}$$

where the integration has to be performed over the occupied volume in the momentum space for both the up-spin and down-spin electrons. It is very difficult to evaluate such integrals, especially near a zone boundary or in cases where the Fermi surface has a complicated shape. We shall not deal with these aspects further.

Let us now proceed to the second approximate method, which is known as the tight-binding approximation.[32] Of all the methods, this one is the most important and has most frequently been regarded as giving the closest approximation in many cases, especially for the transition metals.

The wavefunctions in the tight-binding approximation can be written as

$$\psi_i(\mathbf{r}_j) = N^{-1/2} \sum_l \exp(i\mathbf{K}_i \cdot \mathbf{R}_l)\phi(\mathbf{r}_j - \mathbf{R}_l) \tag{13.130}$$

where the ϕ's are the normalized atomic wavefunctions and the summation is over all the lattice points $\mathbf{R}_l$. For d electrons we have to consider the various degenerate m states, to each of which there corresponds a particular atomic wavefunction ϕ. If the atomic level generating the energy bands is degenerate (e.g., the $3d$ level), (13.130) is replaced by a more complicated function.[33] Once this is done, determinantal wavefunctions can be built up from the one-electron functions of this kind, the various occupied states being different in their $\mathbf{K}$ values only if all the electrons concerned occupy a single band based on a particular atomic function ϕ. The Coulomb and the exchange energies can now be calculated for the single-determinantal wavefunctions and may be used to compare the fully magnetized and the unmagnetized states.

Following Wohlfarth,[8] we can expand the Coulomb energies in terms of two-electron integrals (written in terms of atomic functions) and their degree of overlap. The expression is given by

$$\begin{aligned} C_{ij} = N^{-1} \sum_{m,n,p} \int\int (\mathbf{r}_{12})^{-1} \phi^*(\mathbf{r}_1)\phi(\mathbf{r}_1 - \mathbf{R}_m)\phi^*(\mathbf{r}_2 - \mathbf{R}_n)\phi(\mathbf{r}_2 - \mathbf{R}_p)\, d\tau \\ \cdot \exp i\{\mathbf{K}_i \cdot \mathbf{R}_m - \mathbf{K}_j \cdot (\mathbf{R}_n - \mathbf{R}_p)\} \end{aligned} \tag{13.131}$$

Neglecting all the three- and four-center integrals and also the exponentially decreasing integrals for all but the nearest neighbors, we can reduce (13.131) to the following form:

$$\begin{aligned} C_{ij} = N^{-1}\Big\{ I_0 + \sum_l (\mathbf{R}_l) \Big\} + zN^{-1}\{4I_2(\mathbf{R}) \cos \mathbf{K}_i \cdot \mathbf{R} \\ + 2I_3(\mathbf{R}) \cos \mathbf{K}_i \cdot \mathbf{R} \cos \mathbf{K}_j \cdot \mathbf{R}\} \end{aligned} \tag{13.132}$$

where

$$\begin{aligned} I_0 &= \int\int (\mathbf{r}_{12})^{-1} |\phi(\mathbf{r}_1)|^2 |\phi(\mathbf{r}_2)|^2 \, d\tau(\mathbf{r}_1, \mathbf{r}_2) \\ I_1(\mathbf{R}_l) &= \int\int (\mathbf{r}_{12})^{-1} |\phi(\mathbf{r}_1)|^2 |\phi(\mathbf{r}_2 - \mathbf{R}_l)|^2 \, d\tau(\mathbf{r}_1, \mathbf{r}_2) \\ I_2(\mathbf{R}_l) &= \int\int (\mathbf{r}_{12})^{-1} |\phi(\mathbf{r}_1)|^2 \phi(\mathbf{r}_2)\phi^*(\mathbf{r}_2 - \mathbf{R}) \, d\tau(\mathbf{r}_1, \mathbf{r}_2) \\ I_3(\mathbf{R}_l) &= \int\int (\mathbf{r}_{12})^{-1} \phi^*(\mathbf{r}_1)\phi(\mathbf{r}_1 - \mathbf{R})\phi^*(\mathbf{r}_2 - \mathbf{R})\phi(\mathbf{r}_2) \, d\tau(\mathbf{r}_1, \mathbf{r}_2) \end{aligned} \tag{13.133}$$

and z denotes the number of nearest neighbors. The integral I_0 is independent of the interatomic distance and is likely to be the largest. It has the form of a Coulomb interaction between charge clouds of density ρ_i on the same atom i. Also, I_1 represents an interaction between the charge clouds on different atoms i and j, and I_2 and I_3 denote interactions between the charge clouds that depend on the overlap of the functions on different atoms and therefore fall rapidly with increasing separation and are presumed to vanish except for the nearest neighbors. These integrals can, of course, be evaluated if the atomic function ϕ is known. Similarly, the exchange energy is given by

$$J_{ij}=N^{-1}\sum_{m,n,p}(\mathbf{r}_{12})^{-1}\phi^*(\mathbf{r}_1)\phi^*(\mathbf{r}_2-\mathbf{R}_m)\phi(\mathbf{r}_1-\mathbf{R}_n)\phi(\mathbf{r}_2-\mathbf{R}_p)\,d\tau(\mathbf{r}_1,\mathbf{r}_2)$$

$$\cdot\exp i\{\mathbf{K}_i\cdot\mathbf{R}_p-\mathbf{K}_j\cdot(\mathbf{R}_m-\mathbf{R}_n)\} \tag{13.134}$$

and, with the same simplification as before,

$$J_{ij}=N^{-1}\left[I_0+\sum_l I_1(\mathbf{R}_l)\cos\mathbf{K}_i\cdot\mathbf{R}_l\cos\mathbf{K}_j\cdot\mathbf{R}_l\right]$$

$$+zN^{-1}\left[4I_2(\mathbf{R})\cos\mathbf{K}_i\cdot\mathbf{R}+I_3(\mathbf{R})(1+\cos\mathbf{K}_i\cdot\mathbf{R}\cos\mathbf{K}_j\cdot\mathbf{R})\right] \tag{13.135}$$

where R is the distance between the nearest neighbors.

To obtain the Coulomb and the exchange energies, C_{ij} and J_{ij} are summed over the electron pairs, C_{ij} over all pairs but J_{ij} only over pairs having parallel spins. We now sum over $\mathbf{K}_i$ and $\mathbf{K}_j$. Let us discuss separately two cases, $\zeta=1$ and $\zeta=0$, which represent complete magnetization and no magnetization, respectively.

For $\zeta=1$, all the electron spins are parallel and every state in the Brillouin zone is singly occupied. For this case, if $\mathbf{R}_l$ is a lattice vector, we have

$$\sum_{\text{zone}}\cos(\mathbf{K}\cdot\mathbf{R}_l)=\begin{cases}0 & \text{when } \mathbf{R}_l\neq 0\\ N_e & \text{when } \mathbf{R}_l=0\end{cases} \tag{13.136}$$

where N_e is the number of electrons. It may be noted that $\Sigma_{i,j}I_0$ in C is exactly canceled here by the term $\Sigma_{i,j}I_0\delta_{ij}$ in J. Taking sums over i and j, we now have, for $\zeta=1$,

$$C=\frac{e^2}{2}\sum_{i,j}C_{ij}=\frac{e^2}{2N}(N^2I_0+N^2I_1)$$

$$=\frac{Ne^2}{2}(I_0+I_1) \tag{13.137}$$

$$J=\frac{Ne^2}{2}I_0$$

since there are N^2 pairs of electrons with parallel spins. Similarly, for $\zeta=0$, there being only $2(N/2)^2$ pairs with parallel spins, we have

$$C=\frac{e^2}{2}\sum_{i,j} C_{ij}=\frac{Ne^2}{2}(I_0+I_1)$$

$$J=\frac{e^2}{2}\sum_{i,j} J_{ij}=\frac{Ne^2}{4}I_0 \tag{13.138}$$

Neglecting the integrals, I_3 and I_4, with the assumption that they are small, we have the interaction (magnetic) energies for $\zeta=1$, given by

$$E_M=C-J=\frac{Ne^2}{2}I_1 \tag{13.139}$$

and, for $\zeta=0$,

$$E_M=\frac{Ne^2}{2}\left(\tfrac{1}{2}I_0+I_1\right)$$

Let us now discuss the integrals that are given in (13.133). The form of the integral I_0 shows that the anomalous molecular field arises from the interaction of electrons close to the atom. This anomaly results from the overemphasis on the ionized states in the Bloch approximation. So far, we have neglected the **R**-dependent terms in the Coulomb and exchange energies, C_{ij} and J_{ij}. To include them, it is strictly necessary to calculate first the shape of the occupied part of the Brillouin zone. Assuming this to be spherical with radius k_0, we have

$$\sum_{\text{sphere}} \cos(\mathbf{K}\cdot\mathbf{R})=\frac{V_0}{8\pi^3}\int \cos(\mathbf{K}\cdot\mathbf{R})\,d\tau(\mathbf{K})$$

$$=\frac{Ne}{2}f(k_0R) \tag{13.140}$$

where

$$f(x)=\frac{3\sin x-x\cos x}{x^3}$$

$$=3\left(\frac{\pi}{2}\right)^{1/2}x^{-3/2}J_{3/2}(x) \tag{13.141}$$

V_0 is the volume, and N_e is the number of electrons. The function $f(x)\rightarrow 1$ as $x\rightarrow 0$, and converges rapidly as x increases. Hence the **R**-dependent terms in C and J have the general form

$$\frac{ne^2N_e^2I(R)f^p(k_0R)}{N} \tag{13.142}$$

where n is a numerical factor of the order of unity, $I(R)$ is one of the overlap integrals in (13.133), and $p=0$, 1, or 2, depending on whether the term that has been integrated is constant, depends on either $\mathbf{K}_i$ or $\mathbf{K}_j$, or depends on both. We shall not go into the details of evaluating the $\mathbf{R}$-dependent terms.

We have limited our discussion to the two extreme cases, $\zeta=1$ or $\zeta=0$, since only then can we represent the total wavefunction by a single Slater determinant. It has been shown by Wohlfarth that, after some manipulation, the total energy, as a function of magnetization $E(\zeta)$ involving the integrals $I_2(R)$ and $I_3(R)$, is given by

$$-\tfrac{1}{2}N_e k\theta'\zeta^2\left(1+\tfrac{1}{2}A\zeta^2+\cdots\right)-\left(\frac{ze^2}{4N}\right)N_e^2 I_3(R)\zeta^2 \qquad (13.143)$$

where

$$k\theta'=\tfrac{1}{2}zqe^2 I_2(R)$$

and

$$A\simeq\left(\frac{2}{243}\right)\left(\frac{3\pi^2 qNR^3}{V}\right)^{2/3}$$

corresponding to an effective θ', which increases with magnetization. For nickel, q, the number of electrons per atom (N_e/N), equals 0.6 therefore $A\simeq0.07$.

In conclusion, without going into further detail, we might say that Wohlfarth's collective electron theory of ferromagnetism on the basis of the tight-binding approximation is indeed applicable to real metals and can explain satisfactorily many features of ferromagnetism. It has been suggested from the theory that the magnetic carriers are best described by the tight-binding wavefunctions. These allow, in principle, calculation of the band shape and band width determined by the overlap integrals. The exchange energy $J(\zeta)$ is, however, not strictly proportional to the square of the relative magnetization (Stoner's assumption), $J(\zeta)$ being more generally given by relations of the form (13.143). The importance of correlation correction has been discussed, and it has been shown how this requirement can be satisfied in the absence of a rigorous treatment. The appearance of ferromagnetism is thus closely related to the band structure of the metal and to the magnitude of the interatomic distance.

13.9 SPIN DEGENERACY AND COLLECTIVE ELECTRON FERROMAGNETISM

The collective electron theory of ferromagnetism developed by Stoner and Wohlfarth assumed that the exchange energy responsible for magnetization is adequately represented as a sum of the exchange integrals between pairs of electrons with mutually parallel spins. This is equivalent to using, for the total wavefunction, a determinant built up of one-electron functions. However, for the magnetized states in ferromagnetism, a single-determinantal wavefunction must be regarded as inadequate on account of the so-called spin or exchange degeneracy. This requires us to use the appropriate linear combinations of the degenerate determinants. Here we present a model, due to Lidiard,[34] which shows that the results of neglecting this spin degeneracy and of taking it into account are identical. One great advantage of this model is that it is exactly solvable and it derives the equations for the free energy, magnetization, and so on. This approach may be regarded as a generalization of Stoner's theory of collecting electron ferromagnetism, which is applicable to a system with two energy bands.

As remarked above, it was assumed earlier that the spin-dependent energy of a state can be written as a sum of the exchange integrals for pairs of electrons having parallel spins. This means that the whole system can be described by a wavefunction that is a single determinant, built up of one-electron functions, each one-electron function being simply the product of the orbital part and the spin part. The orbitals may be assumed to be localized on particular atoms, as in the Heisenberg theory, or nonlocalized (itinerant), as in the Stoner theory. When a given state contains one electron in some orbitals, there will be not one but a number of degenerate unperturbed determinantal functions corresponding to the different ways of assigning the spin functions (α or β) to these singly occupied orbitals. This degeneracy is known as the spin degeneracy. Hence the correct wavefunction for calculating the perturbation energy due to the electron interactions should be a linear combination of the degenerate determinantal functions. The general method of calculating the coefficients and the perturbation energy is outlined in Condon and Shortley.[35] It was shown by Dirac[36] that this problem is equivalent to a similar degeneracy problem in which the perturbation part of the Hamiltonian is given by $-2\Sigma_{i<j}J_{ij}\mathbf{S}_i\cdot\mathbf{S}_j$ and is thus a sum of terms, each proportional to the scalar product of the spin vectors of two electrons in singly occupied orbitals. This method of Dirac is very convenient for solving exchange degeneracy problems and therefore will be employed here.

In principle, one can calculate the magnetic and the thermodynamic properties of a system if the partition function can be evaluated exactly. But for large systems such as those arising in the theory of ferromagnetism it is necessary to resort to some approximations in order to evaluate the partition function. At sufficiently high temperatures, one is justified in neglecting the spin degeneracy because in this case the many-determinantal partition function may be shown to reduce to a single-determinantal function. This observation is due to Bloch.[37] This model, due to Lidiard,[34] is specified by the requirement that the available orbitals be divided into two groups such that the exchange integral between any two orbitals depends only on which group they belong to.

Let us suppose that the orbitals of the d electrons, denoted by $\phi_j(\mathbf{r})$, form a set of localized orthonormal functions, and $V(\mathbf{r})$ denotes the potential field due to the inner shell electrons and the nuclei. We also assume that $\phi_j(\mathbf{r})$ are the tight-binding wavefunctions of the Bloch form. The energy of $\phi_j(\mathbf{r})$, denoted by ϵ_j, is given by

$$\epsilon_j = \langle \phi_j(\mathbf{r}) | \mathcal{H} | \phi_j(\mathbf{r}) \rangle \tag{13.144}$$

where

$$\mathcal{H} = \frac{\hbar^2}{2m} \nabla^2 + V(\mathbf{r})$$

This energy ϵ_j is the sum of the kinetic and potential energies. The Coulomb and the exchange integrals, C_{ij} and J_{ij}, are, as usual, written as

$$C_{ij} = \langle \phi_i(\mathbf{r}_1)\phi_j(\mathbf{r}_2) \left| \frac{e^2}{\mathbf{r}_{12}} \right| \phi_i(\mathbf{r}_1)\phi_j(\mathbf{r}_2) \rangle \tag{13.145}$$

and

$$J_{ij} = \langle \phi_i(\mathbf{r}_1)\phi_j(\mathbf{r}_2) \left| \frac{e^2}{\mathbf{r}_{12}} \right| \phi_j(\mathbf{r}_1)\phi_i(\mathbf{r}_2) \rangle \tag{13.146}$$

We have already seen that a state of maximum or minimum multiplicity can be represented accurately by a single determinant in which only electrons with parallel spins contribute to the exchange energy. We now examine how exchange energies depend on the spin configurations for more general states. As discussed in the introduction of this section, for more general states the wavefunction will be a sum of many-determinantal wavefunctions. If we consider only the spin degeneracy, select the same N spatial one-electron functions for each determinant, and permute only the spin functions in going from one determinant to the next, then obviously there will be 2^N determinants which could be included in each sum representing a discrete state of the system. The total wavefunction Ψ for such a general state can be written as

$$\Psi = a_1\psi_1 + a_2\psi_2 + \cdots + a_{2N}\psi_{2N} \tag{13.147}$$

where the coefficients a_p have to be found by perturbation calculations. The total energy of the state described by Ψ is given by

$$E = \langle \Psi | \mathcal{H}_I | \Psi \rangle = \sum_{p,q} a_p^* a_q \langle \Psi_p | \mathcal{H}_I | \Psi_q \rangle \tag{13.148}$$

Now each determinant can be written as

$$\Psi_p = (N!)^{-1/2} \sum_{\mu} (-1)^{\mu} P_{\mu} \Psi_x^p$$

where Ψ_x^p is a simple product of the N one-electron functions, differing from other products (e.g., Ψ_x^q) by a permutation of the spin functions. Substituting for Ψ_p and Ψ_q the above expressions, we see that the total energy E is a sum of terms of the form

$$\langle\Psi_p|\mathcal{H}_I|\Psi_q\rangle = \frac{1}{N!}\sum_{\mu,\nu}(-1)^{\mu}(-1)^{\nu}\langle(P_\nu\Psi_x^q)|\mathcal{H}_I|(P_\mu\Psi_x^p)\rangle \quad (13.149)$$

where

$$\mathcal{H}_I = \sum_{i<j}\frac{e^2}{\mathbf{r}_{ij}} = \mathcal{H}_{ij}$$

so that the right-hand side of (13.149) can be written as

$$\frac{1}{N!}\sum_{i<j}\sum_{\mu,\nu}\langle(-1)^{\nu}(P_\nu\Psi_x^q)|\mathcal{H}_{ij}|(-1)^{\mu}(P_\mu\Psi_x^p)\rangle$$

Now, expanding each product function into spatial and spin parts with P^0 and P^σ as permutation operators for the spatial and spin coordinates, respectively, we get

$$\frac{1}{N!}\sum_{i,j}\sum_{\mu,\nu}(-1)^{\mu+\nu}\langle P_\nu^r O_x^q|\mathcal{H}_{ij}|P_\mu^r O_x^p\rangle\sum_\sigma P_\nu^\sigma \chi_x^{q*} P_\mu^\sigma \chi_x^p$$

If all the one-electron spatial functions forming O_x are different and orthogonal, the above term will be zero except when $P_\nu^r O_x$ differs from $P_\mu^r O_x$ only in the interchange of electrons i and j, on whose coordinates $\mathcal{H}_{ij}$ operates. This requirement reduces the double sum over μ and ν to a single one over μ, say, because we can write $P_{ij}P_\mu$ instead of P_ν, where P_{ij} denotes the interchange of the ith and the jth electrons. Hence we have

$$\frac{1}{N!}\sum_{i,j}\sum_{\mu}(-1)\langle P_\mu^r O_x|\mathcal{H}_{ij}|P_{ij}^r P_\mu^r O_x\rangle$$

$$\cdot\sum_\sigma (P_\mu^\sigma \chi_x^p)(P_{ij}^\sigma P_\mu^\sigma \chi_x^{q*})$$

Now, by evaluating the sum over μ, which gives a factor $N!$, and also by separating in the integral the one-electron spatial functions of all the electrons except the ith and the jth and using their normalization, we can reduce the above expression to

$$-\sum_{i,j}\langle u_i(i)u_j(j)|\mathcal{H}_{ij}|u_j(i)u_i(j)\rangle\sum_\sigma \chi_x^{q*}P_{ij}^\sigma\chi_x^p$$

where the ith and jth electrons are assigned to the spatial states u_i and u_j, respectively, in the product O_x. The bracket above, denoted by $\langle\ \rangle$, is merely a number when evaluated and is, in fact, the exchange integral J_{ij}, so that the term reduces to

$$-\sum_{i,j}\sum_{\sigma}\chi_x^{q*}(J_{ij}P_{ij}^{\sigma})\chi_x^{p}$$

and therefore the total exchange energy, from (13.148), is

$$E_{ex}=-\sum_{p,q}a_p^*a_q\sum_{i,j,\sigma}\chi_x^{q*}(J_{ij}P_{ij}^{\sigma})\chi_x^{p} \tag{13.150}$$

It is to be noted that in the equations given above we have not included the Coulomb part of the energy, which is independent of the spin orientations.

However, to calculate the coefficients denoted by the a's, we have to minimize the total energy (Coulomb plus exchange) with respect to variations of the a's, or, equivalently, a perturbation calculation has to be made. Calculations of this kind for the electrons in atoms and some simple molecules have been performed.[38] In 1926 Dirac[39] showed, however, that there is a much more elegant way of finding the energies in the presence of spin degeneracy that provides a much simpler physical picture. For this reason Dirac's approach is normally adopted in magnetism. We simply write the Dirac operator, which is of the form

$$\mathcal{H}_D=-\frac{1}{2}\sum_{i,j}J_{ij}(1+4\mathbf{s}_i\cdot\mathbf{s}_j) \tag{13.151}$$

Including the Coulomb energy with Dirac's exchange energy (13.151), we can replace the perturbation term in the Hamiltonian by the following operator:[40]

$$\begin{aligned}V=&\sum_{i<j}\left[C_{ij}-\tfrac{1}{2}(1+4\mathbf{s}_i\cdot\mathbf{s}_j)J_{ij}\right]\\&+\sum_{j,\mu}(2C_{j\mu}-J_{j\mu})+\sum_{\mu<\nu}(4C_{\mu\nu}-2J_{\mu\nu})\\&+\sum_{\mu}C_{\mu\mu}-2\mu_B H\sum_i s_{iz}\end{aligned} \tag{13.152}$$

where Greek letters are used to denote the doubly filled orbitals and Roman letters for singly filled orbitals, $\mathbf{s}_i$ is the spin operator for the electron in the ith orbital, and s_{iz} is the z-component of this operator. The Bohr magneton is denoted by μ_B, as usual. The last term in (13.152) represents the effect of the external magnetic field and has been incorporated into the perturbation operator V. Equation (13.152) has been written in a very convenient form for solving the secular problem arising from the spin degeneracy. The possible energies for the various spin states of a particular configuration are the eigenvalues of the matrix of V.

We now divide the whole set of available orbitals into two groups, A and B, such that all pairs of A orbitals have a common exchange integral J_A, all pairs of B orbitals have common J_B, and all pairs with one orbital from A and one from B have the common exchange integral J_{AB}. In this model the spin-dependent part of V is given by

$$\begin{aligned} -2\sum_{i<j} \mathbf{s}_i\cdot\mathbf{s}_j J_{ij} &= -2J_A \sum_{A_i<A_j} \mathbf{s}_{A_i}\cdot\mathbf{s}_{A_j} - 2J_B \sum_{B_i<B_j} \mathbf{s}_{B_i}\cdot\mathbf{s}_{B_j} \\ &\quad -2J_{AB}\sum_{A_i,B_j} \mathbf{s}_{A_i}\cdot\mathbf{s}_{B_j} \\ &= -J_A\Big(\mathbf{s}_A^2 - \sum \mathbf{s}_{A_i}^2\Big) - J_B\Big(\mathbf{s}_B^2 - \sum \mathbf{s}_{B_i}^2\Big) \\ &\quad -2J_{AB}\mathbf{s}_A\cdot\mathbf{s}_B \end{aligned} \tag{13.153}$$

where

$$\mathbf{s}_A = \sum_{A_i} \mathbf{s}_{A_i} \quad \text{and} \quad \mathbf{s}_B = \sum_{B_i} \mathbf{s}_{B_i} \tag{13.154}$$

Since

$$2\mathbf{s}_A\cdot\mathbf{s}_B = \mathbf{s}^2 - \mathbf{s}_A^2 - \mathbf{s}_B^2$$

we see that the secular matrix of V has the eigenvalues

$$\begin{aligned} V &= \sum_{i<j} C_{ij} + \sum_{j,\mu} 2C_{j\mu} + \sum_{\mu<\nu} 4C_{\mu\nu} + \sum_{\mu} C_{\mu\mu} \\ &\quad -\tfrac{1}{4}(J_A N_A^2 + J_B N_B^2 + 2J_{AB}N_A N_B) \\ &\quad + S_A^2(J_{AB}-J_A) + S_B^2(J_{AB}-J_B) - S^2 J_{AB} \\ &\quad -2\mu_B H S_z \\ &= V_0 - 2\mu_B H S_z \end{aligned} \tag{13.155}$$

where N_A and N_B are the total numbers of electrons in A and B orbitals, respectively. Since N_A, N_B, S_A, S_B, and S are very large, of the order of the number of atoms in a magnetic domain, the difference between $S(S+1)$ and S^2, $N_A(N_A-1)$ and N_A^2, and so on, has been ignored in the derivation of (13.155). For a given configuration the number of spin states, with given values of S_A, S_B, S, and S_z, can be obtained by using the branching rule.[40] This number, which is denoted by $g(n, S_A : mS_B)$, is given by

$$g(n,S_A:mS_B) = \left[\binom{n}{\frac{1}{2}n+S_A} - \binom{n}{\frac{1}{2}n+S_A+1}\right]\left[\binom{m}{\frac{1}{2}m+S_B} - \binom{m}{\frac{1}{2}m+S_B+1}\right] \tag{13.156}$$

where

$$\binom{n}{m} = {}^nC_m$$

Now the total energy E of each of the states is obtained by adding to V of (13.155) the sum of the orbital energies for the chosen configuration, $C(n, N_A : mN_B)$. It is now easy to write the partition function Z, from which all the thermodynamic properties of the system must follow:

$$Z = \sum_{m=2S_B}^{N_B} \sum_{n=2S_A}^{N_A} \sum_{c} g(n, S_A : mS_B) e^{-E/kT} \tag{13.157}$$

After some lengthy manipulation we finally get

$$\begin{aligned} \log Z = {} & 2(S_A + S_B)\frac{\mu_B H}{kT} - \frac{V_0}{kT} \\ & + \log\left\{ P_A\left(\tfrac{1}{2}N_A + S_A\right) P_A\left(\tfrac{1}{2}N_A - S_A\right) P_B\left(\tfrac{1}{2}N_B + S_B\right) P_B\left(\tfrac{1}{2}N_B - S_B\right) \right\} \\ & + \log\left\{ 1 - \frac{\lambda_B\left(\tfrac{1}{2}N_B - S_B\right)}{\lambda_B\left(\tfrac{1}{2}N_B + S_B\right)} \right\}\left\{ 1 - \frac{\lambda_A\left(\tfrac{1}{2}N_A - S_A\right)}{\lambda_A\left(\tfrac{1}{2}N_A + S_A\right)} \right\} \end{aligned} \tag{13.158}$$

Here P_A and P_B are the simple Fermi-Dirac partition functions, and the parameter λ is given by

$$\frac{P(\nu)}{P(\nu+1)} = \lambda(\nu)$$

where ν is the number of particles. The last term in (13.158) is vanishingly small except when S_A and S_B tend to zero as at the Curie point. Even then, it is only of the order of $\log N$ and so may be neglected since $\log Z$ must be proportional to N. Except for the last term, (13.158) is just the partition function in the absence of spin degeneracy. From the partition function one obtains the free energy in the usual way. The details of the free energy are given in Lidiard's paper.[34]

Let us now discuss the application of Lidiard's theory to Fe, Co, and Ni. Identifying the A and B groups with $3d$ and $4s$ bands in these three metals and neglecting the B orbitals as being of lesser importance than A, we can write the expression for the magnetic moment as

$$\begin{aligned} \frac{M}{\mu_B} = 2\bar{S}_A = {} & \int_A \eta_A(\epsilon)\left[\exp\frac{1}{kT}\left(\epsilon - \eta_A - J_A\bar{S}_A - \mu_B H\right) + 1\right]^{-1} d\epsilon \\ & - \int_A \eta_A(\epsilon)\left[\exp\frac{1}{kT}\left(\epsilon - \eta_A + J_A\bar{S}_A + \mu_B H\right) + 1\right]^{-1} d\epsilon \end{aligned} \tag{13.159}$$

This equation has been worked out by Stoner[7] for $\eta_A(\epsilon) \sim \epsilon^{1/2}$ and by Wohlfarth[8] for $\eta_A(\epsilon)$ independent of ϵ. In both cases, $N_A J_A/2$ is identified with their parameter $k\theta'$. Lidiard's theory, given above, can also be equally well applied to palladium (platinum) by identifying A and B with $4d$ ($5d$) and $5s$ ($6s$) bands.

13.10 FERROMAGNETISM AND BAND THEORY

This section is devoted to a general critical discussion of the different phenomenological models proposed by Van Vleck, Stoner and Wohlfarth, and Slater and Zener to elucidate the phenomenon of ferromagnetism and its relationship to the exchange mechanism. It is clear from the preceding sections that the exchange forces between electrons provide the coupling between the elementary magnets that is a prerequisite to ferromagnetism. This is well represented by the collective electron ferromagnetism of Stoner, Wohlfarth, and Lidiard. It is fairly evident that accurate solution of the eigenvalue problems for a solid is so complex that it is necessary to resort to some kind of simplifying model for which at least approximate calculations can be made. Some of the important models are as follows:

1. Heisenberg's theory.
2. Collective electron theory of Stoner, Wohlfarth, and Lidiard.
3. Generalized Heisenberg model of Hurwitz and Van Vleck.
4. Band theory and ferromagnetism of Slater.
5. Zener's theory.

Of these five models, we have already discussed the first two in great detail, so it is not necessary to deal with them any further.

We shall now try to appreciate the generalized Heisenberg model of Hurwitz[41] and Van Vleck,[9] which assumes a nonintegral number of spins per atom, with the spins continually being redistributed among the different lattice sites. In this model the motions of the electrons are so constrained that the states corresponding to higher degrees of ionization are excluded. We have already seen that Heisenberg's theory does not offer any explanation of the nonintegral number of spins per atom that is observed experimentally. Let us, for example, take nickel. It has 0.6 electron per atom in the $4s$ band and 9.4 electrons in $3d$ states, so that its configuration can be written as $3d^{9.4}4s^{0.6}$. The electrons in the wide $4s$ conduction band are itinerant and uncorrelated. The question therefore arises, How should the d electrons be treated? In the Stoner-Wohlfarth model we saw that the motions of the $3d$ electrons through the lattice are uncorrelated, meaning that no allowance is made for the fact that the Coulomb repulsions tend to keep the electrons apart. Although in nickel the average configuration is $d^{9.4}$, Hurwitz and Van Vleck propose that

Ni is really a mixture of configurations $d^{10}, d^9, d^8, d^7, \ldots$. Since the work required to tear off an electron increases with the degree of ionization, the energy is appreciably lower if, for instance, two atoms with a combined total of 18 electrons are both in $3d^9$ states rather than one in $3d^8$ and the other in $3d^{10}$. In Van Vleck's model, which neglects the states of higher ionization, the configuration $3d^{9.4}$ for nickel, for example, is considered to be 40% $3d^{10}$ and 60% $3d^9$, and the lattice sites occupied by $3d^9$ and $3d^{10}$ are viewed as being in continuous redistribution. It is important to note that such a continuous redistribution is supported by neutron diffraction experiments. We shall not pursue this idea further since few quantitative calculations have been done so far. This generalized form of the Heisenberg model appears to be capable of explaining many facts that cannot be explained by the classical Heisenberg model.

Let us now consider Slater's method. This method of attacking the full problem and solving it by numerical methods is probably the most complete and quantum-mechanically exact procedure available. The difficulties of handling such a calculation at the present time, however, seem to be enormous. On the other hand, the success of the treatment of the paramagnetic oxygen molecule is noteworthy and constitutes impressive evidence in favor of the molecular orbital method.

Slater[10] advocates the use of a determinantal method with orthogonal one-electron functions or orbitals. If there is a complete orthogonal set of one-electron spin-orbitals $u_i(x)$, x denoting both orbitals and spin, we can pick n functions out of such an orthogonal set and form a determinantal function:

$$(n!)^{-1/2}\det|u_i(x_j)| \tag{13.160}$$

If the n one-electron functions are selected in all possible ways from the complete orthogonal set, and all the possible determinantal functions are built up from them, the result will be a complete orthogonal set of antisymmetric functions. The correct wavefunction of our problem can be rigorously expanded in terms of these functions. The one-electron functions from which the determinants are built up may be the solutions of a periodic potential problem and are the well-known Bloch functions. Alternatively, one may use the Wannier functions, which can be constructed from the Bloch functions. Since we always take a complete orthogonal set, either the Bloch or the Wannier functions form a proper starting point for an exact calculation. If these functions are not available, one can proceed as Mulliken and Parr[42] or Lowdin[43] has done, that is, start with the atomic orbitals and form orthogonal linear combinations of AOs of the Bloch or Wannier type. In their calculations, Mulliken and Lowdin used a single-determinantal wavefunction constructed out of the n Bloch-like energy-band functions whose one-electron energies are the lowest. It is found that a single such determinant describes adequately an actual molecule or a crystal at its observed internuclear distance. However, as the interatomic distance tends to infinity, this single-determinantal solution is not found to behave properly, and its asymptotic energy becomes too high. This situation is described in Figure 13.10. The full curve in the figure represents the diagonal energy of the state given by a single determinant as a function of the internuclear distance. The dotted curve represents the correct ground state energy, which is obtained from linear combination of the determinants. It is apparent from Figure 13.10 that a single determinant is a good approximation to the ground state near the minimum of the curve. The dissociation energy D can be calculated from the single determinant. Although the single-determinantal wavefunction gives good values for the binding energies, because the latter behave incorrectly at large internuclear distances they can lead to serious problems in magnetism. To see this clearly let us analyze the reason why the energy of the single-determinantal wavefunction is too high by an amount A at large distances. Slater argues that this high value results because there is nothing in the wavefunction to keep electrons of opposite spin out of each other's way, and hence there will be ionic states in which the positive and negative ions are found on the lattice sites rather than on the neutral atoms. Thus, in the H_2 molecule problem, the molecular orbital wavefunctions are $(\psi_a \pm \psi_b)$, where a and b denote the H atoms. Since there are two electrons, the determinantal wavefunction, considering the orbital part only, is given by

$$[\psi_a(1)+\psi_b(1)][\psi_a(2)+\psi_b(2)]$$
$$=[\psi_a(1)\psi_b(2)+\psi_b(1)\psi_a(2)]+[\psi_a(1)\psi_a(2)+\psi_b(1)\psi_b(2)] \tag{13.161}$$

where the first term represents the Heitler-London ground state, and the

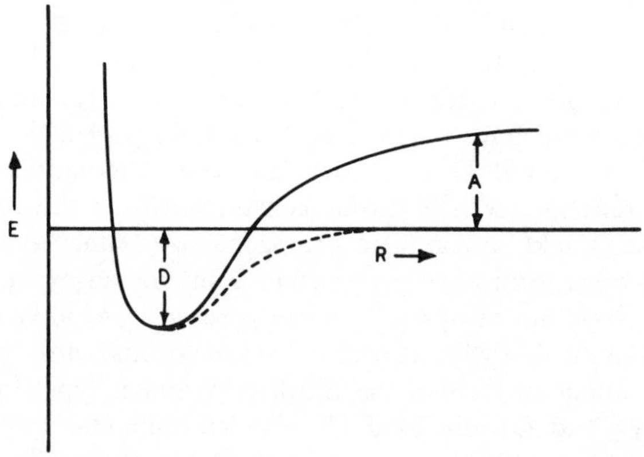

Fig. 13.10 Energy of single determinantal state of H_2 molecule (the full curve). The correct ground state energy is given by the dotted line. R denotes the internuclear distance, and D denotes the dissociation energy. (Taken from Slater.[10])

second term the ionic state with equal coefficients. At large separations there is equal probability that the wavefunction represents an ionic state (the second term). In calculating the diagonal energy of the state, given by the solid line in Figure 13.10, one has to treat the interaction of the two electrons on the same atom; in other words, one has to calculate the energy of the H^- ion. The energy A represents the energy required to form H^+ and H^- ions from the neutral H atoms.

For ferromagnetism it is necessary to compare the energies of states of different magnetic moments; hence one must have a magnetic moment for the ground state, whereas in a nonmagnetic case there should be no magnetic moment. The next question is, How does the energy of a single deteminant, as a function of the internuclear distance, depend on whether or not it represents a magnetic state? It is important to realize that as the internuclear distance decreases the energy band becomes broader, and, as a consequence, an increase in the magnetic moment will increase the energy. This is so because in the nomagnetic state each of the lowest one-electron energy levels is occupied by two electrons, one of each spin, so that there are as many up-spin as down-spin electrons, whereas in the magnetic state electrons with down spin, say, are removed from the low-lying one-electron levels and are placed in higher, unoccupied one-electron levels of the band of up spins. There is thus an increase in energy of the whole system, resulting from increased one-electron energies, the total increase being proportional to the band width, which increases with decreasing internuclear distance. This mechanism was discussed earlier in this chapter. As more and more down-spin electrons go to the upper half-band containing the up-spin electrons, the kinetic energy of the system increases. As more and more electrons are in the up-spin state, however, there is a gradual decrease in exchange energy so that there is an effective competition between these two energies—the kinetic and the exchange. These exchange integrals are always positive, and when the orbitals u_i are Bloch functions, it can easily be seen that at large internuclear separations they do not vanish but become interaction integrals (intra-atomic) between different electrons in the same atom. The result is a set of terms with negative sign, one for each pair of electrons with parallel spin. Since there are more such pairs for a magnetized than for an unmagnetized state, this exchange energy is lower for the magnetized than for the unmagnetized state.

The net result of these two effects is shown in Figure 13.11. A very important idea that can be visualized from this figure follows from the fact that the energy of the magnetized state lies lower than that of the magnetized state at infinite separation. This happens because of the tendency for electrons of opposite spin to come closer together and form ions, as can be seen from the determinantal function given above. On the other hand, electrons of the same spin are prevented from getting too close together because of the Pauli exclusion principle. Hence the greater the magnetization, the fewer are the pairs of electrons of opposite spins to come together and therefore the smaller is the error. If, for example, a magnetized state has all the electrons

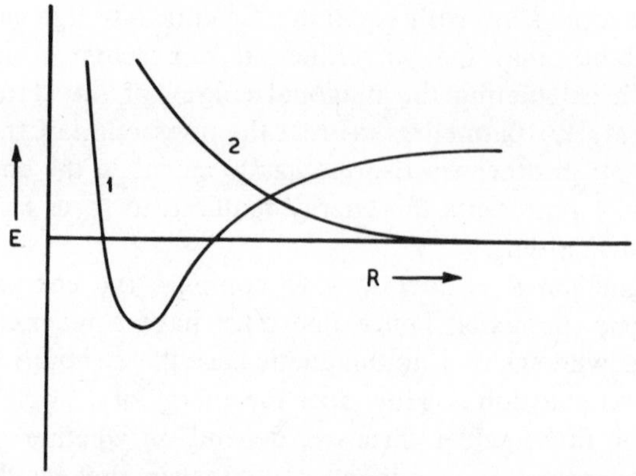

Fig. 13.11 Energy of the molecular orbitals for magnetized and unmagnetized states (single-determinantal). Curve 1 represents the unmagnetized state, and curve 2 the magnetized state. *R* denotes the internuclear distance. (Taken from Slater.[10])

with up-spin filling the upper half-band and there are no electrons with down-spin filling the lower half-band, the exclusion principle will entirely prevent the formation of the ionic states, and consequently the state of maximum magnetization will reach the correct energy at infinity. The most important point, as emphasized by Slater, is the fact that the *intra-atomic* exchange terms get into the energy of the determinantal state, and these are the correct terms to describe this tendency of electrons of opposite spins, in the unmagnetized state, to form positive and negative ions, as discussed earlier.

So far we have considered the behavior of the energy of a single-determinantal function, formed out of the Bloch functions, for both the magnetized and the unmagnetized state. The next question concerns how the energies of the correct wavefunctions are related to the Bloch functions (for example, as indicated in the dotted curve of Figure 13.10). To obtain the relation it is necessary to follow Slater's prescription, which says that we have to use many-determinantal functions and solve a perturbation problem. We may use the Bloch functions or the Wannier functions[44] as the one-electron orbitals and consider all the possible determinants that can be built out of the orbitals in the whole band. All such determinants are connected by a unitary transformation, so that the final result will be the same, regardless of which set we start from. Once these determinants are obtained, the perturbation problem between these states must be solved to get the resulting energy levels for each total spin. The lowest such energy level for each total magnetization will have to lie lower than the diagonal energy of the single-determinantal function of the same magnetization, as indicated in Figures 13.12*a* and 13.12*b*. It is worth noting that, for infinite internuclear distance, the correct solution will have a very high degeneracy. Since the atoms are separated, the energy will be independent of the orientation of the magnetic moments, so

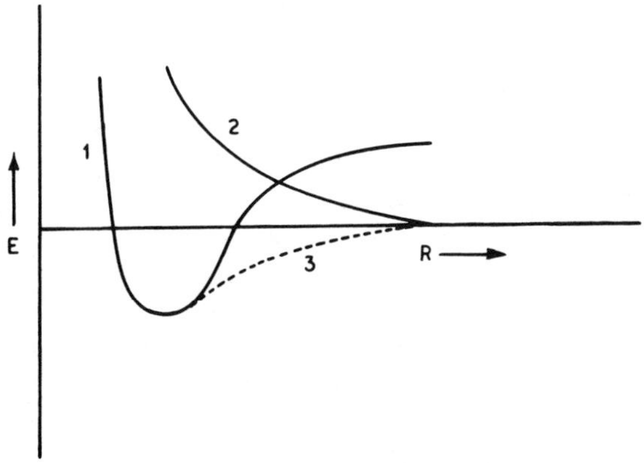

(a)

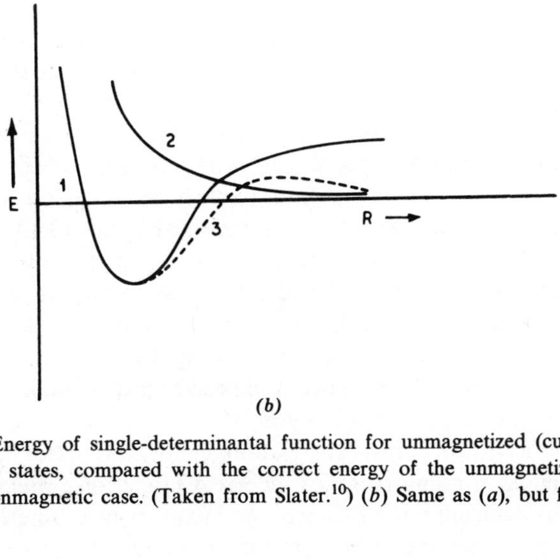

(b)

Fig. 13.12 (*a*) Energy of single-determinantal function for unmagnetized (curve 1) and magnetized (curve 2) states, compared with the correct energy of the unmagnetized state (dotted curve) for the nonmagnetic case. (Taken from Slater.[10]) (*b*) Same as (*a*), but for the ferromagnetic case.

that the energies of the lowest states corresponding to the various total magnetizations will be the same.

The results of such a calculation is described by Figures 13.12*a* and 13.12*b* for the case where the wavefunction of the state of maximum magnetization is correctly given by a single determinant, although the conclusions drawn will not be altered for any general situation, according to Slater. The correct energies of the unmagnetized and the magnetized states as a function of the internuclear distance are given in Figures 13.12*a* and 13.12*b*. In both of the cases shown, it has been indicated that the dotted curve lies below the diagonal energy of the single determinant representing an unmagnetized state but the dotted curve reduces to the same energy at infinity as the magnetized state. In the case of Figure 13.12*a* the unmagnetized state everywhere has lower energy than the magnetized state, and hence the system must be nonmagnetic. In Figure 13.12*b*, on the other hand, the unmagnetized energy lies above the magnetized state at large internuclear distances, and there is a region over which the substance will be ferromagnetic.

Many important conclusions can be drawn from Figures 13.12*a* and 13.12*b*. For the ferromagnetic case the energy difference between the magnetized and unmagnetized states is zero at infinity, increases as the internuclear distance decreases, goes through a maximum, and then decreases to zero and changes sign at smaller internuclear distances. This sort of behavior is exactly what is found in reality. It is important to note that these figures represent the energies of the 3*d* electrons. We now consider the energies of the 4*s* electrons, superimposed on the energies of the 3*d* electrons. Since the 4*s* band is much broader than the 3*d* and its radius is much larger, the result will be a much deeper minimum of the energy curve at large internuclear distances than is shown in Figure 13.12*b*. Thus one expects curves like those in Figures 13.13*a* and 13.13*b* to represent the actual energy of the iron group of elements as a function of the internuclear distance. If the exchange interaction is positive at the minimum of the curve, as in Figure 13.13*a*, the substance will be ferromagnetic. On the other hand, for the earlier elements like titanium and vanadium, since the 3*d* orbital is of larger size, the exchange integral will change sign at a larger internuclear distance and will be negative at the minimum of the curve, as shown in Figure 13.13*b*, so that the substance will be nonmagnetic. Such arguments demonstrate very clearly that, for substances with wide bands, ferromagnetism is impossible. For narrow bands, on the other hand, these arguments tell us only that ferromagnetism is possible; they certainly cannot distinguish between ferromagnetism or antiferromagnetism in the absence of more accurate calculations.

Now, what is the relationship between the energy band treatment and the correlation energy? The term "correlation energy" is used to describe the decrease of energy in going from an incorrect model of a metal, constructed out of a single determinant of Bloch- or Wannier-like functions, in which electrons of opposite spins can be found too close together, to a correct model in which electrons of opposite spins are kept apart. At infinite internuclear distance the correlation energy is given by the difference between the full curve and the dotted curve, which is the amount A indicated in Figure 13.10. If this is really the correlation energy at $R \rightarrow \infty$, we should expect it to constitute a very small correction at the value of R corresponding to the minimum of the curve. This inference agrees well with Löwdin's[43] results for sodium. According to Slater, the correction for correlation should reduce the energy of the unmagnetized state from the diagonal value, coming from the single-determinantal wavefunction, to the correct final value (shown by the dotted line in Figure 13.10 or in Figure 13.12*a* or 13.12*b*). A similar

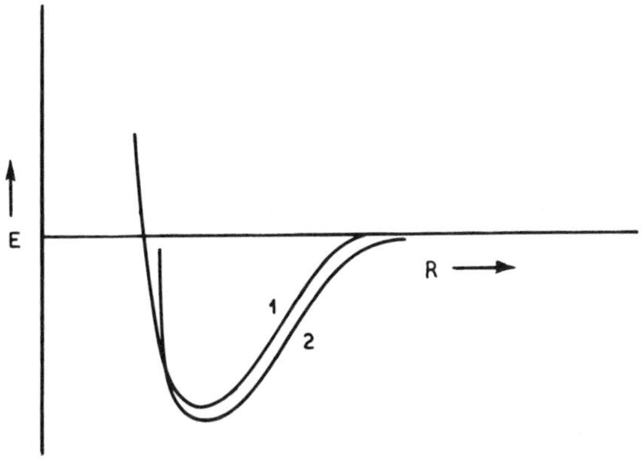

(a)

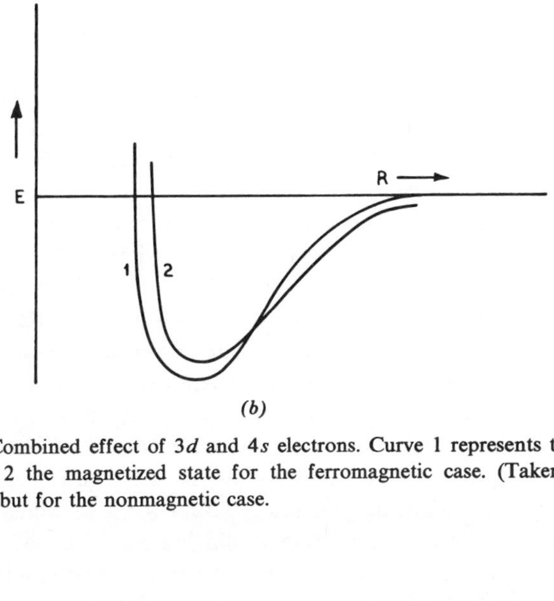

(b)

Fig. 13.13 (*a*) Combined effect of $3d$ and $4s$ electrons. Curve 1 represents the unmagnetized state, and curve 2 the magnetized state for the ferromagnetic case. (Taken from Slater.[10]) (*b*) Same as (*a*), but for the nonmagnetic case.

independent calculation has to be done for the magnetized state as well. From the difference between the final corrected energies for the magnetized and unmagnetized states, which has been calculated as above, we obtain the magnitude of the exchange integral. The sign of the exchange integral determines whether or not the metal is ferromagnetic. We thus see a very important thing, that is, the relationship of the correlation energy with the exchange energy. The details of such a calculation cannot be given at this time since as yet no complete calculation is available.

It is very important to note that the correlation energy originates, not from the attractive interaction of an electron and a hole, but from the repulsive interaction of two charges of like sign. For example, in nickel a band almost filled with d electrons will have some vacancies (the holes), and these holes will tend to repel each other and stay out of each other's way. This kind of calculation is very similar to that for the electrons and holes near the impurity atoms in a lattice of a semiconductor[45] and can be performed by using the Wannier functions. We do not intend to pursue this matter further but simply note, along with Slater, that such a calculation has been done by Meckler[10] on the O_2 molecule. He has considered all the determinantal functions and done the calculation rigorously. He finds that the triplet (^3P) state lies below the singlet, as is to be expected. His calculated dissociation energy and the internuclear distance for the ground state agree very well with experimental values. In conclusion, we must say that this method of calculation, suggested by Slater and performed by Meckler, though in a simple system, gives a correct understanding not only of the correlation energy but also of the exchange interaction. This understanding is absolutely essential for a satisfactory theory of magnetism.

In the last phase of our examination of collective electron ferromagnetism, we shall briefly study Zener's model,[46] where the exchange interactions responsible for ferromagnetism are to be found mainly between the $3d$ and $4s$ electrons rather than between the $3d$ electrons, as ordinarily assumed. According to Zener, the $3d$-$4s$ coupling is so strongly ferromagnetic that it dominates the $3d$-$3d$ interactions, which he assumes to be antiferromagnetic.

We have already seen that, according to Heisenberg, the interaction arising from an exchange of electrons between the d shells of adjacent atoms is known as the direct exchange. This has been the subject of considerable controversy over the sign of the exchange integral J. In the usual sense, a positive J corresponds to ferromagnetic coupling, and a negative J to antiferromagnetic coupling. But in order that a chemical bond may be formed, the sign of J has to be negative. In the classic paper on ferromagnetism, Heisenberg[11] assumed that the direct exchange furnished the ferromagnetic coupling in Fe, Co, and Ni. In accounting for the positive sign of J, which such a coupling demanded, he pointed out that for a sufficiently large value of the principal quantum number one would indeed anticipate a change in J from a negative to a positive sign. The question immediately arises, Why, then, is ferromagnetism absent in the second and third transition elements, which, according to Heisenberg, should be stronger in this regard than the first transition elements? Unfortunately, Heisenberg had no answer.

In 1930, however, Slater[47] answered this puzzling question. He emphasized the importance of the degree of overlap of the d shells and postulated that, as the degree of overlap decreases from a larger to a smaller value, J changes sign from negative to positive. Since the degree of overlap is a continuously decreasing function of the ratio $\tau = R/r$, where R is the internuclear distance and r the radius of the d shell, Slater assumed that J varies in the general manner indicated in Figure 13.14. This single postulate explained not only why ferromagnetism does not occur in the palladium and platinum groups of

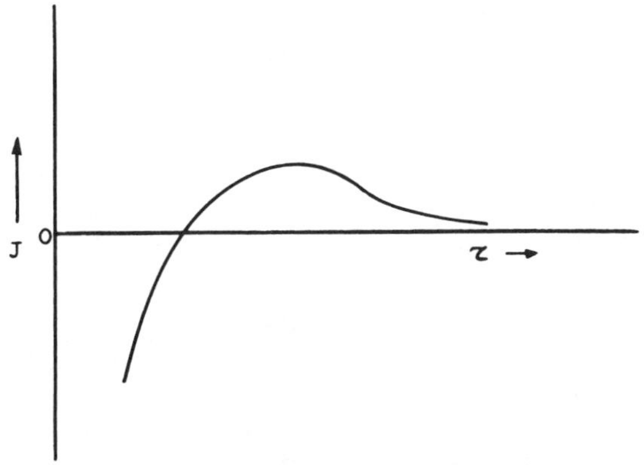

Fig. 13.14 Variation of the exchange integral J postulated by Slater. (Taken from Slater.[47])

metals, where the ratio τ is smaller than in the iron group, but also why only the last elements in the iron group are ferromagnetic.

This postulate of Slater is so satisfactory that it was immediately accepted by all workers in the field who used the localized atomic model for the d electrons. In 1951 Zener[46] proposed his model of $3d$-$4s$ coupling which, when taken into account, eliminated the necessity for a positive J. On the basis of this model it appears more logical to assume that the magnitude of J decreases as the ratio τ increases (Figure 13.15). According to Zener, the integral corresponding to the exchange of a single pair of electrons between two atoms contains two dominant terms, that is,

$$J = J_1 + J_2 \tag{13.162}$$

where J_1 is the mutual energy of the two exchange-charge distributions, and J_2 is the sum of the energies of the two exchange charges in the so-called

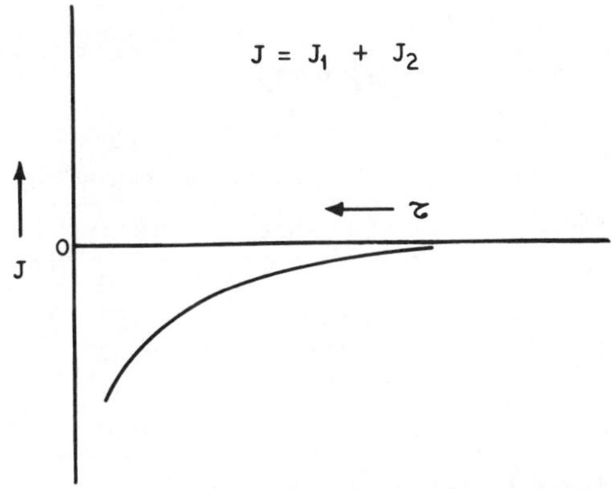

Fig. 13.15 Variation of J with τ. (Taken from Zener.[46])

reduced potential of the two atoms. To obtain the reduced potential one has to take first the actual potential of the two atoms, considering the two exchanging electrons to be absent. This potential is then multiplied by the total exchange charge of one of the exchanging electrons. The first term J_1 is always positive, while the second term J_2 is always negative. The sign of J, therefore, depends on which of the two terms dominates.

Let us look at the theoretical basis of Zener's model. It is well known that Hund's rule of highest multiplicity applies to the coupling of the spin of an outer s electron to the spin of an incomplete d shell of isolated atoms, as well as to the coupling of the spins of the d electrons among themselves. Thus, when the outer s shell contains only one electron, the state of lowest energy will be that in which the spin of this s electron is aligned parallel with the spin of the inner incomplete d shell. In Table 13.1 we consider the lowest configuration in which the outer s shell contains only one electron and compare the two energies where the spin of this electron is parallel and antiparallel, respectively, to that of the inner d shell.

In Zener's model the outer s electrons of the isolated atoms become conduction electrons in the metallic state. The exchange interaction between the conduction s electron and a d electron will therefore be ferromagnetic in nature. Each individual conduction electron tends to align the spins of the incomplete d shell in a direction parallel to its own spin. Usually there are as many conduction electrons with up spin as with down spin, so that their net effect on the spin of the incomplete d electrons is zero. If, with Zener, we think that the conduction electrons with up spin, first of all, align the spins of the d electrons to the up-spin position and then reverse the spins of some of the conduction electrons from the down-spin to the up-spin position, the total energy of the system will be lowered.

If E_{spin} is the spin energy per atom of a metal and S_d and S_c are the mean components per atom of the d electron and the conduction electron spin along the magnetization direction, respectively, then

$$E_{\text{spin}} = \tfrac{1}{2}\alpha S_d^2 - \beta S_d S_c + \gamma S_c^2 \tag{13.163}$$

Table 13.1

Coupling energy ΔE of spin of outer s electron to that of inner d shell electron[a]

Metal	Configuration	ΔE (eV)
Cr	$3d^5(^7S)4s$	0.92
Mn	$3d^6(^5D)4s$	0.77
Fe	$3d^7(^4F)4s$	0.62
Co	$3d^8(^4F)4s$	0.50
Ni	$3d^9(^2F)4s$	0.30

[a]From C. Zener and R. R. Heikes, *Rev. Mod. Phys.* **25**, 191 (1953).

We now minimize this spin energy with respect to S_c; thus

$$\frac{dE_{\text{spin}}}{dS_c} = -\beta S_d + \gamma S_c = 0$$

or

$$S_c = \left(\frac{\beta}{\gamma}\right) S_d \tag{13.164}$$

Inserting this value of S_c in (13.163), we have

$$E_{\text{spin}} = \frac{1}{2}\left(\alpha - \frac{\beta^2}{\gamma}\right) S_d^2 \tag{13.165}$$

Now Zener's theory becomes identical to the Heisenberg-Slater theory if the following correspondence is made:

$$J = \frac{\beta^2}{\gamma} - \alpha \tag{13.166}$$

Relation (13.166) shows that the exchange integral J is the difference between two terms, according to Zener. The term β^2/γ arises from the indirect coupling through the conduction electrons and is always positive and independent of the amount of overlap of the adjacent d shells. The second term, α, arises from the direct exchange between the adjacent d shells. We assume this term to be positive, and its magnitude decreases rapidly with decreasing overlap of the adjacent d shells. The change in sign of the exchange integral J, as we found in Slater's theory, with decreasing amount of overlap thus arises when $\beta^2/\gamma > \alpha$ is satisfied.

The difference between Heisenberg's theory and Zener's theory is that, whereas in the former the ferromagnetic coupling is due to the short-range direct exchange between the adjacent d shells, in the latter it is due to a coupling through the conduction electrons and hence is of long-range type. We can also make a quantitative estimate of the three coupling coefficients, α, β, and γ, appearing in E_{spin}. The value of β for isolated atoms is obtained from spectroscopic data. Comparing the Hartree wavefunction for the s electrons with the cellular wavefunction for the metal shows how β is modified in the metal. One such estimate of β for nickel is 0.48 eV per atom. Similarly, γ can be easily estimated from the treatment of the free-electron gas model of conduction electrons; γ is 2.9 eV per atom for nickel. An estimate is thereby obtained for β^2/γ, neglecting α. The estimate thus obtained for nickel is 25% higher than the observed value of 2.53 eV, which can be ascribed to α coming from the direct exchange. We thus see that in Zener's theory the conduction electrons contribute quite appreciably to the total exchange energy.

REFERENCES

1. J. B. Goodenough, *Magnetism and Chemical Bond*, Interscience, New York, 1963.
2. C. Kittel, *Introduction to Solid State Physics*, 3rd ed, Wiley Eastern Private Limited, New Delhi, India, 1971.
3. D. H. Martin, *Magnetism in Solids*, Iliffe Books Ltd., London, 1967.
4. C. Kittel, *The Quantum Theory of Solids*, John Wiley and Sons, New York, 1963.
5. R. M. Bozorth, *Ferromagnetism*, Van Nostrand, Princeton, N.J., 1951.

5a. M. Sachs, *Solid State Theory*, McGraw-Hill, New York, 1963.

6. P. Weiss, *J. Physi* **6**, 667 (1907); ibid. **1**, 166 (1930). A good survey of the Weiss theory is given in Stoner's *Magnetism and Atomic Structure*, Methuen, London, 1926, p. 75.
7. E. C. Stoner, *Proc. Roy. Soc. (London) A*, **165**, 372 (1938); ibid. **169**, 339 (1939).
8. E. P. Wohlfarth, *Rev. Mod. Phys.* **25**, 211 (1953).
9. J. H. Van Vleck, *Rev. Mod. Phys.* **25**, 220 (1953).
10. J. C. Slater, *Rev. Mod. Phys.* **25**, 199 (1953).
11. W. Heisenberg, *Z. Phys.* **38**, 411 (1926). This effect was simultaneously discovered by Dirac, *Proc. Roy. Soc. (London) A*, **112**, 661 (1926).
12. R. N. Stuart and W. Marshall, *Phys. Rev.* **120**, 353 (1960).
13. H. A. Kramers, *Physica* **1**, 182 (1934).
14. P. A. Anderson, *Solid State Phys.*, **14**, 99 (1963).
15. E. C. Stoner, *Proc. Leeds Phil. Soc.* **2**, 56 (1930); *Phil. Mag.* **10**, 43 (1930).
16. M. L. Neel, *Ann. Phys.* **18**, 5 (1932); **8**, 237 (1937).
17. J. C. Slater, *Phys. Rev.* **49**, 537 (1936).
18. J. McDougall and E. C. Stoner, *Phil. Trans. A*, **237**, 67 (1938).
19. E. C. Stoner, *Proc. Roy. Soc. (London) A*, **152**, 672 (1935).
20. E. C Stoner, *Proc. Leeds Phil. Soc.* **3**, 191 (1936).
21. F. Bloch, *Z Phys* **57**, 545 (1939).
22. E. P. Wohlfarth, *Phil. Mag.* **49**, 703, 1095 (1949).
23. L. Brillouin, *J. Phys.* **3**, 380, 565 (1932); ibid. **4**, 1, 333 (1933).
24. F. Bloch, *Z. Phys.* **52**, 555 (1928); ibid. **57**, 545 (1929).
25. J. C. Slater, *Rev. Mod. Phys.* **6**, 209 (1934).
26. F. Seitz, *Modern Theory of Solids*, McGraw-Hill, New York, 1940.
27. E. Wigner, *Phys. Rev.* **46**, 1002 (1934).
28. E. P. Wohlfarth, *Proc. Leeds Phil. Soc.* **5**, 89 (1948).
29. J. C. Slater, *Phys. Rev.* **49**, 537 (1936).
30. N. F. Mott, *Proc. Phys. Soc.* **47**, 571 (1935).
31. R. Peierls, *Ann. Phys. Leipzig* **4**, 121 (1930).
32. A. Sommerfeld and H. Bethe, *Handbuch der Physik*, 2nd ed., Vol. 24/2, 1933, p. 333.
33. G. C. Fletcher, *Proc. Phys. Soc. A*, **65**, 192 (1952).
34. A. B. Lidiard, *Proc. Phys. Soc. A*, **65**, 885 (1952).
35. E. U. Condon and G. H. Shortley, *The Theory of Atomic Spectra*, Cambridge University Press, London, England, 1963.
36. P. A. M. Dirac, *Proc. Roy. Soc. (London) A*, **123**, 714 (1929).
37. F. Bloch, *Z. Phys.* **74**, 295 (1932).
38. J. C. Slater, *Quantum Theory of Atomic Structure*, Vol. 1, McGraw-Hill, New York, 1960.
39. P. A. M. Dirac, *Proc. Roy. Soc. (London) A*, **112**, 661 (1926).

40. E. M. Corson, *Perturbation Methods in the Quantum Mechanics of n-Electron Systems,* Hafner, New York, 1951.

41. H. Hurwitz, Ph.D. Thesis, Harvard University, 1941.

42. R. S. Mulliken and R. G. Parr, *J. Chem. Phys.* **19**, 1271 (1951) and many other references cited therein.

43. P. O. Löwdin, *J. Chem. Phys.* **18**, 365 (1950); ibid. **19**, 1570, 1579 (1951).

44. G. Wannier, *Phys. Rev.* **52**, 191 (1937).

45. J. C. Slater, *Phys. Rev.* **76**, 1592 (1949).

46. C. Zener, *Phys. Rev.* **81**, 440; ibid. **82**, 403; ibid. **83**, 299 (1951); ibid. **85**, 324 (1952).

47. J. C. Slater, *Phys. Rev.* **36**, 57 (1930).

14

Molecular Field Theory of Antiferromagnetism

In Chapter 13 we studied in great detail the molecular field theory of ferromagnetism, which has been immensely successful in explaining qualitatively and also, to some extent, quantitatively the magnetization of ferromagnetic substances below the Curie temperature. But there are many compounds of the transition metal ions (the list is growing steadily) that are antiferromagnetic instead of ferromagnetic below a certain critical temperature as remarked briefly in Chapter 13. Consider, for example, a lattice in which the magnetic atoms are arranged in a body-centered cubic structure. One may visualise all the ions at the body centers having "spin up" and all the corner atoms having "spin down". The net magnetic moment per unit cell, in this case, will be zero, although the lattice is magnetically ordered. This is a simple case of magnetic structure with two interpenetrating sublattices. Apart from this simple structure, the transition metal compounds, particularly the oxides and the halides, have a wide variety of complicated magnetic structures.[1] These complicated structures possess more than two interpenetrating ordered sublattices, and each of them is spontaneously magnetized in such a direction that the resultant magnetization over all the sublattices is zero. In this chapter we deal with transition metal complexes that are antiferromagnetic below a certain critical temperature.

Materials containing noninteracting magnetic ions possess paramagnetic susceptibility, which increases as the temperature decreases toward absolute zero following the Curie law ($\chi = C/T$). If there is interaction among the magnetic ions, the susceptibility at a particular temperature may be found to be lower than that for the independent ions. It has been experimentally observed that many compounds containing a high density of ions of the transition metals exhibit paramagnetic susceptibilities that pass through a sharp maximum at a critical temperature to decrease with decreasing temperature below this point. In 1936 Neel[1] pointed out that strong negative interactions in some structures are responsible for cooperative ordering of the magnetic moments giving such a behavior. In the ordered state, which sets in at the critical temperature, the magnetic ions are so aligned that they form two or more spontaneously magnetized and interpenetrant sublattices whose magnetizations are differently oriented to give zero resultant magnetization, so that the ordering does not reveal itself directly, as in ferromagnetism. This kind of ordering is known as antiferromagnetic ordering. A great number of materials are known to be antiferromagnetic, including the Mn and Cr metals, many of the rare earth metals, and some compounds of the 3*d* and 4*d* transition elements.

Before considering the details of the molecular fields in magnetic solids, which is the subject matter of this chapter, let us discuss qualitatively the experimental observations on the common magnetic structures found in nature. The effects to which such structures give rise are known as antiferromagnetism. Neel showed that an antiferromagnetic material has magnetic susceptibility, which varies with temperature in a characteristic way. The specific heat of such a material registers a sharp maximum at the temperature at which the order is destroyed by thermal agitation. This characteristic temperature is known as the Neel temperature.

A large number of antiferromagnetic materials of different kinds have been discovered by using the recently available powerful technique of neutron diffraction. Some of these systems are listed in Table 14.1.

Table 14.1

Neel temperature and θ/T_N ratio of some typical antiferromagnetic systems.[a]

Substance	Type of lattice	Ionic magnetic moment (μ_B)	Neel temperature, T_N (°K)	$\frac{-\theta}{T_N}$
$KMnF_3$	Simple Cubic	5.94	88	1.8
CoF_3	Simple Cubic	2.5	460	–
CrF_3	Simple Cubic	3.9	80	1.6
$KFeF_3$	Simple Cubic	5.38	113	–
FeF_2	Body-centered tetragonal	5.6	79	1.5
CoF_2	Body-centered tetragonal	5.13	37	1.3
MnF_2	Body-centered tetragonal	5.7	72	1.6
NiF_2	Body-centered tetragonal	3.5	74	~1.4
CoO	Face-centered cubic	5.1	291	1.1
FeO	Face-centered cubic	4.6	188	1.0
MnO	Face-centered cubic	5.95	118	5.0
NiO	Face-centered cubic	4.6	520	~5.0

[a](Data collected from various sources. See the references at the end of the chapter.)

Measurements of the magnetic susceptibilities of some simple solids, like the fluorides of the transition metals, show a decrease in susceptibility with decreasing temperature below the characteristic temperature, T_N. To interpret this effect, Neel proposed the theory of antiferromagnetism. The ordering process occurring below the Neel temperature T_N more than offsets the decreasing thermal agitation and reduces the response of the individual ions to an applied field, so that an antiferromagnetic crystal shows maximum susceptibility at T_N and is magnetically anisotropic below. Figure 14.1 shows such behavior[4] for MnF_2. Most of the simple compounds of the iron group elements display qualitatively similar behavior. We shall discuss this behavior further later in this chapter. For simple antiferromagnetic systems a Curie-Weiss relation is usually followed, with a negative value for the Weiss constant θ [see (13.1)]. There is considerable variation of the ratio θ / T_N from one material to another, as can be seen from Table 14.1.

Neutron diffraction methods are the most powerful ones available to demonstrate the existence of this order. This technique uses the magnetic scattering of slow neutrons to reveal the magnetic sublattices—a principle very similar to the use of X-rays in crystallography. The typical magnetic structures prevalent in antiferromagnetic materials and the main characteristic features of antiferromagnetic behavior were described above. The temperatures up to which such an antiferromagnetic structure can persist may be well above 100°K, and it can be shown that the exchange effects must be responsible for this kind of magnetic ordering. This section deals with the molecular field representations of the exchange effects in antiferromagnetism.

The MF theory discussed in Chapter 13, which was so successfully applied to the case of ferromagnetism, can be applied equally well to antiferromagnetism. Here, however, the exchange integral, denoted by J, is assumed to be negative in sign, and it is possible to obtain an ordered state such that the nearest-neighbor atoms tend to have their magnetic moments aligned antiparallel. As discussed earlier, the first theory of such a state was developed by Neel in 1936. Neel's theory, which is a generalization of the Weiss molecular field theory, predicts a completely ordered arrangement at a temperature of absolute zero, with the lattice of magnetic atoms divided into two spontaneously magnetized, equivalent sublattices set antiparallel to each other. The spontaneous magnetization decreases with increase of temperature and ultimately vanishes at the transition temperature, that is, the Neel temperature T_N, which is a function of the molecular field constant—λ, say. The susceptibility above the Neel temperature follows the Curie-Weiss law [(Equation 13.1)] as in the case of ferromagnetism but with $\theta = -T_N$. In the region $T < T_N$, the susceptibility decreases with decrease of temperature, the physical reason being that the magnetic moments are locked together by the exchange interactions, which hinder the response to an applied magnetic field. There is also a magnetic specific heat associated with antiferromagnetism, which has the same origin and the same sort of temperature dependence as those for a ferromagnetic system. We shall discuss this aspect also later in this chapter.

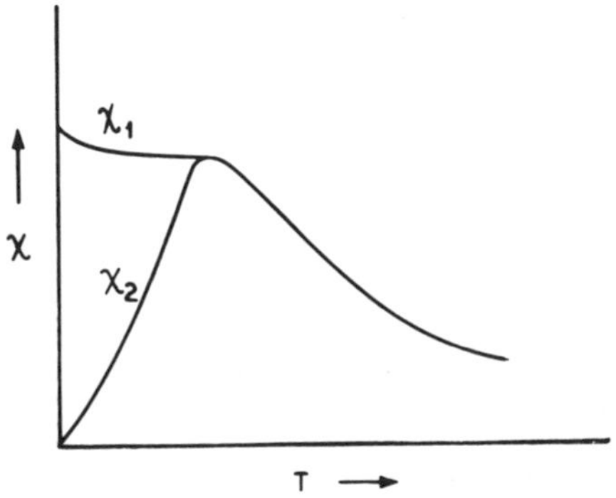

Fig. 14.1 Variation of the mass susceptibility of MnF_2 single crystal with temperature; χ_1 and χ_2 represent the perpendicular and parallel susceptibilities, respectively. (Taken from Foner.[4])

14.1 THE TWO-SUBLATTICE MODEL

The simplest antiferromagnetics are the crystalline ionic compounds of the transition series in which the metallic ions are arranged in such a way that they can be subdivided into two equivalent interpenetrating sublattices, A and B, such that A atoms have only B atoms as the nearest neighbors, and vice versa. Examples of such lattices are MnF_2, FeF_2, and CoF_2, which crystallize in the rutile structure where the body-centered metal ions form sublattice A, say, and the corner metal ions form sublattice B. Now, if the exchange interaction favors an antiparallel alignment of the ionic magnetic moments ($\mathbf{M}_A$ and $\mathbf{M}_B$), sublattices A and B are spontaneously magnetized in opposite directions, as shown schematically in Figure 14.2.

In the MF treatment, let us consider two different molecular fields, $\mathbf{H}_{mA}$ and $\mathbf{H}_{mB}$, acting on sublattices A and B, respectively. Then the effective field acting on the A sublattice is given by

$$\mathbf{H}_{mA} = -\lambda_{AA}\mathbf{M}_A - \lambda_{AB}\mathbf{M}_B \tag{14.1}$$

and similarly the effective field on the B sublattice is given by

$$\mathbf{H}_{mB} = -\lambda_{BA}\mathbf{M}_A - \lambda_{BB}\mathbf{M}_B \tag{14.2}$$

where $\mathbf{M}_A$ and $\mathbf{M}_B$ are the magnetizations of the A and B sublattices,

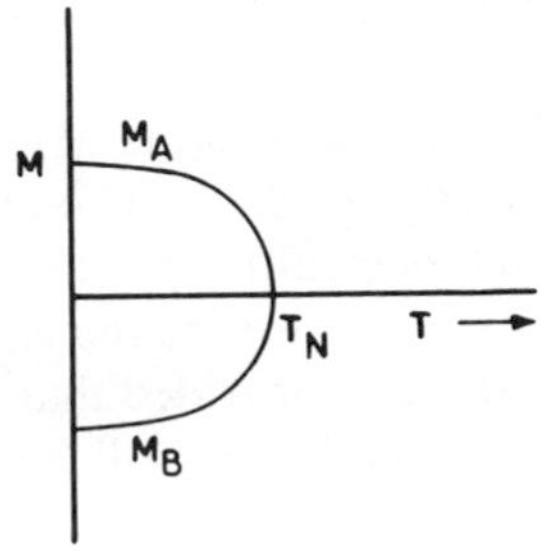

Fig. 14.2 Temperature dependence of the sublattice magnetizations, M_A and M_B, in a two-sublattice antiferromagnet.

respectively. Also, λ_{AB} and λ_{AA} are the molecular field constants for the nearest-neighbor and the next-nearest-neighbor interactions, respectively, for an atom at the A site. Similarly, λ_{BA} and λ_{BB} are related to atoms at the B site. Now, since atoms of the same type occupy both the A and the B lattice sites, we have

$$\lambda_{AA}=\lambda_{BB}=\lambda_{ii} \qquad \text{and} \qquad \lambda_{AB}=\lambda_{BA}$$

In addition, if a field $\mathbf{H}$ is also applied, the effective fields $\mathbf{H}_A$ and $\mathbf{H}_B$ at A and B lattice sites, respectively, are given by

$$\mathbf{H}_A=\mathbf{H}-\lambda_{ii}\mathbf{M}_A-\lambda_{AB}\mathbf{M}_B \tag{14.3}$$

$$\mathbf{H}_B=\mathbf{H}-\lambda_{AB}\mathbf{M}_A-\lambda_{ii}\mathbf{M}_B \tag{14.4}$$

Since the interaction between the nearest neighbors is antiferromagnetic, the molecular field constant λ_{AB} must be positive. In general, it is conceivable that λ_{ii} may be positive, negative, or even zero, depending on the particular situation and material. The expressions for the spontaneous magnetizations $\mathbf{M}_A$ and $\mathbf{M}_B$ as functions of temperature T and those for the susceptibilities above and below the Neel temperature can now be derived. We start from the expression derived in Section 13.2 for the magnetization induced by an externally applied field $\mathbf{H}$ in a system of noninteracting ions:

$$\mathbf{M}=M_0B_S\left(\frac{g\mu_{\mathrm{B}}SH}{kT}\right) \tag{14.5}$$

Here k is the Boltzmann constant, and $g\mu_B S$ denotes the maximum value of the aligned moment of an ion, where g is the Landé g-factor; μ_B, the Bohr magneton; and S, the spin quantum number. As in the case of ferromagnetism, here also we neglect the orbital contribution to the magnetic moment because it is almost completely quenched by the crystal fields in most antiferromagnetics. Also, M_0 is the saturation magnetization which corresponds to complete alignment of the ions and is equal to $Ng\mu_{\mathrm{B}}S$, N being the number of magnetic ions per unit volume, and $B_S(x)$ is the well-known Brillouin function, given by

$$B_S(x)=\frac{2S+1}{2S}\coth\left(\frac{2S+1}{2S}x\right)-\frac{1}{2S}\coth\left(\frac{x}{2S}\right)$$

Because of the existence of two sublattices, A and B, according to the MF hypothesis the magnetizations $\mathbf{M}_A$ and $\mathbf{M}_B$ are given directly by (14.5), where $\mathbf{H}_A$ and $\mathbf{H}_B$ are substituted, respectively, for $\mathbf{H}$. Since the two sublattices are exactly equivalent, however, in the absence of an externally applied magnetic field we can presume that $\mathbf{M}_A$ is equal in magnitude to $\mathbf{M}_B$, and provided that λ_{ii} is positive and much greater than λ_{AB}, $\mathbf{H}_A$ and consequently $\mathbf{M}_A$ will be antiparallel to $\mathbf{M}_B$, that is, $\mathbf{M}_A=-\mathbf{M}_B$.

At thermal equilibrium the magnetizations of the sublattices are therefore given by

$$\mathbf{M}_A = \tfrac{1}{2} N g \mu_B S B_s(x_A) \tag{14.6}$$

Here

$$x_A = \frac{g\mu_B S}{kT} H_A$$

and

$$\mathbf{M}_B = \tfrac{1}{2} N g \mu_B S B_s(x_B) \tag{14.7}$$

where

$$x_B = \frac{g\mu_B S}{kT} H_B$$

Now let us study the situation when $T > T_N$. Although there is no antiferromagnetic ordering above the Neel temperature, a small magnetization is induced by the applied field. For the usual values of the applied field, the saturation effects are negligible, and the Brillouin function can be replaced by the first term of the series expansion x, so that

$$\lim_{x\to 0} B_s(s) = \frac{S+1}{3S} x$$

and (14.6) and (14.7) are reduced to

$$\mathbf{M}_A = \frac{N g^2 \mu_B^2 S(S+1)}{6kT} \mathbf{H}_A$$

and (14.8)

$$\mathbf{M}_B = \frac{N g^2 \mu_B^2 S(S+1)}{6kT} \mathbf{H}_B$$

Since $\mathbf{M}_A$ and $\mathbf{M}_B$ are parallel to $\mathbf{H}$ in the paramagnetic region ($T > T_N$), we have

$$\mathbf{H}_A = \mathbf{H} - \lambda_{ii}\mathbf{M}_A - \lambda_{AB}\mathbf{M}_B$$

and (14.9)

$$\mathbf{H}_B = \mathbf{H} - \lambda_{ii}\mathbf{M}_B - \lambda_{AB}\mathbf{M}_A$$

Substituting these values of $\mathbf{H}_A$ and $\mathbf{H}_B$ in the expressions for $\mathbf{M}_A$ and $\mathbf{M}_B$

given by (14.8), we obtain

$$\mathbf{M}_A = \frac{Ng^2\mu_B^2 S(S+1)}{6kT}(\mathbf{H} - \lambda_{ii}\mathbf{M}_A - \lambda_{AB}\mathbf{M}_B)$$

$$\mathbf{M}_B = \frac{Ng^2\mu_B^2 S(S+1)}{6kT}(\mathbf{H} - \lambda_{ii}\mathbf{M}_B - \lambda_{AB}\mathbf{M}_A) \qquad (14.10)$$

which, when added, give

$$\mathbf{M} = \mathbf{M}_A + \mathbf{M}_B = \frac{Ng^2\mu_B^2 S(S+1)}{6kT}[2\mathbf{H} - (\lambda_{ii} + \lambda_{AB})\mathbf{M}] \qquad (14.11)$$

Therefore the susceptibility χ is given by

$$\chi = \frac{M}{H} = \frac{C}{T+\theta} \qquad (14.12)$$

where

$$C = \frac{Ng^2\mu_B^2 S(S+1)}{3k}$$

and

$$\theta = \frac{C}{2}(\lambda_{ii} + \lambda_{AB})$$

Since the nearest-neighbor interactions are much greater in magnitude than the next-nearest-neighbor interactions, we have $\lambda_{AB} \gg \lambda_{ii}$, and hence θ is positive.

Now let us study the situation below the Neel temperature, that is, $T < T_N$. Below T_N, the two sublattices possess spontaneous magnetizations ($H=0$) of equal magnitude, as seen before. The Neel temperature is defined as the temperature at which the spontaneous magnetization of one of these sublattices vanishes, and T_N can be found by approaching either from the low-temperature side ($T < T_N$) or from the high-temperature side ($T > T_N$). Let us consider the first approach. Putting $H=0$ in (14.10), we have

$$\mathbf{M}_A = \frac{C}{2T}(-\lambda_{ii}\mathbf{M}_A - \lambda_{AB}\mathbf{M}_B)$$

$$\mathbf{M}_B = \frac{C}{2T}(-\lambda_{AB}\mathbf{M}_A - \lambda_{ii}\mathbf{M}_B) \qquad (14.13)$$

Now, inserting $\mathbf{M}_A = -\mathbf{M}_B$ in (14.13), as discussed earlier, we have

$$\mathbf{M}_A = \frac{C}{2T}(\lambda_{AB} - \lambda_{ii})\mathbf{M}_A$$

and (14.14)

$$\mathbf{M}_B = \frac{C}{2T}(\lambda_{AB} - \lambda_{ii})\mathbf{M}_B$$

We now see that each of these equations is identical with the single corresponding equation (13.13) derived in Chapter 13 for a ferromagnetic sample, with $(\lambda_{AB} - \lambda_{ii})$ serving as the molecular field constant in place of γ.

For nonzero values of magnetizations $\mathbf{M}_A$ and $\mathbf{M}_B$ for $T < T_N$, the determinant of the coefficients of $\mathbf{M}_A$ and $\mathbf{M}_B$ must be zero, yielding the Neel temperature T_N as

$$T_N = \frac{C}{2}(\lambda_{AB} - \lambda_{ii}) \tag{14.15}$$

Thus we see that T_N is higher, the stronger is the interaction between the sublattices, as is to be expected on physical grounds. Hence the ratio of the paramagnetic Curie-Weiss temperature θ to the Neel temperature T_N is

$$\frac{\theta}{T_N} = \frac{\lambda_{ii} + \lambda_{AB}}{\lambda_{AB} - \lambda_{ii}} \tag{14.16}$$

Thus, if $\lambda_{ii} = 0$, $\theta = T_N$, as in the case of ferromagnets, whereas if $\lambda_{ii} > 0, \theta > T_N$. Of course, there is an upper limit to the ratio θ / T_N, since, if λ_{ii} becomes too large in comparison to λ_{AB}, the assumed two-sublattice model becomes inadequate. The experimental values show that this ratio lies between 1.5 and 5.0 (with either sign). Measured parameters for some simple antiferromagnetic crystals are given in Table 14.2.

We have seen that, below the Neel temperature, $\mathbf{M}_A$ and $\mathbf{M}_B$ are antiparallel. The crystalline anisotropy is assumed for simplicity to be uniaxial, and it determines the common direction of $\mathbf{M}_A$ and $\mathbf{M}_B$ since these sublattice magnetizations are collinear with the easy direction. It should be remembered that there are always secondary effects in a crystal that make one or other of the crystal axes the preferred direction for the spontaneous magnetization. These are mainly the ordinary magnetic forces between the ions and the various effects due to the spin-orbit coupling which are neglected in the present two-sublattice model. This is justified, however, if the crystalline anisotropy field is small compared to the molecular field and if the applied field is not too large.

Table 14.2

Observed parameters for some simple antiferromagnetics[a]

J_1 and $J_2 (J_2 \ll J_1)$ are the exchange integrals for the nearest neighbors and the next nearest neighbors, respectively. J_1 has been estimated from the observed T_N, which was calculated by the Bethe-Peierls-Weiss method.

Compound	Magnetic sublattice[b]	Spin, S	T_N (°K)	$-\theta$ (°K)	C (μ_B)	$-J_1$ (°K)	$-J_2$ (°K)
$LaFeO_3$	s.c.	5/2	740		4.38	26.3	
$LaCrO_3$	s.c.	3/2	320		1.88	26.9	
$NdFeO_3$	s.c.	5/2	760			27.0	
$HoFeO_3$	s.c.	5/2	700			24.0	
$ErFeO_3$	s.c.	5/2	620			22.0	
$CaMnO_3$	s.c.	3/2	110			9.0	
$KMnF_3$	s.c.	5/2	88		4.74	3.1	
$KFeF_3$	s.c.	2	115		3.38	6.0	
$KCoF_3$	s.c.	3/2	114		3.08	9.6	
$KNiF_3$	s.c.	1	275		2.41	44.5	
CrF_3	s.c.	3/2	80		1.90	6.7	
FeF_3	s.c.	5/2	394			14.0	
CoF_3	s.c.	2	460			23.0	
MoF_3	s.c.	3/2	185			15.6	
CrF_2	b.c.c.	2	53			1.93	
MnF_2	b.c.c.	5/2	68		4.38	1.70	
FeF_2	b.c.c.	5/2	79			2.88	
NiF_2	b.c.c.	1	73			8.1	
MnO	f.c.c.	5/2	116	610	4.40	7.2	3.5
FeO	f.c.c.	2	198	570	6.24	7.8	8.2
CoO	f.c.c.	3/2	292	330	3.45	1.0	20.0
NiO	f.c.c.	1	523	3000	3.64	15.0	95.0
α-MnS	f.c.c.	5/2	154	465	4.23	4.4	4.5
β-MnS	s.c.	5/2	155	982	4.46	10.5	7.2
MnS_2	f.c.c.	5/2	60	590	4.92	5.6	5.9

[a]From J. S. Smart, Magnetism, Vol *3*, 63 (1963) (Eds. Rado and Suhl)
[b]s.c. = simple cubic; b.c.c. = body-centered cubic; f.c.c. = face-centered cubic.

Unfortunately, one cannot verify the predictions of the MF model by simple direct measurements, simply because $\mathbf{M}_A = -\mathbf{M}_B$ and hence $\mathbf{M} = 0$, a result that is in contrast to the one for ferromagnetic and ferrimagnetic substances. One can, however, determine the magnitudes of $\mathbf{M}_A$ and $\mathbf{M}_B$ by the use of neutron diffraction techniques by measuring the intensities of reflection of slow neutron beams from a crystal set at Bragg angles. It has been found that the observed intensity of the spontaneous magnetization very near $T = 0$ agrees approximately with the value $M_0 = \frac{1}{2} N g \mu_B S$ for some systems. This behavior seems to confirm that the orbital moments are, indeed, quenched in these compounds under the action of the asymmetric crystal fields. There are apparent disagreements also. For example, distinct deviations from the predictions of the MF theory have been detected. In particular, the decrease of $\mathbf{M}_A$ and $\mathbf{M}_B$ as $T(T<T_N)$ increases toward T_N is more rapid than is consistent with the low-temperature variation of $\mathbf{M}_A$. This may be due to a neglect of the long-range order. We shall not deal with this aspect.

Let us now derive the expressions for the susceptibility of a two-sublattice model, following the method of Van Vleck.[2] We continue to omit any explicit anisotropic forces and simply assume that $\mathbf{M}_A$ and $\mathbf{M}_B$ must remain in a preferred direction or at least be equally inclined to it. Two important situations to be considered are then (1) the applied field parallel to the preferred axis, and (2) the applied field perpendicular to this axis.

As discussed in Chapter 13.2, Dirac has shown that the exchange interactions are equivalent to an interatomic potential

$$V_{ij} = -\tfrac{1}{2}J(1+4\mathbf{S}_i\cdot\mathbf{S}_j) \tag{14.17}$$

where $\mathbf{S}_i$ and $\mathbf{S}_j$ are the spin angular momenta of atoms i and j, respectively, measured in multiples of $\hbar$, and J is the exchange integral, which is assumed to be negative for antiferromagnetic systems. We now let the subscripts i and j refer to atoms belonging to A and B sublattices, respectively. The effective potential V_i, to which a given atom i in sublattice A is subjected, may be obtained by summing over the various atoms j in sublattice B that are its nearest neighbors. Using the mean values for the positions of the spins of the j atoms and neglecting the additive constant J in (14.17), we may write the effective potential V_i as

$$V_i = -2J\mathbf{S}_i\cdot\sum_j \mathbf{S}_j = -2Jz\mathbf{S}_i\cdot\langle\mathbf{S}_j\rangle \tag{14.18}$$

where z is the number of nearest neighbors. In the absence of an applied field $\mathbf{H}$, we may assume that the mean values of the spins $\mathbf{S}_i$ and $\mathbf{S}_j$ have opposite directions. If the external field is now applied, there will be displacements $\delta\mathbf{S}_i$ and $\delta\mathbf{S}_j$ of the spins, which may be assumed parallel and proportional to $\mathbf{H}$ since we consider only the powder susceptibility and neglect the saturation effects. For small applied fields we may further assume that $\delta\mathbf{S}_i=\delta\mathbf{S}_j=\delta\mathbf{S}$. Thus we have for the two sublattices

$$\langle\mathbf{S}_i\rangle=\mathbf{S}_0+\delta\mathbf{S} \qquad \text{and} \qquad \langle\mathbf{S}_j\rangle=-\mathbf{S}_0+\delta\mathbf{S} \tag{14.19}$$

where $\mathbf{S}_0$ represents the spontaneous antiferromagnetism since it is oppositely directed for the two types of sublattices. The critical temperature above which $\mathbf{S}_0$ vanishes is the Neel temperature T_N. The magnetic moment per unit volume is

$$M=\tfrac{1}{2}Ng\mu_B\left[\langle\mathbf{S}_i\rangle+\langle\mathbf{S}_j\rangle\right]=Ng\mu_B\delta\mathbf{S} \tag{14.20}$$

where μ_B is the Bohr magneton, N is the number of atoms per unit volume, and g is the Landé factor. A given atom i or j also experiences an energy $-g\mu_B\mathbf{H}\cdot\mathbf{S}_i$ or $-g\mu_B\mathbf{H}\cdot\mathbf{S}_j$ due to the applied field $\mathbf{H}$, in addition to the potential (14.18). Hence, using (14.18) and (14.19), we can express the effective fields acting on atoms i and j belonging to sublattices A and B, respectively, as follows:

$$-g\mu_B\mathbf{H}_A^{\text{eff}}\cdot\mathbf{S}_i = -g\mu_B\mathbf{H}\cdot\mathbf{S}_i - 2Jz\mathbf{S}_i\cdot\mathbf{S}_j$$

and

$$-g\mu_B\mathbf{H}_B^{\text{eff}}\cdot\mathbf{S}_j = -g\mu_B\mathbf{H}\cdot\mathbf{S}_j - 2Jz\mathbf{S}_j\cdot\mathbf{S}_i$$

which, after simplification, are given by

$$\mathbf{H}_A^{\text{eff}} = \mathbf{H} + \frac{2Jz}{g\mu_B}(-\mathbf{S}_0 + \delta\mathbf{S})$$

and

$$\mathbf{H}_B^{\text{eff}} = \mathbf{H} + \frac{2Jz}{g\mu_B}(+\mathbf{S}_0 + \delta\mathbf{s})$$

Substituting for $\delta\mathbf{S}$ from (14.20), we find that these equations reduce to

$$\mathbf{H}_A^{\text{eff}} = \mathbf{H} + \frac{2Jz}{g\mu_B}(-\mathbf{S}_0) + \frac{2Jz}{Ng^2\mu_B^2}\mathbf{M}$$

and (14.21)

$$\mathbf{H}_B^{\text{eff}} = \mathbf{H} + \frac{2Jz}{g\mu_B}(+\mathbf{S}_0) + \frac{2Jz}{Ng^2\mu_B^2}\mathbf{M}$$

The magnetic moment $\mathbf{M}$ of atoms subjected to an effective field, as given above, in the direction of the applied field[3] is given by

$$\mathbf{M} = \tfrac{1}{2}g\mu_B NS\left[\cos(\mathbf{H}_A^{\text{eff}},\mathbf{H})B_S(x^A) + \cos(\mathbf{H}_B^{\text{eff}},\mathbf{H})B_S(x^B)\right] \quad (14.22)$$

where $B_S(x)$ is the familiar Brillouin function, and

$$x^A = \frac{g\mu_B S\mathbf{H}_A^{\text{eff}}}{kT}$$

(14.23)

$$x^B = \frac{g\mu_B S\mathbf{H}_B^{\text{eff}}}{kT}$$

For one electron per atom contributing to the moment we have $S=\frac{1}{2}$, and the Brillouin function $B_S(x)$ reduces to tanh x. In the derivation of (14.22) we have considered equal numbers of atoms in both sublattices. If we write the equations given above for the two sublattices separately, we have

$$\begin{aligned}|-\mathbf{S}_0+\delta\mathbf{S}| &= SB_S(x^B)\\ |+\mathbf{S}_0+\delta\mathbf{S}| &= SB_S(x^A)\end{aligned} \tag{14.24}$$

In the absence of the applied field, (14.24) reduce to

$$|\mathbf{S}_0| = SB_S(x_0) \tag{14.25}$$

where

$$x_0 = \frac{2zS|J\mathbf{S}_0|}{kT}$$

Equation (14.25) gives us the antiferromagnetic spontaneous field $\mathbf{S}_0$. The Neel temperature can be found from (14.25) on the assumption of small values for x_0, thus, at $T=T_{\text{N}}$,

$$\begin{aligned}x_0 &= \frac{2|J|zS}{kT_C}|S_0|\\ &= \frac{2|J|zS}{kT_{\text{N}}}\tfrac{1}{3}(S+1)x_0\end{aligned} \tag{14.26}$$

or

$$kT_{\text{N}} = \tfrac{2}{3}|J|zS(S+1) \tag{14.27}$$

since

$$\lim_{x_0\to 0} B_S(x_0) = \frac{S+1}{3S}x_0$$

It is to be noted, however, that (14.26) is identical to the expression obtained for T_C in the Weiss-Heisenberg theory for a ferromagnetic medium, except that $|J|=-J$.

We now study the region $T>T_{\text{N}}$, which is the paramagnetic region. Above the critical temperature we have $S_0=0$ (in 14.21), since the spontaneous magnetization vanishes. The cosines in (14.22) become unity, and again using (14.27), we have, from (14.22),

$$\chi = \frac{M}{H} = \frac{Ng^2\mu_{\text{B}}^2 S(S+1)}{3k(T+T_{\text{N}})} \tag{14.28}$$

with T_{N} as given in (14.26).

When $T<T_N$, there is spontaneous magnetization. In general, the applied field **H** may make any angle (α) with the preferred axis of magnetization, denoted by $\mathbf{S}_0$. But we shall first consider the case where **H** and $\mathbf{S}_0$ are parallel, so that the cosines in (14.22) are ± 1. The alternation in sign expresses the staggering of the internal molecular field, which is now larger than the applied field and hence determines the direction of $\mathbf{H}_{\text{eff}}$. Neglecting the saturation effects and expanding the Brillouin function in (14.22) in Taylor series in **H**, using (14.21), and keeping only the constant term, we have

$$B_S(x^A) = B_S(x_0) + \frac{g\mu_B S}{kT}\left(\mathbf{H} + \frac{2JzM}{Ng^2\mu_B^2}\right)B_S'(x_0)$$

and

$$B_S(x^B) = B_S(x_0) - \frac{g\mu_B S}{kT}\left(\mathbf{H} + \frac{2JzM}{Ng^2\mu_B^2}\right)B_S'(x_0) \tag{14.29}$$

where x^A, x^B, and x_0 are already defined, and $B_S'(x_0)$ is the derivative of $B_S(x_0)$ with respect to x_0. Substituting (14.29) in (14.22), we now get, for the susceptibility $\chi = M/H$,

$$\chi_{\parallel} = \frac{Ng^2\mu_B^2 S^2 B_S'(x_0)}{k\left[T + \{3T_N S/(S+1)\}B_S'(x_0)\right]} \tag{14.30}$$

for **H** parallel to $\mathbf{S}_0$. When only one electron per atom contributes to the spontaneous magnetization, we have $S=\frac{1}{2}$, and hence (14.30) reduces to

$$\chi_{\parallel} = \frac{Ng^2\mu_B^2}{4k\left[T_N + \{T/(1-4S_0^2)\}\right]} \tag{14.31}$$

Van Vleck derived the above relations considering only the nearest neighbors; when the next nearest neighbors are also considered, as we did earlier, we have to replace x_0 in (14.30) by $(\lambda_{AB} - \lambda_{ii})M_S S g\mu_B/kT$, where M_S is the corresponding magnitude of $\mathbf{M}_A$ (or $\mathbf{M}_B$) in zero applied field. The parallel susceptibility thus varies with temperature in the manner illustrated in Figure 14.3. At absolute zero, (14.30) predicts that $\chi_{\parallel}=0$. The physical reason is that, in the approximation of the MF theory, all the atomic moments are either parallel or antiparallel to the applied field at $T=0°$K. Hence the field does not exert any torque on the moments, and the induced magnetization is zero. As T increases, $\chi_{\parallel}$ also increases until, at the Neel temperature, $\chi_{\parallel}(T_N)$ becomes equal to the susceptibility given by (14.28).

We now consider the case where the applied field is perpendicular to the preferred axis of magnetization below the Neel temperature, that is, $\mathbf{H} \perp \mathbf{S}_0$. In this case the field acting on any given atom will consist of two parts: (1)

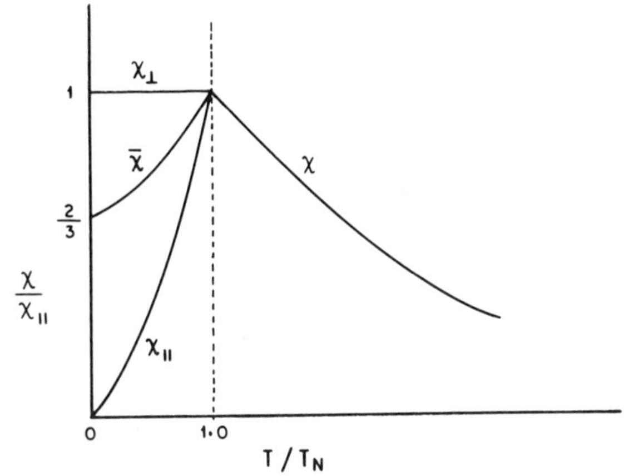

Fig. 14.3 The temperature dependence of susceptibility χ from the molecular field theory of an antiferromagnetic crystal.

the original inner field, $\mathbf{H}_x = \pm 2Jz\mathbf{S}_0/g\mu_B$, directed along, say, the x-axis, and (2) the applied plus the internal induced field, $\mathbf{H}_z = H + 2Jz\delta\mathbf{S}/g\mu_B$, directed along the z-axis. As $\mathbf{H}_z$ is small compared to $\mathbf{H}_x$, we may consider both the cosine factors in (14.22) to be H_z/H_x and replace x^A and x^B by x_0 in the arguments of the Brillouin function. Then the mean moment per unit volume is, by (14.22) and (14.25),

$$\mathbf{M} = \frac{N|\mathbf{H}_z|}{|\mathbf{H}_x|} g\mu_B|\mathbf{S}_0|$$

$$= \frac{Ng^2\mu_B^2(\mathbf{H} + 2Jz\,\delta\mathbf{S}/g\mu_B)}{2|J|z}$$

By using (14.20), eliminating J by means of (14.26), and solving for $\mathbf{M}$, we have, for the perpendicular susceptibility,

$$\chi_\perp = \frac{M}{H} = \frac{Ng^2\mu_B^2 S(S+1)}{6kT_N} \qquad (14.32)$$

which is independent of temperature. This is also illustrated in Figure 14.3.

The materials that are often studied in experiments are polycrystalline or occur in the form of powders, and it is reasonable to assume that the easy directions in the specimen are distributed at random. If the applied field makes an angle α with the easy direction of one crystallite or particle, the susceptibility of the specimen can be calculated by considering the components of the field parallel and perpendicular to $\mathbf{S}_0$. The corresponding contributions to the moment can be seen to be additive and give $M\cos^2(\mathbf{H}, \mathbf{S}_0)$ and $M\sin^2(\mathbf{H}, \mathbf{S}_0)$, respectively. Since the direction of the internal field may obviously be taken to be random in powder, we can replace $\cos^2(\mathbf{H},\mathbf{S}_0)$ and $\sin^2(\mathbf{H},\mathbf{S}_0)$ by the mean values $\frac{1}{3}$ and $\frac{2}{3}$, respectively, in as much as there are always three equivalent directions in a cubic crystal. Hence for $T<T_N$ we can safely write for the susceptibility

$$\bar{\chi}=\tfrac{1}{3}\bar{\chi}_{\parallel}+\tfrac{2}{3}\bar{\chi}_{\perp} \tag{14.33}$$

and

$$\begin{aligned}\chi&=\chi_{\perp} \quad \text{at} \quad T=T_N\\ &=\tfrac{2}{3}\chi_{\perp} \quad \text{at} \quad T=0\end{aligned} \tag{14.34}$$

The predictions of (14.33) along with (14.32) and (14.30) are in fairly good agreement with experiment. The most notable characteristic of an antiferromagnetic is the existence of a critical temperature T_N, at which the susceptibility reaches a maximum, and precisely this behavior is found in many substances,[4] such as MnF_2. The agreement between the theory traced above and the measured properties is not as satisfactory for all antiferromagnetics as it is for MnF_2, however. The ratios of the powder susceptibilities, for example, at $T=0$ and $T=T_N$, are generally decidedly greater than 0.67 [equation (14.34)]. The molecular field model has to be improved, therefore, if it is to cover all antiferromagnetics, quite apart from the neglect of the long-range order implicit in the MF theories. The two-sublattice model is a crude approximation because not all of the crystal structures exhibiting antiferromagnetism can be broken down into only two sublattices in such a way that the nearest neighbors of an ion all occupy sites in the other sublattice. Even in a crystal that can be so divided, if the interactions between ions on the same sublattice as measured by the molecular field constant γ_{ii} are antiferromagnetic and sufficiently strong, the pattern of order will not be as assumed above. In Figure 14.4 is shown this kind of structure (type 2), along with the structure of type 1, which is our two-sublattice model, both being appropriate for a body-centered lattice. In a type 2 magnetic structure the neighboring ions within the same sublattice are antiferromagnetically aligned at the expense of some ferromagnetically paired pairs involving both

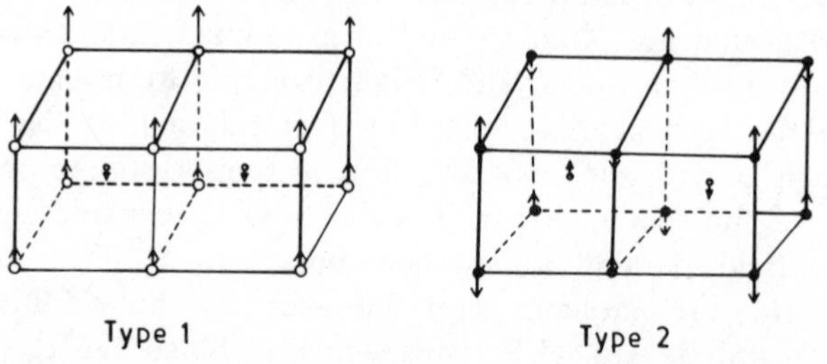

Fig. 14.4 Possible types of magnetic ordering for a b.c.c. lattice. (Taken from Smart.[5])

sublattices. By comparing the energies of the MF approximation, the type 2 structure is seen to be preferred to type 1 if both λ_{ii} and λ_{AB} are negative and $\lambda_{ii} > \frac{1}{2}|\lambda_{AB}|$. When $\lambda_{ii}/\lambda_{AB} = \frac{1}{2}, \theta/T_N$ takes the value -3, from (14.16); and since, in the case of type 2, $|\theta/T_N|$ decreases as the ratio $\lambda_{ii}/\lambda_{AB}$ increases, this is the lowest value allowed in this model. Smart[5] has reviewed the original calculations of Van Vleck and Anderson.

14.2 THE FOUR-SUBLATTICE MODEL

In the preceding section we considered the two-sublattice model in the MF theory of Neel and Van Vleck, which is appropriate for the body-centered cubic (b.c.c.) lattice. Many antiferromagnetic materials, however, crystallize in structures for which the two-sublattice model does not hold good. The most common of these is the face-centered cubic (f.c.c.) lattice, which can be divided into no less than *four* simple-cubic sublattices which have the property that the sublattice does not contain the nearest neighbors. Figure 14.5 describes a f.c.c. structure that has been divided into four sublattices, designated by 1, 2, 3, and 4. It is clear from this figure that each sublattice contains four nearest neighbors in each of the other three sublattices. We now give a detailed treatment of this structure, following the work of Anderson.[6]

As we saw in (14.18), the interaction energy is given by

$$V_{nm} = 2Jz\mathbf{S}_n \cdot \langle \mathbf{S}_m \rangle_{\text{av}} \tag{14.35}$$

where J is the exchange integral, and z is the number of nearest neighbors, which is 4 for a f.c.c. lattice, as discussed above, and ranges from 1 to 4 since there are four sublattices. The effective field, similar to (14.21), is then given by

$$\mathbf{H}_n^{\text{eff}} = \mathbf{H}_0 - \frac{2Jz}{g\mu_B} \sum_{m \neq n} \langle \mathbf{S}_m \rangle_{\text{av}} \tag{14.36}$$

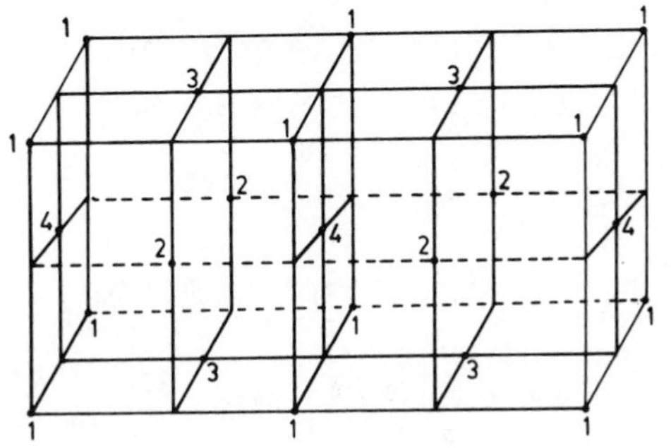

Fig. 14.5 Four sublattices belonging to f.c.c. structure. (Taken from Anderson.[6])

and the magnetization is

$$\mathbf{M}=\tfrac{1}{4}Ng\mu_{\mathrm{B}}\sum_{n}\mathbf{S}_n \tag{14.37}$$

where N is the number of atoms. Similarly to (14.25), we have also in this case

$$\bar{S}_n = SB_S(x_n) \tag{14.38}$$

where

$$x_n=\left(\frac{g\mu_{\mathrm{B}}S}{kT}\right)\mathbf{H}_n^{\mathrm{eff}} \tag{14.39}$$

and $B_S(x)$ is the well-known Brillouin function. If we now first consider the region $T>T_{\mathrm{N}}$, that is, the paramagnetic region, where the effective fields are much smaller than the external field, we can expand the Brillouin function (14.38) for small values of x_n, getting

$$SB_S(x)\cong\tfrac{1}{3}(S+1)x \tag{14.40}$$

Thus, from (14.38), using (14.39), (14.40), and (14.36), we have

$$\mathbf{S}_n=\tfrac{1}{3}S(S+1)\left(\frac{g\mu_{\mathrm{B}}}{kT}\right)\left(\mathbf{H}_0-\frac{2Jz}{g\mu_{\mathrm{B}}}\sum_{m\neq n}\mathbf{S}_m\right) \tag{14.41}$$

We now have to sum over n from 1 to 4 to get the magnetization. Let

$$\mathbf{S}=\sum_{n=1}^{4}\mathbf{S}_n$$

and, putting $\alpha'=\frac{1}{3}S(S+1)(g\mu_{\mathrm{B}}/kT)$ and $\beta'=2Jz/g\mu_{\mathrm{B}}$, we have, for $n=1$ to 4,

$$\begin{aligned}
\mathbf{S}_1&=\alpha'[\mathbf{H}_0-\beta'(\mathbf{S}_2+\mathbf{S}_3+\mathbf{S}_4)]\\
\mathbf{S}_2&=\alpha'[\mathbf{H}_0-\beta'(\mathbf{S}_1+\mathbf{S}_3+\mathbf{S}_4)]\\
\mathbf{S}_3&=\alpha'[\mathbf{H}_0-\beta'(\mathbf{S}_1+\mathbf{S}_2+\mathbf{S}_4)]\\
\mathbf{S}_4&=\alpha'[\mathbf{H}_0-\beta'(\mathbf{S}_1+\mathbf{S}_2+\mathbf{S}_3)]
\end{aligned}$$

Adding, we get

$$\begin{aligned}
\sum_{n=1}^{4}\mathbf{S}_n&=\alpha'[4\mathbf{H}_0-3\beta'(\mathbf{S}_1+\mathbf{S}_2+\mathbf{S}_3+\mathbf{S}_4)]\\
&=\tfrac{1}{3}S(S+1)\left(\frac{g\mu_{\mathrm{B}}}{kT}\right)\left(4\mathbf{H}_0-\frac{6Jz}{g\mu_{\mathrm{B}}}\sum_{n=1}^{4}\mathbf{S}_n\right)
\end{aligned} \tag{14.42}$$

or

$$\sum_{n=1}^{4} \mathbf{S}_n = \frac{\frac{1}{3} S(S+1)(4g\mu_B/kT)\mathbf{H}_0}{1+[2JSz(S+1)/kT]} \tag{14.43}$$

Hence the magnetization, from (14.37) and (14.43), is given by

$$\mathbf{M} = \left(\tfrac{1}{4} Ng\mu_B\right) \frac{\frac{1}{3} S(S+1)(4g\mu_B/kT)\mathbf{H}_0}{1+[2JzS(S+1)/kT]}$$

$$= \frac{\frac{1}{3} S(S+1) Ng^2\mu_B^2/k \cdot \mathbf{H}_0}{T+2JzS(S+1)/k}$$

Thus the susceptibility χ for $T > T_N$ is given by

$$\chi_{T>T_N} = \frac{C}{T+\theta} \tag{14.44}$$

where

$$C = \frac{\frac{1}{3} S(S+1) Ng^2\mu_B^2}{k}$$

and (14.45)

$$\theta = \frac{2JzS(S+1)}{k}$$

To find the Neel temperature ($T = T_N$), we use (14.41) with $\mathbf{H}_0 = 0$, getting

$$\mathbf{S}_n + \tfrac{1}{3} S(S+1)\left(\frac{2Jz}{kT}\right) \sum_{m \neq n} \mathbf{S}_m = 0 \tag{14.46}$$

which can be written in terms of a determinant, given by

$$\begin{vmatrix} 1 & \alpha & \alpha & \alpha \\ \alpha & 1 & \alpha & \alpha \\ \alpha & \alpha & 1 & \alpha \\ \alpha & \alpha & \alpha & 1 \end{vmatrix} = 0 \tag{14.47}$$

where $\alpha = 2JzS(S+1)/3kT$.

Solving the determinant (14.47), we have

$$(1-\alpha)^3(1+3\alpha) = 0$$

Leaving out the impossible root, we have, for $T=T_{\mathrm{N}}$,

$$\alpha=1$$

or

$$\frac{2S(S+1)Jz}{3kT_{\mathrm{N}}}=1$$

or, finally,

$$T_{\mathrm{N}}=\frac{2JzS(S+1)}{3k}=\frac{\theta}{3} \tag{14.48}$$

where θ is already defined by (14.45).

It is now easy to find the susceptibility at $T=T_{\mathrm{N}}$, for which we use (14.44); thus

$$\chi_{T_{\mathrm{N}}}=\frac{\frac{1}{3}S(S+1)Ng^2\mu_{\mathrm{B}}^2}{k[(\theta/3)+\theta]}$$

Substituting the value for θ in this equation, we have

$$\chi_{T_{\mathrm{N}}}=\frac{Ng^2\mu_{\mathrm{B}}^2}{8Jz} \tag{14.49}$$

We now study the behavior of the antiferromagnetic f.c.c. lattice below the Neel temperature, that is, for $T<T_{\mathrm{N}}$. In this case it is impossible to find a unique value for χ because the only restriction on the spins is

$$\sum_n \mathbf{S}_n=0 \tag{14.50}$$

as can be readily seen from (14.46) by putting $T=T_{\mathrm{N}}$. Equation (14.50) shows clearly that an infinite number of possible arrangements of spins are possible, and the actual arrangement can finally be decided by including the next-nearest-neighbor interactions (neglected at the present moment but to be included later) and the anisotropy forces, which always exist in the solid. Anderson[6] therefore assumes two possible arrangements to show what forms of χ are to be expected.

The first arrangement we choose is that in which two spins are parallel and two antiparallel:

$$\begin{matrix}\uparrow & \uparrow & \downarrow & \downarrow\\ \mathbf{S}_1 & \mathbf{S}_2 & \mathbf{S}_3 & \mathbf{S}_4\end{matrix}$$

The spontaneous magnetization can be determined from (14.38), which is

$$\begin{aligned}\mathbf{S}_n &= SB_S(x_n)\\ &= SB_S\left[\frac{g\mu_B S}{kT}\left\{\mathbf{H}_0 - \frac{2Jz}{g\mu_B}\sum_{m\neq n}\langle\mathbf{S}_m\rangle\right\}\right]\end{aligned}$$

But from (14.46) and (14.48) we have

$$\sum_{m\neq n}\mathbf{S}_m = -\mathbf{S}_n$$

Using this in the equation given above, we have

$$\mathbf{S}_n = SB_S\left[\frac{g\mu_B S}{kT}\left\{\mathbf{H}_0 + \frac{2Jz}{g\mu_B}\mathbf{S}_n\right\}\right]$$

Now, putting $\mathbf{H}_0 = 0$, this equation reduces to

$$S_n^0 = SB_S\left[\left(\frac{2SJz}{kT}\right)S_n^0\right] \tag{14.51}$$

Expanding all quantities about their mean values with no external field, we have

$$\begin{aligned}\mathbf{H}_n^{\text{eff}} &= \mathbf{H}_n^0 + \delta\mathbf{H}_n\\ \mathbf{S}_n &= \mathbf{S}_n^0 + \delta\mathbf{S}_n\end{aligned} \tag{14.52}$$

Then

$$\delta\mathbf{H}_n = \mathbf{H}_0 - \left(\frac{2Jz}{g\mu_B}\right)\sum_{m\neq n}\delta\mathbf{S}_m \tag{14.53}$$

We now consider two separate cases: (1) $\mathbf{H}_0$ parallel to $\mathbf{S}_0$, and (2) $\mathbf{H}_0$ perpendicular to $\mathbf{S}_0$.

Case 1. Expanding the Brillouin function in Taylor series, we have

$$\mathbf{S}_n^0 + \delta\mathbf{S}_n = S\left[B_S\left(\frac{g\mu_B S\mathbf{H}_n^0}{kT}\right) + \delta H_n\left(\frac{g\mu_B S}{kT}\right)B_S'\right]$$

whence

$$\begin{aligned}\delta\mathbf{S}_n &= \delta\mathbf{H}_n\left(\frac{g\mu_B S^2}{kT}\right)B_S'\\ &= \left[H_0 - \left(\frac{2Jz}{g\mu_B}\right)\sum_{m\neq n}\delta\mathbf{S}_m\right]\left(\frac{g\mu_B S^2}{kT}\right)B_S'\end{aligned}$$

We now sum up this equation for $n=1$ to 4 and get

$$\sum \delta\mathbf{S}_n=\left(4\mathbf{H}_0-\frac{6Jz}{g\mu_\mathrm{B}}\sum \delta\mathbf{S}_n\right)\left(\frac{g\mu_\mathrm{B}S^2}{kT}\right)B_S'$$

or

$$\sum_n \delta\mathbf{S}_n=\frac{4\mathbf{H}_0(g\mu_\mathrm{B}S^2/kT)B_S'}{1+(6JzS^2/kT)B_S'}$$

The parallel susceptibility is then given by

$$\chi_\parallel=\frac{\frac{1}{4}Ng\mu_\mathrm{B}\sum_n \delta\mathbf{S}_n}{\mathbf{H}_0}=\frac{(Ng^2\mu_\mathrm{B}^2S^2/kT)B_S'}{1+(6JzS^2/kT)B_S'} \tag{14.54}$$

At absolute zero, since

$$B_S'(\infty)=0, \chi_\parallel=0. \tag{14.55}$$

Case 2. Here the spins have to be rotated parallel to the effective field. Since

$$\frac{\delta\mathbf{S}_n}{\mathbf{S}_n}=\frac{\delta\mathbf{M}}{\mathbf{M}}=\frac{\delta\mathbf{H}_n}{\mathbf{H}_n} \tag{14.56}$$

we have

$$\delta\mathbf{S}_n=|S_n^0|\frac{\delta\mathbf{H}_n}{|H_n^0|}$$

$$=|S_n^0|\frac{\mathbf{H}_0-(2Jz/g\mu_\mathrm{B})\sum_{m\neq n}\delta\mathbf{S}_m}{(2Jz/g\mu_\mathrm{B})\sum_{m\neq N}S_m^0}$$

and, since all $\delta\mathbf{S}_n$ are equal, as a first approximation this gives us

$$4\delta\mathbf{S}_n=\frac{g\mu_\mathrm{B}H_0}{2Jz}$$

whence

$$\chi_\perp=\frac{Ng^2\mu_B^2}{8Jz} \tag{14.57}$$

Equation (14.57) shows that $\chi_\perp$ is independent of temperature and is equal to

χ_{T_N} at $T = T_N$ (14.49). Since

$$\chi = \tfrac{1}{3}\chi_{\parallel} + \tfrac{2}{3}\chi_{\perp}$$

at absolute zero, $\chi_{\parallel} = 0$ and hence

$$\chi_{T=0} = \tfrac{2}{3}\chi_{\perp} = \tfrac{2}{3}\chi_{T_N} \tag{14.58}$$

For this particular case, therefore, we get the same result as that given by Van Vleck[2] [see (14.34)].

We now examine the second arrangement, in which two pairs of spins are mutually perpendicular. If $\mathbf{H}_0$ is perpendicular to all four spins, we get essentially the same result as in the previous case, that is, $\chi_{\perp} = \chi_{T_N}$. But if $\mathbf{H}_0$ is parallel to the plane of the spins, we can easily show that

$$\chi_{\parallel} = \chi_{\perp} = \chi_{T_N} = \frac{Ng^2\mu_B^2}{8Jz} \tag{14.59}$$

Thus

$$\chi_{T=0} = \chi_{T_N} \tag{14.60}$$

which does not agree with Van Vleck's result (14.58).

A comparison of Anderson's theory with experiment is given in Table 14.3. It was pointed out by Neel[7] that the large θ / T_N ratio observed for some compounds can be explained by introducing the next-nearest-neighbor interaction λ_{ii}, along with the nearest-neighbor interaction λ_{AB}. Then θ / T_N

Table 14.3

θ / T_N values of some antiferromagnetic compounds[a]

Compound	Type of structure	Magnetic sublattice	Number of sublattices	θ (°K)	T_N (°K)	$\frac{\theta}{T_N}$	$\frac{\lambda_{ii}}{\lambda_{AB}}$	$\frac{\chi_{T=0}}{\chi_{T=T_N}}$
MnF_2	Rutile	b. c. rectangular	2	113	72	1.57	—	0.76
FeF_2	Rutile	b. c. rectangular	2	117	79	1.48	—	0.72
MnO	NaCl	f. c. c.	4	610	122	5.0	0.5	0.67
FeO	NaCl	f. c. c.	4	570	198	2.9	1.07	0.8
MnS	NaCl	f. c. c.	4	528	165	3.2	0.91	—
MnSe	NaCl	f. c. c.	4	~435	~150	~3.0	—	—

[a] P. W. Anderson, *Phys. Rev.* **79**, 350 (1950).

becomes $(1+\lambda_{ii}/\lambda_{AB})/(1-\lambda_{ii}/\lambda_{AB})$. The $\lambda_{ii}/\lambda_{AB}$ ratios (the higher values are mentioned only), computed by Anderson, are also given in Table 14.3. The somewhat larger values of $\lambda_{ii}/\lambda_{AB}$ resulting from Anderson's theory may be due to the superexchange mechanism,[6] which becomes the dominant factor if the separations between the magnetic ions are too large for the direct exchange mechanism to be operative. In the antiferromagnetic crystals (e.g., the monoxides of the iron-group ions: Mn^{2+}, Ni^{2+}, etc.) the ions that interact are separated by more than 4 Å, and the overlap of their actual atomic *d*-shell wavefunctions is therefore negligible. We shall now see how the superexchange mechanism is responsible for producing antiferromagnetism in such systems.

14.3 SUPEREXCHANGE IN ANTIFERROMAGNETICS

By now it is sufficiently clear that in many antiferromagnetic materials the atoms that strongly interact magnetically are definitely separated from each other by intervening nonmagnetic ions. The best example of such a material is MnO, in which it has been possible to actually identify the pattern of the magnetic spins below the Neel point by means of neutron diffraction measurements.[8] It should be remembered that MnO possesses an NaCl type of structure in which the Mn^{2+} ions are situated at the face-centered positions. It has been observed that the 12 nearest neighbors to a particular Mn^{2+} ion are apparently uncorrelated, but if the f. c. c. lattice is regarded as consisting of four separate simple-cubic lattices, it can be seen that each of these lattices is lined up in a perfect antiferromagnetic pattern but there is no correlation amongst them. This situation is illustrated in Figure 14.6, where it is clearly evident that between the Mn^{2+} ions there is an O^{2-} ion. Kramers[9] first showed in 1934 that the exchange interactions between the magnetic ions take place through the agency of the intervening nonmagnetic ions. One has to assume, therefore, that, in addition to the ionic state $Mn^{2+}O^{2-}$, there is a considerable admixture of the state in which at least one *p* electron from the O atom is transferred to the *s* or *d* state of Mn^{2+}, as a result of which oxygen becomes paramagnetic and thus can take part in the magnetic interactions.

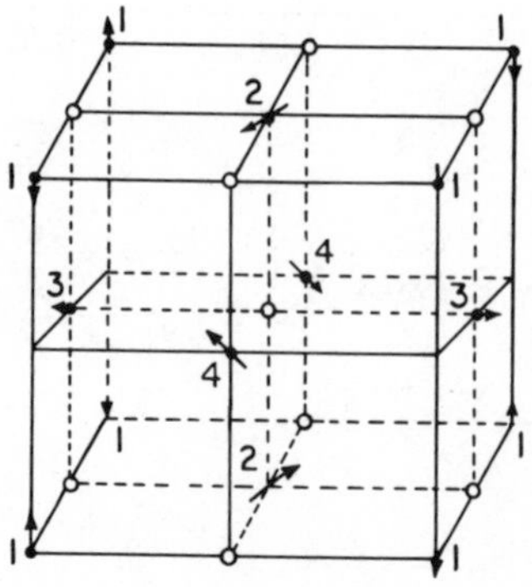

Fig. 14.6 The Mn0 structure. Solid circles represent the Mn^{2+} ions with either up or down spins. The hollow circles represent the O^{2-} ions.

Let us consider the system consisting of two Mn^{2+} ions separated by an O^{2-} ion. If, now, one p electron from O^{2-} goes over to the Mn^{2+} site, O^{2-} becomes O^{-}, which can have exchange interaction with the next Mn^{2+} ion. As a result the excited configuration becomes $Mn^{+}O^{-}Mn^{2+}$. This situation is picture in Figure 14.7. Since one electron is transferred from O^{2-} with down spin, say, to the left-hand Mn^{2+}, there will be an antiferromagnetic exchange between this p electron and one of Mn^{2+}, as a result of which O^{2-} will become O^{-}; since this is now paramagnetic, it will have antiferromagnetic coupling with the next Mn^{2+} ion. This is the way in which the superexchange takes place between two Mn^{2+} ions via the intervening O^{2-} ion. For a clear exposition of this "transfer" mechanism the reader is referred to Van Vleck.[10] The discovery and measurement of the hyperfine interaction of the ligand nuclear spins with the electron spins of the magnetic ion demonstrated conclusively that the ligand wavefunctions are partially magnetic, and to the expected degree[11,12] which tallies with the description given above. A quantitative calculation on these lines is very complicated, however, and leads to uncertainties.

To see the situation more clearly, let us consider the above-mentioned linear three-ion system, that is, two transition metal ions separated by an O ion. We shall, for simplicity, consider only two p electrons with antiparallel spins on the O ion and shall consider a single unpaired electron in the d state of each metal (magnetic) ion. We also consider, by hypothesis, that there is a finite probability that an electron will jump from the O ion into a d state of one of the magnetic ions. Qualitatively, it is now easy to see the superexchange process. In the ground as well as in the excited states there are only two possible spin states: a singlet and a triplet corresponding to the total spins $S=0$ and $S=1$, respectively. Now in the unperturbed, perfectly ionic ground state, the energy of the system is independent of the spin configuration because the paired electrons in the O ion can take no part in the exchange effects and the direct exchange between the metal ions is negligible owing to the large distance (of the order of 4 Å) between them. In the excited state, on the other hand, the energy depends on the spin configuration. The electron transferred from the oxygen to the metal ion will be subject to the usual Hund rules governing the influence of *intraionic* exchange, and it will take a parallel alignment with respect to the unpaired electron already on the ion. In this excited state there will be interionic exchange between the oxygen and the metal ions, and if we assume that these favor antiparallel alignment,

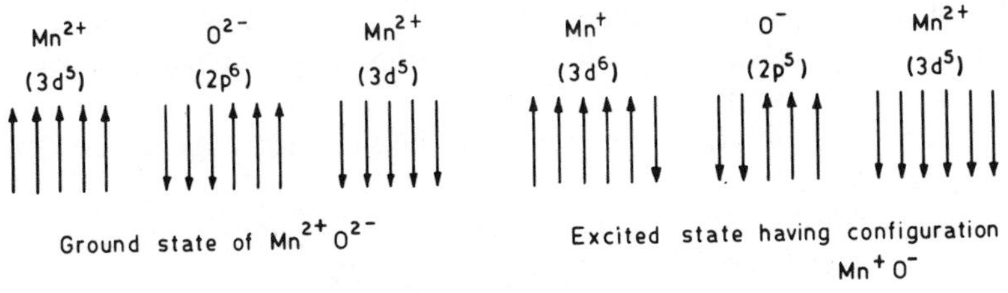

Fig. 14.7 The superexchange mechanism in $Mn^{2+}O^{2-}$.

Table 14.4

Ground and excited states of M–O–M configuration involving four electrons

State:	M	O	O	M	Total spin
Orbitals:	d_1, s	p	p	d_2	
Ground Singlet:	↑	↑	↓	↓	$S=0$
Ground Triplet:	↑	↑	↓	↑	$S=1$
Excited Singlet:	↑↓		↑	↓	$S=0$
Excited Triplet:	↑↓		↑	↑	$S=1$

the spin configuration of lower energy is a triplet state. There are several ways of assigning the spin orientations of the four electrons in the $3d$ and p orbitals of M–O–M configuration, however. If the p electron transferred from the oxygen to one of the metal atoms occupies an s state, the interaction scheme can be pictured as in Table 14.4.

Kramers, in his original paper, proposed a perturbation treatment of the superexchange problem. Before we describe his method, it is important to realize that, if the form of the perturbation causing the distortion from perfect ionicity is spin independent, it can be shown that the triplet and the singlet states will not be mixed by the perturbation. The perturbed wavefunction has two forms when the spin functions are included, one being a combination of the ground and the excited triplet states and the other a combination of the corresponding singlet states. Thus, if s denotes the singlet and t the triplet state, we have

$$\begin{aligned} \psi_s' &= a\psi_{\text{gr}}^s + b\psi_{\text{ex}}^s \\ \psi_t' &= a'\psi_{\text{gr}}^t + b'\psi_{\text{ex}}^t \end{aligned} \tag{14.61}$$

From (14.61) it is evident that the singlet and the triplet states, corresponding to $S=0$ and $S=1$, respectively, are well defined. Now we have to find the energies of ψ_s' and ψ_t', the difference between which gives us the desired exchange energy. It is clear from (14.61) that this difference in energy arises because of the admixed excited state functions since the ground state energy is spin independent. It can be shown that ψ_t' has the lower energy. Although the spin-dependence comes from the admixed excited state, the spatial electron distribution closely resembles that for the unperturbed ionic state (since $a \gg b$). The ionic state can therefore be retained as an approximate picture of the true state. The effect of perturbation induces a strong coupling between the spins of the metal ions, thus producing a splitting between the singlet and triplet configurations.

Let us now write the final expression for the superexchange interaction $H_{tt'}$, given by Kramers:

$$H_{tt'} = \sum_{u,u'} \frac{\langle t|H'|u\rangle\langle u|H_{\text{ex}}|u'\rangle\langle u'|H'|t'\rangle}{(E_u - E_t)(E_{u'} - E_{t'})} \tag{14.62}$$

where t and t' represent the various initially degenerate spin states of the ground state, u and u' refer to the various possible spin states of the excited orbital states, H' denotes the potential representing the transfer of an electron from the O atom to the metal, and H_{ex} is the exchange interaction. In Kramers' formula we have included, following Anderson, the intraionic exchange coupling between the d and s electrons on the Mn^+ ion in the unperturbed Hamiltonian, so that E_u and $E_{u'}$ are not necessarily equal. In this theory the superexchange effect appears to the *third* order in the perturbation theory. Kramers' "ground" state is defined to be a state in which the electronic wavefunctions are orthogonal, all "nonmagnetic" ions contain only the filled shells, and although the interionic exchange is neglected, the *intraionic* exchange is included.

We shall now see how we can evaluate the sum in (14.62), following the method given by Anderson. To calculate the matrix elements in Kramers' expression, Anderson extended the Dirac-Van Vleck vector model described in Chapter 13 of spin coupling by exchange forces to the problem of the interaction of the two configurations, the ground and the excited, discussed above for the four-electron system. As stated above, Kramers' expression is of the correct form if this interaction can be assumed small. The ground configuration G has already been defined as consisting of one electron on each of the two magnetic ions (orbitals d_1 and d_2) and two electrons in the p orbital of the intervening ion. The excited configuration E is that in which one electron from the oxygen p orbital is transferred into an s orbital situated on one of the magnetic ions. This method can easily be extended to more complicated systems. Kramers' expression involves three matrix elements, two of which, designated by $\langle G\|E\rangle$, connect G with E and one, designated by $\langle E\|E\rangle$, involves only the excited configuration. We shall not go into the details of the derivation of the matrix elements, which are given in Anderson's paper.[6] We shall, however, state a few points and then give the final results.

The spin-dependent part of the Hamiltonian can be written in terms of operators referring to the spins of the electrons in the various orbitals. If the orbitals are designated by i, j, and so on, the Dirac-Van Vleck exchange Hamiltonian is given by

$$V = -\frac{1}{2}\sum_{i<j} V_{ij}(1+\boldsymbol{\sigma}_i\cdot\boldsymbol{\sigma}_j) \tag{14.63}$$

where V_{ij} is the matrix element referring to an interchange of orbitals i and j. Dividing this equation into two parts, V_1 and V_2, where V_1 contains all the terms that are included in E_u in our four-electron problem, we have

$$V_1 = \tfrac{1}{2}V_{d_1p_1}(1+\boldsymbol{\sigma}_{d_1}\cdot\boldsymbol{\sigma}_{p_1})$$

and (14.64)

$$V_2 = \tfrac{1}{2}V_{p_2d_2}(1+\boldsymbol{\sigma}_{p_2}\cdot\boldsymbol{\sigma}_{d_2})$$

Now, assuming the orbital states to be

$$\begin{aligned}\psi_t &= a(1)b(2)c(3)d(4)\cdots \\ \psi_u &= a'(1)b(2)c(3)d(4)\cdots\end{aligned} \tag{14.65}$$

where t and u are taken to represent the sets of spins $s_1, s_2, \ldots,$ belonging to various orbits, Anderson, after some clever manipulations, shows the superexchange energy of Kramers to be given by

$$H_{tt'} = \alpha^2 \langle t | V_2' | t' \rangle \tag{14.66}$$

where

$$\alpha = \frac{\langle a | H' | a' \rangle}{(E_u - E_t)_{\text{lowest}}}$$

and $V_2' = \mathcal{P} V_2$, where $\mathcal{P}$ is an appropriate operator that "projects" V_2 on the subspace belonging only to the lowest eigenvalue of V_1. This means that V_2 is modified suitably so as to neglect the higher states because of large denominators, in the cases where V_1 is sufficiently strong. The method of modifying V_2 in such a way as to eliminate the states belonging to higher values of E_u is well known.

Anderson has discussed two cases to make this point clearer. For example, consider the case where $V_{d_1 s}$ is *large and positive*, σ_{d_1}, σ_s are favored, and $\mathcal{P}$ should be taken such that it eliminates all states for which σ_{d_1} and σ_s are antiparallel. The form of $\mathcal{P}$ is then given by

$$\mathcal{P}_1 = \tfrac{1}{4}(3 + \boldsymbol{\sigma}_{d_1} \cdot \boldsymbol{\sigma}_s) \tag{14.67}$$

whence

$$H_{tt'} = \tfrac{1}{4}\alpha^2 \langle t | (3 + \boldsymbol{\sigma}_{d_1} \cdot \boldsymbol{\sigma}_{p_1}) V_{p_2 d_2} (1 + \boldsymbol{\sigma}_{p_2} \cdot \boldsymbol{\sigma}_{d_2}) | t' \rangle \tag{14.68}$$

where we have simply read $\boldsymbol{\sigma}_{p_1}$ for $\boldsymbol{\sigma}_s$, this being the electron making the transition.

Similarly, when $V_{d_1 s}$ is *large and negative*, we have

$$\mathcal{P}_2 = \tfrac{1}{4}(1 - \boldsymbol{\sigma}_{d_1} \cdot \boldsymbol{\sigma}_s) \tag{14.69}$$

whence

$$H_{tt'} = \tfrac{1}{4}\alpha^2 \langle t | V_{pd} (1 - \boldsymbol{\sigma}_{d_1} \cdot \boldsymbol{\sigma}_{p_1})(1 + \boldsymbol{\sigma}_{p_2} \cdot \boldsymbol{\sigma}_{d_2}) | t' \rangle \tag{14.70}$$

where

$$\alpha = \frac{\langle s | H' | p \rangle}{E_{\text{bonding}} - E_{\text{ionic}}} \tag{14.71}$$

We shall not go into the obvious generalization of this method but instead discuss a few important points. The dependence of the coupling (ferromagnetic or antiferromagnetic) on the sign of the internal coupling can be verified in some cases. By Hund's rule it is well known that an electron will be added most easily with parallel spin for ions with less than half-filled shells (e.g., Cr^{2+}) and with antiparallel spin for ions with half-filled or more than half-filled shells, such as Mn^{2+} and Fe^{2+}. Thus it is an observed fact that CrTe is ferromagnetic and MnTe is antiferromagnetic, although the two are very similar in structure and in many other ways.

A few concluding remarks may be helpful. The calculation of the superexchange, as given above, based on a particular excited state, can be extended, in principle, to include other excited states as well. Several cases have been considered where both p electrons are simultaneously transferred, one to each metal ion. This gives rise to a number of exchange couplings, however, depending on the number of transfers involved in the process. Several difficulties arise in making quantitative calculations of this type because of the fact, discussed earlier, that the spin-dependent effects involve rather higher orders in the perturbation theory, and nonorthogonality is also difficult to handle.

Anderson[13] has shown how one can approach the problem in a more general manner. He points out that superexchange involves two aspects: the "potential" exchange, which involves the ferromagnetic exchange, and the "kinetic" exchange involving the antiferromagnetic exchange effect. To perform a quantitative calculation, detailed information about the localized functions, modified by the interaction with the surrounding diamagnetic ions, is required, and this again poses a difficult problem. We saw in Chapter 6 that with the help of the crystal field theory it is possible to classify the energy levels and momenta of ions in crystals where the interaction is represented by an electrostatic field. Even when there is appreciable covalent bonding between the ions, this theory, known as the ligand field theory (Chapter 6), is adequate. When details of the wavefunctions are important, as they are in the superexchange problem, however, more sophisticated theories are needed to handle such a complicated situation.[14] This two-stage approach, advocated by Anderson, has a great advantage in that it provides more opportunity for using experimental findings to help keep the theory on the right track. In particular, nuclear magnetic resonance and electron spin resonance experiments give direct evidence of covalent transfers of electrons. Theoretical estimates for many oxides and halides of the $3d^n$ compounds have been made by Anderson, and the exchange energies thus estimated are within a factor of 2 of those implied by the observed Neel temperatures. This is indeed very creditable in the sense that for no other class of magnetic materials can exchange energies be computed as closely. It is experimentally observed that in most ionic solids the coupling is antiferromagnetic, a situation that may be due to the fact that the "kinetic" exchange predominates over the "potential" exchange for nonorthogonality reasons[13] (localized functions will not normally be orthogonal to each other).

As an application of the MF theory to compounds of the palladium and platinum groups we refer to the work done by De, Desai, and Chakravarty,[25] who have calculated the magnetic susceptibility of Ir^{4+} ($5d^5$) in the antiferromagnetic and paramagnetic phases, wherein the molecular field is calculated self-consistently. It has been shown that the observed and the calculated susceptibilities in the antiferromagnetic phase agree well.

In conclusion it may be pointed out that the molecular field model is very useful to treat the influence of the crystal field on the susceptibility of the antiferromagnetic phase in the compounds of the transition metal ions, as in the rare earth compounds.[21]

14.4 THE CORRELATED EFFECTIVE FIELD THEORY

We shall not describe in some detail the correlated effective field (CEF) theory proposed by Lines,[15,16] which may be regarded as equivalent (though not exactly) to the random phase approximation (RPA) of the Green function theory in the Heisenberg limit, which, of course, assumes the orbital moment to be quenched because of the asymmetric crystal fields existing in the solid. In many-body magnetism it is rather difficult to describe a magnetic solid consisting of interacting magnetic ions when the thermal energy kT, the exchange interaction energy J, and the crystal field splittings are of the same order of magnitude. It is therefore almost impossible to describe such solids in terms of a spin Hamiltonian, especially in systems where the quenching of the orbital angular momentum due to the crystal field is not complete. For instance, there are a number of transition metal oxides (e.g., MnO) where, as we saw in Section 14.3, the *indirect* exchange via an intervening nonmagnetic ion (the superexchange process) is more important than the *direct* exchange (its unimportance is due to the large distance between the two magnetic ions).

The CEF theory of Lines has great mathematical simplicity and, therefore, an advantage, which is due to the reduction of the magnetic interactions to effective local fields. This reduces the many-body problem to the problem of an ensemble of noninteracting effective ions, while maintaining a simple physical picture in terms of single-ion energy levels and their perturbation by the local fields. Using this approach, one can also assess, in a comparatively easier fashion, the exchange effects on the excited orbital levels. The molecular field is, of course, represented very crudely in this model in the sense that all the ions except *one* are replaced by their *ensemble* averages, a substitution that amounts to neglecting all interion spin and orbital static correlations. This constitutes a weakness of the model, as pointed out by Lines himself.

At very low temperatures it is possible to neglect the thermal population of the excited orbital levels, so that one can confine oneself to the ground state; this can then be described by an effective spin Hamiltonian, which was presented in Chapter 8. But for higher temperatures, where the excited levels are also populated to an appreciable extent, this approach is no longer possible, and one needs a formalism that can effectively handle such a situation. In the CEF theory, Lines has proposed such a model, which incorporates the important effects of static spin and orbital correlations in a reasonably sophisticated manner. To do this, Lines introduced the concept of a "correlated effective field," replacing the "molecular field." We shall deal only with the static properties, although a dynamical theory has also been proposed by Lines and Eibschultz.[16] The theory has been formulated for both ordered and disordered phases, as a model for a structural phase transition. In our exposition of the CEF theory we shall follow Lines' papers very closely, since our primary interest is to learn the technique.

The conventional MF theory, it should be noted, violates the "fluctuation" theorem. Lines has shown that static correlations can be included in his model in such a fashion that requiring the resultant theory to obey a rigorous restriction imposed by the fluctuation theorem determines these correlations completely. It is important to recall at this point the limitations of the Weiss MF theory, which assumes that the interactions of a given spin with its neighbors may be replaced by an effective field—the Weiss molecular field, and that there are no correlations between the spins. For describing quantities that are expressed by single-particle operators, such as the magnetization, the MF theory is qualitatively correct. We have seen that it predicts the spontaneous magnetization and gives the Curie-Weiss law, which is qualitatively accurate at sufficiently high temperatures. However, for describing quantities expressed by two-particle operators, such as the exchange interaction or any short-range order, the MF theory is completely incorrect in the paramagnetic region at zero applied field. This is a direct consequence of the assumption of "no correlation" between the spins. This situation has been stated by Lines[15] in general terms as follows: "The effective field susceptibility (e.g., Curie-Weiss law for disordered ferromagnets), when related to statistical correlations through an exact relationship derivable from the fluctuation dissipation theorem, implies non-zero correlations—in direct contradiction of the initial basic assumption. In view of this rather unsatisfactory circumstance, one is tempted to ask whether the contradiction can be put to good use by introducing the concept of correlations into the effective field framework in such a simple way that merely forcing consistency with the exact fluctuation result determines them completely." The CEF theory gives us the right procedure to do this, as we now describe.

To develop the CEF theory of Lines, let us start with the Hamiltonian, described by

$$\mathcal{H} = \sum_i \left[\tfrac{1}{2}\mathbf{p}_i^2 + V(\mathbf{x}_i) - \mathbf{h}_i \mathbf{x}_i \right] - \tfrac{1}{2} \sum_{\substack{i,j \\ i \neq j}} v_{ij} \mathbf{x}_i \mathbf{x}_j \tag{14.72}$$

where $\mathbf{p}_i$ and $\mathbf{x}_i$ are the canonically conjugate momenta and coordinates, respectively, of the particle situated at the ith site of the lattice, $\mathbf{h}_i$ is the applied magnetic field, $V(\mathbf{x}_i)$ is the crystal field potential, and v_{ij} (with $v_{ii} = 0$)

is an interaction potential. In (14.72) it can be seen that $\Sigma_j v_{ij} x_j$ is the effective field experienced by the particle situated at the ith lattice site. We now incorporate this effective field into the equation of motion for the ith particle by using the following approximation: the operator $\mathbf{x}_j$ of the neighbor at the jth site is replaced by its ensemble average $\langle \mathbf{x}_j \rangle$, together with a correlated contribution $A_{ij}(\mathbf{x}_i - \langle \mathbf{x}_i \rangle)$, which is proportional to the instantaneous deviation of the ith particle from its averaged position $\langle \mathbf{x}_i \rangle$.

Let us now find the equation of motion. We have

$$\dot{\mathbf{p}}_i = \frac{i}{\hbar}[\mathcal{H}, \mathbf{p}_i] \tag{14.73}$$

where $\mathcal{H}$ is given by (14.72). We now evaluate $[\mathcal{H}, \mathbf{p}_i]$, one by one, for the four terms appearing in the Hamiltonian given by (14.72). We have

$$[V(\mathbf{x}_i), \mathbf{p}_i] = i\hbar \frac{\partial V(x_i)}{\partial x_i} \tag{14.74}$$

since

$$\begin{aligned} [V(\mathbf{x}), \mathbf{p}]\psi &= (Vp - pV)\psi \\ &= V(p\psi) - (pV)\psi - V(p\psi) \\ &= i\hbar\psi \frac{\partial V}{\partial x} \\ [\mathbf{x}_i, \mathbf{p}_i] &= i\hbar \end{aligned} \tag{14.75}$$

and

$$[v_{kl}\mathbf{x}_k \mathbf{x}_l, \mathbf{p}_i] = v_{kl}[i\hbar\delta_{ki}\mathbf{x}_l + i\hbar\delta_{il}\mathbf{x}_k]$$

Therefore

$$\begin{aligned} \sum_{k,l} [v_{kl}\mathbf{x}_k \mathbf{x}_l, \mathbf{p}_i] &= \sum_l v_{il} i\hbar \mathbf{x}_l + \sum_k v_{ki} i\hbar \mathbf{x}_k \\ &= 2i\hbar \sum_l v_{ij} x_j \end{aligned} \tag{14.76}$$

Substituting (14.74) to (14.76) in (14.73), we get the required equation of motion for the ith particle:

$$\dot{\mathbf{p}}_i = \mathbf{h}_i - \frac{\partial V(\mathbf{x}_i)}{\partial x_i} + \sum_j v_{ij}[\langle \mathbf{x}_j \rangle + A_{ij}(\mathbf{x}_i - \langle \mathbf{x}_i \rangle)] \tag{14.77}$$

where $\mathbf{x}_j$ has been replaced by $\langle \mathbf{x}_j \rangle + A_{ij}(\mathbf{x}_i - \langle \mathbf{x}_i \rangle)$, as discussed earlier.

Again, by using (14.74) to (14.76), it can be easily shown that the equation of motion (14.77) derived above can be exactly reproduced from the effective Hamiltonian, given by

$$\mathcal{H}_i = \tfrac{1}{2}\mathbf{p}_i^2 + V(\mathbf{x}_i) - \mathbf{h}_i\mathbf{x}_i - \sum_j v_{ij}\mathbf{x}_i\left[\langle\mathbf{x}_j\rangle + \tfrac{1}{2}\alpha\mathbf{x}_i - \alpha\langle\mathbf{x}_i\rangle\right] \tag{14.78}$$

where $\alpha = \sum_j v_{ij}A_{ij}/\sum_j v_{ij}$ is the effective correlation parameter, which is dimensionless but temperature-dependent. Now, writing

$$\begin{aligned}\langle\mathbf{x}_j\rangle &= \langle\mathbf{x}_j\rangle_0 + \mathbf{m}_j\\ \langle\mathbf{x}_i\rangle &= \langle\mathbf{x}_i\rangle_0 + \mathbf{m}_i\end{aligned} \tag{14.79}$$

and assuming that $\langle\mathbf{x}_j\rangle_0 = \langle\mathbf{x}_i\rangle_0$, which means that the average is site-independent, we get, by substituting (14.79) in (14.78),

$$\begin{aligned}\mathcal{H}_i = \tfrac{1}{2}\mathbf{p}_i^2 + V(\mathbf{x}_i) - \mathbf{h}_i\mathbf{x}_i - \sum_j v_{ij}\mathbf{x}_i\Big[\langle\mathbf{x}_j\rangle_0 + \mathbf{m}_j + \tfrac{1}{2}\alpha\mathbf{x}_i\\ - \alpha\{\langle\mathbf{x}_i\rangle_0 + \mathbf{m}_i\}\Big]\end{aligned}$$

or

$$\begin{aligned}\mathcal{H}_i = \tfrac{1}{2}\mathbf{p}_i^2 + V(\mathbf{x}_i) - \mathbf{h}_i\mathbf{x}_i - \sum_j v_{ij}x_i\left[\langle\mathbf{x}_j\rangle_0 + \tfrac{1}{2}\alpha\mathbf{x}_i - \alpha\langle\mathbf{x}_i\rangle_0\right]\\ - \sum_j v_{ij}x_i[\mathbf{m}_j - \alpha\mathbf{m}_i]\end{aligned}$$

or

$$\begin{aligned}\mathcal{H}_i = \tfrac{1}{2}\mathbf{p}_i^2 + V(\mathbf{x}_i) - h_i x_i - \sum_j v_{ij}x_i\left[\langle x\rangle_0(1-\alpha) + \tfrac{1}{2}\alpha\mathbf{x}_i\right]\\ - \sum_j v_{ij}\mathbf{x}_i[\mathbf{m}_j - \alpha\mathbf{m}_i]\end{aligned} \tag{14.80}$$

where we have put $\langle\mathbf{x}_j\rangle_0 = \langle\mathbf{x}_i\rangle_0 = \langle x\rangle_0$.

We now Fourier-transform $\mathbf{x}_i$ and v_{ij}; thus

$$\begin{aligned}\mathbf{x}_i &= \frac{1}{\sqrt{N}}\sum_{\mathbf{q}} x(\mathbf{q})e^{i\mathbf{q}\cdot\mathbf{i}}\\ v_{ij} &= \frac{1}{\sqrt{N}}\sum_{\mathbf{q}'} v(\mathbf{q}')e^{i\mathbf{q}'\cdot(\mathbf{i}-\vec{\mathbf{j}})}\end{aligned} \tag{14.81}$$

and hence

$$\sum_j \mathbf{x}_i v_{ij} = \frac{1}{N} \sum_{\mathbf{q},\mathbf{q}',\mathbf{j}} x(\mathbf{q}) v(\mathbf{q}') e^{i(\mathbf{q}+\mathbf{q}')i} e^{-i\mathbf{q}'\cdot\mathbf{j}}$$

$$= \frac{1}{\sqrt{N}} \sum_{\mathbf{q}} x(\mathbf{q}) v(0) e^{i\mathbf{q}\cdot\mathbf{i}} \tag{14.82}$$

where the summation over $\mathbf{j}$ leaves $\mathbf{q}'=0$, $\delta(\mathbf{q}')$. Thus

$$\sum_j v_{ij}\mathbf{x}_i = \mathbf{x}_i v(0) \tag{14.83}$$

Also, similarly,

$$\sum v_{ij}\mathbf{x}_i\mathbf{x}_i = \frac{1}{N\sqrt{N}} \sum_{\substack{\mathbf{j},\mathbf{q},\\ \mathbf{q}',\mathbf{q}''}} x(\mathbf{q}) x(\mathbf{q}') v(\mathbf{q}'') e^{i(\mathbf{q}+\mathbf{q}')i} e^{i\mathbf{q}''(\mathbf{i}-\mathbf{j})}$$

$$= \mathbf{x}_i^2 v(0) \tag{14.84}$$

Substituting (14.83) and (14.84) in (14.80), we get

$$\mathcal{H}_i = \tfrac{1}{2}\mathbf{p}_i^2 + V(\mathbf{x}_i) - \mathbf{h}_i\mathbf{x}_i - v(0)\langle x\rangle_0(1-\alpha)\mathbf{x}_i$$

$$- \tfrac{1}{2}v(0)\alpha\mathbf{x}_i^2 - \sum_j v_{ij}\mathbf{x}_i(\mathbf{m}_j - \alpha\mathbf{m}_i) \tag{14.85}$$

We now divide the effective Hamiltonian given above into field-independent and field-dependent parts:

$$\mathcal{H}_i = \mathrm{v}_i' - \mathbf{x}_i\boldsymbol{\Gamma}_i \tag{14.86}$$

where

$$\mathrm{v}_i' = \tfrac{1}{2}\mathbf{p}_i^2 + V(\mathbf{x}_i) - \tfrac{1}{2}\alpha v(0)\mathbf{x}_i^2 - \mathbf{x}_i\langle x\rangle_0 v(0)(1-\alpha) \tag{14.87a}$$

and

$$\boldsymbol{\Gamma}_i = \mathbf{h}_i + \sum_j v_{ij}(\mathbf{m}_j - \alpha\mathbf{m}_i) \tag{14.87b}$$

Let us now differentiate a function $e^{\beta \mathrm{v}_i'} e^{-\beta\mathcal{H}_i}$ with respect to β $(=1/kT)$:

$$\frac{\partial}{\partial\beta} e^{\beta \mathrm{v}_i'} e^{-\beta\mathcal{H}_i} = e^{\beta \mathrm{v}_i'}\mathrm{v}_i' e^{-\beta\mathcal{H}_i} - e^{\beta \mathrm{v}_i'}\mathcal{H}_i e^{-\beta\mathcal{H}_i}$$

$$= e^{\beta \mathrm{v}_i'}(\mathrm{v}_i' - \mathcal{H}_i) e^{-\beta\mathcal{H}_i}$$

$$= e^{\beta \mathrm{v}_i'}\mathbf{x}_i\boldsymbol{\Gamma}_i e^{-\beta\mathcal{H}_i} \tag{14.88}$$

Therefore

$$\left[e^{\beta V_i'}e^{-\beta \mathcal{H}_i}\right]_0^\beta = \int_0^\beta e^{\lambda V_i'}\mathbf{x}_i\boldsymbol{\Gamma}_i e^{-\lambda \mathcal{H}_i}\,d\lambda$$

or

$$e^{\beta V_i'}e^{-\beta \mathcal{H}_i} = 1 + \int_0^\beta e^{\lambda V_i'}\mathbf{x}_i\boldsymbol{\Gamma}_i e^{-\lambda \mathcal{H}_i}\,d\lambda$$

Hence

$$\begin{aligned} e^{-\beta \mathcal{H}_i} &= e^{-\beta V_i'} + e^{-\beta V_i'}\int_0^\beta e^{\lambda V_i'}\mathbf{x}_i\boldsymbol{\Gamma}_i e^{-\lambda \mathcal{H}_i}\,d\lambda \\ &\simeq e^{-\beta V_i'} + e^{-\beta V_i'}\int_0^\infty e^{\lambda V_i'}\mathbf{x}_i\boldsymbol{\Gamma}_i e^{-\lambda V_i'}\,d\lambda \end{aligned} \tag{14.89}$$

correct to the first order in the applied field $\mathbf{h}_i$ and response $\mathbf{m}_i$. Now

$$\begin{aligned} \langle \mathbf{x}_i \rangle &= \frac{\operatorname{Tr} \mathbf{x}_i e^{-\beta \mathcal{H}_i}}{\operatorname{Tr} e^{-\beta \mathcal{H}_i}} \\ &= \frac{\operatorname{Tr} \mathbf{x}_i \left\{ e^{-\beta V_i'} + e^{-\beta V_i'}\int_0^\beta e^{\lambda V_i'}\mathbf{x}_i\boldsymbol{\Gamma}_i e^{-\lambda V_i'}\,d\lambda \right\}}{\operatorname{Tr}\left\{ e^{-\beta V_i'} + e^{-\beta V_i'}\int_0^\beta e^{\lambda V_i'}\mathbf{x}_i\boldsymbol{\Gamma}_i e^{-\lambda V_i'}\,d\lambda \right\}} \end{aligned} \tag{14.90}$$

using (14.89), which, after some algebra, is given by

$$\langle \mathbf{x}_i \rangle = \frac{Z_N + F_N}{Z_D + F_D} \tag{14.91}$$

where Z_N and F_N denote the zeroth-order and the first-order expressions, respectively, of the numerator, and Z_D and F_D those of the denominator. These expressions are given by

$$\begin{aligned} Z_N &= \operatorname{Tr}\left[\mathbf{x}_i e^{-\beta V_i'}\right] \\ F_N &= \operatorname{Tr}\left[\int_0^\beta e^{\lambda V_i'}\mathbf{x}_i e^{-\lambda V_i'}\mathbf{x}_i\,d\lambda\, e^{-\beta V_i'}\left\{\mathbf{h}_i + \sum_j v_{ij}(\mathbf{m}_j - \alpha\mathbf{m}_i)\right\}\right] \\ Z_D &= \operatorname{Tr} e^{-\beta V_i'} \\ F_D &= \operatorname{tr}\int_0^\beta e^{\lambda V_i'}\mathbf{x}_i\boldsymbol{\Gamma}_i e^{-\lambda V_i'}\,d\lambda\, e^{-\beta V_i'} \end{aligned} \tag{14.92}$$

Now, (14.91) can be approximately written as

$$\langle \mathbf{x}_i \rangle = \frac{Z_N}{Z_D} + \frac{F_N}{Z_D} - \frac{Z_N F_D}{Z_D Z_D} \tag{14.93}$$

where

$$\frac{Z_N}{Z_D}=\frac{\mathrm{Tr}\,\mathbf{x}_i e^{-\beta \mathrm{v}_i'}}{\mathrm{Tr}\,e^{-\beta \mathrm{v}_i'}}=\langle x_i\rangle_0=\langle x\rangle_0$$

$$\frac{F_N}{Z_D}=\left\langle\int_0^\beta e^{\lambda \mathrm{v}_i'}\mathbf{x}_i e^{-\lambda \mathrm{v}_i'}\mathbf{x}_i\,d\lambda\right\rangle_0\left\{\mathbf{h}_i+\sum_j v_{ij}(\mathbf{m}_j-\alpha\mathbf{m}_i)\right\} \tag{14.94}$$

and

$$\frac{Z_N F_D}{Z_D Z_D}=\frac{\left\{\mathrm{Tr}\,\mathbf{x}_i e^{-\beta \mathrm{v}_i'}\right\}\left\{\mathrm{Tr}\int_0^\beta e^{\lambda \mathrm{v}_i'}\mathbf{x}_I\boldsymbol{\Gamma}_i e^{-\lambda \mathrm{v}_i'}d\lambda e^{-\beta \mathrm{v}_i'}\right\}}{\left\{\mathrm{Tr}\,e^{-\beta \mathrm{v}_i'}\right\}\left\{\mathrm{Tr}\,e^{-\beta \mathrm{v}_i'}\right\}}$$

$$=\langle x_i\rangle_0\left\langle\int_0^\beta e^{\lambda \mathrm{v}_i'}\mathbf{x}_i e^{-\lambda \mathrm{v}_i'}d\lambda\right\rangle_0\left\{\mathbf{h}_i+\sum_j v_{ij}(\mathbf{m}_j-\alpha\mathbf{m}_i)\right\}$$

Thus, finally, we have

$$\langle\mathbf{x}_i\rangle=\langle\mathbf{x}\rangle_0+\left\langle\int_0^\beta e^{\lambda \mathrm{v}_i'}\mathbf{x}_i e^{-\lambda \mathrm{v}_i'}\mathbf{x}_i\,d\lambda\right\rangle_0\cdot\boldsymbol{\Gamma}_i$$

$$-\langle\mathbf{x}\rangle_0\left\langle\int_0^\beta e^{\lambda \mathrm{v}_i'}\mathbf{x}_i e^{-\lambda \mathrm{v}_i'}d\lambda\right\rangle_0\cdot\boldsymbol{\Gamma}_i \tag{14.95}$$

Let us define

$$\beta\langle\mathbf{x}:\mathbf{x}\rangle_0=\left\langle\int_0^\beta e^{\lambda \mathrm{v}_i'}\mathbf{x}_i e^{-\lambda \mathrm{v}_i'}\mathbf{x}_i\,d\lambda\right\rangle_0 \tag{14.96}$$

and evaluate the traces in which v_i' is diagonal:

$$\mathrm{Tr}\int_0^\beta e^{\lambda \mathrm{v}_i'}\mathbf{x}_i e^{-\lambda \mathrm{v}_i'}d\lambda e^{-\beta \mathrm{v}_i'}=\sum_n\int_0^\beta e^{\lambda E_i^n}(\mathbf{x}_i)_{nn}e^{-\lambda E_i^n}d\lambda e^{-\beta E_i^n}$$

$$\mathrm{Tr}\,e^{-\beta \mathrm{v}_i'}=\sum_n e^{-\beta E_i^n}$$

$$\mathrm{Tr}\int_0^\beta e^{\lambda \mathrm{v}_i'}\mathbf{x}_i e^{-\lambda \mathrm{v}_i'}d\lambda e^{-\beta \mathrm{v}_i'}=\sum_n\int_0^\beta(\mathbf{x}_i)_{nn}e^{-\beta E_i^n}d\lambda$$

$$=\beta\sum_n(\mathbf{x}_i)_{nn}e^{-\beta E_i^n}$$

Therefore

$$\left\langle\int_0^\beta e^{\lambda \mathrm{v}_i'}\mathbf{x}_i e^{-\lambda \mathrm{v}_i'}d\lambda\right\rangle_0=\beta\langle\mathbf{x}_i\rangle_0=\beta\langle\mathbf{x}\rangle_0$$

Thus

$$\langle \mathbf{x}_i \rangle = \langle \mathbf{x} \rangle_0 \rangle + \boldsymbol{\Gamma}_i \beta \left[\langle \mathbf{x} : \mathbf{x} \rangle_0 - \langle \mathbf{x} \rangle_0^2 \right]$$

or

$$\mathbf{m}_i = \beta \left[\langle \mathbf{x} : \mathbf{x} \rangle_0 - \langle \mathbf{x} \rangle_0^2 \right] \left\{ \mathbf{h}_i + \sum_j v_{ij} (\mathbf{m}_j - \alpha \mathbf{m}_i) \right\} \tag{14.97}$$

We now Fourier-transform $\mathbf{m}_i$, $\mathbf{h}_i$, and v_{ij} by using the following relations:

$$\begin{aligned} \mathbf{m}_i &= \frac{1}{\sqrt{N}} \sum_{\mathbf{q}} \mathbf{m}(\mathbf{q}) e^{i\mathbf{q}\cdot\vec{\mathbf{i}}} \\ \mathbf{h}_i &= \frac{1}{\sqrt{N}} \sum_{\mathbf{q}} \mathbf{h}(\mathbf{q}) e^{i\mathbf{q}\cdot\vec{\mathbf{i}}} \\ v_{ij} &= \frac{1}{\sqrt{N}} \sum_{\mathbf{q}'} v(\mathbf{q}') e^{i\mathbf{q}'\cdot(\vec{\mathbf{i}} - \vec{\mathbf{j}})} \end{aligned} \tag{14.98}$$

where N is the number of particles in the lattice. By using (14.98), we obtain

$$\begin{aligned} \sum_j v_{ij} \mathbf{m}_j &= \frac{1}{N} \sum_{\mathbf{q},\mathbf{q}',\vec{\mathbf{j}}} \mathbf{m}(\mathbf{q}) v(\mathbf{q}') e^{i\mathbf{q}\cdot\vec{\mathbf{j}}} e^{i\mathbf{q}'\cdot(\vec{\mathbf{i}} - \vec{\mathbf{j}})} \\ &= \frac{1}{N} \sum_{\mathbf{q},\mathbf{q}',\vec{\mathbf{j}}} \mathbf{m}(\mathbf{q}) v(\mathbf{q}') e^{i(\mathbf{q}-\mathbf{q}')\cdot\vec{\mathbf{j}}} e^{i\mathbf{q}'\cdot\vec{\mathbf{i}}} \\ &= \frac{1}{\sqrt{N}} \sum_{q} \mathbf{m}(\mathbf{q}) v(\mathbf{q}) e^{i\mathbf{q}\cdot\vec{\mathbf{i}}} \end{aligned} \tag{14.99}$$

[$\because$ summation over $\mathbf{j}$ gives $\sqrt{N}\,\delta(\mathbf{q}-\mathbf{q}')$].

Similarly,

$$\sum_j v_{ij} \mathbf{m}_i = \frac{1}{\sqrt{N}} \sum_{\mathbf{q}} \mathbf{m}(\mathbf{q}) v(0) e^{i\mathbf{q}\cdot\vec{\mathbf{i}}}$$

Substituting (14.98) and (14.99) in (14.97), we get

$$\begin{aligned} \mathbf{m}(\mathbf{q}) = \beta \left[\langle \mathbf{x} : \mathbf{x} \rangle_0 - \langle \mathbf{x} \rangle_0^2 \right] \{ &\mathbf{h}(\mathbf{q}) + \mathbf{m}(\mathbf{q}) v(\mathbf{q}) \\ &- \alpha v(0) m(\mathbf{q}) \} \end{aligned} \tag{14.100}$$

where

$$\mathbf{m}(\mathbf{q}) = \frac{1}{\sqrt{N}} \sum_i \mathbf{m}_i e^{-i\mathbf{q}\cdot\vec{\mathbf{i}}}$$

From (14.100) we can easily write the wave-vector-dependent susceptibility, $\chi(\mathbf{q})$. We have

$$\mathbf{m}(\mathbf{q})=\frac{1}{\tau}\mathbf{h}(\mathbf{q})+\frac{1}{\tau}\mathbf{m}(\mathbf{q})v(\mathbf{q})-\alpha v(0)\mathbf{m}(\mathbf{q})$$

or

$$\mathbf{m}(\mathbf{q})\left\{1-\frac{1}{\tau}v(\mathbf{q})+\frac{1}{\tau}\alpha v(0)\right\}=\frac{1}{\tau}\mathbf{h}(\mathbf{q})$$

or

$$\left[\chi(\mathbf{q})\right]^{-1}=\frac{\mathbf{h}(\mathbf{q})}{\mathbf{m}(\mathbf{q})}=\tau+\alpha v(0)-v(\mathbf{q}) \tag{14.101}$$

where

$$\tau=\frac{kT}{\left[\langle\mathbf{x}:\mathbf{x}\rangle_0-\langle\mathbf{x}\rangle_0^2\right]}$$

The correlation parameter α may now be determined by requiring the susceptibility $\chi(\mathbf{q})$ of (14.101) to obey an exact fluctuation-dissipation relation, which we shall derive. We can calculate the susceptibility directly from the Hamiltonian given by (14.72) as

$$\chi(\mathbf{q})=\frac{\partial}{\partial h(\mathbf{q})}\frac{\mathrm{Tr}\left[x(\mathbf{q})\rho\right]}{\mathrm{Tr}\,\rho} \tag{14.102}$$

the susceptibility being the derivative of response, $\mathbf{x}$, with respect to the applied stress, $\mathbf{h}$. In (14.102), $x(\mathbf{q})$ is the Fourier transform of the displacement $\mathbf{x}_i$ with respect to the lattice, and $\rho/\mathrm{Tr}\,\rho$ is the normalized density matrix, where

$$\rho=e^{-\beta\mathcal{H}} \tag{14.103}$$

Here

$$\mathcal{H}=\mathcal{H}_0-\mathcal{H}'$$

$$\mathcal{H}_0=\frac{1}{2}\sum_i\mathbf{p}_i^2+\sum_i V(\mathbf{x}_i)-\frac{1}{2}\sum_{i,j}v_{ij}\mathbf{x}_i\mathbf{x}_j$$

$$\mathcal{H}'=\sum_q h(\mathbf{q})x(-\mathbf{q})$$

Exactly as before, we have

$$\frac{\partial}{\partial\beta}e^{\beta\mathcal{H}_0}e^{-\beta\mathcal{H}}=e^{\beta\mathcal{H}_0}(\mathcal{H}_0-\mathcal{H})e^{-\beta\mathcal{H}}$$
$$=e^{\beta\mathcal{H}_0}\sum_{\mathbf{q}}\mathbf{h}(\mathbf{q})x(-\mathbf{q})e^{-\beta\mathcal{H}}$$

Hence

$$e^{\beta\mathcal{H}_0}e^{-\beta\mathcal{H}}=1+\int_0^{\beta}e^{\lambda\mathcal{H}_0}\sum_{\mathbf{q}}\mathbf{h}(\mathbf{q})x(-\mathbf{q})e^{-\lambda\mathcal{H}}d\lambda$$

To the first order, we have, therefore,

$$e^{-\beta\mathcal{H}}=e^{-\beta\mathcal{H}_0}+e^{-\beta\mathcal{H}_0}\int_0^{\beta}e^{\lambda\mathcal{H}_0}\sum_{\mathbf{q}}h(\mathbf{q})x(-\mathbf{q})e^{-\lambda\mathcal{H}_0}d\lambda \tag{14.104}$$

Hence

$$\frac{\partial}{\partial h(\mathbf{q})}e^{-\beta\mathcal{H}}=e^{-\beta\mathcal{H}_0}\int_0^{\beta}e^{\lambda\mathcal{H}_0}x(-\mathbf{q})e^{-\lambda\mathcal{H}_0}d\lambda$$
$$=\frac{\partial}{\partial h(\mathbf{q})}\cdot\rho \tag{14.105}$$

Now, from (14.102), we have

$$\chi=\frac{\partial}{\partial h(\mathbf{q})}\frac{\mathrm{Tr}[x(\mathbf{q})\rho]}{\mathrm{Tr}\,\rho}$$
$$=\frac{\mathrm{Tr}\,x(\mathbf{q})[\partial\rho/\partial h(\mathbf{q})]}{\mathrm{Tr}\,\rho}-\frac{\mathrm{Tr}\,x(\mathbf{q})\rho\cdot\mathrm{tr}[\partial\rho/\partial h(\mathbf{q})]}{(\mathrm{Tr}\,\rho)^2}$$
$$=\chi_1+\chi_2 \tag{14.106}$$

The first term of (14.104) can be simplified by using the well-known cyclic permutation of trace, and we have

$$\chi_1=\frac{\mathrm{Tr}\,x(\mathbf{q})e^{-\beta\mathcal{H}_0}\int_0^{\beta}e^{\lambda\mathcal{H}_0}x(-\mathbf{q})e^{-\lambda\mathcal{H}_0}d\lambda}{\mathrm{Tr}\,\rho}$$
$$=\frac{\mathrm{Tr}\int_0^{\beta}e^{\lambda\mathcal{H}_0}x(-\mathbf{q})e^{-\lambda\mathcal{H}_0}x(\mathbf{q})e^{-\beta\mathcal{H}_0}d\lambda}{\mathrm{Tr}\,\rho}$$
$$=\beta\langle x(-\mathbf{q}):x(\mathbf{q})\rangle_0 \tag{14.107}$$

using (14.96), the previous definition. Similarly,

$$\chi_2 = \frac{\langle x(\mathbf{q})\rangle_0 \operatorname{Tr} \int_0^\beta e^{\lambda \mathcal{H}_0} x(-\mathbf{q}) e^{-\lambda \mathcal{H}_0} d\lambda\, e^{-\beta \mathcal{H}_0}}{\operatorname{Tr} \rho}$$

$$= \beta \langle x(\mathbf{q})\rangle_0 \langle x(-\mathbf{q})\rangle_0 \tag{14.108}$$

Thus

$$\sum_{\mathbf{q}} \chi(\mathbf{q}) = \beta \Big[\sum_{\mathbf{q}} \langle x(-\mathbf{q}) : x(\mathbf{q})\rangle_0 - \langle x(\mathbf{q})_0 < x(-\mathbf{q})\rangle_0 \Big] \tag{14.109}$$

Noting that

$$x(\mathbf{q}) = \frac{1}{\sqrt{N}} \sum_i \mathbf{x}_i e^{-i\mathbf{q}\cdot\vec{\mathbf{i}}}$$

and

$$x(-\mathbf{q}) = \frac{1}{\sqrt{N}} \sum_j \mathbf{x}_j e^{i\mathbf{q}\cdot\vec{\mathbf{j}}} \tag{14.110}$$

we have

$$\sum_{\mathbf{q}} x(\mathbf{q}) x(-\mathbf{q}) = N \mathbf{x}_i \mathbf{x}_i$$

We have, therefore, from (14.109)

$$\sum_{\mathbf{q}} \chi(\mathbf{q}) = \beta [\langle \mathbf{x}_i : \mathbf{x}_i \rangle_0 - \langle \mathbf{x}_i \rangle_0 \langle \mathbf{x}_i \rangle_0]$$

$$= \frac{N}{kT} [\langle \mathbf{x} : \mathbf{x} \rangle_0 - \langle \mathbf{x} \rangle_0 \langle \mathbf{x} \rangle_0]$$

$$= \frac{N}{\tau} \tag{14.111}$$

whereas from our previous result (14.101), we had

$$\sum_{\mathbf{q}} \chi(\mathbf{q}) = \sum_{\mathbf{q}} \frac{1}{\tau + \alpha v(0) - v(\mathbf{q})} \tag{14.112}$$

Comparing (14.111) with (14.112), we have finally

$$\frac{N}{\tau} = \sum_{\mathbf{q}} [\tau + \alpha v(0) - v(\mathbf{q})]^{-1} \tag{14.113}$$

We thus obtain a self-consistent equation for the correlation parameter α in the lowest order of the correlated effective field theory of Lines, given by (14.113). We therefore see that, by forcing a consistency with the fluctuation-dissipation theorem of statistical mechanics, we can obtain the correlation parameter α uniquely, of course in the lowest order. However, obtaining higher orders is just a logical extension of the procedure described above.

One great advantage of this theory is that it is defined for all temperatures and for the arbitrary local potential $V(\mathbf{x}_i)$, and therefore it can be related to the existing statistical approximations derived under more restrictive circumstances. Two important points should be made here. For the local potential $V(\mathbf{x}_i)$ of a quasi-harmonic form, the Hamiltonian (14.72) is adequate for approximation by self-consistent linearization of the normal mode equations of motion.[17] In the opposite limit of extreme anharmonicity, however, $V(\mathbf{x}_i)$ can be represented in the form of a narrow double well, and the ensemble average $\langle \mathbf{x}:\mathbf{x}\rangle_0$ is independent of temperature and consequently becomes an Ising problem.

Now that the theory has been set out in general terms, we deal with magnetism in the ordered phase. For this purpose we shall follow Lines's paper of 1974.[15]

First of all, we write the effective Hamiltonian in the presence of a static magnetic field (infinitesimally small), $\mathbf{h}_i$, applied in the direction ζ. If the susceptibility is taken to be diagonal in the coordinate system (x,y,z), the ensemble averages are nonvanishing only in these directions. We have already seen that the correlated field has the form

$$\mathbf{h}^{\gamma}_{\text{corr}} = 2\sum_j J^{\gamma}_{ij}\left[\langle \mathbf{S}^{\gamma}_j\rangle + A^{\gamma}_{ij}(\mathbf{S}^{\gamma}_i - \langle \mathbf{S}^{\gamma}_i\rangle)\right] \tag{14.114}$$

where A_{ij} are the temperature-dependent static correlation parameters of as yet unspecified form, J_{ij} is the Heisenberg exchange integral, and γ specifies the three directions of the coordinate axes. We do not need to know the detailed forms of A_{ij} because by using

$$\sum_j A^{\gamma}_{ij}J^{\gamma}_{ij} = \alpha^{\gamma}\sum_j J^{\gamma}_{ij} \tag{14.115}$$

to define the correlation parameters α^{γ} $(\gamma = x,y,z)$, (14.114) can be expressed in terms of α^{γ} alone, that is,

$$\mathbf{h}^{\gamma}_{\text{corr}} = 2\sum_j J^{\gamma}_{ij}\left[\langle \mathbf{S}^{\gamma}_j\rangle + \alpha^{\gamma}(\mathbf{S}^{\gamma}_i - \langle \mathbf{S}^{\gamma}_i\rangle)\right] \tag{14.116}$$

where we assume all lattice sites to be perfectly equivalent. It is evident from (14.116) that the correlation parameters α^{γ} will tend to be positive for ferromagnets and negative for antiferromagnets. We can now write the effective Hamiltonian in the presence of an applied field, which is given by

$$\mathcal{H}_{\text{eff}}^i = \mathcal{H}_i - \mathbf{h}_i^\zeta \boldsymbol{\mu}_i^\zeta - \sum_{\gamma, j} J_{ij}^\gamma \alpha^\gamma (\mathbf{S}_i^\gamma)^2$$
$$-2\sum_{j,\gamma} J_{ij}^\gamma \mathbf{S}_i^\gamma (\langle \mathbf{S}_j^\gamma \rangle - \alpha^\gamma \langle \mathbf{S}_i^\gamma \rangle) \tag{14.117}$$

where $\boldsymbol{\mu}$ signifies the magnetic moment.

Equation (14.117) is a very important relation in the sense that the many-body problem has been reduced to a single-body form. This artifice, of course, has introduced, in general, three temperature-dependent parameters: α^x, α^y, α^z. However, two of these, or even all three, may be equal by symmetry reasons, depending on the particular problem. It has been shown by Lines that even in the most general situations all these parameters can be determined in an unambiguous fashion by using the fluctuation theorem.

In (14.117) the averages $\langle \mathbf{S}_j^\gamma \rangle$ consist of two parts in the ordered phase: nonzero average $\langle \mathbf{S}_j^\gamma \rangle_0$ in the absence of the field, and some induced moment due to the field. Thus we write

$$\langle \mathbf{S}_i^\gamma \rangle_h = \langle \mathbf{S}_i^\gamma \rangle_0 + \mathbf{m}_j^\gamma \tag{14.118}$$

where $\mathbf{m}_j^\gamma$ is in the first order in $\mathbf{h}$.

We now divide the Hamiltonian (14.117) into two parts:

$$\mathcal{H}_{\text{eff}}^i = (\mathcal{H}_{\text{eff}}^i)_0 + \mathcal{H}_i' \tag{14.119}$$

where

$$(\mathcal{H}_{\text{eff}}^i)_0 = \mathcal{H}_i - \sum_{j,\gamma} J_{ij}^\gamma \alpha^\gamma (\mathbf{S}_i^\gamma)^2 - 2\sum_{j,\gamma} J_{ij}^\gamma \mathbf{S}_i^\gamma (\langle \mathbf{S}_j^\gamma \rangle_0 - \alpha^\gamma \langle \mathbf{S}_i^\gamma \rangle_0) \tag{14.120}$$

and

$$\mathcal{H}_i' = -\mathbf{h}_i^\zeta \boldsymbol{\mu}_i^\zeta - 2\sum_{j,\gamma} J_{ij}^\gamma \mathbf{S}_i^\gamma (\mathbf{m}_j^\gamma - \alpha^\gamma \mathbf{m}_i^\gamma) \tag{14.121}$$

Equation (14.120) represents the correlated effective Hamiltonian for the ith magnetic ion. Its eigenfunctions $\phi_{in}^0(\alpha)$ and the eigenvalues $E_{in}^0(\alpha)$ can be calculated as functions of α^γ for a particular case.

We now consider the perturbation on $\phi_{in}^0(\alpha)$ by $\mathcal{H}'$, given by (14.121). The resulting perturbed eigenvalues and eigenfunctions, to the first order of smallness in $\mathbf{h}_i^\gamma$, $\langle \mathbf{S}_j^\gamma \rangle$, and $\langle \mathbf{S}_i^\gamma \rangle$, are given by

$$E_{in}(\alpha) = E_{in}^0(\alpha) + (\mathcal{H}_i')_{nn}$$
$$\phi_{in}(\alpha) = \phi_{in}^0(\alpha) + \sum_{m \neq n} \frac{(\mathcal{H}_i')_{mn} \phi_{im}^0(\alpha)}{E_{in}^0(\alpha) - E_{im}^0(\alpha)} \tag{14.122}$$

where

$$(\mathcal{H}_i')_{mn} = \langle \phi_{im}^0(\alpha) | \mathcal{H}_i' | \phi_{in}^0(\alpha) \rangle$$

For convenience we drop the superscript temporarily in the following. Let us find the average value of an operator, to the first order in **h**, which is given by

$$\langle \mathbf{A} \rangle = \frac{\sum_n A_{nn} e^{-E_n \beta}}{\sum_n e^{-E_n \beta}} \tag{14.123}$$

Using (14.122), we notice that

$$\begin{aligned}\sum e^{-E_n\beta} &= \sum_n e^{-E_n^0\beta} e^{-\beta'_{nn}} = \sum_n e^{-E_n^0\beta}(1-\beta\mathcal{H}'_{nn}) \\ &= \sum_n e^{-E_n^0\beta} - \beta \sum_n \mathcal{H}'_{nn} e^{-E_n^0\beta} \\ &= S - S_1 \end{aligned} \tag{14.124}$$

where

$$S = \sum_n e^{-E_n^0\beta}$$

and

$$S_1 = \beta \sum_n \mathcal{H}'_{nn} e^{-E_n^0\beta}$$

therefore, writing D for denominator of (14.123), we see that

$$\frac{1}{D} \simeq \frac{1}{S} + \frac{S_1}{S^2} \tag{14.125}$$

Similarly, using (14.122), we can easily show that the numerator of (14.123) can be written as

$$\begin{aligned}\mathbf{A}_{nn} e^{-E_n\beta} &= \left\{ \mathbf{A}_{nn} + \sum_{m\neq n} \frac{\mathcal{H}'_{nm}\mathbf{A}_{mn} + \mathbf{A}_{nm}\mathcal{H}'_{mn}}{E_n - E_m} \right\} e^{-E_n^0\beta}(1-\mathcal{H}'_{nn}\beta) \\ &\simeq \mathbf{A}_{nn} e^{-E_n^0\beta} + \sum_{m\neq n} \left(\frac{\mathcal{H}'_{nm}\mathbf{A}_{mn} + \mathbf{A}_{nm}\mathcal{H}'_{mn}}{E_n - E_m} \right) e^{-E_n^0\beta} \\ &\quad - \beta \mathbf{A}_{nn} \mathcal{H}'_{nn} e^{-E_n^0\beta} \end{aligned} \tag{14.126}$$

Now, substituting (14.125) and (14.126) in (14.123), we have

$$\langle \mathbf{A}\rangle_h = \sum_n \frac{\mathbf{A}_{nn} e^{-E_n^0\beta}}{S} + \sum_{m\neq n} \frac{\left(\dfrac{\mathcal{H}'_{nm}\mathbf{A}_{mn} + \mathbf{A}_{nm}\mathcal{H}'_{mn}}{E_n - E_m}\right)}{S} e^{-E_n^0\beta}$$
$$-\beta\sum_n \frac{\mathbf{A}_{nn}\mathcal{H}'_{nn}e^{-E_n^0\beta}}{S} + \beta\sum_n \frac{\mathbf{A}_{nn}e^{-E_n^0\beta}}{S}\sum_m \frac{\mathcal{H}'_{mm}e^{-E_m^0\beta}}{S} \quad (14.127)$$

Evaluating the averages with the zero-field basis, we rewrite (14.127) as

$$\langle \mathbf{A}\rangle_h = -\beta\langle \mathbf{A}:\mathcal{H}'\rangle_0 + \langle \mathbf{A}\rangle_0 + \beta\langle \mathbf{A}\rangle_0\langle \mathcal{H}'\rangle_0 \quad (14.128)$$

where

$$\langle \mathbf{A}:\mathcal{H}'\rangle_0 = \sum_n \frac{\mathbf{A}_{nn}\mathcal{H}'_{nn}e^{-E_n^0\beta}}{S} + \beta\frac{\dfrac{(\mathbf{A}_{nm}\mathcal{H}'_{mn} + \mathcal{H}'_{nm}\mathbf{A}_{mn})e^{-E_n^0\beta}}{E_m - E_n}}{S}$$

$$\langle \mathbf{A}\rangle_0 = \frac{\sum_n \mathbf{A}_{nn}e^{-E_n^0\beta}}{S}$$

$$S = \sum_n e^{-E_n^0\beta}$$

and

$$\mathcal{H}' = -\mathbf{h}_i^\zeta\boldsymbol{\mu}_i^\zeta - 2\sum_j J_{ij}^\zeta \mathbf{S}_i^\zeta(\mathbf{m}_j^\zeta - \alpha^\zeta\mathbf{m}_i^\zeta)$$

We now substitute for $\mathcal{H}'$ from (14.121) in (14.128) and get

$$\langle \mathbf{A}\rangle_h = \beta h_i^\zeta\langle \mathbf{A}:\boldsymbol{\mu}_i^\zeta\rangle_0 - \beta\mathbf{h}_i\langle \mathbf{A}\rangle_0\langle \boldsymbol{\mu}_i^\zeta\rangle_0 + \langle \mathbf{A}\rangle_0$$
$$+2\beta\langle \mathbf{A}:\mathbf{S}_i^\gamma\rangle_0\left\{\sum_{j,\gamma} J_{ij}^\gamma(\mathbf{m}_j^\gamma - \alpha^\gamma\mathbf{m}_i^\gamma)\right\}$$
$$-2\beta\langle \mathbf{A}\rangle_0\langle \mathbf{S}_i^\gamma\rangle_0\left\{\sum_{j,\gamma} J_{ij}^\gamma(\mathbf{m}_j^\gamma - \alpha^\gamma\mathbf{m}_i^\gamma)\right\} \quad (14.129)$$

In place of **A** we now substitute $\mathbf{S}_i^\zeta$ and $\mathbf{L}_i^\zeta$. Also we write

$$\langle \mathbf{S}_i^\zeta\rangle = \langle \mathbf{S}_i^\zeta\rangle_0 + \mathbf{m}_i^\zeta$$
$$\langle \mathbf{L}_i^\zeta\rangle = \langle \mathbf{L}_i^\zeta\rangle_0 + \mathbf{m}_i'^\zeta \quad (14.130)$$

where $\mathbf{m}_i^\zeta$ is the induced moment due to the spin part, and $\mathbf{m}_i'^\zeta$ is the induced moment due to the orbital part. After these substitutions we get

$$\begin{aligned}\mathbf{m}_i'^{\xi} &= \beta \mathbf{h}_i^{\xi}\left[\langle \mathbf{L}_i^{\xi} : \boldsymbol{\mu}_i^{\xi}\rangle_0 - \langle \mathbf{L}_i^{\xi}\rangle_0 \langle \boldsymbol{\mu}_i^{\xi}\rangle_0\right] \\ &\quad + 2\beta\left[\langle \mathbf{L}_i^{\xi} : \mathbf{S}_i^{\xi}\rangle_0 - \langle \mathbf{L}_i^{\xi}\rangle_0 \langle \mathbf{S}_i^{\xi}\rangle_0\right]\cdot \boldsymbol{\Gamma} \\ 2\mathbf{m}_i^{\xi} &= \beta \mathbf{h}_i^{\xi}\left[\langle 2\mathbf{S}_i^{\xi} : \boldsymbol{\mu}_i^{\xi}\rangle_0 - \langle 2\mathbf{S}_i^{\xi}\rangle_0 \langle \boldsymbol{\mu}_i^{\xi}\rangle_0\right] \\ &\quad + 2\beta\left[\langle 2\mathbf{S}_i^{\xi} : \mathbf{S}_i^{\xi}\rangle - \langle 2\mathbf{S}_i^{\xi}\rangle_0 \langle \mathbf{S}_i^{\xi}\rangle_0\right]\cdot \boldsymbol{\Gamma}\end{aligned} \tag{14.131}$$

where

$$\boldsymbol{\Gamma} = \sum_j J_{ij}^{\xi}\left(\mathbf{m}_j^{\xi} - \alpha^{\xi}\mathbf{m}_i^{\xi}\right)$$

Now we make a Fourier transform of (14.131). After some lengthy manipulation we get for the Fourier transform of the susceptibility

$$\begin{aligned}\chi^{\xi}(\mathbf{q}) &= \frac{\mu_B\left[\mathbf{m}'^{\xi}(\mathbf{q}) + 2\mathbf{m}^{\xi}(\mathbf{q})\right]}{\mathbf{h}^{\xi}(\mathbf{q})} \\ &= \mu_B\left[\langle \boldsymbol{\mu}_i^{\xi} : \boldsymbol{\mu}_i^{\xi}\rangle_0 - \langle \boldsymbol{\mu}_i^{\xi}\rangle_0 \langle \boldsymbol{\mu}_i^{\xi}\rangle_0\right] \\ &\quad + 2\mu_B\left[\langle \boldsymbol{\mu}_i^{\xi} : \mathbf{S}_i^{\xi}\rangle_0 - \langle \boldsymbol{\mu}_i^{\xi}\rangle_0 \langle \mathbf{S}_i^{\xi}\rangle_0\right] \\ &\quad \cdot \frac{\left[\langle \mathbf{S}_i^{\xi} : \boldsymbol{\mu}_i^{\xi}\rangle_0 - \langle \mathbf{S}_i^{\xi}\rangle_0 \langle \boldsymbol{\mu}_i^{\xi}\rangle_0\right]\left[J^{\xi}(\mathbf{q}) - \alpha^{\xi} J^{\xi}(0)\right]}{kT - 2\left[J^{\xi}(\mathbf{q}) - \alpha^{\xi} J(0)\right]\left[\langle \mathbf{S}_i^{\xi} : \mathbf{S}_i^{\xi}\rangle_0 - \langle \mathbf{S}_i^{\xi}\rangle_0 \langle \mathbf{S}_i^{\xi}\rangle_0\right]}\end{aligned} \tag{14.132}$$

In deriving (14.132) we have made use of the fact that α, the zero-field averages, and the colon brackets are site-independent. This is definitely true in the case of ferromagnetism and is valid also for antiferromagnetism, since all the moments are either parallel or antiparallel to a unique axis. In such cases, since we always have terms of the form $\langle \boldsymbol{\mu}_i : \boldsymbol{\mu}_i\rangle_0$ or $\langle \boldsymbol{\mu}_i\rangle_0 \langle \boldsymbol{\mu}_i\rangle_0$, the site-index terms in (14.132) are site independent. Equation (14.132) can be written as

$$\begin{aligned}kT\chi^{\xi}(\mathbf{q}) &= \left[\langle \boldsymbol{\mu}_i^{\xi} : \boldsymbol{\mu}_i^{\xi}\rangle_0 - \langle \boldsymbol{\mu}_i^{\xi}\rangle_0 \langle \boldsymbol{\mu}_i^{\xi}\rangle_0\right] \\ &\quad + U(\mathbf{q})\end{aligned} \tag{14.133}$$

where $U(\mathbf{q})$ is the second term in (14.132) without the factor μ_B, and $\mathbf{q}$ is the wave vector. Summing over $\mathbf{q}$ in (14.133), we get

$$\begin{aligned}kT\chi^{\xi} &= \sum_{\mathbf{q}} kT\chi^{\xi}(\mathbf{q}) \\ &= N\left[\langle \boldsymbol{\mu}_i^{\xi} : \boldsymbol{\mu}_i^{\xi}\rangle_0 - \langle \boldsymbol{\mu}_i^{\xi}\rangle_0 \langle \boldsymbol{\mu}_i^{\xi}\rangle_0\right] + \sum_{\mathbf{q}} U(\mathbf{q})\end{aligned} \tag{14.134}$$

From the fluctuation theorem we have the result

$$kT\chi^{\xi}=N\left[\langle\mu_i^{\xi}:\mu_i^{\xi}\rangle_0-\langle\mu_i^{\xi}\rangle_0\langle\mu_i^{\xi}\rangle_0\right] \tag{14.135}$$

which has been proved by Lines[15] for the paramagnetic case. Desai and Chakravarty[18] have shown that (14.135) is valid also for the ordered phase. It therefore follows that

$$\sum_{\mathbf{q}} U(\mathbf{q})=0 \tag{14.136}$$

Equation (14.136) essentially determines α^{γ}. Thus the α^{γ} are not free parameters but have to be determined self-consistently. Using (14.136), it can be shown that

$$\alpha^{\gamma}=\frac{\sum_{\mathbf{q}}\eta^{\gamma}(\mathbf{q})J^{\gamma}(\mathbf{q})}{\sum_{\mathbf{q}}\eta^{\gamma}(\mathbf{q})J^{\gamma}(0)} \tag{14.137}$$

where

$$\eta^{\gamma}(\mathbf{q})=kT-2\left[J^{\gamma}(\mathbf{q})-\alpha^{\gamma}J^{\gamma}(0)\right]a$$

and

$$a=\left[\langle\mathbf{S}_i^{\gamma}:\mathbf{S}_i^{\gamma}\rangle_0-\langle\mathbf{S}_i^{\gamma}\rangle_0\langle\mathbf{S}_i^{\gamma}\rangle_0\right] \tag{14.138}$$

This self-consistent equation was derived by Lines. Because of the temperature appearing in (14.138), α^{γ} becomes temperature dependent. The sum over $\mathbf{q}$ is over the first Brillouin zone.

Desai and Chakravarty[18] have derived an alternative formula for self-consistency, using (14.136) but rearranging the terms suitably. Their expression is

$$\alpha^{\gamma}=\frac{\sum_{\mathbf{q}}\dfrac{J^{\gamma}(\mathbf{q})kT}{kT-2\left[J^{\gamma}(\mathbf{q})-\alpha^{\gamma}J^{\gamma}(0)\right]a}}{\sum_{\mathbf{q}}J(0)} \tag{14.139}$$

where a is given by (14.138). Equation (14.139) is computationally very convenient to use since there is essentially only one integral over $\mathbf{q}$.

We now give the procedure for the calculation of susceptibility. Equation (14.120) gives the CEF Hamiltonian. After some α, $0\leqslant|\alpha|\leqslant 1$, and J are assumed, the Hamiltonian is diagonalized. The eigenvalues and the eigenfunctions are used, and the averages and the colon brackets are evaluated [by means of (14.128)] for a particular temperature. Then new α^{γ} are determined, using (14.139). This procedure is repeated until self-consistency is achieved. The susceptibility is then calculated using (14.132). This process is repeated for other temperatures.

14.5 APPLICATION OF THE CORRELATED EFFECTIVE FIELD THEORY

To show the usefulness and importance of the correlated effective field theory proposed by Lines, we now apply[19] the theory to investigate the magnetic properties of the Re^{4+} complexes.

Ter Haar and Lines[20] used a molecular field method due to Anderson[6] to study the stability of various antiferromagnetic patterns in f.c.c. lattices. They included third- and fourth-neighbor isotropic exchange interaction, in addition to the anisotropic first- and second-neighbor interactions J_1 and J_2, and derived expressions for T_N and θ in terms of J_1 and J_2. If we assume only the nearest and the next-nearest isotropic interactions, T_N and θ for type 1 order (see Figure 14.8) are given by[21]

$$T_N = \frac{S(S+1)}{3k}(-8J_1 + 12J_2)$$
$$\theta = \frac{2S(S+1)}{3k}(12J_1 + 6J_2) \tag{14.140}$$

Furthermore, in order that type 1 order (Figure 14.8) be stable with respect to the other types, the second-neighbor interaction should be zero or ferromagnetic.[22] As discussed in Ref. 20, the anisotropies in the exchange integrals also stabilize type 1 order.

Note that the CEF Hamiltonian (14.120) contains the crystal field terms. This can be written as an effective spin Hamiltonian, given by

$$\mathcal{H}_i = D(S_i^z)^2 \tag{14.141}$$

To write the other terms we note that in type 1 order the magnetic moments

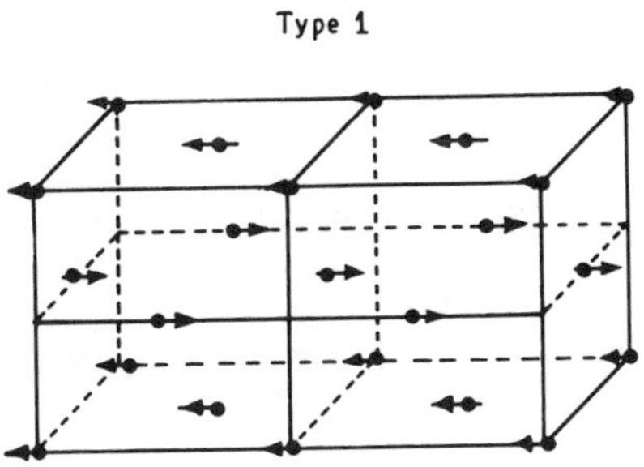

Fig. 14.8 Antiferromagnetic ordering of Type 1.

are ferromagnetically aligned in xy planes (Figure 14.8), and as one goes along the z-axis the direction of the alignment alternates. Thus each ion has four nearest neighbors parallel and eight nearest neighbors antiparallel to itself. Of six next nearest neighbors four are parallel and two antiparallel. With no loss of generality, we take the x-axis as the axis of alignment. Then $\langle S^y \rangle_0 = \langle S^z \rangle_0 = 0$. Furthermore, we retain only the isotropic exchange integrals between the first and second neighbors (J_1 and J_2, respectively), neglecting the anisotropies, since they are small. Inclusion of the second-neighbor exchange integral assures the stability of type 1 order. The anisotropy in the correlation parameters α^γ, which are determined self-consistently, simulates anisotropic exchange interaction to some extent. Thus the complete CEF Hamiltonian for our problem can be written as

$$\mathcal{H}_{\text{CEF}} = D(S^z)^2 - (12J_1 + 6J_2)\{\alpha^x(S^x)^2 + \alpha^y(S^y)^2 + \alpha^z(S^z)^2\} + \langle S^x \rangle_0 \{(8J_1 - 12J_2) + 12(2J_1 + J_2)\alpha^x\} S^x \tag{14.142}$$

No experimental value of the zero-field splitting factor D is available, but this is expected to be small since analysis of the optical spectra[23] shows that

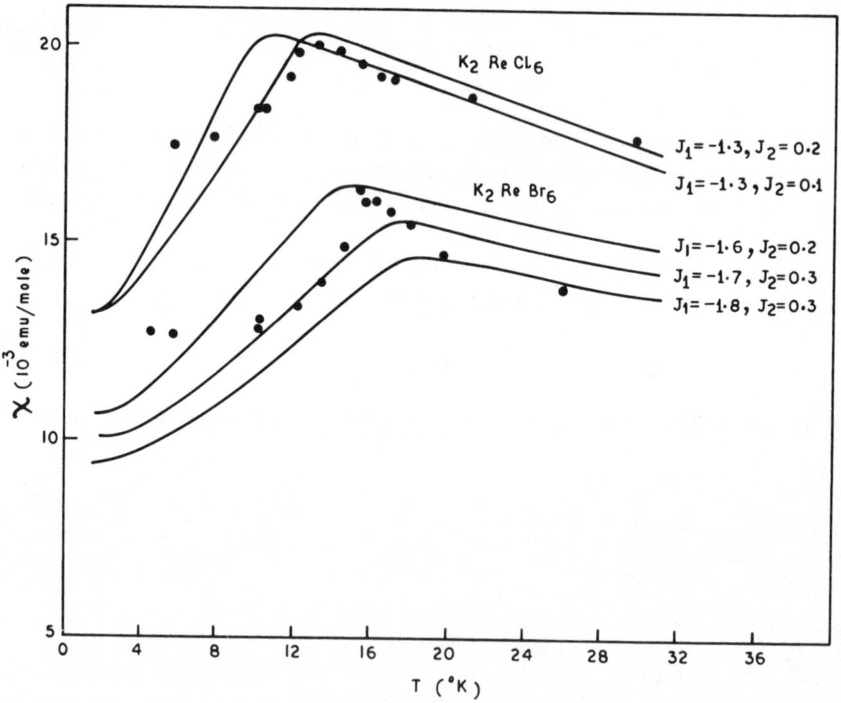

Fig. 14.9 Variation of susceptibility with temperature in K_2ReCl_6 and K_2ReBr_6. J_1 and J_2 represent the nearest- and next-nearest neighbor exchange integrals. The solid circles represent the experimental values.

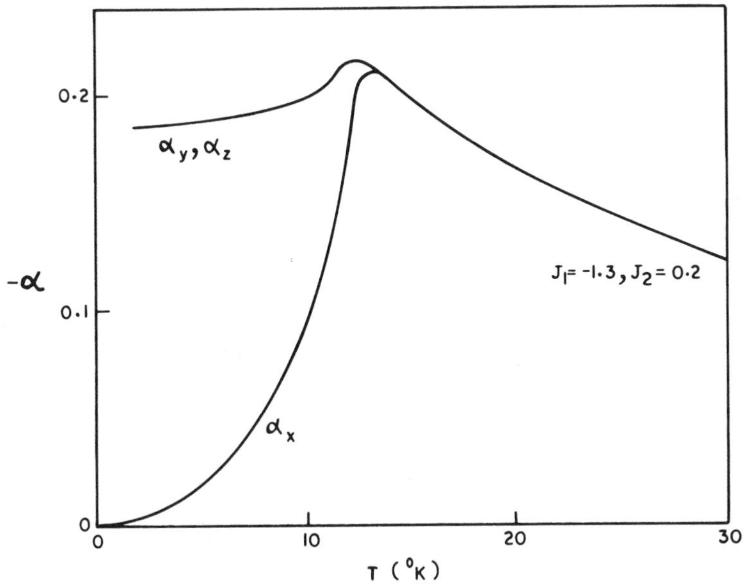

Fig. 14.10 Variation of α with T (see the text).

the cluster $(ReCl_6)^{2-}$ is an almost perfect octahedron down to liquid helium temperature. The susceptibility is calculated as described earlier, and is shown in Figure 14.9, along with the experimental points.[24] As there is fair agreement with the experimental values, the estimates of the exchange integrals are rather reasonable. In Figure 14.10 the correlation parameters are plotted as functions of temperature for particular J_1 and J_2.

In conclusion, we may say that the CEF theory gives a good account of these antiferromagnetic materials. For any realistic theory of antiferromagnetism in insulators it is necessary to know the J_1 and J_2 integrals and a tractable model Hamiltonian. Toward this end the CEF theory proves to be a good model.

REFERENCES

1. L. Neel, *Ann. Phys.* **5**, 256 (1936); J. S. Smart, *Magnetism*, Vol. 3, G. T. Rado and H. Suhl (Eds.), Academic Press, New York, 1963, p. 63.
2. J. H. Van Vleck, *J. Chem. Phys.* **9**, 85 (1941).
3. J. H. Van Vleck, *The Theory of Electric and Magnetic Susceptibilities*, Oxford University Press, London, England, 1932, p. 257.
4. S. Foner, *J. Phys. Radium* **20**, 336 (1959).
5. J. S. Smart, *Rev. Mod. Phys.* **25**, 327 (1953).
6. P. W. Anderson, *Phys. Rev.* **79**, 350, 705 (1950).
7. L. Neel, *Ann. Phys.* **3**, 137 (1948).

8. C. G. Shull and J. S. Smart, *Phys. Rev.* **76**, 1256 (1949).
9. H. A. Kramers, *Physica* **1**, 182 (1934).
10. J. H. Van Vleck, *J. Phys. Radium* **12**, 262 (1951).
11. M. Tinkham, *Proc. Roy. Soc. (London) A*, **236**, 535, 549 (1956).
12. R. G. Shulman and V. Jaccarino, *Phys. Rev.* **103**, 1126 (1956); ibid. **108**, 1219 (1957).
13. P. W. Anderson, *Solid State Phys.* **14**, 99 (1963). J. S. Smart, *Magnetism*, Vol. 3, G. T. Rado and H. Suhl (Eds.), Academic Press, New York, 1963, p. 63.
14. S. Sugano, *J. Appl. Phys.*, Suppl., **33**, 303 (1962).
15. M. E. Lines, *J. Phys. Chem. Solids* **33**, 269 (1972); *Phys. Rev. B* **9**, 950, 3927 (1974).
16. M. E. Lines and M. Eibschultz, *Phys. Rev.* **B11**, 4583 (1975).
17. N. Boccara and G. Sarma, *Physics* **1**, 219 (1965); E. Pytte and J. Feder, *Phys. Rev.* **187**, 1077 (1969); N. R. Werthamer, *Phys. Rev. B*, **1**, 572 (1970); N. S. Gillis and T. R. Koehler, *Phys. Rev. B*, **4**, 3971 (1971); ibid. **5**, 1925 (1972); M. Kohen and T. L. Einstein, *Phys. Rev. B*, **7**, 1932 (1973).
18. V. P. Desai and A. S. Chakravarty, unpublished work (1976).
19. Ibha Chatterjee and V. P. Desai, to be communicated to *Phys. Rev.* (1980).
20. D. ter Haar and M. E. Lines, *Phil. Trans. Roy. Soc. (London) A*, **255**, 1 (1962).
21. J. S. Smart, *Effective Field Theories of Magnetism*, Saunders, Philadelphia, 1966.
22. J. B. Goodenough, *Magnetism and Chemical Bond*, Interscience, New York, 1962.
23. L. Press, K. Rossler, and H. J. Schenk, *J. Inorg. Nucl. Chem.* **36**, 317 (1974).
24. R. H. Busey and E. Sonder, *J. Chem. Phys.* **36**, 93 (1962).
25. Ibha De, V. P. Desai, and A. S. Chakravarty, *Phys. Rev. B*, **10**, 1113 (1974).

15

Magnetism in Disordered Systems

So far, our attention has been confined to the crystalline solids, where the group-theoretical concepts of the translational and rotational symmetries are applicable. In recent years, however, there has been rapidly increasing interest in the magnetism of amorphous solids.[1,27] This interest is mainly due, of course, to the expected and semiquantitatively realized applications in technical devices, for example, in magnetic "bubble" storage chips and magnetic Metglas ribbons. Theoretical interest in amorphous systems, on the other hand, is also sustained by the challenge that these and other strongly disordered systems present to physicists' understanding of the origin and basic mechanism of the special properties of the systems after the pioneering work by Gubanov[28] in 1960.

15.1 AMORPHOUS MAGNETISM IN SOLIDS

The term "amorphous magnetism" can be used both in a narrow and in a wider sense. In the wider sense it comprises all kinds of disordered solid state magnetic systems, that is, (1) the spin glasses, (2) the crystalline magnetic alloys, and (3) the amorphous magnets in the conventional narrow sense. Among these, we shall study only the first one since much attention has been paid to the spin glass systems.

Before discussing these systems, let us try to understand what we mean by the "crystalline magnetic alloys" (case 2 above). These are the simplest disordered spin systems, where there is no structural disorder at all except in regard to composition. Also, the magnetic ground state is generally simple (ferro-, antiferro-, or ferri-magnetic), although exceptions may, of course, exist, as mentioned below. In this context two important points should be made:

1. If the exchange integrals and/or the spin quantum numbers of the different magnetic components of the alloy differ considerably, the magnetic excitation spectra can be very complex.[29] This is true, for instance, of antiferromagnetic system $KMn_{0.25}Ni_{0.75}F_3$, which has been studied both theoretically[29] and experimentally.[30]
2. The origin of the so-called Invar anomalies, for example, the extremely large-volume magnetostriction in the f.c.c. Fe_xNi_{1-x} crystalline alloys near $x \sim 0.75$,[31] is still a problem to the physicist. This Invar problem may ultimately be connected to the spin glass problem, which we shall discuss in this chapter.

We now briefly consider the amorphous magnets (case 3 above). In comparison to the spin glass problem, the amorphous magnetic problem is a bit simpler. Structurally, the amorphous systems correspond to a "moment" photograph of a fluid.[32] Furthermore, in amorphous ferromagnets the distribution of the exchange integrals should be small enough and should have a positive mean value so that an almost perfect spin alignment can be expected. The alignment will be disturbed only by the magnetic dipole interaction and by local magnetic anisotropies of a uniaxial character. These are present for spin quantum numbers greater than $\frac{1}{2}$ because of the lack of structural order (symmetry). We do not expect very drastic effects except in cases where these local anisotropies are very strong, as they seem to be for the amorphous $TbFe_2$ alloys.[33]

One of the most interesting examples of research in this field comes from the experimental work of Mook, Wakabayashi, and Pau[34] on amorphous phosphate glass CoP. Drastic effects might also show up in the phase transition of amorphous ferromagnets; for example, one might imagine that the random anisotropies can become effectively enhanced near the phase transition. In fact, here also it is not clear whether a conventional second-order phase transition occurs.

Many of the amorphous magnets are, in fact, binary or ternary alloys with two or more magnetic components and are ordered mostly ferrimagnetically, for example, the Gd-Co amorphous system[35] and other rare earth-transition metal systems. The amorphous ferrimagnets can often be "compensated" at a certain temperature T_N, where the macroscopic average of the magnetic moment vanishes; but since there can be no exact compensation on a microscopic scale, in contrast to the crystalline compensated ferrimagnets, the term "antiferromagnet" should be used here.

Finally, we should mention that the strange "sperimagnetic" configuration observed in amorphous $DyCo_3$ is due to the random anisotropies acting preferentially on the Dy spin; the Dy spin direction varies strongly with position but with mean direction opposite to that of the aligned Co spin. However, our present knowledge of the possible spin configurations in amorphous magnetic systems is far from complete, as in the case of spin glass, which we describe now.

15.2 THE SPIN GLASSES

During the past two decades there has been a great deal of activity, both theoretical and experimental, in regard to the magnetic alloys, in which a magnetic species of atoms is randomly combined with a nonmagnetic metal, for example, a solid solution of gold and iron. For the most part, two important approaches of modern solid state physics are employed to understand these apparently simple magnetic alloys. The first approach is to try to understand the nature of the amorphous magnetic structure and the resulting magnetism, while the second approach is to investigate the critical phenomena and "percolation" in these random magnetic alloys, which have been named the "spin glasses." A spin glass has been defined[1] as a random, metallic, magnetic system in a nonmagnetic host characterized by a random freezing of moments without long-range order at a rather well-defined temperature T_f.

To appreciate the problem of spin glasses, we begin[1] by considering a crystal of pure gold and hypothetically dissolve a single atom of iron into the crystal structure. Naively one expects in this situation that Fe^{2+} should retain its free-ion magnetic moment, given by $g\sqrt{S(S+1)}=4.9$, as seems to be reasonable from the standpoint of an insulating crystal. If one approaches from the metal point of view, however, there exists the possibility of a strong interaction between the $3d$ electrons of Fe^{2+} with the conduction s electrons of the host metal. It is well known that the magnetic coupling of the s and d electrons for the isolated impurities gives rise to the so-called Kondo effect, which causes a rapidly varying magnitude of the Fe spin. This process may be described in terms of a spin fluctuation frequency ω_{SF}, given by $\hbar\omega_{SF}=kT_K$, where T_K is the Kondo temperature and k is the Boltzmann constant. In addition to the magnetic fluctuations of the magnetic moment at finite temperatures, its orientation or direction will also continually change because of thermal disorder. The Kondo effect is nicely demonstrated in Figure 15.1, where the low-temperature electrical resistivity has been plotted against temperature for various types of metals.

In practice, however, it is very difficult to investigate the effects of "single" magnetic impurities experimentally because of measurement sensitivity. To be able to detect the effects, one needs a sufficient number of these impurities. It is also important to note that there exists a long-range interaction between the (localized) impurities via the conduction electrons, which destroys the Kondo state (see Figure 15.1). This strong long-range interaction, which is investigated by the Ruderman-Kittel-Kasuya-Yoshida (RKKY) theory, leads us to the idea of a spin glass. The very important property of RKKY which gives spin glasses their unique nature is the oscillating character of the exchange

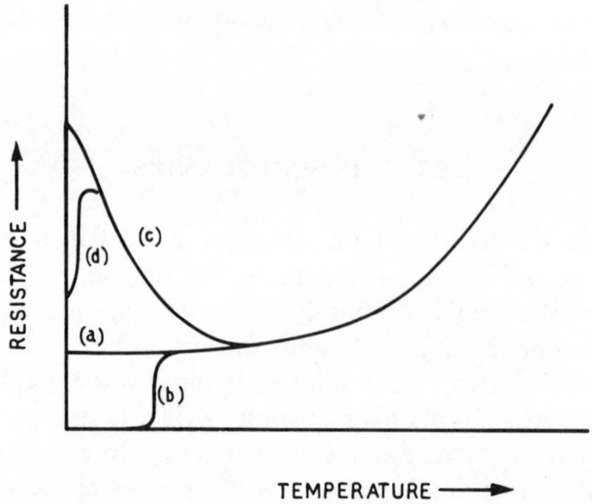

Fig. 15.1 Variation of the electrical resistance with temperature for (*a*) normal metal, (*b*) the superconductor, (*c*) the Kondo metal, and (*d*) the Kondo metal with impurity-impurity interactions. (Taken from Mydosh.[1])

interactions which fall off as $1/\mathbf{r}^3$. This is so because the randomly located spins then have interactions of essentially random sign, so that no ferromagnetic or (with less confidence) long-range ordered antiferromagnetic state is particularly favored energetically. This is the peculiar characteristic of "spin glass," to use a very apt term invented by Brian R. Coles. He used this term to describe the entire class of magnetic alloys of moderate dilution in which the magnetic atoms are far enough apart so that the magnetic structure no longer resembles that of the pure metal, but close enough so that their exchange interactions dominate other energies such as the Kondo effect and other free-electron interactions. The RKKY theory tells us that the effective coupling between the magnetic moments can be either ferromagnetic or antiferromagnetic, depending on the separation between two impurities. Now, since the impurities are randomly located within the crystal, the magnetic interactions are also randomly distributed. Thus the term "spin glass" is used in analogy with a real glass or an amorphous solid where the atoms are randomly distributed without any order or regular structure.

Now, what happens when the alloy is cooled down to liquid helium temperature? As the freezing temperature T_f, which is rather well defined, is reached, there is a sudden loss of rotational freedom because of the magnetic moments being locked or frozen in random orientations which effectively means that $\Sigma_i \boldsymbol{\mu}_i = 0$. Since the freezing temperature is quite sharp, it is reasonable to assume that a new type of phase transition does take place at T_f. The situation is very similar to that for a real glass. As the temperature is lowered through its melting point, one observes a slowing down of the atomic or molecular mobility and consequently an increase in viscosity; as a result, the static amorphous structure appears. For spin glass we have a long-range interaction via the conduction electrons, which causes the freezing process, and hence a cooperative behavior would be expected.

If now we increase the concentration of Fe impurities in our Au lattice, the probability of each Fe atom having similar atoms as nearest neighbors also increases. Hence at a few atomic percent of impurity it is reasonable to think that local groups of directly coupled Fe spins will appear and give rise to "giant magnetic moments." These regions are called the "mictomagnetic" clusters. These clusters will increase in size if a short-range crystallographic order is also present to some extent. Figure 15.2 shows the mictomagnetic clusters within the spin glass matrix for about 10% of the sites occupied with spins. These clusters give rise to the superparamagnetism. Because of the intrinsic anistropy of the clusters, as can be seen from Figure 15.2, the axes of the up-spin and down-spin orientations become a random directional variable throughout the spin glass matrix.

As the concentration of Fe impurities is increased more and more, the probability that each impurity will have at least one nearest-neighbor impurity greatly increases and approaches unity; therefore a "percolation" limit will be reached for long-range ferromagnetic ordering. This means that, within the crystal, long chains of varying thickness, containing ferromagnetically coupled spins, will zigzag from one end of the alloy to the other. For a f.c.c.

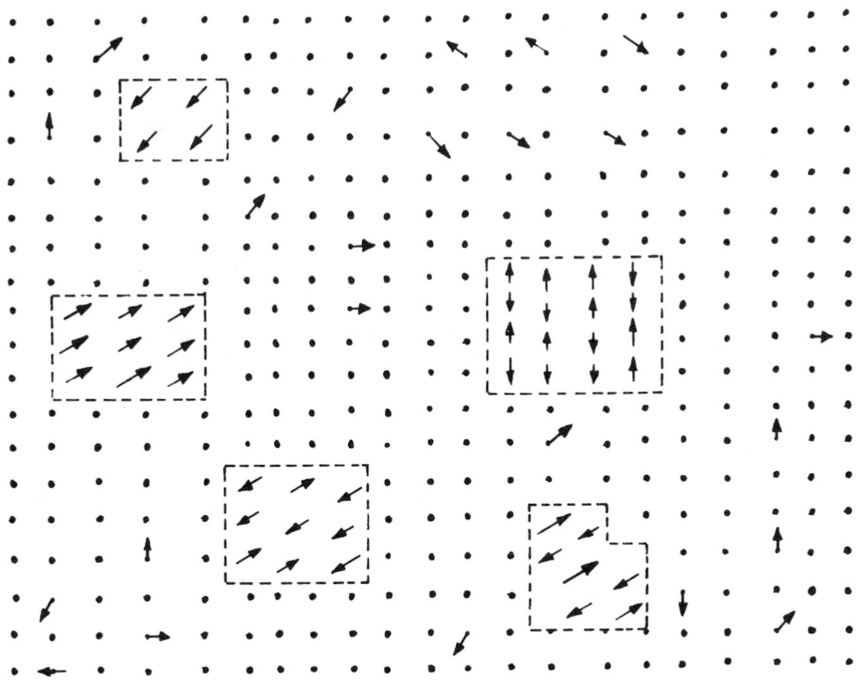

Fig. 15.2 Mictomagnetic clusters with approximately 10 atomic percent impurities in spin glass matrix (schematic). (Taken from Mydosh.[1])

lattice, such as gold, approximately 15% replacement by Fe atoms is required to reach the percolation limit. Above this percolation limit there will be ferromagnetic ordering, since the number of impurities greatly increases.

Let us now briefly discuss the experimental characteristics of the spin glasses. Experimentally one observes in a spin glass a sharp, cusp like maximum in the very low (or even zero) field susceptibility at the freezing temperature T_f, which becomes greatly broadened for external fields of only a few hundred oersteds. Figure 15.3 represents this effect for two Au-Fe alloys. At the same T_f, the Mössbauer spectrum shows a sudden splitting from a single line to six ordered components. For $T<T_f$ there appear, in the magnetization of remanence, irreversibility and relaxation.

We shall not elaborate on these aspects any further. The electrical resistivity, as well as the specific heat, has been measured in the spin glass systems. From these measurements it has been found that a great deal of short-range magnetic clustering is already present at $T \gg T_f$. Such measurements establish beyond doubt that a majority of the Fe spins participate in a local type of correlation. Furthermore, spin glass can be strongly magnetized by applying magnetic fields, but below T_f there is large hysteresis, and aftereffects with a characteristic time of several hours appear.

All these observations can be qualitatively understood by means of a simple model. As the temperature is lowered, many of the randomly located, freely rotating spins come together, by means of some correlation, into clusters that can then rotate as a whole. The remaining isolated spins are uncorrelated but serve to transmit interactions between the clusters. At $T=T_f$ these independent isolated spins freeze out in random directions; these spins are, however, difficult to detect experimentally because of their very low concentration. However, their freezing reacts on the largest of the clusters,

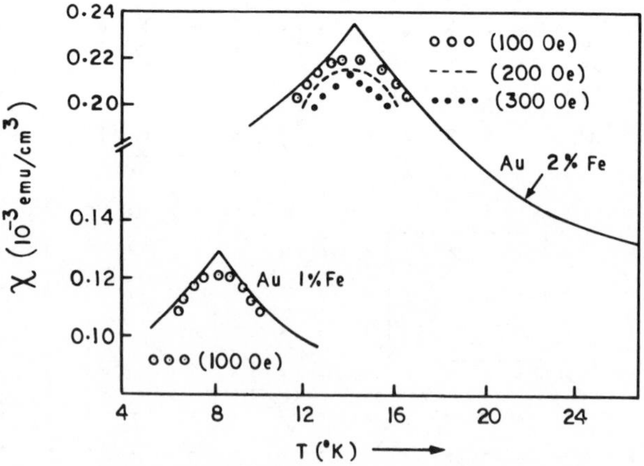

Fig. 15.3 Variation of χ with T in a spin glass with different external fields. Solid line represents the zero-field limit susceptibility. (After Mydosh.[1])

and they become rigid. Thus the freezing of a spin glass is more subtle in nature than the ordinary phase transitions in solids.

The application of the term "spin glass" has been extended to encompass very many, diverse magnetic materials. Let us now survey these systems, starting with the noble metal-transition metal alloys, that is, Au-(Fe, Mn, Cr), Ag-Mn, and Cu-Mn. We should also mention here Zn-Mn and $Al_{30}Fe_{70}$ alloys, as well as the giant moment ferromagnetic "dilute" systems, that is, Pd-(Fe, Co, Mn) and Pt-(Fe, Co), which exhibit spin glass ordering. Next come the rare earth spin glass systems which we do not elaborate[1] here.

Such a large collection of systems (both binary and pseudo-binary alloys) demonstrates the importance of the RKKY indirect exchange. Since the rare earth $4f$ electrons are highly localized, the magnetic couplings must be transmitted via the conduction electrons in order to show a cooperative freezing in a dilute alloy. The similar properties of the $4f$ spin glasses and the $3d$ ones further emphasize the generality of the freezing process, for, barring complications due to the crystal field, any metallic system having the local moments randomly distributed should be a spin glass.

Another interesting class of spin glass systems consists of the rare earth impurities in superconducting hosts. The Kondo theory predicts a strongly temperature-dependent pair-breaking parameter which, for $T_K \ll T_C$, where T_K is the Kondo temperature, causes the normal state to stabilize at lower temperatures.[2] Two important examples of such behavior occur in $La_{1-x}Ce_x$-Al_2 and $La_{0.8}Th_{0.2}$ with about one atomic percent of Ce. It would be interesting to increase the Ce concentration until the interactions become dominant, and then test for the appearance of spin glass freezing and its effect on the superconductivity. Another important aspect to study would be the coexistence of superconductivity and magnetism. Experimental evidence suggests that this magnetism is of the spin glass type. It seems, therefore, that the frozen spin glass and the superconducting states can simultaneously exist, whereas long-range magnetic order and superconductivity cannot.

Three different classes of crystalline metallic systems that show spin glass behavior are as follows:

1. The Ti-V sesquioxides, $(Ti_{1-x}V_x)_2O_3$, in the conducting phase, and with a local moment associated with the V atoms randomly distributed, fit nicely into the spin glass framework.[3] Specific heat studies have also been made on such systems.[4]
2. The two-dimensional intercalated layer compounds, for example, $NbSe_2$, are also of great interest because of their highly anisotropic superconductivity.[5] Moreover, the coexistence of superconductivity and magnetism becomes an important question. All these effects may be simply described using the random freezing concepts of a spin glass.
3. Lastly, the recently discovered Chevrel-phase ternary Mo chalcogenides, for example, $(M)Mo_6(S \text{ or } Se)_8$, exhibit a very high critical field superconductivity.[6] In the above formula M represents a variety of metals, or even

a combination of two metals, one of which may be magnetic, for example, $M = Sn_{1-x}Eu_x$. Mössbauer studies on such a system have revealed a hyperfine splitting associated with a spin glass freezing at temperatures far below the superconducting transition.[7] The unique feature here is that the localized Eu magnetic impurities can interact among themselves and yet do not cause any significant pair breaking because of the small overlap of the very different magnetic and superconducting electrons.

So far, we have considered the crystalline (even single-crystal) metallic systems with a random distribution of magnetic spins that interact in "conflict." Now we can relax the condition of lattice order and study the amorphous magnets, neglecting the long-range ordered amorphous ferromagnets. A large number of such disordered systems have been studied in thin-film form and also in bulk form by Korn.[8] We shall not go into the details of these systems, however. One more interesting type of system consists of insulating random substitutional substances and real glasses, both of which may be doped with magnetic elements. Since these materials do not possess the strong RKKY interaction, it is not proper to group them as typical spin glasses. According to Coey,[9] however, we should use the term "superparamagnets" to describe their behavior.

In this section we have reviewed briefly a very interesting and rapidly developing aspect of the solid known as spin glass. We have defined, following Mydosh,[1] what we mean by a "spin glass" and have limited the usage of this concept to metallic alloys where long-range magnetic interactions are present. It is this long-range interaction (short-range interaction may also be present, though to a much smaller extent) that produces the random freezing of the spin moments at a rather well-defined temperature T_f. We have also examined a variety of recent experiments performed on spin glass alloys with respect to three ranges of temperature: $T > T_f$, $T \simeq T_f$, and $T < T_f$.

In the interpretation of the experimental behavior, a phenomenological model depending on the dynamically growing clusters has been proposed by Mydosh.[10] It has been observed, even at $T \gg T_f$, that some local correlations among the randomly separated spins also exist. The growth of magnetic clusters continues until $T \simeq T_f$, when a few of the largest clusters freeze out in random orientations. Such a freezing process is dramatically observed in "direct" magnetic measurements (e.g., the susceptibility), but not in "indirect" ones (e.g., the specific heat).[10] The spin glass behavior may be theoretically described by the methods of critical phenomena, where local correlations must be incorporated into the calculations. The first significant theoretical approach to understand the spin glass behavior was due to Edwards and Anderson[11] but has been subsequently extended by several workers, notably Binder.[12] A different theoretical approach based on fluctuations has been proposed by Riess.[13]

We now discuss the theoretical approaches that have been made by a number of workers to understand the spin glass properties, which apparently are a consequence of the interactions between the magnetic moments. There are several possible interactions, the most important one being the Rudermann-Kittel-Kasuya-Yosida (RKKY) interaction, where the spin system is described by a Heisenberg or Ising Hamiltonian ($b = g\mu_B H_0$, where H_0 is an external field):

$$\mathcal{H} = -\sum_{i<j} J_{ij}\mathbf{S}_i \cdot \mathbf{S}_j - b\sum_i S_i^z \tag{15.1}$$

where the exchange interaction $J_{ij} = J(|\mathbf{R}_i - \mathbf{R}_j|)$ and can be written as

$$J(R) = \frac{J' \cos 2K_F R}{R^3} \tag{15.2}$$

where K_F is the Fermi momentum of the host conduction electrons. Equation (15.2) is oscillatory in nature and therefore is long range, yielding ferro- and antiferromagnetically coupled spins. Whether this oscillatory nature of the exchange interaction is essential for spin glass is, however, not yet clear. Amorphous dilute magnetic alloys, in which the RKKY interaction is exponentially damped, also show spin glass properties.[8]

It can be shown that the R^{-3} dependence of the RKKY interaction leads to some kind of corresponding states for low concentrations of the magnetic impurities,[14] that is, $c \to 0$. Thus the specific heat C_H, which is a function of T, H_0, and c, "scales" as $C_H = cf_1(T/c, H_0/c)$; the magnetization, as $M(T, H_0, c) = cf_2(T/c, H_0/c)$; and so on. This leads to $T_f \propto c$, in agreement with experimental results, and one can also write

$$\begin{aligned} C_H &= cf_1\left(\frac{T}{T_f}, \frac{b}{kT_f}\right) \\ M &= cf_2\left(\frac{T}{T_f}, \frac{b}{kT_f}\right) \end{aligned} \tag{15.3}$$

Fischer[15] has given more general relations (15.3) with $T_f(c)$, extending the theory of Edwards and Anderson.[11]

For more concentrated spin glasses $T_f \propto c^m$ with $0.55 < m < 0.75$, at least for the Au-Fe system described earlier. In such concentrations one should expect direct exchange interactions to become important. For Au-Fe with ferromagnetically coupled nearest-neighbor pairs,[16] the direct interaction leads to an abrupt transition of $T_f(c)$[17] at roughly the percolation limit. Apart from small ferromagnetic clusters above T_f, clusters with more or less random spin directions are indicated by the specific heat measurements.

For a fixed impurity configuration we have, for $T > T_f$,

$$\langle \mathbf{S}_i \rangle = 0 \tag{15.4}$$

since the internal field felt by a spin averages to zero if the (thermal) average is taken over a sufficiently long time, whereas, for $T<T_f$, we have, for spin glasses,

$$\overline{\langle \mathbf{S}_i\rangle \cdot \langle \mathbf{S}_i\rangle} \neq 0 \tag{15.5}$$

since the internal field at the lattice site i no longer averages to zero. This is so because, for spin glasses, there exists a well-defined temperature T_f below which all spins assume a fixed local axis. However, if the local axes are randomly distributed (as is assumed for spin glasses, since the total magnetization is zero), the configuration average $\langle \overline{\mathbf{S}}_i\rangle = 0$, for all temperatures.

Let us now consider the static susceptibility for a spin glass. For N spins it follows from the free energy that

$$\chi = \frac{\chi_0}{NS(S+1)} \sum_{i,j} \left[\langle \overline{\mathbf{S}_i \mathbf{S}_j} \rangle - \langle \overline{\mathbf{S}}_i \rangle \langle \overline{\mathbf{S}}_j \rangle \right] \tag{15.6}$$

where $\chi_0 = NS(S+1)g^2\mu_B^2/3kT$.

We now consider a simple model[11] in which the *random* distribution of impurity sites is replaced by a *symmetric* distribution of exchange couplings J_{ij} with $P(J_{ij}) = P(-J_{ij})$, where $P(J_{ij})$ is the probability of finding a coupling constant and is given by

$$P(J_{ij}) \propto e^{-J_{ij}^2/2(\Delta J)^2} \tag{15.7}$$

where the interaction J_{ij} between a given spin and its z nearest neighbors has a Gaussian distribution. The Hamiltonian that has been used by Edwards and Anderson to treat the spin glasses is the familiar Heisenberg Hamiltonian with the external field given by (15.1).

As a first approximation, (15.1) has been treated in the mean field approximation (MFA). There one should calculate the free energy, given by

$$\begin{aligned} F &= \langle F(J_{ij}) \rangle = \langle -kT \ln Z(J_{ij}) \rangle \\ &= \langle -kT \ln \operatorname{Tr} e^{-\mathcal{H}/kT} \rangle \end{aligned} \tag{15.8}$$

for a fixed set J_{ij}, the brackets denoting an average with the Gaussian distribution (15.7), and Z representing the partition function. This procedure is very difficult to carry through even in MFA, and therefore Edwards and Anderson use the "replica trick." Instead of calculating $\langle \ln Z(J_{ij}) \rangle$, one calculates $\langle Z^n(J_{ij}) \rangle$, where n is an arbitrary positive integer, and then computes F by analytic continuation to $n=0$, that is,

$$F = -kT \langle \ln Z(J_{ij}) \rangle = -kT \lim_{n\to 0} \frac{1}{n} \left[\langle Z^n(J_{ij}) \rangle - 1 \right] \tag{15.9}$$

where the partition function Z, which can be derived from (15.1) with $H=0$, is given by

$$Z(J_{ij}) = \prod_{i,j,\mu} e^{J_{ij}S_i^\mu S_j^\mu/kT} \tag{15.10}$$

Here μ runs from 1 to m if $\mathbf{S}$ is an m-component spin ($m=1$ in the Ising model, whereas $m=3$ in the Heisenberg model).

Let us now return to the problem of static susceptibility (15.6). For a system with no correlation between different sites, it follows that

$$\langle \overline{\mathbf{S}_i\mathbf{S}_j} \rangle = S(S+1)\delta_{ij} \tag{15.11}$$

and

$$\langle \overline{\mathbf{S}_i} \rangle \langle \overline{\mathbf{S}_j} \rangle = \langle \overline{\mathbf{S}_i} \rangle^2 \delta_{ij}$$

since each spin has a neighboring spin j in the upward or downward direction with the same probability. For a classical Heisenberg or an Ising model with $S^2=1$, we have, finally,

$$\chi = \chi^0(1-q) \tag{15.12}$$

with

$$q \equiv \langle \overline{\mathbf{S}_i} \rangle^2$$

where q is the relevant spin glass parameter. From (15.4) it follows that $q(T)=0$ for $T>T_f$. In this approach, which is essentially due to Edwards and Anderson, the short-range correlations are neglected.

Without going into more details of the Edwards-Anderson theory, let us discuss its merits. This theory explains the sharp cusp in the static susceptibility $\chi(T)$ by the concept of a sudden freezing of the impurity spins in random directions. From the configuration-averaged free energy, a self-consistent equation for the parameter $q(T)$ is derived under the assumption of a symmetric distribution $P(J_{ij})$. The susceptibility and the specific heat have also been calculated for a finite external field. As expected, the cusp in $\chi(T)$ and $C_H(T)$ is rounded off if the field energy $b \simeq kT_f$. There is a large discrepancy between the theory and experiment for specific heat, which shows no indication of a cusp. As compared to susceptibility, specific heat is more sensitive to the formation of random clusters. Also, a MFA yields worse results for $C_M(T)$ than for $\chi(T)$, near a phase transition. The disagreement with experiment seems to be due to the MFA and not to the model.

The Edwards-Anderson theory has been extended by several authors.[15,18] In a theory based on the formation of clusters and proposed by Smith,[19] a magnetic cluster is defined as a connected group of spins that are coupled together by exchange interactions with a magnitude greater than the thermal energy. With decreasing temperature these clusters grow until the percolation limit is reached at T_f. For $T>T_f$, (15.4) holds since the clusters are still able to rotate and the internal field at each lattice site averages to zero. For $T<T_f$, $\overline{\langle \mathbf{S}_i \rangle} \neq 0$ for a macroscopic contribution of all spins frozen into an infinite cluster. Detailed calculations have been performed for a Bethe lattice. The number of "loose" spins below T_f decreases exponentially, leading to an exponential temperature dependence of $\chi(T)$ for $T \to 0$. At T_f a sharp cusp is obtained, and for $T>T_f$ and a small impurity concentration, one obtains (as in the Edwards-Anderson model) the susceptibility of the free spins.

Thouless and his co-workers[20] have studied a spherical model with infinite-range random interaction that can be solved exactly, but it is almost equivalent to the Edwards-Anderson model in the mean field approximation. Although this treatment removes the inconsistences at $T \to 0$, the discrepancies with experiment remain, and it is very difficult to generalize this theory to a finite number of nearest neighbors, where fluctuations have to be included.

Recently another approach has been proposed by Mattis and Luthinger,[21] whereby one starts from a random binary alloy (AB). The exchange integrals between AA pairs, AB pairs, and BB pairs are denoted by J_{AA}, J_{AB} and J_{BB}, respectively. In terms of occupation numbers, C_i ($C_i=1$ if the ith site is occupied by an A-type atom and zero otherwise), (15.1) can be written as

$$J_{ij} = J_{AA} C_i C_j + J_{AB}\left[C_i(1-C_i) + C_j(1-C_j) \right] + J_{BB}\left[(1-C_i)(1-C_j) \right] \tag{15.13}$$

Assuming $J_{AA} = J_{BB} = -J_{AB} = J$ and defining $\epsilon_i = 2C_i - 1(\pm 1)$, one can show that (15.13) reduces to

$$\mathcal{H} = -J \sum_{i \neq j} \epsilon_i \epsilon_j \mathbf{S}_i \cdot \mathbf{S}_j - b \sum_i S_i^z \tag{15.14}$$

where $\epsilon_i = \pm 1$. If one now chooses $\epsilon_i \mathbf{S}_i = \mathbf{S}_i'$, (15.14) is transformed to an ordinary m-vector model in a random field. However, since the Mattis-Luthinger model does not improve the agreement with experiment obtained with the Edwards-Anderson model, we do not discuss it any farther.

On the other hand, the concept of "superparamagnetic clusters" has been utilized by Kueller and others[22] to derive the magnetization of a spin glass system in thermal equilibrium, which is given by

$$m = N \int_0^\infty d\mu\, P_\mu \mu L\left(\frac{\mu H}{kT} \right)$$

$$\simeq \frac{N}{3kT} \bar{\mu}^2 H \tag{15.15}$$

where $L(x) \simeq x/3$ is the well-known Langevin function, and

$$\overline{\mu^2} = \int_0^\infty d\mu \, \mu^2 P_\mu \tag{15.16}$$

In (15.15), N denotes the superparamagnetic grains, which are described by the distribution function of their magnetic moments P_μ. Here, no interaction between the particles is, of course, considered. If magnetic anisotropy is also present within the superparamagnetic cluster, then (if the time of observation is not large enough) the magnetic anisotropy energy $K\mu$ will prevent the particles from following the field. This is known as the blocking of the grains. In this case one can show by using an Arrhenius formula that

$$\frac{1}{\tau_\mu} = \nu_a e^{-K\mu/kT} \tag{15.17}$$

where τ_μ is the time required for the orientation of a magnetic moment μ, and ν_a is an attempt frequency. Now, clearly, only the moments for which $\tau_\mu < t$, the time of observation, should be included in (15.17), and hence

$$\chi = \frac{N}{3kT} \int_0^a d\mu \mu^2 P_\mu \tag{15.18}$$

where $a = kT/K \ln \nu_a t$. Clearly, χ will have a maximum at $kT \simeq K\sqrt{\overline{\mu^2}} / \ln \nu_a t$. According to Wohlfarth,[23] the magnetic behavior of spin glass is very similar to "rock magnetism" or is entirely due to chemical clustering of the magnetic atoms within the host material. On the basis of this idea, Thouless and Fourier[24] have developed a theory where the spins of the spin glass are thought to be grouped into clusters, each of which can be characterized as a superparamagnetic particle in the above sense. This treatment is quite successful in explaining the absence of long-range order, the occurrence of short-range order, the strong magnetic field dependence of susceptibility, the absence of singularities in specific heat, the ultrasonic attenuation, and so on, as well as all the irreversible behaviors like the remanent magnetization and relaxation.

There are some deficiencies of this model, however. For instance, the maximum of χ at the effective blocking temperature is rounded, not sharp, and it is also asymmetric around T_f in contrast with experiment. Grouping the spins into "clouds," where the spins are strongly locked together within a cloud but there is no interaction between neighboring clouds, is also unrealistic and has little basis in the microscopic picture of the interactions in real spin glass materials. To be realistic, therefore, one has to take into account the interactions between these clusters and has to consider the RKKY, the superexchange, and also the dipolar interactions.

To improve the agreement with experiment, Binder[25] has proposed a model that combines the essential ingredients of the Edwards-Anderson theory and the superparamagnetic cluster concept described above. In this model the spins are grouped into clusters such that all the spins within a cluster are locked together by interactions much larger than kT_f. At temperatures near T_f and for small applied fields, one can neglect the internal degrees of freedom of the clusters, meaning that variations of the magnetic moments $\{\boldsymbol{\mu}_i\}$ of these clusters with temperature and field, as well as the magnetic excitations, are not considered. Thus the short-range-order effects and also the low-temperature behavior of the spin glass, for which the internal degrees of freedom may be important, are dispensed with. By generalizing the Edwards-Anderson model of interacting clusters described above, Binder has shown that the zero-field susceptibility and the remanent magnetic moment are given by

$$\chi = \left(\frac{N_c \bar{\mu}^2}{kT} \right) (1 - \langle \langle \sigma \rangle^2 \rangle) \qquad (15.19)$$

and

$$\langle m(t) \rangle = N_c \bar{\mu} \langle \langle \sigma(t) \rangle \rangle$$

where

$$\bar{\mu} = \int_0^\infty \mu_i P_\mu \, d\mu$$

Although this model has some good features, one is not sure whether it can explain all the important observations in a spin glass.

We do not go into the details of the calculations of the different models proposed so far. The reader is referred to the appropriate references already cited.

In conclusion, it can be said that, despite considerable effort, the situation in regard to the spin glass theory is still very unsatisfactory, since many important basic questions about spin glasses remain unanswered. According to Fischer, these areas of uncertainty are as follows:

1. Is there a phase transition at T_f? If so, what type is it? Whereas the cusp of $\chi(T)$ can be explained by the sudden freezing of a macroscopic number of spins, it remains unclear why no anomalous behavior is seen for the specific heat at T_f. Also, it is not understood under what conditions one obtains a sharp or a gradual transition.
2. It is not clear whether percolation is important at T_f.
3. Remanence seems to be essential for spin glasses, but the nature of the energy barriers involved is obscure. Is an additional anisotropic interaction essential?
4. Well below T_f, one expects collective excitations. Antiferromagnetic excitations have therefore been proposed. However, the temperature dependence

of the specific heat and the resistivity calculated with these excitations are in disagreement with experiment. Therefore a domain structure has been invoked by Edwards and Anderson[26] which separates two of the many degenerate ground states. For a Heisenberg model with $P(J_{ij}) = P(-J_{ij})$, this degeneracy is at least of the order 2^N, where N is the number of spins. Thus there is still no explanation of the linear temperature dependence of the specific heat for $T \to 0$, despite the fact that the Edwards-Anderson theory predicts such behavior.

Hence no explicit analytic theory exists that reproduces the various static and dynamic properties of spin glasses with fair accuracy.

REFERENCES

1. J. A. Mydosh, *AIP Conference Proceedings*, Vol. 24, p. 131, 1975; *J. Magn. Magn. Mater.* (1978). In the introduction to this Chapter I have followed these two articles very closely.
2. M. B. Maple, *Appl. Phys.* **9**, 179 (1976).
3. Y. Miyako, *Physica*, **86–88B**, 869 (1977).
4. M. E. Sjostrand and P. H. Keesom, *Phys. Rev. B*, **7**, 3558 (1973).
5. R. C. Morris et al., *Phys. Rev. B*, **5**, 895 (1972).
6. O. Fischer, *Low Temperature Physics-LT14*, Vol. 5, M. Krusins and M. Vuorio (Eds.), North Holland, Amsterdam, 1975, p. 172.
7. J. Bolz et al., *J. Low Temp. Phys.* **28**, 61 (1977).
8. D. Korn, *Z. Phys.* **187**, 463 (1965); *ibid.* **214**, 136 (1968); *ibid.* **238**, 275 (1970).
9. J. M. D. Coey, *J. Appl. Phys.* **49**, 1646 (1978).
10. J. A. Mydosh, *Amorphous Magnetism*, Vol. II, R. A. Levy and R. Hasegawa, (Eds.), Plenum Press, New York, 1977, p. 73.
11. S. F. Edwards and P. W. Anderson, *J. Phys., F: Met. Phys.* **5**, 965 (1975).
12. K. Binder, *Z. Phys. B*, **26**, 339 (1977).
13. I. Riess, *Commun. Phys.* **2**, 37 (1977).
14. J. Souletie and R. Fourier, *J. Low Temp. Phys.* **1**, 95 (1969).
15. K. H. Fischer, *Solid State Commun.* **18**, 1515 (1976).
16. C. E. Violet and R. J. Borg, *Phys. Rev.* **149**, 540 (1966).
17. A. P. Murani et al., *J. Phys. F*, **6**, 425 (1976).
18. D. Sherrington and B. W. Southern, *J. Phys. F*, **5**, L49 (1975); D. Sherrington and S. Kirkpatrick, *Phys. Rev. Lett.* **35**, 1792 (1975).
19. D. A. Smith, *J. Phys. F*, **4**, L266 (1974).
20. J. M. Kosterlitz, D. J. Thouless, and R. C. Jones, *Phys. Rev. Lett.* **36**, 1217 (1976); D. J. Thouless, P. W. Anderson, and R. G. Palmer, *Phil. Mag.* (in press).
21. D. C. Mattis, *Phys. Lett.* **56A**, 421 (1976); J. M. Luttinger, *Phys. Rev. Lett.* **37**, 778 (1976).
22. E. Kueller, *Handbuch der Physik*, Vol. **18**, Part 2, S. Flugge (Ed.), Springer, Berlin, 1966; L. Neel, *Ann. Geophys.* **5**, 99 (1949).
23. E. P. Wohlfarth, *Physica,* **86–88B + C**, 852 (1977).
24. J. L. Thouless and R. Fourier, *J. Phys.* **35**, C4, 229 (1974); *Physica,* **86–88B + C**, 873 (1977).
25. K. Binder, *Fest Körperprobleme*, Vol XVII, 55 (1977).

26. S. F. Edwards and P. W. Anderson, *J. Phys. F*, **6**, 1927 (1976).

27. H. O. Hooper and A. M. de Graaf (Eds.), *Amorphous Magnetism*, Plenum Press, New York, 1973; R. A. Levy and R. Hasegawa (Eds.), *Proceedings of the Second International Symposium on Amorphous Magnetism at Troy, N.Y.*, Plenum Press, New York, 1977.

28. A. Gubanov, *Sov. Phys. Solid State* **2**, 468 (1960).

29. U. Krey, *Phys. Cond. Matter* **18**, 17 (1974); H. J. Schlichting and U. Krey, *Z. Phys. B*, **20**, 375 (1975); U. Krey and H. J. Schlichting, *Z. Phys. B*, **21**, 157 (1975); *ibid.*, **24**, 417 (1976).

30. J. Als Nielsen, R. J. Birgeneau, R. J. Guggenheim, and G. Shirane, *Phys. Rev. B*, **12**, 4963 (1975).

31. Y. Nakamura, *IEEE Mag.* **12**, 278 (1976).

32. R. J. Cargill III, *Solid State Phys.*, Vol. 30, F. Seitz and D. Turnbull (Eds.), Academic Press, New York, 1975.

33. J. J. Rhyne, S. J. Pickart, and H. A. Alperin, *Phys. Rev. Lett.* **29**, 1562 (1972); R. Harris, M. Plischke, and M. J. Zuckermann, *Phys. Rev. Lett.* **3**, 160 (1973).

34. H. A. Mook, N. Wakabayashi, and D. Pau, *Phys. Rev. Lett.* **34**, 1029 (1975).

35. L. J. Tao, R. J. Gamleino, S. Kirkpatrick, J. J. Cuomo, and H. Lilienthal, AIP Conference Proceedings, Vol. 18, p. 641 (1974).

16

The Cooperative Phenomenon

A theory of many-body systems is confronted with the difficulty of describing both the internal properties of a system and its response to external perturbation. In a solid we are interested in knowing such characteristics of a system as its energy spectrum, that is, the ground state energy and the excited state energies, the average values of the dynamical variables and their distributions, and the transition probabilities and their thermodynamic and kinetic characteristics. To obtain information about these, we start, of course, with a model of a given system with known characteristics of the individual particles, the way in which they interact, and the external conditions.

Many-body theory provides us with an efficient method of solving a given problem in terms of some approximation and deals effectively with the interactions between the particles. For the perfect gases, however, there is no interaction between the particles, and their properties can be calculated by elementary methods. It is just the interaction between the particles that makes the difference between a solid and a perfect gas.

The difficulty in accounting for the particle interactions in solids stems from the fact that the Schrödinger equation for a solid cannot be separated into individual equations containing each variable because of the interparticle interactions. This nonseparable Schrödinger equation is far more difficult to solve, from the mathematical point of view, than the Schrödinger equation for particles in the presence of an external perturbation. Since there is a continuous field of force between any particular particle and the other particles in the system, the concept of particle states (i.e., the wavefunction, energy, etc. of a particle) has no precise meaning, and one should think of the states of the system as a whole—in other words, such a system represents a truly cooperative phenomenon.

The many-body theory depending on the field-theoretical methods utilizes the method of second quantization, which is essentially very simple and compact. Moreover, the method of second quantization contains all that is essential for a field-theoretical description of a many-body system. In this method the concept of a particle is of minor importance, and it is replaced by elementary excitations, which are the quanta of a continuous field, denoted by $\psi(x)$ and $\psi^{\dagger}(x)$, the particle annihilation and creation operators (for the quasi-particles), respectively. These operators create and annihilate the field quanta. In this formalism the interparticle interactions are represented in terms of particle creation and annihilation operators. It is also important to note that the number of particles in an interaction process is not generally conserved, and also that the process of excitation of a system can be represented by the quasi-particles. By the introduction of the idea of quasi-particles it is possible to describe the interactions in a solid in the single-particle form, which can then be easily solved to obtain the relevant properties of the system. Since the field-theoretical methods provide a natural basis for such an approach in the many-body theory, we shall deal with the quantum field theory briefly in the next section before discussing the second quantization formalism.

16.1 A BRIEF OUTLINE OF THE QUANTUM FIELD THEORY[1]

It is well known that in the Schrödinger representation the wavefunctions are time dependent, whereas the operators are independent of time. If $\Psi_S(q_i, t)$ represent a wavefunction that depends on time and $\mathcal{H}$ is the total Hamiltonian (time independent), consisting of the kinetic and potential energies, then, in the Schrödinger representation we have,

$$\mathcal{H}\Psi_S(q_i,t) = i\hbar\frac{\partial\Psi_S(q_i,t)}{\partial t} \tag{16.1}$$

where $q_i = [q_1, q_2, \ldots, q_N]$ is the total number of coordinates of a system consisting of N particles, and q_i represents both space and spin coordinates.

The form of the wavefunction that satisfies (16.1) can easily be shown to be

$$\Psi_S(q_i,t) = e^{-i\mathcal{H}(t-t_0)/\hbar}\Psi_S(q_i,t_0) \tag{16.2}$$

Alternatively, we can also assume that the wavefunctions are time independent and the operators time dependent—just the reverse of the Schrödinger picture given above. This representation, which is known as the Heisenberg representation, has some advantages, which will be obvious shortly. The unitary transformation that transforms the Schrödinger representation to the Heisenberg representation is given by

$$\Psi_H = e^{i\mathcal{H}(t-t_0)/\hbar}\Psi_S(q_i,t) \tag{16.3}$$

From (16.3) we can easily show that

$$\frac{\partial}{\partial t}\Psi_H = 0 \tag{16.4}$$

To prove (16.4), we proceed as follows

$$\frac{\partial}{\partial t}\Psi_{\mathrm{H}}=\frac{\partial}{\partial t}\left[e^{i\mathcal{H}t/\hbar}\Psi_{\mathrm{S}}(q_i,t)\right]$$
$$=\frac{i}{\hbar}\mathcal{H}e^{i\mathcal{H}t/\hbar}\Psi_{\mathrm{S}}(q_i,t)+e^{i\mathcal{H}t/\hbar}\frac{\partial}{\partial t}\Psi_{\mathrm{S}}(q_i,t)$$

But from (16.1)

$$\frac{\partial}{\partial t}\Psi_{\mathrm{S}}(q_i,t)=-\frac{i}{\hbar}\mathcal{H}\Psi_{\mathrm{S}}(q_i,t)$$

Hence

$$\frac{\partial}{\partial t}\Psi_{\mathrm{H}}=\frac{i\mathcal{H}}{\hbar}e^{i\mathcal{H}t/\hbar}\Psi_{\mathrm{S}}(q_i,t)-\frac{i}{\hbar}e^{i\mathcal{H}t/\hbar}\mathcal{H}\Psi_{\mathrm{S}}(q_i,t)$$
$$=0\qquad (\because\ \mathcal{H}\text{ and }e^{i\mathcal{H}t/\hbar}\text{ commute})$$

To preserve the observable quantities unaltered by the transformation, we must have, in the Heisenberg representation for all operators,

$$\mathbf{Q}_{\mathrm{H}}=e^{i\mathcal{H}(t-t_0)/\hbar}\mathbf{Q}_{\mathrm{S}}e^{-i\mathcal{H}(t-t_0)/\hbar}\tag{16.5}$$

It is important to note that in the Schrödinger representation the Hamiltonian remains unchanged, and the operators that commute with the Hamiltonian are time independent. In the Heisenberg representation all operators have a time dependence, which can be easily shown by differentiating (16.5). If we do this, we find that

$$\frac{\partial}{\partial t}\mathbf{Q}_{\mathrm{H}}=\frac{\partial}{\partial t}\left[e^{i\mathcal{H}t/\hbar}\mathbf{Q}_{\mathrm{S}}e^{-i\mathcal{H}t/\hbar}\right]$$
$$=\frac{i\mathcal{H}}{\hbar}e^{i\mathcal{H}t/\hbar}\mathbf{Q}_{\mathrm{S}}e^{-i\mathcal{H}t/\hbar}+e^{i\mathcal{H}t/\hbar}\mathbf{Q}_{\mathrm{S}}\left(\frac{-i\mathcal{H}}{\hbar}\right)e^{-i\mathcal{H}t/\hbar}$$
$$=\frac{i}{\hbar}\mathcal{H}e^{i\mathcal{H}t/\hbar}\mathbf{Q}_{\mathrm{S}}e^{-i\mathcal{H}t/\hbar}-\frac{i}{\hbar}e^{i\mathcal{H}t/\hbar}\mathbf{Q}_{\mathrm{S}}e^{-i\mathcal{H}t/\hbar}\mathcal{H}$$
$$=\frac{i}{\hbar}\left[\mathcal{H}\mathbf{Q}_{\mathrm{H}}-\mathbf{Q}_{\mathrm{H}}\mathcal{H}\right]$$
$$=\frac{i}{\hbar}\left[\mathcal{H},\mathbf{Q}_{\mathrm{H}}\right]$$

We have, therefore,

$$\dot{\mathbf{Q}}_{\mathrm{H}}=\frac{i}{\hbar}\left[\mathcal{H},\mathbf{Q}_{\mathrm{H}}\right]\tag{16.6}$$

In addition to the Schrödinger and Heisenberg representations, which have been described above, there is another representation that is very useful in the many-body theory. This is known as the interaction (or Dirac) representation. To obtain this representation, let us write the total Hamiltonian as

$$\mathcal{H} = \mathcal{H}_0 + \mathcal{H}_{\text{int}} \tag{16.7}$$

where $\mathcal{H}_0$ is the free Hamiltonian without interaction, and $\mathcal{H}_{\text{int}}$ represents the interaction between the particles. In the Schrödinger representation we now remove from the wavefunction the time development due to $\mathcal{H}_0$. It is important to note that in the interaction representation, the time dependence of the problem is divided between the operators and the wavefunctions. The representation is defined by

$$|\Psi_I(t)\rangle = e^{i\mathcal{H}_0 t/\hbar}|\Psi_S(t)\rangle \tag{16.8}$$

and

$$\mathbf{Q}_I(t) = e^{i\mathcal{H}_0 t/\hbar}\mathbf{Q}_S e^{-i\mathcal{H}_0 t/\hbar} \tag{16.9}$$

where, for simplicity, we have omitted writing the coordinates. In (16.7), the Hamiltonian operator $\mathcal{H}$ has been divided into the sum of two terms, $\mathcal{H}_0$ and $\mathcal{H}_{\text{int}}$. If $\mathcal{H}_{\text{int}}$ vanishes, the interaction and the Heisenberg representations become exactly identical. In the presence of $\mathcal{H}_{\text{int}}$, the time dependence of the field operators is governed by the zero-order Hamiltonian $\mathcal{H}_0$. The wavefunction Ψ_I is time dependent solely as a result of the interaction $\mathcal{H}_{\text{int}}$ being present.

Let us now differentiate $\Psi_I(t)$, given by (16.8), with respect to time:

$$\Psi_I(t) = e^{i\mathcal{H}_0 t/\hbar}\Psi_S(t) \tag{16.10}$$

$$\begin{aligned}
\therefore \frac{\partial}{\partial t}\Psi_I(t) &= e^{i\mathcal{H}_0 t/\hbar}\frac{\partial}{\partial t}\Psi_S(t) + \left(\frac{\partial}{\partial t}e^{i\mathcal{H}_0 t/\hbar}\right)\Psi_S(t) \\
&= e^{i\mathcal{H}_0 t/\hbar}\left[-\frac{i}{\hbar}\mathcal{H}\Psi_S(t)\right] + e^{i\mathcal{H}_0 t/\hbar}\left(\frac{i\mathcal{H}_0 t}{\hbar}\right)\Psi_S(t) \\
&= -\frac{i}{\hbar}e^{i\mathcal{H}_0 t/\hbar}(\mathcal{H} - \mathcal{H}_0)\Psi_S(t) \\
&= -\frac{i}{\hbar}e^{i\mathcal{H}_0 t/\hbar}\mathcal{H}_{\text{int}}\Psi_S(t) \\
&= -\frac{i}{\hbar}e^{i\mathcal{H}_0 t/\hbar}\mathcal{H}_{\text{int}}e^{-i\mathcal{H}_0 t/\hbar}\Psi_I(t)
\end{aligned}$$

Therefore from (16.10)

$$\frac{\partial}{\partial t}\Psi_I(t) = -\frac{i}{\hbar}\mathcal{H}^I_{\text{int}}(t)\Psi_I(t) \tag{16.11}$$

where

$$\mathcal{H}_{\text{int}}^{I}(t)=e^{i\mathcal{H}_0 t/\hbar}\mathcal{H}_{\text{int}}e^{-i\mathcal{H}_0 t/\hbar} \tag{16.12}$$

One should note that the functions $\Psi_I(t)$ determine the interaction representation. Every operator, $\mathbf{O}(t)$, say, in this representation is obtained from the corresponding Schrödinger operator by using (16.12), as in the case of $\mathcal{H}_{\text{int}}^{I}(t)$, and therefore $\mathbf{O}(t)$ satisfies the following equation:

$$\frac{\partial \mathbf{O}(t)}{\partial t}=\frac{i}{\hbar}\left[\mathcal{H}_0, \mathbf{O}(t)\right] \tag{16.13}$$

This means that we have an equation similar to that for a Heisenberg operator for a system of noninteracting particles. We therefore conclude from the above that in the interaction representation all operators have the same form as the Heisenberg operators for the corresponding noninteracting system, while the wavefunctions satisfy a Schrödinger equation with the Hamiltonian $\mathcal{H}_{\text{int}}^{I}(t)$.

Let us now determine the time dependence of $\Psi_I(t)$ in the interaction representation. Since the operators $\mathcal{H}_{\text{int}}^{I}(t)$ do not commute at different times, it is not possible to write the solution of (16.11) in the form

$$\Psi_I(t)=\exp\left\{-i\int_{t_0}^{t}\mathcal{H}_{\text{int}}^{I}(t')\,dt'\right\}\Psi_I(t_0)$$

Now, if we suppose that the value of Ψ_I at t_0 is known, then, at a later time $t>t_0$, if we also suppose that $\Psi_I(t)$ can be obtained from $\Psi_I(t_0)$ by a unitary matrix $U(t,t_0)$, we can write

$$\Psi_I(t)=U(t,t_0)\Psi_I(t_0) \tag{16.14}$$

Note that $U(t,t_0)\to S$, the scattering matrix, in the limit $(t_0,t)\to(-\infty,+\infty)$. To determine $\Psi_I(t)$, we must know $U(t,t_0)$, which can be found from (16.14) as follows:

$$\frac{\partial}{\partial t}\Psi_I(t)=\frac{\partial}{\partial t}\{U(t,t_0)\}\Psi_I(t_0)$$

By using (16.11), the above relation reduces to

$$-\frac{i}{\hbar}\mathcal{H}_{\text{int}}^{I}(t)\Psi_I(t)=\frac{\partial}{\partial t}\{U(t,t_0)\}\Psi_I(t_0)$$

which, with the help of (16.14), becomes

$$-\frac{i}{\hbar}\mathcal{H}_{\text{int}}^{I}(t)\,U(t,t_0)\Psi_I(t_0)=\frac{\partial}{\partial t}\{U(t,t_0)\}\Psi_I(t_0)$$

Now, removing $\Psi_I(t_0)$ from both sides, we have, finally,

$$\frac{\partial}{\partial t} U(t,t_0) = -\frac{i}{\hbar} \mathcal{H}^I_{\text{int}}(t) U(t,t_0) \tag{16.15}$$

Since $U(t_0,t_0)=1$, we get from (16.15)

$$U(t,t_0) = 1 - \frac{i}{\hbar} \int_{t_0}^{t} \mathcal{H}^I_{\text{int}}(t') U(t',t_0)\, dt' \tag{16.16}$$

Let us now examine whether we can retrieve (16.15) by differentiating (16.16). We get (16.15) by using the relation

$$\frac{\partial}{\partial x} \int_a^x f(x')\, dx' = f(x)$$

which follows from

$$\frac{\partial}{\partial x} \int_a^x f(x')\, dx' = \lim_{h\to 0} \frac{\int_a^{x+h} f(x')\, dx' - \int_a^x f(x')\, dx'}{h}$$

$$= \lim_{h\to 0} \frac{\int_x^{x+h} f(x')\, dx'}{h} = \frac{f(x)h}{h} = f(x)$$

It is important to note that (16.16) cannot be solved exactly, and therefore recourse must be had to an approximate solution given by the Neumann-Liouville iterative procedure. The solution of (16.16) in the zeroth approximation is

$$U^{(0)}(t,t_0) = 1$$

In the first approximation the solution is

$$U^{(1)}(t,t_0) = 1 - \frac{i}{\hbar} \int_{t_0}^{t} \mathcal{H}^I_{\text{int}}(t_1)\, dt_1$$

and in the second approximation it is

$$U^{(2)}(t,t_0) = 1 - \frac{i}{\hbar} \int_{t_0}^{t} \mathcal{H}^I_{\text{int}}(t_1)\, dt_1 + \left(\frac{-i}{\hbar}\right)^2 \int_{t_0}^{t} dt_1 \int_{t_0}^{t_1} dt_2$$

$$\cdot \mathcal{H}^I_{\text{int}}(t_1) \mathcal{H}^I_{\text{int}}(t_2)\, dt_1\, dt_2$$

Finally, in the nth approximation, we have the series

$$U(t,t_0)=1-\frac{i}{\hbar}\int_{t_0}^{t}\mathcal{H}_{\text{int}}^{I}(t_1)\,dt_1+\cdots$$
$$+\left(\frac{-i}{\hbar}\right)^n\int_{t_0}^{t}\mathcal{H}_{\text{int}}^{I}(t_1)\,dt_1+\cdots+\int_{t_0}^{t_{n-1}}\mathcal{H}_{\text{int}}^{I}(t_n)\,dt_n$$
$$+\cdots \qquad (16.17)$$

where $t_0 \leqslant t_1 \leqslant t_2 \leqslant \cdots \leqslant t_n$. Note the characteristic of series (16.17) that the operators $\mathcal{H}_{\text{int}}^{I}$ taken at later times always appear to the left of the operators taken at earlier times, since the inequality

$$t > t_1 > t_2 > \cdots > t_n > t_0$$

always holds.

Equation (16.17) can be made more symmetric. Let us consider the nth term of the series

$$\left(-\frac{i}{\hbar}\right)^n\int_{t>t_1>\cdots>t_0}\mathcal{H}_{\text{int}}^{I}(t_1)\mathcal{H}_{\text{int}}^{I}(t_2)\cdots\mathcal{H}_{\text{int}}^{I}(t_n)\,dt_1\,dt_2\cdots dt_n \qquad (16.18)$$

Obviously, this expression does not change if we subject the variables of integration to any permutation

$$t_1, t_2, t_3, \ldots, t_n \rightarrow t_{x_1}, t_{x_2}, t_{x_3}, \ldots, t_{x_n}$$

Permuting the variables $t_1, \ldots, t_n$, adding the resulting expressions, and dividing by $n!$, the total number of permutations, we can extend the range of integration of each variable from t_0 to t_1. For this purpose it is important that the operators $\mathcal{H}_{\text{int}}^{I}$ in the integral (16.18) should always be arranged from left to right in the order of decreasing times. If T denotes this ordering operation, the nth term of (16.17) can be expressed as

$$U^{(n)}(t,t_0)=\left(\frac{-i}{\hbar}\right)^n\frac{1}{n!}\int_{t_0}^{t}\cdots\int_{t_0}^{t}T\left[\mathcal{H}_{\text{int}}^{I}(t_1)\cdots\mathcal{H}_{\text{int}}^{I}(t_n)\right]dt_1\cdots dt_n$$

Thus (16.17) can conveniently be written as

$$U(t,t_0)=T\exp\left\{\frac{-i}{\hbar}\int_{t_0}^{t}\mathcal{H}_{\text{int}}^{I}(t')\,dt'\right\} \qquad (16.19)$$

It is easy to see that, by simply expanding the exponential series in (16.19) and then using the definition of the chronological or ordering operation, we can retrieve (16.17). The operator $U(t, t_0)$ has the property

$$U(t_2, t_1)U(t_1, t_0) = U(t_2, t_0) \tag{16.20}$$

where $t_2 > t_1 > t_0$.

Let us now generalize the definition of T-ordering for Fermi and Bose operators, which we shall use later. If we have a set of Fermi operators $A(t_1), B(t_2), C(t_3)$, the T-product of the operators will be given by writing them from left to right, multiplied by the factor $(-1)^P$, where P is the number of permutations needed to obtain the chronological product from $A(t_1)B(t_2)C(t_3)\ldots$. If $F_1(t_1)$ and $F_2(t_2)$ are Fermi operators, while $B_1(t_3)$ and $B_2(t_4)$ are Bose operators, we get

$$\begin{aligned} T[F_1(t_1)F_2(t_2)] &= \begin{cases} F_1(t_1)F_2(t_2) & \text{for} \quad t_1 > t_2 \\ -F_2(t_2)F_1(t_1) & \text{for} \quad t_2 > t_1 \end{cases} \\ T[B_1(t_3)B_2(t_4)] &= \begin{cases} B_1(t_3)B_2(t_4) & \text{for} \quad t_3 > t_4 \\ B_2(t_4)B_1(t_3) & \text{for} \quad t_4 > t_3 \end{cases} \end{aligned} \tag{16.21}$$

It can be shown that the operation of T-ordering is the same both for operators in the interaction representation and for those in the Heisenberg representation. Let us write the chronological products of a number of Heisenberg operators averaged over the ground state of the system Φ_{H}^0, that is,

$$\langle \Phi_{\mathrm{H}}^{0*} T[\mathbf{A}(t_1)\mathbf{B}(t_2)\mathbf{C}(t_3)\cdots]\Phi_{\mathrm{H}}^0\rangle \tag{16.22}$$

where we have

$$t_1 > t_2 > t_3 > \cdots$$

Now, going over to the operators in the interaction representation by using

$$\mathbf{O}(t) = U^{-1}(t)O(t)U(t) \tag{16.23}$$

we get

$$\begin{aligned} &\langle \Phi_{\mathrm{H}}^{0*} U^{-1}(t_1)\mathbf{A}(t_1)U(t_1)U^{-1}(t_2)\mathbf{B}(t_2)U(t_2)\cdots\Phi_{\mathrm{H}}^0\rangle \\ &\qquad = \langle \Phi_{\mathrm{H}}^{0*} U^{-1}(\infty)U(\infty, t_1)\mathbf{A}(t_1)U(t_1, t_2)\mathbf{B}(t_2)\cdots\Phi_{\mathrm{H}}^0\rangle \\ &\qquad = \langle \Phi_{\mathrm{H}}^{0*} U^{-1}(\infty)T[\mathbf{A}(t_1)\mathbf{B}(t_2)\mathbf{C}(t_3)\cdots U(\infty)]\Phi_{\mathrm{H}}^0\rangle \end{aligned} \tag{16.24}$$

from (16.22). We now have to determine the quantity

$$\Phi_{\mathrm{H}}^{0*} U^{-1}(\infty) = [U(\infty)\Phi_{\mathrm{H}}^0]^*$$

and therefore the ground state function has to be operated upon by $U(\infty)$. We now use the concept of adiabatically turned on interaction, which means that at time $t=-\infty$ there is no interaction between the particles, and that afterwards the interaction is switched on infinitely slowly. Thus we have

$$\Phi_{\mathrm{H}}^{0}=\Phi_{i}(-\infty)$$

and

$$U(\infty)\Phi_{\mathrm{H}}^{0}=\Phi_{i}(\infty)$$

Using these, we find that (16.22) finally reduces to

$$\langle\Phi_{\mathrm{H}}^{0}{}^{*}T[\mathbf{A}(t_1)\mathbf{B}(t_2)\mathbf{C}(t_3)\cdots]\Phi_{\mathrm{H}}^{0}\rangle=\frac{\langle\Phi_{\mathrm{H}}^{0}{}^{*}T[\mathbf{A}(t_1)\mathbf{B}(t_2)\mathbf{C}(t_3)\cdots U(\infty)]\Phi_{\mathrm{H}}^{0}\rangle}{\langle\Phi_{\mathrm{H}}^{0}{}^{*}U(\infty)\Phi_{\mathrm{H}}^{0}\rangle} \tag{16.25}$$

It is important to note that (16.25) is valid only for averages over the ground state of the system, which is nondegenerate. For degenerate states the system can make a transition from one state to another as a result of collisions between the particles. Thus, in averaging over an excited state, (16.24) is valid but (16.25) is not.

In this section we have dealt briefly with the quantum field theory, which will be very useful in connection with Green function formalism, discussed in Chapter 17.

16.2 SECOND QUANTIZATION FORMALISM

To appreciate the second quantization formalism we deal, first of all, with the quantization of the one-dimensional harmonic oscillator. The quantum-mechanical problem of the harmonic oscillator is that of a particle of mass m constrained to move along an axis subject to a restoring force proportional to its displacement from a point located on that axis. If $\mathbf{q}$ is the positional coordinate of the particle on the axis, taking the center of force as the origin, and $\mathbf{p}$ is its momentum, m its mass, and $-m\omega^2 q$ the restoring force, the Hamiltonian of the harmonic oscillator is given by

$$\mathcal{H}=\frac{1}{2m}(p^2+m^2\omega^2q^2) \tag{16.26}$$

where $\mathbf{q}$ and $\mathbf{p}$ are connected by the commutation relation

$$[\mathbf{q},\mathbf{p}]=i\hbar \tag{16.27}$$

The equation of motion of the particle is then derived from the Hamiltonian (16.26), and it is easy to show that the motion is an oscillatory, sinusoidal motion of angular frequency ω about the origin. Study of the harmonic oscillator is of great importance in quantum theory since a Hamiltonian of type (16.26) enters into all problems involving quantized oscillations; for example, one encounters it in quantum field theory, as well as in the theory of molecular and crystalline vibrations.

To reduce (16.26) to a simpler form, let us make the following substitutions:

$$\mathcal{H} = H\hbar\omega \tag{16.28}$$

$$\mathbf{q} = \left(\frac{\hbar}{m\omega}\right)^{1/2}\mathbf{Q} \tag{16.29}$$

$$\mathbf{p} = (m\hbar\omega)^{1/2}\mathbf{P} \tag{16.30}$$

As a result of this, (16.26) reduces to

$$H = \tfrac{1}{2}(P^2 + Q^2) \tag{16.31}$$

where the Hermitian operators $\mathbf{P}$ and $\mathbf{Q}$ satisfy the following commutation relations:

$$[\mathbf{Q}, \mathbf{P}] = i \tag{16.32}$$

We now have to find the eigenvalues and the eigenvectors of (16.31). To diagonalize this Hamiltonian, we now transform it to the new variables a and its Hermitian conjugate $a^\dagger$, such that the commutator of a and $a^\dagger$ is equal to one. To accomplish this transformation, we define new variables $\mathbf{Q}$ and $(-i\mathbf{P})$, which have the desired commutation relation

$$[\mathbf{Q}, (-i\mathbf{P})] = 1 \tag{16.33}$$

We define the new variables a and $a^\dagger$ in terms of $\mathbf{Q}$ and $(-i\mathbf{P})$ as

$$\mathbf{a} = \alpha\mathbf{Q} + \beta(-i\mathbf{P}) \tag{16.34}$$

and

$$\mathbf{a}^\dagger = \alpha\mathbf{Q} - \beta(-i\mathbf{P}) \tag{16.35}$$

where the real constants α and β are to be chosen properly so that

$$[\mathbf{a}, \mathbf{a}^\dagger] = 1 \tag{16.36}$$

By substituting (16.34) and (16.35) in (16.36), we get, after some algebra,

$$\alpha\beta = -\tfrac{1}{2}$$

Also, to make the $\mathbf{a}^2$ and $(\mathbf{a}^\dagger)^2$ terms in the Hamiltonian (16.31) vanish, we must have $\alpha^2=\beta^2$. Thus we find the important result

$$\alpha=-\beta=\frac{1}{\sqrt{2}} \tag{16.37}$$

Now, substituting these values of α and β in (16.34) and (16.35), we can show that

$$\mathbf{Q}=\frac{1}{\sqrt{2}}(\mathbf{a}+\mathbf{a}^\dagger) \tag{16.38}$$

and

$$\mathbf{P}=\frac{1}{\sqrt{2}\,i}(\mathbf{a}-\mathbf{a}^\dagger) \tag{16.39}$$

Inserting these expressions for **P** and **Q** in our simplified Hamiltonian (16.28), we finally get

$$\mathcal{H}=\left(\mathbf{a}^\dagger\mathbf{a}+\tfrac{1}{2}\right)\hbar\omega \tag{16.40}$$

with

$$\left[\mathbf{a},\mathbf{a}^\dagger\right]=1 \tag{16.41}$$

The important feature of this Hamiltonian is that only the product $\mathbf{a}^\dagger\mathbf{a}$ appears. Using (16.40) and (16.41), we can easily show that

$$\mathcal{H}(\mathbf{a}|n\rangle)=(E_n-\hbar\omega)(\mathbf{a}|n\rangle) \tag{16.42}$$

This means that, if $|n\rangle$ is an eigenvector of $\mathcal{H}$, that is,

$$\mathcal{H}|n\rangle=E_n|n\rangle$$

then $(\mathbf{a}|n\rangle)$ is also an eigenvector with the eigenvalue $(E_n-\hbar\omega)$. Also,

$$\begin{aligned}\mathcal{H}|n-1\rangle&=E_{n-1}|n-1\rangle\\&=(E_n-\hbar\omega)|n-1\rangle\end{aligned} \tag{16.43}$$

Comparing (16.42) and (16.43), we see that the eigenvector $\mathbf{a}|n\rangle$ must equal the eigenvector $|n-1\rangle$ times a constant. This constant can be found as follows. Let

$$\mathbf{a}|n\rangle=c|n-1\rangle \tag{16.44}$$

Taking the complex conjugate of (16.44), we get

$$\langle n|\mathbf{a}^{\dagger} = \langle n-1|c^* \tag{16.45}$$

Now, combining (16.44) and (16.45) gives

$$\langle n|\mathbf{a}^{\dagger}\mathbf{a}|n\rangle = \langle n-1|c^*c|n-1\rangle$$

or

$$c^*c = n$$

or

$$c = (n)^{1/2} \tag{16.46}$$

Similarly, let

$$\mathbf{a}^{\dagger}|n\rangle = d|n+1\rangle \tag{16.47}$$

Then the complex conjugate of (16.47) is

$$\langle n|\mathbf{a} = \langle n+1|d^* \tag{16.48}$$

Combining (16.47) and (16.48), as before, we easily get

$$d = (1+n)^{1/2} \tag{16.49}$$

Inserting these values of the constants c and d in (16.44) and (16.47), respectively, we obtain two very basic, important relations:

$$\begin{aligned} \mathbf{a}|n\rangle &= (n)^{1/2}|n-1\rangle \\ \mathbf{a}^{\dagger}|n\rangle &= (n+1)^{1/2}|n+1\rangle \end{aligned} \tag{16.50}$$

where $n = 0, 1, 2, \ldots, \infty$. From (16.50) it follows that

$$\mathbf{a}^{\dagger}\mathbf{a}|n\rangle = n|n\rangle \tag{16.51}$$

Hence we see that the energy eigenvalues of the harmonic oscillator Hamiltonian (16.40) are

$$E_n = \left(n + \tfrac{1}{2}\right)\hbar\omega \tag{16.52}$$

where $n = 0, 1, 2, \ldots, \infty$.

The operators $\mathbf{a}^{\dagger}$ and $\mathbf{a}$, in terms of which the harmonic oscillator has been expressed, are known as the creation and annihilation operators, respectively. The creation operator $\mathbf{a}^{\dagger}$ changes the oscillator from its original state to a state higher in energy by one unit, that is, it creates one quantum of energy. Similarly, the annihilation operator **a** annihilates one quantum of energy and thereby changes the original state to a state lower in energy by one unit.

We thus see that the Hamiltonian of the harmonic oscillator can be written as the product of **a** and its Hermitian conjugate $\mathbf{a}^\dagger$, where the commutator of **a** and $\mathbf{a}^\dagger$ is equal to one [see (16.41)].

For N uncoupled harmonic oscillators, the analysis given above is unchanged. All the variables—**Q**, **P**, **a**, and $\mathbf{a}^\dagger$—will have subscripts i, where $i=1,2,\ldots,N$, and the Hamiltonian will be summed over i from 1 to N. All the commutation relations between operators with different values of i vanish. The eigenvector of the system is the product of the N single oscillator eigenvectors.

The spins, situated at regular intervals on the lattice of a solid, oscillate like harmonic oscillators. The spins are, of course, coupled by the exchange interactions. Hence the low-lying energy states of the spin systems are wavelike in character because any spin deviation propagates through the entire lattice in the form of a wave. These waves are commonly known as spin waves. This concept was originally shown by Bloch for ferromagnets. The energy of a spin wave is quantized, and the unit of energy of a spin wave is called a magnon. Therefore for the magnetic properties of a solid we shall be interested in the properties of the ferromagnetic and antiferromagnetic magnons. Before we consider the spin waves in a magnetic system, however, we discuss the second quantization formalism outlined above in connection with the harmonic oscillator problem in a more general fashion, following the method given by Schrieffer.[3]

The Hamiltonian for n identical particles in the Schrödinger representation is given by

$$H(\mathbf{x}_1,\ldots,\mathbf{x}_n)=\sum_i \frac{\mathbf{p}_i^2}{2m}+\sum V(\mathbf{x}_i)+\frac{1}{2}\sum_{i\neq j} V_2(\mathbf{x}_i,\mathbf{x}_j) \tag{16.53}$$

where $\mathbf{x}_i$ labels the position and spin of particle i, and three-body potentials and higher interactions are excluded. The many-body Schrödinger equation is, therefore,

$$H(\mathbf{x}_1,\ldots,\mathbf{x}_n)\Psi(\mathbf{x}_1,\ldots,\mathbf{x}_n,t)=i\hbar\frac{\partial\Psi(\mathbf{x}_1,\ldots,\mathbf{x}_n,t)}{\partial t} \tag{16.54}$$

Let us now introduce a complete set of n-particle wavefunctions, denoted by Φ. These are constructed as a properly symmetrized product of one-particle wavefunctions $u_k(\mathbf{x})$, which form a complete orthonormal set. The orthonormality and completeness are given, respectively, by

$$\int u_{k'}^*(\mathbf{x})u_k(\mathbf{x})\,d\mathbf{x}=\delta_{kk'} \tag{16.55}$$

and

$$\sum_k u_k^*(\mathbf{x}')u_k(\mathbf{x})=\delta(\mathbf{x}-\mathbf{x}') \tag{16.56}$$

The function Φ is then given by

$$\Phi=\mathcal{Y}u_{k_1}(\mathbf{x}_1)u_{k_2}(\mathbf{x}_2)\cdots u_{k_n}(\mathbf{x}_n) \tag{16.57}$$

where $\mathcal{Y}=(1/n!)\Sigma P$ for particles obeying Bose statistics and $\mathcal{Y}=(1/n!)\Sigma(-1)^P P$ for those obeying Fermi statistics. The summation is over all $n!$ possible permutations of the coordinates $\mathbf{x}_1,\ldots,\mathbf{x}_n$, and P is the order of permutations. Instead of labeling Φ by the quantum numbers $k_1,k_2,\ldots,k_n$, we may specify a state by the number of particles occupying it. This is known as the occupation number representation. Let this occupation number be n_k for the kth state. Then the set of numbers $n_1,n_2,\ldots,n_k$ uniquely specifies the symmetrized state $\Phi_{n_1,n_2,\ldots,n_k}$. For the state consisting of n particles we therefore have $\Sigma_k n_k=n$. For Fermi statistics the occupation number n_k is restricted to the values zero and one, whereas for Bose statistics it can have any possible integer value, including zero. The functions

$$\sqrt{n!}\;\Phi_{n_1,\ldots,n_k,\ldots}(\mathbf{x}_1,\ldots,\mathbf{x}_n)$$

form a complete orthonormal set of n-particle functions for fermions, and the functions

$$\left(\frac{n!}{n_1!n_2!\cdots}\right)^{1/2}\Phi_{n_1,\ldots,n_k,\ldots}(\mathbf{x}_1,\ldots,\mathbf{x}_n)$$

do the same for bosons. The orthonormality condition is

$$\int\Phi_{n_1',n_2',\ldots}(\mathbf{x}_1,\ldots,\mathbf{x}_n)\Phi_{n_1,n_2,\ldots}(\mathbf{x}_1,\ldots,\mathbf{x}_n)\,d\mathbf{x}_1\cdots d\mathbf{x}_n=\delta_{n_1'n_1}\delta_{n_2'n_2}\cdots \tag{16.58}$$

In general, the total wavefunction in the Schrödinger representation may be expanded in the complete set of the $\Phi_{n_1,\ldots,n_k}$, thus:

$$\Psi(\mathbf{x}_1,\ldots,\mathbf{x}_n,t)=\sum c(n_1,\ldots,n_k,t)\Phi_{n_1,\ldots,n_k}(\mathbf{x}_1,\ldots,\mathbf{x}_n) \tag{16.59}$$

where the coefficients $c(n_1,\ldots,n_k,t)$ are in the occupation number representation whose norm gives the probability of finding n_k particles in the kth state.

For the particles obeying Bose statistics we have from (16.50)

$$\begin{aligned} a_k^\dagger\Phi_{n_1,\ldots,n_k}(\mathbf{x}_1,\ldots,\mathbf{x}_n)&=(n_k+1)^{1/2}\Phi_{n_1,\ldots,n_k+1,\ldots}(\mathbf{x}_1,\ldots,\mathbf{x}_{n+1})\\ \mathbf{a}_k\Phi_{n_1,\ldots,n_k}(\mathbf{x}_1,\ldots,\mathbf{x}_n)&=(n_k)^{1/2}\Phi_{n_1,\ldots,n_k-1,\ldots}(\mathbf{x}_1,\ldots,\mathbf{x}_{n-1})\end{aligned} \tag{16.60}$$

where, as in the case of the harmonic oscillator, the operator $a_k^\dagger$ (the creation operator) creates an additional particle in the kth state, and $\mathbf{a}_\mathbf{k}$ (the annihilation operator) destroys a particle in the **k**th state. If, however, no particle is present in Φ, the operator $\mathbf{a}_\mathbf{k}$ gives zero.

If we define a new operator, $N_k = \mathbf{a}_\mathbf{k}^\dagger \mathbf{a}_\mathbf{k}$, it follows from (16.51) that

$$N_k \Phi_{n_1,\dots,n_k}(\mathbf{x}_1,\dots,\mathbf{x}_n) = n_k \Phi_{n_1,\dots,n_k}(\mathbf{x}_1 \cdots \mathbf{x}_n) \tag{16.61}$$

Thus N_k may be interpreted as the number operator, which gives us the number of particles in the kth state. If N is the total number of particles in the system, then N is given by

$$N = \sum_\mathbf{k} N_\mathbf{k} = \sum_k \mathbf{a}_\mathbf{k}^\dagger \mathbf{a}_\mathbf{k} \tag{16.62}$$

The commutation relations between $\mathbf{a}_\mathbf{k}^\dagger$ and $\mathbf{a}_\mathbf{k}$ are given by

$$\begin{gathered} \left[\mathbf{a}_\mathbf{k}, \mathbf{a}_{\mathbf{k}'}^\dagger\right] = \delta_{kk'} \\ \left[\mathbf{a}_\mathbf{k}, \mathbf{a}_{\mathbf{k}'}\right] = \left[\mathbf{a}_\mathbf{k}^\dagger, \mathbf{a}_{\mathbf{k}'}^\dagger\right] = 0 \end{gathered} \tag{16.63}$$

We are now in a position to express the Schrödinger Hamiltonian (16.53) in the occupation number representation, and it can be seen that (16.53) becomes

$$\begin{aligned} H = & \sum_{k,k'} \langle \mathbf{k}'|H_1|\mathbf{k}\rangle \mathbf{a}_{\mathbf{k}'}^\dagger \mathbf{a}_\mathbf{k} \\ & + \frac{1}{2} \sum_{\mathbf{k}_1', \mathbf{k}_2', \mathbf{k}_1, \mathbf{k}_2} \langle \mathbf{k}_1'\mathbf{k}_2'| V_2 |\mathbf{k}_1 \mathbf{k}_2\rangle \mathbf{a}_{\mathbf{k}_1'}^\dagger \mathbf{a}_{\mathbf{k}_2'}^\dagger \mathbf{a}_{\mathbf{k}_1} \mathbf{a}_{\mathbf{k}_2} \end{aligned} \tag{16.64}$$

where

$$\langle \mathbf{k}'|H_1|\mathbf{k}\rangle = \int u_{\mathbf{k}'}^*(\mathbf{x}) \left[\frac{\mathbf{p}^2}{2m} + V_1(\mathbf{x}) \right] u_\mathbf{k}(\mathbf{x})\, d\mathbf{x}$$

and

$$\langle \mathbf{k}_1'\mathbf{k}_2'| V_2 |\mathbf{k}_1 \mathbf{k}_2\rangle = \int u_{\mathbf{k}_1'}^*(\mathbf{x}_1) u_{\mathbf{k}_2'}^*(\mathbf{x}_2) V_2(\mathbf{x}_1, \mathbf{x}_2) u_{\mathbf{k}_1}(\mathbf{x}_1) u_{\mathbf{k}_2}(\mathbf{x}_2)\, d\mathbf{x}_1\, d\mathbf{x}_2$$

We shall not give the proof of (16.64), preferring to leave it as an exercise for the reader. The proof is straightforward though lengthy.

We now define some new operators that are independent of the state **k** but dependent on the variable **x**, thus:

$$\Psi(\mathbf{x}) = \sum_\mathbf{k} u_\mathbf{k}(\mathbf{x}) \mathbf{a}_\mathbf{k}$$

and

$$\Psi^{\dagger}(\mathbf{x})=\sum_{\mathbf{k}} u_{\mathbf{k}}^{*}(\mathbf{x})\mathbf{a}_{\mathbf{k}}^{\dagger} \tag{16.65}$$

The Ψ's are known as the field operators; they satisfy the following commutation relations:

$$\begin{aligned}
\left[\Psi(\mathbf{x}),\Psi^{\dagger}(\mathbf{x}')\right] &= \sum_{\mathbf{k},\mathbf{k}'} u_{\mathbf{k}}(\mathbf{x})u_{\mathbf{k}'}^{*}(\mathbf{x}')\left[\mathbf{a}_{\mathbf{k}},\mathbf{a}_{\mathbf{k}'}^{\dagger}\right]\\
&=\delta(\mathbf{x}-\mathbf{x}')\\
\left[\Psi(\mathbf{x}),\Psi(\mathbf{x}')\right] &= \left[\Psi(\mathbf{x}),\Psi(\mathbf{x}')\right]^{\dagger}=0
\end{aligned} \tag{16.66}$$

In terms of the field operators, the density of particles in space is given by

$$\rho(\mathbf{x})=\Psi^{\dagger}(\mathbf{x})\Psi(\mathbf{x}) \tag{16.67}$$

from which the number operator is given by

$$\begin{aligned}
N=\int\rho(\mathbf{x})\,d\mathbf{x} &= \sum_{k,k'} a_{k'}^{\dagger}a_{k}\int u_{\mathbf{k}'}^{*}(\mathbf{x})u_{\mathbf{k}}(\mathbf{x})\,d\mathbf{x}\\
&=\sum_{k}\mathbf{a}_{\mathbf{k}}^{\dagger}\mathbf{a}_{\mathbf{k}}
\end{aligned} \tag{16.68}$$

The Fourier transform of ρ is given by

$$\rho_{\mathbf{q}}=\int e^{i\mathbf{q}\mathbf{x}}\rho(\mathbf{x})\,d\mathbf{x}=\sum_{k,k'} a_{\mathbf{k}'}^{\dagger}a_{\mathbf{k}}\int e^{i\mathbf{q}x}u_{\mathbf{k}'}^{*}(\mathbf{x})u_{\mathbf{k}}(\mathbf{x})\,d\mathbf{x} \tag{16.69}$$

When $u_{\mathbf{k}}(\mathbf{x})$ are the plane waves $e^{i\mathbf{k}\mathbf{x}}$, and when we normalize them in a box of unit volume, we get $\mathbf{k}'=\mathbf{k}-\mathbf{q}$ ($\mathbf{q}$, the wave vector). Hence (16.69) reduces to

$$\rho_{\mathbf{q}}=\sum_{k'} a_{\mathbf{k}'}^{\dagger}a_{\mathbf{k}'+\mathbf{q}} \tag{16.70}$$

We can also write the Hamiltonian (16.53) in terms of the field operators $\Psi(\mathbf{x})$; thus:

$$\begin{aligned}
H=&\int\Psi^{\dagger}(\mathbf{x})H_{1}(\mathbf{x})\Psi(\mathbf{x})\,d\mathbf{x}\\
&+\frac{1}{2}\int\Psi^{\dagger}(\mathbf{x})\Psi^{\dagger}(\mathbf{x}')V_{2}(\mathbf{x},\mathbf{x}')\Psi(\mathbf{x}')\Psi(\mathbf{x})\,d\mathbf{x}\,d\mathbf{x}'
\end{aligned} \tag{16.71}$$

where, in the second term, the order of the field operators has to be preserved in order to omit the $i=j$ term in the two-body potential V_2. If $V_2(\mathbf{x},\mathbf{x}')$ is translationally invariant, then $V_2(\mathbf{x},\mathbf{x}')=V_2(\mathbf{x}-\mathbf{x}')$, and (16.71) can be written as

$$H=\int \Psi^\dagger(\mathbf{x})H_1(\mathbf{x})\Psi(\mathbf{x})\,d\mathbf{x}+\frac{1}{2}\sum_{\mathbf{q}} V_2(\mathbf{q})T(\rho_{\mathbf{q}}^\dagger\rho_{\mathbf{q}}) \tag{16.72}$$

where T is the normally ordered product such that all $\Psi^\dagger$ are placed to the left, and all Ψ to the right, in the product. We omit the proof of (16.72) since it is simple. A hint for finding the proof is as follows: Expand $V_2(\mathbf{x}-\mathbf{x}')$ as its Fourier transform, and then use (16.69).

The n-body interaction can now be simply written in the occupation number representation as

$$V_\lambda=\frac{1}{\lambda!}\int \Psi^\dagger(\mathbf{x}_1)\cdots\Psi^\dagger(\mathbf{x}_\lambda)V_\lambda(\mathbf{x}_1,\ldots,\mathbf{x}_\lambda)\Psi(\mathbf{x}_\lambda)\cdots\Psi(\mathbf{x}_1)\,d\mathbf{x}_1\cdots d\mathbf{x}_\lambda \tag{16.73}$$

Suppressing the variables $\mathbf{x}_1\cdots\mathbf{x}_n$ in Φ, we can represent Φ simply by

$$\Phi_{n_1,\ldots,n_i,\ldots}(\mathbf{x}_1,\cdots,\mathbf{x}_n)=|n_1,\ldots,n_i\cdots\rangle \tag{16.74}$$

where $|n_1\cdots n_i\ldots\rangle$ can be very well considered to be a vector in the Hilbert space which has components $\langle\mathbf{x}_1,\ldots,\mathbf{x}_n|n_1,\ldots,n_i\cdots\rangle$ along a complete set of position eigenvectors $|\mathbf{x}_1,\ldots,\mathbf{x}_n\rangle$. These components are identified with the function $\Phi_{n_1,\ldots,n_i\cdots}(\mathbf{x}_1,\ldots,\mathbf{x}_n)$.

The many-body Schrödinger equation may now be easily expressed in the occupation number formalism as

$$H\Psi(t)=i\hbar\frac{\partial\Psi(t)}{\partial t} \tag{16.75}$$

where

$$\Psi(t)=\sum_{n_1,n_2,\ldots} A(n_1,n_2,\ldots,t)|n_1,n_2,\ldots\rangle$$

and

$$H=H_1+V_2+\cdots+V_\lambda+\cdots$$

Thus (16.75) is the second quantized form of the many-body Schrödinger equation.

Let us now describe the second quantization for fermions. For particles obeying Fermi statistics, we denote the creation and annihilation operators by $\mathbf{C}_{\mathbf{k}}^\dagger$ and $\mathbf{C}_{\mathbf{k}}$ respectively, instead of $\mathbf{a}_{\mathbf{k}}^\dagger$ and $\mathbf{a}_{\mathbf{k}}$, which were used for bosons.

Equations (16.60) should now be written as

$$\mathbf{C}_k^\dagger \Phi_{n_1,\ldots,n_k\ldots}(\mathbf{x}_1,\ldots,\mathbf{x}_n) = (n_k+1)^{1/2}\Phi_{n_1,\ldots,n_k+1\cdots}(\mathbf{x}_1,\ldots,\mathbf{x}_{n+1})$$
$$\mathbf{C}_k \Phi_{n_1,\ldots,n_k\ldots}(\mathbf{x}_1,\ldots,\mathbf{x}_n) = (n_k)^{1/2}\Phi_{n_1,\ldots,n_k-1\ldots}(\mathbf{x}_1,\ldots,\mathbf{x}_{n-1}) \tag{16.76}$$

Since these operators are for fermions, they satisfy the following anticommutation relations:

$$\{\mathbf{C}_k^\dagger, \mathbf{C}_{k'}\} = \delta_{kk'}$$
$$\{\mathbf{C}_k, \mathbf{C}_{k'}\} = \{\mathbf{C}_k^\dagger, \mathbf{C}_{k'}^\dagger\} = 0 \tag{16.77}$$

where

$$\{\mathbf{A},\mathbf{B}\} = \mathbf{AB} + \mathbf{BA}$$

We can easily see that the above choice of commutation relations restricts the occupation number of the states k to zero or one, as demanded by Fermi statistics. We thus have, for $\mathbf{C}_k$ and $\mathbf{C}_k^\dagger$,

$$\mathbf{C}_k \Phi_{\cdots n_k \cdots} \begin{cases} = \Phi_{\cdots n_k - 1 \cdots} & \text{when} \quad n_k = 1 \\ = 0 & \text{when} \quad n_k = 0 \end{cases}$$

and (16.78)

$$\mathbf{C}_k^\dagger \Phi_{\cdots n_k \cdots} \begin{cases} = \Phi_{\cdots n_k + 1 \cdots} & \text{when} \quad n_k = 0 \\ = 0 & \text{when} \quad n_k = 1 \end{cases}$$

The Hamiltonian for fermions is given in the second quantization formalism by

$$H = \sum_{k',k,s} \langle k's|H_1|ks\rangle \mathbf{C}_{k's}^\dagger \mathbf{C}_{ks} + \frac{1}{2} \sum_{i,j,k,l,s,s'} \langle ij|V_2|lk\rangle \mathbf{C}_{is}^\dagger \mathbf{C}_{js'}^\dagger \mathbf{C}_{ks'} \mathbf{C}_{ls} \tag{16.79}$$

where s and s' label the spin of the particle. The order of $\mathbf{C}_{ks}$ with regard to the matrixelement indices is the inverse of the $\mathbf{C}_{ks}^\dagger$. The field operators are now defined as

$$\psi(\mathbf{x}) = \sum_k u_k(\mathbf{x})\mathbf{C}_k$$
$$\psi^\dagger(\mathbf{x}) = \sum_k u_k(\mathbf{x})\mathbf{C}_k^\dagger \tag{16.80}$$

which satisfy the following commutation relations:

$$\begin{aligned} \{\psi(\mathbf{x}),\psi^{\dagger}(\mathbf{x}')\} &= \delta(\mathbf{x}-\mathbf{x}') \\ \{\psi(\mathbf{x}),\psi(\mathbf{x}')\} &= \{\psi^{\dagger}(\mathbf{x}),\psi^{\dagger}(\mathbf{x}')\} = 0 \end{aligned} \tag{16.81}$$

The many-body interaction for fermions can now be written as

$$\begin{aligned} V_\lambda = \frac{1}{\lambda!} \sum_{k'_\lambda s'_\lambda \cdots k'_1 s'_1 \cdots k_1 s_1} &\langle k'_\lambda \cdots k'_1 | V_\lambda | k_\lambda \cdots k_1 \rangle \\ &\cdot \mathbf{C}^{\dagger}_{k'_\lambda s'_\lambda} \cdots \mathbf{C}^{\dagger}_{k'_1 s'_1} \mathbf{C}_{k_1 s_1} \cdots \mathbf{C}_{k_\lambda s_\lambda} \end{aligned} \tag{16.82}$$

where the ordering is not entirely arbitrary because of the sign changes arising from the anticommutation of the fermion operators.

With this brief outline of the second quantization formalism, we proceed to the famous Holstein-Primakoff transformation in the next section.

16.3 HOLSTEIN-PRIMAKOFF TRANSFORMATION

Before we deal with the spin-wave theory of ferromagnetism and antiferromagnetism, we discuss in this section a very important transformation derived by Holstein and Primakoff[4] in 1940. The Holstein-Primakoff (H-P) transformation is a transformation from the spin operators $\mathbf{S}_i$ to the spin-deviation creation and annihilation operators $\mathbf{a}_i^{\dagger}$ and $\mathbf{a}_i$, which have the following commutation relations [see (17.63)]:

$$\begin{aligned} \left[\mathbf{a}_i, \mathbf{a}_j^{\dagger}\right] &= \delta_{ij} \\ \left[\mathbf{a}_i, \mathbf{a}_j\right] &= \left[\mathbf{a}_i^{\dagger}, \mathbf{a}_j^{\dagger}\right] = 0 \end{aligned} \tag{16.83}$$

The spin-deviation occupation number n_i, which is the eigenvalue of $\mathbf{a}_i^{\dagger}\mathbf{a}_i$, is defined by the relation

$$n_i = S - S_{iz} \tag{16.84}$$

It is very important to note that n_i is the reduction in the z-component of the ith spin from its maximum possible value of S. Thus the operator $\mathbf{a}_i$ creates a quantum of spin deviation, that is, it reduces S_{iz} by one unit. In Section 16.2 we saw that the operators $\mathbf{a}_i^{\dagger}$ and $\mathbf{a}_i$ have the harmonic oscillator matrix elements.

To derive the H-P transformation, we use

$$\mathbf{S}_i^{\pm} = \mathbf{S}_{ix} \pm i\mathbf{S}_{iy} \tag{16.85}$$

as the two independent spin operators, simply because they have the convenient raising and lowering properties. We already know from the theory of atomic spectra[5] that

$$\mathbf{S}_i^{\pm}|S_{iz}\rangle=[S(S+1)-S_{iz}(S_{iz}\pm 1)]^{1/2}|S_{iz}\pm 1\rangle \tag{16.86}$$

Now, taking the first relation from (16.86), we have, dropping for the moment the subscript,

$$\begin{aligned}\mathbf{S}^{+}|S_z\rangle&=[S(S+1)-S_z(S_z+1)]^{1/2}|S_z+1\rangle\\&=(S^2+S+SS_z-SS_z-S_z^2-S_z)^{1/2}|S_z+1\rangle\\&=(S+S_z+1)^{1/2}(S-S_z)^{1/2}|S_z+1\rangle\\&=(2S)^{1/2}\left(1-\frac{S-S_z-1}{2S}\right)^{1/2}(S-S_z)^{1/2}|S_z+1\rangle\end{aligned} \tag{16.87}$$

or, restoring the subscript i, we get, using (16.84),

$$\mathbf{S}_i^{+}|n_i\rangle=(2S)^{1/2}\left(1-\frac{n_i-1}{2S}\right)^{1/2}(n_i)^{1/2}|n_i-1\rangle \tag{16.88}$$

Note that (16.88) follows from (16.87) simply by correspondence. The fact that $\mathbf{S}^{+}$ operating on S_z gives rise to a state having (S_z+1) means that the occupation number n_i decreases by one unit, becoming (n_i-1). On the left-hand side of (16.88), we had originally n_i spin deviations, say.

Now, using (16.50), we find that (16.88) reduces to

$$\mathbf{S}_i^{+}|n_i\rangle=(2S)^{1/2}\left(1-\frac{n_i-1}{2S}\right)^{1/2}\mathbf{a}_i|n_i\rangle \tag{16.89}$$

whence we have

$$\mathbf{S}_i^{+}=(2S)^{1/2}\left(1-\frac{\mathbf{a}_i^{\dagger}\mathbf{a}_i}{2S}\right)^{1/2}\mathbf{a}_i \tag{16.90}$$

In (16.90) we have written $\mathbf{a}_i^{\dagger}\mathbf{a}_i$ in place of (n_i-1) because, when $\mathbf{a}_i$ operates on n_i, we get the state (n_i-1) and also from (16.51)

$$\mathbf{a}_i^{\dagger}\mathbf{a}_i|n_i-1\rangle=(n_i-1)|n_i-1\rangle$$

Thus, in (16.90), we have replaced the eigenvalue (n_i-1) by the corresponding (number) operator $\mathbf{a}_i^{\dagger}\mathbf{a}_i$, as is possible since $n_i/2S\ll 1$, that is, $[1-(n_i/2S)]^{1/2}$ converges.

Similarly, taking the second relation from (16.86), we get

$$\mathbf{S}_i^- |n_i\rangle = (2S)^{1/2}(n_i+1)^{1/2}\left(1-\frac{n_i}{2S}\right)^{1/2}|n_i+1\rangle$$
$$= (2S)^{1/2}\mathbf{a}_i^\dagger\left(1-\frac{n_i}{2S}\right)^{1/2}|n_i\rangle$$

or

$$\mathbf{S}_i^- = (2S)^{1/2}\mathbf{a}_i^\dagger\left(1-\frac{n_i}{2S}\right)^{1/2} \tag{16.91}$$

Thus we have derived the H-P transformation, that is, the relationship of the spin operators with the spin-deviation creation and annihilation operators, $\mathbf{a}_i^\dagger$ and $\mathbf{a}_i$. Collecting the latter from (16.84), (16.90), and (16.91), we have

$$\begin{aligned}\mathbf{S}_{iz} &= S-\mathbf{a}_i^\dagger\mathbf{a}_i\\ \mathbf{S}_i^+ &= (2S)^{1/2}\left(1-\frac{\mathbf{a}_i^\dagger\mathbf{a}_i}{2S}\right)^{1/2}\mathbf{a}_i\\ \mathbf{S}_i^- &= (2S)^{1/2}\mathbf{a}_i^\dagger\left(1-\frac{\mathbf{a}_i^\dagger\mathbf{a}_i}{2S}\right)^{1/2}\end{aligned} \tag{16.92}$$

Holstein and Primakoff's pioneering work was the first successful attempt to derive the variation of magnetization of a ferromagnetic sample with temperature and external field. To derive the expression for magnetization, Holstein and Primakoff diagonalized the Hamiltonian consisting of the exchange, the Zeeman, and the dipole-dipole interactions by means of three successive transformations, the first of which is outlined in (16.92). The Hamiltonian that has to be diagonalized is written as

$$\begin{aligned}\mathcal{H} = &-\sum_{i,j=1}^{N} J(\mathbf{R}_{ij})\mathbf{S}_i\cdot\mathbf{S}_j - \sum_{i=1}^{N} 2\mu_B H\mathbf{S}_{iz}\\ &+\frac{1}{2}\sum_{i,j=1}^{N}{}' \left(\frac{4\mu_B^2}{\mathbf{R}_{ij}^5}\right)\left[\mathbf{R}_{ij}^2(\mathbf{S}_i\cdot\mathbf{S}_j)-3(\mathbf{R}_{ij}\cdot\mathbf{S}_i)(\mathbf{R}_{ij}\cdot\mathbf{S}_j)\right]\end{aligned} \tag{16.93}$$

where the prime over summation denotes $j\neq i$, $\mathbf{R}_{ij}$ $(=|\mathbf{R}_i-\mathbf{R}_j|)$ is the relative position vector of spins $\mathbf{S}_i$ and $\mathbf{S}_j$, and N is the total number of atoms. It is convenient to include the z-demagnetizing factor in the Zeeman term by defining

$$\mathbf{H}=\mathbf{H}_0-4\pi N_z\mathbf{M} \tag{16.94}$$

where $\mathbf{H}_0$ is the applied field, and $-4\pi N_z M$ is the demagnetizing field in the z-direction, which is the direction of the applied field.

In (16.93) the first term is the familiar Heisenberg exchange energy, expressed in terms of the atomic spin operators, and the third term arises from the magnetic dipole-dipole interaction between the electrons on different atoms.

Now, using the first transformation given by (16.92), we find that the first term of the Hamiltonian (16.93) reduces to

$$\begin{aligned}\mathcal{H}_{\text{ex}} = &-\sum_{i,j} J_{ij}(\mathbf{R}_{ij})\Bigg\{(S-\mathbf{a}_i^\dagger\mathbf{a}_i)(S-\mathbf{a}_j^\dagger\mathbf{a}_j)\\ &+\tfrac{1}{2}(2S)\left(1-\frac{\mathbf{a}_i^\dagger\mathbf{a}_i}{2S}\right)^{1/2} a_i a_j^\dagger\left(1-\frac{\mathbf{a}_j^\dagger\mathbf{a}_j}{2S}\right)^{1/2}\\ &+\tfrac{1}{2}(2S)\mathbf{a}_i^\dagger\left(1-\frac{\mathbf{a}_i^\dagger\mathbf{a}_i}{2S}\right)^{1/2}\left(1-\frac{\mathbf{a}_j^\dagger\mathbf{a}_j}{2S}\right)^{1/2}\mathbf{a}_j\Bigg\}\\ = &-\sum_{i,j} J_{ij}(\mathbf{R}_{ij})\Bigg\{S^2-2S\mathbf{a}_i^\dagger\mathbf{a}_i+\mathbf{a}_i^\dagger\mathbf{a}_i\mathbf{a}_j^\dagger\mathbf{a}_j\\ &+(2S)\mathbf{a}_i^\dagger\left(1-\frac{\mathbf{a}_i^\dagger\mathbf{a}_i}{2S}\right)^{1/2}\left(1-\frac{\mathbf{a}_j^\dagger\mathbf{a}_j}{2S}\right)^{1/2}\mathbf{a}_j\Bigg\}\end{aligned}\tag{16.95}$$

by using the commutation relations (16.89).

The dipolar and the Zeeman terms in (16.93) similarly reduce to

$$\begin{aligned}\mathcal{H}_{\text{dipole}} = &\frac{1}{2}\sum_{i,j}\frac{4\mu_B^2}{\mathbf{R}_{ij}^3}\Bigg\{(S^2-2S\mathbf{a}_i^\dagger\mathbf{a}_i+\mathbf{a}_i^\dagger\mathbf{a}_i\mathbf{a}_j^\dagger\mathbf{a}_j)\\ &+(2S)\mathbf{a}_i^\dagger\left(1-\frac{\mathbf{a}_i^\dagger\mathbf{a}_i}{2S}\right)^{1/2}\left(1-\frac{\mathbf{a}_j^\dagger\mathbf{a}_j}{2S}\right)^{1/2}\mathbf{a}_j\Bigg\}\\ &+\frac{1}{2}\sum_{i,j}\frac{4\mu_B^2}{\mathbf{R}_{ij}^5}\Bigg[\{-3(\mathbf{R}_{ij}^z)^2\}(S^2-2S\mathbf{a}_i^\dagger\mathbf{a}_i+\mathbf{a}_i^\dagger\mathbf{a}_i\mathbf{a}_j^\dagger\mathbf{a}_j)\\ &+\mathbf{R}_{ij}^z\mathbf{R}_{ij}^+(-\tfrac{3}{2})2(S-\mathbf{a}_i^\dagger\mathbf{a}_i)(2S)^{1/2}\mathbf{a}_j^\dagger\left(1-\frac{\mathbf{a}_j^\dagger\mathbf{a}_j}{2S}\right)^{1/2}\\ &+(\mathbf{R}_{ij}^+)^2(-\tfrac{3}{4})(2S)\mathbf{a}_i^\dagger\left(1-\frac{\mathbf{a}_i^\dagger\mathbf{a}_i}{2S}\right)^{1/2}\mathbf{a}_j^\dagger\left(1-\frac{\mathbf{a}_j^\dagger\mathbf{a}_j}{2S}\right)^{1/2}\\ &+\mathbf{R}_{ij}^z\mathbf{R}_{ij}^-(-\tfrac{3}{2})\cdot 2\cdot(S-\mathbf{a}_i^\dagger\mathbf{a}_i)(2S)^{1/2}\left(1-\frac{\mathbf{a}_j^\dagger\mathbf{a}_j}{2S}\right)^{1/2}\mathbf{a}_j\\ &+(\mathbf{R}_{ij}^-)^2(-\tfrac{3}{4})(2S)\left(1-\frac{\mathbf{a}_i^\dagger\mathbf{a}_i}{2S}\right)^{1/2}\mathbf{a}_i\left(1-\frac{\mathbf{a}_j^\dagger\mathbf{a}_j}{2S}\right)^{1/2}\mathbf{a}_j\\ &+2(\mathbf{R}_{ij}^+\mathbf{R}_{ij}^-)(-\tfrac{3}{4})(2S)\mathbf{a}_i^\dagger\left(1-\frac{\mathbf{a}_i^\dagger\mathbf{a}_i}{2S}\right)^{1/2}\left(1-\frac{\mathbf{a}_j^\dagger\mathbf{a}_j}{2S}\right)^{1/2}\mathbf{a}_j\Bigg]\end{aligned}\tag{16.96}$$

where

$$\mathbf{R}_{ij}^{\pm} = \mathbf{R}_{ij}^{x} \pm \mathbf{R}_{ij}^{y}$$

and

$$\mathcal{H}_{\text{Zeeman}} = -\sum_i 2\mu_B H\left(S - \mathbf{a}_i^{\dagger}\mathbf{a}_i\right) \tag{16.97}$$

To reduce expressions (16.95) to (16.97) to a much simpler, and thus solvable, form, Holstein and Primakoff made the following approximations: If the magnetization is near saturation, we expect that

$$\frac{1}{2}\,\frac{M_0 - M(T,H)}{M_0} \ll 1$$

or

$$\frac{S - S_z}{2S} \ll 1$$

or

$$\frac{\langle\langle \mathbf{a}_i^{\dagger}\mathbf{a}_i \rangle\rangle}{2S} \ll 1$$

Hence

(i) $$\left(1 - \frac{\mathbf{a}_i^{\dagger}\mathbf{a}_i}{2S}\right)^{1/2} \simeq 1$$

(ii) $$\mathbf{a}_i^{\dagger}\mathbf{a}_i\mathbf{a}_j^{\dagger}\mathbf{a}_j = n_i n_j \simeq 0$$

which effectively meanS that there is no correlation in the location of the different spin deviations, that is,

$$\begin{aligned}\langle\langle n_i n_j \rangle\rangle_{\text{av}} &\approx \langle\langle n_i \rangle\rangle_{\text{av}} \langle\langle n_j \rangle\rangle_{\text{av}} \\ &\approx (2S)\langle\langle n_i \rangle\rangle_{\text{av}} \frac{\langle\langle n_j \rangle\rangle_{\text{av}}}{2S} \\ &\ll 1\end{aligned}$$

since S is a very large number.

(iii) $$(2S)^{1/2}\mathbf{a}_i^{\dagger}\mathbf{a}_i\mathbf{a}_j = (2S)^{1/2} n_i \mathbf{a}_j^{\dagger} \ll 1$$

These are the terms that cause the system to make transitions between the states of different total spin. They are nonvanishing only for transitions taking place near the atoms on which spin deviations are already present.

(iv)
$$\sum_{i,j} f(\mathbf{R}_{ij})\mathbf{x}_{ij}\mathbf{z}_{ij}\mathbf{a}_j^\dagger = \sum_j \mathbf{a}_j^\dagger \sum_h f(h)\mathbf{x}_h\mathbf{z}_h$$
$$=0$$

for the type of crystals considered.

With these approximations (16.95), (16.96), and (16.97) take much simpler forms, which, when added together, give us the total Hamiltonian in a tractable form:

$$\begin{aligned}\mathcal{H}=\mathcal{C}&-\sum_{i,j} 2SJ_{ij}(\mathbf{R}_{ij})(\mathbf{a}_i^\dagger\mathbf{a}_j-\mathbf{a}_i^\dagger\mathbf{a}_i)\\ &+\frac{1}{2}\sum_{i,j}\frac{4\mu_B^2}{\mathbf{R}_{ij}^3}(2S)(\mathbf{a}_i^\dagger\mathbf{a}_j-\mathbf{a}_i^\dagger\mathbf{a}_i)\\ &+\frac{1}{2}\sum_{i,j}\frac{4\mu_B^2}{\mathbf{R}_{ij}^5}(-3)(2S)\Big[-z_{ij}^2\mathbf{a}_i^\dagger\mathbf{a}_i+\tfrac{1}{2}(\mathbf{x}_{ij}^2+\mathbf{y}_{ij}^2)\\ &\cdot\mathbf{a}_i^\dagger\mathbf{a}_j+\tfrac{1}{4}\{(\mathbf{R}_{ij}^+)^2\mathbf{a}_i^\dagger\mathbf{a}_j^\dagger+(\mathbf{R}_{ij}^-)^2\mathbf{a}_i\mathbf{a}_j\}\Big]\\ &+\sum_i 2\mu_B H\mathbf{a}_i^\dagger\mathbf{a}_i\end{aligned} \tag{16.98}$$

where

$$\begin{aligned}\mathcal{C}=&-\sum_{i,j} J_{Ij}(\mathbf{R}_{ij})S^2-2\mu_B HNS\\ &+\frac{1}{2}\sum_{i,j}\frac{4\mu_B^2}{\mathbf{R}_{ij}^3}S^2\left[1-\frac{3z_{ij}^2}{\mathbf{R}_{ij}^2}\right]\\ &\mathbf{R}_{ij}=|\mathbf{R}_i-\mathbf{R}_j|\end{aligned} \tag{16.98a}$$

and

$$\mathbf{R}_{ij}^{\pm}=\mathbf{x}_{ij}\pm i\mathbf{y}_{ij}$$

The constant $\mathcal{C}$ denotes the energy when all the magnetic atoms point along the direction of the applied magnetic field. The first term in $\mathcal{C}$ is the exchange energy; the second, the interaction between the atoms and the magnetic field; and the third, the dipolar interaction. If the magnetic dipolar interactions are omitted from (16.98), approximations (i)–(iv) lead to the same energy levels as those obtained by Bloch[6] in his derivation of the $T^{3/2}$ law for atoms with spin $S=\frac{1}{2}$. Bloch's method has been extended by Möller[7] to the case of $S>\frac{1}{2}$. We shall deal with Bloch's treatment in connection with the spin-wave theory in Section 16.4.

From the form of the Hamiltonian (16.98), it is very clear that the spin deviations are not localized on any one atom but are propagated through the entire crystal. Also, it is apparent from the terms in (16.98) that the Hamiltonian is not yet reduced to the desired diagonal form, since the subscripts of $\mathbf{a}^\dagger$ and $\mathbf{a}$ are different. The standard transformation to diagonalize such terms is the well-known Fourier transformation. The physical reason for making the Fourier transformation is as follows. The Hamiltonian (16.98) contains factors like $\mathbf{a}_i^\dagger\mathbf{a}_j$, which flips the ith spin down while flipping the jth spin up, so that this factor relates to two different lattice sites, i and j. Physically, the Fourier transforms (16.99) given below take us from the operator $\mathbf{a}_i$, related to a single site i, to the operators $\mathbf{a}_\lambda$, which are collective operators in the sense that $\mathbf{a}_\lambda$ relate to all the lattice sites, i, j, and so on, in the sample. However, if only terms such as $\mathbf{a}_i^\dagger\mathbf{a}_i$, $\mathbf{a}_i^\dagger\mathbf{a}_i^\dagger$, and $\mathbf{a}_i\mathbf{a}_i$ appear, the Fourier transformation is not needed.

The propagation of the spin deviations is essentially of a wavelike character, which can be visualized by the introduction of new variables (the Fourier transformation), defined by the following relations:

$$\begin{aligned}\mathbf{a}_l &= N^{-1/2}\sum_\lambda e^{(-i\mathbf{K}_\lambda\cdot\mathbf{R}_l)}\mathbf{a}_\lambda\\ \mathbf{a}_l^\dagger &= N^{-1/2}\sum_\lambda e^{(i\mathbf{K}_\lambda\cdot\mathbf{R}_l)}\mathbf{a}_\lambda^\dagger\end{aligned}\tag{16.99}$$

and the inverse transformations:

$$\begin{aligned}\mathbf{a}_\lambda &= N^{-1/2}\sum_l e^{(i\mathbf{K}_\lambda\cdot\mathbf{R}_l)}\mathbf{a}_l\\ \mathbf{a}_\lambda^\dagger &= N^{-1/2}\sum_l e^{(-i\mathbf{K}_\lambda\cdot\mathbf{R}_l)}\mathbf{a}_l^\dagger\end{aligned}\tag{16.100}$$

where N is the total number of spins in the system. One can show easily that these new operators obey the proper commutation relation:

$$\begin{aligned}\left[\mathbf{a}_\lambda,\mathbf{a}_\mu^\dagger\right] &= \frac{1}{N}\sum_l\left(e^{(i\mathbf{K}_\lambda\cdot\mathbf{R}_l)}\mathbf{a}_l,\, e^{(-i\mathbf{K}_\mu\cdot\mathbf{R}_l)}\mathbf{a}_l^\dagger\right)\\ &= \frac{1}{N}\sum_l e^{i(\mathbf{K}_\lambda-\mathbf{K}_\mu)\cdot\mathbf{R}_l}\\ &= \delta_{\mathbf{K}_\lambda,\mathbf{K}_\mu} = \delta_{\lambda\mu}.\end{aligned}\tag{16.101}$$

In (16.99) and (16.100), $\mathbf{R}_l$ denotes the vector from an arbitrary origin to the lth atom, whose magnitude measures the corresponding distance in units of the lattice constant; $\mathbf{K}_\lambda$ is a reduced wave vector. The periodicity of the lattice demands that

$$K_\lambda^i = \frac{2\pi\lambda_i}{G_i}, \qquad i = x, y, z \tag{16.102}$$

where λ_i may have any integer values between $-\frac{1}{2}G_i$ and $(\frac{1}{2}G_i - 1)$, the G_i's being the ratios of the lengths of the specimen and the lattice constants.

The replacement of the $\mathbf{a}_l$ and $\mathbf{a}_\lambda$ constitutes the first step in the evaluation of the eigenvalues of the Hamiltonian (16.98) (the second transformation). To carry out this replacement, one has to evaluate sums of the type

$$\begin{aligned}\sum_{l,m} f(R_{lm})\mathbf{a}_l^\dagger \mathbf{a}_m &= \frac{1}{N}\sum_{l,m,\lambda,\lambda'} f(R_{lm})\mathbf{a}_{\lambda'}^\dagger \mathbf{a}_\lambda\, e^{i(\mathbf{K}_{\lambda'}\mathbf{R}_l - \mathbf{K}_\lambda \mathbf{R}_m)}\\ &= \frac{1}{N}\sum_{\lambda,\lambda',l,h} f(R_h)\mathbf{a}_{\lambda'}^\dagger \mathbf{a}_\lambda\, e^{i(\mathbf{K}_{\lambda'}\mathbf{R}_l - \mathbf{K}_\lambda(\mathbf{R}_l - \mathbf{R}_h))}\\ &= \frac{1}{N}\sum_{\lambda,\lambda',l,h} f(R_h)\mathbf{a}_{\lambda'}^\dagger \mathbf{a}_\lambda\, e^{i(\mathbf{K}_{\lambda'} - \mathbf{K}_\lambda)\mathbf{R}_l}\cdot e^{i\mathbf{K}_\lambda\cdot\mathbf{R}_h}\end{aligned} \tag{16.103}$$

where we have introduced $\mathbf{R}_h = \mathbf{R}_l - \mathbf{R}_m = \mathbf{R}_{lm}$. We can now carry out the summations over l and λ', provided that we neglect the contributions arising from the surface of the specimen, which are negligible compared to the exchange forces. The summation over l gives the factor $N\delta_{\lambda'\lambda}$, and we have

$$\sum_{l,m} f(R_{lm})\mathbf{a}_l^\dagger \mathbf{a}_m = \sum_{\lambda,h} f(\mathbf{R}_h)e^{i\mathbf{K}_\lambda\cdot\mathbf{R}_h}\mathbf{a}_\lambda^\dagger \mathbf{a}_\lambda$$

and

$$\sum_{l,m} J_{lm}(\mathbf{R}_{lm})\mathbf{a}_l^\dagger \mathbf{a}_l = \sum_h J(\mathbf{R}_h)\sum_\lambda \mathbf{a}_\lambda^\dagger \mathbf{a}_\lambda \tag{16.104}$$

Thus

$$\sum_{l,m}\frac{1}{\mathbf{R}_{lm}^3}(\mathbf{a}_l^\dagger \mathbf{a}_m - \mathbf{a}_l^\dagger \mathbf{a}_l) = \sum_{\lambda,h}\frac{1}{a^3\mathbf{R}_h^3}(e^{i\mathbf{K}_\lambda\cdot\mathbf{R}_h} - 1)\mathbf{a}_\lambda^\dagger \mathbf{a}_\lambda$$

Before applying these approximations, let us rewrite (16.98) in the following form for our convenience:

$$\begin{aligned}\mathcal{H} = \mathcal{C} &- \sum_{l,m} 2SJ_{lm}(\mathbf{a}_l^\dagger \mathbf{a}_m - \mathbf{a}_l^\dagger \mathbf{a}_l) + \sum_{l,m}\frac{4\mu_B^2 S}{R_{lm}^3}(\mathbf{a}_l^\dagger \mathbf{a}_m - \mathbf{a}_l^\dagger \mathbf{a}_l)\\ &- \sum_{l,m}\frac{6\mu_B^2 S}{\mathbf{R}_{lm}^5}\left[(x^2+y^2)\mathbf{a}_l^\dagger \mathbf{a}_m - 2z_{lm}^3 \mathbf{a}_l^\dagger \mathbf{a}_l\right]\\ &- \frac{1}{2}\sum_{l,m}\frac{6\mu_B^2 S}{R_{lm}^5}\left[(\mathbf{R}_{lm}^+)^2 \mathbf{a}_l^\dagger \mathbf{a}_m^\dagger + (R_{lm}^-)^2 \mathbf{a}_l \mathbf{a}_m\right]\\ &+ \frac{1}{2}\sum_l 2\mu_B H \mathbf{a}_l^\dagger \mathbf{a}_l\end{aligned} \tag{16.105}$$

where

$$\mathcal{C} = -\sum_{l,m} J_{lm} S^2 - 2\mu_B HNS$$

$$-\sum_{l,m} \mu_B S\left(-\frac{2\mu_B S}{\mathbf{R}_{lm}^3}\right)\left(1-\frac{3z_{lm}^2}{\mathbf{R}_{lm}^2}\right)$$

We now apply the Fourier transforms (16.104) to the terms in (16.105) given below:

$$\frac{4\mu_B^2 S}{\mathbf{R}_{lm}^3}(\mathbf{a}_l^\dagger \mathbf{a}_m - \mathbf{a}_l^\dagger \mathbf{a}_l) - \frac{6\mu_B^2 S}{\mathbf{R}_{lm}^5}\left[(x^2+y^2)\mathbf{a}_l^\dagger \mathbf{a}_m - 2z_{lm}^2 \mathbf{a}_l^\dagger \mathbf{a}_l\right]$$

$$= \frac{4\mu_B^2 S}{\mathbf{R}_{lm}^3}\left[(e^{i\mathbf{K}_\lambda \cdot \mathbf{R}_h} - 1)\sum_\lambda \mathbf{a}_\lambda^\dagger \mathbf{a}_\lambda\right] - \frac{6\mu_B^2 S}{\mathbf{R}_{lm}^5}\left[(x^2+y^2)e^{i\mathbf{K}_\lambda \cdot \mathbf{R}_h}\sum_\lambda \mathbf{a}_\lambda^\dagger \mathbf{a}_\lambda - 2z_{lm}^2 \sum_\lambda \mathbf{a}_\lambda^\dagger \mathbf{a}_\lambda\right]$$

$$= \left[\frac{4\mu_B^2 S}{\mathbf{R}_{lm}^3} e^{i\mathbf{K}_\lambda \cdot \mathbf{R}_h} - \frac{4\mu_B^2 S}{\mathbf{R}_{lm}^3} - \frac{6\mu_B^2 S}{\mathbf{R}_{lm}^5}(\mathbf{R}_{lm}^2 - z_{lm}^2)e^{i\mathbf{K}_\lambda \cdot \mathbf{R}_h} + \frac{6\mu_B^2 S}{\mathbf{R}_{lm}^5}\cdot 2z_{lm}^2\right]\sum_\lambda \mathbf{a}_\lambda^\dagger \mathbf{a}_\lambda$$

$$= \frac{2\mu_B^2 S}{a^3 \mathbf{R}_{lm}^3}\left[2e^{i\mathbf{K}_\lambda \cdot \mathbf{R}_h} - 2 - 3e^{i\mathbf{K}_\lambda \cdot \mathbf{R}_h} + \frac{3z_{lm}^2}{\mathbf{R}_{lm}^2} e^{i\mathbf{K}_\lambda \cdot \mathbf{R}_h} + \frac{2}{\mathbf{R}_{lm}^2}(3z_{lm}^2)\right]\sum_\lambda \mathbf{a}_\lambda^\dagger \mathbf{a}_\lambda$$

$$= \frac{2\mu_B^2 S}{a^3 \mathbf{R}_{lm}^3}\left[-e^{i\mathbf{K}_\lambda \cdot \mathbf{R}_h} - 2 + \frac{3z_{lm}^2}{\mathbf{R}_{lm}^2} e^{i\mathbf{K}_\lambda \cdot \mathbf{R}_h} + \frac{2}{\mathbf{R}_{lm}^2}(2z_{lm}^2)\right]\sum_\lambda \mathbf{a}_\lambda^\dagger \mathbf{a}_\lambda$$

$$= \frac{2\mu_B^2 S}{a^3 \mathbf{R}_{lm}^3}\left[-e^{i\mathbf{K}_\lambda \cdot \mathbf{R}_h}\left(1-\frac{3z_{lm}^2}{\mathbf{R}_{lm}^2}\right) - 2\left(1-\frac{3z_{lm}^2}{\mathbf{R}_{lm}^2}\right)\right]\sum_\lambda \mathbf{a}_\lambda^\dagger \mathbf{a}_\lambda$$

$$= \frac{2\mu_B^2 S}{a^3 \mathbf{R}_{lm}^3}\left[\left(1-\frac{3z_{lm}^2}{\mathbf{R}_{lm}^2}\right) - \left(1-\frac{3z_{lm}^2}{\mathbf{R}_{lm}^2}\right)e^{i\mathbf{K}_\lambda \cdot \mathbf{R}_h} - 3\left(1-\frac{3z_{lm}^2}{\mathbf{R}_{lm}^2}\right)\right]\sum_\lambda \mathbf{a}_\lambda^\dagger \mathbf{a}_\lambda$$

$$= \sum_h\left[\left\{-\frac{6\mu_B^2 S}{a^3 \mathbf{R}_h^3}\left(1-\frac{3z_h^2}{\mathbf{R}_h^2}\right)\right\} + \frac{2\mu_B^2 S}{a^3 \mathbf{R}_h^3}\left(1-\frac{3z_h^2}{\mathbf{R}_h^2}\right)(1-e^{i\mathbf{K}_\lambda \cdot \mathbf{R}_h})\right]\sum_\lambda \mathbf{a}_\lambda^\dagger \mathbf{a}_\lambda \tag{16.106}$$

Also

$$\sum_{l,m} -\frac{6\mu_B^2 S}{\mathbf{R}_{lm}^5}\left[(\mathbf{R}_{lm}^-)^2 \mathbf{a}_l \mathbf{a}_m\right]$$

$$= \sum_{l,m} -\frac{6\mu_B^2 S}{\mathbf{R}_{lm}^5}(x_h^2 - y_h^2 - 2ix_h y_h)\mathbf{a}_l \mathbf{a}_m$$

$$= \sum_{l,m} -\frac{6\mu_B^2 S}{\mathbf{R}_{lm}^5}(x_h^2 - y_h^2 - 2ix_h y_h)\frac{1}{N}\sum_{\lambda,\lambda'} e^{-i\mathbf{K}_\lambda\cdot\mathbf{R}_l}\mathbf{a}_\lambda \cdot e^{-i\mathbf{K}_{\lambda'}\cdot\mathbf{R}_m}\mathbf{a}_{\lambda'}$$

$$= \sum_{l,m,\lambda,\lambda'} -\frac{6\mu_B^2 S}{\mathbf{R}_{lm}^5}(x_h^2 - y_h^2 - 2ix_h y_h)\frac{1}{N}\sum_{\lambda,\lambda'} e^{-i\mathbf{K}_\lambda\cdot\mathbf{R}_l - i\mathbf{K}_{\lambda'}\cdot(\mathbf{R}_l - \mathbf{R}_h)}\mathbf{a}_\lambda \mathbf{a}_{\lambda'}$$

$$= \sum_{l,m,\lambda,\lambda'} -\frac{6\mu_B^2 S}{\mathbf{R}_{lm}^5}(x_h^2 - y_h^2 - 2ix_h y_h)\frac{1}{N}\sum_{\lambda,\lambda'} e^{-i(\mathbf{K}_\lambda + \mathbf{K}_{\lambda'})\cdot\mathbf{R}_l} e^{i\mathbf{K}_{\lambda'}\cdot\mathbf{R}_h}\mathbf{a}_\lambda \mathbf{a}_{\lambda'}$$

$$= \sum_{h,\lambda,\lambda'} -\frac{6\mu_B^2 S}{\mathbf{R}_h^5}(x_h^2 - y_h^2 - 2ix_h y_h)\frac{1}{N}N\delta_{\lambda',-\lambda} e^{i\mathbf{K}_{\lambda'}\cdot\mathbf{R}_h}\mathbf{a}_\lambda \mathbf{a}_{\lambda'}$$

$$= \sum_{h,\lambda} -\frac{6\mu_B^2 S}{\mathbf{R}_h^5}(x_h^2 - y_h^2 - 2ix_h y_h) e^{i\mathbf{K}_\lambda\cdot\mathbf{R}_h}\mathbf{a}_\lambda \mathbf{a}_{-\lambda}$$

$$= \left[\sum_{h,\lambda} -\frac{6\mu_B^2 S}{a^3\mathbf{R}_h^3}\frac{x_h^2 - y_h^2 - 2ix_h y_h}{\mathbf{R}_h^2} + \sum_{h,\lambda}\frac{6\mu_B^2 S}{a^3\mathbf{R}_h^3}\frac{x_h^2 - y_h^2 - 2ix_h y_h}{\mathbf{R}_h^2}(1 - e^{i\mathbf{K}_\lambda\cdot\mathbf{R}_h})\right]\mathbf{a}_\lambda \mathbf{a}_{-\lambda} \tag{16.107}$$

Treating in this way all the summations in (16.105), we finally get

$$\mathcal{H} = \mathcal{C} + \sum_\lambda A_\lambda \mathbf{a}_\lambda^\dagger \mathbf{a}_\lambda + \tfrac{1}{2}(B_\lambda \mathbf{a}_\lambda \mathbf{a}_{-\lambda} + B_\lambda^\dagger \mathbf{a}_\lambda^\dagger \mathbf{a}_{-\lambda}^\dagger) \tag{16.108}$$

where

$$A_\lambda = -\sum_h \frac{6\mu_B^2 S}{a^3\mathbf{R}_h^3}\left(1 - \frac{3z_h^2}{\mathbf{R}_h^2}\right) + \frac{2\mu_B^2 S}{a^3\mathbf{R}_h^3}\left(1 - \frac{3z^2}{\mathbf{R}_h^2}\right)(1 - e^{i\mathbf{K}_\lambda\cdot\mathbf{R}_h}) + \sum_h 2SJ_h(\mathbf{R}_h)(1 - e^{i\mathbf{K}_\lambda\cdot\mathbf{R}_h}) + 2\mu_B H \tag{16.109}$$

$$B_\lambda = -\sum_h \frac{6\mu_B^2 S}{a^3\mathbf{R}_h^3}\frac{x_h^2 - y_h^2 - 2ix_h y_h}{\mathbf{R}_h^2} + \sum \frac{6\mu_B^2 S}{a^3\mathbf{R}_h^3}\frac{x_h^2 - y_h^2 - 2ix_h y_h}{\mathbf{R}_h^2}(1 - e^{i\mathbf{K}_\lambda\cdot\mathbf{R}_h}) \tag{16.110}$$

and a is the lattice constant.

The sums over h in A_λ and B_λ can be evaluated[8] with little trouble. Evaluating the sums of h in (16.109) and (16.110), we get for $\mathbf{K}_\lambda \ll 1$

$$A_\lambda = A_{-\lambda} = 2SJ\mathbf{K}_\lambda^2 + 4\pi\mu_B M_0 \sin^2\theta_\lambda + 2\mu_B H \tag{16.111}$$

$$B_\lambda = B_{-\lambda} = 4\pi\mu_B M_0 \sin^2\theta_\lambda e^{-2i\phi_\lambda} \tag{16.112}$$

Here $J \equiv J_h(\mathbf{R}_h)$, where $\mathbf{R}_h$ is the distance between the nearest neighbors; θ_λ and ϕ_λ are the polar angles of $\mathbf{K}_\lambda$ with the polar axis parallel to the applied magnetic field.

It is important to note that in (16.108) we have retained only the bilinear terms, neglecting the higher order ones. For a simple cubic lattice and for J different from zero for the nearest neighbors only, in the long-wavelength limit $\mathbf{K}_\lambda \ll 1$, the expressions for A_λ and B_λ are given in (16.111) and (16.112).

It should be noticed that, though the first term in (16.108) is in the diagonal form, in the second term the operators appear as two creation operators or two annihilation operators for the particles λ and $-\lambda$. Before performing the third Holstein-Primakoff transformation to remove these nondiagonal terms, we make the important observation that the coefficients A_λ are much larger than the coefficients B_λ, so that the B_λ terms can be neglected, leaving the Hamiltonian (16.108) in the diagonal form.

Excluding the $\mathcal{C}$ term, which is constant, and replacing the **a**'s by **b**'s, for convenience, we can rewrite the Hamiltonian (16.108) in a more symmetric form;

$$\mathcal{H} = \sum{}^{+} \mathcal{H}_\lambda = A_\lambda(\mathbf{b}_\lambda^\dagger \mathbf{b}_\lambda + \mathbf{b}_{-\lambda}^\dagger \mathbf{b}_{-\lambda}) + (B_\lambda \mathbf{b}_\lambda \mathbf{b}_{-\lambda} + B_\lambda^* \mathbf{b}_{-\lambda}^\dagger \mathbf{b}_\lambda^\dagger) \tag{16.113}$$

where plus indicates summation over only half of the $\mathbf{K}_\lambda$-space. The Hamiltonian (16.113) is of the bilinear form, and the necessary transformation to diagonalize such bilinear Hamiltonians is popularly known as the Bogolieubov transformation.

Our principal aim now is to define a set of new operators, $\mathbf{c}_\lambda$ and $\mathbf{c}_{-\lambda}$, say, and their Hermitian conjugates, $\mathbf{c}_\lambda^\dagger$ and $\mathbf{c}_{-\lambda}^\dagger$, such that our Hamiltonian (16.113) is reduced to the diagonal form. These new operators should be such that they can be expressed as linear combinations of the old operators (the **b**'s). To derive this canonical transformation, let us consider the Heisenberg equation of motion:

$$i\hbar \frac{d\mathbf{b}_\lambda}{dt} = [\mathbf{b}_\lambda, \mathcal{H}_\lambda] \tag{16.114}$$

Substituting (16.113) in (16.114), we can easily show that

$$i\hbar \frac{d\mathbf{b}_\lambda}{dt} = A_\lambda \mathbf{b}_\lambda + B_\lambda^* \mathbf{b}_{-\lambda}^\dagger \tag{16.115}$$

and

$$i\hbar\frac{d\mathbf{b}_{-\lambda}^{\dagger}}{dt} = -A_{\lambda}\mathbf{b}_{-\lambda}^{\dagger} - B_{\lambda}\mathbf{b}_{\lambda} \tag{16.116}$$

where we have used the well-known commutation relations

$$\left[\mathbf{b}_i, \mathbf{b}_j^{\dagger}\right] = \delta_{ij} \tag{16.117}$$

Now, assuming the time dependence of the form $e^{-i\omega_{\lambda}t}$ for $\mathbf{b}_{\lambda}$ and $\mathbf{b}_{-\lambda}^{\dagger}$, we can easily show that the secular equation that results from (16.115) and (16.116) is given by

$$\begin{vmatrix} A_{\lambda} - \hbar\omega_{\lambda} & B_{\lambda}^{*} \\ B_{\lambda} & A_{\lambda} + \hbar\omega_{\lambda} \end{vmatrix} = 0$$

or

$$\hbar\omega_{\lambda} = \left[A_{\lambda}^2 - |B_{\lambda}|^2\right]^{1/2} \tag{16.118}$$

It is apparent from (16.115) and (16.116) that $\mathbf{b}_{\lambda}$ and $\mathbf{b}_{-\lambda}^{\dagger}$ are coupled together. Thus the new operators $\mathbf{c}_{\lambda}$ and $\mathbf{c}_{-\lambda}^{\dagger}$ should be a linear combination of $\mathbf{b}_{\lambda}$ and $\mathbf{b}_{-\lambda}^{\dagger}$. It can be shown that the following transformations diagonalize (16.115) and (16.116):

$$\begin{aligned} \mathbf{b}_{\lambda} &= u_{\lambda}\mathbf{c}_{\lambda} - v_{\lambda}\mathbf{c}_{-\lambda}^{\dagger} \\ \mathbf{b}_{-\lambda}^{\dagger} &= u_{-\lambda}\mathbf{c}_{-\lambda} - v_{-\lambda}\mathbf{c}_{\lambda}^{\dagger} \end{aligned} \tag{16.119}$$

The inverse transformation is

$$\begin{aligned} \mathbf{c}_{\lambda} &= u_{-\lambda}^{*}\mathbf{b}_{\lambda} + v_{-\lambda}\mathbf{b}_{-\lambda}^{\dagger} \\ \mathbf{c}_{-\lambda} &= u_{\lambda}\mathbf{b}_{-\lambda}^{\dagger} + v_{\lambda}^{*}\mathbf{b}_{\lambda} \end{aligned} \tag{16.120}$$

The complex coefficients in this transformation must now be suitably chosen so that not only are the commutation relations $[\mathbf{c}_i, \mathbf{c}_j^{\dagger}] = \delta_{ij}$ satisfied, but also we should have

$$\begin{aligned} i\hbar\frac{d\mathbf{c}_{\lambda}}{dt} &= \hbar\omega_{\lambda}\mathbf{c}_{\lambda} \\ i\hbar\frac{d\mathbf{c}_{-\lambda}}{dt} &= \hbar\omega_{\lambda}\mathbf{c}_{-\lambda} \end{aligned} \tag{16.121}$$

This means that the nondiagonal terms in the Hamiltonian must vanish. Assuming u_{λ} real and $(u_{\lambda}^2 - |v_{\lambda}|^2) = 1$, we see that the proper commutation relations are satisfied. Let us, therefore, write u_{λ} and v_{λ} in terms of two new variables, μ_λ and ν_λ, given by

$$u_\lambda = \cosh \mu_\lambda$$
$$v_\lambda = e^{i\nu_\lambda} \sinh \mu_\lambda \tag{16.122}$$

where μ_λ and ν_λ are to be chosen properly so that our bilinear Hamiltonian is diagonal. This is our final goal, that is, we must have

$$\mathcal{H} = \hbar\omega_\lambda \mathbf{c}_\lambda^\dagger \mathbf{c}_\lambda + \hbar\omega_\lambda \mathbf{c}_{-\lambda}^\dagger \mathbf{c}_{-\lambda}$$

Assuming harmonic time dependence of the **c**'s and using (16.120), and then equating the coefficients of $\mathbf{b}_\lambda$ and $\mathbf{b}_{-\lambda}^\dagger$, we get

$$(\hbar\omega_\lambda - A_\lambda)\cosh \mu_\lambda = -B_\lambda e^{i\nu_\lambda} \sinh \mu_\lambda \tag{16.123}$$

where we have used (16.122). We now note that the phase $e^{-2i\phi_\lambda}$ of B_λ in (16.112) must cancel $e^{i\nu_\lambda}$, in order to satisfy the imaginary part of the equation. We thus have $\nu_\lambda = 2\phi_\lambda$, whereas the real part of (16.123) determines μ_λ and is given by

$$\tanh \mu_\lambda = \frac{A_\lambda - \hbar\omega_\lambda}{|B_\lambda|}$$

which can be recast into a more convenient form by eliminating $\hbar\omega_\lambda$, using (16.118). Thus

$$\tanh 2\mu_\lambda = \frac{|B_\lambda|}{A_\lambda} \tag{16.124}$$

We have now derived all the important and necessary equations, so that, by using these, the Hamiltonian can be brought to the required diagonal form. It can be shown with the help of a little algebra that our Hamiltonian (16.113) finally reduces to the following form:

$$\mathcal{H}_\lambda = \mathcal{C} + \frac{1}{2}\sum_\lambda \left[(A_\lambda^2 - |B_\lambda|^2)^{1/2} - A_\lambda \right] + \sum_\lambda (A_\lambda^2 - |B_\lambda|^2)^{1/2} \mathbf{c}_\lambda^\dagger \mathbf{c}_\lambda \tag{16.125}$$

as a result of the Bogolieubov transformation. In (16.125) the sum over λ goes over the whole of $\mathbf{K}_\lambda$-space, as in (16.113).

The eigenvalues E of $\mathcal{H}_\lambda$ are now easy to write. The eigenvalues N_λ of $\mathbf{c}_\lambda^\dagger \mathbf{c}_\lambda$ are $0, 1, 2, 3, \ldots$. Hence

$$E = E_{N_\lambda} = \mathcal{C} + \sum_\lambda \left[\tfrac{1}{2}(A_\lambda^2 - |B_\lambda|^2)^{1/2} - \tfrac{1}{2}A_\lambda \right] + \sum_\lambda (A_\lambda^2 - |B_\lambda|^2)^{1/2} N_\lambda \tag{16.126}$$

Once the Hamiltonian is diagonalized and the energies (16.126) are known, we can easily calculate the magnetization from the partition function Z, using

$$M=\frac{kT}{V}\frac{\partial}{\partial H}\log Z \tag{16.127}$$

where

$$Z=\sum e^{-E/kT}$$
$$=e^{\mathcal{C}/kT}\exp\left[\frac{-\sum_{\lambda}\{(A_{\lambda}^{2}-|B_{\lambda}|^{2})^{1/2}-A_{\lambda}\}}{2kT}\right]\exp\left[\frac{-\sum_{\lambda}(A_{\lambda}^{2}-|B_{\lambda}|^{2})^{1/2}N_{\lambda}}{kT}\right] \tag{16.128}$$

Now, writing

$$x_{\lambda}=-\frac{(A_{\lambda}^{2}-|B_{\lambda}|^{2})^{1/2}}{kT}$$

we see that

$$\sum_{N_{\lambda}}\exp\left(\sum_{\lambda}x_{\lambda}N_{\lambda}\right)=1+e^{x_{\lambda_1}N_{\lambda_1}}+e^{x_{\lambda_2}N_{\lambda_2}}+\cdots$$
$$=1+e^{x_{\lambda_1}}+(e^{x_{\lambda_1}})^{2}+(e^{x_{\lambda_1}})^{3}+\cdots$$
$$=(1-e^{x_{\lambda_1}})^{-1}$$

and similarly for the other exponential terms.

Thus

$$\sum_{N_{\lambda}}\exp\left(\sum_{\lambda}x_{\lambda}N_{\lambda}\right)=\prod_{\lambda}\left[1-\exp\left\{-\frac{(A_{\lambda}^{2}-|B_{\lambda}|^{2})^{1/2}}{kT}\right\}\right]^{-1} \tag{16.129}$$

Using (16.129), we thus have

$$Z=e^{-\mathcal{C}/kT}\exp\left[-\frac{\sum_{\lambda}\{(A_{\lambda}^{2}-|B_{\lambda}|^{2})^{1/2}-A_{\lambda}\}}{2kT}\right]$$
$$\cdot\prod_{\lambda}\left[1-\exp\left\{-\frac{(A_{\lambda}^{2}-|B_{\lambda}|^{2})^{1/2}}{kT}\right\}\right]^{-1}$$
$$=e^{-\mathcal{C}/kT}e^{-A}\prod_{\lambda}(1-e^{-B/kT})^{-1} \tag{16.130}$$

where

$$A = \frac{\left(A_\lambda^2 - |B_\lambda|^2\right)^{1/2} - A_\lambda}{2kT}$$

and

$$B = \left(A_\lambda^2 - |B_\lambda|^2\right)^{1/2}$$

From (16.130)

$$\log Z = -\frac{\mathcal{C}}{kT} - A - \prod_\lambda \log(1 - e^{-B/kT})$$

Therefore

$$\frac{\partial}{\partial H}\log Z = -\frac{1}{kT}\frac{\partial \mathcal{C}}{\partial H} - \frac{\partial A}{\partial H} - \prod_\lambda \frac{\partial}{\partial H}\log(1 - e^{-B/kT}) \qquad (16.131)$$

Now, from (16.98a), we have

$$\frac{\partial \mathcal{C}}{\partial H} = -2\mu_{\mathbf{B}} SN$$

and

$$\frac{\partial A}{\partial H} = \frac{\mu_{\mathbf{B}}}{kT}\left[\frac{A_\lambda}{\left(A_\lambda^2 - |B_\lambda|^2\right)^{1/2}} - 1\right]$$

$$\prod_\lambda \frac{\partial}{\partial H}\log(1 - e^{-B/kT}) = \frac{1}{e^{-B/kT} - 1}\cdot\frac{1}{kT}\cdot\frac{2\mu_{\mathbf{B}} A_\lambda}{\left(A_\lambda^2 - |B_\lambda|^2\right)^{1/2}}$$

Inserting (16.131) in (16.127), we get finally

$$\begin{aligned}\mathbf{M}(T,H) &= \frac{2\mu_B SN}{V} - \frac{G_x G_y G_z}{(2\pi)^3 V}\int \frac{A(\mathbf{K})}{P}\left(\frac{1}{e^{P/kT} - 1}\right) d\mathbf{K} \\ &\quad - \frac{G_x G_y G_z}{(2\pi)^3 V}\int\left[\frac{A(\mathbf{K})}{P} - 1\right] d\mathbf{K} \\ &= M_0 - M_T(T,H) - M_\beta(H)\end{aligned} \qquad (16.132)$$

where

$$\sum_\lambda f(\mathbf{K}_\lambda) = \lambda_x\lambda_y\lambda_z = \frac{G_x G_y G_z}{(2\pi)^3}\int f(\mathbf{K})\, d\mathbf{K}$$

$$P = \left[A(\mathbf{K})^2 - |B(\mathbf{K})|^2\right]^{1/2}$$

and we have also replaced A_λ, B_λ, and so on, by $A(\mathbf{K})$, $B(\mathbf{K})$, and so on, respectively. In (16.132), M_0 gives us the saturation magnetization (at $T=0$), which corresponds to the situation when all the spins are parallel to each other in the direction of the applied magnetic field; M_T is the deviation of M from M_0 arising because of the temperature, the magnetic field, and the dipole-dipole interaction; and M_β is the deviation of M from M_0 due to the dipole-dipole interaction and the magnetic field. The existence of M_β shows that, even at $T=0$, complete saturation is possible only for $H \gg 4\pi M_0$.

The famous $T^{3/2}$ law of Bloch is obtained from (16.132) by neglecting the magnetic interactions and by setting $H=0$. In this case

$$A(\mathbf{K}) = A(-\mathbf{K}) = 2SJK^2$$

and

$$B(\mathbf{K}) = B(-\mathbf{K}) = 0$$

Hence

$$M(T)^{\text{Bloch}} = M_0 - M_T^{\text{Bloch}} = M_0 - \text{const.} \int \frac{d\mathbf{K}}{e^{2SJK^2/kT} - 1} \tag{16.133}$$

Evaluating the integral in (16.133), we get

$$\begin{aligned} M(T)^{\text{Bloch}} &= M_0 - G_x G_y G_z V^{-1} 2\mu_B (2\pi)^{-3} \left(\frac{k}{2ST} \right)^{3/2} 2\pi \cdot 2 \cdot (1 \cdot 157) T^{3/2} \\ &= M_0 - \left[1 - \text{const. } T^{3/2} \right] \end{aligned} \tag{16.134}$$

where the integration is extended over all **K**-space. The exchange energy is given by

$$\sum_h 2SJ_h(\mathbf{R}_h)(1 - e^{i\mathbf{K}\cdot\mathbf{R}_h}) \simeq 2SJK^2 \tag{16.135}$$

This is permissible provided that $kT/2SJ \ll 1$, a condition that is not as well fulfilled as is desirable. For example, for iron at room temperature this ratio ≈ 0.3.

We do not intend to become involved in the evaluations of the other integrals appearing in (16.132) but would like to point out that the contribution of M_β is very small compared to the first two terms and hence can be neglected in future considerations. In evaluating the integral appearing in the second term of (16.132), it has been found, after rather lengthy manipulations, that the "intrinsic susceptibility," $\partial\mathbf{M}/\partial\mathbf{H}$, is given by

$$\chi=\frac{\partial M}{\partial H}=\left(\frac{M_0-M(T,0)}{M_0}\right)\left(\frac{1}{16\cdot 2^{1/2}\cdot 1.157}\right)\left(\frac{4\pi\mu_B M_0}{kT}\right)^{1/2}$$

$$\left[\left(\frac{4\pi M_0}{H}\right)^{1/2}+\sin^{-1}\left(\frac{4\pi M_0}{H+4\pi M_0}\right)^{1/2}\right] \qquad (16.136)$$

where the constant J has been eliminated by using the fact that its numerical value can be directly obtained from an experimental measurement of $[M_0-M(T,0)]$, instead of trying to calculate it from theoretical considerations—a rather difficult procedure. It is evident from (16.136) that χ has an inverse square-root dependence of H for $H\gg 4\pi M_0$, and this seems to be a general feature of the exchange interaction model subject to the approximations adopted here. Also, it should be noted that the susceptibility is proportional to the absolute temperature.

Let us now go to the experiments and see how the theoretical result (16.136) compares to the real situation. For iron and nickel at $T=287$ °K and $H=4000$ oersteds, (16.136) gives

$$\chi(\text{Fe})=1.7\times10^{-4}, \qquad \chi(\text{Ni})=1.2\times10^{-4} \qquad (16.137)$$

taking $[M_0-M(T,0)]/M_0$ and $4\pi M_0$ for iron to be 0.018 and 21,800, respectively. For nickel, they are 0.040 and 6400. Akulov's experimental χ values[9] for iron and nickel are 1.5×10^{-5} and 1.2×10^{-5}, respectively. The discrepancy is thus of the order of a factor of 10. Akulov's χ values are also independent of H. We thus come to the conclusion that the agreement with experiment is not good, though the magnetic dipole-dipole interactions are taken into consideration in addition to the exchange interaction. We have, of course, already discussed in Chapter 13 on molecular fields the fact that the magnetic dipolar interactions are not significant compared to the other interactions. The importance of Holstein and Primakoff's work lies in the fact that they were the first to show how the spin deviations can be written in terms of the creation and annihilation operators. These transformations ultimately diagonalize a complicated Hamiltonian like (16.93). It should also be noted that Bloch's $T^{3/2}$ law can easily be obtained by switching off the magnetic dipole-dipole interactions.

In the next section we discuss the spin-wave theory, which deals effectively with the elementary excitations in magnetic systems.

16.4 SPIN WAVES IN MAGNETIC SOLIDS

In solid state physics the concept of spin waves was established by Bloch[6] in 1930. Bloch's ideas remained of only theoretical interest for a long time, until experimentors, as a result of their investigations, demonstrated unmistakably that spin waves exist in magnetic materials. It is a common experimental observation that the magnetization M of a ferromagnet decreases when its temperature is raised from absolute zero. Bloch hit upon the spin-wave theory in order to explain this experimental observation. At absolute zero the ionic magnetic moments are all oriented parallel to the applied magnetic field. Now, if the temperature is raised from absolute zero, it is expected that there will be some misalignment due to the thermal activation, as a result of which the total magnetization will decrease from its value at absolute zero. The decrease in magnetization at very low temperatures can be represented by the simple relation

$$M = M_0 \cdot (1 - \alpha T^n) \tag{16.138}$$

where M_0 is the magnetization at 0 °K, and $n = \frac{3}{2}$. This is known as the famous $T^{3/2}$ law of Bloch [see (16.134)].

The misalignment of the spins (or the magnetic moments) due to the thermal disturbance is not a static matter, however, since the spins will continually be changing their orientations. This dynamic state of affairs gives rise to a complicated, haphazard pattern so far as the spin distributions are concerned. This complicated picture may be simplified by considering it as a superposition of a set of single misorientations, each of which travels from one ion to the next throughout the lattice. This propagation of a single misorientation through the lattice is known as a spin wave.

If the magnetic moment of an ion is due to one electron only, there are two possible values of S_z, that is, $\pm\frac{1}{2}\hbar$ $(S = \frac{1}{2}\hbar)$ corresponding to the up-spin and down-spin directions. The propagation of a spin wave will, in this case, be the propagation of a single spin reversal. If more than one electron per atom contributes to the magnetic moment, $S > \frac{1}{2}\hbar$ and consequently the spin wave will be the propagation of a spin deviation from the maximum value of S_z [see, e.g., (16.84)]. Since the spins are connected by the Heisenberg exchange interaction, any spin deviation may be considered to be associated with all the ions in the crystal, and it will form a collective excitation. It should also be noted that the spin waves are the lowest type of magnetic excitation. The spin wave may be considered as a quasi-particle, and, by analogy with phonons (quanta of lattice wave) and photons (quanta of light wave), it is often referred to as a magnon.

In the presence of an external magnetic field $\mathbf{H}$ in the z-direction, the Hamiltonian is written as

$$\mathcal{H} = -\sum_{l,m} J(\mathbf{l}-\mathbf{m})\mathbf{S}_l \cdot \mathbf{S}_m - g\mu_B H \sum_l S_l^z \tag{16.139}$$

where $\mathbf{S}_l$ is the spin at the lattice site $\mathbf{l}$ of magnitude S in units of $\hbar$, g is the Landé g-factor, μ_B is the Bohr magneton, $J(\mathbf{l}-\mathbf{m})$ is the exchange integral, which is positive, and $J(\mathbf{l}-\mathbf{m})=J$ when $\mathbf{l}$ and $\mathbf{m}$ are nearest neighbors and zero otherwise. Using (16.92), we find that the Hamiltonian (16.139) is reduced to

$$\begin{aligned}\mathcal{H} = -J\sum_{l,m}\Bigg[& S\mathbf{a}_l^\dagger\left(1-\frac{\mathbf{a}_l^\dagger\mathbf{a}_l}{2S}\right)^{1/2}\left(1-\frac{\mathbf{a}_m^\dagger\mathbf{a}_m}{2S}\right)^{1/2}\mathbf{a}_m \\ & + S\left(1-\frac{\mathbf{a}_l^\dagger\mathbf{a}_l}{2S}\right)\mathbf{a}_l\mathbf{a}_m^\dagger\left(1-\frac{\mathbf{a}_m^\dagger\mathbf{a}_m}{2S}\right)^{1/2} \\ & + S^2 - S\mathbf{a}_l^\dagger\mathbf{a}_l - S\mathbf{a}_m^\dagger\mathbf{a}_m + \mathbf{a}_l^\dagger\mathbf{a}_l\mathbf{a}_m^\dagger\mathbf{a}_m\Bigg] \\ & - g\mu_B H\sum_l (S-\mathbf{a}_l^\dagger\mathbf{a}_l)\end{aligned} \tag{16.140}$$

As in the preceding section, we make the following assumption on the ground that the average value of the spin deviation is small compared to $2S$;

$$\frac{\langle\mathbf{a}_l^\dagger\mathbf{a}_l\rangle}{2S} \ll 1$$

so that

$$\left(\frac{1-\mathbf{a}_l^\dagger\mathbf{a}_l}{2S}\right)^{1/2} \simeq 1$$

We also neglect the terms $\mathbf{a}_l^\dagger\mathbf{a}_l\mathbf{a}_m^\dagger\mathbf{a}_m$ and thereby neglect the interaction of the spin waves (linear spin-wave theory).

As a result of the approximation given above, the Hamiltonian (16.140) simplifies to

$$\mathcal{H} = \mathcal{C} - J\sum_{l,m} S(\mathbf{a}_l^\dagger\mathbf{a}_m + \mathbf{a}_l\mathbf{a}_m^\dagger - \mathbf{a}_l^\dagger\mathbf{a}_l - \mathbf{a}_m^\dagger\mathbf{a}_m) + g\mu_B H\sum_l \mathbf{a}_l^\dagger\mathbf{a}_l \tag{16.141}$$

where

$$\mathcal{C} = -\tfrac{1}{2}JNzS^2 - g\mu_B HNS$$

Here z is the number of nearest neighbors, N is the total number of spins, and $\mathcal{C}$ obviously gives the ground state energy in the external field. As discussed in Section 16.3, the spin deviations are not localized to a particular lattice site but are propagated throughout the lattice. This is best seen by introducing the spin-wave variables by the Fourier transforms (16.99), as is possible because of the translational invariance of the lattice. Using the Fourier transforms (16.99), we find that the Hamiltonian (16.141) takes the form

$$\mathcal{H} = \mathcal{C} + \sum_{\mathbf{K}} \left[2JzS(1-\gamma_{\mathbf{K}}) + g\mu_{\mathrm{B}}H \right] \mathbf{a}_{\mathbf{K}}^{\dagger}\mathbf{a}_{\mathbf{K}} \tag{16.142}$$

where

$$\gamma_{\mathbf{K}} = \frac{1}{z} \sum_{\mathbf{K}} e^{i\mathbf{K}(l-m)} \tag{16.143}$$

The eigenvalues of (16.142) are obviously

$$E_{n_K} = \mathcal{C} + \sum_{K} \omega_{\mathbf{K}} n_{\mathbf{K}} \tag{16.144}$$

where

$$\omega_{\mathbf{K}} = 2JzS(1-\gamma_{\mathbf{K}}) + g\mu_{\mathrm{B}}H \tag{16.145}$$

Equation (16.144) gives the dispersion relation for magnons. In the long-wavelength approximation, which is valid at very low temperatures, we have

$$\mathbf{K} \cdot \mathbf{h} \ll 1$$

so that

$$\begin{aligned} z(1-\gamma_K) &= \frac{1}{2} \sum_h (\mathbf{K} \cdot \mathbf{h})^2 \qquad \text{up to the 2nd power in } \mathbf{K} \\ &= \frac{1}{2} \sum_h \mathbf{K}^2 h^2 \cos^2\theta_{\mathbf{K},h} \\ &= \tfrac{1}{2}\mathbf{K}^2 a^2 \end{aligned} \tag{16.146}$$

where

$$\cos^2\theta_{\mathbf{K},h} = \tfrac{1}{3} \qquad \text{(the mean value)}$$

and

$$\sum_h h^2 = 6a^2$$

a being the lattice constant. Using (16.146) in (16.145), we have

$$\omega_{\mathbf{K}} = 2JS\mathbf{K}^2 a^2 + g\mu_{\mathrm{B}}H \tag{16.147}$$

We now calculate the spontaneous magnetization for $H=0$. At absolute zero all the spins are in the z-direction, so the magnetization $M_S(0)$ per site is $g\mu_B$. At higher temperature the magnetization $M_S(T)$ is given by the Bose distribution function, that is,

$$M_S(0)-M_S(T)=g\mu_B\sum_{\mathbf{K}}\langle n_{\mathbf{K}}\rangle$$

$$=g\mu_B\sum_{\mathbf{K}}\frac{1}{e^{\omega_{\mathbf{K}}/kT}-1} \tag{16.148}$$

where k is the Boltzmann constant. Since most of the contribution comes from low values of $\mathbf{K}$, the sum in (16.148) can be replaced by an integral over the whole of $\mathbf{K}$-space. Thus

$$M_S(0)-M_S(T)=g\mu_B\cdot\frac{v}{(2\pi)^3}\int_{-\infty}^{\infty}\frac{d^3\mathbf{K}}{e^{D\mathbf{K}^2/kT}-1} \tag{16.149}$$

where

$$D=2SJa^2$$

and

$$v=\begin{cases} a^3 & \text{for a simple cubic (s.c.) lattice} \\ \frac{1}{2}a^3 & \text{for a body-centered cubic (b.c.c.) lattice} \\ \frac{1}{4}a^3 & \text{for a face-centered cubic (f.c.c) lattice} \end{cases}$$

v being the volume of the primitive cell. Performing the integration in (16.149), we get

$$M_S(T)=M_S(0)\left[1-0.1187\alpha\left(\frac{kT}{2SJ}\right)^{3/2}\right] \tag{16.150}$$

where $\alpha=1$, $\frac{1}{2}$, and $\frac{1}{4}$ for s.c., b.c.c, and f.c.c. lattices, respectively. This the $T^{3/2}$ law of Bloch.

The most important result of the linear spin-wave theory described above, as applied to a ferromagnet that has a Hamiltonian of the ordinary exchange type, is that the energy of a spin wave of wave vector $\mathbf{K}$ is $E_{\mathbf{K}}=DK^2$, in the limit of small $\mathbf{K}$. This energy is assumed to be the same for all spin waves, no matter how many are present. It is important to note, however, that the spin-wave states are no longer eigenstates of the Hamiltonian when more than one spin is reversed. Various theoretical treatments have evolved in order to determine the limitations of the linear spin-wave approximation; of these, the Holstein-Primakoff approach has been studied in detail.

In connection with the H-P approach, we were forced to use quite a few approximations in order to diagonalize the Hamiltonian, as a result of which there was poor agreement with experiment. To develop the theory properly, the above calculation should be carried through using the complete expression

$$\left(1-\frac{\mathbf{a}_i^\dagger \mathbf{a}_i}{2S}\right)^{1/2}$$

rather than just the first term, in the expansion. If this could be done exactly, the effect of the interactions between the spin waves would be properly taken into account. Unfortunately, mathematical difficulties occur, and not all of the various attempts agree.

It is more convenient, therefore, to take account of the more than one spin deviation that will be present above 0 °K by using a linear theory in which $\omega_{\mathbf{K}}$ is temperature dependent. This will lead to correction terms in the expression for magnetization. The spin-wave energy must therefore be renormalized to take account of the interactions between the spin waves. This problem has been resolved by Dyson[10] in an extensive and rigorous treatment of the localized Heisenberg model. He concludes that the linear theory is a good approximation up to temperatures of the order of one third of the Curie temperature, and the spin-wave interactions may now be taken into account if the expression for $\omega_{\mathbf{K}}$ includes a temperature-dependent correction term.

Let us discuss Dyson's theory very briefly. Generalizing Bloch's definition, Dyson defined a spin-wave state with more than one spin wave. Let $|a\rangle$ denote any set of nonnegative integers a_λ, one attached to each reciprocal lattice vector λ. Then the spin-wave state $|a\rangle$ containing a_λ spin waves with wave vectors λ is defined by

$$|a\rangle=\prod_{\lambda}\left[(2S)^{-(1/2)a_\lambda}(a_\lambda!)^{-1/2}(\mathbf{S}_\lambda^+)^{a_\lambda}\right]|0\rangle \tag{16.151}$$

The order of factors in this product is immaterial since they all commute. The states are neither orthogonal nor normal, and the number of states (which is infinite) is clearly redundant since there are only $(2S+1)^N$ independent states for the system. However, the redundancy removes the mathematical ambiguity of Holstein and Primakoff. When $\Sigma_\lambda a_\lambda \ll N$, the states are approximately orthonormal.

The nonorthogonality of states produces an interaction between the spin waves that we call the kinematical interaction. The physical reason for this is that a spin reversal greater than $2S$ cannot be attached to any site. The result is a statistical hindrance to any dense packing of many spin waves within a limited volume.

There is another spin-wave interaction, which arises from the fact that the Hamiltonian (16.139) is not diagonal in the states (16.151), that is,

$$\mathcal{H}|a\rangle = \sum_b Q_{ab}|b\rangle \tag{16.152}$$

This interaction Q_{ab} is called the dynamical interaction.

Introducing an ideal spin-wave model (for which $S=\infty$, so that the states are orthonormal) and an effective Hamiltonian for it, one can relate the behavior of the real spin system to that of the ideal spin system. Thus Dyson separated the two types of interactions and showed that the kinematical interaction is not important because of the cancellation of a term from this interaction by a term from the dynamical interaction. As a result, he obtained a term proportional to T^4 and due to the spin-wave interaction, besides the terms of half-odd integral powers of T. This fact raised the validity of the spin-wave theory to $\frac{1}{4}T_C$. For future comparison we write his results fully:

$$M(T) = \left(\frac{m}{S}\right)\left(S - a_0\theta^{3/2} - a_1\theta^{5/2} - a_2\theta^{7/2} - a_3 S^{-1}\theta^4\right) + O(T^{9/2}) \tag{16.153}$$

Here m is the moment per site, equal to $g\mu_B S$, and

$$a_0 = \zeta\left(\tfrac{3}{2}\right)$$
$$a_1 = \tfrac{3}{4}\pi\nu\zeta\left(\tfrac{5}{2}\right)$$
$$a_2 = \pi^2\omega\nu^2\zeta\left(\tfrac{7}{2}\right)$$
$$a_3 = \tfrac{3}{2}\pi\nu\zeta\left(\tfrac{3}{2}\right)\zeta\left(\tfrac{5}{2}\right)Q$$

where

$$\nu = \left\{\begin{matrix} 1 \\ \frac{3}{4}\,2^{2/3} \\ 2^{1/3} \end{matrix}\right\} \quad \text{and} \quad \omega = \left\{\begin{matrix} 33/32 \\ 281/288 \\ 15/16 \end{matrix}\right\}$$

$$Q = 1 + \tfrac{4}{3}(GS-1)^{-1} + \alpha(3S)^{-1}$$

$$G = \left\{\begin{matrix} 10 \\ 16 \\ 24 \end{matrix}\right\}, \qquad \alpha = \left\{\begin{matrix} 0.53 \\ 0.39 \\ 0.34 \end{matrix}\right\}$$

for the s.c., b.c.c., and f.c.c. lattices, respectively.

For an s.c. lattice

$$a_3 = \tfrac{3}{2}\pi\nu\zeta\left(\tfrac{3}{2}\right)\zeta\left(\tfrac{5}{2}\right)\left[1 + \frac{0.31}{5} + O\left(\frac{1}{S^2}\right)\right]$$

Here $\zeta(n)$ is the well-known Riemann zeta function, and

$$\theta = \frac{3kT}{4\pi JzvS} = \frac{T}{2\pi T_{\mathrm{C}}} \tag{16.154}$$

where

$$T_{\mathrm{C}} = \frac{2}{3}\frac{JzvS}{k}$$

Later Oguchi[11] derived a similar result by taking into account the latter terms in the Holstein-Primakoff expansion; we do not describe Oguchi's work.

16.4.1 Spin Waves in Antiferromagnets

For an antiferromagnetic sample, let us consider the Hamiltonian

$$\mathcal{H} = J\sum_{j,\delta} \mathbf{S}_j \cdot \mathbf{S}_{j+\delta} - 2\mu_0 H_A \sum_j S_{jz}^A + 2\mu_0 H_A \sum_j S_{jz}^B \tag{16.155}$$

where J is the nearest-neighbor exchange integral and is negative for an antiferromagnet, and $\mu_0 = g\mu_{\mathrm{B}}S$. We neglect the effect of the next-nearest-neighbor interactions, although they may be important in real antiferromagnets. Exactly as in the molecular field approximation for antiferromagnets described in Chapter 14, we also assume here that the spin structure of the crystal may be divided into two interpenetrating sublattices, A and B, with the property that all the nearest neighbors of an atom A lie on B, and vice versa. This simple two-sublattice model does not conform to all types of antiferromagnetic structures, however, because the spin structures in antiferromagnetics may be very complicated. As discussed earlier, for an antiferromagnetic material having a f.c.c. structure, a minimum of *four* sublattices is required to describe the spin structure. The quantity H_A is positive and is a fictitious magnetic field, which approximates the effect of the crystal anisotropy energy, with the property of tending, for positive μ_0, to align the spins on A sublattice in the $+z$-direction and those on B in the $-z$-direction. We thus introduce H_A mainly to stabilize the spin arrays along a preferred axis, the z-axis.

Using the Holstein-Primakoff transformations described earlier, we transform the Hamiltonian (16.155) to the magnon variables:

$$\mathcal{H} = -2NzJS^2 - 4N\mu_0 H_A S + \mathcal{H}_0 + \mathcal{H}_1 \tag{16.156}$$

where the term bilinear in magnon variables is

$$\mathcal{H}_0 = 2JzS\sum_{\mathbf{K}}\left[\gamma_{\mathbf{K}}(c_{\mathbf{K}}^{\dagger}d_{\mathbf{K}}^{\dagger} + c_{\mathbf{K}}d_{\mathbf{K}}) + (c_{\mathbf{K}}^{\dagger}c_{\mathbf{K}} + d_{\mathbf{K}}^{\dagger}d_{\mathbf{K}})\right]$$

$$+ 2\mu_0 H_A \sum_{\mathbf{K}} (c_{\mathbf{K}}^{\dagger}c_{\mathbf{K}} + d_{\mathbf{K}}^{\dagger}d_{\mathbf{K}}) \tag{16.157}$$

with

$$\gamma_{\mathbf{K}} = \frac{1}{z} \sum_{\boldsymbol{\delta}} e^{i\mathbf{K}\cdot\boldsymbol{\delta}} = \gamma_{-\mathbf{K}}$$

where we assume a center of symmetry. The term $\mathcal{H}_1$ in (16.156) contains fourth and higher order terms in magnon operators, and it may be neglected when the excitation is low.

We now look for a transformation that will diagonalize $\mathcal{H}_0$. This can be done by using the Bogolieubov transformation, which we described in detail earlier. After diagonalization the magnon frequencies $\omega_{\mathbf{K}}$ are given by

$$\omega_{\mathbf{K}}^2 = (\omega_0 + \omega_A)^2 - \omega_0^2 \gamma_{\mathbf{K}}^2 \tag{16.157a}$$

where

$$\omega_0 = 2JzS \qquad \text{and} \qquad \omega_A = 2\mu_0 H_A$$

Equation (16.157a) gives the dispersion relation for the antiferromagnetic magnons. Using these, we can write the bilinear Hamiltonian (16.157), with $\omega_{\mathbf{K}}$ positive, as

$$\mathcal{H}_0 = -N(\omega_0 + \omega_A) + \sum_{\mathbf{K}} \omega_{\mathbf{K}} \left(\alpha_{\mathbf{K}}^\dagger \alpha_{\mathbf{K}} + \beta_{\mathbf{K}}^\dagger \beta_{\mathbf{K}} + 1 \right) \tag{16.158}$$

There are thus two degenerate modes for each **K**, one associated with the α operators and the other with the β operators. The total Hamiltonian (16.156) is therefore

$$\mathcal{H} = -2NzJS(S+1) - 4\pi\mu_0 H_A \left(S + \tfrac{1}{2}\right) + \sum_{\mathbf{K}} \omega_{\mathbf{K}} \left(n_{\mathbf{K}} + \tfrac{1}{2}\right) + \mathcal{H}_1$$

where each value of **K** is to be counted twice because of the double degeneracy, and $n_{\mathbf{K}}$ is a positive integer.

If we neglect ω_A and assume that $Ka \ll 1$, then $(1-\gamma_{\mathbf{K}}^2)^{1/2} \simeq 3^{-1/2} Ka$, for a simple cubic lattice, and we have finally

$$\omega_{\mathbf{K}} \simeq 4(3)^{1/2} JSKa \tag{16.159}$$

This is the dispersion law for antiferromagnets in the long-wavelength limit, provided that $\omega_{\mathbf{K}}/\omega_0 \gg 1$.

We shall now describe Anderson's work[12] on the determination of the ground state energy and wavefunction of an antiferromagnet, using the spin-wave theory and following the work of Klein and Smith.[13] This method is important in the sense that it has been applied for general lattices like linear, square, and also simple cubic, and for arbitrary spin quantum number S. The results obtained should be valid to the order $1/S$ or perhaps even $1/zS$, where z is the number of nearest neighbors. The ground state energy in all cases lies between the rigorous limits that have been derived on the variational principle,[15] $-\frac{1}{2}NJz(S^2)$ and $-\frac{1}{2}NJz(S^2)[1+(1/zS)]$. For the linear chain with $S=\frac{1}{2}$ there is good agreement with the rigorous result of Bethe.[16]

The Heisenberg Hamiltonian for an antiferromagnet is taken to be

$$\mathcal{H}=J\sum_{\langle j,k\rangle}\mathbf{S}_j\cdot\mathbf{S}_k \tag{16.160}$$

where J is the exchange integral, $\Sigma_{\langle j,k\rangle}$ denotes a sum over all pairs of j and k of the neighboring atoms, and j and k each represent a set of D numbers, where D is the dimensionality—in other words, j and k are the D-dimensional vectors. We are concerned with the nearest-neighbor interaction and also with lattices that can be divided into two interpenetrant sublattices with the neighbors of one all on the other, and vice versa. Although we are imposing all these restrictions, we should bear in mind that the procedure is perfectly general.

As already discussed in great detail in Chapter 13 on molecular field theory, the basic assumption in deriving the spin waves is that the state of an antiferromagnet is not greatly different from the classical ground state, in which all the spins of one sublattice point in one direction, say the $+z$-direction, whereas all the spins of the other sublattices are in the opposite direction. This is, of course, an assumption that cannot be justified by direct reasoning but will be indirectly justified by the results of the theory. We therefore assume that

$$S_{zj}\sim+S,\qquad S_{zk}\sim-S \tag{16.161}$$

Henceforth, let us assume that the atoms labeled j are always on the sublattice A, say, while those labeled k are on the sublattice B. For a linear chain lattice, for example, this would mean that j is odd and k is even. Now

$$S_z^2=S_c^2-\left(S_x^2+S_y^2\right) \tag{16.162}$$

where S_c is the classical total spin of an atom with spin quantum number S;

$$S_c=\left[S(S+1)\right]^{1/2} \tag{16.163}$$

Assuming that (16.161) is correct, we have from (16.162)

$$S_{zj} = \left[S_c^2 - \left(S_{xj}^2 + S_{yj}^2 \right) \right]^{1/2}$$
$$= S_c \left(1 - \frac{S_{xj}^2 + S_{yj}^2}{S_c^2} \right)^{1/2}$$
$$\approx S_c - \frac{S_{xj}^2 + S_{yj}^2}{2S_c}$$

and (16.164)

$$S_{zk} \approx -S_c + \frac{S_{xk}^2 + S_{yk}^2}{2S_c}$$

after the binomial expansion. Inserting (16.164) in (16.160), we then have

$$\mathcal{H} = -\tfrac{1}{2} JNzS_c^2 + \tfrac{1}{2} Jz \left\{ \sum_j \left(S_{xj}^2 + S_{yj}^2 \right) + \sum_k \left(S_{xk}^2 + S_{yk}^2 \right) \right\}$$
$$+ J \sum_{\langle j,k \rangle} \left(S_{xj} S_{xk} + S_{yj} S_{yk} \right) \tag{16.165}$$

Expression (16.165) is of course valid in the first order of smallness of S_x^2 and S_y^2. Let us introduce two sets of spin waves, one pair for each sublattice:

$$\left.\begin{aligned} \mathbf{S}_{xj} &= \left(\frac{2S}{N} \right)^{1/2} \sum_{\boldsymbol{\lambda}} e^{i\boldsymbol{\lambda}\cdot\mathbf{j}} Q_{\boldsymbol{\lambda}} \\ \mathbf{S}_{yj} &= \left(\frac{2S}{N} \right)^{1/2} \sum_{\boldsymbol{\lambda}} e^{-i\boldsymbol{\lambda}\cdot\mathbf{j}} P_{\boldsymbol{\lambda}} \\ \mathbf{S}_{xk} &= \left(\frac{2S}{N} \right)^{1/2} \sum_{\boldsymbol{\lambda}} e^{-i\boldsymbol{\lambda}\cdot\mathbf{k}} R_{\boldsymbol{\lambda}} \\ \mathbf{S}_{yk} &= \left(\frac{2S}{N} \right)^{1/2} \sum_{\boldsymbol{\lambda}} e^{i\boldsymbol{\lambda}\cdot\mathbf{k}} S_{\boldsymbol{\lambda}} \end{aligned}\right\} \tag{16.166}$$

where the wave number $\boldsymbol{\lambda}$ runs only over $N/2$ values from $-\pi$ to π, giving $2N$ coordinates in all. For example, for a linear chain of length N,

$$\boldsymbol{\lambda} = \frac{2\pi n}{N}, \qquad n = -\tfrac{1}{2}N + 2, \ldots, -2, 0, 2, \ldots, \tfrac{1}{2}N.$$

Now, substituting (16.166) into the Hamiltonian (16.165), we obtain, after some lengthy but straightforward algebra,

$$\mathcal{H} = -DJNS_c^2 + DJS\sum_{\lambda}\Big[(2N_{1\lambda}+1)(1-\gamma_\lambda^2)^{1/2} + (2n_{2\lambda}+1)(1-\gamma_\lambda^2)^{1/2}\Big] \tag{16.167}$$

For the ground state all $n_\lambda = 0$; therefore

$$E_g = -DJNS_c^2 + 2DJS\sum_{\lambda}(1-\gamma_\lambda^2)^{1/2} \tag{16.168}$$

It is evident from (16.167) that the frequencies of the spin waves fall into two identical branches:

$$\omega = DJS(1-\gamma_\lambda^2)^{1/2} \tag{16.169}$$

Thus

$$\omega_\lambda \sim DSJ\boldsymbol{\lambda}, \qquad \boldsymbol{\lambda}\to 0 \tag{16.170}$$

since, for $\boldsymbol{\lambda}\to 0$,

$$\gamma_\lambda \sim 1 - \tfrac{1}{2}\boldsymbol{\lambda}^2$$

As one finds, this dispersion law is quite different from the ferromagnetic case, where $\omega \sim \boldsymbol{\lambda}^2$. The decrease in the magnetization of a sublattice and the specific heat will vary as T^3 at low temperatures, if this dispersion law is correct, in contrast to the $T^{3/2}$ variation of a ferromagnet. This should be observable by experiment.

The energy of the ground state has been calculated from (16.168) by evaluating the sum over λ and is given by

$$E_g \simeq -\tfrac{1}{2}NzJS^2\left(1+\frac{1}{2Sz}\right) \tag{16.171}$$

For most known antiferromagnets this correction is fairly small.[15] The respective energies of the linear chain, the square lattice, and the simple cubic lattices are given by

$$\begin{aligned} E_g(D=1) &= -NJS^2\left(1+\frac{0.363}{S}\right) \\ E_g(D=2) &= -2NJS^2\left(1+\frac{0.158}{S}\right) \\ E_g(D=3) &= -3NJS^2\left(1+\frac{0.097}{S}\right) \end{aligned} \tag{16.172}$$

after evaluating the respective integrals.[12] It is interesting that $E_g(D=1)$ agrees nicely with the case $S=\frac{1}{2}$ for the linear chain computed by Bethe.[16] Unfortunately, no comparison can be made of (16.172) with experiments because the latter are not yet available. However, what is important here is that the apparent success of the spin-wave theory in describing the ground state of an antiferromagnet lends some confidence regarding its validity as a description of the higher lying states (the excited states) as well. We have, however, not gone into the details of Anderson's calculation, which the reader can find in Ref. 12.

Kubo[17] has proposed a method of calculation in which the spin-wave theory, in a modified form, is applied to antiferromagnets. The interaction between the different modes of the spin waves, which is due to higher order terms in the customary spin-wave treatment, has been taken into account by Kubo in a self-consistent manner. It has been shown that the interaction terms cause a change in the frequency spectrum of the spin waves which is equivalent to the existence of an anisotropic field.

Let us try to appreciate the difficulty encountered in the spin-wave theory because of the neglect of the interaction terms. This difficulty is directly revealed in the fact that the fluctuation of the magnetization (sublattice magnetization in the case of antiferromagnets) diverges abnormally for both ferromagnets and antiferromagnets and is of the order of $N^{4/D}$, where N is the total number of magnetic atoms or ions, and D the dimensionality of the lattice, provided that no anisotropic field is acting on the spins. This abnormal fluctuation might be thought to be correlated with the free rotation of the resultant spin moment in the absence of the anisotropic field. This view is inconsistent, however, with the usual picture of the spins, primarily aligned by strong exchange forces, giving rise to the resultant moment, which moves in a relatively weak anisotropic field and also in an external field, if any. This picture requires the resultant moment to be a well-defined quantity with a normal fluctuation of the order of N. The very fact that the spin-wave theory does not give such a picture constitutes the difficulty that the theory poses.

To remove this difficulty, Kubo has proposed a variational method. He chooses a system of spin waves with a certain frequency spectrum which represents the best approximation of the spin system. These waves are then determined in a self-consistent manner, taking into account the interaction terms. This variational method gives a finite frequency for the spin wave of infinite wave length in the absence of an anisotropic field. We shall not go into the details of this calculation.

16.4.2 Spin Waves in Ferrimagnets

Just like ferromagnets, ferrimagnetic materials also possess spin-wave excitations. Since the ferrimagnetic structures are much more complicated, there will be several spin-wave dispersion curves corresponding to the optical and the acoustic branches of the spectrum. At low temperatures, the spin waves of the lowest acoustic branch will be excited. We shall not go into the details of these waves, which have been discussed in an excellent article by Van Kranendonk and Van Vleck.[18] It is well known that the ferrimagnets are insulators, for example, yttrium iron garnet and Fe_3O_4, which are available as good single crystals. Many experiments have been done on them, and the results have been interpreted without much difficulty.

16.4.3 Spin Waves in Metals

Let us look at the spin waves in metals, the existence of which was shown by Herring and Kittel.[19] They demonstrated that the magnetization of the metal responded to a magnetic field in such a way as to indicate the possibility of spin-wave-like excitations. Theoretical investigations have been done by Izuyama,[20] Izuyama and Kubo,[21] and Mattis.[22] These investigations utilized Stoner's collective electron model, which we have described. We have already seen that the electrons are divided between "up" and "down" bands whose energies are shifted relative to one another by exchange interaction. The displacement of the two half-bands means that the wave vector for the electron in the highest filled state will be different for the up and the down spins (Figure 16.1). Let us consider the excitation of an electron from one band to another.

It is apparent from Figure 16.1 that, whereas an electron by reversing its spin can go from the down to the up band without any change of wave vector, the reverse transition takes place only when the wave vector changes by at least Δq_F. Both these processes are single-particle excitations, however. The excitation corresponding to a down→up transition with no change of wave vector is assumed to be a spin wave in the itinerant electron model. The wave vector of this spin wave is equal to **K**. The Hamiltonian has been assumed to be

$$\mathcal{H} = \sum_{q,\sigma} E_q \mathbf{C}^\dagger_{q,\sigma}\mathbf{C}_{q,\sigma} + \frac{1}{2}\sum_{\mathbf{K},q,\sigma,q',\sigma'} J(\mathbf{K})\mathbf{C}^\dagger_{q+\mathbf{K},\sigma}\mathbf{C}^\dagger_{q'-\mathbf{K},\sigma'}\mathbf{C}_{q'\sigma'}\mathbf{C}_{q,\sigma} \qquad (16.173)$$

where $\mathbf{C}^\dagger$ and $\mathbf{C}$ are the Fermi creation and annihilation operators for Bloch waves of wave vector $\mathbf{q}$ and spin $\boldsymbol{\sigma}$. The first term of (16.173) represents the kinetic energy, and the second term represents the interaction. $J(\mathbf{K})$ is the Fourier transform of the interaction potential. One now investigates the equation of motion of the quantity

$$\mathbf{S}^-_{\mathbf{q}}(\mathbf{K}) = \mathbf{C}^\dagger_{q+\mathbf{K},\uparrow}\mathbf{C}_{q,\downarrow} \qquad (16.174)$$

which excites an electron in the state q with down spin into a state $(q+K)$ with up spin. The equation of motion of $\mathbf{S}^-$ is given by

$$i\hbar\dot{\mathbf{S}}^-_{\mathbf{q}}(\mathbf{K}) = \left[S_{\mathbf{q}}(\mathbf{K}), \mathcal{H}\right] \qquad (16.175)$$

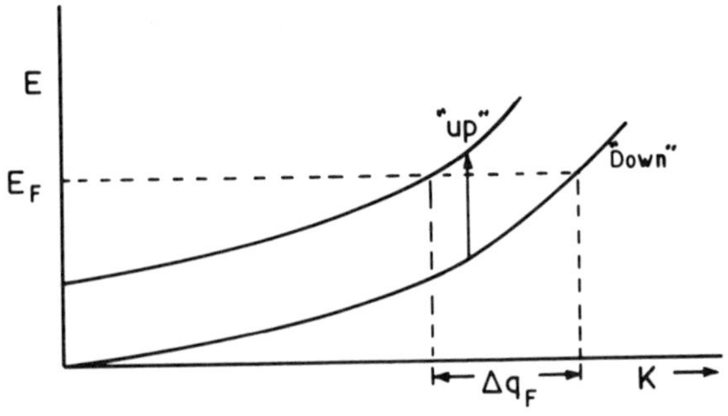

Fig. 16.1 Figure shows the displacement of the up electron band relative to the down electron band. Since both bands are filled to the same energy level, E_F, the Fermi energy, the transition from the down to up electron band can take place without a change of the wave vector, as shown by the arrow, whereas transitions from up to down must be accompanied by a change of the wave vector of at least Δq_F. (Taken from phillips and Rosenberg.[23])

We shall not go into the details of the calculation, which are given by Walker[24] and Mattis.[22] In the calculations of Izuyama[20] and Izuyama and Kubo[21] a Green function method was used.

The most important interactions that should be taken into account are the magnon-magnon, the magnon-electron, and the magnon-phonon interactions. Assuming the form of the Hamiltonian given by (16.173) and including these interactions, we can show that the spin-wave energy $E_{\mathbf{K}}$ at a temperature T is of the form

$$E_{\mathbf{K}}(T)=E_{\mathbf{K}}(0)+\sum_{\mathbf{K}'} J(\mathbf{K},\mathbf{K}')\langle n_{\mathbf{K}}'\rangle+\sum_{\mathbf{q}} J(\mathbf{K},\mathbf{q})\langle \delta N_{\mathbf{q}}\rangle+\sum_{\mathbf{l}} J(\mathbf{K},\mathbf{l})\langle \phi_{\mathbf{l}}\rangle \tag{16.176}$$

where $\mathbf{q}$ and $\mathbf{l}$ are the wave vectors of the electrons and phonons, respectively; $\delta N_{\mathbf{q}}$ is the deviation of the occupation number of the electrons from its value at absolute zero; $\phi_{\mathbf{l}}$ is the number of phonons of wave vector $\mathbf{l}$; and $\langle\ \rangle$ denotes a thermal average. We can assume that $E_{\mathbf{K}}\approx\mathbf{K}^2$ for small K and that the phonon energy is proportional to $\mathbf{l}$ for small $\mathbf{l}$. The interaction constant J will then be of the form

$$\begin{aligned} J(\mathbf{K},\mathbf{K}')&\simeq\text{const}\,\mathbf{K}^2\mathbf{K}'^2\\ J(\mathbf{K},\mathbf{q})&\simeq f(\mathbf{q})\mathbf{K}^2\\ J(\mathbf{K},\mathbf{l})&\simeq\text{const}\,\mathbf{K}^2|\mathbf{l}| \end{aligned} \tag{16.177}$$

where $f(\mathbf{q})$ is a smooth, arbitrary function of $\mathbf{q}$. Choosing appropriate distribution functions for $n_{\mathbf{K}}$, δN_q, and $\phi_{\mathbf{l}}$ and integrating over all $\mathbf{K}$, $\mathbf{q}$, and $\mathbf{l}$, respectively, we get

$$\frac{E_{\mathbf{K}}(T)}{\mathbf{K}^2} = D_0 + (\text{magnon-magnon})\, T^{5/2} + (\text{magnon-electron})\, T^2 + (\text{magnon-phonon})\, T^4 \tag{16.178}$$

We now go through the magnetization experiments that have a bearing on the spin-wave theory.[12] It is difficult to estimate the spontaneous magnetization of even a single crystal, since the magnetic domains existing in the sample reduce the macroscopic magnetization. The effect of the domains can be made nullified if the measurements are done in a high magnetic field. Unfortunately, this high field will itself tend to align the elementary dipoles so that the magnetization of the sample will be greater than the zero-field value, which should be compared with the theory. It is essential, therefore, that some extrapolation technique be employed to obtain the zero-field value of magnetization. It should be noted, however, that at low temperatures the variation of magnetization with field (irrespective of the strength of the field) is small and linear, so that the extrapolation to the zero-field value is quite straightforward and cannot lead to much inaccuracy. At high temperatures, on the other hand, especially near the Curie point, this problem is very serious. In the high-temperature ranges, therefore, one performs magnetocaloric experiments along with the magnetization measurements. For the details of experiment, the reader should consult Weiss and Forrer,[25] Oliver and Sucksmith,[26] and Foner.[27]

It has been found, from measurements on the variation of magnetization of Fe, Ni, and Co with temperature, that at very low temperatures the experimental data fit approximately a $(1-aT^{3/2})$ law predicted by Bloch, whereas at liquid air temperature the data fit a $(1-bT^2)$ curve quite well.[24] Since at very low temperatures the changes in magnetization are very small, it is difficult, however, to verify the $T^{3/2}$ relation convincingly. For rare earths also there is a good fit with the $T^{3/2}$ relation, though additional correction terms are necessary.

The experimental values on ^{59}Co powder[28] in the range 1 to 550 °K and those[29] on ^{57}Fe, obtained by the nuclear magnetic resonance technique, fit a relation having the form $(1-aT^{3/2}-bT^3)$. Assuming that the spin-wave theory is also applicable to nonmetallic ferrites in the long-wavelength region, one finds[30] that the magnetization fits a relation of the form $(1-aT^{3/2}-bT^{5/2})$ for yttrium iron garnet (YFeG) in the range 4.2 to 50 °K. In this connection the reader should also see Mercereau and Feynman[31] and Walker.[24]

In conclusion, it can be said, as a result of recent magnetization experiments, that the prediction of the spin-wave theory of magnetization of the form

$$M = M_0(1 - aT^{3/2}) + \text{correction terms} \tag{16.179}$$

holds good for many systems.

Let us now discuss the specific heat of the magnons. When the temperature of a ferromagnet is increased, there is excitation of the spin waves, as a result of which the magnetic contribution to specific heat also increases.

To derive an expression for the specific heat of magnons,[32] let us assume that $H=0$ and $ka \ll 1$; then, neglecting the magnon-magnon interactions as a first approximation, we have, as shown before,

$$\omega_{\mathbf{K}} = (2SJa^2)\mathbf{K}^2 \tag{16.180}$$

The internal energy per unit volume of the magnon gas in thermal equilibrium at temperature T is given by

$$\begin{aligned} U &= \sum_{\mathbf{K}} \omega_{\mathbf{K}} \langle n_{\mathbf{K}} \rangle_T \\ &= \frac{1}{(2\pi)^3} \int d^3\mathbf{K}\, D\mathbf{K}^2 \frac{1}{e^{D\mathbf{K}^2/kT} - 1} \end{aligned} \tag{16.181}$$

where $D = 2SJa^2$.

Performing the integration in (16.181) in a straightforward manner, we get the specific heat per unit volume of the magnons:

$$C_M = \frac{dU}{dT} = 0.113k\left(\frac{kT}{D}\right)^{3/2} \tag{16.182}$$

We thus see that the specific heat of magnons is proportional to $T^{3/2}$, whereas that of phonons[32] is proportional to T^3. Hence a plot of $CT^{-3/2}$ versus $T^{3/2}$ should be a straight line, with the intercept at $T=0$ giving the magnon contribution and the slope giving the phonon contribution. Figure 16.2 shows the specific heat of YFeG plotted as $CT^{-3/2}$ against $T^{3/2}$ to demonstrate the $T^{3/2}$ dependence of the spin-wave contribution to the specific heat. Thus, for the nonmetallic ferromagnets, C should be of the form

$$C = C_p T^3 + C_m T^{3/2} \tag{16.183}$$

where C_p is the specific heat for phonons, and C_m that for magnons. This relation was first demonstrated by Kouvel[33] from experimental measurements on magnetite in the temperature range 1 to 4 °K.

By virtue of the neutrons possessing a spin of $\frac{1}{2}$ and a magnetic moment, they will interact with the spins situated periodically on the lattice, as a result

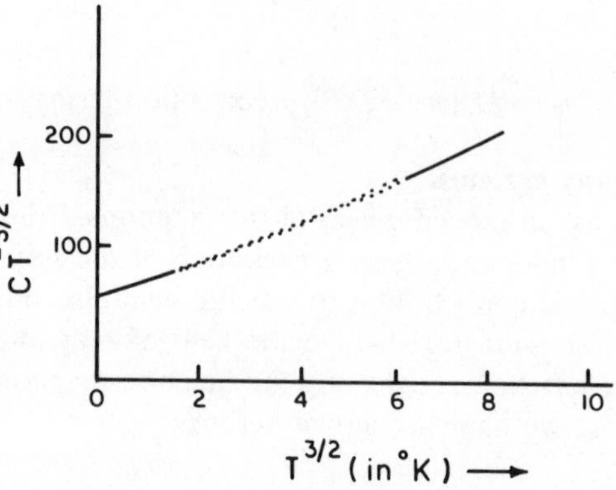

Fig. 16.2 $CT^{-3/2}$ (in erg cm^{-3} /degree$^{-5/2}$) versus $T^{3/2}$ for YFeG, showing the temperature dependence of the spin-wave contribution to the specific heat. (Taken from Shinozaki.[34])

of which the Bragg reflections are enhanced. Neutron scattering experiments provide vital information about the magnetic ordering of the lattice. There are two types of neutron scattering: elastic and inelastic. The elastic type of scattering preserves the energy, and the spin of the neutron remains unchanged, whereas in inelastic scattering the energy and spin are exchanged between the neutrons and ions in the lattice. It is important to note that only in inelastic scattering can one observe the effect of the spin waves. Lowde[35] has written a review on neutron scattering processes, which should be consulted for detailed information. Brockhouse[36] obtained the spin-wave dispersion curve from the inelastic neutron scattering on Fe_3O_4.

Spin waves are directly observed by the resonance method, in which they are selectively excited by radio-frequency resonant absorption. Kittel[37] first suggested such experiments, which were first performed by Seavey and Tannenwald.[38] The experimental arrangements are exactly similar to those for paramagnetic resonance. The sample is placed in a microwave cavity in such a position that the alternating homogeneous magnetic field, $h\,e^{i\omega t}$, is at right angles to an externally applied direct current (d.c.) field, **H**, assumed to be along the z-axis; $h\,e^{i\omega t}$ couples only the transverse components (x and y) of magnetization. As in paramagnetic resonance, the static field **H** causes the spins to precess about the z-axis with frequency γH_z, γ being the gyromagnetic ratio; and when this is equal to the radio frequency ω, energy is absorbed from the radio-frequency field, and consequently the precession angle increases. When all the spins precess in phase, the spin waves of the wave vector $\mathbf{K}=0$ become excited. This procedure[39] makes it possible to determine the dispersion relation for small **K**. This resonance method has to be used[40,41] to determine the temperature variation of the constant D. The experiments in Fe, Ni, and Co showed[34] that approximately a $T^{3/2}$ law is followed, which can be represented quite closely as T^2. From the experimental measurements on the transition metals one gets the impression that the itinerant model holds good.

In conclusion, we might comment that, in spite of a large amount of experimental and theoretical work, the magnetic properties of solids are by no means completely understood. In particular, very little is understood regarding the elementary excitations and the nature of the excited states in magnetic solids. Our understanding of the precise nature of the exchange interaction in metals is still incomplete, and there is thus much scope for research on the effect of temperature on the spin-wave interactions. More experiments are necessary to determine the effect of external perturbations on the exchange interactions.

REFERENCES

1. For the details of quantum field theory, see D. A. Kirzhnits, *Field Theoretical Methods in Many-Body Systems*, Pergamon Press, Oxford, 1967; English translation by A. J. Meadows.
2. A. A. Abrikosov, L. P. Gorkov, and I. E. Dzyaloshinski, *Methods of Quantum Field Theory in Statistical Physics*, Prentice-Hall, Englewood Cliffs, N.J., 1963; English translation by A. A. Silverman.
3. J. R. Schrieffer, *Theory of Superconductivity*, Benjamin, New York, 1964.
4. T. Holstein and H. Primakoff, *Phys. Rev.* **58**, 1098 (1940).
5. E. U. Condon and G. H. Shortley, *The theory of Atomic Spectra*, Cambridge University Press, London, 1953.
6. F. Bloch, *Phys* **61**, 206 (1930).
7. Von Chr. Möller, *Phys* **82**, 559 (1963).
8. Appendix II of Ref. 1.
9. See Ref. 18 of Ref. 1.
10. F. J. Dyson, *Phys. Rev.* **102**, 1217, 1230 (1956).
11. T. Oguchi, *Phys. Rev.* **117**, 117 (1960).
12. P. W. Anderson, *Phys. Rev.* **86**, 694 (1952).
13. M. J. Klein and R. S. Smith, *Phys. Rev.* **80**, 1111 (1951).
14. L. Hulthen, *Proc. Roy. Acad. Sci. Amsterdam* **39**, 190 (1936).
15. P. W. Anderson, *Phys. Rev.* **83**, 1260 (1951).
16. H. A. Bethe, *Z. Phys* **21**, 205 (1931).
17. R. Kubo, *Rev. Mod. Phys.* **25**, 344 (1953).
18. J. Van Kranendonk and J. H. Van Vleck, *Rev. Mod. Phys.* **30**, 1 (1958).
19. C. Herring and C. Kittel, *Phys. Rev.* **81**, 869 (1951).
20. T. Izuyama, *Prog. Theor. Phys. Japan* **23**, 969 (1960).
21. T. Izuyama and R. Kubo, *J. Appl. Phys.* **35**, 1074 (1964).
22. D. C. Mattis, *Phys. Rev.* **132**, 2521 (1963).
23. T. G. Phillips and H. M. Rosenberg, *Rep. Prog. Phys.* **29**, Part 1, 285 (1966).
24. L. R. Walker, *Magnetism*, Vol. 1, G. T. Rado and H. Suhl (Eds.), Academic Press, New York, 1963, p. 299.
25. P. Weiss and R. Forrer, *Ann. Phys. Paris* **5**, 153 (1926); *ibid.* **12**, 279 (1929).

26. D. J. Oliver and W. Sucksmith, *Proc. Roy. Soc. (London) A*, **219**, 1 (1953).
27. S. Foner, *J. Sci. Instrum.* **27**, 548 (1956).
28. V. Jaccarino, *Bull. Am. Phys. Soc.* **4**, 461 (1959).
29. I. J. Budnick, L. J. Bruner, R. J. Blume, and E. L. Boyd, *J. Appl. Phys.* **32**, 120S (1961).
30. I. H. Solt, *J. Appl. Phys.* **33**, 1189 (1962).
31. J. Mercereau and R. P. Feynman, *Phys. Rev.* **105**, 390 (1956).
32. C. Kittel, *Quantum Theory of Solids*, John Wiley and Sons, New York, 1963.
33. J. S. Kouvel, *Phys. Rev.* **102**, 1489 (1956).
34. S. S. Shinozaki, *Phys. Rev.* **122**, 388 (1961).
35. R. D. Lowde, *J. Appl. Phys.* **36**, 884 (1965).
36. B. N. Brockhouse, *Phys. Rev.* **106**, 859 (1957).
37. C. Kittel, *Phys. Rev.* **110**, 1295 (1958).
38. M. H. Seavey and P. E. Tannenwald, *Phys. Rev. Lett.* **1**, 168 (1958).
39. J. H. E. Griffiths, *Nature* **158**, 670 (1946).
40. P. E. Tannenwald and R. Weber, *J. Phys. Chem. Solids* **24**, 1357 (1964).
41. T. G. Philips and H. M. Rosenberg, *Proceedings of the International Conference on Magnetism, Nottingham*, Institute of Physics and Physical Society, London, 1964, p. 306.

17

Green Function Formalism

There is a great deal of similarity between the quantum field theory and the theories based on statistical mechanics as far as the many-body aspect is concerned. As we discussed earlier, the real many-body problem begins with a system of particles that interact with one another. Here we can no longer consider the particles to act independently but must take into account the enormously complicated influence that each particle has on the behavior of all the others. Almost all physical systems, except gases, are of this type —molecules in a liquid, electrons in a solid, protons and neutrons in a nucleus, and so on. The many-body problem is thus the study of the effects of interactions between the particles on the behavior of a many-particle system. For example, we are interested in the influence that the interactions have on the ground and the excited state energies, on the thermodynamic properties, and on the electrical and magnetic properties.

One of the most successful early methods of approaching the many-body problem, which is still in use today, is the canonical transformation techniques, which will be described shortly. This technique involves transforming the Schrödinger equation to a new set of coordinates in which the interaction term becomes small. The main difficulty with this approach is that it is not as systematic as one would like and therefore is difficult to apply. Because of this lack of a systematic method there was little advancement of the many-body theory up to the 1950s. The situation has changed greatly since then; by a series of very important papers it was shown that the methods of quantum field theory (described in Chapter 16), which was successfully applied to elementary particle physics, provide a very powerful and unified way of attacking the many-body problem.

The starting point for attacking the many-body problem is to express the Hamiltonian describing the system in terms of the creation and annihilation operators, which were discussed earlier. In both quantum field theory and statistical mechanics one is concerned with the averages of the quantum-mechanical operators, but in the quantum field theory one usually considers averages over the ground state of the system ($T=0$), whereas in statistical mechanics one is interested in the ensemble averages ($T\neq 0$).

Statistical mechanics is concerned with systems whose energy levels are very dense, so that the distance between the energy levels tends to zero as the volume tends to infinity. Hence the spectrum is practically continuous, and the perturbation energy is always larger than the distance between the levels. One should therefore use the perturbation theory for the continuous spectrum. There has been a great advancement in perturbation theory due to the use of the diagram technique involving connected diagrams only.[1–5]

In recent years the so-called Green functions or suitable modifications of these functions have been applied in quantum field theory to statistical problems. The Green functions are especially useful for summing over the restricted classes of perturbation theory diagrams, and are very powerful when combined with spectral representations, as we shall show. In quantum field theory the Green functions are the so-called propagators. This name is based on the idea that, in order to find the important physical properties of a system, it is essential to know, not the detailed behavior of each particle in the system, but rather just the average behavior of one or two typical particles. The quantities that describe this average behavior are called the one-particle and two-particle propagators, respectively.

Let us consider the one-particle propagator first. Obviously, it is impossible to describe in detail the enormously complicated motion of a particle moving from one point to another in an interacting system. Nevertheless, we can describe the motion in an "average" way by giving the probability (the probability amplitude) for the motion. Thus the one-particle propagator, $G(\mathbf{r}_2,t_2;\,\mathbf{r}_1,t_1)$, is defined as the probability (amplitude) that a single particle in a system will move or "propagate" to point $\mathbf{r}_2$ at time t_2 if it starts from point $\mathbf{r}_1$ at time t_1. The two-particle propagator is defined in a similar way. From these propagators one obtains the physical properties of the many-body system.

To describe the canonical transformation, discussed earlier, for solving the many-body problem, let us start from N noninteracting particles. If these particles have masses $m_1,m_2,\ldots,m_N$ and are placed in a time-independent external force field $F(\mathbf{r})$, with associated potential $V(\mathbf{r})$, our problem is to discover their behavior. When these particles are noninteracting, the Schrödinger equation for the N-particle system separates into the single-particle Schrödinger equations, given by

$$\mathcal{H}_i\phi_{\mathbf{K}_i}(\mathbf{r}_i)=E_{\mathbf{K}_i}\phi_{\mathbf{K}_i}(\mathbf{r}_i),\qquad i=1,\ldots,N \tag{17.1}$$

where

$$\mathcal{H}_i=\frac{\mathbf{p}_i^2}{2m}+V(\mathbf{r}_i)$$

The total energy is just the sum of the single-particle energies, $E=\Sigma_i E_{K_i}$. On the other hand, when the particles do interact with each other, we have to solve N coupled equations, given by

$$F(\mathbf{r}_i)+\sum_{j=1}^{N} F(\mathbf{r}_i,\mathbf{r}_j)=m_i\frac{d^2\mathbf{r}_i}{dt^2}, \qquad i=1,\ldots,N \tag{17.2}$$

where $F(\mathbf{r}_i,\mathbf{r}_j)$ denotes the interaction between two particles. The nonseparable Schrödinger equation is

$$\left[\sum_{i=1}^{N}\left\{\frac{\mathbf{p}_i^2}{2m}+V(\mathbf{r}_i)\right\}+\frac{1}{2}\sum_{\substack{i,j=1\\ i\neq j}}^{N} V(\mathbf{r}_i,\mathbf{r}_j)\right]\Psi(\mathbf{r}_1,\ldots,\mathbf{r}_N)=E\Psi(\mathbf{r}_1,\ldots,\mathbf{r}_N) \tag{17.3}$$

where $V(\mathbf{r}_i,\mathbf{r}_j)$ is the two-particle interaction potential. Let us now distinguish between the "weak" and "strong" interactions. An interaction is weak if it causes just a small perturbation to the solutions of the noninteracting case. This means that it can be handled by the ordinary finite-order perturbation theory. If the interaction cannot be handled in this way, it is referred to as strong. Most of the interactions we have to deal with, in the case of solids, are of the strong type. For example, the Coulomb interaction between two electrons in a metal has the form

$$V(\mathbf{r}_i,\mathbf{r}_j)=\frac{e^2}{|\mathbf{r}_i-\mathbf{r}_j|} \tag{17.4}$$

If the energy of the ground state of the system is calculated using (17.4), one obtains

$$E_0=E_0^{(0)}+E_0^{(1)}+\text{infinity} \tag{17.5}$$

that is, infinity to all orders of the perturbation theory above the first. This impasse is resolved by means of a transformation that changes to a new set of coordinates in which (17.2) becomes approximately decoupled. We do not go into the details of this transformation, which transforms a system of interacting real particles into a system of approximately noninteracting fictitious bodies; the reader is referred to the work of Ziman.[6]

Let us now consider the elementary excitations in solids, with which we shall be mostly concerned. To understand what an elementary excitation is and how it is related to the fictitious bodies, let us look at the quanta of vibration, the phonons. In connection with the quantization of harmonic oscillators, we have already seen that the quantized energy of a harmonic oscillator is given by

$$E_q'=\hbar\omega_q\left(n_q+\tfrac{1}{2}\right) \tag{17.6}$$

which can be regarded as a set of n_q quanta, each having energy $\hbar\omega_q$, together with a ground state of energy $\frac{1}{2}\hbar\omega_q$. These quanta of the sound wave are called phonons. Like photons (the quanta of light waves), they behave very much like particles, that is, particles in the quantum-mechanical sense. For a given n_q there is only one quantized sound wave of wave number q, but there are many (i.e., n_q) phonons of wave number q. Hence it is more appropriate to call a phonon a quantum or a particle of sound.

The energy $\hbar\omega_q$ is just the minimum unit of excitation energy above the zero-point energy, $\frac{1}{2}\hbar\omega_q$. Since the phonon carries this minimum unit, it is referred to as an elementary excitation. The compound excitations are divided[7] into two general categories: the "collective excitations" and the "quasi-particles," although this practice is not a universal one. For example, Ter Haar[8] prefers to refer to all elementary excitations as quasi-particles. The collective excitations are the quanta associated with collective motions of the macroscopic groups of particles in the system, that is, with motions of the system as a whole. The collective excitations show no resemblance to the real particles, whereas the quasi-particles resemble the real particles quite closely. As a single particle moves through the system, it pushes or pulls the particles in its neighborhood and thus becomes surrounded by a cloud of agitated particles. The real particle plus its cloud constitutes the quasi-particle. Since the particle cloud screens the real particle, thus greatly reducing the force field, a quasi-particle interacts only weakly with other quasi-particles and hence can be treated as approximately independent of them.

We should like to develop a systematic method of obtaining the energy of the ground state and the energies and lifetimes of the elementary excitations. As discussed earlier, the quantum field theory, which was previously restricted to elementary particle physics, provides just the method we are looking for. It gives a unified way of finding the ground state energy, the excitation energies, the lifetimes, and also other properties of the system.

Green functions or propagators play the most important part in the field-theoretic treatment of the many-body problem. There are different types of Green functions: one-particle, two-particle,...,n-particle, advanced, retarded, causal, zero-temperature, finite-temperature, real-time, imaginary-time, and so on. The Green functions enjoy popularity because they yield, in a direct way, the most important physical properties of a system, have a simple physical interpretation, and can be calculated in a systematic way.

Let us take the single-particle propagator $G(\mathbf{r}_2,t_2;\,\mathbf{r}_1,t_1)$, for example. This gives the probability amplitude that, if we put a particle into the interacting system at point $\mathbf{r}_1$ at time t_1 and let it collide with the other particles for a while (i.e., propagate through the system), it will be observed at point $\mathbf{r}_2$ at time t_2. The propagator G yields directly the energies and lifetimes of the quasi-particles. It also gives the momentum distribution, the spin, and the particle density and can be used as well to calculate the ground state energy. All these properties may be obtained at finite temperatures by using the finite-temperature version[9] of G.

The two-particle propagator G_2 gives the probability amplitude for observing one particle at $(\mathbf{r}_3, t_3)$ and another at $(\mathbf{r}_4, t_4)$ if the first particle was put into the system at $(\mathbf{r}_1, t_1)$ and the second at $(\mathbf{r}_2, t_2)$. Also, G_2 gives directly the energies and lifetimes of the collective excitations, as well as the magnetic susceptibility, the electrical conductivity, and a host of other nonequilibrium properties at all temperatures.

In addition, there is a quantity called the vacuum amplitude, which, though it plays a less central role in the many-body theory, is of considerable importance. The zero-temperature vacuum amplitude can be used to calculate the ground state energy; its finite-temperature version yields the grand partition function, from which all the equilibrium properties of the system may be determined.

Now the question is, How are the Green functions calculated? One method is to expand the propagator (or the vacuum amplitude) in an infinite perturbation series and evaluate the series approximately. The usual method of summing all terms up to the second or third order is not sufficient for the Green function since the series coverges very slowly. In some cases it may so happen that all the terms in the series are infinite, as discussed earlier [equation (17.5)]. One then has to sum over certain types of terms up to infinite order, a procedure known as selective summation. Of course, a new technique is required to carry out this selective summation in the infinite-order perturbation theory. This technique is well known as the method of Feynman diagrams,[10] which will not be discussed in this book.

In another method, the analytic method, one has to solve the coupled differential equations that the Green functions satisfy. This means that the single-particle propagator G obeys a differential equation involving the unknown two-particle propagator G_2, which again obeys an equation involving G_3, and so on. As a result one has to deal with an infinite hierarchy of coupled nonlinear differential equations. In reality, one has to decouple these coupled equations at some convenient stage by using a suitable cut-off and then solve the decoupled equations. We shall go through this procedure[11] later in this chapter.

17.1 SINGLE-PARTICLE GREEN FUNCTION[12]

In the method of quantum field theory, the one-particle Green function is one of the most important quantities characterizing the microscopic properties of a system. Before we consider the double-time temperature-dependent Green functions, which will be used for the magnetic properties of solids, it is essential to deal with the general properties of the Green function. We accomplish this by making use of the single-particle Green function. For simplicity, let us drop the spin indices, α and β, since, in the absence of ferromagnetism and of an external magnetic field, the single-particle Green functions $G_{\alpha\beta}$ must be of the form

$$G_{\alpha\beta} = G\delta_{\alpha\beta} \tag{17.7}$$

and here we consider only this case.

Let us introduce the single-particle Green function, and choose a metal as a model for our considerations. Let $|0\rangle$ denote the ground state of the metal. If we now place an electron in a Bloch state $\mathbf{K}$ at $t=0$, we produce the state

$$\mathbf{a}_{\mathbf{K}}^{\dagger}|0\rangle = |\mathbf{K},0\rangle \tag{17.8}$$

This added electron will interact with the rest of the system and undergo scattering as well as various inelastic processes. Thus, we expect $|\mathbf{K},0\rangle$ to be not an exact eigenstate, but rather a form of decaying state, as it develops in time. We now ask, What will be the form of the state $|\mathbf{K},t\rangle$ at a later time t? From quantum mechanics we have

$$|\mathbf{K},t\rangle = e^{I\mathcal{H}t}|\mathbf{K},0\rangle \tag{17.9}$$

The probability that the system, after a time t, will still be in the same state as at $t=0$ is written $\langle\mathbf{K},0|\mathbf{K},t\rangle$; this is closely related to the one-electron Green function, which we define as

$$G(\mathbf{K},t) = \begin{cases} -i\langle\mathbf{K},0|\mathbf{K},t\rangle & \text{for} \quad t>0 \\ 0 & \text{for} \quad t<0 \end{cases} \tag{17.10}$$

By introducing explicitly the Heisenberg operators for the creation and annihilation of an electron in the Bloch state $\mathbf{K}$, we can write (17.10) as

$$G(\mathbf{K},t) = \begin{cases} -i\langle 0|\mathbf{a}_{\mathbf{K}}(t)\mathbf{a}_{\mathbf{K}}^{\dagger}(0)|0\rangle & \text{for} \quad t>0 \\ 0 & \text{for} \quad t<0 \end{cases} \tag{17.11}$$

This can be simplified in the following way:

$$\begin{aligned} G(\mathbf{K},t) &= -i\langle 0|\mathbf{a}_{\mathbf{K}}(t)a_{\mathbf{K}}^{\dagger}(0)|0\rangle \\ &= -i\sum_n \langle 0|e^{i\mathcal{H}t}\mathbf{a}_{\mathbf{K}}(0)e^{-i\mathcal{H}t}|n\rangle\langle n|\mathbf{a}_{\mathbf{K}}^{\dagger}|0\rangle \\ &= -i\sum_n |\langle n|\mathbf{a}_{\mathbf{K}}^{\dagger}|0\rangle|^2 e^{-i[E_n(N+1)-E_0(N)]t} \\ &= -i\sum_n |\langle n|\mathbf{a}_{\mathbf{K}}^{\dagger}|0\rangle|^2 e^{-i\mathcal{E}_n t} \quad \text{for} \quad t>0 \end{aligned} \tag{17.12}$$

where $\mathcal{E}_n = E_n(N+1) - E_0(N)$. It is possible to rearrange $\mathcal{E}_n$ as

$$\begin{aligned} \mathcal{E}_n &= [E_n(N+1) - E_0(N+1)] + E_0(N+1) - E_0(N) \\ &= \omega_n + \mu \end{aligned} \tag{17.13}$$

where ω_n is the excitation energy, and μ is the chemical potential. Note that the smallest value of $\mathcal{E}_n$ is μ.

Taking the Fourier transform of (17.12) in the energy space, we get

$$G(\mathbf{K},\omega)=\int_{-\infty}^{\infty} dt\, e^{i\omega t} G(\mathbf{K},t)=\sum_n \frac{|\langle n|\mathbf{a}_\mathbf{K}^\dagger|0\rangle|^2}{\omega-\mathcal{E}_n+i\delta} \tag{17.14}$$

where δ is a positive infinitesimal which ensures that

$$G(\mathbf{K},t)=0 \qquad \text{for} \quad t<0.$$

Let us introduce the hole now, instead of an electron. The probability amplitude related to the hole Green function is given by

$$\langle 0|\mathbf{a}_\mathbf{K}^\dagger(t)\mathbf{a}_\mathbf{K}(0)|0\rangle$$

Using the negative time axis for holes, we can define the hole Green function as

$$G(\mathbf{K},t)=\begin{cases} 0 & \text{for} \quad t>0 \\ i\langle 0|\mathbf{a}_\mathbf{K}^\dagger(0)\mathbf{a}_\mathbf{K}(t)|0\rangle & \text{for} \quad t<0 \end{cases} \tag{17.15}$$

Proceeding exactly as for the electron Green function, we have the Fourier transform of the hole Green function:

$$G(\mathbf{K},\omega)=-\sum_n \frac{|\langle n|\mathbf{a}_\mathbf{K}|0\rangle|^2}{\omega-\mathcal{E}_n-i\delta} \tag{17.16}$$

Now, combining the particle and hole Green functions, we get the time-ordered or causal Green function, defined by

$$\begin{aligned} G_c(\mathbf{K},t) &= -i\langle 0|T\{\mathbf{a}_\mathbf{K}(t)\mathbf{a}_\mathbf{K}^\dagger(0)\}|0\rangle \\ &= \begin{cases} -i\langle 0|\mathbf{a}_\mathbf{K}(t)\mathbf{a}_\mathbf{K}^\dagger(0)|0\rangle & \text{for} \quad t>0 \\ i\langle 0|\mathbf{a}_\mathbf{K}^\dagger(0)\mathbf{a}_\mathbf{K}(t)|0\rangle & \text{for} \quad t<0 \end{cases} \end{aligned} \tag{17.17}$$

The time-ordering symbol T means that we should order the operators chronologically, so that the ones with earlier times are placed to the right. Also, we should note that a negative sign has to be introduced for fermions arising from anticommutation.

The so-called spectral representation of the Green function can be obtained by combining (17.14) and (17.16). We should note that, for electrons, $\mathcal{E}(\mathbf{K})>\mu$ and, for holes, $\mathcal{E}(\mathbf{K})<\mu$. It is convenient to shift the energy scale and measure the energies from the Fermi energy, so that the particles have positive energies and the holes have negative energies.

From (17.13) we can define a function

$$A^{+}(\mathbf{K},\omega)=\sum_{n}|\langle n|\mathbf{a}_{\mathbf{K}}^{\dagger}|0\rangle|^{2}\delta(\omega-\omega_{n}) \tag{17.18}$$

which gives the probability of finding the $(N+1)$-particle system excited to an energy in the interval ω and $(\omega+d\omega)$. Similarly, for the holes

$$A^{-}(\mathbf{K},\omega)=\sum_{n}|\langle n|\mathbf{a}_{\mathbf{K}}|0\rangle|^{2}\delta(\omega-\omega_{n}) \tag{17.19}$$

where ω_n is negative for holes. Adding (17.14) and (17.16), we have

$$\begin{aligned} G(\mathbf{K},\omega) &= \int_{0}^{\infty} d\omega' \frac{A^{+}(\mathbf{K},\omega')}{\omega-\omega'+i\delta}+\int_{-\infty}^{0} d\omega' \frac{A^{-}(\mathbf{K},\omega')}{\omega-\omega'-i\delta} \\ &= \int_{-\infty}^{\infty} d\omega' \frac{A(\mathbf{K},\omega')}{\omega-\omega'+i\omega\delta} \end{aligned} \tag{17.20}$$

where

$$A(\mathbf{K},\omega)=A^{+}(\mathbf{K},\omega)+A^{-}(\mathbf{K},\omega) \tag{17.21}$$

is the total spectral weight function. Equation (17.20) is a very important relation, known as the spectral representation of $G(\mathbf{K},\omega)$, which is exact and from which various analytic properties can be deduced. Note that $A(\mathbf{K},\omega)$ is a positive real quantity and fulfills the condition

$$\int_{-\infty}^{\infty} A(\mathbf{K},\omega)\,d\omega=1$$

which is known as the sum rule.

The spectral weight function $A(\mathbf{K},\omega)$ is the general distribution function with regard to both the momentum and the energy of the system, $A^{+}(\mathbf{K},\omega)$ being related to processes involving the addition of an electron, whereas $A^{-}(\mathbf{K},\omega)$ is related to processes involving the holes. Note that integration over the momenta gives the energy distribution in the system, and integration over the frequencies gives the momentum distribution.

In actual cases the Green function G is determined directly, rather than first calculating the spectral weight function A. Applying the well-known relation

$$\frac{1}{x\pm i\epsilon}=P\frac{1}{x}\mp i\pi\delta(x) \tag{17.22}$$

to (17.20), we get

$$\operatorname{Im} G(\mathbf{K},\omega)=\begin{cases} -\pi A(\mathbf{K},\omega), & \omega>0 \\ +\pi A(\mathbf{K},\omega), & \omega<0 \end{cases} \tag{17.23}$$

Using (17.23) in (17.20), we obtain a dispersion relation connecting the real and imaginary parts of G:

$$\operatorname{Re} G(\mathbf{K},\omega) = -\frac{1}{\pi}\int_0^{\infty}\frac{\operatorname{Im} G(\mathbf{K},\omega')\,d\omega'}{\omega-\omega'} + \frac{1}{\pi}\int_{-\infty}^{0}\frac{\operatorname{Im} G(\mathbf{K},\omega')\,d\omega'}{\omega-\omega'} \tag{17.24}$$

We shall not go into further details of the single-particle Green function but instead indicate how the present treatment can be adapted to more general situations.

The space-time Green function can be defined with the help of the Heisenberg operators for the wave fields (see Chapter 16 for the quantum field theory):

$$\begin{aligned} G(\mathbf{x},t;\,\mathbf{x}',t') &= -i\langle 0|T\{\psi(\mathbf{x},t)\psi^{\dagger}(\mathbf{x}',t')\}|0\rangle \\ &= -i\langle 0|\psi(\mathbf{x},t)\psi^{\dagger}(\mathbf{x}'t')|0\rangle, \qquad t>t' \\ &= -i\langle 0|\psi(\mathbf{x},t)\psi^{\dagger}(\mathbf{x}',t')|0\rangle, \qquad t<t' \end{aligned} \tag{17.25}$$

Analogously, the n-particle Green function can be defined by

$$\begin{aligned} &G(\mathbf{x}_1 t_1,\ldots,\mathbf{x}_n t_n, \mathbf{x}_1' t_1',\ldots,\mathbf{x}_n' t_n') \\ &\quad = (-i)^n\langle 0|T\{\psi(\mathbf{x}_1 t_1)\cdots\psi(\mathbf{x}_n t_n)\psi^{\dagger}(\mathbf{x}_n' t_n')\cdots\psi^{\dagger}(\mathbf{x}_1 t_1)\}|0\rangle \end{aligned} \tag{17.26}$$

The time-ordered Green functions are very useful in connection with the perturbation theory. The retarded and the advanced Green functions are often used, and examples of these will be given later.

Note that the Green functions introduced here are defined as the expectation values with regard to the ground state of the system. The appropriate generalization to finite temperatures is obtained by replacing them with the appropriate statisical averages for the temperature T, that is,

$$\langle 0|\cdots|0\rangle \to \sum_S \rho_S\langle S|\cdots|S\rangle = \langle\cdots\rangle_T \tag{17.27}$$

where S denotes the state of the system, and ρ_S represents the probability that state S will be realized.

To apply the above results let us now consider a gas of free fermions. The time-dependent Green functions for the fermions are given by

$$G(\mathbf{K},t) = \begin{cases} -i\langle \mathbf{a}_{\mathbf{K}}\mathbf{a}_{\mathbf{K}}^{\dagger}\rangle e^{-i\epsilon(\mathbf{K})t}, & t>0 \\ i\langle \mathbf{a}_{\mathbf{K}}^{\dagger}\mathbf{a}_{\mathbf{K}}\rangle e^{-i\epsilon(\mathbf{K})t}, & t<0 \end{cases} \tag{17.28}$$

or

$$G(\mathbf{K},t)=\begin{cases} -i\left(1-n_{\mathbf{K}}^{0}\right)e^{-i\epsilon(\mathbf{K})t}, & t>0 \\ in_{\mathbf{K}}^{0}\,e^{-i\epsilon(\mathbf{K})t}, & t<0 \end{cases} \tag{17.29}$$

Equation (17.29) expresses the fact that, for positive times, the Green function describes the "electrons" outside the Fermi sea and, for negative times, the "holes" in the Fermi sea.

Taking the Fourier transform to the energy space, we get

$$G(\mathbf{K},\omega)=\frac{1}{\omega-\epsilon(\mathbf{K})} \tag{17.30}$$

where we must add the prescription of how to treat the singularity. It is easy to show that, if we choose

$$G(\mathbf{K},\omega)=\frac{1}{\omega-\epsilon(\mathbf{K})+i\omega\delta} \tag{17.31}$$

we get back (17.28) and (17.29). This means that we have to add small imaginary parts to the single-particle energies.

If a particle is decaying with time, which means that the amplitude $|\mathbf{K},t\rangle$ decays with time, we have

$$G(\mathbf{K},t)=\begin{cases} -i\langle\mathbf{K},0|\mathbf{K},t\rangle\approx e^{-\Gamma t}e^{-i\epsilon(\mathbf{K})t}, & t>0 \\ 0, & t<0 \end{cases} \tag{17.32}$$

The Fourier transform of (17.32) is, as before,

$$G(\mathbf{K},\omega)=\frac{1}{\omega-\epsilon(\mathbf{K})+i\Gamma} \tag{17.33}$$

For the imaginary part we obtain the expected Lorentzian form for the spectral function:

$$A(\mathbf{K},\omega)=\frac{1}{\pi}|\operatorname{Im}G(\mathbf{K},\omega)|=\frac{1}{\pi}\frac{\Gamma}{\left[\omega-\epsilon(\mathbf{K})\right]^{2}+\Gamma^{2}} \tag{17.34}$$

Let us now briefly describe the calculation of the expectation values of the ground state properties of a system from the Green functions. We have already discussed the fact that the Green functions contain information about the excitations of the system. However, taking the appropriate limits, we can also calculate the ground state properties.

For example, the average density of particles of a system is given by

$$\begin{aligned} \rho(\mathbf{x})&=\langle 0|\psi^{\dagger}(\mathbf{x},t)\psi(\mathbf{x},t)|0\rangle \\ &=-i\underset{\substack{\mathbf{x}'\to\mathbf{x},\\ t'\to t+0}}{G}(\mathbf{x},t;\mathbf{x}',t') \end{aligned} \tag{17.35}$$

The momentum distribution is given by

$$
\begin{aligned}
N(\mathbf{K}) &= \langle 0|\mathbf{a}_{\mathbf{K}}^{\dagger}\mathbf{a}_{\mathbf{K}}|0\rangle \\
&= -i \underset{t\to -0}{G}(\mathbf{K},t) \\
&= -i\int_{-\infty}^{\infty}\frac{d\omega}{2\pi}G(\mathbf{K},\omega)e^{i\omega\delta} = \int_{-\infty}^{0}A(\mathbf{K},\omega)\,d\omega
\end{aligned}
\tag{17.36}
$$

To get the energy distribution of electrons we have to sum $A(\mathbf{K},\omega)$ over all momenta, that is,

$$
N(\omega) = \sum_{\mathbf{K}} A(\mathbf{K},\omega) \tag{17.37}
$$

The kinetic energy can be expressed in terms of the single-particle Green function:

$$
\langle 0|\int \psi^{\dagger}(\mathbf{x},t)\left(-\frac{\nabla^2}{2m}\right)\psi(\mathbf{x},t)\,d^3\mathbf{x}|0\rangle = \int d^3\mathbf{x}\left[i\frac{\nabla_x^2}{2m}G(\mathbf{x},t;\,\mathbf{x}'t')\right]_{\substack{\mathbf{x}'\to\mathbf{x},\\ t'\to t+0}} \tag{17.38}
$$

If the particles of a system interact via a two-body interaction, then the interaction energy can be expressed in terms of the two-particle Green function, that is,

$$
\begin{aligned}
&\langle|\tfrac{1}{2}\int \psi^{\dagger}(\mathbf{x},t)\psi^{\dagger}(\mathbf{x}',t)\,V(\mathbf{x}-\mathbf{x}')\psi(\mathbf{x},t')\psi(\mathbf{x}',t')\,d^3\mathbf{x}\,d^3\mathbf{x}'|0\rangle \\
&\qquad = -\frac{1}{2}\int d^3\mathbf{x}\,d^3\mathbf{x}'\,V(\mathbf{x}-\mathbf{x}')\underset{\substack{t'\to t,\\ t'>t}}{G}(\mathbf{x}t,\mathbf{x}'t;\,\mathbf{x}t',\mathbf{x}'t')
\end{aligned}
\tag{17.39}
$$

Let us now consider the interaction between our system and external probes like neutron beams, charged particles, or electromagnetic radiations. In such experiments we can assume that the probe goes from the initial state $|\rho\rangle$ to the final state $|\rho'\rangle$, while the system makes a transition from $|S\rangle$ to $|S'\rangle$. If we consider the interaction to be weak, we can treat it in the first order (linear response). The transition probability, per unit time, is then given by

$$
\frac{dW(\rho\to\rho')}{dt} = \frac{2\pi}{\hbar^2}\sum_{S'}|\langle S'|\mathcal{H}_{\text{int}}^{\rho'\rho}|S\rangle|^2\,\delta(\omega-\omega_{S'S}) \tag{17.40}
$$

where $\omega_{S'S} = \omega_{S'} - \omega_S$ denotes the excitation energy of the system, and $\omega = \omega_p - \omega_p'$ denotes the energy change of the probe (positive change for loss of energy in the process). Also, $\mathcal{H}_{\text{int}}$ denotes the interaction between the probe and the system, and $\langle S'|\mathcal{H}_{\text{int}}^{\rho'\rho}|S\rangle$ denotes the matrix element of the system. In reality, we do not have a single initial state but rather a statistical distribution over the initial states, so that we have to replace (17.40) by

$$\frac{dW(\rho\to\rho')}{dt}=\frac{2\pi}{\hbar^2}\sum_S \rho_S\left\{\sum_{S'}|\langle S'|\mathcal{H}_{\text{int}}^{\rho'\rho}|S\rangle|^2\right\}\delta(\omega-\omega_{S'S}) \qquad (17.41)$$

This expression can be simplified considerably by using the formula

$$\delta(\omega)=\frac{1}{2\pi}\int_{-\infty}^{\infty} e^{i\omega t}\,dt \qquad (17.42)$$

which is a very important relation. Inserting (17.42) in (17.41), we get

$$\sum_{S,S'}\rho_S\langle S|\mathcal{H}_{\text{int}}^{*\rho\rho'}|S'\rangle\langle S'|\mathcal{H}_{\text{int}}^{\rho'\rho}|S\rangle\,\delta(\omega-\omega_{S'S})$$

$$=\frac{1}{2\pi}\int_{-\infty}^{\infty} dt\,e^{i\omega t}\left\{\sum_{S,S'}\rho_S\langle S|\mathcal{H}_{\text{int}}^{*\rho'\rho}|S'\rangle\langle S'|\mathcal{H}_{\text{int}}^{\rho'\rho}|S\rangle\right\}e^{-i\omega_{S'}t}e^{i\omega_S t}$$

$$=\frac{1}{2\pi}\int_{-\infty}^{\infty} dt\,e^{i\omega t}\left\{\sum_{S,S'}\rho_S\langle S|e^{i\mathcal{H}t}\mathcal{H}_{\text{int}}^{*\rho\rho'}e^{-i\mathcal{H}t}|S'\rangle\langle S'|\mathcal{H}_{\text{int}}^{\rho'\rho}|S\rangle\right\}$$

$$=\frac{1}{2\pi}\int_{-\infty}^{\infty} dt\,e^{i\omega t}\sum_S\rho_S\langle S|\mathcal{H}_{\text{int}}^{*\rho\rho'}(t)\mathcal{H}_{\text{int}}^{\rho'\rho}(0)|S\rangle$$

$$=\frac{1}{2\pi}\int_{-\infty}^{\infty} dt\,e^{i\omega t}\langle|\mathcal{H}_{\text{int}}^{*\rho\rho'}(t)\mathcal{H}_{\text{int}}^{\rho'\rho}(0)|\rangle_T \qquad (17.43)$$

where $\mathcal{H}$ denotes the total Hamiltonian of the scattering system, and we have used

$$\sum_{S'}|S'\rangle\langle S'|=1$$

and the definition of a Heisenberg operator (see Chapter 16). The transition probability, per unit time, is then given by

$$\frac{dW(\rho-\rho')}{dt}=\frac{1}{\hbar^2}\int_{-\infty}^{\infty} dt\,e^{i\omega t}\langle|\mathcal{H}_{\text{int}}^{*\rho\rho'}(t)\mathcal{H}_{\text{int}}^{\rho'\rho}(0)|\rangle \qquad (17.44)$$

This shows that the scattering depends on a certain time-correlation function of the interaction operator, taken with itself at two different times. Note that the transition probability itself is, except for a constant, just the Fourier transform of the time-correlation function. Without going into further detail, we note from (17.44) that only two times occur, rather than the full number of times contained in the corresponding Green function. For this reason reference is often made to the double-time Green functions, which we shall describe now.

17.2 METHOD OF THE DOUBLE-TIME TEMPERATURE-DEPENDENT GREEN FUNCTION

We shall now formally develop the method of the double-time temperature-dependent Green function, following the excellent article by Zubarev.[13] The Green functions in statistical mechanics are the appropriate generalizations of the correlation functions, which we shall see later. They are useful in calculating the average of dynamical quantities, and they have great advantages when equations are framed and solved.

The retarded and advanced Green functions $G_r(t,t')$ and $G_a(t,t')$ are defined as

$$\begin{aligned} G_r(t,t') &\equiv \langle\langle \mathbf{A}(t);\mathbf{B}(t')\rangle\rangle_r \\ &= -i\theta(t-t')\langle[\mathbf{A}(t),\mathbf{B}(t')]\rangle \end{aligned} \tag{17.45}$$

and

$$\begin{aligned} G_a(t,t') &\equiv \langle\langle \mathbf{A}(t);\mathbf{B}(t')\rangle\rangle_a \\ &= i\theta(t'-t)\langle[\mathbf{A}(t),\mathbf{B}(t')]\rangle \end{aligned} \tag{17.46}$$

where $\langle\langle\cdots\rangle\rangle_{r,a}$ are the abbreviated notations for the corresponding Green functions, and $\langle\cdots\rangle$ denotes averaging over a grand canonical ensemble. This is appropriate since the number of particles is not constant. The process $\langle\cdots\rangle$ in quantum mechanics is defined by

$$\langle\cdots\rangle = Z^{-1}\mathrm{Tr}(e^{-\mathcal{H}/\theta}\cdots) \tag{17.47}$$

where

$$Z = \mathrm{Tr}(e^{-\mathcal{H}/\theta}) = e^{-\Omega/\theta}$$

Here $\theta = kT$, k being the Boltzmann constant and T the absolute temperature; Z is the grand partition function; and Ω is the thermodynamic potential. The operator $\mathcal{H}$ is the "generalized Hamiltonian," given by

$$\mathcal{H} = \mathbf{H} - \mu\mathbf{N}$$

where $\mathbf{H}$ is the time-independent Hamiltonian, $\mathbf{N}$ the operator for the total number of particles, and μ the chemical potential (renormalized). In (17.45) and (17.46) $\mathbf{A}(t), \mathbf{B}(t')$ are operators in the Heisenberg representation, which can be expressed as the product of the quantized field operators (see Chaper 16), that is,

$$\mathbf{A}(t) = e^{i\mathcal{H}t}\mathbf{A}(0)e^{-i\mathcal{H}t}; \qquad \hbar = 1 \tag{17.48}$$

and $\theta(t)$ is the step function, that is,

$$\theta(t)=\begin{cases}0, & t<0\\ 1 & t>0\end{cases} \tag{17.49}$$

Also, $[\mathbf{A},\mathbf{B}]$ is a commutator or anticommutator, that is,

$$[\mathbf{A},\mathbf{B}]=\mathbf{AB}-\eta\mathbf{BA}, \qquad \eta=\pm 1 \tag{17.50}$$

The sign of η is positive if **A** and **B** are both Bose operators and negative if they are Fermi operators. In general, **A** and **B** are neither Bose nor Fermi operators since products of operators can satisfy more complicated commutation relations. The sign of η is chosen by considering what is most convenient for the problem.

Using (17.50), we write (17.45) and (17.46) as

$$G_r(t,t')=-i\theta(t-t')[\langle\mathbf{A}(t)\mathbf{B}(t')\rangle-\eta\langle\mathbf{B}(t')\mathbf{A}(t)\rangle] \tag{17.51}$$

$$G_a(t,t')=i\theta(t-t')[\langle\mathbf{A}(t)\mathbf{B}(t')-\eta\langle\mathbf{B}(t')\mathbf{A}(t)\rangle] \tag{17.52}$$

We note from (17.49) and (17.51) that $G_r(t,t')\neq 0$ when $t'<t$, $G_r(t,t')=0$ when $t'>t$, and $G_r(t,t')$ is not defined when $t=t'$, because of the discontinuity of $\theta(t)$ at $t=0$. Similar considerations apply to $G_a(t,t')$.

The important difference from the field-theoretic Green function is that, instead of averaging over the ground (vacuum) state of the system, we have averaged over a grand canonical ensemble. The Green functions $G_{r,a}(t,t')$ depend on time and temperature as well. In the limit $T\to 0$, our Green functions go over to the field-theoretic Green functions.

One important property of $G_{r,a}(t,t')$ is that they depend only on the difference $(t-t')$ in the case of statistical equilibrium. This property is utilized in finding the "spectral representation" of the Green function, which is, in fact, a Fourier integral representation. This fact is verified as follows.

Using (17.51), (17.47), (17.48), and (17.45), we get

$$G_a(t,t')=-i\theta(t-t')\langle\mathbf{A}(t)\mathbf{B}(t')\rangle-i\eta\theta(t'-t)\langle\mathbf{B}(t')\mathbf{A}(t)\rangle \tag{17.53}$$

But

$$\begin{aligned}\mathbf{A}(t)\mathbf{B}(t')&=e^{i\mathcal{H}t}\mathbf{A}(0)e^{-i\mathcal{H}t}e^{i\mathcal{H}t'}\mathbf{B}(0)e^{-i\mathcal{H}t'}\\ &=e^{i\mathcal{H}t}e^{-i\mathcal{H}t'}\mathbf{A}(0)e^{-i\mathcal{H}t}e^{i\mathcal{H}t'}\mathbf{B}(0)e^{-i\mathcal{H}t'}e^{i\mathcal{H}t'}\\ &=e^{i\mathcal{H}(t-t')}\mathbf{A}(0)e^{-i\mathcal{H}(t-t')}\mathbf{B}(0)\end{aligned} \tag{17.54}$$

Similarly,

$$\mathbf{B}(t')\mathbf{A}(t)=e^{i\mathcal{H}(t'-t)}\mathbf{B}(0)e^{-i\mathcal{H}(t'-t)}\mathbf{A}(0) \tag{17.55}$$

Hence, using (17.54) and (17.55) in (17.53), we have

$$G_a(t,t') = -i\theta(t-t')Z^{-1}\mathrm{Tr}\left[e^{\mathcal{H}[i(t-t')-\beta]}A(0)e^{-i\mathcal{H}(t-t')}\mathbf{B}(0)\right]$$
$$-i\eta\theta(t'-t)Z^{-1}\mathrm{Tr}\left[e^{\mathcal{H}[i(t'-t)-\beta]}\mathbf{B}(0)e^{-i\mathcal{H}(t'-t)}\mathbf{A}(0)\right] \quad (17.56)$$

where

$$\beta = \frac{1}{kT} = \frac{1}{\theta}$$

Equation (17.56) shows that

$$G_a(t,t') = G_a(t-t')$$

Similarly,

$$G_r(t,t') = G_r(t-t')$$

The averages over the statistical ensemble of the product of operators in the Heisenberg representation of the kind

$$\begin{aligned} F_{\mathbf{BA}}(t,t') &= \langle \mathbf{B}(t')\mathbf{A}(t)\rangle \\ F_{\mathbf{AB}}(t,t') &= \langle \mathbf{A}(t)\mathbf{B}(t')\rangle \end{aligned} \quad (17.57)$$

are of importance in statistical mechanics. These are called the time-correlation functions. When the times are different ($t \neq t'$), these averages yield the time-correlation functions, which are essential for transport processes. Just like the Green functions in statistical equilibrium, these time-correlation functions also depend on $(t-t')$, that is,

$$\begin{aligned} F_{\mathbf{BA}}(t,t') &= F_{\mathbf{BA}}(t-t') \\ F_{\mathbf{AB}}(t,t') &= F_{\mathbf{AB}}(t-t') \end{aligned} \quad (17.58)$$

as can be verified easily. This fact helps in finding a spectral representation for them that relates them to the Green functions.

We now derive the spectral representation.[14] It has been pointed out that the Green function and the time-correlation functions depend on $(t-t')$ only. This is utilized in finding a Fourier integral for each, which is called the spectral representation. We first set up the representation for the time-correlation function and then that for the Green functions and find the relation between the two.

If we use for the matrix elements a representation in which $\mathcal{H}$ is diagonal, then

$$\langle \phi_\mu | \mathcal{H} | \phi_\nu \rangle = \delta_{\mu\nu} E_\nu$$

which means that

$$\mathcal{H}|\phi_\nu\rangle = E_\nu|\phi_\nu\rangle \tag{17.59}$$

We now write explicitly the statistical averaging operation in the definition of the time-correlation functions given by (17.58), that is,

$$\begin{aligned} F_{\mathbf{BA}}(t,t') &= \langle \mathbf{B}(t')\mathbf{A}(t)\rangle \\ &= Z^{-1}\sum_\nu \langle\phi_\nu|\mathbf{B}(t')\mathbf{A}(t)|\phi_\nu\rangle e^{-E_\nu/\theta} \\ &= Z^{-1}\sum_\mu\sum_\nu \langle\phi_\nu|\mathbf{B}(t')|\phi_\mu\rangle\langle\phi_\mu|\mathbf{A}(t)|\phi_\nu\rangle e^{-E_\nu/\theta} \end{aligned} \tag{17.60}$$

where, in writing the last step, use is made of the completeness of $|\phi_\nu\rangle$. Now, using (17.43) in (17.60), we get

$$\begin{aligned} \langle\mathbf{B}(t')\mathbf{A}(t)\rangle &= Z^{-1}\sum_{\mu,\nu}\langle\phi_\nu|e^{i\mathcal{H}t'}\mathbf{B}(0)e^{-i\mathcal{H}t'}|\phi_\mu\rangle\langle\phi_\mu|e^{i\mathcal{H}t}\mathbf{A}(0)e^{-i\mathcal{H}t}|\phi_\nu\rangle e^{-E_\nu/\theta} \\ &= Z^{-1}\sum_{\mu,\nu}\langle\phi_\nu|e^{iE_\nu t'}\mathbf{B}(0)e^{-iE_\mu t'}|\phi_\mu\rangle\langle\phi_\mu|e^{iE_\mu t}\mathbf{A}(0)e^{-iE_\nu t}|\phi_\nu\rangle e^{-E_\nu/\theta} \\ &= Z^{-1}\sum_{\mu,\nu}\langle\phi_\nu|\mathbf{B}(0)|\phi_\mu\rangle\langle\phi_\mu|\mathbf{A}(0)|\phi_\nu\rangle e^{-i(E_\nu-E_\mu)(t-t')}e^{-E_\nu/\theta} \end{aligned} \tag{17.61}$$

Similarly,

$$\begin{aligned} \langle\mathbf{A}(t)\mathbf{B}(t')\rangle = Z^{-1}\sum_{\mu,\nu}\langle\phi_\nu|\mathbf{A}(0)|\phi_\mu\rangle\langle\phi_\mu\mathbf{B}(0)|\phi_\nu\rangle \\ \cdot e^{-E_\nu/\theta}e^{-i(E_\nu-E_\mu)(t-t')} \end{aligned} \tag{17.62}$$

Now, changing the summation indices μ,ν in (17.62), we have

$$\begin{aligned} \langle\mathbf{A}(t)\mathbf{B}(t')\rangle &= Z^{-1}\sum_{\mu,\nu}\langle\phi_\mu|\mathbf{A}(0)|\phi_\nu\rangle\langle\phi_\nu|\mathbf{B}(0)|\phi_\mu\rangle \\ &\quad\cdot e^{-E_\mu/\theta}e^{-i(E_\mu-E_\nu)(t'-t)} \\ &= Z^{-1}\sum_{\mu,\nu}\langle\quad\rangle\langle\quad\rangle e^{-E_\mu/\theta}e^{-i(E_\nu-E_\mu)(t-t')} \\ &= Z^{-1}\sum_{\mu,\nu}\langle\quad\rangle\langle\quad\rangle e^{-E_\nu/\theta}e^{\omega/\theta}e^{-i(E_\nu-E_\mu)(t-t')} \\ &= \int_{-\infty}^{\infty} J(\omega)e^{\omega/\theta}e^{-i\omega(t-t')}\,d\omega \end{aligned} \tag{17.63}$$

where

$$\omega - E_\nu + E_\mu = 0$$

and

$$J(\omega) = Z^{-1} \sum_{\mu,\nu} \langle \phi_\nu | \mathbf{B}(0) | \phi_\mu \rangle \langle \phi_\mu | \mathbf{A}(0) | \phi_\mu \rangle \cdot e^{-E_\nu/\theta} \delta(\omega - E_\nu + E_\mu) \tag{17.64}$$

Similarly,

$$F_{\mathbf{BA}}(t-t') = \langle \mathbf{B}(t')\mathbf{A}(t) \rangle = \int_{-\infty}^{\infty} J(\omega) e^{i\omega(t-t')} d\omega. \tag{17.65}$$

where $J(\omega)$ is the same as in (17.64). Note that (17.63) and (17.65) are the required spectral representations for the time-correlation functions where $J(\omega)$ is the spectral intensity of the function $F_{\mathbf{BA}}(t)$ and really its Fourier transform.

Let us now consider the spectral representations of $G_r(t-t')$ and $G_a(t-t')$. These are obtained by means of (17.63) and (17.65). Let $G_r(E)$ be the Fourier transform of $G_r(t-t')$, that is,

$$G_r(t-t') = \int_{-\infty}^{\infty} dE\, G_r(E) e^{-iE(t-t')} \tag{17.66}$$

or

$$G_r(E) = \frac{1}{2\pi} \int_{-\infty}^{\infty} G_r(t-t') e^{iE(t-t')} dt \tag{17.67}$$

Substituting into (17.67) expression (17.51), we have

$$G_r(E) = \frac{1}{2\pi i} \int_{-\infty}^{\infty} dt \left[e^{iE(t-t')} \theta(t-t') \cdot \{ \langle \mathbf{A}(t)\mathbf{B}(t') \rangle - \eta \langle \mathbf{B}(t')\mathbf{A}(t) \rangle \} \right] \tag{17.68}$$

Now, inserting the time-correlation functions (17.63) and (17.65) in (17.68), we obtain

$$G_r(E) = \frac{1}{2\pi i} \int_{-\infty}^{\infty} dt \Bigg[e^{iE(t-t')} \theta(t-t') \cdot \left\{ \int_{-\infty}^{\infty} J(\omega) e^{\omega/\theta} e^{-i\omega(t-t')} d\omega - \eta \int_{-\infty}^{\infty} J(\omega) e^{-i\omega(t-t')} d\omega \right\} \Bigg]$$

$$= \int_{-\infty}^{\infty} d\omega\, J(\omega)(e^{\omega/\theta} - \eta) \frac{1}{2\pi i} \int_{-\infty}^{\infty} dt\, e^{iE(t-t')} e^{-i\omega(t-t')} \theta(t-t')$$

Now, puting $t-t'=t$, the time difference, we finally get

$$G_r(E)=\int_{-\infty}^{\infty} d\omega J(\omega)(e^{\omega/\theta}-\eta)\frac{1}{2\pi i}\int_{-\infty}^{\infty} dt\, e^{i(E-\omega)t}\theta(t) \qquad (17.69)$$

Here the discontinuous function $\theta(t)$ can be written in the form

$$\theta(t)=\int_{-\infty}^{t} e^{\mathcal{E}t}\delta(t)\,dt, \qquad \mathcal{E}\to 0\ (\mathcal{E}>0)$$

where

$$\delta(t)=\frac{1}{2\pi}\int_{-\infty}^{\infty} e^{-ixt}\,dx$$

Therefore

$$\begin{aligned}\theta(t)&=\frac{1}{2\pi}\int_{-\infty}^{t} e^{\mathcal{E}t}\,dt\int_{-\infty}^{\infty} e^{-ixt}\,dx\\ &=\frac{1}{2\pi}\int_{-\infty}^{\infty}\int_{-\infty}^{t} e^{(\mathcal{E}-ix)t}\,dt\,dx\\ &=\frac{1}{2\pi}\lim_{\mathcal{E}\to 0}\int_{-\infty}^{\infty}\frac{1}{\mathcal{E}-ix}e^{(\mathcal{E}-ix)t}\,dx\\ &=\frac{i}{2\pi}\lim_{\mathcal{E}\to 0}\int_{-\infty}^{\infty}\frac{e^{-ixt}}{x+i\mathcal{E}}\,dx \qquad (17.70)\end{aligned}$$

It can easily be verified that the function defined by (17.70) has the properties of the discontinuous θ function. We shall consider x as a complex variable and assume that the integral (17.70) is taken over the contour depicted in Figure 17.1. The integrand has a pole in the lower half-plane at $x=-i\mathcal{E}$. Putting $x=x_1+ix_2$, we have

$$e^{-ixt}=e^{-i(x_1+ix_2)t}=e^{-ix_1t}e^{x_2t}$$

Now, if $t>0$, x_2 must be negative in order for the integral to be nonvanishing.

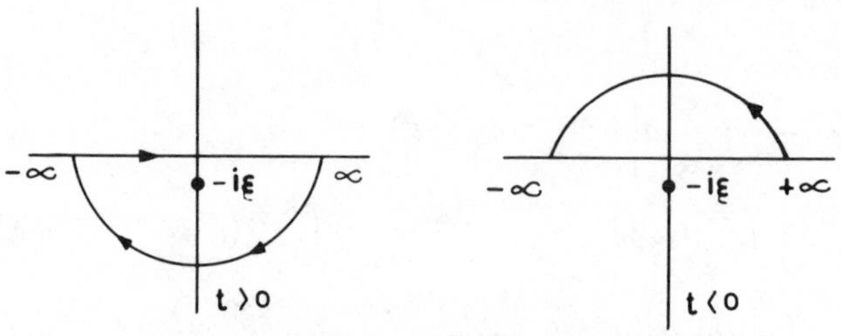

Fig. 17.1 Path of integration in a complex plane.

Also, since the pole is in the lower half-plane, the value of the integral is unity. When $t<0$, x_2 must be positive, and therefore the contour must be in the upper half-plane; but since the pole is not situated in the upper half-plane, the value of the integrand vanishes.

Using (17.70), we have then

$$\frac{1}{2\pi}\int_{-\infty}^{\infty} dt\, e^{i(E-\omega)t}\theta(t)$$

$$=\frac{1}{2\pi}\int_{-\infty}^{\infty} dt\, e^{i(E-\omega)t}\frac{i}{2\pi}\int_{-\infty}^{\infty}\frac{e^{-ixt}}{x+i\mathfrak{S}}\,dx$$

$$=\frac{i}{2\pi}\int_{-\infty}^{\infty}\frac{dx}{x+i\mathfrak{S}}\frac{1}{2\pi}\int_{-\infty}^{\infty} e^{-i(x-E+\omega)t}\,dt$$

$$=\frac{i}{2\pi}\int_{-\infty}^{\infty}\frac{dx}{x+i\mathfrak{S}}\delta(x-E+\omega)\qquad\left[\because \delta(x)=\frac{1}{2\pi}\int_{-\infty}^{\infty} e^{-ixt}\,dt\right]$$

$$=\frac{i}{2\pi}\frac{1}{E-\omega+i\mathfrak{S}}$$

Equation (17.69) then becomes

$$G_r(E)=\frac{1}{2\pi}\int_{-\infty}^{\infty}(e^{\omega/\theta}-\eta)J(\omega)\frac{d\omega}{E-\omega+i\mathfrak{S}}\tag{17.71}$$

Similarly,

$$G_a(E)=\frac{1}{2\pi}\int_{-\infty}^{\infty}(e^{\omega/\theta}-\eta)J(\omega)\frac{d\omega}{E-\omega-i\mathfrak{S}}\tag{17.72}$$

Combining (17.71) and (17.72), we have

$$G_{r,a}(E)=\frac{1}{2\pi}\int_{-\infty}^{\infty}(e^{\omega/\theta}-\eta)J(\omega)\frac{d\omega}{E-\omega\pm i\mathfrak{S}}\tag{17.73}$$

where the plus sign corresponds to the index r, and the minus sign to a. If E is assumed to be complex, (17.73) can be continued analytically in the complex E plane; thus

$$\frac{1}{2\pi}\int_{-\infty}^{\infty}(e^{\omega/\theta}-\eta)J(\omega)\frac{d\omega}{E-\omega}=\begin{cases} G_r(E), & \operatorname{Im} E>0\\ G_a(E), & \operatorname{Im} E<0\end{cases}\tag{17.74}$$

The left-hand side of (17.74) is a Cauchy-type integral. Also $G_r(E)$ and $G_a(E)$ can be considered to be two branches of the same analytic function, $G(E)$, defined in the whole complex plane with a singularity on the real axis:

$$G(E)=\begin{cases} G_r(E), & \operatorname{Im} E>0\\ G_a(E), & \operatorname{Im} E<0\end{cases}\tag{17.75}$$

For the sake of completeness, let us give the proof of (17.75). The analyticity of $G(E)$ follows from a theorem proved by Bogolyubov and Parasyuk[14] in the theory of dispersion relations. We have already seen that

$$G_r(E)=\frac{1}{2\pi}\int_{-\infty}^{\infty} G_r(t)e^{iEt}\,dt$$

where $G_r(t)=0$ when $t<0$. Let us show that the function $G_r(E)$ can be analytically continued into the region of complex E. Let E have a nonvanishing imaginary part:

$$E=\operatorname{Re}E+i\operatorname{Im}E=\alpha+i\gamma, \qquad \gamma>0$$

We then have

$$G_r(\alpha+i\gamma)=\frac{1}{2\pi}\int_{0}^{\infty} G_r(t)e^{i\alpha t}e^{-\gamma t}\,dt, \qquad \gamma>0$$

In this equation, $e^{-\gamma t}$ plays the role of a cut-off factor which makes the integral of $G_r(E)$ and its derivatives with respect to E convergent under sufficiently general assumptions about the function $G_r(t)$. The function $G_r(E)$ can thus be analytically continued in the upper half-plane. One can similarly prove that $G_a(E)$ can be analytically continued into the lower half-plane. If a cut is made the real axis, the function

$$G(E)=\begin{cases} G_r(E), & \operatorname{Im}E>0 \\ G_a(E), & \operatorname{Im}E<0 \end{cases}$$

can be considered to be two branches of $G(E)$. This completes the proof of (17.75).

We now find the relation between $J(\omega)$ in (17.64) and $G(E)$ in (17.75). From (17.73) we have

$$G(E)=\frac{1}{2\pi}\int_{-\infty}^{\infty}(E^{\omega/\theta}-\eta)J(\omega)\frac{d\omega}{E-\omega} \tag{17.76}$$

Now, interchanging E and ω, we obtain

$$G(\omega)=\frac{1}{2\pi}\int_{-\infty}^{\infty}(e^{E/\theta}-\eta)J(E)\frac{dE}{\omega-E} \tag{17.77}$$

Therefore

$$G(\omega+i\mathfrak{S})=\frac{1}{2\pi}\int_{-\infty}^{\infty}(e^{E/\theta}-\eta)J(E)\frac{dE}{\omega+i\mathfrak{S}-E} \tag{17.78}$$

Similarly,

$$G(\omega - i\mathfrak{S}) = \frac{1}{2\pi}\int_{-\infty}^{\infty}(e^{E/\theta} - \eta)J(E)\frac{dE}{\omega - i\mathfrak{S} - E} \tag{17.79}$$

Combining the two, we get

$$G(\omega + i\mathfrak{S}) - G(\omega - i\mathfrak{S})$$
$$= \frac{1}{2\pi}\int_{-\infty}^{\infty}(e^{E/\theta} - \eta)J(E)\left(\frac{1}{\omega - E + i\mathfrak{S}} - \frac{1}{\omega - E - i\mathfrak{S}}\right)dE \tag{17.80}$$

Now, using the δ-function representation

$$\delta(x) = \frac{1}{2\pi i}\left(\frac{1}{x - i\mathfrak{S}} - \frac{1}{x + i\mathfrak{S}}\right) \tag{17.81}$$

we obtain

$$G(\omega + i\mathfrak{S}) - G(\omega - i\mathfrak{S})$$
$$= -\frac{2\pi i}{2\pi}\int_{-\infty}^{\infty}(e^{E/\theta} - \eta)J(E)\delta(\omega - E)\,dE$$
$$= -i(e^{\omega/\theta} - \eta)J(\omega) \tag{17.82}$$

Thus we get the most important relation between the time-correlation function and $G(E)$. We thus see that from a knowledge of $G(E)$ we can obtain $J(\omega)$ and thus the correlation functions given by (17.63) and (17.65). From (17.65) we have

$$\langle \mathbf{B}(t')\mathbf{A}(t)\rangle = \int_{-\infty}^{\infty} J(\omega)e^{-i\omega(t-t')}\,d\omega \tag{17.83}$$

But $J(\omega)$ from (17.82) is

$$J(\omega) = -\frac{1}{i}\,\frac{G(\omega + i\mathfrak{S}) - G(\omega - i\mathfrak{S})}{e^{\omega/\theta} - \eta}$$

Inserting this in (17.83), we finally get

$$\langle \mathbf{B}(t')\mathbf{A}(t)\rangle = i\lim_{\mathfrak{S}\to +0}\int_{-\infty}^{\infty}\frac{G(\omega + i\mathfrak{S}) - G(\omega - i\mathfrak{S})}{e^{\omega/\theta} - \eta}e^{-i\omega(t-t')}\,d\omega \tag{17.84}$$

Relation (17.84) is a very important one and will be used later.

We mention now a few very useful relations for the Green functions. We use the well-known identity

$$\frac{1}{E-\omega \pm i\mathcal{E}} = P\frac{1}{E-\omega} \mp i\pi\,\delta(E-\omega) \tag{17.85}$$

in (17.71) and (17.72), where $\mathcal{E} \to 0$, $\mathcal{E} > 0$, and P denotes the principal value of the integral, and obtain

$$\begin{aligned} G_r(E) &= \frac{1}{2\pi} P\int_{-\infty}^{\infty} (e^{\omega/\theta}-\eta)J(\omega)\frac{d\omega}{E-\omega} - \frac{i}{2}(e^{E/\theta}-\eta)J(E) \\ G_a(E) &= \frac{1}{2\pi} P\int_{-\infty}^{\infty} (e^{\omega/\theta}-\eta)J(\omega)\frac{d\omega}{E-\omega} + \frac{i}{2}(e^{E/\theta}-\eta)J(E) \end{aligned} \tag{17.86}$$

where $(E-\omega)$ is considered to be a real quantity. From (17.86) we obtain a relation between the real and the imaginary parts of the Green functions:

$$\begin{aligned} \operatorname{Re} G_r(E) &= \frac{P}{\pi}\int_{-\infty}^{\infty} \frac{\operatorname{Im} G_r(\omega)}{\omega - E}\,d\omega \\ \operatorname{Re} G_a(E) &= -\frac{P}{\pi}\int_{-\infty}^{\infty} \frac{\operatorname{Im} G_a(\omega)}{\omega - E}\,d\omega \end{aligned} \tag{17.87}$$

Equations (17.87) have the form of dispersion relations.

We should like to point out some salient features of $J(E)$, appearing in (17.86). First, $J(E)$ may have the poles on the real axis. If such poles exist at E_i, $F_{\mathbf{BA}}$ oscillates with frequencies (as $\hbar = 1$)E_i. It is evident from the definition of quantum-mechanical average and the definition of $J(E)$ that the frequencies E_i at $T=0$ are the exact energy eignevalues of the system and hence give the stationary states of the system. For $T \neq 0$, if E_i at all exist, they depend on the temperature and chemical potential and their interpretation is not exact. However they characterize undamped motion and give the "temperature-dependent energy levels." In general, $J(E)$ can have singularities of complicated nature and therefore cannot be reduced to poles, and consequently $F_{\mathbf{BA}}$ will be a damped function of time. Hence we get no stationary states for $T=0$ and no ground state average for $T \neq 0$.

These facts define the "quasi-particle" picture of the low-lying branches of the energy spectrum of a many-body system and its limits of applicability. The spin-wave theory described in Chapter 16 is based on such a concept. Because of damping, in general, the number of quasi-particles is not a constant of motion. Only on the assumption of sufficiently weak damping is the quasi-particle picture valid, and it describes the quasi-stationary states of the system. One can determine the limits of its applicability because the damping constants can be determined from $J(E)$.

We now derive the equation of motion of the Green function. To use relation (17.84) we have to find the Green function. This is achieved by solving the equation of motion for the Green function. It is known that $\mathbf{A}(t)$ and $\mathbf{B}(t)$ satisfy equations of the form

$$i\frac{d\mathbf{A}}{dt} = [\mathbf{A}, \mathcal{H}] \tag{17.88}$$

We now differentiate the Green functions (17.45) and (17.46) (here G denotes both G_r and G_a) and obtain

$$\begin{aligned}
i\frac{dG}{dt} &= i\frac{d}{dt}\langle\langle \mathbf{A}(t); \mathbf{B}(t')\rangle\rangle \\
&= \frac{d}{dt}\langle[\theta(t-t')\mathbf{A}(t)\mathbf{B}(t') + \eta\theta(t'-t)\mathbf{B}(t')\mathbf{A}(t)]\rangle \\
&= \frac{d}{dt}\theta(t-t')\langle[\mathbf{A}(t)\mathbf{B}(t') - \eta\mathbf{B}(t')\mathbf{A}(t)]\rangle \\
&\quad + \theta(t-t')\left\langle\left[\frac{d\mathbf{A}(t)}{dt}\mathbf{B}(t') - \eta\mathbf{B}(t')\frac{d\mathbf{A}(t)}{dt}\right]\right\rangle \\
&= \frac{d}{dt}\theta(t-t')\langle[\mathbf{A}(t); \mathbf{B}(t')]\rangle + \left\langle T\frac{d\mathbf{A}(t)}{dt}; \mathbf{B}(t')\right\rangle \\
&= \frac{d}{dt}\theta(t-t')\langle[\mathbf{A}(t); \mathbf{B}(t')]\rangle + \left\langle\left\langle i\frac{d\mathbf{A}(t)}{dt}; \mathbf{B}(t')\right\rangle\right\rangle
\end{aligned} \tag{17.89}$$

since

$$\frac{d\theta(t)}{dt} = -\frac{d\theta(-t)}{dt}$$

Now

$$\theta(t) = \int_{-\infty}^{t} \delta(t)\, dt$$

Therefore

$$\frac{d\theta(t)}{dt} = \delta(t)$$

Hence, using (17.88), we get

$$\begin{aligned}
i\frac{dG}{dt} &= \delta(t-t')\langle[\mathbf{A}(t), \mathbf{B}(t')]\rangle \\
&\quad + \langle\langle[\mathbf{A}(t)\mathcal{H}(t) - \mathcal{H}(t)\mathbf{A}(t)]; \mathbf{B}(t')\rangle\rangle
\end{aligned} \tag{17.90}$$

or

$$i\frac{dG}{dt}=\delta(t-t')\langle[\mathbf{A},\mathbf{B}]_{-\eta}\rangle+\langle\langle[\mathbf{A},\mathcal{H}]_{-};\mathbf{B}\rangle\rangle \tag{17.91}$$

Now, taking the Fourier transform of (17.91) and using

$$\delta(t)=\frac{1}{2\pi}\int_{-\infty}^{\infty}e^{-iEt}\,dt$$

we have

$$EG(E)=\frac{1}{2\pi}\langle[\mathbf{A}(t),\mathbf{B}(t')]\rangle+\langle\langle[\mathbf{A}(t),\mathcal{H}(t)]_{-};\mathbf{B}(t')\rangle\rangle \tag{17.92}$$

where $[\ \ ]_{-}$ means a commutator. For future use we write (17.92) in the form

$$E\langle\langle\mathbf{A},\mathbf{B}\rangle\rangle_{E}=\frac{1}{2\pi}\langle[\mathbf{A},\mathbf{B}]_{-}\rangle+\langle\langle[\mathbf{A},\mathcal{H}]_{-};\mathbf{B}\rangle\rangle_{E} \tag{17.93}$$

Equation (17.92) follows since the Fourier transform of the δ function is $1/2\pi$, and

$$G(t)=\int G(E)e^{-iEt}\,dE$$

from which it can be shown that

$$\begin{aligned}\frac{\partial G}{\partial t}&=-iE\times\text{Fourier transform of } G(t)\\&=-iEG(E)\end{aligned}$$

This means that the Fourier transform of a time derivative is equal to $-iE$ times the Fourier transform of the quantity itself.

Equations (17.84) and (17.93) are the basic relations that we shall often require.

If $\langle\langle\mathbf{A};\mathbf{B}\rangle\rangle$ denotes the Fourier transform of the Green function involving the operators **A** and **B** ($\langle\langle\mathbf{A};\mathbf{B}\rangle\rangle$ will thus be a function of E), it satisfies the equation of motion (17.93), where the double angular brackets $\langle\langle\cdots\rangle\rangle$ indicate the Fourier transforms of the Green functions. The single brackets $\langle\cdots\rangle$ indicate the thermal average over a canonical ensemble, that is,

$$\langle F\rangle=\frac{\operatorname{Tr}e^{-\beta\mathcal{H}}F}{\operatorname{Tr}e^{-\beta\mathcal{H}}}$$

where $\beta=1/kT$, and $\mathcal{H}$ is the Hamiltonian of the system considered. From the analytical properties of the Green functions it follows that the correlation function $\langle \mathbf{B}(t')\mathbf{A}(t)\rangle$ can be obtained from the equation

$$\langle \mathbf{B}(t')\mathbf{A}(t)\rangle = \lim_{\mathcal{E}\to 0} i\int_{-\infty}^{\infty} \frac{\langle\langle A;B\rangle\rangle_{E=\hbar\omega+i\mathcal{E}} - \langle\langle A;B\rangle\rangle_{E=\hbar\omega-i\mathcal{E}}}{e^{\beta\hbar\omega}-1} e^{-i\omega(t-t')}\,d\omega \tag{17.94}$$

which is nothing but (17.84), given earlier. This form will be used later when we deal with the work of Tahir Kheli and ter Haar.

From (17.93) it is seen that the second term on the right-hand side is a higher order Green function than the original one. We can construct for the higher order Green functions, equations of the form (17.93), which entails still higher order functions. We thus obtain a hierarchy of coupled equations. There is, as yet, no method to find the exact solution of such a set of coupled equations, and therefore we are forced to solve them by using the approximate method of decoupling the chain of equations, that is, reducing it to a finite set of equations which then can be solved. It is to be noted that the success of the Green function method lies in the handling of the coupled chain of equations.

In the theories of ferromagnetism based on the Green function method, various decoupling schemes are used. Hence we describe these theories in the subsequent sections on different coupling schemes.

17.3 GREEN FUNCTION THEORIES OF FERROMAGNETISM IN RANDOM PHASE DECOUPLING: SPIN-1/2 CASE

Bogolyubov and Tyablikov,[15] who first applied the Green function method to ferromagnetism for the spin-1/2 case, evoked the random phase decoupling scheme to reduce the Green functions to the lowest order. The decoupling procedure is to postulate that

$$\langle\langle n_g(t)\mathbf{b}_p(t);\mathbf{b}_f^\dagger(t')\rangle\rangle = \langle n_g\rangle\langle\langle \mathbf{b}_p(t);\mathbf{b}_f^\dagger(t')\rangle\rangle \tag{17.95}$$

which means that there is no correlation between n_g and $\mathbf{b}_p$ and also that they have no phase relationship. Hence this is known as random phase decoupling.

Let the ferromagnet studied here consist of N atoms of a ferromagnetic element that form a regular simple cubic lattice. We shall assume, moreover, that each atom has only a single ferromagnetic electron and that we may neglect the interaction between these electrons and the conduction electrons. Furthermore, we assume that there is an isotropic exchange interaction between the electrons, and that the magnetic field is directed along the z-direction. Under these circumstances the system considered is described by a Heisenberg Hamiltonian, which can be written in the form

$$\mathcal{H} = -g\mu_b H\sum_g \mathbf{S}_g^Z - \tfrac{1}{2}\sum_{g,f} J(\mathbf{g}-\mathbf{f})\mathbf{S}_g\cdot\mathbf{S}_f \tag{17.96}$$

where the symbols have the usual meaning, and the $S_f^\alpha(\alpha=x,y,z)$ are the well-known Pauli spin matrices, given by

$$S_f^x=\begin{pmatrix}0&1\\1&0\end{pmatrix},\qquad S_f^y=\begin{pmatrix}0&-i\\i&0\end{pmatrix},\qquad S_f^z=\begin{pmatrix}1&0\\0&-1\end{pmatrix}\tag{17.97}$$

where f indicates the lattice site on which the spin is situated. We now introduce a set of operators $\mathbf{b}_f$ and $\mathbf{b}_f^\dagger$, given by the equations

$$\mathbf{b}_f=\begin{pmatrix}0&1\\0&0\end{pmatrix}\quad\text{and}\quad \mathbf{b}_f^\dagger=\begin{pmatrix}0&0\\1&0\end{pmatrix}\tag{17.98}$$

The Pauli spin operators (17.97) can now be expressed in terms of the spin operators (17.98) as follows:

$$\begin{aligned}\mathbf{S}_f^x&=\mathbf{b}_f^\dagger+\mathbf{b}_f\\ \mathbf{S}_f^y&=i(\mathbf{b}_f^\dagger-\mathbf{b}_f)\\ \mathbf{S}_f^z&=1-2\mathbf{b}_f^\dagger\mathbf{b}_f\end{aligned}\tag{17.99}$$

It can be easily proved that the $\mathbf{b}_f$'s and $\mathbf{b}_f^\dagger$'s satisfy the following commutation relations:

$$\left.\begin{aligned}\mathbf{b}_f\mathbf{b}_f^\dagger+\mathbf{b}_f^\dagger\mathbf{b}_f&=1\\ \mathbf{b}_f^2=\mathbf{b}_f^{\dagger 2}&=0\\ \mathbf{b}_f\mathbf{b}_g^\dagger-\mathbf{b}_g^\dagger\mathbf{b}_f&=0\\ \mathbf{b}_f\mathbf{b}_g-\mathbf{b}_g\mathbf{b}_f&=0\\ \mathbf{b}_f^\dagger\mathbf{b}_g^\dagger-\mathbf{b}_g^\dagger\mathbf{b}_f^\dagger&=0\end{aligned}\right\}\tag{17.100}$$

It is important to note that the Pauli operators are of the Fermi type for the same lattice site, and of the Bose type for different sites. With the help of (17.99), our Hamiltonian reduces to

$$\begin{aligned}\mathcal{H}=&-N\left[g\mu_B H+\tfrac{1}{2}J(0)\right]+\left[2g\mu_B H+2J(0)\right]\sum_f \mathbf{b}_f^\dagger\mathbf{b}_f\\ &-\sum_{\mathbf{g},\mathbf{f}}2J(\mathbf{g}-\mathbf{f})\mathbf{b}_g^\dagger\mathbf{b}_f^\dagger-\sum_{g,f}2J(\mathbf{g}-\mathbf{f})n_g n_f\end{aligned}\tag{17.101}$$

where $J(0)$ is the zeroth component of the Fourier transform

$$J(K)=\sum_f J(\mathbf{f})e^{-iK\cdot\mathbf{f}}$$

and

$$n_g = \mathbf{b}_g^\dagger \mathbf{b}_g$$

Now, because of the translational invariance of the lattice and the equivalence of all lattice sites, the average $\langle n_g \rangle$ is independent of the site index, that is,

$$\langle n_g \rangle = \bar{n} \tag{17.102}$$

It is also well known from (17.88) that

$$i\frac{d\mathbf{A}}{dt} = -[\mathcal{H}, \mathbf{A}]$$

Hence, from (17.101), we have

$$\begin{aligned} i\frac{d\mathbf{b}_g}{dt} = &-\left[\sum_g \{2g\mu_B H + 2J(0)\}\mathbf{b}_g^\dagger \mathbf{b}_g, \mathbf{b}_g\right] \\ &+ \sum_{g,f}\left[2J(\mathbf{g}-\mathbf{f})\mathbf{b}_g^\dagger \mathbf{b}_f, \mathbf{b}_g\right] \\ &+ \sum_{g,f}\left[2J(\mathbf{g}-\mathbf{f}) n_g n_f, \mathbf{b}_g\right] \end{aligned} \tag{17.103}$$

Using the commutation relations (17.100), we can reduce (17.103) to

$$\begin{aligned} i\frac{d\mathbf{b}_g}{dt} = &[2g\mu_B H + 2J(0)]\mathbf{b}_g - \sum_p 2J(\mathbf{g}-\mathbf{p})\mathbf{b}_p \\ &+ \sum_p 4J(\mathbf{g}-\mathbf{p})(n_g b_p - \mathbf{b}_g n_p) \end{aligned} \tag{17.104}$$

Similarly,

$$i\frac{dn_g}{dt} = -2\sum_p J(\mathbf{g}-\mathbf{p})\left(\mathbf{b}_g^\dagger \mathbf{b}_p - \mathbf{b}_p^\dagger \mathbf{b}_g\right) \tag{17.105}$$

We now introduce the Green function ($n=1$):

$$\begin{aligned} G_{\mathbf{g},\mathbf{f}}(t-t') &\equiv \langle\langle \mathbf{b}_\mathbf{g}(t); \mathbf{b}_\mathbf{f}^\dagger(t')\rangle\rangle \\ G_{\mathbf{g}_1,\mathbf{g}_2,\mathbf{f}}(t-t') &\equiv \langle\langle n_{\mathbf{g}_1}(t)\mathbf{b}_{\mathbf{g}_2}(t); \mathbf{b}_\mathbf{f}^\dagger(t')\rangle\rangle \end{aligned} \tag{17.106}$$

Using the equation of motion for the Green function (17.89), we get

$$\begin{aligned} i\frac{dG_{g,f}}{dt} = &(1-2\bar{n})\delta(t-t')\delta_{\mathbf{gf}} + [2g\mu_B H + 2J(0)]G_{\mathbf{g},\mathbf{f}} \\ &- \sum_p 2J(\mathbf{g}-\mathbf{p})G_{\mathbf{p},\mathbf{f}} + \sum_p 4J(\mathbf{g}-\mathbf{p})(G_{\mathbf{gp},\mathbf{f}} - G_{\mathbf{pg},\mathbf{f}}) \end{aligned} \tag{17.107}$$

The first term on the right-hand side appears because

$$\langle [\mathbf{b}_f, \mathbf{b}_f^\dagger] \rangle = \langle 1 - 2\bar{n}_f \rangle_{\text{av}}$$
$$= (1 - 2\bar{n})\,\delta(t - t')\,\delta_{\mathbf{gf}}$$

Now random phase decoupling is evoked, that is,

$$\begin{aligned} G_{\mathbf{gp},\mathbf{f}}(t - t') &= \langle\langle n_{\mathbf{g}}(t)\mathbf{b}_{\mathbf{p}}(t); \mathbf{b}_{\mathbf{f}}^\dagger(t')\rangle\rangle \\ &= \langle n_g \rangle \langle\langle \mathbf{b}_p(t); \mathbf{b}_{\mathbf{f}}^\dagger(t')\rangle\rangle \\ &= \bar{n} G_{\mathbf{p},\mathbf{f}}(t - t') \end{aligned} \tag{17.108}$$

Equation (17.107) then becomes

$$\begin{aligned} i\frac{dG_{g,f}}{dt} &= \left[2g\mu_{\mathrm{B}}H + (1 - 2\bar{n})2J(0)\right] G_{\mathbf{g},\mathbf{f}} \\ &\quad - \sum_p (1 - 2\bar{n})2J(\mathbf{g} - \mathbf{p})\,G_{\mathbf{p},\mathbf{f}} + (1 - 2\bar{n})\,\delta(t - t')\,\delta_{\mathbf{gf}} \end{aligned} \tag{17.109}$$

which no longer contains higher order Green functions. The method of decoupling (17.108) corresponds to the method of approximate second quantization,[16] improved thermodynamically, for the higher temperature region. Actually, in the method of Ref. 16, the Pauli operators were assumed to be approximately Bose operators, and the last term in the Hamiltonian (17.101) was neglected. This, however, corresponds to the low-temperature region well below the Curie temperature. In this approximation one must put $\bar{n} = 0$ in (17.109). When the temperature rises however, the role of the terms containing a factor $\bar{n}$ increases, and they must then be taken into account.

$$G_{\mathbf{g},\mathbf{f}}(t) = \int_{-\infty}^{\infty} G_{\mathbf{g},\mathbf{f}}(E)\,e^{-iEt}\,dE$$

we get from (17.109)

$$\begin{aligned} EG_{\mathbf{g},\mathbf{f}} &= \frac{1}{2\pi}(1 - 2\bar{n})\,\delta_{\mathbf{gf}} + \left[2g\mu_{\mathrm{B}}H + (1 - 2\bar{n})2J(0)\right] G_{\mathbf{g},\mathbf{f}} \\ &\quad - \sum_p (1 - 2\bar{n})2J(\mathbf{g} - \mathbf{p})\,G_{\mathbf{p},\mathbf{f}} \end{aligned} \tag{17.110}$$

since

$$\delta(t - t') = \frac{1}{2\pi}\int_{-\infty}^{\infty} e^{-ix(t - t')}\,dx$$

or, for $t = t'$,

$$\delta(0) = \frac{1}{2\pi}$$

This $1/2\pi$ appears in the first term of (17.110). We then take the following Fourier transforms, based on the translational invariance of the lattice:

$$G_{\mathbf{g},\mathbf{p}}(E)=\frac{1}{N}\sum_{\mathbf{K}}e^{i\mathbf{K}(\mathbf{g}-\mathbf{f})}G_{\mathbf{K}}(E)$$

$$\delta_{\mathbf{gf}}=\frac{1}{N}\sum_{\mathbf{K}}e^{i\mathbf{K}(\mathbf{g}-\mathbf{f})} \tag{17.111}$$

$$J(\mathbf{g}-\mathbf{f})=\frac{1}{N}\sum_{\mathbf{K}}e^{i\mathbf{K}(\mathbf{g}-\mathbf{f})}J(\mathbf{K})$$

where **K** is the wave vector and, because of the periodic boundary conditions, lies in the first Brillouin zone. Substituting (17.111) into (17.110), we have

$$G_{\mathbf{K}}(E)=\frac{1}{2\pi}\frac{1-2\bar{n}}{E-E_{\mathbf{K}}} \tag{17.112}$$

where

$$E_{\mathbf{K}}=2g\mu_{\mathrm{B}}H+(1-2\bar{n})2\left[J(0)-J(\mathbf{K})\right]$$

and (17.113)

$$J(\mathbf{K})=\sum_{f}J(\mathbf{f})e^{i\mathbf{f}\cdot\mathbf{K}}$$

Let us find the spectral intensity $I_{\mathbf{K}}(\omega)$ by using (17.82) and (17.84) (note that we have written $I_{\mathbf{K}}(\omega)$ for $J(\omega)$ to avoid confusion):

$$\begin{aligned}
I_{\mathbf{K}}(\omega)&=i\frac{G_{\mathbf{K}}(\omega+i\mathcal{E})-G_{\mathbf{K}}(\omega-i\mathcal{E})}{e^{\omega/kT}-1}\\
&=\frac{i}{e^{\omega/kT}-1}\left(\frac{1-2\bar{n}}{2\pi}\right)\left(\frac{1}{\omega+i\mathcal{E}-E_{\mathbf{K}}}-\frac{1}{\omega-i\mathcal{E}-E_{\mathbf{K}}}\right)\\
&=\frac{i}{e^{\omega/kT}-1}\left(\frac{1-2\bar{n}}{2\pi}\right)\left[-2\pi i\,\delta(\omega-E_{\mathbf{K}})\right]\\
&=\frac{1-2\bar{n}}{e^{E_{\mathbf{K}}/kT}-1}\delta(\omega-E_{\mathbf{K}})
\end{aligned} \tag{17.114}$$

by using (17.81). Hence

$$\begin{aligned}
I_{\mathbf{g},\mathbf{f}}(\omega)&=\frac{1}{N}\sum_{\mathbf{K}}e^{i(\mathbf{g}-\mathbf{f})\cdot\mathbf{K}}I_{\mathbf{K}}(\omega)\\
&=\frac{1}{N}\sum_{\mathbf{K}}\frac{1-2\bar{n}}{e^{E_{\mathbf{K}}/\theta}-1}e^{i(\mathbf{g}-\mathbf{f})\cdot\mathbf{K}}\delta(\omega-E_{\mathbf{K}})
\end{aligned} \tag{17.115}$$

Also, from (17.65)

$$\langle \mathbf{b}_f^\dagger(t')\mathbf{b}_g(t)\rangle = \int_{-\infty}^{\infty} I_{\mathbf{g},\mathbf{f}}(\omega)e^{-i\omega(t-t')}\,d\omega$$

$$= \frac{1}{N}\sum_{\mathbf{K}} e^{i(\mathbf{g}-\mathbf{f})\mathbf{K}}e^{-iE_{\mathbf{K}}(t-t')}\frac{1-2\bar{n}}{e^{E_{\mathbf{K}}/\theta}-1} \tag{17.116}$$

Now, substituting $t'=t$, $f=g$ and changing from a sum over $\mathbf{K}$ to an integral, we get for $\bar{n}$

$$\bar{n}(t) = \langle \mathbf{b}_g^\dagger(t)\mathbf{b}_g(t)\rangle = \frac{1}{N}\sum_{\mathbf{K}}\frac{1-2\bar{n}}{e^{E_{\mathbf{K}}/\theta}-1}$$

or

$$\frac{\bar{n}}{1-2\bar{n}} = \frac{v}{(2\pi)^3}\int\frac{d^3\mathbf{K}}{e^{E_{\mathbf{K}}/\theta}-1} \tag{17.117}$$

where $v = V/N$ is the volume of the elementary cell, that is, the volume per site.

We now introduce the relative magnetization, defined as the magnetization per site. Noting that the z-component of the spin is connected to the operator $n = \mathbf{b}^\dagger\mathbf{b}$, we get the relative magnetization:

$$\sigma = \langle S_g^z\rangle = \langle 1-2n_g\rangle = 1-2\bar{n} \tag{17.118}$$

Now let

$$\frac{\bar{n}}{1-2\bar{n}} = \alpha \qquad \text{or} \qquad 2\alpha = \frac{2\bar{n}}{1-2\bar{n}}$$

Therefore

$$2\alpha+1 = \frac{1}{1-2\bar{n}} = \frac{1}{\sigma} = \frac{v}{(2\pi)^3}\int d^3\mathbf{K}\left(\frac{2}{e^{E_{\mathbf{K}}/\theta}-1}+1\right)$$

$$= \frac{v}{(2\pi)^3}\int\frac{e^{E_{\mathbf{K}}/\theta}+1}{e^{E_{\mathbf{K}}/\theta}-1}\,d^3\mathbf{K}$$

Hence

$$\frac{1}{\sigma} = \frac{v}{(2\pi)^3}\int \coth\left(\tfrac{1}{2}\beta E_{\mathbf{K}}\right)d^3\mathbf{K} \tag{17.119}$$

where

$$\beta = \frac{1}{\theta} = \frac{1}{kT}$$

We note that $G_{\mathbf{K}}(E)$ in (17.112) has a pole at $E_{\mathbf{K}}$. This gives the spin-wave energy. The expression for it, given by the spin-wave theory,[16] is

$$E_{\mathbf{K}}=2g\mu_{\mathrm{B}}H+2[J(0)-J(\mathbf{K})]$$

which differs from (17.113) in terms involving the wave vector being proportional to the relative magnetization, $\sigma=1-2\bar{n}$. Thus we get the "magnetization renormalized" spin-wave energy. Only when $\bar{n}=0$ is the agreement with the spin-wave theory exact.

We now calculate the magnetization in different temperature ranges and show that (1) for $T\ll T_{\mathrm{C}}$ we get the spin-wave result, (2) for $T\lesssim T_{\mathrm{C}}$, the results of the molecular field theory, and (3) for $T\gg T_{\mathrm{C}}$, the results of Opechowski.

The following dimensionless quantities are introduced:

$$L=\frac{g\mu_{\mathrm{B}}H}{J(0)}, \qquad \tau=\frac{1}{\beta J(0)}, \qquad \mathcal{E}(\mathbf{K})=1-\frac{J(\mathbf{K})}{J(0)} \tag{17.120}$$

From (17.113) we have

$$\begin{aligned}\tfrac{1}{2}\beta E_{\mathbf{K}}&=g\mu_{\mathrm{B}}H\beta+(1-2\bar{n})\beta J(0)\left[1-\frac{J(\mathbf{K})}{J(0)}\right]\\ &=\frac{L}{\tau}+\frac{\sigma}{\tau}[1-\mathcal{E}(\mathbf{K})]\end{aligned}$$

Therefore, from (17.119),

$$\frac{1}{\sigma}=\frac{v}{(2\pi)^3}\int\coth\frac{L+\sigma\mathcal{E}(\mathbf{K})}{\tau}\,d^3\mathbf{K} \tag{17.121}$$

Case 1 $T\ll T_{\mathrm{C}}$, $\tau\to 0$, $L+\sigma\mathcal{E}(\mathbf{K})\gg\tau$

We use the expansion

$$\begin{aligned}\coth x&=(1+e^{-2x})(1-e^{-2x})^{-1}\\ &=1+2\sum_{l=1}^{\infty}e^{-2lx}\end{aligned}$$

Equation (17.121) then assumes the form

$$\frac{1}{\sigma}=1+F\left(\frac{\tau}{\sigma},\frac{L}{\tau}\right) \tag{17.122}$$

where

$$F=\sum_{l=1}^{\infty}\frac{2v}{(2\pi)^3}\int d^3\mathbf{K}\exp\left[-2l\frac{L+\sigma\mathcal{E}(\mathbf{K})}{\tau}\right]$$

Since the main contribution comes from the small wave vector, we can assume that

$$\mathcal{E}(\mathbf{K}) = \gamma \mathbf{K}^2$$

where γ, for a simple cubic lattice, is given by

$$\gamma = \frac{1}{6J(0)} \sum_f f^2 J(f) \tag{17.123}$$

so that

$$F\left(\frac{\tau}{\sigma}, \frac{L}{\tau}\right) = \frac{2v}{(2\pi)^3} \sum_{l=1}^{\infty} \int d^3\mathbf{K} \exp\left(-2l \frac{L + \sigma\gamma\mathbf{K}^2}{\tau}\right)$$

The integration is a simple Γ-integral, as can be seen by putting

$$x = \frac{2l\gamma\sigma}{\tau} \mathbf{K}^2$$

and we get

$$F\left(\frac{\tau}{\sigma}, \frac{L}{\tau}\right) = 2v \left(\frac{\tau}{8\pi\gamma\sigma}\right)^{3/2} Z_{3/2}\left(\frac{2L}{\tau}\right) \tag{17.124}$$

where

$$Z_p(x) = \sum_{l=1}^{\infty} l^{-p} e^{-lx}$$

We then write (17.122) as

$$\sigma = 1 - 2v \left(\frac{\tau}{8\pi\gamma\sigma}\right)^{3/2} Z_{3/2}\left(\frac{2L}{\tau}\right)\sigma \tag{17.125}$$

We now solve (17.125) by iteration and put as the zeroth approximation $\sigma_0 = 1$. We get up to terms of order $\tau^{3/2}$:

$$\sigma \simeq 1 - 2v \left(\frac{\tau}{8\pi\gamma\sigma}\right)^{3/2} Z_{3/2}\left(\frac{2L}{\tau}\right) \tag{17.126}$$

which is the same as Bloch's $T^{3/2}$ law for $L = 0$. We have already derived this law from the spin-wave theory given in Chapter 16.

Now, to obtain an expression of σ in a power series of τ, we expand $\mathcal{E}(\mathbf{K})$ in powers of $\mathbf{K}$ and evaluate F as a function of the ratio τ/σ:

$$F\left(\frac{\tau}{\sigma},\frac{L}{\tau}\right)=a_1\left(\frac{L}{\tau}\right)\left(\frac{\tau}{\sigma}\right)^{3/2}+a_2\left(\frac{L}{\tau}\right)\left(\frac{\tau}{\sigma}\right)^{5/2}$$
$$+a_3\left(\frac{L}{\tau}\right)\left(\frac{\tau}{\sigma}\right)^{7/2}+\cdots \qquad (17.127)$$

where $a_1, a_2, a_3, \cdots$ are functions of L/τ. We then solve (18.122) for σ and obtain

$$\sigma=1-\sum_{j>3} A_j\tau^{j/2} \qquad (17.128)$$

where the A_j's are the functions of L/τ, which depends also on the lattice structure. In particular,

$$A_3=\frac{2v}{(8\pi\gamma)^{3/2}}Z_{3/2}\left(\frac{L}{\tau}\right), \qquad A_4=0 \qquad (17.129)$$

For a simple cubic lattice and the nearest-neighbor interaction, the coefficients A_j's are as follows:

$$A_3=2\left(\frac{3}{4\pi}\right)^{3/2}Z_{3/2}, \qquad A_4=0$$
$$A_5=\frac{3\pi}{2}\left(\frac{3}{4\pi}\right)^{5/2}Z_{5/2}, \qquad A_6=2\left(\frac{3}{4\pi}\right)^{3}Z_3$$
$$A_7=\frac{33\pi^2}{16}\left(\frac{3}{4\pi}\right)^{7/2}Z_{7/2} \qquad (17.130)$$
$$A_8=6\pi\left(\frac{3}{4\pi}\right)^{4}Z_{3/2}Z_{5/2}$$

We wish to calculate these coefficients in the next section in connection with the ferromagetism for the general spin S case. Hence we have simply written the coefficients in (17.130).

We see from (17.126) to (17.130) that for Case 1 the first two terms in the expansion for the relative magnetization give Bloch's results, while the latter terms represent corrections to them. It is of interest to compare these with the results of Dyson,[17] who found, from expansion (17.128), $A_4=A_6=0$, and, for the s.c. lattice, $A_8=10\pi(\frac{3}{4\pi})^4 Z_{3/2}Z_{5/2}$ (in the present notation). It may be thought that the discrepancy between these results and those of Dyson are connected with the decoupling procedure. Nevertheless, the agreement with Dyson's results is very satisfactory, apart from one or two correction terms.

Case 2. $T \ll T_C$, $L=0$

We turn now to the case of high temperatures and no external field ($T \ll T_C$, $L=0$). We note, first of all, that the system considered here has a phase transition point above which the spontaneous magnetization vanishes. As $\tau \to \infty$, $\sigma \ll 1$ and $\sigma \mathcal{E}(\mathbf{K})/\tau \to 0$. Since $\coth \zeta = 1/\zeta$ as $\zeta \to 0$, we get from (17.121)

$$\frac{1}{\sigma} \approx \frac{\tau}{\sigma} \frac{v}{(2\pi)^3} \int \frac{d^3\mathbf{K}}{\mathcal{E}(\mathbf{K})}$$

It is evident from this equation that no solution is possible for sufficiently high temperatures. To obtain the temperature dependence of relative magnetization, we assume that $\sigma \to 0$ as $\tau \to \tau_C (\tau < \tau_C)$. We call τ_C the Curie temperature. For small x,

$$\begin{aligned}
\coth x &= \left(1 + \frac{x^2}{2!} + \frac{x^4}{4!} + \cdots\right) \Big/ \left(x + \frac{x^3}{3!} + \frac{x^5}{5!} + \cdots\right) \\
&= \left(1 + \frac{x^2}{2!} + \frac{x^4}{4!} + \cdots\right) \frac{1}{x} \left[1 + \left(\frac{x^2}{3!} + \frac{x^4}{5!} + \cdots\right)\right]^{-1} \\
&= \frac{1}{x} + \frac{x}{3} - \frac{x^3}{45} + \frac{2x^5}{945} - \cdots, \qquad x^2 < \pi^2 \\
&= \frac{1}{x} + \sum_{l=1}^{\infty} \frac{2^{2l} B_{2l}}{(2l)!} x^{2l-1}
\end{aligned} \tag{17.131}$$

where B_{2l} are the Bernoulli members ($B_2 = \frac{1}{6}$, $B_4 = -\frac{1}{30}, \cdots$). Substituting (17.131) into (17.121) and introducing the Curie temperature:

$$\frac{1}{\tau_C} = \frac{v}{(2\pi)^3} \int \frac{d^3\mathbf{K}}{\mathcal{E}(\mathbf{K})} \tag{17.132}$$

we get

$$\frac{1}{\sigma} = \frac{\tau}{\tau_C} \cdot \frac{1}{\sigma} + \sum_{l=1}^{\infty} \frac{2^{2l} B_{2l}}{(2l)!} \left(\frac{\sigma}{\tau}\right)^{2l-1} C_{2l-1}$$

where

$$C_n = \frac{v}{(2\pi)^3} \int \mathcal{E}^n(\mathbf{K})\, d^3\mathbf{K}$$

and $C_1 = 1$. We thus get

$$\frac{\sigma}{\tau} = \left[\frac{3}{\tau}\left(1 - \frac{\tau}{\tau_C}\right)\right]^{1/2} \left[1 + 3 \sum_{l=1}^{\infty} \frac{2^{2l} B_{2l}}{(2l)!} C_{2l-1} \left(\frac{\sigma}{\tau}\right)^{2l-2}\right]^{-1/2} \tag{17.133}$$

To obtain a solution of (17.133) by iteration we put, in the zeroth approximation,

$$\sigma=\tau\left[\frac{3}{\tau}\left(1-\frac{\tau}{\tau_C}\right)\right]^{1/2} \tag{17.134}$$

which corresponds to the usual molecular field approximation. It is seen from (17.133) and (17.134) that there is nonvanishing spontaneous magnetization σ for $\tau<\tau_C$, and that it vanishes at $\tau=\tau_C$, thus predicting correctly the phase transition. The Curie temperatures can be calculated from (17.132) in the nearest-neighbor approximation. The results for the three lattices are as follows:

$$\tau_C=\begin{cases}0.66 & \text{for the s.c. lattice}\\ 0.725 & \text{for the b.c.c. lattice}\\ 0.745 & \text{for the f.c.c. lattice}\end{cases}$$

Case 3. $T>T_C$, $L\neq 0$

For comparison with Opechowski's[18] result, we write $\coth(x)$ in the following form:

$$\begin{aligned}\coth\left(\frac{L+\sigma\mathcal{E}(\mathbf{K})}{\tau}\right)&=\frac{1+\tanh(L/\tau)\tanh[\sigma\mathcal{E}(\mathbf{K})/\tau]}{\tanh(L/\tau)+\tanh[\sigma\mathcal{E}(\mathbf{K})/\tau]}\\ &=\frac{1+t_0t_1}{t_0+t_1},\qquad \text{say}\\ &=(1+t_0t_1)\frac{1}{t_0}\left(1+\frac{t_1}{t_0}\right)^{-1}\\ &=\frac{1}{t_0}\left[1+(1-t_0^2)\sum_{l=1}^{\infty}(-1)^l\left(\frac{t_1}{t_0}\right)^l\right]\end{aligned} \tag{17.135}$$

Hence we get from (17.121)

$$\frac{1}{\sigma}=\frac{1}{t_0}\left\{1+(1-t_0^2)\sum_{l=1}^{\infty}\frac{(-1)^l}{t_0^l}\cdot\frac{v}{(2\pi)^3}\int\left[\tanh\frac{\sigma\mathcal{E}(\mathbf{K})}{\tau}\right]^l d^3\mathbf{K}\right\} \tag{17.136}$$

If we now use the expansion of tanh $(\sigma\mathcal{E}(\mathbf{K})/\tau)$ in powers of its argument and restrict ourselves to terms up to $(\sigma/\tau)^2$, we get the following expression for σ:

$$\frac{1}{\sigma}=\frac{1}{t_0}\left\{1-\frac{1-t_0^2}{t_0}\frac{\sigma}{\tau}+\frac{1-t_0^2}{t_0^2}\left(\frac{\sigma}{\tau}\right)^2C_2\right\} \tag{17.137}$$

where C_2 is defined as before. Up to terms in τ^{-2}, we get

$$\sigma = t_0 + t_0(1-t_0^2)\frac{1}{\tau} + t_0(1-t_0^2)(2-C_2-2t_0^2)\frac{1}{\tau^2}$$

In the nearest-neighbor approximation

$$C_2 = \frac{1+\nu}{\nu}$$

ν being the number of nearest neighbors. We then obtain

$$\sigma = t_0 + t_0(1-t_0^2)\frac{1}{\tau} + t_0(1-t_0^2)\left(\frac{\nu-1}{\nu} - 2t_0^2\right)\frac{1}{\tau^2} \tag{17.138}$$

We have written (17.138) in this form for comparison with Opechowski's result, which was obtained by thermodynamic perturbation theory. Opechowski's result, of course, differs from (17.138) by a quantity of the order $1/\nu\tau^2$. Here also the agreement is very satisfactory.

Equation (17.121) thus gives very satisfactory results for the magnetization at all temperatures and also describes the phase transition from the ferromagnetic to the paramagnetic phase. Thus it is very remarkable that the Green function method offers a well-defined possibility to investigate the phase changes. Once we know the correlation function $\langle \mathbf{b}_f^\dagger \mathbf{b}_g \rangle$, it should be easy to evaluate other thermodynamic quantities, for instance, the average energy. We must note that the solution studied here exists for arbitrary fields only if the exchange integral is positive.

17.4 FERROMAGNETISM: GENERAL SPIN S CASE

Tahir-Kheli and ter Haar[19] extended the random phase decoupling method to derive the magnetization for the general spin S case and compared these values with Dyson's results at low temperature, with the spherical model results just below the Curie temperature, and with the molecular field theory results for the high-temperature region.

The following properties of the spin operators are used:

$$\mathbf{S}_l^{\pm} = \mathbf{S}_l^x \pm i\mathbf{S}_l^y$$

$$\left[\mathbf{S}_l^+ \mathbf{S}_g^-\right] = 2\hbar \mathbf{S}_g^z \delta_{lg}$$

$$\left[\mathbf{S}_l^{\pm}, \mathbf{S}_g^z\right] = \mp \hbar \mathbf{S}_g^{\pm} \delta_{lg} \tag{17.139}$$

$$\mathbf{S}_l^- \mathbf{S}_l^+ = \hbar^2 S(S+1) - \hbar \mathbf{S}_l^z - (\mathbf{S}_l^z)^2$$

A generalization of the last equation in (17.139) is

$$(\mathbf{S}_l^-)^n(\mathbf{S}_l^+)^n = (\mathbf{S}_l^-)^{n-1}\left[\hbar^2 S(S+1) - \hbar\mathbf{S}_l^z - (\mathbf{S}_l^z)^2\right](\mathbf{S}_l^+)^{n-1}$$
$$= \hbar^2 S(S+1)(\mathbf{S}_l^-)^{n-1}(\mathbf{S}_l^+)^{n-1} - (\mathbf{S}_l^-)^{n-2}\hbar\mathbf{S}_l^-\mathbf{S}_l^z(S_l^+)^{n-1}$$
$$- (\mathbf{S}_l^-)^{n-2}\mathbf{S}_l^-(\mathbf{S}_l^z)^2(\mathbf{S}_l^+)^{n-1}$$

Now,

$$\left[\mathbf{S}_l^-, \mathbf{S}_l^z\right] = \mathbf{S}_l^-\mathbf{S}_l^z - \mathbf{S}_l^z\mathbf{S}_l^- = \hbar\mathbf{S}_l^-$$

or

$$\mathbf{S}_l^-\mathbf{S}_l^z = (\mathbf{S}_l^z + \hbar)\mathbf{S}_l^-$$

Similarly,

$$\mathbf{S}_l^-(\mathbf{S}_l^z)^2 = (\mathbf{S}_l^z + \hbar)^2\mathbf{S}_l^-$$

Thus

$$(\mathbf{S}_l^-)^n(\mathbf{S}_l^+)^n = \hbar^2 S(S+1)(\mathbf{S}_l^-)^{n-1}(\mathbf{S}_l^+)^{n-1} - (\mathbf{S}_l^-)^{n-2}\hbar(\mathbf{S}_l^z + \hbar)\mathbf{S}_l^-$$
$$\cdot(\mathbf{S}_l^+)^{n-1} - (\mathbf{S}_l^-)^{n-2}(\mathbf{S}_l^z + \hbar)^2\mathbf{S}_l^-(\mathbf{S}_l^+)^{n-1}$$
$$= \hbar^2 S(S+1)(\mathbf{S}_l^-)^{n-1}(\mathbf{S}_l^+)^{n-1}$$
$$- (\mathbf{S}_l^-)^{n-2}\left[(\mathbf{S}_l^z)^2 + 3\hbar\mathbf{S}_l^z + 2\hbar^2\right]\mathbf{S}_l^-(\mathbf{S}_l^+)^{n-1}$$

and

$$(\mathbf{S}_l^-)^{n-3}(\mathbf{S}_l^-)\left[(\mathbf{S}_l^z)^2 + 3\hbar\mathbf{S}_l^z + 2\hbar^2\right](\mathbf{S}_l^-)$$
$$= (\mathbf{S}_l^-)^{n-3}\left[(\mathbf{S}_l^z + \hbar)^2 + 3\hbar(\mathbf{S}_l^z + \hbar) + 2\hbar^2\right](\mathbf{S}_l^-)$$
$$= (\mathbf{S}_l^-)^{n-3}\left[(\mathbf{S}_l^z)^2 + 5\hbar\mathbf{S}_l^z + 6\hbar^2\right](\mathbf{S}_l^-)^2$$

Thus the power of $(\mathbf{S}_l^-)^{n-3}$ is decreasing while $(\mathbf{S}_l^-)^2$ on the right-hand-side corner is increasing. Proceeding up to $(\mathbf{S}_l^-)^2$, we find that the expression given above reduces to

$$\left[(\mathbf{S}_l^z)^2 + (2n-1)\hbar\mathbf{S}_l^z + n(n-1)\hbar^2\right](\mathbf{S}_l^-)^{n-1}$$

Hence

$$(\mathbf{S}_l^-)^n(\mathbf{S}_l^+)^n = \left[\hbar^2 S(S+1) - n(n-1)\hbar^2 - (2n-1)\hbar\mathbf{S}_l^z\right.$$
$$\left.- (\mathbf{S}_l^z)^2\right](\mathbf{S}_l^-)^{n-1}(\mathbf{S}_l^+)^{n-1}$$

from which we finally get

$$(\mathbf{S}_l^-)^n(\mathbf{S}_l^+)^n = \prod_{p=1}^{n} \left[S(S+1)\hbar^2 - (n-p)(n-p+1)\hbar^2 \right.$$
$$\left. - (2n-2p+1)\hbar \mathbf{S}_l^z - (\mathbf{S}_l^z)^2 \right] \tag{17.140}$$

The following operator equation is also required:

$$\prod_{r=-s}^{+s} (S^z - rh) + 0 \tag{17.141}$$

where r takes integral or half-integral values according to whether S is integral or half-integral. Finally,

$$\begin{aligned}
&\left[(\mathbf{S}_l^+), (\mathbf{S}_l^-)^n(\mathbf{S}_l^+)^{n-1}\right]_- \\
&\quad = \mathbf{S}_l^+(\mathbf{S}_l^-)^n(\mathbf{S}_l^+)^{n-1} - (\mathbf{S}_l^-)^n(\mathbf{S}_l^+)^{n-1}(\mathbf{S}_l^+) \\
&\quad = \mathbf{S}_l^+\mathbf{S}_l^-(\mathbf{S}_l^-)^{n-1}(\mathbf{S}_l^+)^{n-1} - (\mathbf{S}_l^-)^n(\mathbf{S}_l^+)^n \\
&\quad = \left[\mathbf{S}_l^-\mathbf{S}_l^+ + 2\hbar\mathbf{S}_l^z\right](\mathbf{S}_l^-)^{n-1}(\mathbf{S}_l^+)^{n-1} \\
&\qquad - \left[\hbar^2 S(S+1) - n(n-1)\hbar^2 - (2n-1)\hbar\mathbf{S}_l^z - (\mathbf{S}_l^z)^2\right] \\
&\qquad \cdot (\mathbf{S}_l^-)^{n-1}(\mathbf{S}_l^+)^{n-1} \\
&\quad = \left[2n\hbar\mathbf{S}_l^z + (n^2-n)\hbar^2\right](\mathbf{S}_l^-)^{n-1}(\mathbf{S}_l^+)^{n-1} \\
&\quad = \left[2n\hbar\mathbf{S}_l^z + (n^2-n)\hbar^2\right] \prod_{p=1}^{n-1} \left[\hbar^2 S(S+1) - (n-p-1)(n-p)\hbar^2 \right. \\
&\qquad \left. - (2n-2p-1)\hbar\mathbf{S}_l^z - (\mathbf{S}_l^z)^2\right]
\end{aligned} \tag{17.142}$$

where we have used (17.140).

Our basic Hamiltonian is

$$\mathcal{H} = -g\mu_B H \sum_f \mathbf{S}_f^z - \sum_{f,m} J(\mathbf{f}-\mathbf{m})\mathbf{S}_f \cdot \mathbf{S}_m \tag{17.143}$$

To find the correlation function of the form $\langle(\mathbf{S}_l^-)^n(\mathbf{S}_l^+)^n\rangle$ we will study the Green function $\langle\langle \mathbf{S}_g^+ ; (\mathbf{S}_l^-)^n(\mathbf{S}_l^+)^{n-1}\rangle\rangle$, where n is a positive integer.

The equation of motion is, by (17.93),

$$E\langle\langle \mathbf{S}_g^+ ; (\mathbf{S}_l^-)^n(\mathbf{S}_l^+)^{n-1}\rangle\rangle = \frac{1}{2\pi}\langle\left[\mathbf{S}_g^+, (\mathbf{S}_l^-)^n(\mathbf{S}_l^+)^{n-1}\right]_-\rangle$$
$$+ \langle\langle\left[\mathbf{S}_g^+, \mathcal{H}\right]_- ; (\mathbf{S}_l^-)^n(\mathbf{S}_l^+)^{n-1}\rangle\rangle \tag{17.144}$$

By using the spin commutation relations, it can be easily shown that

$$\left[\mathbf{S}_g^+,\mathcal{H}\right]=g\mu_{\mathrm{B}}\hbar H\mathbf{S}_g^+-2\hbar\sum_f J(\mathbf{g}-\mathbf{f})(\mathbf{S}_g^z\mathbf{S}_f^+-\mathbf{S}_f^z\mathbf{S}_g^+) \tag{17.145}$$

Using (17.142) and (17.145) in (17.144), we get

$$\begin{aligned}
&E\langle\langle\mathbf{S}_g^+;(\mathbf{S}_l^-)^n(\mathbf{S}_l^+)^{n-1}\rangle\rangle\\
&=\frac{\delta_{gl}}{2\pi}\cdot\langle\left[2n\hbar\mathbf{S}_l^z+(n^2-n)\hbar^2\right]\prod_{p=1}^{n-1}\left[S(S+1)\hbar^2\right.\\
&\left.-(n-p-1)(n-p)\hbar^2-(2n-2p-1)\hbar\mathbf{S}_l^z-(\mathbf{S}_l^z)^2\right]\\
&+g\mu_{\mathrm{B}}H\langle\langle\mathbf{S}_g^+;(\mathbf{S}_l^-)^n(\mathbf{S}_l^+)^{n-1}\rangle\rangle\\
&-2\sum_f J(\mathbf{g}-\mathbf{f})\langle\langle(\mathbf{S}_g^z\mathbf{S}_f^+-\mathbf{S}_f^z\mathbf{S}_g^+);(\mathbf{S}_l^-)^n(\mathbf{S}_l^+)^{n-1}\rangle\rangle
\end{aligned} \tag{17.146}$$

Note that (17.146) involves higher order Green functions in the last term, and we decouple it by using the random phase approximation, that is,

$$\begin{aligned}
\langle\langle\mathbf{S}_g^z\mathbf{S}_f^+;(\mathbf{S}_l^-)^n(\mathbf{S}_l^+)^{n-1}\rangle\rangle&=\langle\mathbf{S}_g^z\rangle\langle\langle\mathbf{S}_f^+;(\mathbf{S}_l^-)^n(\mathbf{S}_l^+)^{n-1}\rangle\rangle\\
&=\langle\mathbf{S}_g^z\rangle\langle\langle\mathbf{S}_f^+;(\mathbf{S}_l^-)^n(\mathbf{S}_l^+)^{n-1}\rangle\rangle
\end{aligned} \tag{17.147}$$

The last step follows because the translational invariance makes any average $\langle\rangle$ independent of the site index. Taking the Fourier transforms yields

$$\begin{aligned}
\langle\langle\mathbf{S}_g^+;(\mathbf{S}_l^-)^n(\mathbf{S}_l^+)^{n-1}\rangle\rangle&=\frac{1}{N}\sum_{\mathbf{K}}G_{\mathbf{K}}(E)e^{i\mathbf{K}(\mathbf{g}-\mathbf{l})}\\
\delta_{gl}&=\frac{1}{N}\sum_{\mathbf{K}}e^{i\mathbf{K}(\mathbf{g}-\mathbf{l})}\\
J(\mathbf{g}-\mathbf{f})&=\frac{1}{N}\sum_{\mathbf{K}}e^{-i\mathbf{K}(\mathbf{g}-\mathbf{f})}J(\mathbf{K})
\end{aligned} \tag{17.148}$$

Using (17.147) and (17.148), we find that (17.146) reduces to

$$\begin{aligned}
(E-g\mu_BH)G_{\mathbf{K}}(E)&=\frac{1}{2\pi}\langle\left[2n\hbar\mathbf{S}_z+\hbar^2(n^2-n)\right]\\
&\prod_{p=1}^{n-1}\left[S(S+1)\hbar^2-(n-p-1)(n-p)\hbar^2-(2n-2p-1)\hbar\mathbf{S}^z\right.\\
&\left.-(\mathbf{S}^z)^2\right]\rangle+2\langle\mathbf{S}^z\rangle\left[J(0)-J(\mathbf{K})\right]G_{\mathbf{K}}(E)
\end{aligned}$$

Thus we get

$$2\pi\left(E-E_{\mathbf{K}}^{(S)}\right)G_{\mathbf{K}}(E)=\langle\left[2n\hbar S^{z}+\hbar^{2}(n^{2}-n)\right]$$
$$\prod_{p=1}^{n-1}\left[S(S+1)\hbar^{2}-(n-p-1)(n-p)\hbar^{2}-(2n-2p-1)\hbar S^{z}-(S^{z})^{2}\right]\rangle \tag{17.149}$$

where

$$E_{\mathbf{K}}^{(S)}=g\mu_{B}H+2\hbar\langle S^{z}\rangle\left[J(0)-J(\mathbf{K})\right] \tag{17.150}$$

is the magnetization-renormalized energy eigenvalue.

From (17.94) we get

$$\langle(\mathbf{S}_{l}^{-})^{n}(\mathbf{S}_{l}^{+})^{n}\rangle=\lim_{\epsilon\to 0}\frac{1}{N}i\sum_{\mathbf{K}}(e^{\beta E}-1)^{-1}e^{-1(E/\hbar)(t-t')}.\left[G_{\mathbf{K}}(E)_{E+i\epsilon}-G_{\mathbf{K}}(E)_{E-i\epsilon}\right]$$

$$=i\frac{1}{N}\lim_{\epsilon\to 0}\sum_{\mathbf{K}}(e^{\beta E}-1)^{-1}e^{-i(E/\hbar)(t-t')}.\frac{A}{2\pi}\left[\frac{1}{E-E_{\mathbf{K}}^{(S)}+i\epsilon}-\frac{1}{E-E_{\mathbf{K}}^{(S)}-i\epsilon}\right]$$

$$=i\frac{1}{N}\sum_{\mathbf{K}}(e^{\beta E}-1)^{-1}e^{-i(E/\hbar)(t-t')}\frac{A}{2\pi}\lim_{\epsilon\to 0}\left[\frac{1}{E-E_{\mathbf{K}}^{(S)}+i\epsilon}-\frac{1}{E-E_{\mathbf{K}}^{(S)}-i\epsilon}\right]$$

$$=\sum_{\mathbf{K}}\frac{1}{N}(e^{\beta E}-1)^{-1}\frac{A}{2\pi}i(-2\pi i)\,\delta\left(E-E_{\mathbf{K}}^{(S)}\right),\quad\text{for}\quad t=t'$$

$$=\frac{1}{N}\sum_{\mathbf{K}}\frac{A}{e^{\beta E_{\mathbf{K}}^{(S)}}-1}$$

$$=A\phi(S) \tag{17.151}$$

where A denotes the entire quantity on the right-hand side of (17.149), $\beta=1/kT$, and

$$\phi(S)=\frac{1}{N}\sum_{\mathbf{K}}(e^{\beta E_{\mathbf{K}}(S)}-1)^{-1} \tag{17.152}$$

We note that, if we use (17.140), we have on both sides of (17.151) a sum of averages of powers of $\mathbf{S}^z$. This yields $2S$ independent simultaneous linear equations in $\langle\mathbf{S}^z\rangle,\langle(\mathbf{S}^z)^2\rangle,\ldots,\langle(\mathbf{S}^z)^{2S}\rangle$ by putting $n=1,2,\ldots,2S$ in (17.151). For $n>2S$ the equations are not independent because of (17.141).

We now go to individual cases. For $S=\frac{1}{2}$ we get from (17.149), for $n=1$ and $p=0$,

$$\langle 2\hbar\mathbf{S}^z\rangle\phi\left(\tfrac{1}{2}\right)=\langle\mathbf{S}^-\mathbf{S}^+\rangle=\langle\tfrac{3}{4}\hbar^2-\hbar\mathbf{S}^z-(\mathbf{S}^z)^2\rangle \tag{17.153}$$

Also, from (17.141),

$$(\mathbf{S}^z - \tfrac{1}{2}\hbar)(\mathbf{S}^z + \tfrac{1}{2}\hbar) = 0$$

or

$$S_z^2 = \tfrac{1}{4}\hbar^2$$

Thus

$$\langle 2\hbar S^z \rangle \phi(\tfrac{1}{2}) = \langle \tfrac{1}{2}\hbar^2 - \hbar S^z \rangle \tag{17.154}$$

or

$$\langle S^z \rangle_{S=1/2} = \frac{\hbar}{2\left[1 + 2\phi(\tfrac{1}{2})\right]}$$

Similarly, for $S = 1(n = 1$ and $p = 0)$,

$$\langle 2\hbar \mathbf{S}^z \rangle \phi(1) = \langle \mathbf{S}^- \mathbf{S}^+ \rangle = \langle 2\hbar^2 - \hbar S^s - (S^z)^2 \rangle \tag{17.155}$$

Also, from (17.149), for $n = 2$ and $p = 1$,

$$\begin{aligned}
&\langle [4\hbar \mathbf{S}^z + 2\hbar^2][S(S+1)\hbar^2 - \hbar \mathbf{S}^z - (\mathbf{S}^z)^2] \rangle \phi(1) \\
&\quad = \langle (\mathbf{S}^-)^2 (\mathbf{S}^+)^2 \rangle \\
&\quad = \langle [S(S+1)\hbar^2 - 2\hbar^2 - 3\hbar S^z - (S^z)^2][S(S+1)\hbar^2 - \hbar S^z - (S^z)^2] \rangle \\
&\quad = \langle [-6\hbar^3 S^z + \hbar^2 (S^z)^2 + 4\hbar (S^z)^3 + (S^z)^4] \rangle \qquad \text{(from 17.140)}
\end{aligned}$$

Now, simplyfying the left-hand side of this equation, we get

$$\begin{aligned}
&\langle 4\hbar^4 + 6\hbar^3 S^z - 6\hbar^2 (S^z)^2 - 4\hbar (S^z)^3 \rangle \phi(1) \\
&\quad = \langle -6\hbar^3 S^z + \hbar^2 (S^z)^2 + 4\hbar (S^z)^3 + (S^z)^4 \rangle
\end{aligned} \tag{17.156}$$

But, by (17.141),

$$(S^z - \hbar) S^z (S^z + \hbar) = 0$$

or

$$(S^z)^3 = \hbar^2 S^z \tag{17.157}$$

Now, eliminating $(S^z)^2$, $(S^z)^3$, and $(S^z)^4$ from (17.155) to (17.157), we obtain

$$\langle S^z \rangle_{S=1} = \hbar \frac{1 + 2\phi(1)}{1 + 3\phi(1) + 3[\phi(1)]^2} \tag{17.158}$$

The results for higher values of S are obtained in exactly similar fashion, and we get

$$\begin{aligned}
\langle S^z\rangle_{S=3/2} &= \hbar\frac{\frac{3}{2}+5\phi(\frac{3}{2})+5\left[\phi(\frac{3}{2})\right]^2}{\left[1+\phi(\frac{3}{2})\right]^4-\left[\phi(\frac{3}{2})\right]^4}\\
\langle S^z\rangle_{S=2} &= \hbar\frac{2+9\phi(2)+15[\phi(2)]^2+10[\phi(2)]^3}{[1+\phi(2)]^5-[\phi(2)]^5}\\
\langle S^z\rangle_{S=5/2} &= \hbar\frac{\frac{5}{2}+14\phi(\frac{5}{2})+\frac{63}{2}\left[\phi(\frac{5}{2})\right]^2+35\left[\phi(\frac{5}{2})\right]^3+\frac{35}{2}\left[\phi(\frac{5}{2})\right]^4}{\left[1+\phi(\frac{5}{2})\right]^6-\left[\phi(\frac{5}{2})\right]^6}\\
\langle S^z\rangle_{S=3} &= \hbar\frac{3+20\phi(3)+56[\phi(3)]^2+84[\phi(3)]^3+70[\phi(3)]^4+28[\phi(3)]^5}{[1+\phi(3)]^7-[\phi(3)]^7}
\end{aligned} \tag{17.159}$$

Now our aim is to obtain the series expansion for $\phi(3)$, which is required for the calculation of magnetization. Introducing

$$\frac{1}{\tau}=\beta J(0)\hbar^2 \text{ where } \beta=\frac{1}{kT}$$

We have from (17.150) and (17.152)

$$\phi(S)=\frac{1}{N}\sum_{\mathbf{K}}\left[\exp\left(\frac{2\eta(\mathbf{K})\cdot\langle S^z\rangle}{\hbar\tau}\right)-1\right]^{-1} \tag{17.160}$$

where

$$\eta(\mathbf{K})=1-\frac{J(\mathbf{K})}{J(0)},\quad \text{for}\quad H=0 \tag{17.161}$$

In the nearest-neighbor approximation for the s.c., the b.c.c., and the f.c.c. lattices, respectively, it can be shown that

$$\begin{aligned}
J(\mathbf{K})_{\text{s.c.}} &= \tfrac{1}{3}J(0)\left[\cos(K_x a)+\cos(K_y a)+\cos(K_z a)\right]\\
J(\mathbf{K})_{\text{b.c.c.}} &= J(0)\left[\cos(\tfrac{1}{2}K_x a)\cos(\tfrac{1}{2}K_y a)\cos(\tfrac{1}{2}K_z a)\right]\\
J(\mathbf{K})_{\text{f.c.c.}} &= \tfrac{1}{3}J(0)\left[\cos(\tfrac{1}{2}K_x a)\cos(\tfrac{1}{2}K_y a)+\cos(\tfrac{1}{2}K_y a)\cos(\tfrac{1}{2}K_z a)\right.\\
&\quad\left.+\cos(\tfrac{1}{2}K_z a)\cos(\tfrac{1}{2}K_x a)\right]
\end{aligned} \tag{17.162}$$

where a is the lattice constant.

These results are obvious if one remembers that the nearest-neighbor distance is μa; here μ is given by $\mu^2=6/z$, where z is the number of nearest neighbors; 6 for s.c., 8 for b.c.c., and 12 for f.c.c. lattices, respectively.

For the low-temperature region the sum over the first Brillouin zone of the **K**-space is, as usual, replaced by an integral over the whole of **K**-space, and we obtain from (17.160)

$$\phi(S)=\frac{v}{(2\pi)^3}\int d^3\mathbf{K}\sum_{\mathbf{K}=1}^{\infty}\exp\left[-\frac{2r\eta(\mathbf{K})\langle S^z\rangle}{\hbar\tau}\right] \tag{17.163}$$

We expand $\eta(\mathbf{K})$ in powers of **K**, as **K** is small in magnitude, and then evaluate (17.163). We illustrate the process for the s.c. lattice only. From (17.162)

$$\eta(\mathbf{K})=1-\frac{J(\mathbf{K})}{J(0)}$$
$$=\frac{1}{3}\left(\frac{K_x^2+K_y^2+K_z^2}{2}a^2-\frac{K_x^4+K_y^4+K_z^4}{24}a^4+\frac{K_x^6+K_y^6+K_z^6}{720}a^6-\cdots\right)$$

Therefore

$$\exp\left[-\frac{2r\langle S^z\rangle}{\hbar\tau}\eta(\mathbf{K})\right]=\exp\left[-\frac{r\langle S^z\rangle}{\hbar\tau}\frac{K^2a^2}{3}\right]$$
$$\cdot\left[1+\frac{r\langle S^z\rangle}{\hbar\tau}\cdot\frac{K_x^4+K_y^4+K_z^4}{36}a^4-\frac{r\langle S^z\rangle}{\hbar\tau}\cdot\frac{K_x^6+K_y^6+K_z^6}{1080}a^6\right.$$
$$\left.+\frac{r^2\langle S^z\rangle^2}{2(36)^2\hbar^2\tau^2}a^8\left(K_x^8+K_y^8+K_z^8+2K_x^4K_y^4+2K_y^4K_z^4+2K_z^4K_x^4\right)\right] \tag{17.164}$$

after some lengthy manipulation.

Now, let us evaluate the integral in (17.163) for the terms in (17.164).

Integral for first term:

$$\int_{-\infty}^{\infty}\exp\left(-\frac{r\langle S^z\rangle}{\hbar\tau}\frac{K^2a^2}{3}\right)d^3\mathbf{K}=\left(\frac{3\pi\hbar\tau}{r\langle S^z\rangle a^2}\right)^{3/2}=\frac{\pi^{3/2}}{a^3}\left(\frac{3\hbar\tau}{\langle S^z\rangle}\right)^{3/2}\frac{1}{r^{3/2}} \tag{A}$$

Integral for second term:

$$\int_{-\infty}^{\infty}\exp\left[-\frac{r\langle S^z\rangle K_x^2a^2}{3\hbar\tau}\right]a^4\frac{r\langle S^z\rangle}{36h\tau}K_x^4\,dK_x$$
$$\cdot\int_{-\infty}^{\infty}\exp\left[-\frac{r\langle S^z\rangle K_y^2a^2}{3\hbar\tau}\right]dK_y\int_{-\infty}^{\infty}\exp\left[-\frac{r\langle S^z\rangle K_z^2a^2}{3\hbar\tau}\right]dk_z$$
$$=\left(\frac{3\hbar\tau}{r\langle S^z\rangle a^2}\right)^{5/2}\frac{r\langle S^z\rangle a^4}{36\hbar\tau}\left(\frac{3\hbar\tau}{r\langle S^z\rangle a^2}\right)^{1/2}\left(\frac{3\hbar\tau}{r\langle S^z\rangle a^2}\right)^{1/2}\tfrac{3}{4}\pi^{3/2}$$
$$=\frac{\pi^{3/2}}{16}\left(\frac{3\hbar\tau}{\langle S^z\rangle}\right)^{5/2}\frac{1}{a^3}\frac{1}{r^{5/2}}$$

From (17.164) we see that there are three such integrals. Therefore, second term of (17.164) gives

$$\frac{3\pi^{3/2}}{16}\left(\frac{3\hbar\tau}{\langle S^z\rangle}\right)^{5/2}\frac{1}{a^3}\cdot\frac{1}{r^{5/2}} \tag{B}$$

Integral for third term:

$$\int_{-\infty}^{\infty}\exp\left[-\frac{r\langle S^z\rangle}{3\hbar\tau}K_x^2a^2\right]\frac{r\langle S^z\rangle}{1080\tau}a^6K_x^6\,dK_x$$

$$\cdot\int_{-\infty}^{\infty}\exp\left[-\frac{r\langle S^z\rangle}{3\hbar\tau}K_y^2a^2\right]dK_y\int_{-\infty}^{\infty}\exp\left[-\frac{r\langle S^z\rangle K_z^2a^2}{3\hbar\tau}\right]dK_z$$

$$=\left(\frac{3\hbar\tau}{r\langle S^z\rangle a^2}\right)^{7/2}\frac{r\langle S^z\rangle a^6}{1080\hbar\tau}\left(\frac{3\hbar\tau}{r\langle S^z\rangle a^2}\right)^{1/2}\left(\frac{3\hbar\tau}{r\langle S^z\rangle a^2}\right)^{1/2}\cdot\left(\tfrac{5}{3}\cdot\tfrac{3}{2}\cdot\tfrac{1}{2}\cdot\pi^{3/2}\right)$$

$$=\frac{5\cdot 3^2\cdot\pi^{3/2}}{2^3\cdot 1080}\left(\frac{3\hbar\tau}{\langle S^z\rangle}\right)^{7/2}\frac{1}{a^3}\cdot\frac{1}{r^{7/2}}.$$

From (17.164) we see that there are three such integrals. Hence the third term gives (with a minus sign)

$$\frac{5\cdot 3^3\cdot\pi^{3/2}}{2^3\cdot 1080}\cdot\frac{1}{a^3}\left(\frac{3\hbar\tau}{\langle S^z\rangle}\right)^{7/2}\frac{1}{r^{7/2}} \tag{C}$$

For the fourth term in (17.164) there will be two types of integrals.

First type of integral for the fourth term:

$$\int_{-\infty}^{\infty}\exp\left[-\frac{r\langle S^z\rangle K_x^2a^2}{3\hbar\tau}\right]\frac{r^2\langle S^z\rangle^2}{2\cdot(36)^2\hbar^2\tau^2}K_x^8\,dK_x$$

$$\cdot\int_{-\infty}^{\infty}\exp\left[-\frac{r\langle S^z\rangle K_y^2a^2}{3\hbar\tau}\right]dK_y\int_{-\infty}^{\infty}\exp\left[-\frac{r\langle S^z\rangle K_z^2a^2}{3\hbar\tau}\right]dK_z$$

$$=\frac{7\cdot5\cdot3\cdot{}^3\cdot\pi^{3/2}}{2^5\cdot(36)^2}\frac{1}{a^3}\left(\frac{3\hbar\tau}{\langle S^z\rangle}\right)^{7/2}\frac{1}{r^{7/2}}.$$

There are three such integrals, giving

$$\frac{7\cdot5\cdot3^4\cdot\pi^{3/2}}{2^5\cdot(36)^2}\cdot\frac{1}{a^3}\left(\frac{3\hbar\tau}{\langle S^z\rangle}\right)^{7/2}\frac{1}{r^{7/2}} \tag{D}$$

Second type of integral for fourth term:

$$\frac{6\cdot3^4\cdot\pi^{3/2}}{2^5\cdot(36)^2}\frac{1}{a^3}\left(\frac{3\hbar\tau}{\langle S^z\rangle}\right)^{7/2}\frac{1}{r^{7/2}} \tag{E}$$

From (C), (D), and (E), we see that the coefficient of

$$\frac{\pi^{3/2}}{a^3}\cdot\left(\frac{3\hbar\tau}{\langle S^z\rangle}\right)^{7/2}\frac{1}{r^{7/2}}=\frac{7\cdot5\cdot3^4}{2^5\cdot(36)^2}+\frac{6\cdot3^4}{2^5\cdot(36)^2}-\frac{5\cdot3^3}{2^3\cdot1080}=\frac{33}{32}\cdot\frac{1}{2^4} \tag{F}$$

Therefore, the contribution of the third and fourth terms in (17.164) is

$$\frac{33}{32}\cdot\frac{1}{2^4}\cdot\frac{\pi^{3/2}}{a^3}\left(\frac{3\hbar\tau}{\langle S^z\rangle}\right)^{7/2}\frac{1}{r^{7/2}}$$

Now, collecting results (A), (B), and (F), we have for (17.163) ($v=a^3$ for a s.c. lattice)

$$\phi(S)=\frac{a^3}{(2\pi)^3}\sum_{r=1}^{\infty}\left[\frac{\pi^{3/2}}{a^3}\left(\frac{3\hbar\tau}{\langle S^z\rangle}\right)^{3/2}\cdot\frac{1}{r^{3/2}}+\frac{3\pi^{3/2}}{16a^3}\cdot\left(\frac{3\hbar\tau}{\langle S^z\rangle}\right)^{5/2}\frac{1}{r^{5/2}}\right.$$
$$\left.+\frac{33}{32}\cdot\frac{1}{2^4}\cdot\frac{\pi^{3/2}}{a^3}\left(\frac{3\hbar\tau}{\langle S^z\rangle}\right)^{7/2}\cdot\frac{1}{r^{7/2}}+\cdots\right.$$

Noting that

$$\sum_{r=1}^{\infty}\frac{1}{r^p}=\zeta(p)$$

where $\zeta(p)$ is the well-known Riemann zeta function, we get for the s.c. lattice

$$\phi(S)=\zeta(\tfrac{3}{2})\left(\frac{3\hbar\tau}{4\pi\langle S^z\rangle}\right)^{3/2}+\frac{3\pi}{4}\zeta(\tfrac{5}{2})\left(\frac{3\hbar\tau}{4\pi\langle S^z\rangle}\right)^{5/2}$$
$$+\frac{33\pi^2}{32}\cdot\zeta(\tfrac{7}{2})\left(\frac{3\hbar\tau}{4\pi\langle S^z\rangle}\right)^{7/2}+\cdots$$

Similarly, proceeding for b.c.c. and f.c.c. lattices, which are more complicated and noting that

$$v=\begin{bmatrix}\frac{1}{2}a^3\\ \frac{1}{4}a^3\end{bmatrix}\qquad\text{for}\begin{bmatrix}\text{b.c.c.}\\ \text{f.c.c.}\end{bmatrix}\text{lattices}$$

we get

$$\phi(S)=a_0\left(\frac{3\hbar\tau}{4\pi\langle S^z\rangle}\right)^{3/2}+a_1\left(\frac{3\hbar\tau}{4\pi\langle S^z\rangle}\right)^{5/2}+a_2\left(\frac{3\hbar\tau}{4\pi\langle S^z\rangle}\right)^{7/2}+\cdots \tag{17.165}$$

where

$$a_0=\zeta\left(\tfrac{3}{2}\right), \qquad a_1=\frac{3\pi}{4}\nu\zeta\left(\tfrac{5}{2}\right), \qquad a_2=\pi^2\omega\nu\zeta\left(\tfrac{7}{2}\right)$$

and

$$\nu=\begin{bmatrix} & 1 \\ \frac{3}{4} & 2^{2/3} \\ & 2^{1/3} \end{bmatrix}; \qquad \omega=\begin{bmatrix} 33/32 \\ 281/288 \\ 15/16 \end{bmatrix}$$

where ν and ω are listed for s.c., b.c.c., and f.c.c. lattices, respectively.

The next temperature range to be discussed is that just below the Curie temperature, where $\langle S^z\rangle/\hbar\tau\to 0$. From (17.160) we have

$$\begin{aligned}\phi(S)&=\frac{1}{N}\sum_{\mathbf{K}}\left[\exp\frac{2\eta(\mathbf{K})\langle S^z\rangle}{\hbar\tau}-1\right]^{-1}\\ &=\frac{1}{N}\sum_{\mathbf{K}}\frac{\hbar\tau}{2\eta(\mathbf{K})\langle S^z\rangle}\end{aligned}$$

or

$$2\phi(S)=\frac{1}{N}\sum_{\mathbf{K}}\frac{\hbar\tau}{\eta(\mathbf{K})\langle S^z\rangle} \tag{17.166}$$

Now,

$$\begin{aligned}\coth\frac{\eta(\mathbf{K})\rangle S^z\rangle}{\hbar\tau}-1&=\frac{\hbar\tau}{\eta(\mathbf{K})\langle S^z\rangle}-1=\frac{\hbar\tau-\eta(\mathbf{K})\langle S^z\rangle}{\eta(\mathbf{K})\langle S^z\rangle}\\ &\approx\frac{\hbar\tau}{\eta(\mathbf{K})\langle S^z\rangle} \qquad (\because \hbar\tau\rangle\rangle\langle S^z\rangle)\end{aligned} \tag{17.167}$$

Substituting (17.167) in (17.166), we get

$$1+2\phi(S)=\frac{1}{N}\sum_{\mathbf{K}}\coth\left[\frac{\eta(\mathbf{K})\langle S^z\rangle}{\hbar\tau}\right]$$

Using (17.131), we obtain

$$1+2\phi(S)=\frac{1}{N}\sum_{\mathbf{K}}\left[\frac{\hbar\tau}{\eta(\mathbf{K})\langle S^z\rangle}+\frac{\eta(\mathbf{K})\langle S^z\rangle}{3\hbar\tau}-\frac{1}{45}\left(\frac{\eta(\mathbf{K})\langle S^z\rangle}{\hbar\tau}\right)^3+\cdots\right. \tag{17.168}$$

Now we restrict our integration to the first Brillouin zone. Using (17.162), we evaluate

$$F(n)=\frac{1}{N}\sum_{\mathbf{K}}[\eta(\mathbf{K})]^n \tag{17.169}$$

This we do for the s.c. lattice. We have

$$F(1)=\frac{a^3}{(2\pi)^3}\iiint_{-\pi/a}^{\pi/a}\left[1-\tfrac{1}{3}(\cos K_xa+\cos K_ya+\cos K_za)\right]dK_x\,dK_y\,dK_z$$

$$=\frac{a^3}{(2\pi)^3}\left[\frac{(2\pi)^3}{a^3}-0\right]=1$$

$$F(2)=\frac{a^3}{(2\pi)^3}\iiint_{-\pi/a}^{\pi/a}\left[1-\tfrac{1}{3}(\cos K_xa+\cos K_ya+\cos K_za)\right]^2dK_x\,dK_y\,dK_z$$

$$=\frac{a^3}{(2\pi)^3}\iiint_{-\pi/a}^{\pi/a}\left[1+\tfrac{1}{9}(\cos^2K_xa+\cos^2K_ya+\cos^2K_za)\right]dK_x\,dK_y\,dK_z$$

$$=\frac{a^3}{(2\pi)^3}\left[\frac{(2\pi)^3}{a^3}+\tfrac{1}{9}\cdot3\cdot\tfrac{1}{2}\cdot\frac{2\pi}{a}\left(\frac{2\pi}{a}\right)^2\right]\qquad\text{(other integrals vanish)}$$

$$=1+\tfrac{1}{6}=1+\frac{1}{z}=\frac{z+1}{z}\qquad(\because z=6\text{ for s.c. lattice})$$

Similarly,

$$F(3)=\frac{a^3}{(2\pi)^3}\iiint_{-\pi/a}^{\pi/a}\left[1-\tfrac{1}{3}(\cos K_xa+\cos K_ya+\cos K_za)\right]^3dK_x\,dK_y\,dK_z$$

$$=\frac{a^3}{(2\pi)^3}\iiint_{-\pi/a}^{\pi/a}\left[1+\tfrac{3}{9}(\cos^2K_xa+\cos^2K_ya+\cos^2K_za)\right]dK_x\,dK_y\,dK_z$$

$$=\frac{a^3}{(2\pi)^3}\left[1+\tfrac{3}{9}\cdot3\cdot\tfrac{1}{2}\right]\left(\frac{2\pi}{a}\right)^3\qquad\text{(other integrals vanish)}$$

$$=\frac{z+3}{z}$$

Note that these results are same for all three lattices, so that

$$F(1)=1,\qquad F(2)=\frac{z+1}{z},\qquad F(3)=\frac{z+3}{z} \tag{17.170}$$

The sum $F(-1)$ is difficult to evaluate, but it has been derived by Watson.[20] The values are as follows:

$$F(-1)=\begin{cases}1.51638\text{ for the s.c. lattice}\\1.39320\text{ for the b.c.c. lattice}\\1.34466\text{ for the f.c.c. lattice}\end{cases} \tag{17.171}$$

Finally, for the high temperature region and $\mathbf{H}\neq 0$, we take expansion (17.135), that is,

$$[1+2\phi(S)]_{\tau>>1} \simeq 2[\phi(S)]_{\tau>>1} = \frac{1}{N}\sum_{\mathbf{K}} \coth\left[\tfrac{1}{2}g\mu_{\mathrm{B}}H\beta + \frac{\eta(\mathbf{K})\langle S^z\rangle}{\hbar\tau}\right]$$

$$= \frac{1}{N}\sum_{\mathbf{K}} \frac{1}{t_0}\left[1+(1-t_0^2)\sum_{n=1}^{\infty}(-1)^n\left(\frac{t_1}{t_0}\right)^n\right]$$

$$= \frac{1}{N}\sum_{\mathbf{K}}\left(1-\frac{t_1}{t_0}+\frac{t_1^2}{t_0^2}-\frac{t_1^3}{t_0^3}+\cdots\right) \tag{17.172}$$

where

$$t_0 = \tanh\left(\tfrac{1}{2}g\mu_{\mathrm{B}}H\beta\right), \qquad t_1 = \tanh\left(\frac{\eta(\mathbf{K})\langle S^z\rangle}{\hbar\tau}\right)$$

Note that, for $\tau>>1$, $t_0^2<<1$; hence $1-t_0^2\simeq 1$. Now, expanding the hyperbolic tangents t_0, t_1, and using (17.170), we get from (17.172)

$$[\phi(S)]_{\tau>>1} = \frac{1}{2t_0}\left[1-\frac{\langle S^z\rangle}{\hbar\tau t_0}F(1)+\left(\frac{\langle S^z\rangle}{\hbar\tau t_0}\right)^2 F(2)+O\left(\frac{1}{\tau^3}\right)\right] \tag{17.173}$$

where, since $\tau>>1$, we have neglected $1/\tau^3$ and also the higher terms, which are $<<1$.

We now use (17.165), (17.168), and (17.173) to study the behavior of the ferromagnet in the three temperature ranges.

1. The Low-Temperature Region

From (17.165) we find $\phi(S)$ to be small, and hence we can expand (17.154), (17.158), and similar equations for $\langle S^z\rangle$ in powers of $\phi(S)$ and thus of τ. From (17.154)

$$\langle S^z\rangle_{S=1/2} = \left[\tfrac{1}{2}-\phi\left(\tfrac{1}{2}\right)+2\left[\phi\left(\tfrac{1}{2}\right)\right]^2+O\left(\left[\phi\left(\tfrac{1}{2}\right)\right]^3\right)\right] \tag{17.174}$$

$$\langle S^z\rangle_{S\geqslant 1} = \left[S-\phi(S)+O\left([\phi(S)]^3\right)\right] \tag{17.175}$$

Substituting (17.165) in (17.174), we get

$$\langle S^z\rangle_{S=1/2} = \left[\tfrac{1}{2}-a_0\left(\frac{3\hbar\tau}{4\pi\nu\langle S^z\rangle}\right)^{3/2}-a_1\left(\frac{3\hbar\tau}{4\pi\nu\langle S^z\rangle}\right)^{5/2}\right.$$

$$\left.-a_2\left(\frac{3\hbar\tau}{4\pi\nu\langle S^z\rangle}\right)^{7/2}+\cdots+2a_0\left(\frac{3\hbar\tau}{4\pi\nu\langle S^z\rangle}\right)^3+4a_0a_1\left(\frac{3\hbar\tau}{4\pi\nu\langle S^z\rangle}\right)^3+\cdots\right]$$

(17.176)

This is solved by iteration. In the zeroth approximation we take $\langle S^z\rangle_{S=1/2}=\frac{1}{2}$, and in the first approximation

$$\begin{aligned}\langle S^z\rangle_{S=1/2}&=\tfrac{1}{2}-a_0\left(\frac{3\hbar\tau}{2\pi\nu}\right)^{3/2}-a_1\left(\frac{3\hbar\tau}{2\pi\nu}\right)^{5/2}\\&=\tfrac{1}{2}\left[1-2a_0\left(\frac{3\hbar\tau}{2\pi\nu}\right)^{3/2}-2a_1\left(\frac{3\hbar\tau}{2\pi\nu}\right)^{5/2}\right]\end{aligned}\tag{17.177}$$

Putting (17.177) on the right-hand side of (17.174), we obtain after expansion, collecting terms of equal powers of τ,

$$\begin{aligned}\langle S^z\rangle_{S=1/2}=\Bigg[&\tfrac{1}{2}-a_0\left(\frac{3\hbar\tau}{2\pi\nu}\right)^{3/2}-a_1\left(\frac{3\hbar\tau}{2\pi\nu}\right)^{5/2}-a_2\left(\frac{3\hbar\tau}{2\pi\nu}\right)^{7/2}\\&-a_0^2\left(\frac{3\hbar\tau}{2\pi\nu}\right)^{3}-4a_0a_1\left(\frac{3\hbar\tau}{2\pi\nu}\right)^{4}-\cdots\Bigg]\end{aligned}\tag{17.178}$$

Similarly, from (17.165) and (17.175), we get

$$\begin{aligned}\langle S^z\rangle_{S\geqslant 1}=\Bigg[&S-a_0\left(\frac{3\hbar\tau}{4\pi\nu S}\right)^{3/2}-a_1\left(\frac{3\hbar\tau}{4\pi\nu S}\right)^{5/2}-a_2\left(\frac{3\hbar\tau}{4\pi\nu S}\right)^{7/2}\\&-\cdots-\tfrac{3}{25}a_0^2\left(\frac{3\hbar\tau}{4\pi\nu S}\right)^{3}-\tfrac{4}{5}a_0a_1\left(\frac{3\hbar\tau}{4\pi\nu S}\right)^{4}-\cdots\Bigg]\end{aligned}\tag{17.179}$$

The results, when compared with Dyson's,[17] show that the terms proportional to τ°, $\tau^{3/2}$, $\tau^{5/2}$, and $\tau^{7/2}$ are the same as those of Dyson, but that there is a spurious τ^3 term in (17.178) and (17.179), which is absent from Dyson's results. The coefficient of τ^4 term also differs. Tahir-Kheli and ter Haar[19] pointed out that the correct spin-wave energy should be

$$E_{\mathbf{K}}^{(S)}=\alpha+\gamma\tau^{5/2}+\cdots\tag{17.180}$$

while that given by $E_{\mathbf{K}}^{(S)}$ in (17.150) [c.f. (17.181)] is of the form

$$E_{\mathbf{K}}^{(S)}=\alpha+\beta\tau^{3/2}+\cdots\tag{17.181}$$

Tahir-Kheli and ter Haar[21] showed in a second paper how to obtain the correct spin-wave energy by the Green function method. This, of course, we shall not discuss.

2. Region with Temperatures Just below the Curie Temperature

From (17.168) we see that, just below the Curie temperature, $\phi(S)\rangle 1$, and hence we expand $\langle S^z\rangle$ in inverse powers of $\phi(S)$. It follows from (17.151) and (17.141) that the expansion of $\langle S^z\rangle$ starts as

$$\langle S^z\rangle=\frac{S(S+1)}{3\phi(S)}+O([\phi(S)]^{-2})\tag{17.182}$$

Hence the Curie temperature is the temperature for which

$$3\phi(S)\langle S^Z\rangle = S(S+1), \qquad \langle S^z\rangle = 0 \tag{17.183}$$

But from (17.168)

$$1+2\phi(S) = \frac{1}{N}\sum_{\mathbf{K}} \frac{\hbar\tau}{\eta(\mathbf{K})\langle S^z\rangle}$$

Therefore

$$\frac{\langle S^z\rangle}{\hbar\tau} + \frac{2\phi(S)\langle S^z\rangle}{\hbar\tau} = \frac{1}{N}\sum_{\mathbf{K}} [\eta(\mathbf{K})]^{-1} = F(-1).$$

Therefore from (17.183)

$$\frac{2}{3}\cdot\frac{S(S+1)}{\hbar\tau_C} = F(-1)$$

or

$$\frac{2}{3}\,\frac{S(S+1)J(0)}{kT_C} = F(-1) \qquad \left[\because \tau = \frac{kT}{J(0)}\right]$$

In other words, the Curie temperature is given by

$$T_C = \frac{\theta_C}{F(-1)} \tag{17.184}$$

where θ_C is the Curie temperature, given by the molecular field theory [22]

$$k\theta_C = \tfrac{2}{3}S(S+1)J(0) \tag{17.185}$$

Note that the Curie temperature (17.184) is exactly the one given by the spherical model.[23] Expanding in powers of $(\tau_C-\tau)^2$ and writing $\sigma = \langle S^Z\rangle/S$, we find, retaining terms up to $(\tau_C-\tau)^2$, that

$$\sigma^2 = C_S\left(1-\frac{\tau}{\tau_C}\right) + D_S\left(1-\frac{\tau}{\tau_C}\right)^2 \tag{17.186}$$

where the coefficients C_S and D_S depend on the S values and the lattice structure. Tahir-Kheli and ter Haar have given the values for the three cubic lattices, and we shall not discuss them any more.

3. High-Temperature Region, $\tau \gg 1$

Here we assume that $\mathbf{H}\neq 0$. Now (17.173) can be used to find $\langle S^z\rangle$ and thus the susceptibility χ per atom. In the limit $\mathbf{H}\to 0, \beta\to 0, \phi(S)\gg 1$, so that we can use an expansion like (17.182). Up to terms of order T^{-2} we have

$$\chi=\frac{g^2\mu_B^2S(S+1)}{3kT}\left[1+\frac{\theta_C}{T}+\frac{z-1}{z}\left(\frac{\theta_C}{T}\right)^2+O\left(\frac{1}{T^3}\right)\right] \qquad (17.187)$$

Apart from the factor $(z-1)/z$ in the third term, (17.187) is identical up to T^{-3} with the result of the molecular field theory, that is, the Curie-Weiss law:[22]

$$\chi_{\text{mol. fields}}=\frac{g^2\mu_B^2S(S+1)}{3k(T-\theta_C)} \qquad (17.188)$$

We thus see from the work of Tahir-Kheli and ter Haar that the Green function method can be successfully employed to obtain the magnetization of a ferromagnet for the general spin S case for all ranges of temperature. This method is thus very powerful indeed.

17.5 GREEN FUNCTION THEORIES IN SYMMETRIC-TYPE DECOUPLING: SPIN S CASE

In an attempt to improve the low-temperature results, mainly to eliminate the spurious T^3 term, as seen at the end of Section 17.4, Callen[24] postulated a new type of decoupling, known as the symmetric type, and thus partially succeeded in his aim. We now discuss Callen's work with more emphasis on the decoupling procedure and the method of solution than on working out the details of the calculation, which are similar to those given in Section 17.4.

As a necessity for a better decoupling than the random phase type, Callen pointed out that the theories based on random phase decoupling give the quasi-particle energy renormalized by the magnetization at all temperatures, whereas from physical consideration[25] this energy should be renormalized by the thermodynamic energy at low temperatures and by the magnetization at high temperatures.

The Hamiltonian, as usual, is

$$\mathcal{H}=-g\mu_BH\sum_g S_g^z-\sum_{f,g}J(\mathbf{g}-\mathbf{f})\mathbf{S}_g\cdot\mathbf{S}_f \qquad (17.189)$$

We now consider the Green function of the form

$$G_t^a(\mathbf{g},\mathbf{l})\equiv\langle\langle \mathbf{S}_g^{\dagger}(t);e^{a\mathbf{S}_l^z}\mathbf{S}_l^-\rangle\rangle \qquad (17.190)$$

where a is an adjustable parameter.

Similarly, the equation of motion, by (17.93), is given by

$$EG_E^a(\mathbf{g},\mathbf{l}) = \frac{\theta(a)}{2\pi}\delta_{gl} + g\mu_B H G_E^a(\mathbf{g},\mathbf{l})$$

$$-2\sum_f J(\mathbf{g}-\mathbf{f})\langle\langle \mathbf{S}_g^z S_f^+ - \mathbf{S}_f^z \mathbf{S}_g^+ ; e^{a\mathbf{S}_l^z} S_l^- \rangle\rangle \qquad (17.191)$$

where

$$\theta(a) = \langle [\mathbf{S}^+, e^{a\mathbf{S}^z}\mathbf{S}^-] \rangle \qquad (17.192)$$

The symmetric decoupling procedure is best described for spin 1/2. We can write $\mathbf{S}_g^z$ in either of two forms:

$$\mathbf{S}_g^z = S - \mathbf{S}_g^- \mathbf{S}_g^+ \qquad (17.193)$$

or

$$\mathbf{S}_g^z = \tfrac{1}{2}(\mathbf{S}_g^+ \mathbf{S}_g^- - \mathbf{S}_g^- \mathbf{S}_g^+) \qquad (17.194)$$

Multiplying (17.193) by α and (17.194) by $(1-\alpha)$ and adding, we have

$$\mathbf{S}_g^z = \alpha S + \tfrac{1}{2}\alpha(1-\alpha)\mathbf{S}_g^+ \mathbf{S}_g^- - \tfrac{1}{2}(1+\alpha)\mathbf{S}_g^- \mathbf{S}_g^+ \qquad (17.195)$$

The Green function $\langle\langle \mathbf{S}_g^- \mathbf{S}_g^+ \mathbf{S}_f^+ ; \mathbf{B} \rangle\rangle$ is now decoupled in the following symmetric form:

$$\langle\langle \mathbf{S}_g^- \mathbf{S}_g^+ \mathbf{S}_f^+ ; \mathbf{B} \rangle\rangle \underset{g\neq f}{\rightarrow} \langle \mathbf{S}_g^- \mathbf{S}_g^+ \rangle \langle\langle \mathbf{S}_f^+ ; \mathbf{B} \rangle\rangle + \langle \mathbf{S}_g^- \mathbf{S}_f^+ \rangle \langle\langle \mathbf{S}_g^+ ; \mathbf{B} \rangle\rangle$$

$$+ \langle \mathbf{S}_g^+ \mathbf{S}_f^+ \rangle \langle\langle \mathbf{S}_g^- ; \mathbf{B} \rangle\rangle$$

$$= \langle \mathbf{S}_g^- \mathbf{S}_g^+ \rangle \langle\langle \mathbf{S}_f^+ ; \mathbf{B} \rangle\rangle + \langle \mathbf{S}_g^- \mathbf{S}_f^+ \rangle \langle\langle \mathbf{S}_g^+ ; \mathbf{B} \rangle\rangle \qquad (17.196)$$

since the last term vanishes because $\langle \mathbf{S}_g^+ \mathbf{S}_f^+ \rangle$ is nondiagonal in the total z-component of the spin. Similarly,

$$\langle\langle \mathbf{S}_g^+ \mathbf{S}_g^- \mathbf{S}_f^+ ; \mathbf{B} \rangle\rangle \underset{g\neq f}{\rightarrow} \langle \mathbf{S}_g^+ \mathbf{S}_g^- \rangle \langle\langle \mathbf{S}_f^+ ; \mathbf{B} \rangle\rangle + \langle \mathbf{S}_g^- \mathbf{S}_f^+ \rangle \langle\langle \mathbf{S}_g^+ ; \mathbf{B} \rangle\rangle \qquad (17.197)$$

Thus (17.195) with (17.196) and (17.197) gives

$$\begin{aligned}
\langle\langle \mathbf{S}_g^z\mathbf{S}_f^+;\mathbf{B}\rangle\rangle &= \langle\langle\{\alpha S+\tfrac{1}{2}(1-\alpha)\mathbf{S}_g^+\mathbf{S}_g^- -\tfrac{1}{2}(1+\alpha)\mathbf{S}_g^-\mathbf{S}_g^+\}\mathbf{S}_f^+;\mathbf{B}\rangle\rangle\\
&= \langle\langle \alpha S\mathbf{S}_f^+ +\tfrac{1}{2}(1-\alpha)\mathbf{S}_g^+\mathbf{S}_g^-\mathbf{S}_f^+ -\tfrac{1}{2}(1+\alpha)\mathbf{S}_g^-\mathbf{S}_g^+\mathbf{S}_f^+;\mathbf{B}\rangle\rangle\\
&= \alpha S\langle\langle \mathbf{S}_f^+;\mathbf{B}\rangle\rangle +\tfrac{1}{2}(1-\alpha)\{\langle \mathbf{S}_g^+\mathbf{S}_g^-\rangle\langle\langle \mathbf{S}_f^+;\mathbf{B}\rangle\rangle\\
&\quad +\langle \mathbf{S}_g^-\mathbf{S}_f^+\rangle\langle\langle \mathbf{S}_g^+;\mathbf{B}\rangle\rangle\} -\tfrac{1}{2}(1+\alpha)\{\langle \mathbf{S}_g^-\mathbf{S}_g^+\rangle\langle\langle \mathbf{S}_f^+;\mathbf{B}\rangle\rangle\\
&\quad +\langle \mathbf{S}_g^-\mathbf{S}_f^+\rangle\langle\langle \mathbf{S}_g^+;\mathbf{B}\rangle\rangle\}\\
&= \alpha S\langle\langle \mathbf{S}_f^+;\mathbf{B}\rangle\rangle +\langle\langle \mathbf{S}_f^+;\mathbf{B}\rangle\rangle\{\tfrac{1}{2}(1-\alpha)\langle \mathbf{S}_g^+\mathbf{S}_g^-\rangle\\
&\quad -\tfrac{1}{2}(1+\alpha)\langle \mathbf{S}_g^-\mathbf{S}_g^+\rangle\} +\langle\langle \mathbf{S}_g^+;\mathbf{B}\rangle\rangle\{\tfrac{1}{2}(1-\alpha)\langle \mathbf{S}_g^-\mathbf{S}_f^+\rangle\\
&\quad -\tfrac{1}{2}(1+\alpha)\langle \mathbf{S}_g^-\mathbf{S}_f^+\rangle\}\\
&= \alpha S\langle\langle \mathbf{S}_f^+;\mathbf{B}\rangle\rangle +\langle\langle \mathbf{S}_f^+;\mathbf{B}\rangle\rangle\{\tfrac{1}{2}(1-\alpha)\langle \mathbf{S}_g^+\mathbf{S}_g^-\rangle\\
&\quad -\tfrac{1}{2}(1+\alpha)\langle \mathbf{S}_g^-\mathbf{S}_g^+\rangle\} +\langle\langle \mathbf{S}_g^+;\mathbf{B}\rangle\rangle\left\{\langle \mathbf{S}_g^-\mathbf{S}_f^+\rangle\left(\frac{1}{2}-\frac{\alpha}{2}-\frac{1}{2}-\frac{\alpha}{2}\right)\right\}\\
&= \langle\langle \mathbf{S}_f^+;\mathbf{B}\rangle\rangle\{\alpha S-\tfrac{1}{2}(1+\alpha)\langle \mathbf{S}_g^-\mathbf{S}_g^+\rangle +\tfrac{1}{2}(1-\alpha)\langle \mathbf{S}_g^+\mathbf{S}_g^-\rangle\}\\
&\quad -\alpha\langle \mathbf{S}_g^-\mathbf{S}_f^+\rangle\langle\langle \mathbf{S}_g^+;\mathbf{B}\rangle\rangle\\
&= \langle S^z\rangle\langle\langle \mathbf{S}_f^+;\mathbf{B}\rangle\rangle -\alpha\langle \mathbf{S}_g^-\mathbf{S}_f^+\rangle\langle\langle \mathbf{S}_g^+;\mathbf{B}\rangle\rangle
\end{aligned}\tag{17.198}$$

where $\langle \mathbf{S}^z\rangle=\alpha S+\tfrac{1}{2}(1-\alpha)\mathbf{S}_g^+\mathbf{S}_g^- -\tfrac{1}{2}(1+\alpha)\mathbf{S}_g^-\mathbf{S}_g^+$

If $\alpha=1$, the result corresponds to decoupling on the basis of (17.193). If $\alpha=0$, the decoupling is on the basis of (17.194). But if $\alpha=-1$, the decoupling rests on

$$\mathbf{S}_g^z=-S+\mathbf{S}_g^+\mathbf{S}_g^-$$

Therefore it is evident that a physical criterion is required at this stage. The operator $\mathbf{S}^-\mathbf{S}^+$ in (17.193) represents the deviation of S^z from $+S$. It is this operator $\mathbf{S}^-\mathbf{S}^+$ that is treated approximately when decoupling is on the basis of (17.193). Hence it is reasonable to use (17.193) when the deviation from $S^z=S$ is small, that is, when $\langle S^z\rangle\approx S$, at low temperatures. Similarly, the operator $\tfrac{1}{2}(\mathbf{S}^+\mathbf{S}^- -\mathbf{S}^-\mathbf{S}^+)$ in (17.194) represents the deviation from zero, and (17.194) can then be reasonably used for decoupling when $\langle S^z\rangle\simeq 0$ at high temperatures.

Both of the facts stated above are contained in the choice

$$\alpha=\frac{\langle S^z\rangle}{S},\qquad S=\tfrac{1}{2}\tag{17.199}$$

Equation (17.195) then becomes

$$S_g^z = \langle S^z \rangle + \left[\frac{S - \langle S^z \rangle}{2S} \mathbf{S}_g^+ \mathbf{S}_g^- - \frac{S + \langle S^z \rangle}{2S} \mathbf{S}_g^- \mathbf{S}_g^+ \right] \tag{17.200}$$

The operator in the brackets, which is to be decoupled, represents the deviation of S^z from $\langle S^z \rangle$ and should be (self-consistently) small in all temperature regions.

Inserting in (17.198) the value of α given by (17.199), we obtain for $S = \frac{1}{2}$

$$\langle\langle \mathbf{S}_g^z \mathbf{S}_f^+ ; \mathbf{B} \rangle\rangle \underset{g \neq f}{\rightarrow} \langle \mathbf{S}^z \rangle \langle\langle \mathbf{S}_f^+ ; \mathbf{B} \rangle\rangle - \frac{\langle S^z \rangle}{S} \langle \mathbf{S}_g^- \mathbf{S}_f^+ \rangle \langle\langle \mathbf{S}_g^+ ; \mathbf{B} \rangle\rangle \tag{17.201}$$

This is the decoupling scheme for $S = \frac{1}{2}$. We now generalize it for higher spin. The analog of (17.193), for general spin S, is

$$S_g^z = S(S+1) - (S_g^z)^2 - \mathbf{S}_g^- \mathbf{S}_g^+ \tag{17.202}$$

whereas (17.194) remains the same. Relation (17.202) follows from (17.140) simply by putting $n = 1$ and $p = 1$. Decoupling as before, and neglecting the fluctuations of $(S^z)^2$, we find that

$$\langle\langle \mathbf{S}_g^z \mathbf{S}_f^+ ; \mathbf{B} \rangle\rangle \underset{g \neq f}{\rightarrow} \langle S^z \rangle \langle\langle \mathbf{S}_f^+ ; \mathbf{B} \rangle\rangle - \alpha \langle \mathbf{S}_g^- \mathbf{S}_f^+ \rangle \langle\langle \mathbf{S}_g^+ ; \mathbf{B} \rangle\rangle \tag{17.203}$$

Unfortunately $\mathbf{S}_g^- \mathbf{S}_g^+$ is *not* the only operator treated approximately in decoupling (17.202), and the interpretation of the deviation from $S^z = +S$ is *no longer* valid. Therefore the choice of α is not as evident as for $S = \frac{1}{2}$. However, it is determined by the following criteria:

1. For $S = \frac{1}{2}$, α should reduce to our previous result, that is $\alpha = \langle S^z \rangle / S$.
2. For $\langle S^z \rangle = 0$, α should vanish. This follows from the fact that (17.194) retains its interpretation for arbitrary S.
3. For $\langle S^z \rangle \simeq S$, we expect $\langle S^z \rangle \simeq (S - n)$, where n is a deviation of the order of unity rather than of the order of S. This implies that $\alpha \langle \mathbf{S}_g^- \mathbf{S}_f^+ \rangle$ should be of the order of unity, rather than of the order of S, at low temperatures. Now $(1/2S)\langle \mathbf{S}_g^- \mathbf{S}_g^+ \rangle$ is the spin deviation in the lowest order (because, as we saw in the spin-wave theory, $\mathbf{S}^-$ and $\mathbf{S}^+$ correspond to $(2S)^{1/2}\mathbf{a}^\dagger$ and $(2S)^{1/2}\mathbf{a}$, respectively). Similarly $\langle \mathbf{S}_g^- \mathbf{S}_f^+ \rangle$ will be of the order of $2S$ if $\mathbf{f}$ and $\mathbf{g}$ are closely coupled. Therefore we choose

$$\alpha = \frac{1}{2S} \cdot \frac{\langle S^z \rangle}{S} \tag{17.204}$$

This clearly satisfies all the criteria. Equations (17.203) and (17.204) are the basic decoupling approximations of the theory.

Inserting these approximations in (17.191), we have

$$\begin{aligned}EG_E^a(\mathbf{g},\mathbf{l}) &= \frac{\theta(a)}{2\pi}\delta_{gl} + g\mu_B H G_E^a(\mathbf{g},\mathbf{l}) \\ &\quad -\langle S^z\rangle \sum_f J(\mathbf{g}-\mathbf{f})\left[G_E^a(\mathbf{f},\mathbf{l}) - G_E^a(\mathbf{g},\mathbf{l})\right] \\ &\quad + \frac{\langle S^z\rangle}{S^2}\sum_f J(\mathbf{g}-\mathbf{f})\left[\langle \mathbf{S}_g^- \mathbf{S}_f^+\rangle G_E^a(\mathbf{g},\mathbf{l}) - \langle \mathbf{S}_f^- \mathbf{S}_g^+\rangle G_E^a(\mathbf{f},\mathbf{l})\right]\end{aligned} \tag{17.205}$$

Now, introducing the familiar Fourier transforms, $G_E^a(\mathbf{K})$, $J(\mathbf{K})$, and

$$\Psi(\mathbf{K},a) = \sum_{\mathbf{g}-\mathbf{l}} e^{-i(\mathbf{g}-\mathbf{l})\cdot\mathbf{K}}\langle e^{aS_l^z}\mathbf{S}_l^- \mathbf{S}_g^+\rangle \tag{17.206}$$

we get, as usual,

$$G_E^a(\mathbf{K}) = \frac{\theta(a)}{2\pi\left[E - E(\mathbf{K})\right]} \tag{17.207}$$

where

$$\begin{aligned}E(\mathbf{K}) &= g\mu_B H + 2\langle S^z\rangle\left[J(0) - J(\mathbf{K})\right] \\ &\quad + \frac{\langle S^z\rangle}{NS^2}\sum_{\mathbf{K}'}\left[J(\mathbf{K}') - J(\mathbf{K}'-\mathbf{K})\right]\Psi(\mathbf{K}',0)\end{aligned} \tag{17.208}$$

The correlation function, which can be determined from $G_E^a(\mathbf{K})$ by (17.84), is $\Psi(\mathbf{K},a)$ in (17.206), that is,

$$\Psi(\mathbf{K},a) = \frac{\theta(a)}{e^{E(\mathbf{K})/kT} - 1} \tag{17.209}$$

Equation (17.209), together with (17.192) defining $\theta(a)$, and (17.208) giving $E(\mathbf{K})$, are the basic equations of the theory.

Let us now indicate the method of solution. Instead of the laborious procedure of Tahir-Kheli and ter Haar,[19], given in Section 17.4, Callen constructs a differential equation exploiting the functional dependence of

$$\Psi(a) = \frac{1}{N}\sum_{\mathbf{K}}\Psi(\mathbf{K},a)$$

and $\theta(a)$ on the parameter a, and this replaces, to our great advantage, the $2S$ simultaneous equations in the method of Tahir-Kheli and ter Haar. Since

$$\theta(a) = \langle[\mathbf{S}^+, e^{aS^z}\mathbf{S}^-]\rangle \tag{17.210}$$

we take the identity

$$[\mathbf{S}^+,(S^z)^n]=[(S^z-1)^n-(S^z)^n]\mathbf{S}^+$$

which is obtained easily from the commutation relation $[\mathbf{S}^+,\mathbf{S}^z]_-=-\mathbf{S}^+$. Thus

$$\begin{aligned}[\mathbf{S}^+,(e^{\mathbf{S}^z})^a]_-&=[e^{(\mathbf{S}^z-1)a}-e^{a\mathbf{S}^z}]\mathbf{S}^+\\&=(e^{-a}-1)e^{a\mathbf{S}^z}\mathbf{S}^+\end{aligned}\tag{17.211}$$

Therefore

$$\begin{aligned}\theta(a)&=\langle[\mathbf{S}^+,e^{a\mathbf{S}^z}\mathbf{S}^-]\rangle\\&=\langle\mathbf{S}^+e^{a\mathbf{S}^z}\mathbf{S}^--e^{a\mathbf{S}^z}\mathbf{S}^-\mathbf{S}^+\rangle\\&=\langle\mathbf{S}^+e^{a\mathbf{S}^z}\mathbf{S}^--e^{a\mathbf{S}^z}\mathbf{S}^-\mathbf{S}^++e^{a\mathbf{S}^z}\mathbf{S}^+\mathbf{S}^--e^{a\mathbf{S}^z}\mathbf{S}^+\mathbf{S}^-\rangle\\&=\langle e^{a\mathbf{S}^z}(\mathbf{S}^+\mathbf{S}^--\mathbf{S}^-\mathbf{S}^+)+(\mathbf{S}^+e^{a\mathbf{S}^z}-e^{a\mathbf{S}^z}\mathbf{S}^+)\mathbf{S}^-\rangle\\&=\langle e^{a\mathbf{S}^z}[\mathbf{S}^+,\mathbf{S}^-]_-+[\mathbf{S}^+,e^{a\mathbf{S}^z}]_-\mathbf{S}^-\rangle\\&=2\langle e^{a\mathbf{S}^z}\mathbf{S}^z\rangle+(e^{-a}-1)\langle e^{a\mathbf{S}^z}\mathbf{S}^+\mathbf{S}^-\rangle\\&=2\langle e^{a\mathbf{S}^z}\mathbf{S}^z\rangle+(e^{-a}-1)\langle e^{a\mathbf{S}^z}\{S(S+1)-(S^z)^2+S^z\}\rangle\\&=S(S+1)(e^{-a}-1)\langle e^{a\mathbf{S}^z}\rangle+\langle e^{a\mathbf{S}^z}\mathbf{S}^z\rangle(1+e^{-a})\\&\quad-(e^{-a}-1)\langle e^{a\mathbf{S}^z}(\mathbf{S}^z)^2\rangle\end{aligned}\tag{17.212}$$

Now, introducing

$$\Omega(a)=\langle e^{a\mathbf{S}^z}\rangle\tag{17.213}$$

and

$$D=\frac{d}{da}\tag{17.214}$$

we have, from (17.212),

$$\theta(a)=S(S+1)(e^{-a}-1)\theta+(e^{-a}+1)D\Omega-(e^{-a}-1)D^2\Omega\tag{17.215}$$

and

$$\begin{aligned}\Psi(a)&=\frac{1}{N}\sum_{\mathbf{K}}\Psi(\mathbf{K},a)\\&=\frac{1}{N}\sum_{\mathbf{K}}\sum_{\mathbf{g}-\mathbf{l}}e^{-i(\mathbf{g}-\mathbf{l})\cdot\mathbf{K}}\langle e^{aS_l^z}S_l^-S_g^+\rangle\\&=\langle e^{aS^z}S^-S^+\rangle\\&=S(S+1)\Omega-D\Omega-D^2\Omega\end{aligned}\tag{17.216}$$

Also, (17.209) can be rewritten as

$$\Psi(\mathbf{K},a)=\phi(\mathbf{K})\theta(a) \tag{17.217}$$

where

$$\phi(\mathbf{K})=\left[e^{E(\mathbf{K})/kT}-1\right]^{-1}$$

Now, summing over $\mathbf{K}$, we have

$$\Psi(a)=\Phi\theta(a) \tag{17.218}$$

where

$$\begin{aligned}\Phi&=\frac{1}{N}\sum_{\mathbf{K}}\phi(\mathbf{K})\\&=\frac{1}{N}\sum_{\mathbf{K}}\left[e^{E(\mathbf{K})/kT}-1\right]^{-1}\end{aligned}$$

This Φ is the same as the $\Phi(S)$ in (17.160). Relation (17.218) is recast, as follows, into a differential equation with the help of (17.215) and (17.216). Since

$$\Psi(a)=\Phi\theta(a)$$

using (17.215) and (17.216), we see that

$$\begin{aligned}S(S+1)\Omega-D\Omega-D^2\Omega=\Phi\big[&S(S+1)(e^{-a}-1)\Omega\\&+(1-e^{-a})D\Omega-(e^{-a}-1)D^2\Omega\big]\end{aligned}$$

or

$$D^2\Omega\left[1-(e^{-a}-1)\Phi\right]+D\Omega\left[1+(1+e^{-a})\Phi\right]-S(S+1)\Omega\left[1-(e^{-a}-1)\Phi\right]=0$$

Now, multiplying throughout by e^a and rearranging, we obtain

$$D^2\Omega+\frac{(1+\phi)e^a+\phi}{(1+\phi)e^a-\phi}D\Omega-S(S+1)\Omega=0 \tag{17.219}$$

This equation is the analog of the set of $2S$ coupled equations of Tahir-Kheli and ter Haar. Solution of (17.219) with the boundary condition $\Omega(0)=1$ and

$$\prod_{r=-S}^{+S}(S^z-r)=0$$

gives

$$\begin{aligned}\langle S^z\rangle&=D\Omega(a)_{a=0}\\&=\frac{(S-\Phi)(1+\Phi)^{2S+1}+(S+1+\Phi)\Phi^{2S+1}}{(1-\phi)^{2S+1}-\phi^{2S+1}}\end{aligned} \tag{17.220}$$

This should be compared with (17.154), (17.158), and (17.159) for $\langle S^z \rangle$ in Section 17.4.

We do not wish to proceed further with the theory and will only mention the results. The coupled set of equations (17.208) giving $E(\mathbf{K})$, and (17.218) giving $\Phi(\mathbf{K})$, is solved self-consistently for $\langle S^z \rangle$, and for the nearest-neighbor approximation we get the following results:

$$E(\mathbf{K}) = g\mu_B H + 2SR[J(0) - J(\mathbf{K})] \tag{17.221}$$

where R is the normalization factor, given by

$$R = \frac{\langle S^z \rangle}{S}\left[1 + \frac{\langle S^z \rangle}{S^2}\frac{1}{NJ(0)}\sum_{\mathbf{K}} J(\mathbf{K})\left[e^{E(\mathbf{K})/kT} - 1\right]^{-1}\right] \tag{17.222}$$

The second term in the bracket is a function of T, $\langle S^z \rangle$, and $\mathbf{H}$, as is shown by Bloch.[26] Thus the quasi-particle energy is renormalized by a factor depending only on the temperature (for $\mathbf{H}=0$) and independent of the wave vector $\mathbf{K}$.

The low-temperature expansion for the magnetization is given in terms of

$$\tau = \frac{3kT}{4\pi vzJS}$$

as

$$\begin{aligned}\langle S^z \rangle = {} & S - \zeta\left(\tfrac{3}{2}\right)\tau^{3/2} - \tfrac{3}{4}\pi v\zeta\left(\tfrac{5}{2}\right)\tau^{5/2} - \pi^2\omega v^2\zeta\left(\tfrac{7}{2}\right)\tau^{7/2} \\ & - \cdots - \frac{3}{2S}\pi v\zeta\left(\tfrac{3}{2}\right)\zeta\left(\tfrac{5}{2}\right)\tau^4 + \cdots \\ & + (2S+1)\zeta^{2S+1}\left(\tfrac{3}{2}\right)\tau^{3S+(3/2)} + (2S+1)^2\tfrac{3}{4}\pi v\zeta^{2S}\left(\tfrac{3}{2}\right)\zeta\left(\tfrac{5}{2}\right)\tau^{3S+(5/2)} + \cdots\end{aligned} \tag{17.223}$$

where the symbols have the usual meaning.

We see exact agreement with Dyson's[17] results regarding τ°, $\tau^{3/2}$, $\tau^{5/2}$, and $\tau^{7/2}$ terms. Equation (17.223) has no τ^3 term. The coefficient of τ^4 given by Dyson is

$$-\frac{3}{2S}\pi v\zeta\left(\tfrac{3}{2}\right)\zeta\left(\tfrac{5}{2}\right)Q\tau^4$$

where

$$Q = 1 + \tfrac{4}{3}(GS-1)^{-1} + \alpha(3S)^{-1}$$

G and α being constants. Thus the τ^4 term here corresponds to the leading term ($Q=1$) of Dyson.

The spurious terms, $\tau^{3S+(3/2)}$ and $\tau^{3S+(5/2)}$, in (17.223) result from the decoupling approximation. For $S=\frac{1}{2}$, they make the incorrect contribution of

τ^3 and τ^4 terms. For $S=1$, the first spurious contribution is a $\tau^{9/2}$ term, and with higher S values it moves to a higher power. Thus the spurious terms are of importance only for $S=\frac{1}{2}$, at low temperatures.

The Curie temperature is given by

$$\frac{kT_C}{J}=\frac{2z(S+1)}{9[F(-1)]^2}[(4S+1)F(-1)-(S+1)] \tag{17.224}$$

The Curie temperature is, however, extremely sensitive to the decoupling parameter α. If the chosen value of α is multiplied by $(\langle S^z\rangle/S)^\epsilon$, where ϵ is a small positive quantity, the Curie temperatures become identical with those of Tahir-Kheli and ter Haar, though the low- and high-temperature results remain unchanged.

Finally, we write the high-temperature expansion as obtained by Callen giving the susceptibility as

$$\chi=\frac{g^2\mu_B^2S(S+1)}{3kT}\left[1+\frac{T_M}{T}+\left(1-\frac{2S-1}{3Sz}\right)\left(\frac{T_M}{T}\right)^2+O\left(\frac{1}{T^3}\right)\right] \tag{17.225}$$

where T_M is the Curie temperature of the molecular field theory and is given by

$$T_M=\frac{2}{3k}zJS(S+1) \tag{17.226}$$

This agrees with Opechowski's expansion[18] up to T^{-2}. In conclusion, we mention the work of Haas,[27] which includes in the Hamiltonian the classical magnetic dipole-dipole interaction

$$\sum_{i,j}\frac{g^2\mu_B^2}{\mathbf{r}_{ij}^5}\left[\mathbf{r}_{ij}^2\mathbf{S}_i\cdot\mathbf{S}_j-3(\mathbf{S}_i\cdot\mathbf{r}_{ij})(\mathbf{S}_j\cdot\mathbf{r}_{ij})\right] \tag{17.227}$$

This becomes important in the long-wavelength limit, and by using Callen's symmetric decoupling procedure, Haas has found the "renormalization factor" for long-wavelength magnons. The factor is similar to Callen's but is not independent of the wave vector **K**. Haas's results explain well the parallel pumping experiment of Le-Craw and Walker[28] on "curvature renormalization," which we shall not discuss.

17.6 GREEN FUNCTION THEORIES IN HIGHER ORDER DECOUPLING

The occurrence of the spurious T^3 term in the theories of random phase decoupling for general spin S and in Callen's work for $S=\frac{1}{2}$ only, and the incorrect coefficient of the T^4 term in low-temperature results, are significant indications of the drawbacks of the decoupling schemes discussed above, which are restricted to the Green function of the lowest order. Hewson and ter Haar[29] showed that the error in such a decoupling is of the order T^3 in the spontaneous magnetization. This also indicates that the kinematical interaction, which is strongest for $S=\frac{1}{2}$, is inadequately accounted for by these decoupling procedures. Also, the correlation function $\langle \mathbf{S}_i^- \mathbf{S}_j^- \mathbf{S}_k^+ \mathbf{S}_l^- \rangle$, which should, by definition, be zero for $S=\frac{1}{2}$, is not properly given by the theories presented above.

Morita and Tanaka[30] undertook to resolve these difficulties by setting up the equation of motion for the higher order Green function and then decoupling it in the next higher order, and they finally succeeded in their aim. We give below a very brief sketch of their decoupling procedure and then present the important results.

To show the elimination of the T^3 term, Morita and Tanaka considered the case of spin $S=\frac{1}{2}$. The Hamiltonian is taken as

$$\mathcal{H} = -\mathbf{H}\sum_f S_f^z - \sum_{f,g} J_{fg}\mathbf{S}_f \cdot \mathbf{S}_g \tag{17.228}$$

where $g\mu_B$ is set equal to unity so that $\mathbf{H}$ is in energy units. The usual transformation to $\mathbf{S}^+$ and $\mathbf{S}^-$ gives

$$\mathcal{H} = [H+J(0)]\sum_f \mathbf{S}_f^- \mathbf{S}_f^+ - \sum_{f,g} J_{fg}[\mathbf{S}_f^- \mathbf{S}_g^+ + \mathbf{S}_f^- \mathbf{S}_g^- \mathbf{S}_g^+ \mathbf{S}_f^+] \tag{17.229}$$

where the ground state energy is dropped for simplicity. For $S=\frac{1}{2}$, the following commutation relations hold:

$$\mathbf{S}_f^+ \mathbf{S}_g^- - \mathbf{S}_g^- \mathbf{S}_f^+ = \delta_{fg}[1-2\mathbf{S}_f^- \mathbf{S}_f^+] \tag{17.230}$$

$$\mathbf{S}_f^+ \mathbf{S}_g^+ - \mathbf{S}_g^+ \mathbf{S}_f^+ = 0 = \mathbf{S}_f^- \mathbf{S}_g^- - \mathbf{S}_g^- \mathbf{S}_f^-, \qquad g \neq f \tag{17.231}$$

$$[\mathbf{S}_f^-]^2 = 0 = [\mathbf{S}_f^+]^2 \tag{17.232}$$

Note that (17.232) is the kinematical interaction. The deviation from the Boson commutation relation in (17.230) for $f=g$ is also the kinematical interaction. We now introduce the Green function $\langle\langle \mathbf{S}_g^+(t); \mathbf{S}_f^-(0)\rangle\rangle$, and so the equation of motion for its Fourier tranform is given [by using (17.93)] by

$$E\langle\langle \mathbf{S}_g^+; \mathbf{S}_f^- \rangle\rangle = \frac{-1}{2\pi}[1-2\langle \mathbf{S}_f^- \mathbf{S}_f^+ \rangle]\delta_{fg} + [H+J(0)]\langle\langle \mathbf{S}_g^+; \mathbf{S}_f^- \rangle\rangle$$
$$- \sum_j J_{jg}\langle\langle \mathbf{S}_j^+; \mathbf{S}_f^- \rangle\rangle + 2\sum_J J_{jg}[\langle\langle \mathbf{S}_g^- \mathbf{S}_g^+ \mathbf{S}_j^+; \mathbf{S}_f^- \rangle\rangle - \langle\langle \mathbf{S}_j^- \mathbf{S}_j^+ \mathbf{S}_g^+; \mathbf{S}_f^- \rangle\rangle] \tag{17.233}$$

The correlation function $2\langle \mathbf{S}_f^- \mathbf{S}_f^+ \rangle$ in the first term on the right-hand side of (17.233) is due to the kinematical interaction and contributes a T^3 term to the spontaneous magnetization. We are to find another term in (17.233) that is also of kinematic origin and whose contribution exactly cancels that due to $2\langle \mathbf{S}_f^- \mathbf{S}_f^+ \rangle$. Such a kinematical interaction appears in the higher order Green function in two ways. When $m=n$, $\langle\langle \mathbf{S}_l^- \mathbf{S}_m^+ \mathbf{S}_n^+ ; \mathbf{S}_f^- \rangle\rangle$ vanishes because of the kinematical interaction. This indicates a strong correlation between $\mathbf{S}_m^+$ and $\mathbf{S}_n^+$. There is another correlation between $\mathbf{S}_l^-$ and $\mathbf{S}_f^-$ due to the same interaction. This correlation, as shown by Hewson and ter Haar,[29] is destroyed when any decoupling is imposed. For instance, a symmetric decoupling

$$\langle\langle \mathbf{S}_l^- \mathbf{S}_m^+ \mathbf{S}_n^+ ; \mathbf{S}_f^- \rangle\rangle \rightarrow \langle \mathbf{S}_l^- \mathbf{S}_m^+ \rangle \langle\langle \mathbf{S}_n^+ ; \mathbf{S}_f^- \rangle\rangle + \langle \mathbf{S}_l^- \mathbf{S}_n^+ \rangle \langle\langle \mathbf{S}_m^+ ; \mathbf{S}_f^- \rangle\rangle$$

clearly destroys the correlation between $\mathbf{S}_l^-$ and $\mathbf{S}_f^-$. Hence $\langle\langle \mathbf{S}_l^- \mathbf{S}_m^+ \mathbf{S}_n^+ ; \mathbf{S}_f^- \rangle\rangle$ has to be more accurately evaluated. Its equation of motion is, for $m \neq n$,

$$\begin{aligned} E\langle\langle \mathbf{S}_l^- \mathbf{S}_m^+ \mathbf{S}_n^+ ; \mathbf{S}_f^- \rangle\rangle = &-\frac{1}{2\pi}\{\delta_{mf}[\langle \mathbf{S}_l^- \mathbf{S}_n^+ \rangle - 2\langle \mathbf{S}_l^- \mathbf{S}_m^- \mathbf{S}_m^+ \mathbf{S}_n^+ \rangle] \\ &+ \delta_{nf}[\langle \mathbf{S}_l^- \mathbf{S}_m^+ \rangle - 2\langle \mathbf{S}_l^- \mathbf{S}_n^- \mathbf{S}_n^+ \mathbf{S}_m^+ \rangle]\} \\ &+ [H + J(0) - 2J_{mn}]\langle\langle \mathbf{S}_l^- \mathbf{S}_m^+ \mathbf{S}_n^+ ; \mathbf{S}_f^- \rangle\rangle \\ &+ \sum_j [J_{ji}\langle\langle \mathbf{S}_j^- \mathbf{S}_m^+ \mathbf{S}_n^+ ; \mathbf{S}_f^- \rangle\rangle - J_{jm}\langle\langle \mathbf{S}_l^- \mathbf{S}_j^+ \mathbf{S}_n^+ ; \mathbf{S}_f^- \rangle\rangle \\ &- J_{jn}\langle\langle \mathbf{S}_l^- \mathbf{S}_m^+ \mathbf{S}_j^+ ; \mathbf{S}_f^- \rangle\rangle] \\ &+ [\text{terms of higher order Green functions of} \\ &\quad \text{the type } \langle\langle \mathbf{S}^- \mathbf{S}^- \mathbf{S}^+ \mathbf{S}^+ \mathbf{S}^+ ; \mathbf{S}_f^- \rangle\rangle] \end{aligned} \tag{17.234}$$

It is shown by Morita and Tanaka[30] that $\langle \mathbf{S}_f^- \mathbf{S}_g^+ \rangle$ and $\langle \mathbf{S}_f^- \mathbf{S}_l^- \mathbf{S}_m^+ \mathbf{S}_n^+ \rangle$ are of the first and second orders in fugacity $e^{-\beta H}$. In (17.234), $\langle \mathbf{S}_l^- \mathbf{S}_n^+ \rangle$'s appear in the inhomogeneous term. Therefore, $\langle\langle \mathbf{S}_l^- \mathbf{S}_m^+ \mathbf{S}_n^+ ; \mathbf{S}_f^- \rangle\rangle$ is of the first order in fugacity. If an equation is set up for the higher order Green function $\langle\langle \mathbf{S}^- \mathbf{S}^- \mathbf{S}^+ \mathbf{S}^+ \mathbf{S}^+ ; \mathbf{S}_f^- \rangle\rangle$, a correlation function of the type $\langle \mathbf{S}_f^- \mathbf{S}_l^- \mathbf{S}_m^+ \mathbf{S}_n^+ \rangle$ occurs in the inhomogeneous term, and hence it is of the second order in fugacity. For spin-wave interaction we require $\langle\langle \mathbf{S}_l^- \mathbf{S}_m^+ \mathbf{S}_n^+ ; \mathbf{S}_f^- \rangle\rangle$ in the first order of fugacity. To that order, the last term in (17.234) can be neglected. In (17.234) the fact that $\langle\langle \mathbf{S}_l^- \mathbf{S}_m^+ \mathbf{S}_m^+ ; \mathbf{S}_f^- \rangle\rangle = 0$ is explicitly taken into account.

The equation of motion is then solved by constructing the eigenfunctions and eigenvalues appropriate to the problem, and the correlation functions are found. The latter diverge, and the divergence seems to arise because of the neglect of higher order Green functions. How this divergence arises and how it is eliminated are discussed by using an example of a two-spin-1/2 problem.

The spontaneous magnetization is calculated from the correlation function $\langle \mathbf{S}_f^- \mathbf{S}_g^+ \rangle$ by setting $f=g$, and it is shown that there is no T^3 term. Thus the major drawback of the Green function theory seems to be eliminated.

In a second paper,[31] Morita and Tanaka applied the method to the general spin S case and obtained exactly the same coefficient for the T^4 term as did Dyson.

17.7 CONCLUDING REMARKS

After reviewing the work of Morita and Tanaka, it can be stated that the Green function theories of ferromagnetism are at present the only theories that predict perfectly the behavior of a ferromagnet throughout the temperature range.

For the reader who wishes to gain an idea of the achievement of the Green function theories without becoming involved with the details of calculation, Table 17.1 lists the Curie temperatures (kT_C/J) from Callen's[24] paper. Morita and Tanaka have not calculated the Curie temperatures.

For each type of lattice the first column (1) gives the values of the Curie temperature obtained by Brown and Luttinger.[32] These were calculated by the Kramers-Opechowski method of finding the roots to the inverse of the susceptibility up to the fourth order of approximation.

Table 17.1

Curie temperatures (kT_C/J) for cubic lattices with nearest-neighbor interaction

	Type of lattice								
	s. c.			b. c. c.			f. c. c.		
S	(1)[a]	(2)[b]	(3)[c]	(1)[a]	(2)[b]	(3)[c]	(1)[a]	(2)[b]	(3)[c]
$\frac{1}{2}$	1.9	2.0	2.7	2.4	2.9	3.7	4.2	4.5	5.6
1	5.4	5.3	6.5	7.8	7.7	9.1	12.7	11.9	13.9
$\frac{3}{2}$	10.6	9.9	11.7	15.4	14.4	16.6	24.7	22.3	25.5
2	17.5	15.8	18.5	25.2	23.0	26.2	40.0	35.7	40.1
$\frac{5}{2}$	25.8	23.1	26.8	37.1	33.5	37.9	58.7	52.1	58.3
3	35.7	31.7	36.4	51.2	45.9	51.6	80.9	71.4	79.5

[a] H.A. Brown and J.M. Luttinger, *Phys. Rev.* **100**, 685 (1955).
[b] R.A. Tahir-Kheli and D. ter Haar, *Phys. Rev.* **127**, 88 (1962).
[c] H.B. Callen, *Phys. Rev.* **130**, 890 (1963).

The second column (2) gives the results of Tahir-Kheli and ter Haar,[19] calculated from the formula (z = number of nearest neighbors)

$$\frac{kT_C}{J}=\frac{2}{3}\frac{S(S+1)z}{F(-1)} \tag{17.235}$$

where

$$F(-1)=\begin{cases}1.516 & \text{for the s.c. lattice}\\ 1.393 & \text{for the b.c.c. lattice}\\ 1.345 & \text{for the f.c.c. lattice}\end{cases} \tag{17.236}$$

The third column (3) gives the values obtained by Callen. They were calculated from the formula

$$\frac{kT_C}{J}=\frac{2z(S+1)}{9F^2(-1)}\left[(4S+1)F(-1)-(S+1)\right] \tag{17.237}$$

where $F(-1)$ has the same values as in (17.236).

It is evident from Table 17.1 that the values of Brown and Luttinger lie between those of Tahir-Kheli and ter Haar and those of Callen. With increasing spin values the agreement with the latter author improves and the agreement with the former authors becomes poorer.

REFERENCES

1. L. Van Hove, *Physica* **21**, 901 (1955); *ibid.* **22**, 343 (1956).
2. J. Goldstone, *Proc. Roy. Soc. (London) A*, **239**, 267 (1957).
3. N. M. Hugenholtz, *Physica* **23**, 481, 533 (1957).
4. F. J. Dyson, *Phys. Rev.* **102**, 1217 (1956).
5. C. Bloch and C. De Dommicis, *Nucl. Phys.* **7**, 459 (1958).
6. J. M. Ziman, *Electrons and Phonons*, Clarendon Press, Oxford, England, 1962.
7. D. Pines, *The Many-Body Problem*, Benjamin, New York, 1961.
8. D. Ter Haar, *Cont. Phys.* **1**, 112 (1960).
9. T. Matsubara, *Prog. Theor. Phys. (Japan)* **14**, 351 (1955).
10. A. A. Abrikosov, L. P. Gorkov, and I. E. Dzyaloshinski, *Methods of Quantum Field Theory in Statistical Physics*, Prentice Hall, Englewood Cliffs, N. J., 1963.
11. V. Bonch Bruevich and S. Tyablikov, *The Green-Function Method in Statistical Mechanics*, North Holland, Amsterdam, 1962.
12. This section is based on the lecture notes prepared by S. Lundqvist, International Course on the Theory of Condensed Matter, Trieste, Italy, Many Body Theory, Part II, 3 October-16 December, 1966.
13. D. N. Zubarev, *Usp. Fiz. Nauk* **71**, 71 (1960) [English Translation: *Sov. Phys.—Usp.* **3**, 320 (1960)].
14. N. N. Bogolyubov and O. S. Parasyuk, *Dokl. Akad. Nauk. SSSR* **107**, 717 (1956).

15. N. N. Bogolyubov and S. V. Tyablikov, *Dokl. Akad. Nauk SSSR* **126**, 53 (1957) (English Translation: *Sov. Phys.—Dokl.* **4**, 604 (1959); V. L. Bonch-Breuvich and S. V. Tyablikov, *The Green Function Method in Statistical Mechanics*, North-Holland, Amsterdam, 1962.
16. N. N. Bogolyubov and S. V. Tyablikov, *JETP* **19** 251 (1949).
17. F. J. Dyson, *Phys. Rev.* **102**, 1217, 1230 (1956).
18. W. Opechowski, *Physica* **4**, 181, 715 (1937).
19. R. A. Tahir-Kheli and D. ter Haar, *Phys. Rev.* **127**, 88 (1962).
20. G. N. Watson, *Quart. J. Math.* **10**, 266 (1939); M. Tikson, *J. Natl. Bur. Stand.* **50** 177 (1953).
21. R. A. Tahir-Kheli and D. ter Haar, *Phys. Rev.* **127**, 88 (1962).
22. J. H. Van Vleck, *J. Chem. Phys.* **9**, 85 (1941).
23. M. Lax, *Phys. Rev.* **97**, 629 (1955).
24. H. B. Callen, *Phys. Rev.* **130**, 890 (1963).
25. F. Keffer and F. London, *J. Appl. Phys.* **32**, 25 (1961).
26. M. Bloch, *Phys. Rev. Lett.* **9**, 286 (1962).
27. C. W. Haas, *Phys. Rev.* **132**, 228 (1964).
28. See Ref. 27.
29. A. C. Hewson and D. ter Haar, *Phys. Lett.*, **6**, 136 (1963).
30. T. Morita and T. Tanaka, *Phys. Rev.* **137A**, 648 (1965).
31. T. Morita and T. Tanaka, *Phys. Rev.* **138A**, 1395 (1965).
32. H. A. Brown and J. M. Luttinger, *Phys. Rev.* **100** 685 (1955).

18

Green Function Theories in Antiferromagnetism

We saw in Chapter 17 that Bogolyubov and Tyablikov first employed the double-time temperature-dependent Green functions in an approximate treatment of statistical problems in ferromagnetism. With the help of the Green functions they obtained a formula for the magnetization of the Heisenberg ferromagnet that is a reasonable approximation over the entire temperature range. We also saw that the exact treatment of the problem involves the solution of an infinite set of coupled equations in the Green functions, and that approximate solutions are obtained by making the set of equations finite by using a decoupling approximation. In reality, people have tried to invent a suitable decoupling that will isolate from the rest just a few equations, which can then be solved without much difficulty. But the moment these equations are isolated from the rest, the major portion of the many-body aspect is lost. Hence it would be more advantageous for us if the decoupling could be done at a later stage, but unfortunately mathematical difficulties arise and it becomes progressively more difficult to obtain a good solution of the problem.

It is very surprising that, in contrast to the wealth of literature on the Green function approach to ferromagnetism, comparatively little material has been published concerning the analogous antiferromagnetic problem. In spite of the fact that the two problems are quite closely related and similar questions of decoupling procedure arise for both, the inequivalence of the sublattices in the presence of external magnetic fields and the great number of possible spin structures add to the formidable difficulties of treating an antiferromagnetic problem completely. In 1960 Fu-Cho Pu[4] first considered the problem of the antiferromagnetic structure in some generality, but he treated only the spin-1/2 case.

We shall present here a treatment due to Lines,[5] who use the Green function technique to treat the problem of a general antiferromagnetic structure with arbitrary spin S, with Heisenberg exchange interaction between any or all pairs of spins in the lattice and in the presence of an external magnetic field. Lines's theory is developed on the basis that there will be a single preferred direction of antiferromagnetic alignment in the ordered state of the spin system and also that each of the two ferromagnetic sublattices will be translationally invariant.

In Section 17.2, we discussed in detail the method of the double-time temperature-dependent Green function, following the excellent article by Zubarev.[3] For the sake of completeness we shall mention here also a few important relations before we proceed with the actual formulation and solution of the problem. We shall follow very closely the papers by Lines[5] because they are extremely well written.

We saw in (17.45) that from two Heisenberg operators $\mathbf{A}(t)$ and $\mathbf{B}(t')$ we can construct the double-time temperature-dependent retarded Green function, which is defined as

$$\langle\langle \mathbf{A}(t); \mathbf{B}(t')\rangle\rangle = -i\theta(t-t')\langle[\mathbf{A}(t), \mathbf{B}(t')]_{-}\rangle \tag{18.1}$$

where the square brackets denotes a commutator, and the single-pointed brackets represent a thermal average over a canonical ensemble. Also, $\theta(t-t')$ is the step function, such that

$$\theta(t-t') = \begin{cases} 1 & \text{when} \quad (t-t')>0 \\ 0 & \text{when} \quad (t-t')<0 \end{cases} \tag{18.2}$$

The point to note here is that, if our Hamiltonian is time independent, the retarded Green function (18.1) is a function of $(t-t')$ and as such may be Fourier-transformed with respect to $(t-t')$. The Fourier transform is a function of E $(=\hbar\omega)$ and is denoted by $\langle\langle \mathbf{A}; \mathbf{B}\rangle\rangle_E$, which satisfies the following equation of motion [see (17.9)]:

$$E\langle\langle \mathbf{A}; \mathbf{B}\rangle\rangle_E = \frac{1}{2\pi}\langle[\mathbf{A}, \mathbf{B}]_{-}\rangle + \langle\langle[\mathbf{A}, \mathcal{H}]_{-}; \mathbf{B}\rangle\rangle_E \tag{18.3}$$

Another relation that we shall use has already been given as (17.94), that is,

$$\langle \mathbf{B}(t')\mathbf{A}(t)\rangle = \lim_{\epsilon\to+0} i\int_{-\infty}^{\infty} \frac{\langle\langle \mathbf{A}; \mathbf{B}\rangle\rangle_{\omega+i\epsilon} - \langle\langle \mathbf{A}; \mathbf{B}\rangle\rangle_{\omega-i\epsilon}}{e^{\frac{\omega}{kT}}-1} \cdot e^{-i\omega(t-t')}\,d\omega \tag{18.4}$$

which gives a relation between the Green function and the correlated function $\langle \mathbf{B}(t')\mathbf{A}(t)\rangle$, and where $\hbar = 1$.

The Heisenberg Hamiltonian, as usual, is written as

$$\mathcal{H}=\sum_{\langle i,j\rangle} 2J_{ij}\mathbf{S}_i\cdot\mathbf{S}_j \tag{18.5}$$

where $\Sigma_{\langle i,j\rangle}$ runs over all pairs of spins in the lattice, and the other symbols are as defined in Chapter 17.

To derive the required equation of motion we replace $\mathbf{A}$ by $\mathbf{S}_g^+$ and substitute (18.5) in (18.3), thus getting

$$E\langle\langle \mathbf{S}_g^+;\mathbf{B}\rangle\rangle=\frac{1}{2\pi}\langle[\mathbf{S}_g^+,\mathbf{B}]_-\rangle+\sum_{i,j}2J_{ij}\langle\langle[\mathbf{S}_g^+,\mathbf{S}_i\cdot\mathbf{S}_j]_-;\mathbf{B}\rangle\rangle_E \tag{18.6}$$

But

$$\begin{aligned}[\mathbf{S}_g^+,\mathbf{S}_j\cdot\mathbf{S}_g]&=\tfrac{1}{2}[\mathbf{S}_g^+,\mathbf{S}_j^+\mathbf{S}_g^-]+\tfrac{1}{2}[\mathbf{S}_g^+,\mathbf{S}_j^-\mathbf{S}_g^+]\\&\quad+[\mathbf{S}_g^+,\mathbf{S}_j^z\mathbf{S}_g^z]\end{aligned} \tag{18.7}$$

By using the well-known spin commutation relations, we can easily show that

$$\tfrac{1}{2}[\mathbf{S}_g^+,\mathbf{S}_j^+\mathbf{S}_g^-]=\mathbf{S}_j^+\mathbf{S}_g^z \tag{A}$$

$$\tfrac{1}{2}[\mathbf{S}_g^+,\mathbf{S}_j^-\mathbf{S}_g^+]=\mathbf{S}_g^z\delta_{gj}\mathbf{S}_g^+ \tag{B}$$

and

$$[\mathbf{S}_g^+,\mathbf{S}_j^z\mathbf{S}_g^z]=-\mathbf{S}_j^z\mathbf{S}_g^+-\mathbf{S}_g^+\delta_{jg}\mathbf{S}_g^z \tag{C}$$

Adding (A), (B), and (C), we have

$$[\mathbf{S}_g^+,\mathbf{S}_j\mathbf{S}_g]=\mathbf{S}_j^+\mathbf{S}_g^z+\mathbf{S}_g^z\delta_{gj}\mathbf{S}_g^+-\mathbf{S}_j^z\mathbf{S}_g^+-\mathbf{S}_g^+\delta_{jg}\mathbf{S}_g^z$$

if $j\neq g$, as indeed is the case since otherwise the exchange integral would be zero. We have, therefore, from (18.7)

$$[\mathbf{S}_g^+,\mathbf{S}_j\cdot\mathbf{S}_g]=\mathbf{S}_g^z\mathbf{S}_j^+-\mathbf{S}_g^+\mathbf{S}_j^z \tag{18.8}$$

Thus, using the Hamiltonian (18.5) together with (18.8), we may write the equation of motion of the Green function in the form

$$\begin{aligned}E\langle\langle \mathbf{S}_g^+;\mathbf{B}\rangle\rangle&=\frac{1}{2\pi}\langle[\mathbf{S}_g^+;f(\mathbf{S}_h^z)\mathbf{S}_h^-]_-\rangle\\&\quad+\sum_{j-g}2J_{jg}\langle\langle(\mathbf{S}_g^z\mathbf{S}_j^+-\mathbf{S}_g^+\mathbf{S}_j^z);\mathbf{B}\rangle\rangle\\&=\frac{1}{2\pi}F\delta_{gh}+\sum_{j-g}2J_{jg}\langle\langle(\mathbf{S}_g^z\mathbf{S}_j^+-\mathbf{S}_g^+\mathbf{S}_j^z);\mathbf{B}\rangle\rangle\end{aligned} \tag{18.9}$$

where

$$F = \langle [\mathbf{S}_h^+; f(\mathbf{S}_h^z)\mathbf{S}_h^-]_- \rangle \qquad \text{and} \qquad J_{jj} = 0 \tag{18.10}$$

In (18.10), $f(\mathbf{S}_h^z)$ is an arbitrary function of S^z at site h. Now, using the "Tyablikov" decoupling procedure,[2] we write

$$\langle\langle \mathbf{S}_g^z \mathbf{S}_j^+; \mathbf{B} \rangle\rangle = \langle \mathbf{S}_g^z \rangle \langle\langle \mathbf{S}_j^+; \mathbf{B} \rangle\rangle$$

and

$$\langle\langle (\mathbf{S}_g^+ \mathbf{S}_j^z; \mathbf{B} \rangle\rangle = \langle \mathbf{S}_j^z \rangle \langle\langle \mathbf{S}_g^+; \mathbf{B} \rangle\rangle, \qquad g \neq j \tag{18.11}$$

which gives, on substitution into (18.9),

$$E\langle\langle \mathbf{S}_g^+; \mathbf{B} \rangle\rangle = \frac{F\delta_{gh}}{2\pi} + \sum_{j-g} 2J_{jg}[\langle \mathbf{S}_g^z \rangle \langle\langle \mathbf{S}_j^+; \mathbf{B} \rangle\rangle - \langle \mathbf{S}_j^z \rangle \langle\langle \mathbf{S}_g^+; \mathbf{B} \rangle\rangle] \tag{18.12}$$

As discussed earlier, we now assume a unique direction of spin alignment and then assume the lattice to be split into two sublattices, the "up" and the "down" sublattice with average values of spin per site of $\bar{S}$ and $-\bar{S}$, respectively. If we now suppose that the two sublattices are translationally invariant, we may Fourier-transform with respect to the reciprocal sublattices as follows. When g and h are situated on the same sublattice, the Fourier transform of the left-hand side of (18.12) is given by

$$\langle\langle \mathbf{S}_g^+; \mathbf{B}(\mathbf{h}) \rangle\rangle = \frac{2}{N} \sum_{\mathbf{K}} G_{1\mathbf{K}} e^{i\mathbf{K}(\mathbf{g}-\mathbf{h})}$$

$$G_{1\mathbf{K}} = \sum_{\mathbf{g}-\mathbf{h}} \langle\langle \mathbf{S}_g^+; \mathbf{B}(\mathbf{h}) \rangle\rangle e^{-i\mathbf{K}(\mathbf{g}-\mathbf{h})} \tag{18.13}$$

Similarly, when **g** and **h** belong to different sublattices, the Fourier transform of the left-hand side of (18.10) is given by

$$\langle\langle \mathbf{S}_g^+; \mathbf{B}(\mathbf{h}) \rangle\rangle = \frac{2}{N} \sum_{\mathbf{K}} G_{2\mathbf{K}} e^{i\mathbf{K}(\mathbf{g}-\mathbf{h})}$$

$$G_{2\mathbf{K}} = \sum_{\mathbf{g}-\mathbf{h}} \langle\langle \mathbf{S}_g^+; \mathbf{B}(\mathbf{h}) \rangle\rangle e^{-i\mathbf{K}(\mathbf{g}-\mathbf{h})} \tag{18.14}$$

In (18.13) and (18.14), N is the total number of spins in the lattice, and **K** is a reciprocal lattice vector that runs over $\frac{1}{2}N$ points in the first Brillouin zone of the reciprocal sublattice. Taking the Fourier transform of the right-hand side of (18.12) and assuming **g** and **h** on the same sublattice or on different sublattices, we obtain

$$\left(E-\mu\bar{S}\right)G_{1\mathbf{K}}=\frac{F}{2\pi}+\lambda\bar{S}G_{2\mathbf{K}} \tag{18.15}$$

$$\left(E+\mu\bar{S}\right)G_{2\mathbf{K}}=-\lambda\bar{S}G_{1\mathbf{K}} \tag{18.16}$$

where

$$\mu=\sum_{j-g}^{s}2J_{jg}\left[e^{i\mathbf{K}(\mathbf{j}-\mathbf{g})}-1\right]+\sum_{j-g}^{d}2J_{jg} \tag{18.17}$$

$$\lambda=\sum_{j-g}^{d}2J_{jg}e^{i\mathbf{K}(\mathbf{j}-\mathbf{g})} \tag{18.18}$$

where Σ_{j-g}^{s} holds for all values for which j and g are on the same sublattice, and Σ_{j-g}^{d} for all values for which j and g are on different sublattices. Substituting (18.16) in (18.15), we get

$$\left(E-\mu\bar{S}\right)G_{1\mathbf{K}}=\frac{F}{2\pi}-\frac{\lambda^2\bar{S}^2}{E+\mu\bar{S}}G_{1\mathbf{K}}$$

or

$$4\pi G_{1\mathbf{K}}=\frac{2F\left(E+\mu\bar{S}\right)}{\left(E+E_0\bar{S}\right)\left(E-E_0\bar{S}\right)} \tag{18.19}$$

Now, putting $\mu=E_0A$ in (18.19), we have

$$4\pi G_{1\mathbf{K}}=\frac{F\left[\left(E-E_0\bar{S}\right)(1-A)+\left(E+E_0\bar{S}\right)(1+A)\right]}{\left(E+E_0\bar{S}\right)\left(E-E_0\bar{S}\right)}$$

$$=\frac{(1-A)F}{E+E_0\bar{S}}+\frac{(1+A)F}{E-E_0\bar{S}} \tag{18.20}$$

where

$$A=\frac{\mu}{E_0}=\frac{\mu}{\left(\mu^2-\lambda^2\right)^{1/2}} \tag{18.21}$$

and

$$E_0=\left(\mu^2-\lambda^2\right)^{1/2} \tag{18.22}$$

Using (18.4), (18.13), and (18.20) and the identity

$$\lim_{\epsilon\to+0}\left\{\frac{1}{\omega+i\epsilon-E_{\mathbf{K}}}-\frac{1}{\omega-i\epsilon-E_{\mathbf{K}}}\right\}=-2\pi i\delta(\omega-E_{\mathbf{K}}) \qquad (18.23)$$

we obtain, for the limit $(t-t')\to 0$,

$$\langle \mathbf{B}\mathbf{S}_g^+\rangle=\frac{F}{N}\sum_{\mathbf{K}}\left[A\coth\left(\frac{E_0\bar{S}}{2kT}\right)-1\right]e^{i\mathbf{K}(\mathbf{g}-\mathbf{h})} \qquad (18.24)$$

When $g=h$, we get finally

$$\langle f(S_h^z)S_h^-S_h^+\rangle=\tfrac{1}{2}F\left[\langle A\coth\left(\frac{E_0\bar{S}}{2kT}\right)\rangle_{\mathbf{K}}-1\right] \qquad (18.25)$$

where $\langle\cdots\rangle_{\mathbf{K}}$ indicates an average for $\mathbf{K}$ over $\frac{1}{2}N$ values in the first Brillouin zone of the reciprocal lattice.

We have already discussed in connection with Tahir-Kheli and ter Haar's work, how to obtain $\bar{S}$ as a function of temperature from (18.25). For $S=\frac{1}{2}$ we get from (18.25)

$$\frac{1}{\bar{S}}=2\langle A\coth\left(\frac{E_0\bar{S}}{2kT}\right)\rangle_{\mathbf{K}} \qquad (18.26)$$

The solution for general spin S has been obtained by Callen,[24] as we have seen, by writing $f(\mathbf{S}^z)=e^{a\mathbf{S}^z}$ and using the functional dependence of F and $\langle f(S^z)S^-S^+\rangle$ on the parameter a. Callen's result [equation (52) of Ref. 24] can be expressed as

$$\frac{2\bar{S}+x}{2\bar{S}+1}=\frac{(x+1)^{2S+1}+(x-1)^{2S+1}}{(x+1)^{2S+1}-(x-1)^{2S+1}} \qquad (18.27)$$

where

$$x=\langle A\coth\left(\frac{E_0\bar{S}}{2kT}\right)\rangle_{\mathbf{K}} \qquad (18.28)$$

From (18.25) the Neel temperature is obtained for the general spin S. As $T\to T_{\mathrm{N}}, \bar{S}\to 0$ and we have

$$\frac{S(S+1)}{3kT_{\mathrm{N}}}=\langle\frac{A}{E_0}\rangle_{\mathbf{K}}=\langle\frac{\mu}{\mu^2-\lambda^2}\rangle_{\mathbf{K}} \qquad (18.29)$$

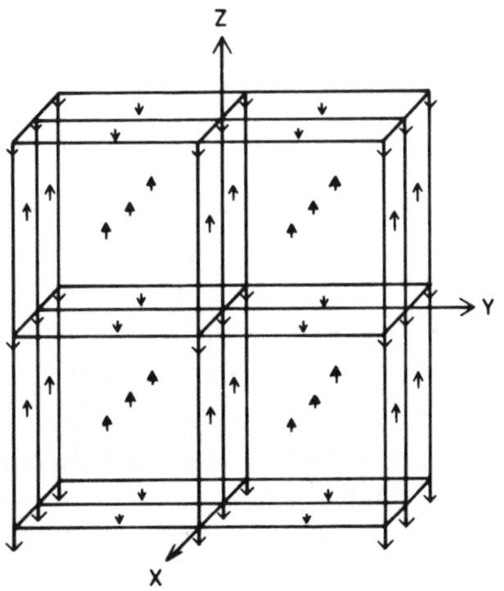

Fig. 18.1 Spin structure in type 1 order for a f.c.c. lattice. (Taken from Lines.[6])

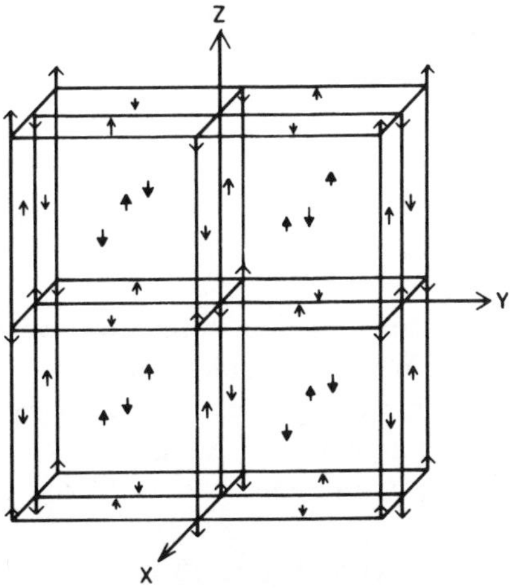

Fig. 18.2 Spin structure in type 2 order for a f.c.c. lattice. (Taken from Lines.[6])

Lines has shown, as an example, how to use (18.20) to obtain, for the first time, an expression of Neel temperature T_N for the face-centered cubic (f. c. c.) antiferromagnetic orders. Following ter Haar and Lines, [6,7] we shall consider order of types 1, 2, and 3 of the f. c. c. lattice, including the nearest- and the next-nearest-neighbors exchange interaction only, which we shall denote by J_1 and J_2, respectively.

Before evaluating μ and λ by using (18.17) and (18.18), let us see order of types 1, 2, and 3 of an f. c. c. lattice, given in Figures 18.1, 18.2, and 18.3,

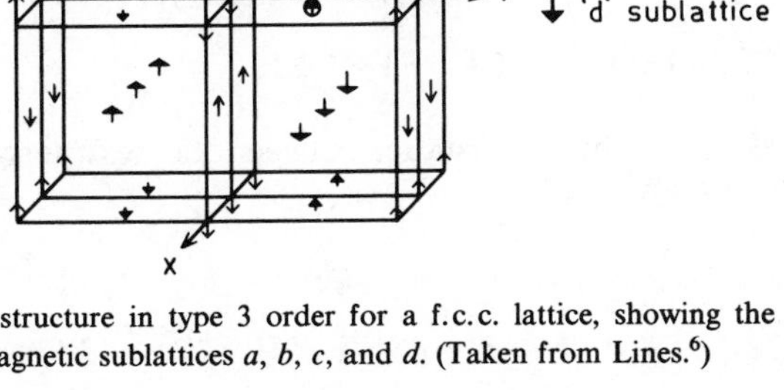

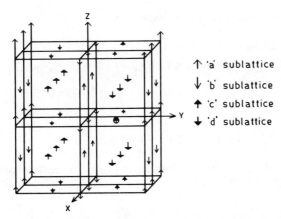

Fig. 18.3 Spin structure in type 3 order for a f.c.c. lattice, showing the four translationally invariant ferromagnetic sublattices a, b, c, and d. (Taken from Lines.[6])

respectively. From (18.17) and (18.18) we have

$$\mu = \sum_{j-g}^{s} 2J_{jg}\left[e^{i\mathbf{K}(\mathbf{j}-\mathbf{g})} - 1\right] + \sum_{j-g}^{d} 2J_{jg}$$

$$\lambda = \sum_{j-g} 2J_{jj}e^{i\mathbf{K}(\mathbf{j}-\mathbf{g})}$$

To carry out the summation we fix g at the origin (0,0,0) and then sum over j. We also assume, with no loss of generality, that the spin at g, that is, at the origin is "up" spin. Then the sum Σ^s is over "up"-spin nearest neighbors (n.n) and next nearest neighbors (n.n.n) (only J_1 and J_2 are considered). Similarly, the sum Σ^d is over "down"-spin n.n.'s and n.n.n.'s.

Type 1 Order (Figure 18.1)

The coordinates of the same spin n.n.'s and n.n.n.'s are as follows:

$$\text{n.n.}\quad J_1\!:\left(\frac{a}{2},\frac{a}{2},0\right),\quad \left(-\frac{a}{2},\frac{a}{2},0\right),\quad \left(-\frac{a}{2},\frac{a}{2},0\right),\quad \left(\frac{a}{2},-\frac{a}{2},0\right)$$

$$\text{n.n.n.}\quad J_2\!: (a,0,0),\quad (-a,0,0),\quad (0,a,0),\quad (0,-a,0),\quad (0,0,a),\quad (0,0,-a)$$

Similarly, the coordinates of the different spins n.n.'s and n.n.n.'s are as follows:

$$\text{n.n.}\quad J_1\left(\frac{a}{2},0,\frac{a}{2}\right),\quad \left(-\frac{a}{2},0,\frac{a}{2}\right),\quad \left(\frac{a}{2},0,-\frac{a}{2}\right),\quad \left(-\frac{a}{2},0,-\frac{a}{2}\right),$$
$$\left(0,\frac{a}{2},\frac{a}{2}\right),\quad \left(0,-\frac{a}{2},\frac{a}{2}\right),\quad \left(0,\frac{a}{2},-\frac{a}{2}\right),\quad \left(0,-\frac{a}{2},-\frac{a}{2}\right)$$

n.n.n. J_2 none.

Using the above coordinates, we now have

$$\begin{aligned}\mu = {} & 2J_1\left[e^{i(a/2)(K_x+K_y)} - 1\right] + 2J_1\left[e^{-i(a/2)(K_x+K_y)} - 1\right] \\ & + 2J_1\left[e^{i(a/2)(K_x-K_y)} - 1\right] + 2J_1\left[e^{-i(a/2)(K_x-K_y)} - 1\right] \\ & + 2J_2\left[e^{iaK_x} - 1\right] + 2J_2\left[e^{-iaK_x} - 1\right] + 2J_2\left[e^{iaK_y} - 1\right] \\ & + 2J_2\left[e^{-iaK_y} - 1\right] + 2J_2\left[e^{iaK_z} - 1\right] + 2J_2\left[e^{-iaK_z} - 1\right] \\ & + 2J_1(1+1+1+1+1+1+1+1)\end{aligned} \tag{18.30}$$

where we have taken a to be the distance between the next nearest neighbors. We now simplify (18.30) thus:

$$\begin{aligned}\mu = {} & 2J_1\left[e^{ip} + e^{-ip} + e^{iq} + e^{-iq}\right] - 8J_1 \\ & + 2J_2\left[e^{i\alpha} + e^{-i\alpha} + e^{i\beta} + e^{-i\beta} + e^{i\gamma} + e^{-i\gamma}\right] - 12J_2 + 16J_1\end{aligned}$$

where

$$p=\frac{a}{2}(K_x+K_y)$$

$$q=\frac{a}{2}(K_x-K_y)$$

$$\alpha=aK_x$$

$$\beta=aK_y$$

$$\gamma=aK_z$$

Simplifying, we have

$$\mu=2J_1[2\cos p+2\cos q]+8J_1$$
$$+2J_2[2\cos\alpha+2\cos\beta+2\cos\gamma]-12J_2$$

After restoring the expressions for p, q, and so on and after a little algebra, we get finally

$$\mu=8J_1(c_1c_2+1)+8J_2(c_1^2+c_2^2+c_3^2-3) \tag{18.31}$$

where

$$c_1=\cos\left(\frac{a}{2}K_x\right), \qquad c_2=\cos\left(\frac{a}{2}K_y\right), \qquad c_3=\cos\left(\frac{a}{2}K_z\right)$$

Similarly,

$$\lambda=2J_1\left[e^{i(a/2)(K_x+K_z)}+e^{i(a/2)(K_x+K_z)}+e^{i(a/2)(K_x-K_z)}+e^{i(a/2)(-K_x-K_z)}\right.$$
$$\left.+e^{i(a/2)(K_y+K_z)}+e^{i(a/2)(-K_y+K_z)}+e^{i(a/2)(K_y-K_z)}+e^{i(a/2)(-K_y-K_z)}\right]$$
$$=4J_1\left[\cos\frac{a}{2}(K_x+K_z)+\cos\frac{a}{2}(K_x-K_z)+\cos\frac{a}{2}(K_y+K_z)\right.$$
$$\left.+\cos\frac{a}{2}(K_y-K_z)\right.$$
$$=8J_1(c_1c_3+c_2c_3) \tag{18.32}$$

after proceeding exactly as before.

Type 2 Order (Figure 18.2)

For type 2 order, we have the following coordinates for n.n.'s and n.n.n.'s having the same and different spins:

Same spins

$$\text{n.n.}\quad J_1\text{: } \left(-\frac{a}{2},\frac{a}{2},0\right),\ \left(-\frac{a}{2},-\frac{a}{2},0\right),\ \left(0,\frac{a}{2},\frac{a}{2}\right),$$
$$\left(0,-\frac{a}{2},-\frac{a}{2}\right),\ \left(\frac{a}{2},0,\frac{a}{2}\right),\ \left(-\frac{a}{2},0,-\frac{a}{2}\right)$$

n.n.n. J_2: none.

Different spins

$$\text{n.n.}\quad J_1\text{: } \left(\frac{a}{2},\frac{a}{2},0\right),\ \left(-\frac{a}{2},-\frac{a}{2},0\right),\ \left(0,-\frac{a}{2},\frac{a}{2}\right),$$
$$\left(0,\frac{a}{2},\frac{a}{2}\right),\ \left(-\frac{a}{2},0,\frac{a}{2}\right),\ \left(\frac{a}{2},0,-\frac{a}{2}\right)$$
$$\text{n.n.n.}\quad J_2\text{: } (a,0,0),\ \ (-a,0,0),\ \ (0,a,0),\ \ (0,-a,0),$$
$$(0,0,a),\ \ (0,0,-a)$$

Using these coordinates, we have, as before,

$$\begin{aligned}\mu=&2J_1\left[e^{i(a/2)(-K_x+K_y)}-1\right]+2J_1\left[e^{i(a/2)(K_x-K_y)}-1\right]\\&+2J_1\left[e^{i(a/2)(K_y+K_z)}-1\right]+2J_1\left[e^{i(a/2)(-K_y-K_z)}-1\right]\\&+2J_1\left[e^{i(a/2)(K_x+K_z)}-1\right]\\&+2J_1\left[e^{i(a/2)(-K_x-K_z)}-1\right]+2J_1[6]+2J_2[6]\end{aligned}$$

which, after simplification, gives

$$\begin{aligned}\mu=&4J_1\left[\cos\frac{a}{2}(K_x-K_y)+\cos\frac{a}{2}(K_y+K_z)+\cos\frac{a}{2}(K_x+K_z)\right]\\&+12J_2\end{aligned}$$

Similarly,

$$\begin{aligned}\lambda=&2J_1\left[e^{i(a/2)(K_x+K_y)}+e^{i(a/2)(-K_x-K_y)}+e^{i(a/2)(-K_y+K_z)}\right.\\&\left.+e^{i(a/2)(K_y-K_z)}+e^{i(a/2)(-K_x+K_z)}+e^{i(a/2)(K_x-K_z)}\right]\\&+2J_2\left[e^{iaK_x}+e^{-iaK_x}+e^{iaK_y}+e^{-iaK_y}+e^{iaK_z}+e^{-iaK_z}\right]\\=&4J_1\left[\cos\frac{a}{2}(K_x+K_y)+\cos\frac{a}{2}(K_y-K_z)+\cos\frac{a}{2}(K_x-K_z)\right]\\&+4J_2(\cos aK_x+\cos aK_y+\cos aK_z)\end{aligned}$$

From these relations of μ and λ, we can show, exactly as before, that

$$\mu+\lambda=8J_1[c_1c_2+c_2c_3+c_3c_1]+8J_2[c_1^2+c_2^2+c_3^2] \tag{18.33}$$
$$\mu-\lambda=8J_1[s_1s_2+s_2s_3+s_3s_1]+8J_2[s_1^2+s_2^2+s_3^2] \tag{18.34}$$

where

$$s_1=\sin\left(\frac{a}{2}K_x\right),\qquad s_2=\sin\left(\frac{a}{2}K_y\right),\qquad s_3=\sin\left(\frac{a}{2}K_z\right)$$

Type 3 Order (Figure 18.3)

A glance at Figure 18.3 shows that type 3 order is an example of spin structure for which the "up" and "down" sublattices are not translationally invariant. Also, to consider this kind of ordering in the Green function approximation, it is essential to subdivide the lattice into *four* ferromagnetic sublattices, each of which is translationally invariant. We thus have to introduce *four* Green functions $G_{i\mathbf{K}}(i=1, 2, 3, \text{and } 4)$, in place of the *two* that are sufficient to describe order of types of 1 and 2. Consequently the equation of motion (18.9) gives rise to four equations in $G_{i\mathbf{K}}$, which have to be solved.

In the four ferromagnetic sublattices of type 3 order, a and c are up spins while b and d are down spins. Also, in the division of the sublattices, the periodicity of the magnetic structure in the x- and z-directions is *only one* crystallographic translation, whereas in the y-direction there are *two* crystallographic translations.

To derive the required relations we take the circled atom in Figure 18.3 as the origin. The coordinates and the sublattice designations of the nearest and the next nearest neighbors are given below. To avoid confusion with a, the dimension of the unit cell, we denote the sublattices by capital letters A, B, C, and D, which corresponds to a, b, c, and d, respectively, in Figure 18.3.

The coordinates of the sublattices for the nearest neighbors and the next nearest neighbors with one atom on the a sublattice as origin are as follows:

$$\text{n.n.}\quad \overset{C}{\left(\frac{a}{2},\frac{a}{2},0\right)},\quad \overset{C}{\left(-\frac{a}{2},\frac{a}{2},0\right)},\quad \overset{D}{\left(-\frac{a}{2},-\frac{a}{2},0\right)},\quad \overset{D}{\left(\frac{a}{2},-\frac{a}{2},0\right)}$$

$$\overset{B}{\left(\frac{a}{2},0,\frac{a}{2}\right)},\quad \overset{D}{\left(0,\frac{a}{2},\frac{a}{2}\right)},\quad \overset{B}{\left(-\frac{a}{2},0,\frac{a}{2}\right)},\quad \overset{C}{\left(0,-\frac{a}{2},\frac{a}{2}\right)}$$

$$\overset{B}{\left(\frac{a}{2},0,-\frac{a}{2}\right)},\quad \overset{D}{\left(0,\frac{a}{2},-\frac{a}{2}\right)},\quad \overset{B}{\left(-\frac{a}{2},0,-\frac{a}{2}\right)},\quad \overset{C}{\left(0,-\frac{a}{2},-\frac{a}{2}\right)}$$

$$\text{n.n.n.}\quad \overset{A}{(a,0,0)},\ \overset{A}{(-a,0,0)},\ \overset{B}{(0,a,0)},\ \overset{B}{(0,-a,0)},\ \overset{A}{(0,0,a)},\ \overset{A}{(0,0,-a)}$$

Now, we have from (18.12)

$$E\langle\langle \mathbf{S}_g^+;\mathbf{B}\rangle\rangle_E=\frac{F\delta_{gh}}{2\pi}+\sum_{j-g}2J_{jg}\left[\langle \mathbf{S}_g^z\rangle\langle\langle \mathbf{S}_j^+;\mathbf{B}\rangle\rangle - \langle \mathbf{S}_j^z\rangle\langle\langle \mathbf{S}_g^+;\mathbf{B}\rangle\rangle\right] \tag{18.35}$$

where

$$\langle\langle \mathbf{S}_g^+;\mathbf{B}\rangle\rangle_E=\frac{4}{N}\sum_{\mathbf{K}} G_{1\mathbf{K}}e^{i\mathbf{K}(\mathbf{g}-\mathbf{h})} \tag{18.36}$$

$$G_{1\mathbf{K}}=\sum_{g-h}\langle\langle \mathbf{S}_g^+;\mathbf{B}\rangle\rangle_E e^{-i\mathbf{K}(\mathbf{g}-\mathbf{h})} \tag{18.37}$$

Multiplying both sides of (18.35) by $e^{-i\mathbf{K}(\mathbf{g}-\mathbf{h})}$ and summing over $(\mathbf{g}-\mathbf{h})$, we get

$$\begin{aligned} E\sum_{g-h}\langle\langle \mathbf{S}_g^+;\mathbf{B}\rangle\rangle e^{-i\mathbf{K}(\mathbf{g}-\mathbf{h})}=&\frac{F}{2\pi}\sum\delta_{gh}e^{-i\mathbf{K}(\mathbf{g}-\mathbf{h})} \qquad \text{(I)}\\ &+\sum_{j-g}\sum_{g-h}2J_{jg}\langle \mathbf{S}_g^z\rangle e^{i\mathbf{K}(\mathbf{j}-\mathbf{g})}\langle\langle \mathbf{S}_j^+;\mathbf{B}\rangle\rangle e^{-i\mathbf{K}(\mathbf{j}-\mathbf{h})} \qquad \text{(II)}\\ &-\sum_{j-g}\sum_{g-h}2J_{jg}\langle \mathbf{S}_j^z\rangle\langle\langle \mathbf{S}_g^+;\mathbf{B}\rangle\rangle e^{-i\mathbf{K}(\mathbf{g}-\mathbf{h})} \qquad \text{(III)} \end{aligned} \tag{18.38}$$

We now use relation (18.38) to derive the equations for $G_{i\mathbf{K}}(i=1,2,3,4)$. We take h to be on a sublattice and g on a, b, c, and d sublattices and derive the four equations. Note that both (II) and (III) break up into four sums, depending on whether j is on the a, b, c, or d sublattice. The summations over $(\mathbf{j}-\mathbf{h})$ in (II) and $(\mathbf{g}-\mathbf{h})$ in (III) give the $G_{i\mathbf{K}}$'s. To show the process we shall derive only the first relation. The other three relations can be derived similarly. Also note that the average value of spin per site for the a and c sublattices is $\bar{S}$, while that for the b and d sublattices is $-\bar{S}$, as discussed earlier. We have, for the first relation,

$$\begin{aligned} EG_{1\mathbf{K}}=&\frac{F}{2\pi}+\bar{S}\sum_{j-g}^{a}2J_{jg}e^{i\mathbf{K}(\mathbf{j}-\mathbf{g})}G_{1\mathbf{K}}\\ &+\bar{S}\sum_{j-g}^{b}2J_{jg}e^{i\mathbf{K}(\mathbf{j}-\mathbf{g})}G_{2\mathbf{K}}\\ &+\bar{S}\sum_{j-g}^{c}2J_{jg}e^{i\mathbf{K}(\mathbf{j}-\mathbf{g})}G_{3\mathbf{K}}\\ &+\bar{S}\sum_{j-g}^{d}2J_{jg}e^{i\mathbf{K}(\mathbf{j}-\mathbf{g})}G_{4\mathbf{K}}\\ &-\bar{S}\sum_{j-g}^{a}2J_{jg}G_{1\mathbf{K}}\sum_{j-g}^{b}2J_{jg}G_{1\mathbf{K}}\\ &-\bar{S}\sum_{j-g}^{c}2J_{jg}G_{1\mathbf{K}}+\bar{S}\sum_{j-g}^{d}2J_{jg}G_{1\mathbf{K}} \end{aligned} \tag{18.39}$$

Using the coordinates given in (18.39), we obtain

$$EG_{1\mathbf{K}}=\frac{F}{2\pi}+2J_2\bar{S}(e^{iaK_x}+e^{-iaK_x}+e^{iaK_z}+e^{-iaK_z})G_{1\mathbf{K}}$$
$$+\bar{S}\big[2J_1\{e^{i(a/2)(K_x+K_z)}+e^{i(a/2)(K_x-K_z)}+e^{i(a/2)(-K_x+K_z)}+e^{i(a/2)(-K_x-K_z)}\}$$
$$+2J_2\{e^{iaK_y}+e^{-iaK_y}\}\big]G_{2\mathbf{K}}$$
$$+2J_1\bar{S}\big[e^{i(a/2)(K_x+K_y)}+e^{i(a/2)(-K_x+K_y)}+e^{i(a/2)(-K_y+K_z)}+e^{i(a/2)(-K_y-K_z)}\big]G_{3\mathbf{K}}$$
$$+2J_1\bar{S}\big[e^{i(a/2)(-K_x-K_y)}+e^{i(a/2)(K_x-K_y)}+e^{i(a/2)(K_y+K_z)}+e^{i(a/2)(K_y-K_z)}\big]G_{4\mathbf{K}}$$
$$-8J_2\bar{S}G_{1\mathbf{K}}+\bar{S}(8J_1+4J_2)G_{1\mathbf{K}}-8J_1\bar{S}G_{1\mathbf{K}}+8J_1\bar{S}G_{1\mathbf{K}}$$

which, after simplification, gives

$$(E-\alpha')G_{1\mathbf{K}}-\beta'G_{2\mathbf{K}}-\gamma'G_{3\mathbf{K}}-\delta'G_{4\mathbf{K}}=\frac{F}{2\pi} \tag{18.40}$$

where

$$\alpha'=\alpha\bar{S}=\bar{S}\big[8J_1-12J_2+8J_2(c_1^2+c_3^2)\big]$$
$$\beta'=\beta\bar{S}=\bar{S}\big[8J_1c_1c_3-4J_2+8J_2c_2^2\big]$$
$$\gamma'=\gamma\bar{S}=\bar{S}\big[4J_1(e^{i(a/2)K_y}c_1+e^{-i(a/2)K_y}c_3)\big]$$
$$\delta'=\delta\bar{S}=\bar{S}\big[4J_1(e^{-i(a/2)K_y}c_1+e^{i(a/2)K_y}c_3)\big]$$

Similarly, to derive the second relation we have to use the following coordinates with the b sublattice as origin:

n.n.
$$\overset{D}{\left(\frac{a}{2},\frac{a}{2},0\right)},\quad \overset{D}{\left(\frac{a}{2},-\frac{a}{2},0\right)},\quad \overset{C}{\left(-\frac{a}{2},-\frac{a}{2},0\right)},\quad \overset{C}{\left(\frac{a}{2},-\frac{a}{2},0\right)}$$
$$\overset{A}{\left(\frac{a}{2},0,\frac{a}{2}\right)},\quad \overset{A}{\left(-\frac{a}{2},0,\frac{a}{2}\right)},\quad \overset{C}{\left(0,\frac{a}{2},\frac{a}{2}\right)},\quad \overset{D}{\left(0,-\frac{a}{2},\frac{a}{2}\right)}$$
$$\overset{A}{\left(\frac{a}{2},0,-\frac{a}{2}\right)},\quad \overset{A}{\left(-\frac{a}{2},0,-\frac{a}{2}\right)},\quad \overset{C}{\left(0,\frac{a}{2},-\frac{a}{2}\right)},\quad \overset{D}{\left(0,-\frac{a}{2},-\frac{a}{2}\right)}$$

n.n.n.
$$\overset{B}{(a,0,0)},\ \overset{B}{(-a,0,0)},\ \overset{A}{(0,a,0)},\ \overset{A}{(0,-a,0)}\ \overset{B}{(0,0,a)},\ \overset{B}{(0,0,-a)}$$

Using these coordinates, we find that the second relation is given by

$$\beta'G_{1\mathbf{K}}+(E+\alpha')G_{2\mathbf{K}}+\delta'G_{3\mathbf{K}}+\gamma'G_{4\mathbf{K}}=0 \tag{18.41}$$

For the third relation the following coordinates, with the c sublattice as origin, are used:

$$\text{n.n.}\ \overset{B}{\left(\frac{a}{2},\frac{a}{2},0\right)},\quad \overset{B}{\left(-\frac{a}{2},\frac{a}{2},0\right)},\quad \overset{A}{\left(-\frac{a}{2},-\frac{a}{2},0\right)},\quad \overset{A}{\left(\frac{a}{2},-\frac{a}{2},0\right)}$$

$$\overset{D}{\left(\frac{a}{2},0,\frac{a}{2}\right)},\quad \overset{D}{\left(-\frac{a}{2},0,\frac{a}{2}\right)},\quad \overset{A}{\left(0,\frac{a}{2},\frac{a}{2}\right)},\quad \overset{B}{\left(0,-\frac{a}{2},\frac{a}{2}\right)}$$

$$\overset{D}{\left(\frac{a}{2},0,-\frac{a}{2}\right)},\quad \overset{D}{\left(-\frac{a}{2},0,-\frac{a}{2}\right)},\quad \overset{A}{\left(0,\frac{a}{2},-\frac{a}{2}\right)},\quad \overset{B}{\left(0,-\frac{a}{2},-\frac{a}{2}\right)}$$

$$\text{n.n.n.}\ \overset{C}{(a,0,0)},\overset{C}{(-a,0,0)},\overset{D}{(0,a,0)},\overset{D}{(0,-a,0)},\overset{C}{(0,0,a)},\overset{C}{(0,0,-a)}$$

The third relation is given by

$$-\delta' G_{1\mathbf{K}}-\gamma' G_{2\mathbf{K}}+(E-\alpha')G_{3\mathbf{K}}-\beta'_{4\mathbf{K}}=0 \qquad (18.42)$$

For the fourth relation the following coordinates, with the d sublattice as origin, are used:

$$\text{n.n.}\ \overset{A}{\left(\frac{a}{2},\frac{a}{2},0\right)},\quad \overset{A}{\left(-\frac{a}{2},\frac{a}{2},0,\right)},\quad \overset{B}{\left(-\frac{a}{2},-\frac{a}{2},0\right)},\quad \overset{B}{\left(\frac{a}{2},-\frac{a}{2},0\right)}$$

$$\overset{C}{\left(\frac{a}{2},0,\frac{a}{2}\right)},\quad \overset{C}{\left(-\frac{a}{2},0,\frac{a}{2}\right)},\quad \overset{B}{\left(0,\frac{a}{2},\frac{a}{2}\right)},\quad \overset{A}{\left(0,-\frac{a}{2},\frac{a}{2}\right)}$$

$$\overset{C}{\left(\frac{a}{2},0,-\frac{a}{2}\right)},\quad \overset{C}{\left(-\frac{a}{2},0,-\frac{a}{2}\right)},\quad \overset{B}{\left(0,\frac{a}{2},-\frac{a}{2}\right)},\quad \overset{A}{\left(0,-\frac{a}{2},-\frac{a}{2}\right)}$$

$$\text{n.n.n.}\ \overset{D}{(a,0,0)},\overset{D}{(-a,0,0)},\overset{C}{(0,a,0)},\overset{C}{(0,-a,0)},\ \overset{D}{(0,0,a)},\ \overset{D}{(0,0,-a)}$$

The fourth relation is

$$\gamma' G_{1\mathbf{K}}+\delta' G_{2\mathbf{K}}+\beta' G_{3\mathbf{K}}+(E+\alpha')G_{4\mathbf{K}}=0 \qquad (18.43)$$

We thus get the following matrix for $G_{i\mathbf{K}}$:

$$\begin{bmatrix} E-\alpha' & -\beta' & -\gamma' & -\delta' \\ \beta' & E+\alpha' & \delta' & \gamma' \\ -\delta' & -\gamma' & E-\alpha' & -\beta' \\ \gamma' & \delta' & \beta' & E+\alpha' \end{bmatrix}\begin{bmatrix} G_{1\mathbf{K}} \\ G_{2\mathbf{K}} \\ G_{3\mathbf{K}} \\ G_{4\mathbf{K}} \end{bmatrix}=\begin{bmatrix} F/2\pi \\ 0 \\ 0 \\ 0 \end{bmatrix} \qquad (18.44)$$

It would be a good exercise to derive relations (18.41) to (18.43) by following the procedure outlined in the derivation of (18.40).

To derive the expression for parallel susceptibility we now consider an antiferromagnet placed in an external magnetic field **H** in such a way that the field is applied parallel to the z-direction of the antiferromagnetic spin alignment. The Hamiltonian is given by

$$\mathcal{H} = \sum_{\langle i,j \rangle} 2J_{ij}\mathbf{S}_i \cdot \mathbf{S}_j - g\mu_{\mathrm{B}} H \sum_i S_i^z \tag{18.45}$$

where μ_{B} is the Bohr magneton. We also have, from (17.93),

$$E\langle\langle \mathbf{A}; \mathbf{B} \rangle\rangle_E = \frac{1}{2\pi}\langle [\mathbf{A}, \mathbf{B}]_- \rangle + \langle\langle [\mathbf{A}, \mathcal{H}]_-; \mathbf{B} \rangle\rangle_E \tag{18.46}$$

Putting $\mathbf{A} = \mathbf{S}_g^+$, we have from (18.46)

$$E\langle\langle \mathbf{S}_g^+; \mathbf{B} \rangle\rangle_E = \frac{1}{2\pi}\langle [\mathbf{S}_g^+; \mathbf{B}]_- \rangle + \langle\langle [\mathbf{S}_g^+, \mathcal{H}]_-; \mathbf{B} \rangle\rangle_E \tag{18.47}$$

Also, from (18.45)

$$\begin{aligned} \mathcal{H} = & \sum_{g,j} 2J_{gj}\left(\tfrac{1}{2}\mathbf{S}_g^+\mathbf{S}_j^- + \tfrac{1}{2}\mathbf{S}_g^-\mathbf{S}_j^+ + \mathbf{S}_g^z\mathbf{S}_j^z\right) \\ & + \sum_{\substack{i,j \\ i \neq g}} 2J_{ij} - g\mu_{\mathrm{B}} H \mathbf{S}_g^z - g\mu_{\mathrm{B}} H \sum_{j \neq g} \mathbf{S}_j^z \end{aligned} \tag{18.48}$$

Using the following spin commutation relations:

$$\begin{aligned} &[\mathbf{S}_x, \mathbf{S}_y] = i\mathbf{S}_z \qquad [\mathbf{S}^+, \mathbf{S}^z] = -\mathbf{S}^+ \\ &[\mathbf{S}_x, \mathbf{S}_z] = -i\mathbf{S}_y \qquad [\mathbf{S}^+, \mathbf{S}^-] = 2\mathbf{S}_z \\ &[\mathbf{S}_y, \mathbf{S}_z] = i\mathbf{S}_x \qquad [\mathbf{S}_l^+, \mathbf{S}_g^-] = 2\mathbf{S}_g^z \delta_{lg} \end{aligned} \tag{18.49}$$

we can show that

$$[\mathbf{S}_g^+, \mathcal{H}]_- = \sum_{\langle g,j \rangle} 2J_{gj}(\mathbf{S}_g^z\mathbf{S}_j^+ - \mathbf{S}_g^+\mathbf{S}_j^z) + g\mu_{\mathrm{B}} H \mathbf{S}_g^+ \tag{18.50}$$

Now, inserting (18.50) in (18.47), we have

$$\begin{aligned} E\langle\langle \mathbf{S}_g^+; \mathbf{B} \rangle\rangle_E = & \frac{1}{2\pi}\langle [\mathbf{S}_g^+, B]_- \rangle \sum_{j,g} 2J_{jg}[\langle\langle \mathbf{S}_g^z\mathbf{S}_j^+; \mathbf{B} \rangle\rangle \\ & - \langle\langle \mathbf{S}_g^+\mathbf{S}_j^z; B \rangle\rangle] + g\mu_{\mathrm{B}} H \langle\langle \mathbf{S}_g^+; \mathbf{B} \rangle\rangle \end{aligned} \tag{18.51}$$

Use of the Tyablikov decoupling, as before, reduces (18.51) to

$$(E-g\mu_B H)\langle\langle \mathbf{S}_g^+;\mathbf{B}\rangle\rangle_E=\frac{F\delta_{gh}}{2\pi}+\sum_{j-g}2J_{jg}\cdot\left[\langle \mathbf{S}_g^z\rangle\langle\langle \mathbf{S}_j^+;\mathbf{B}\rangle\rangle-\langle \mathbf{S}_j^z\rangle\langle\langle \mathbf{S}_g^+;\mathbf{B}\rangle\rangle\right] \tag{18.52}$$

where

$$F=\langle[\mathbf{S}_h^+;f(\mathbf{S}_h^z)\mathbf{S}_h^-]_-\rangle$$

In the two-sublattice model the lattice can be divided into two translationally invariant ferromagnetic sublattices, and we may, as usual, Fourier-transform the Green functions with respect to the reciprocal sublattices and consequently have two Green functions, $G_{1\mathbf{K}}$ and $G_{2\mathbf{K}}$, as we saw in (18.13) and (18.14). These can be used again, but with a little modification because of the presence of the magnetic field. It is important to note that, when an external field is present, the average spin per site on the up sublattice (which we now denote as $\bar{S}_u$) is *no longer equal and opposite* to the average spin per site on the down sublattice (denoted as $\bar{S}_d$), and consequently the equation of motion will have these two averages, $\bar{S}_u$ and $\bar{S}_d$, in place of a single $\bar{S}$, as we saw in (18.15) and (18.16). We now have to incorporate this difference in (18.13) and (18.14), that is, we have to consider two cases:

Case 1: h is on the up sublattice and
g is also on the up sublattice
Case 2: h is on the up sublattice and
g is on the down sublattice

Let us first consider Case 1.

We multiply (18.52) by $e^{-i\mathbf{K}(\mathbf{g}-\mathbf{h})}$ and sum over $(\mathbf{g}-\mathbf{h})$ we then have for the left-hand side of (18.52)

$$(E-g\mu_B H)G_{1\mathbf{K}}$$

The second term on the right-hand side of (18.52) is then given by

$$\sum_{j-g}\sum_{g-h}2J_{jg}\langle \mathbf{S}_g^z\rangle\langle\langle \mathbf{S}_j^+;\mathbf{B}\rangle\rangle e^{-i\mathbf{K}(\mathbf{g}-\mathbf{h})}$$

$$=\sum_{j-g}\sum_{g-h}2J_{jg}\langle \mathbf{S}_g^z\rangle\langle\langle \mathbf{S}_j^+;\mathbf{B}\rangle\rangle e^{-i\mathbf{K}(\mathbf{j}-\mathbf{h})}e^{i\mathbf{K}(\mathbf{j}-\mathbf{g})}$$

$$=\sum_{j-g}^{j\ \mathrm{up}}2J_{jg}\bar{S}_u G_{1\mathbf{K}}e^{i\mathbf{K}(\mathbf{j}-\mathbf{g})}$$

$$+\sum_{j-g}^{j\ \mathrm{down}}2J_{jg}\bar{S}_u G_{2\mathbf{K}}e^{i\mathbf{K}(\mathbf{j}-\mathbf{g})}$$

Similarly, the third term on the right-hand side of (18.52) is given by

$$-\sum_{j-g}\sum_{g-h} 2J_{jg}\langle \mathbf{S}_j^z\rangle\langle\langle \mathbf{S}_g^+;\mathbf{B}\rangle\rangle e^{-i\mathbf{K}(\mathbf{g}-\mathbf{h})} = -\sum_{j-g} 2J_{jg}\langle \mathbf{S}_j^z\rangle G_{1\mathbf{K}}$$

$$= -\sum_{j-g}^{j\text{ up}} 2J_{jg}\bar{S}_u G_{1\mathbf{K}} - \sum_{j-g}^{j\text{ down}} 2J_{jg}\bar{S}_d G_{1\mathbf{K}}$$

Collecting all the above terms, we finally have

$$(E-g\mu_{\mathrm{B}}H)G_{1\mathbf{K}} = \frac{F_u}{2\pi} + \left\{\sum_{j-g}^{j\text{ up}} 2J_{jg}\bar{S}_u\left[e^{i\mathbf{K}(\mathbf{j}-\mathbf{g})}-1\right]\right.$$

$$\left.-\sum_{j-g}^{j\text{ down}} 2J_{jg}\bar{S}_d\right\}G_{1\mathbf{K}} + \sum_{j-g}^{j\text{ down}} 2J_{jg}\bar{S}_u \exp^{i\mathbf{K}(\mathbf{j}-\mathbf{g})} G_{2\mathbf{K}}$$

(18.53)

Similarly, for Case 2 we have

$$(E-g\mu_{\mathrm{B}}H)G_{2\mathbf{K}} = \sum_{j-g}^{j\text{ up}} 2J_{jg}\bar{S}_d e^{i\mathbf{K}(\mathbf{j}-\mathbf{g})} G_{1\mathbf{K}}$$

$$+\sum_{j-g}^{j\text{ down}} 2J_{jg}\bar{S}_d\left[e^{i\mathbf{K}(\mathbf{j}-\mathbf{g})}-1\right]G_{2\mathbf{K}}$$

$$-\sum_{j-g}^{j\text{ up}} 2J_{jg}\bar{S}_u G_{2\mathbf{K}} \tag{18.54}$$

Equations (18.53) and (18.54) can be written as

$$\left(E-g\mu_{\mathrm{B}}H-\mu_1\bar{S}_u-\mu_2\bar{S}_d\right)G_{1\mathbf{K}} = \frac{F}{2\pi}+\lambda\bar{S}_u G_{2\mathbf{K}} \tag{18.55}$$

and

$$\left(E-g\mu_{\mathrm{B}}H-\mu_1\bar{S}_d-\mu_2\bar{S}_u\right)G_{2\mathbf{K}} = \lambda\bar{S}_d G_{1\mathbf{K}} \tag{18.56}$$

where

$$F_u = \langle\left[S_h^+, f(S_h^z)S_h^-\right]_-\rangle \tag{18.57}$$

and

$$\mu_1 = \sum_{j-g}^{s} 2J_{jg}\left[e^{i\mathbf{K}(\mathbf{j}-\mathbf{g})} - 1\right] \tag{18.58}$$

$$\mu_2 = -\sum_{j-g}^{d} 2J_{jg} \tag{18.59}$$

$$\lambda = \sum_{j-g}^{d} 2J_{jg} e^{i\mathbf{K}(\mathbf{j}-\mathbf{g})} \tag{18.60}$$

Now, writing

$$\bar{S}_u = \bar{S} + \delta\bar{S}, \qquad \bar{S}_d = -\bar{S} + \delta\bar{S}$$

we solve for $G_{1\mathbf{K}}$ and $G_{2\mathbf{K}}$, given by (18.55) and (18.56), and then, using identity (18.25), obtain after some lengthy manipulation, the correlation function in the form

$$\langle f(\mathbf{S}_h^z)\mathbf{S}_h^- \mathbf{S}_h^+ \rangle = \tfrac{1}{2} F_{\substack{u\\d}} \left\langle \frac{\sinh\alpha \mp A'\sinh\beta}{\cosh\alpha - \cosh\beta} - 1 \right\rangle_{\mathbf{K}} \tag{18.61}$$

where

$$\alpha = \frac{1}{kT}\left[g\mu_{\mathrm{B}}H + \delta\bar{S}(\mu_1 + \mu_2)\right] \tag{18.62}$$

$$\beta = \frac{4}{kT}\left[(\mu_1 - \mu_2)^2(\bar{S})^2 - \lambda^2(\bar{S})^2 + \lambda^2(\delta\bar{S})^2\right]^{1/2} \tag{18.63}$$

$$A' = \frac{(\mu_1 - \mu_2)\bar{S}}{\beta kT} \tag{18.64}$$

Here u and the minus sign (in $\langle \cdots \rangle$) correspond to the case where h is on the up sublattice; d and the plus sign (in $\langle \cdots \rangle$), to the case when h is on the down sublattice. We also note that, for $\delta\bar{S} \to 0$, we have $A' \to A$ and $\beta \to E_0\bar{S}/kT$, with A and E_0 given by (18.21) and (18.22).

To obtain the expression for susceptibility in the ordered state, for the simplest case of $S = \frac{1}{2}$, we put $f(S^z) = 1$ and get from (18.61)

$$\bar{S} + \delta\bar{S} = \left[2\left\langle \frac{\sinh\alpha - A'\sinh\beta}{\cosh\alpha - \cosh\beta} \right\rangle_{\mathbf{K}} \right]^{-1} \tag{18.65}$$

and

$$-\bar{S} + \delta\bar{S} = \left[2\left\langle \frac{\sinh\alpha + A'\sinh\beta}{\cosh\alpha - \cosh\beta} \right\rangle_{\mathbf{K}} \right]^{-1} \tag{18.66}$$

Let us now consider the case where $g\mu_B H/kT \to 0$ and obtain the zero-field susceptibility. In this limit we have

$$\alpha = \frac{\delta\bar{S}(\mu_1+\mu_2)}{kT}$$

$$\beta_0 = \frac{1}{kT}\left[(\mu_1-\mu_2)^2(\bar{S})^2 - \lambda^2(\bar{S})^2\right]^{1/2}$$

$$= \frac{E_0\bar{S}}{kT}$$

from (18.22), since $\mu_1 - \mu_2 = \mu$. Equations (18.61) may be combined in this limit to give, after some manipulation,

$$\delta\bar{S} = \frac{C}{2(C^2-B^2)} \tag{18.67}$$

where

$$B = \langle A \coth\left(\tfrac{1}{2}\beta_0\right)\rangle_{\mathbf{K}}$$

$$C = \left\langle -\tfrac{1}{2}\alpha \operatorname{cosech}^2\left(\frac{E_0\bar{S}}{2kT}\right)\right\rangle_{\mathbf{K}}$$

Neglecting C^2, as is justified in the limit $H \to 0$ right up to the Neel point, we can show that

$$\delta\bar{S} = (\bar{S})^2\left\langle \alpha \operatorname{cosech}^2\left(\frac{E_0\bar{S}}{2kT}\right)\right\rangle_{\mathbf{K}} \tag{18.68}$$

and hence, from (18.62),

$$\chi_{\|} = \frac{Ng\mu_B\,\delta\bar{S}}{H}$$

$$= \frac{N_g^2\mu_B^2(\bar{S})^2\langle \operatorname{cosech}^2(E_0\bar{S}/2kT)\rangle_{\mathbf{K}}}{kT - (\bar{S})^2\langle(\mu_1+\mu_2)\operatorname{cosech}^2(E_0\bar{S}/2kT)\rangle_{\mathbf{K}}} \tag{18.69}$$

If there is no interaction between the spins on the same sublattice, we have $\mu_1 = 0$ and hence

$$\chi_{\|} = \frac{Ng^2\mu_B^2(\bar{S})^2 R}{kT - \mu_2(\bar{S})^2 R} \tag{18.70}$$

where

$$R=\left\langle \operatorname{cosech}^2\left(\frac{E_0\bar{S}}{2kT}\right)\right\rangle_{\mathbf{K}}$$

Equation (18.70) was obtained previously by Ginzburg and Fain.[7] For very low temperatures,

$$\operatorname{cosech}^2\left(\frac{E_0\bar{S}}{2kT}\right)=\left(\frac{e^{x}-e^{-x}}{2}\right)^{-2}$$

For T very small, $e^{-x}\rightarrow 0$, where $x=E_0\bar{S}/2kT$, and therefore

$$\operatorname{cosech}^2\left(\frac{E_0\bar{S}}{2kT}\right)=4e^{-E_0\bar{S}/kT}$$

which is very small when $\bar{S}$ is very small compared to kT. Thus both (18.69) and (18.70) reduce to

$$\chi_{\|}=\frac{Ng^2\mu_{\mathrm{B}}^2(\bar{S})^2R}{kT} \tag{18.71}$$

For spin S, Lines[5] has shown that

$$\chi_{\|}=\frac{Ng^2\mu_B^2F(B)\langle\operatorname{cosech}^2(E_0\bar{S}/2kT)\rangle_{\mathbf{K}}}{4kT-F(B)\langle(\mu_1+\mu_2)\operatorname{cosech}^2(E_0\bar{S}/2kT)\rangle_{\mathbf{K}}} \tag{18.72}$$

where $F(B)$ is given by

$$1-F(B)=\frac{4(2S+1)^2(B^2-1)^{2S}}{\left[(B+1)^{2S+1}-(B-1)^{2S+1}\right]^2}$$

For very low temperatures (18.72) simplifies to

$$\chi_{\|}=\frac{Ng^2\mu_B^2F(B)R}{4kT} \tag{18.73}$$

which is the spin-wave result given by Ziman.[8]

For temperatures above the Neel point, Lines[5] has shown that

$$\chi_{\|}=\frac{Ng^2\mu_{\mathrm{B}}^2}{\tau}\left(1+\frac{C_1}{\tau}+\frac{C_2}{\tau^2}+\cdots\right) \tag{18.74}$$

where

$$\tau=\frac{3kT}{S(S+1)}$$

$$C_1=\langle(\mu_1+\mu_2)\rangle_{\mathbf{K}}$$

$$C_2=2\langle(\mu_1+\mu_2)\rangle_{\mathbf{K}}^2-\langle(\mu_1+\mu_2)^2+\lambda^2\rangle_{\mathbf{K}}$$

As Lines points out, the coefficients C_i should be independent of the type of spin arrangement that occurs below the Neel temperature, since exact high-temperature expansions have this property.[9] For the f.c.c. lattice, Lines has shown that

$$C_1=-(24J_1+12J_2)$$
$$C_2=528J_1^2+576J_1J_2+120J_2^2 \tag{18.75}$$

At the transition temperature it can be shown that

$$(\chi_{\|})_{T_{\mathrm{N}}}=\frac{Ng^2\mu_{\mathrm{B}}^2}{\sum\limits_{j-g}^{d}4J_{jg}} \tag{18.76}$$

which is the molecular field result (see Chapter 14) obtained by Van Vleck.[10] This result shows clearly that at the Neel point the susceptibility depends on interactions between spins of opposite sublattices.

The perpendicular susceptibility is complicated, and therefore we shall not deal with this aspect any further but instead quote the final results obtained by Lines. He has shown that

$$\chi_{\perp}=\frac{Ng^2\mu_{\mathrm{B}}^2(\bar{S})^2}{\sum\limits_{j-g}^{d}4J_{jg}\langle S_g^{y_1}S_j^{y_2}+S_g^{z_1}S_j^{z_2}\rangle} \tag{18.77}$$

which is the result obtained also by Kanamori and Tachiki.[11] In the very low-temperature region, (18.77) reduces to

$$\chi_{\perp}=\frac{Ng^2\mu_{\mathrm{B}}^2}{\sum\limits_{j-g}4J_{jg}(1+\Delta)} \tag{18.78}$$

where

$$\Delta(\bar{S})^2=\langle S_g^{y_1}S_j^{y_2}\rangle \tag{18.79}$$

which differs from the molecular field result simply by the term Δ in the denominator. At $T=0$ we have

$$(\chi_\perp)_{T=0}=\frac{Ng^2\mu_B^2}{24J\left[1+\left(0.13/\bar{S}\right)\right]} \tag{18.80}$$

which, to the first order in $1/S$, is the result obtained by Kubo[12] in his "second approximation" of the spin-wave theory.

In dealing with the two-sublattice model we saw that the Green function formalism is adequate to describe the magnetization of an antiferromagnetic sample over the entire range of temperature. Derivation of the perpendicular susceptibility is very complicated, however, compared to the parallel susceptibility. Let us examine the difficulties encountered with $\chi_\perp$. If we introduce an external field H along the z-direction, which we choose to be perpendicular to the preferred direction, y, of antiferromagnetic spin alignment, then, when the field is switched on, each sublattice (in the two-sublattice model, of course) rotates through an angle ϕ toward the z-direction. Defining two new sets of orthogonal coordinates, one for each sublattice, with respect to the equilibrium positions ϕ of the sublattices, we have the following transformations connecting the new coordinates (x_1,y_1,z_1) for sublattice 1, and (x_2,y_2,z_2) for sublattice 2 (shown in Figure 18.4), with the original (x,y,z):

$$\begin{bmatrix} x_1 \\ y_1 \\ z_1 \end{bmatrix} = \begin{bmatrix} 1 & 0 & 0 \\ 0 & \sin\phi & -\cos\phi \\ 0 & \cos\phi & \sin\phi \end{bmatrix} \begin{bmatrix} x \\ y \\ z \end{bmatrix}$$

and (18.81)

$$\begin{bmatrix} x_2 \\ y_2 \\ z_2 \end{bmatrix} = \begin{bmatrix} 1 & 0 & 0 \\ 0 & \sin\phi & \cos\phi \\ 0 & -\cos\phi & \sin\phi \end{bmatrix} \begin{bmatrix} x \\ y \\ z \end{bmatrix}$$

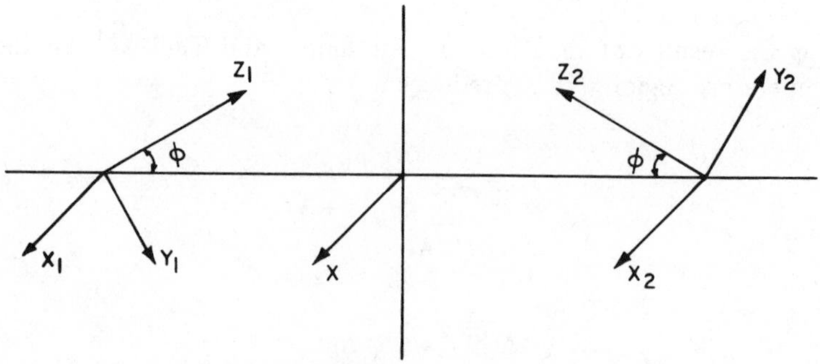

Fig. 18.4 The orthogonal coordinate systems of the two sublattices. (Taken from Lines.[6])

In these new coordinates the Hamiltonian (18.45) may be recast, giving

$$
\begin{aligned}
\mathcal{H} = &\sum_{\langle i,i'\rangle} 2J_{ii'}\mathbf{S}_i\cdot\mathbf{S}_{i'} + \sum_{\langle j,j'\rangle} 2J_{jj'}\mathbf{S}_j\cdot\mathbf{S}_{j'} \\
&+ \sum_{\langle i,j\rangle} 2J_{ij}\{ S_i^{x_1}S_j^{x_2} + \sin(2\phi)[S_i^{z_1}S_j^{y_2} - S_i^{y_1}S_j^{z_2}] \\
&- \cos(2\phi)[S_i^{y_1}S_j^{y_2} + S_i^{z_1}S_j^{z_2}]\} \\
= &\, g\mu_B H\sum_i (-\cos\phi S_i^{y_1} + \sin\phi S_i^{z_1}) \\
&- g\mu_B H\sum_j (\cos\phi S_j^{y_2} + \sin\phi S_j^{z_2})
\end{aligned}
\tag{18.82}
$$

where the suffix i refers to spins on sublattice 1, and suffix j to spins on sublattice 2. To set up the corresponding equations of motion is consequently very involved, and we do not present them here. With the external field on, the equilibrium value of ϕ is the value that minimizes the free energy of the system the condition for which is

$$
\frac{\partial}{\partial\phi}\left[kT\log(\mathrm{Tr}\,e^{-\mathcal{H}/kT})\right] = 0 \tag{18.83}
$$

This reduces to

$$
\left\langle \frac{\partial\mathcal{H}}{\partial\phi} \right\rangle = 0 \tag{18.84}
$$

From (18.82) it is evident that this condition requires a knowledge of the ϕ dependence of the correlation functions, which cannot be evaluated by using the simple decoupling procedure adopted by Lines. These are the difficulties that makes it impossible, unfortunately, to obtain a general functional dependence of $\chi_\perp$ on temperature.

We now consider the four-sublattice model. In (18.44) we derived the Green functions $G_{i\mathbf{K}}$ for the four-sublattice model. As long as no external magnetic field is present, finding the correlation function $\langle f(S_h^z)S_h^-S_h^+\rangle$ is no problem. This can be obtained in the usual way shown before, and is given by

$$
\langle f(S_h^z)S_h^-S_h^+\rangle = \frac{F}{4}\left\langle \frac{\alpha - A}{E_1}\coth\left(\frac{E_1\bar{S}}{2kT}\right) + \frac{\alpha + A}{E_2}\coth\left(\frac{E_2\bar{S}}{2kT}\right) - 2\right\rangle_{\mathbf{K}} \tag{18.85}
$$

where

$$A=\frac{2\alpha\gamma\delta-\beta(\gamma^2+\delta^2)}{\Delta^{1/2}}$$

$$E_1^2=\alpha^2-\beta^2-\Delta^{1/2}$$

$$E_2^2=\alpha^2-\beta^2+\Delta^{1/2}$$

$$\Delta=(\gamma^2-\delta^2)^2-4\alpha\beta(\gamma^2+\delta^2)+4\gamma\delta(\alpha^2+\beta^2)$$

The Neel temperature can be obtained quite easily by using (18.85) and is given by

$$\frac{S(S+1)}{3kT_{\mathrm{N}}}=\left\langle\frac{\alpha-A}{E_1^2}+\frac{\alpha+A}{E_2^2}\right\rangle_{\mathbf{K}}$$

$$=\left\langle\frac{\alpha(\alpha^2-\beta^2)+\beta(\gamma^2+\delta^2)-2\alpha\gamma\delta}{\left[(\alpha+\beta)^2-(\gamma+\delta)^2\right]\left[(\alpha-\beta)^2+(\gamma-\delta)^2\right]}\right\rangle_{\mathbf{K}} \tag{18.86}$$

The Neel temperatures for f. c. c. lattices having order of types 2 and 3 are plotted as a function of J_2/J_1 by Lines in Ref. 5.

When the magnetic field is included, the problem of obtaining the magnetization in the four-sublattice model becomes very complicated. Fortunately, Chakravarty and Basu[13] have recently found an analytic solution of the parallel magnetization by using the well-known Descarte method, which we describe now. The problem of perpendicular magnetization in the four sublattice model is very complicated but has been solved recently.[13]

Let us now consider the presence of a uniform magnetic field applied parallel to the z-direction of the antiferromagnetic spin alignment. In this case the average value of spin per site for the a and c sublattices is $\bar{S}_u=\bar{S}+\delta\bar{S}$, while that for the b and d sublattices is $\bar{S}_d=-\bar{S}+\delta\bar{S}$. Then, instead of the zero-field equation (18.35), we start from (18.52) and, proceeding as before, obtain equations corresponding to the zero-field equations (18.40), (18.41), (18.42), and (18.43). The present set of equations gives the following matrix relation:

$$\begin{bmatrix} E-\mu_1 & -\bar{S}_u\beta & -\bar{S}_u\gamma & -\bar{S}_u\delta \\ -\bar{S}_d\beta & E-\mu_2 & -\bar{S}_d\delta & -\bar{S}_d\gamma \\ -\bar{S}_u\delta & -\bar{S}_u\gamma & E-\mu_1 & -\bar{S}_u\beta \\ -\bar{S}_d\gamma & -\bar{S}_d\delta & -\bar{S}_d\beta & E-\mu_2 \end{bmatrix}\begin{bmatrix} G_{1\mathbf{K}} \\ G_{2\mathbf{K}} \\ G_{3\mathbf{K}} \\ G_{4\mathbf{K}} \end{bmatrix}=\begin{bmatrix} F/2\pi \\ 0 \\ 0 \\ 0 \end{bmatrix} \tag{18.87}$$

where

$$\mu_1 = \bar{S}_u\left[8J_2(c_1^2 + c_3^2) - 8J_1 - 16J_2\right] - \bar{S}_d(16J_1 + 4J_2) + g\mu_B H$$
$$\mu_2 = \bar{S}_d\left[8J_2(c_1^2 + c_3^2) - 8J_1 - 16J_2\right] - \bar{S}_u(16J_1 + 4J_2) + g\mu_B H$$

From (18.87) we obtain

$$G_{1\mathbf{K}} = \frac{F}{2\pi} \cdot \frac{\begin{vmatrix} E-\mu_2 & -\bar{S}_d\delta & -\bar{S}_d\gamma \\ -\bar{S}_u\gamma & E-\mu_1 & -\bar{S}_u\beta \\ -\bar{S}_d\delta & -\bar{S}_d\beta & E-\mu_2 \end{vmatrix}}{\begin{vmatrix} E-\mu_1 & -\bar{S}_u\beta & -\bar{S}_u\gamma & -\bar{S}_u\delta \\ -\bar{S}_d\beta & E-\mu_2 & -\bar{S}_d\delta & -\bar{S}_d\gamma \\ -\bar{S}_u\delta & -\bar{S}_u\gamma & E-\mu_1 & -\bar{S}_u\beta \\ -\bar{S}_d\gamma & -\bar{S}_d\delta & -\bar{S}_d\beta & E-\mu_2 \end{vmatrix}} \tag{18.88}$$

which reduces to

$$G_{1\mathbf{K}} = \frac{F}{2\pi} \cdot \frac{E^3 + c'E^2 + d'E + e'}{E^4 + bE^3 + cE^2 + dE + e} \tag{18.89}$$

where

$$c' = -(\mu_1 + 2\mu_2)$$
$$d' = 2\mu_1\mu_2 + \mu_2^2 - \bar{S}_u\bar{S}_d\beta^2 - \bar{S}_u\bar{S}_d\gamma\delta - \bar{S}_d^2\gamma\delta$$
$$e' = -\mu_1\mu_2^2 + \bar{S}_u\bar{S}_d\beta^2\mu_2 - \bar{S}_u\bar{S}_d^2\beta\gamma^2 + \bar{S}_u\bar{S}_d\gamma\delta\mu_2 + \mu_1\bar{S}_d^2\gamma\delta - \bar{S}_u\bar{S}_d^2\beta\delta^2$$
$$b = -2(\mu_1 + \mu_2)$$
$$c = (\mu_1 + \mu_2)^2 + 2\mu_1\mu_2 - (\bar{S}_u + \bar{S}_d)^2\gamma\delta - 2\bar{S}_u\bar{S}_d\beta^2$$
$$d = 2\gamma\delta(\mu_1\bar{S}_d^2 + \mu_2\bar{S}_u^2) - 2\bar{S}_u\bar{S}_d\beta(\gamma^2 + \delta^2)(\bar{S}_u + \bar{S}_d) + 2(\mu_1 + \mu_2)(\bar{S}_u\bar{S}_d\beta^2 - \mu_1\mu_2 + \bar{S}_u\bar{S}_d\gamma\delta)$$
$$e = -\gamma\delta(\bar{S}_u\mu_2 + \bar{S}_d\mu_1)^2 - \bar{S}_u^2\bar{S}_d^2(\gamma^2 - \delta^2)^2 + 2\bar{S}_u\bar{S}_d\beta(\bar{S}_u\mu_2 + \bar{S}_d\mu_1)(\gamma^2 + \delta^2) - 4\bar{S}_u^2\bar{S}_d^2\beta^2\gamma\delta + (\bar{S}_u\bar{S}_d\beta^2 - \mu_1\mu_2)^2$$

First we consider the temperature region below T_N. In evaluating the zero-field susceptibility, we are concerned with the limiting condition: $H \to 0$, $\delta\bar{S} \to 0$. In this limit we can neglect products and higher powers of H and $\delta\bar{S}$ in comparison with $\bar{S}$, which is nonzero at $T<T_N$. Then the coefficients $c', d', \ldots, e$ given above turn out to be linear in $\bar{S}$, H, and $\delta\bar{S}$:

$$\begin{aligned}
c' &= \alpha\bar{S} - 3p\,\delta\bar{S} - 3g\mu_B H\\
d' &= \left[(\beta^2-\alpha^2)\bar{S} + (2\gamma\delta - 2p\alpha)\,\delta\bar{S} - 2\alpha g\mu_B H\right]\bar{S}\\
e' &= \left\{\left[\alpha(\beta^2-\alpha^2) + 2\alpha\gamma\delta - \beta(\gamma^2+\delta^2)\right]\bar{S}\right.\\
&\quad \left. + \left[p(\alpha^2-\beta^2) - 2\alpha\gamma\delta + \beta(\gamma^2+\delta^2)\right]\delta\bar{S} + (\alpha^2-\beta^2)g\mu_B H\right\}\bar{S}^2
\end{aligned} \tag{18.90}$$

$$\begin{aligned}
b &= -4p\,\delta\bar{S} - 4g\mu_B H\\
c &= 2(\beta^2-\alpha^2)\bar{S}^2\\
d &= \left\{\left[4p(\alpha^2-\beta^2) + 4\beta(\gamma^2+\delta^2) - 8\alpha\gamma\delta\right]\delta\bar{S} + 4(\alpha^2-\beta^2)g\mu_B H\right\}\bar{S}^2\\
e &= \left[4\alpha\beta(\gamma^2+\delta^2) - 4\gamma\delta(\alpha^2+\beta^2) + (\alpha^2-\beta^2)^2 - (\gamma^2-\delta^2)^2\right]\bar{S}^4
\end{aligned}$$

where

$$p = 4\left[2J_2(c_1^2 + c_3^2) - 6J_1 - 5J_2\right]$$

which, like α, β, γ, and δ, depends on $\mathbf{K}$, J_1, and J_2. Our primary task is to express (18.89) in the form

$$G_{1\mathbf{K}} = \frac{F}{2\pi}\left[\frac{\mathcal{K}_1}{E-\theta_1} + \frac{\mathcal{K}_2}{E-\theta_2} + \frac{\mathcal{K}_3}{E-\theta_3} + \frac{\mathcal{K}_4}{E-\theta_4}\right] \tag{18.91}$$

Thus we require the biquadratic polynomial, $E^4 + bE^3 + cE^2 + dE + e$, to be factorized into $(E-\theta_1)(E-\theta_2)(E-\theta_3)(E-\theta_4)$. This factorization can be done by using the well-known Descarte method. At every step of the process we can conveniently utilize the usual limiting condition to achieve simplifications. Thus we obtain

$$\left.\begin{aligned}
\theta_1 &= E_1\bar{S} + \left(\frac{\alpha\gamma\delta}{\Delta^{1/2}} - A + p\right)\delta\bar{S} + g\mu_B H\\
\theta_2 &= -E_2\bar{S} + (-\alpha\gamma\delta\Delta^{-1/2} + A + p)\,\delta\bar{S} + g\mu_B H\\
\theta_3 &= E_2\bar{S} + (-\alpha\gamma\delta\Delta^{-1/2} + A + p)\,\delta\bar{S} + g\mu_B H\\
\theta_4 &= -E_1\bar{S} + (\alpha\gamma\delta\Delta^{-1/2} - A + p)\,\delta\bar{S} + g\mu_B H
\end{aligned}\right\} \tag{18.92}$$

where A, Δ, E_1, and E_2 are the same quantities that appeared earlier in (18.85) for the two-sublattice model.

The constants $\mathcal{K}_i (i=1,2,3,4)$ in (18.91) are given by

$$\mathcal{K}_i = \frac{\theta_i^3 + c'\theta_i^2 + d'\theta_i + e'}{(\theta_i - \theta_j)(\theta_i - \theta_k)(\theta_i - \theta_l)} \tag{18.93}$$

with no two of i, j, k, l being identical.

By using (18.4), (18.36), and (18.91), we obtain the following expression for the correlation function (at $t = t'$):

$$\langle f(S_h^z) S_h^- S_h^+ \rangle = F \left\langle \frac{\mathcal{K}_1}{e^{\theta_1/kT} - 1} + \frac{\mathcal{K}_2}{e^{\theta_2/kT} - 1} + \frac{\mathcal{K}_3}{e^{\theta_3/kT} - 1} + \frac{\mathcal{K}_4}{e^{\theta_4/kT} - 1} \right\rangle_{\mathbf{K}} \tag{18.94}$$

Finally we use (18.92) and (18.93) and utilize the usual limiting condition to express (18.94) as a linear combination of $\delta\bar{S}$ and H:

$$\begin{aligned} \langle f(S_h^z) S_h^- S_h^+ \rangle = \frac{F}{4} \Bigg[& \left\langle \frac{\alpha - A}{E_1} \coth\left(\frac{\bar{S} E_1}{2kT}\right) + \frac{\alpha + A}{E_2} \coth\left(\frac{\bar{S} E_2}{2kT}\right) \right\rangle_{\mathbf{K}} \\ & -2 + \left\{ \left\langle b_1 \coth\left(\frac{\bar{S} E_1}{2kT}\right) - b_2 \coth\left(\frac{\bar{S} E_2}{2kT}\right) \right\rangle_{\mathbf{K}} \cdot \left(\frac{1}{\bar{S}}\right) \right. \\ & \left. - \frac{1}{kT} \left\langle a_1 \operatorname{cosech}^2\left(\frac{\bar{S} E_1}{2kT}\right) + a_2 \operatorname{cosech}^2\left(\frac{\bar{S} E_2}{2kT}\right) \right\rangle_{\mathbf{K}} \right\} (\delta\bar{S}) \\ & - \left\{ \frac{1}{2kT} \left\langle \operatorname{cosech}^2\left(\frac{\bar{S} E_1}{2kT}\right) + \operatorname{cosech}^2\left(\frac{\bar{S} E_2}{2kT}\right) \right\rangle_{\mathbf{K}} \right\} \cdot g\mu_{\mathrm{B}} H \Bigg] \\ = \frac{F}{4} \big(& P_0 - 2 + P_1 \delta\bar{S} - P_2 g\mu_{\mathrm{B}} H \big), \quad \text{say} \end{aligned} \tag{18.95}$$

where we have put

$$a_1 = \tfrac{1}{2}(\alpha\gamma\delta\Delta^{-1/2} - A + p)$$

$$a_2 = \tfrac{1}{2}(-\alpha\gamma\delta\Delta^{-1/2} + A + p)$$

$$b_1 = \alpha\gamma\delta\Delta^{-1/2}/E_1, \qquad b_2 = \alpha\gamma\delta\Delta^{-1/2}/E_2$$

We note that, for $H=0$, $\delta\bar{S}=0$, and the present result (18.95) reduces to the zero-field result (18.85).

For the simplest case, $S=\frac{1}{2}$, we put $f(S_h^z)=1$ in (18.95) and obtain the following result:

$$1 = (\bar{S} + \delta\bar{S})(P_0 + P_1 \delta\bar{S} - P_2 g\mu_{\mathrm{B}} H) \tag{18.96}$$

For $H=0$, $\delta\bar{S}=0$, this yields

$$\frac{1}{\bar{S}}=P_0=\left\langle \frac{\alpha-A}{E_1}\coth\left(\frac{\bar{S}E_1}{2kT}\right)+\frac{\alpha+A}{E_2}\coth\left(\frac{\bar{S}E_2}{2kT}\right)\right\rangle_{\mathbf{K}} \tag{18.97}$$

Using (18.97) in (18.96), we get in the zero-field limit,

$$\delta\bar{S}\left(P_1\bar{S}^2+1\right)=P_2\bar{S}^2\cdot g\mu_{\mathrm{B}}H$$

and hence

$$\chi_{\parallel}=\frac{NG\mu_{\mathrm{B}}\,\delta\bar{S}}{H}=\frac{Ng^2\mu_{\mathrm{B}}^2P_2\bar{S}^2}{1+P_1\bar{S}^2}=\left(Ng^2\mu_{\mathrm{B}}^2\right)$$

$$\frac{\dfrac{\bar{S}^2}{2kT}\left\langle \operatorname{cosech}^2\left(\dfrac{\bar{S}E_1}{2kT}\right)+\operatorname{cosech}^2\left(\dfrac{\bar{S}E_2}{2kT}\right)\right\rangle_{\mathbf{K}}}{\left[1+\bar{S}\left\langle b_1\coth\left(\dfrac{\bar{S}E_1}{2kT}\right)-b_2\coth\left(\dfrac{\bar{S}E_2}{2kT}\right)\right\rangle_{\mathbf{K}} -\dfrac{\bar{S}^2}{kT}\left\langle a_1\operatorname{cosech}^2\left(\dfrac{\bar{S}E_1}{2kT}\right)+a_2\operatorname{cosech}^2\left(\dfrac{\bar{S}E_2}{2kT}\right)\right\rangle_{\mathbf{K}}\right]} \tag{18.98}$$

Here $\bar{S}$ can be computed at a given T by solving (18.97) self-consistently.

For the general spin case, we have to resort to Callen's result (18.27):

$$\frac{2\bar{S}_h^z+x}{2S+1}=\frac{(x+1)^{2S+1}+(x-1)^{2S+1}}{(x+1)^{2S+1}-(x-1)^{2S+1}} \tag{18.99}$$

where, for the present case [see (18.95)],

$$S_h^z=\bar{S}+\delta\bar{S}$$

and

$$x=\tfrac{1}{2}\left(P_0+P_1\,\delta\bar{S}-P_2\,g\mu_{\mathrm{B}}H\right) \tag{18.100}$$

At zero-field ($H=0$, $\delta\bar{S}=0$), (18.99) gives

$$\frac{2\bar{S}+P_0'}{2S+1}=\frac{(P_0'+1)^{2S+1}+(P_0'-1)^{2S+1}}{(P_0'+1)^{2S+1}-(P_0'-1)^{2S+1}} \tag{18.101}$$

with

$$P_0' = \tfrac{1}{2}P_0 = \tfrac{1}{2}\left\langle \frac{\alpha - A}{E_1}\coth\left(\frac{\bar{S}E_1}{2kT}\right) + \frac{\alpha + A}{E_2}\coth\left(\frac{\bar{S}E_2}{2kT}\right)\right\rangle_{\mathbf{K}}$$

From (18.99) we get, after some algebra, the following expression for the zero-field susceptibility for the general spin S:

$$\chi_{\parallel} = Ng^2\mu_B^2 \cdot \frac{F(P_0')\cdot P_2\bar{S}^2}{1 + F(P_0')P_1\bar{S}^2}$$

$$= (Ng^2\mu_B^2)\cdot \frac{F(P_0')\left[\dfrac{\bar{S}^2}{2kT}\left\langle 2\operatorname{cosech}^2\left(\dfrac{\bar{S}E_1}{2kT}\right) + \operatorname{cosech}^2\left(\dfrac{\bar{S}E_2}{2kT}\right)\right\rangle_{\mathbf{K}}\right]}{1 + F(P_0')\left[\begin{array}{l} \bar{S}\left\langle b_1\coth\left(\dfrac{\bar{S}E_1}{2kT}\right) - b_2\coth\left(\dfrac{\bar{S}E_2}{2kT}\right)\right\rangle_{\mathbf{K}} \\ -\dfrac{\bar{S}^2}{kT}\left\langle a_1\operatorname{cosech}^2\left(\dfrac{\bar{S}E_1}{2kT}\right) + a_2\operatorname{cosech}^2\left(\dfrac{\bar{S}E_2}{2kT}\right)\right\rangle_{\mathbf{K}} \end{array}\right]} \tag{18.102}$$

where

$$F(P_0') = \frac{\bar{S}(\bar{S} + P_0') - S(S+1)}{\bar{S}^2(1 - P_0'^2)}$$

Here $\bar{S}$ can be self-consistently evaluated from (18.101). For $S = \frac{1}{2}$ we have, from (18.97), $P_0' = 1/2\bar{S}$. Using this value we find $F(P_0')$ reducing to unity. Thus (18.102) reduces to (18.98) for $S = \frac{1}{2}$, as expected.

From (18.102) we see that, at $T = 0$, $\chi_{\parallel}$ reduces to zero, as it should.

At the transition temperature ($T = T_N$), $\bar{S} = 0$. In this limit ($T \to T_N$, $\bar{S} \to 0$), we may write

$$\coth\left(\frac{\bar{S}E_1}{2kT}\right) \to \frac{2kT_N}{E_1\bar{S}}$$

$$\operatorname{cosech}\left(\frac{\bar{S}E_1}{2kT}\right) \to \frac{2kT_N}{E_1\bar{S}},$$

and so on. Then, after some lengthy algebra and using the earlier result (18.86), we obtain from (18.102) the following expression for $\chi_{\|}$ at $T = T_N$:

$$(\chi_{\|})_{T_N} = Ng^2\mu_B^2 \cdot \frac{\left\langle \dfrac{1}{E_1^2} + \dfrac{1}{E_2^2} \right\rangle_{\mathbf{K}}}{\dfrac{2S(S+1)}{2kT_N} - \left\langle \dfrac{p-A}{E_1^2} + \dfrac{p+A}{E_2^2} \right\rangle_{\mathbf{K}}} \tag{18.104}$$

Finally, we consider the temperature region above T_N. Here, in the paramagnetic phase, $\bar{S}$ vanishes, and, unlike the situation in the ordered phase ($T < T_N$), it is not permissible to use the approximation given in (18.90) for the coefficients in (18.89). Fortunately, for the present case with $\bar{S} = 0$, the determinants in (18.88) become quite simple; the larger one can be exactly factorized without much labor, and the remaining task of writing (18.88) in the form (18.91) is trivial. We simply write the result:

$$G_{1\mathbf{K}} = \frac{F}{2\pi} \cdot \frac{1}{4} \left[\frac{1}{E - g\mu_B H - (p - \gamma - \delta + \beta)\delta\bar{S}} + \frac{1}{E - g\mu_B H - (p + \gamma + \delta + \beta)\delta\bar{S}} + \frac{1}{E - g\mu_B H - \{p - i(\gamma - \delta) - \beta\}\delta\bar{S}} + \frac{1}{E - g\mu_B H - \{p + i(\gamma - \delta) - \beta\}\delta\bar{S}} \right] \tag{18.105}$$

It is important to note that, since $\gamma^* = \delta$, $i(\gamma - \delta)$ is a real quantity.

From (18.105) we find the following correlation function in the zero-field limit;

$$\langle \mathbf{S}^- \mathbf{S}^+ \rangle = F\frac{kT}{4} \left[\left\langle \frac{1}{g\mu_B H + (p - \gamma - \delta + \beta)\delta\bar{S}} + \frac{1}{g\mu_B H + (p + \gamma + \delta + \beta)\delta\bar{S}} + \frac{1}{g\mu_B H + \{p - i(\gamma - \delta) - \beta\}\delta\bar{S}} + \frac{1}{g\mu_B H + \{p + i(\gamma - \delta) - \beta\}\delta\bar{S}} \right\rangle_{\mathbf{K}} \right] \tag{18.106}$$

which immediately gives an implicit equation for the zero-field susceptibility:

$$\frac{S(S+1)}{3\eta}=\frac{kT}{4}\left[\left\langle\frac{1}{1+\eta(p-\gamma-\delta+\beta)}+\frac{1}{1+\eta(p+\gamma+\delta+\beta)}\right.\right.$$
$$\left.\left.+\frac{1}{1+\eta\{p-i(\gamma-\delta)-\beta\}}+\frac{1}{1+\eta\{p+i(\gamma-\delta)-\beta\}}\right\rangle_{\mathbf{K}}\right] \tag{18.107}$$

where

$$\eta=\frac{\chi_{\|}}{Ng^2\mu_{\mathrm{B}}^2}$$

we can solve (18.107) self-consistently to obtain $\chi_{\|}$ as a function of T. From this equation we can also express $\chi_{\|}$ as a polynomial series in $1/T$. At sufficiently high temperatures only the first few terms of the series will be important. Thus we obtain

$$\chi_{\|}=\frac{Ng^2\mu_{\mathrm{B}}^2}{\tau}\left(1+\frac{c_1}{\tau}+\frac{c_2}{\tau^2}+\cdots\right) \tag{18.108}$$

where

$$\tau=\frac{3kT}{S(S+1)}$$
$$c_1=\langle p\rangle_{\mathbf{K}}$$
$$=4\left[\langle 2J_2(c_1^2+c_3^2)-6J_1-5J_2\rangle_{\mathbf{K}}\right]$$
$$c_2=\langle p^2-\beta^2-2\gamma\delta\rangle_{\mathbf{K}}$$

Finally, it will be interesting to compare results (18.97), (18.98), (18.102), (18.104), and (18.108) for the four-sublattice model with the corresponding results for the two-sublattice model.

REFERENCES

1. N. N. Bogolyubov and S. V. Tyablikov, *Sov. Phys. Dokl.* **4**, 589 (1959).
2.. S. V. Tyablikov, *Ukrain. Mat. Zh.* **11**, 287 (1959).
3. D. N. Zubarev, *Usp. Fiz. Nauk* **71**, 71 (1960) [English translation: *Sov. Phys. Usp.* **3**, 320 (1960)].
4. Fu-Cho Pu, *Dokl. Akad. Nauk SSSR* **130**, 1244 (1966); *ibid.* **131**, 546 (1960) [English translation: *Sov. Phys. Dokl.* **5**, 128, 321 (1960)].
5. M. E. Lines, *Phys. Rev.*, **135**, 1336 (1964); *ibid.* **131**, 540 (1963).

6. D. ter Haar and M. E. Lines, *Phil. Trans. Roy. Soc. A*, **254**, 521 (1962).
7. V. L. Ginzburg and V. M. Fain, *Sov. Phys.—JETP*, **12**, 923 (1961).
8. J. M. Ziman, *Proc. Phys. Soc. (London) A*, **65**, 540, 548 (1952).
9. H. A. Brown and J. M. Luttinger, *Phys. Rev.* **100**, 685 (1955).
10. J. H. Van Vleck, *J. Chem. Phys* **9**, 85 (1941).
11. J. Kanamori and M. Tachiki, *J. Phys. Soc. Japan* **17**, 1385 (1962).
12. R. Kubo, *Phys. Rev.* **87** 568 (1952).
13. A. S. Chakravarty and S. Basu, *Phys. Rev. B.*, 1 Sept. Issue (1980).

Appendix I

Electron Configurations of the Neutral Atoms of the Transition Metals

Table I.1

Atoms of the iron group

Atom	1*s*	2*s*	2*p*	3*s*	3*p*	3*d*	4*s*	4*p*
Ar	2	2	6	2	6			
Sc	Ar core					1	2	
Ti						2	2	
V						3	2	
Cr						5	1	
Mn						5	2	
Fe						6	2	
Co						7	2	
Ni						8	2	

Table I.2

Atoms of the palladium group

Atom		4*d*	5*s*	5*p*
Y	Kr core	1	2	
Zr		2	2	
Nb		4	1	
Mo		5	1	
Tc		5	2	
Ru		7	1	
Rh		8	1	
Pd		10		
Ag		10	1	

Table I.3

Atoms of the platinum group

Atom		5d	5f	6s
Lu		1		2
Hf	Lu^{3+} core	2		2
Ta		3		2
W		4		2
Re		5		2
Os		6		2
Ir		9		
Pt		9		1
Au		10		1

Ar core $\equiv 1s^2\, 2s^2\, 2p^6\, 3s^2\, 3p^6$

Kr core $\equiv 1s^2\, 2s^2\, 2p^6\, 3s^2\, 3p^6\, 3d^{10}\, 4s^2\, 4p^6$

Lu^{3+} core $\equiv$ [Kr core] $4d^{10}\, 4f^{14}\, 5s^2\, 5p^6$

Appendix II

Electron Configurations and Ground Terms

Table II.1

Neutral atoms of the iron group

Atom:	Sc	Ti	V	Cr	Mn	Fe	Co	Ni	Cu
Configuration:	$3d^1 4s^2$	$3d^2 4s^2$	$3d^3 4s^2$	$3d^5 4s$	$3d^5 4s^2$	$3d^6 4s^2$	$3d^7 4s^2$	$3d^9 4s^1$	$3d^{10} 4s^1$
Ground term:	^2D	^3F	^4F	^7S	^6S	^5D	^4F	^3D	^2S

Table II.2

Doubly ionized ions of the iron group

Ion:	Sc^{2+}	Ti^{2+}	V^{2+}	Cr^{2+}	Mn^{2+}	Fe^{2+}	Co^{2+}	Ni^{2+}	Cu^{2+}
Configuration:	$3d^1$	$3d^2$	$3d^3$	$3d^4$	$3d^5$	$3d^6$	$3d^7$	$3d^8$	$3d^9$
Ground term:	^2D	^3F	^4F	^5D	^6S	^5D	^4F	^3F	^2D

Table II.3

Triply ionized ions of the palladium group

Ion:	Zr^{3+}	Nb^{3+}	Mo^{3+}	Tc	Ru^{3+}	Rh^{3+}	Pd^{3+}	Ag^{3+}
Configuration:	$4d^1$	$4d^2$	$4d^3$	—	$4d^5$	$4d^6$	$4d^7$	$4d^8$
Ground term:	^2D	^3F	^4F	—	^6S	^5D	^4F	^3F

Appendix III

Racah Parameters B and C (in cm^{-1}) of the Iron Group Atoms[a]

First rows of Tables III.1*a* and III.1*b* refer to the neutral atom, M.

Table III.1*a*

B parameters

B	Ti	V	Cr	Mn	Fe	Co	Ni	Cu
I	560	578	790	720	806	798	1025	—
II	682	659	710	873	869	878	1037	1216
III	718	766	830	960	1058	1115	1084	1238
IV	—	861	1030	1140	—	—	—	—
V	—	—	1039	—	1144	—	—	—

Table III.1*b*

C parameters

C	Ti	V	Cr	Mn	Fe	Co	Ni	Cu
I	1840	2273	2520	3087	3506	4167	4226	—
II	2481	2417	2790	3130	3638	3828	4314	4745
III	2629	2855	3430	3325	3091	4366	4831	4659
IV	—	4165	3850	3675	—	—	—	—
V	—	—	4238	—	4459	—	—	—

[a]From J. S. Griffith, *The Theory of Transition Metal Ions*, Cambridge University Press, London, 1961.

Appendix IV

The Spin-Orbit Parameters ζ for $3d^n$ Ions

Configuration	Ion	ζ (calculated)[a] (cm^{-1})	ζ (experimental[b] (cm^{-1})
$3d^1$	Sc^{2+}	86	79
	Ti^{3+}	159	154
	V^{4+}	255	248
$3d^2$	Ti^{2+}	61	60
	V^{3+}	106	104
	Cr^{4+}	163	164
$3d^3$	V^{2+}	57	55
	Cr^{3+}	91	91
	Mn^{4+}	135	134
$3d^4$	Cr^{2+}	59	58
	Mn^{3+}	87	88
	Fe^{4+}	125	129
$3d^6$	Mn^{+}	−64	−64
	Fe^{2+}	−114	−103
	Co^{3+}	−145	—
	Ni^{4+}	−197	—
$3d^7$	Fe^{+}	−115	−119
	Co^{2+}	−189	−178
	Ni^{3+}	−272	—
$3d^8$	Co^{+}	−228	−228
	Ni^{2+}	−343	−324
$3d^9$	Cu^{2+}	−830	−830

[a]From M. Blume and R. E. Watson, *Proc. Roy. Soc.* (*London*) *A*, **271**, 565 (1963).
[b]From T. M. Dunn, *Trans. Faraday Soc.* **57**, 1441 (1961).

Appendix V

Character Tables and Standard Basis Sets for the Most Frequently Used Finite Groups

$$D_{3d}=D_3\times C_i,\qquad D_{4h}=D_4\times C_i,\qquad O_h=O\times C_i$$

where C_i denotes the inversion at the center.

Table V.1

Character table for D_3

D_3'	E	R	C_3 C_3^2R	C_3^2 C_3R	$3C_2'$	$3C_2'R$	Bases
A_1	1	1	1	1	1	1	x_0
A_2	1	1	1	1	-1	-1	a_0
E	2	2	-1	-1	0	0	$(a_+,a_-),(\theta_+,\theta_-),(x_+,x_-)$
Γ_4	2	-2	1	-1	0	0	$(\alpha,\beta);(\|\frac{3}{2},\frac{1}{2}\rangle,-\|\frac{3}{2}-\frac{1}{2}\rangle)$
Γ_5	1	-1	-1	1	i	$-$i	$\frac{1}{\sqrt{2}}[\|\frac{3}{2}\frac{3}{2}\rangle-i\|\frac{3}{2}-\frac{3}{2}\rangle]$
Γ_6	1	-1	-1	1	$-$i	i	$\frac{1}{\sqrt{2}}[\|\frac{3}{2}\frac{3}{2}\rangle+i\|\frac{3}{2}-\frac{3}{2}\rangle]$

Notes

1. We use Mulliken's notation for the ordinary irreducible representation and Bethe's labels for the augmented irreducible representation.
2. Γ_5 and Γ_6 together form Kramers' pair.
3. The basis set has the axis of quantization along the (1 1 1) axis.

4. Here α, β are spin-1/2 functions, that is, $\alpha = |\frac{1}{2}\,\frac{1}{2}\rangle$, $\beta = |\frac{1}{2}\, -\frac{1}{2}\rangle$. Also,

$$a_+ = Y_1^1(\theta,\phi)$$
$$a_- = Y_1^{-1}(\theta,\phi)$$
$$a_0 = Y_1^0(\theta,\phi)$$

$$x_+ = -\sqrt{\frac{2}{3}}\, Y_2^{-2} - \frac{1}{\sqrt{3}}\, Y_2^1 \qquad x_- = \sqrt{\frac{2}{3}}\, Y_2^2 - \frac{1}{\sqrt{3}}\, Y_2^{-1}$$

$$x_0 = Y_2^0$$

$$\theta_+ = -\frac{1}{\sqrt{3}}\, Y_2^{-2} + \sqrt{\frac{2}{3}}\, Y_2^1 \qquad \theta_- = \frac{1}{\sqrt{3}}\, Y_2^2 + \sqrt{\frac{2}{3}}\, Y_2^{-1}$$

$Y_l^m = Y_l^m(\theta,\phi)$ are standard spherical harmonics. $|jm\rangle$ are bases of the full three-dimensional rotation group R_3.

We follow the phases of the Japanese authors, Ref. 39 of Chapter X.

Table V.2

Character table for D_4

D_4	E	R	C_4 C_4^3R	C_4^3 C_4R	C_4^2 C_4^2R	$2C_2'$ $2C_2'R$	$2C_2''$ $2C_2''R$	Bases
A_1	1	1	1	1	1	1	1	θ
A_2	1	1	1	1	1	-1	-1	z
B_1	1	1	-1	-1	1	1	-1	ζ
B_2	1	1	-1	-1	1	-1	1	ϵ
E	2	2	0	0	-2	0	0	(x,y); (ξ,η)
Γ_6	2	-2	$\sqrt{2}$	$-\sqrt{2}$	0	0	0	(α,β)
Γ_7	2	-2	$-\sqrt{2}$	$\sqrt{2}$	0	0	0	$A_2 \times \Gamma_6$

$$\xi = \frac{1}{\sqrt{2}}\, i(Y_2^1 + Y_2^{-1})$$

$$\eta = -\frac{1}{\sqrt{2}}(Y_2^1 - Y_2^{-1})$$

$$\zeta = -\frac{1}{\sqrt{2}}\, i(Y_2^2 - Y_2^{-2})$$

$$\theta = Y_2^0$$

$$\epsilon = \frac{1}{\sqrt{2}}(Y_2^2 + Y_2^{-2})$$

Functions are quantized along the z-(C_4) axis.

Table V.3

Character table for the octahedral and tetrahedral double groups.

T_d'	E	R	$8C_3$	$8C_3R$	$6C_2$	$6S_4$	$6S_4R$	$12\sigma_d$
O'	E	R	$4C_3$ $4C_3^2R$	$4C_3^2$ $4C_3R$	$3C_4^2$ $3C_4^2R$	$3C_4$ $3C_4^3R$	$3C_4^3$ $3C_4R$	$3C_2'$ $3C_2'R$
A_1	1	1	1	1	1	1	1	1
A_2	1	1	1	1	1	-1	-1	-1
E	2	2	-1	-1	2	0	0	0
T_1	3	3	0	0	-1	1	1	-1
T_2	3	3	0	0	-1	-1	-1	1
Γ_6	2	-2	1	-1	0	$\sqrt{2}$	$-\sqrt{2}$	0
Γ_7	2	-2	1	-1	0	$-\sqrt{2}$	$\sqrt{2}$	0
Γ_8	4	-4	-1	1	0	0	0	0

Basis sets for O'

A_1: $(x^2+y^2+z^2)$; $(\xi^2+\eta^2+\zeta^2)$
A_2: xyz
E: (θ,ϵ)
T_1: (x,y,z)
T_2: (ξ,η,ζ)
Γ_6: (α,β)
Γ_7: $A_2\times\Gamma_6$
Γ_8: $(|\frac{3}{2}\,\frac{3}{2}\rangle, |\frac{3}{2}\,\frac{1}{2}\rangle, |\frac{3}{2}\,-\frac{1}{2}\rangle, |\frac{3}{2}\,-\frac{3}{2}\rangle)$

Table V.4

Behavior of functions belonging to irreducible representations of O and O' under group-theoretical operations denoted by C_4^z, C_4^x, and $C_3^{(111)}$, where C_4^z denotes a fourfold rotation about the z-axis of an octahedron (Figure 5.2), etc. Remember that the group operations rotate the functions and not the coordinate system.

O O'	C_4^z	C_4^x	$C_3^{(111)}$
A_1 a_1	a_1	a_1	a_1
A_2 a_2	$-a_2$	$-a_2$	a_2
E θ	θ	$-\frac{1}{2}\theta-\frac{\sqrt{3}}{2}\epsilon$	$-\frac{1}{2}\theta+\frac{\sqrt{3}}{2}\epsilon$
ϵ	$-\epsilon$	$-\frac{\sqrt{3}}{2}\theta+\frac{1}{2}\epsilon$	$-\frac{\sqrt{3}}{2}\theta-\frac{1}{2}\epsilon$
T_1 x	y	x	y
y	$-x$	z	z
z	z	$-y$	x
T_2 ξ	$-\eta$	$-\xi$	η
η	ξ	$-\zeta$	ζ
ζ	$-\zeta$	η	ξ

Table V.5

$\Gamma_1 \times \Gamma_2 = \sum_i \Gamma_i$ for the O' group

Γ_1 \ Γ_2	A_1	A_2	E	T_1	T_2	Γ_6	Γ_7	Γ_8
A_1	A_1	A_2	E	T_1	T_2	Γ_6	Γ_7	Γ_8
A_2		A_1	E	T_2	T_1	Γ_7	Γ_6	Γ_8
E			A_1+A_2+E	T_1+T_2	T_1+T_2	Γ_8	Γ_8	$\Gamma_6+\Gamma_7+\Gamma_8$
T_1				A_1+E $+T_1+T_2$	A_2+E $+T_1+T_2$	$\Gamma_6+\Gamma_8$	$\Gamma_7+\Gamma_8$	$\Gamma_6+\Gamma_7+2\Gamma_8$
T_2					A_1+E $+T_1+T_2$	$\Gamma_7+\Gamma_8$	$\Gamma_6+\Gamma_8$	$\Gamma_6+\Gamma_7+2\Gamma_8$
Γ_6						A_1+T_1	A_2+T_2	$E+T_1+T_2$
Γ_7							A_1+T_1	$E+T_1+T_2$
Γ_8								A_1+A_2+E $+2T_1+2T_2$

Appendix VI

Clebsch-Gordan Coefficients for O_h Symmetry Group $\langle \Gamma_1 \Gamma_2 ab \mid \Gamma_1 \Gamma_2 \Gamma c \rangle$

$A_2 \times A_2$	A_1 a_1
$a_2 \times a_2$	-1

$A_2 \times E$	E	
	θ	ϵ
$a_2\ \theta$	0	-1
$a_2\ \epsilon$	1	0

$A_2 \times T_1$	T_2		
	ξ	η	ζ
$a_2\ x$	1	0	0
$a_2\ y$	0	1	0
$a_2\ z$	0	0	-1

$A_2 \times T_2$	T_1		
	x	y	z
$a_2\ \xi$	-1	0	0
$a_2\ \eta$	0	-1	0
$a_2\ \zeta$	0	0	-1

$E \times E$		A_1	A_2	E	
		a_1	a_2	θ	ϵ
θ	θ	$1/\sqrt{2}$	0	$-1/\sqrt{2}$	0
θ	ϵ	0	$1/\sqrt{2}$	0	$1/\sqrt{2}$
ϵ	θ	0	$-1/\sqrt{2}$	0	$1/\sqrt{2}$
ϵ	ϵ	$1\sqrt{2}$	0	$1\sqrt{2}$	0

$E \times T_1$		T_1			T_2		
		x	y	z	ξ	η	ζ
θ	x	$-\frac{1}{2}$	0	0	$\sqrt{3}/2$	0	0
θ	y	0	$-\frac{1}{2}$	0	0	$-\sqrt{3}/2$	0
θ	z	0	0	1	0	0	0
ϵ	x	$\sqrt{3}/2$	0	0	$\frac{1}{2}$	0	0
ϵ	y	0	$-\sqrt{3}/2$	0	0	$\frac{1}{2}$	0
ϵ	z	0	0	0	0	0	-1

$E \times T_2$		T_1			T_2		
		x	y	z	ξ	η	ζ
θ	ξ	$-\sqrt{3}/2$	0	0	$-\frac{1}{2}$	0	0
θ	η	0	$-\sqrt{3}/2$	0	0	$-\frac{1}{2}$	0
θ	ζ	0	0	0	0	0	1
ϵ	ξ	$-\frac{1}{2}$	0	0	$\sqrt{3}/2$	0	0
ϵ	η	0	$-\frac{1}{2}$	0	0	$-\sqrt{3}/2$	0
ϵ	ζ	0	0	1	0	0	0

	A_1	E		T_1			T_2		
$T_1 \times T_1$	a_1	θ	ϵ	x	y	z	ξ	η	ζ
$x \quad x$	$-1/\sqrt{3}$	$1/\sqrt{6}$	$-1/\sqrt{2}$	0	0	0	0	0	0
$x \quad y$	0	0	0	0	0	$-1/\sqrt{2}$	0	0	$-1/\sqrt{2}$
$x \quad z$	0	0	0	0	$1/\sqrt{2}$	0	0	$-1/\sqrt{2}$	0
$y \quad x$	0	0	0	0	0	$1/\sqrt{2}$	0	0	$-1/\sqrt{2}$
$y \quad y$	$-1/\sqrt{3}$	$1/\sqrt{6}$	$1/\sqrt{2}$	0	0	0	0	0	0
$y \quad z$	0	0	0	$-1/\sqrt{2}$	0	0	$-1/\sqrt{2}$	0	0
$z \quad x$	0	0	0	0	$-1/\sqrt{2}$	0	0	$-1/\sqrt{2}$	0
$z \quad y$	0	0	0	$1/\sqrt{2}$	0	0	$-1/\sqrt{2}$	0	0
$z \quad z$	$-1/\sqrt{3}$	$-2/\sqrt{6}$	0	0	0	0	0	0	0

	A_2	E		T_1			T_2		
$T_1 \times T_2$	a_2	θ	ϵ	x	y	z	ξ	η	ζ
$x \quad \xi$	$-/1\sqrt{3}$	$-1/\sqrt{2}$	$-1/\sqrt{6}$	0	0	0	0	0	0
$x \quad \eta$	0	0	0	0	0	$1/\sqrt{2}$	0	0	$-1/\sqrt{2}$
$x \quad \zeta$	0	0	0	0	$1/\sqrt{2}$	0	0	$1/\sqrt{2}$	0
$y \quad \xi$	0	0	0	0	0	$1/\sqrt{2}$	0	0	$1/\sqrt{2}$
$y \quad \eta$	$-1/\sqrt{3}$	$1/\sqrt{2}$	$-1/\sqrt{6}$	0	0	0	0	0	0
$y \quad \zeta$	0	0	0	$1/\sqrt{2}$	0	0	$-1/\sqrt{2}$	0	0
$z \quad \xi$	0	0	0	0	$1/\sqrt{2}$	0	0	$-1/\sqrt{2}$	0
$z \quad \eta$	0	0	0	$1/\sqrt{2}$	0	0	$1/\sqrt{2}$	0	0
$z \quad \zeta$	$-1/\sqrt{3}$	0	$2/\sqrt{6}$	0	0	0	0	0	0

	A_1	E		T_1			T_2		
$T_2 \times T_2$	a_1	θ	ϵ	x	y	z	ξ	η	ζ
ξ ξ	$1/\sqrt{3}$	$-1/\sqrt{6}$	$1/\sqrt{2}$	0	0	0	0	0	0
ξ η	0	0	0	0	0	$1/\sqrt{2}$	0	0	$1/\sqrt{2}$
ξ ζ	0	0	0	0	$-1/\sqrt{2}$	0	0	$-1/\sqrt{2}$	0
η ξ	0	0	0	0	0	$-1/\sqrt{2}$	0	0	$1/\sqrt{2}$
η η	$1\sqrt{3}$	$-1\sqrt{6}$	$-1/\sqrt{2}$	0	0	0	0	0	0
η ζ	0	0	0	$1/\sqrt{2}$	0	0	$1/\sqrt{2}$	0	0
ζ ξ	0	0	0	0	$1/\sqrt{2}$	0	0	$1/\sqrt{2}$	0
ζ η	0	0	0	$-1/\sqrt{2}$	0	0	$1/\sqrt{2}$	0	0
ζ ζ	$1/\sqrt{3}$	$2/\sqrt{6}$	0	0	0	0	0	0	0

Appendix VII

The Wavefunctions of t_2^m and e^n

The wavefunctions of t_2^m and e^n are constructed by using the tables of Clebsch-Gordan coefficients for the octahedral group given in Appendix VI; $\eta^2=\eta^+\eta^-$, etc. The orbitals ξ, η, ζ, θ, and ϵ were given in Appendix V.

Configuration	Wavefunctions
t_2^1	$\lvert {}^2T_2\tfrac{1}{2}\xi\rangle = \lvert \overset{+}{\xi} \rangle$
	$\lvert {}^2T_2\tfrac{1}{2}\eta\rangle = \lvert \overset{+}{\eta} \rangle$
	$\lvert {}^2T_2\tfrac{1}{2}\zeta\rangle = \lvert \overset{+}{\zeta} \rangle$
t_2^2	${}^3T_1 \quad 1x = \lvert \overset{+}{\zeta}\ \overset{+}{\eta} \rangle$
	$1y = \lvert \overset{+}{\xi}\ \overset{+}{\zeta} \rangle$
	$1z = \lvert \overset{+}{\eta}\ \overset{+}{\xi} \rangle$
	${}^1A_1 \quad a_1 = \frac{1}{\sqrt{3}}\lvert \xi^2+\eta^2+\zeta^2\rangle$
	${}^1E \quad \theta = \frac{1}{\sqrt{6}}\lvert \xi^2+\eta^2-2\zeta^2\rangle$
	$\epsilon = \frac{1}{\sqrt{2}}\lvert \eta^2-\xi^2\rangle$
	${}^1T_2 \quad \xi = -\frac{1}{\sqrt{2}}\lvert \eta^+\zeta^-+\zeta^+\eta^-\rangle$
	$\eta = -\frac{1}{\sqrt{2}}\lvert \xi^+\zeta^-+\zeta^+\xi^-\rangle$
	$\zeta = -\frac{1}{\sqrt{2}}\lvert \xi^+\eta^-+\eta^+\xi^-\rangle$
t_2^3	${}^4A_2 \quad \tfrac{3}{2}a_2 = -\lvert \xi^+\eta^+\zeta^+\rangle$
	${}^2E \quad \tfrac{1}{2}\theta = \frac{1}{\sqrt{2}}\lvert \xi^+\eta^-\zeta^+-\xi^-\eta^+\zeta^+\rangle$

$$\tfrac{1}{2}\epsilon = \frac{1}{\sqrt{6}}|2\xi^+\eta^+\zeta^- - \xi^+\eta^-\zeta^+ - \xi^-\eta^+\zeta^+\rangle$$

$$^2T_1 \quad \tfrac{1}{2}x = \frac{1}{\sqrt{2}}|\xi^+\eta^2 - \xi^+\zeta^2\rangle$$

$$\tfrac{1}{2}y = \frac{1}{\sqrt{2}}|\eta^+\zeta^2 - \xi^2\eta^+\rangle$$

$$\tfrac{1}{2}z = \frac{1}{\sqrt{2}}|\xi^2\zeta^+ - \eta^2\zeta^+\rangle$$

$$^2T_2 \quad \tfrac{1}{2}\xi = \frac{1}{\sqrt{2}}|\xi^+\zeta^2 + \xi^+\eta^2\rangle$$

$$\tfrac{1}{2}\eta = \frac{1}{\sqrt{2}}|\eta^+\zeta^2 + \xi^2\eta^+\rangle$$

$$\tfrac{1}{2}\zeta = \frac{1}{\sqrt{2}}|\xi^2\zeta^+ + \eta^2\zeta^+\rangle$$

t_2^4

$$^3T_1 \quad 1x = |\xi^2\eta^+\zeta^+\rangle$$

$$1y = -|\xi^+\eta^2\zeta^+\rangle$$

$$1z = |\xi^+\eta^+\zeta^2\rangle$$

$$^1A_1 \quad a_1 = \frac{1}{\sqrt{3}}|\xi^2\eta^2 + \eta^2\zeta^2 + \zeta^2\xi^2\rangle$$

$$^1E \quad \theta = \frac{1}{\sqrt{6}}|\eta^2\zeta^2 + \zeta^2\xi^2 - 2\xi^2\eta^2\rangle$$

$$\epsilon = \frac{1}{\sqrt{2}}|\zeta^2\xi^2 - \eta^2\zeta^2\rangle$$

$$^1T_2 \quad \xi = \frac{1}{\sqrt{2}}|\xi^2\eta^+\zeta^- - \xi^2\eta^-\zeta^+\rangle$$

$$\eta = \frac{1}{\sqrt{2}}|\xi^+\eta^2\zeta^- - \xi^-\eta^2\zeta^+\rangle$$

$$\zeta = \frac{1}{\sqrt{2}}|\xi^+\eta^-\zeta^2 - \xi^-\eta^+\zeta^2\rangle$$

t_2^5

$$^2T_2 \quad \tfrac{1}{2}\xi = |\xi^+\eta^2\zeta^2\rangle$$

$$\tfrac{1}{2}\eta = |\xi^2\eta^+\zeta^2\rangle$$

$$\tfrac{1}{2}\zeta = |\xi^2\eta^2\zeta^+\rangle$$

t_2^6

$$^1A_1 \quad a_1 = |\xi^2\eta^2\zeta^2\rangle$$

e^2

$$^3A_2 \quad 1a_2 = |\theta^+\epsilon^+\rangle$$

$$^1A_1 \quad a_1 = \frac{1}{\sqrt{2}}|\theta^2 + \epsilon^2\rangle$$

$$^1E \quad \theta = \frac{1}{\sqrt{2}}|\epsilon^2 - \theta^2\rangle$$

$$\epsilon = \frac{1}{\sqrt{2}}|\theta^+\epsilon^- - \theta^-\epsilon^+\rangle$$

e^3

$$^2E \quad \tfrac{1}{2}\theta = |\theta^+\epsilon^2\rangle$$

$$\tfrac{1}{2}\epsilon = |\theta^2\epsilon^+\rangle$$

e^4

$$^1A_1 \quad a_1 = |\theta^2\epsilon^2\rangle$$

Appendix VIII

Strong-Field Wavefunctions of the Ground States and a Few First Excited States of d^n Configurations

These transform according to the irreducible representations of the O_h symmetry group, which can be easily constructed by using Appendices VI and VII.

Table VIII.1

Configuration d^2

$(t_{2g})^2\,{}^3T_{1g}$

$${}^3T_{1g} \quad 1x = \zeta^+\eta^+$$
$$1y = \xi^+\zeta^+$$
$$1z = \xi^+\eta^+$$

$(t_{2g}\cdot e_g)\;{}^3T_{1g}$

$${}^3T_{1g} \quad 1x = -\frac{\sqrt{3}}{2}\xi^+\theta^+ - \frac{1}{2}\xi^+\epsilon^+$$
$$1y = \frac{\sqrt{3}}{2}\eta^+\theta^+ - \frac{1}{2}\eta^+\epsilon^+$$
$$1z = \zeta^+\epsilon^+$$

$(t_{2g}\cdot e_g)\,{}^3T_{2g}$

$${}^3T_{2g} \quad 1\xi = -\frac{1}{2}\xi^+\theta^+ + \frac{\sqrt{3}}{2}\xi^+\epsilon^+$$
$$1\eta = -\frac{1}{2}\eta^+\theta^+ - \frac{\sqrt{3}}{2}\eta^+\epsilon^+$$
$$1\zeta = \zeta^+\theta^+$$

Table VIII.2

Configuration d^3

(t_{2g}^3) $^4A_{2g}$:

$$^4A_{2g} \quad \frac{3}{2}a_2 = -\xi^+\eta^+\zeta^+$$

$(t_{2g}^2 \cdot e_g)^4 A_{1g}$:

$$^4A_{1g} \quad \frac{3}{2}x = -\frac{1}{2}\zeta^+\eta^+\theta + \frac{\sqrt{3}}{2}\zeta^+\eta^+\epsilon^+$$

$$\frac{3}{2}y = -\frac{1}{2}\xi^+\zeta^+\theta^+ - \frac{\sqrt{3}}{2}\xi^+\zeta^+\epsilon^+$$

$$\frac{3}{2}z = \eta^+\xi^+\theta^+$$

$(t_{2g}^2 \cdot e_g)$ $^4T_{2g}$:

$$^4T_{2g} \quad \frac{3}{2}\xi = -\frac{\sqrt{3}}{2}\zeta^+\eta^+\theta^+ - \frac{1}{2}\zeta^+\eta^+\epsilon^+$$

$$\frac{3}{2}\eta = \frac{\sqrt{3}}{2}\xi^+\zeta^+\theta^+ - \frac{1}{2}\xi^+\zeta^+\epsilon^+$$

$$\frac{3}{2}\zeta = \eta^+\xi^+\epsilon^+$$

Table VIII.3

Configuration d^4

(t_{2g}^4) $^3T_1$:

$$^3T_1 \quad 1x = \xi^2\eta^+\zeta^+$$

$$1y = -\xi^+\eta^2\zeta^+$$

$$1z = \xi^+\eta^+\zeta^2$$

$(t_{2g}^3 \cdot e_g)$ $^3A_1$:

$$^3A_1 \quad 1a_1 = \frac{1}{2}\xi^+\eta^-\zeta^+\theta^+ - \frac{1}{2}\xi^-\eta^+\zeta^+\theta^+$$

$$+\frac{1}{\sqrt{3}}\xi^+\eta^+\zeta^-\epsilon^+ - \frac{1}{2\sqrt{3}}\xi^+\eta^-\zeta^+\epsilon^+$$

$$-\frac{1}{2\sqrt{3}}\xi^-\eta^+\zeta^+\epsilon^+$$

Table VIII.3 (continued)

$\left(t_{2g}^3 \cdot e_g\right)^3 A_2$:

$$^3A_2 \quad 1a_2 = \frac{1}{2}\xi^+\eta^-\zeta^+\epsilon^+ - \frac{1}{2}\xi^-\eta^+\zeta^+\epsilon^+ - \frac{1}{\sqrt{3}}\xi^+\eta^+\zeta^-\theta^+ + \frac{1}{2\sqrt{3}}\xi^+\eta^-\zeta^+\theta^+ + \frac{1}{2\sqrt{3}}\xi^-\eta^+\zeta^+\theta^+$$

$\left(t_{2g}^3 \cdot e_g\right) 1^3E$:

$$^3E \quad 1\theta = -\frac{1}{\sqrt{6}}[\xi^-\eta^+\zeta^+\epsilon^+ + \xi^+\eta^-\zeta^+\epsilon^+ - 2\xi^+\eta^+\zeta^-\epsilon^+]$$

$$1\epsilon = \frac{1}{\sqrt{6}}[\xi^-\eta^+\zeta^+\theta^+ + \xi^+\eta^-\zeta^+\theta^+ - 2\xi^+\eta^+\zeta^-\theta^+]$$

$\left(t_{2g}^3 \cdot e_g\right) 2^3E$:

$$^3E \quad 1\theta = -\frac{1}{2}\xi^+\eta^-\zeta^+\theta^+ + \frac{1}{2}\xi^-\eta^+\zeta^+\theta^+ + \frac{1}{\sqrt{3}}\xi^+\eta^+\zeta^-\epsilon^+ - \frac{1}{2\sqrt{3}}\xi^+\eta^-\zeta^+\epsilon^+ - \frac{1}{2\sqrt{3}}\xi^-\eta^+\zeta^+\epsilon^+$$

$$1\epsilon = \frac{1}{2}\xi^+\eta^-\zeta^+\epsilon^+ - \frac{1}{2}\xi^-\eta^+\zeta^+\epsilon^+ + \frac{1}{\sqrt{3}}\xi^+\eta^+\zeta^-\theta^+ - \frac{1}{2\sqrt{3}}\xi^+\eta^-\zeta^+\theta^+ - \frac{1}{2\sqrt{3}}\xi^-\eta^+\zeta^+\theta^+$$

$\left(t_{2g}^3 \cdot e_g\right) 1^3T_1$:

$$^3T_1 \quad 1x = -\frac{1}{2\sqrt{2}}\xi^+\eta^2\theta^+ + \frac{1}{2\sqrt{2}}\xi^+\zeta^2\theta^+ + \frac{\sqrt{3}}{2\sqrt{2}}\xi^+\eta^2\epsilon^+ - \frac{\sqrt{3}}{2\sqrt{2}}\xi^+\zeta^2\epsilon^+$$

$$1y = -\frac{1}{2\sqrt{2}}\eta^+\zeta^2\theta^+ + \frac{1}{2\sqrt{2}}\xi^2\eta^+\theta^+ - \frac{\sqrt{3}}{2\sqrt{2}}\eta^+\zeta^2\epsilon^+ + \frac{\sqrt{3}}{2\sqrt{2}}\xi^2\eta^+\epsilon^+$$

$$1z = \frac{1}{\sqrt{2}}\xi^2\eta^+\theta^+ - \frac{1}{\sqrt{2}}\eta^2\zeta^+\theta^+$$

$\left(t_{2g}^3 \cdot e_g\right) 2^3T_1$:

$$^3T_1 \quad 1x = -\frac{\sqrt{3}}{2\sqrt{2}}\xi^+\zeta^2\theta^+ - \frac{\sqrt{3}}{2\sqrt{2}}\xi^+\eta^2\theta^+ - \frac{1}{2\sqrt{2}}\xi^+\zeta^2\epsilon^+ - \frac{1}{2\sqrt{2}}\xi^+\eta^2\epsilon^+$$

$$1y = \frac{\sqrt{3}}{2\sqrt{2}}\eta^+\zeta^2\theta^+ + \frac{\sqrt{3}}{2\sqrt{2}}\xi^2\eta^+\theta^+ - \frac{1}{2\sqrt{2}}\eta^+\zeta^2\epsilon^+ - \frac{1}{2\sqrt{2}}\xi^2\eta^+\epsilon^+$$

$$1z = \frac{1}{\sqrt{2}}\xi^2\zeta^+\epsilon^+ + \frac{1}{\sqrt{2}}\eta^2\zeta^+\epsilon^+$$

$\left(t_{2g}^3 \cdot e_g\right) 1^3T_2$:

$$^3T_2 \quad 1\xi = -\frac{\sqrt{3}}{2\sqrt{2}}\xi^+\eta^2\theta^+ + \frac{\sqrt{3}}{2\sqrt{2}}\xi^+\zeta^2\theta^+ - \frac{1}{2\sqrt{2}}\xi^+\eta^2\epsilon^+ + \frac{1}{2\sqrt{2}}\xi^+\zeta^2\epsilon^+$$

$$1\eta = \frac{\sqrt{3}}{2\sqrt{2}}\eta^+\zeta^2\theta^+ - \frac{\sqrt{3}}{2\sqrt{2}}\xi^2\eta^+\theta^+ - \frac{1}{2\sqrt{2}}\eta^+\zeta^2\epsilon^+ + \frac{1}{2\sqrt{2}}\xi^2\eta^+\epsilon^+$$

$$1\zeta = \frac{1}{\sqrt{2}}\xi^2\zeta^+\epsilon^+ - \frac{1}{\sqrt{2}}\eta^2\zeta^+\epsilon^+$$

Table VIII.3 (continued)

$(t_{2g}^3 \cdot e_g) 2^3T_2$:

$$^3T_2 \quad 1\xi = -\frac{1}{2\sqrt{2}}\xi^+\zeta^2\theta^+ - \frac{1}{2\sqrt{2}}\xi^+\eta^2\theta^+ + \frac{\sqrt{3}}{2\sqrt{2}}\xi^+\zeta^2\epsilon^+ + \frac{\sqrt{3}}{2\sqrt{2}}\xi^+\eta^2\epsilon^+$$

$$1\eta = -\frac{1}{2\sqrt{2}}\eta^+\zeta^2\theta^+ - \frac{1}{2\sqrt{2}}\xi^2\eta^+\theta^+ - \frac{\sqrt{3}}{2\sqrt{2}}\eta^+\zeta^2\epsilon^+ - \frac{\sqrt{3}}{2\sqrt{2}}\xi^2\eta^+\epsilon^+$$

$$1\zeta = \frac{1}{\sqrt{2}}\xi^2\zeta^+\theta^+ + \frac{1}{\sqrt{2}}\eta^2\zeta^+\theta^+$$

$(t_{2g}^3 \cdot e_g)^5E$:

$$^5E \quad 2\theta = -\xi^+\eta^+\zeta^+\epsilon^+$$

$$2\epsilon = \xi^+\eta^+\zeta^+\theta^+$$

$(t_{2g}^2 \cdot e_g^2)^5T_2$:

$$^5T_2 \quad 2\xi = \zeta^+\eta^+\theta^+\epsilon^+$$

$$2\eta = \xi^+\zeta^+\theta^+\epsilon^+$$

$$2\zeta = \eta^+\xi^+\theta^+\epsilon^+$$

Table VIII.4

Configuration d^5

$(t_{2g}^5)^2T_2$:

$$^2T_2 \quad \frac{1}{2}\xi = \xi^+\eta^2\zeta^2$$

$$\frac{1}{2}\eta = \xi^2\eta^+\zeta^2$$

$$\frac{1}{2}\zeta = \xi^2\eta^2\zeta^+$$

$(t_{2g}^4 \cdot e_g)^2A_1$:

$$^2A_1 \quad \frac{1}{2}a_1 = \frac{1}{2\sqrt{3}}\eta^2\zeta^2\theta^+ + \frac{1}{2\sqrt{3}}\zeta^2\xi^2\theta^+ - \frac{1}{\sqrt{3}}\xi^2\eta^2\theta^+$$

$$+\frac{1}{2}\zeta^2\xi^2\epsilon^+ - \frac{1}{2}\eta^2\zeta^2\epsilon^+$$

$(t_{2g}^4 \cdot e_g)^2A_2$:

$$^2A_2 \quad \frac{1}{2}a_2 = \frac{1}{2\sqrt{3}}\eta^2\zeta^2\epsilon^+ + \frac{1}{2\sqrt{3}}\zeta^2\xi^2\epsilon^+ - \frac{1}{\sqrt{3}}\xi^2\eta^2\epsilon^+$$

$$-\frac{1}{2}\zeta^2\xi^2\theta^+ + \frac{1}{2}\eta^2\zeta^2\theta^+$$

$(t_{2g}^4 \cdot e_g)1^2E$:

$$^2E \quad \tfrac{1}{2}\theta = \frac{1}{\sqrt{3}}(\xi^2\eta^2\theta^+ + \eta^2\zeta^2\theta^+ + \zeta^2\xi^2\theta^+)$$

$$\tfrac{1}{2}\epsilon = \frac{1}{\sqrt{3}}(\xi^2\eta^2\epsilon^+ + \eta^2\zeta^2\theta^+ + \zeta^2\xi^2\epsilon^+)$$

$(t_{2g}^4 \cdot e_g)2^2E$:

$$^2E \quad \tfrac{1}{2}\theta = -\frac{1}{2\sqrt{3}}\eta^2\zeta^2\theta^+ - \frac{1}{2\sqrt{3}}\zeta^2\xi^2\theta^+ + \frac{1}{\sqrt{3}}\xi^2\eta^2\theta^+ + \frac{1}{2}\zeta^2\xi^2\epsilon^+ - \frac{1}{2}\eta^2\zeta^2\epsilon^+$$

$$\tfrac{1}{2}\epsilon = \frac{1}{2\sqrt{3}}\eta^2\zeta^2\epsilon^+ + \frac{1}{2\sqrt{3}}\zeta^2\xi^2\epsilon^+ - \frac{1}{\sqrt{3}}\xi^2\eta^2\epsilon^+ + \frac{1}{2}\zeta^2\xi^2\theta^+ - \frac{1}{2}\eta^2\zeta^2\theta^+$$

$(t_{2g}^4 \cdot e_g)1^2T_1$:

$$^2T_1 \quad \tfrac{1}{2}x = -\frac{\sqrt{3}}{2\sqrt{2}}\xi^2\eta^+\zeta^-\theta^+ + \frac{\sqrt{3}}{2\sqrt{2}}\xi^2\eta^-\zeta^+\theta^+ - \frac{1}{2\sqrt{2}}\xi^2\eta^-\zeta^+\epsilon^+ + \frac{1}{\sqrt{2}}\xi^2\eta^-\zeta^+\epsilon^+$$

$$\tfrac{1}{2}y = \frac{\sqrt{3}}{2\sqrt{2}}\xi^+\eta^2\zeta^-\theta^+ - \frac{\sqrt{3}}{2\sqrt{2}}\xi^-\eta^2\zeta^+\theta^+ - \frac{1}{2\sqrt{2}}\xi^+\eta^2\zeta^-\epsilon^+ + \frac{1}{2\sqrt{2}}\xi^-\eta^2\zeta^+\epsilon^+$$

$$\tfrac{1}{2}z = \frac{1}{\sqrt{2}}\xi^+\eta^-\zeta^2\epsilon^+ - \frac{1}{\sqrt{2}}\xi^-\eta^+\zeta^2\epsilon^+$$

$(t_{2g}^4 \cdot e_g)2^2T_1$:

$$^2T_1 \quad \tfrac{1}{2}x = -\frac{1}{2\sqrt{2}}\xi^2\eta^-\zeta^+\theta^+ + \frac{1}{2\sqrt{2}}\xi^2\eta^+\zeta^-\theta^+ + \frac{\sqrt{3}}{2\sqrt{2}}\xi^2\eta^-\zeta^+\epsilon^+ - \frac{\sqrt{3}}{2\sqrt{2}}\xi^2\eta^+\zeta^-\epsilon^+$$

$$\tfrac{1}{2}y = \frac{1}{2\sqrt{2}}\xi^-\eta^2\zeta^+\theta^+ - \frac{1}{2\sqrt{2}}\xi^+\eta^2\zeta^-\theta^+ + \frac{\sqrt{3}}{2\sqrt{2}}\xi^-\eta^2\zeta^+\epsilon^+ - \frac{\sqrt{3}}{2\sqrt{2}}\xi^+\eta^2\zeta^-\epsilon^+$$

$$\tfrac{1}{2}z = \frac{1}{\sqrt{2}}\xi^-\eta^+\zeta^2\theta^+ - \frac{1}{\sqrt{2}}\xi^+\eta^-\zeta^2\theta^+$$

$(t_{2g}^4 \cdot e_g)1^2T_2$:

$$^2T_2 \quad \tfrac{1}{2}\xi = -\frac{1}{2\sqrt{2}}\xi^2\eta^+\zeta^-\theta^+ + \frac{1}{2\sqrt{2}}\xi^2\eta^-\zeta^+\theta^+ + \frac{\sqrt{3}}{2\sqrt{2}}\xi^2\eta^+\zeta^-\epsilon^+ - \frac{\sqrt{3}}{2\sqrt{2}}\xi^2\eta^-\zeta^+\epsilon^+$$

Table VIII.4 (continued)

$$\frac{1}{2}\eta = -\frac{1}{2\sqrt{2}}\xi^{+}\eta^{2}\zeta^{-}\theta^{+} + \frac{1}{2\sqrt{2}}\xi^{-}\eta^{2}\zeta^{+}\theta^{+} - \frac{\sqrt{3}}{2\sqrt{2}}\xi^{+}\eta^{2}\zeta^{-}\epsilon^{+}$$

$$+\frac{\sqrt{3}}{2\sqrt{2}}\xi^{-}\eta^{2}\zeta^{+}\epsilon^{+}$$

$$\frac{1}{2}\zeta = \frac{1}{\sqrt{2}}\xi^{+}\eta^{-}\zeta^{2}\theta^{+} - \frac{1}{\sqrt{2}}\xi^{-}\eta^{+}\zeta^{2}\theta^{+}$$

$(t_{2g}^{4}\cdot e_{g})2^{2}T_{2}$:

$^{2}T_{2}$ $$\frac{1}{2}\xi = -\frac{\sqrt{3}}{2\sqrt{2}}\xi^{2}\eta^{-}\zeta^{+}\theta^{+} + \frac{\sqrt{3}}{2\sqrt{2}}\xi^{2}\eta^{+}\zeta^{-}\theta^{+} - \frac{1}{2\sqrt{2}}\xi^{2}\eta^{-}\zeta^{+}\epsilon^{+}$$

$$+\frac{1}{2\sqrt{2}}\xi^{2}\eta^{+}\zeta^{-}\epsilon^{+}$$

$$\frac{1}{2}\eta = -\frac{\sqrt{3}}{2\sqrt{2}}\xi^{-}\eta^{2}\zeta^{+}\theta^{+} + \frac{\sqrt{3}}{2\sqrt{2}}\xi^{+}\eta^{2}\zeta^{-}\theta^{+} + \frac{1}{2\sqrt{2}}\xi^{-}\eta^{2}\zeta^{+}\epsilon^{+}$$

$$-\frac{1}{2\sqrt{2}}\xi^{+}\eta^{2}\zeta^{-}\epsilon^{+}$$

$$\frac{1}{2}\zeta = \frac{1}{\sqrt{2}}\xi^{-}\eta^{+}\zeta^{2}\epsilon^{+} - \frac{1}{\sqrt{2}}\xi^{+}\eta^{-}\zeta^{2}\epsilon^{+}$$

Table VIII.5

Configuration d^{6}

$(t_{2g}^{6})^{1}A_{1}$:

$^{1}A_{1}$ $a_{1} = \xi^{2}\eta^{2}\zeta^{2}$

$(t_{2g}^{5}\cdot e_{g})^{1}T_{1}$:

$^{1}T_{1}$ $$x = -\frac{\sqrt{3}}{2\sqrt{2}}\xi^{-}\eta^{2}\zeta^{2}\theta^{+} + \frac{\sqrt{3}}{2\sqrt{2}}\xi^{+}\eta^{2}\zeta^{2}\theta^{-}$$

$$-\frac{1}{2\sqrt{2}}\xi^{-}\eta^{2}\zeta^{2}\epsilon^{+} + \frac{1}{2\sqrt{2}}\xi^{+}\eta^{2}\zeta^{2}\epsilon^{-}$$

$$y = \frac{\sqrt{3}}{2\sqrt{2}}\xi^{2}\eta^{-}\zeta^{2}\theta^{+} - \frac{\sqrt{3}}{2\sqrt{2}}\xi^{2}\eta^{+}\zeta^{2}\theta^{-}$$

$$-\frac{1}{2\sqrt{2}}\xi^{2}\eta^{-}\zeta^{2}\epsilon^{+} + \frac{1}{2\sqrt{2}}\xi^{2}\eta^{+}\zeta^{2}\epsilon^{-}$$

$$z = \frac{1}{\sqrt{2}}\xi^{2}\eta^{2}\zeta^{-}\epsilon^{+} - \frac{1}{\sqrt{2}}\xi^{2}\eta^{2}\zeta^{+}\epsilon^{-}$$

Table VIII.5 (continued)

$\left(t_{2g}^{5}\cdot e_{g}\right)^{1}T_{2}$:

$$^{1}T_{2}\quad \xi=-\frac{1}{2\sqrt{2}}\xi^{-}\eta^{2}\zeta^{2}\theta^{+}+\frac{1}{2\sqrt{2}}\xi^{+}\eta^{2}\zeta^{2}\theta^{-}+\frac{\sqrt{3}}{2\sqrt{2}}\xi^{-}\eta^{2}\zeta^{2}\epsilon^{+}-\frac{\sqrt{3}}{2\sqrt{2}}\xi^{+}\eta^{2}\zeta^{2}-\epsilon^{-}$$

$$\eta=-\frac{1}{2\sqrt{2}}\xi^{2}\eta^{-}\zeta^{2}\theta^{+}+\frac{1}{2\sqrt{2}}\xi^{2}\eta^{+}\zeta^{2}\theta^{-}-\frac{\sqrt{3}}{2\sqrt{2}}\xi^{2}\eta^{-}\zeta^{2}\epsilon^{+}+\frac{\sqrt{3}}{2\sqrt{2}}\xi^{2}\eta^{+}\zeta^{2}\epsilon^{-}$$

$$\zeta=\frac{1}{\sqrt{2}}\xi^{2}\eta^{2}\zeta^{-}\theta^{+}-\frac{1}{\sqrt{2}}\xi^{2}\eta^{2}\zeta^{+}\theta^{-}$$

$\left(t_{2g}^{4}\cdot e_{g}^{2}\right)^{5}T_{2g}$:

$$^{5}T_{2g}\quad 2\xi=\xi^{2}\eta^{+}\zeta^{+}\theta^{+}\epsilon^{+}$$

$$2\eta=-\xi^{+}\eta^{2}\zeta^{+}\theta^{+}\epsilon^{+}$$

$$2\zeta=\xi^{+}\eta^{+}\zeta^{2}\theta^{+}\epsilon^{+}$$

$\left(t_{2g}^{3}\cdot e_{g}^{3}\right)^{5}E_{g}$:

$$^{5}E_{g}\quad 2\theta=-\xi^{+}\eta^{+}\zeta^{+}\theta^{2}\epsilon^{+}$$

$$2\epsilon=\xi^{+}\eta^{+}\zeta^{+}\theta^{+}\epsilon^{2}$$

Table VIII.6

Configuration d^{7}

$\left(t_{2g}^{6}\cdot e_{g}\right)^{2}E$:

$$^{2}E\quad \frac{1}{2}\theta=\xi^{2}\eta^{2}\zeta^{2}\theta^{+}$$

$$\frac{1}{2}\epsilon=\xi^{2}\eta^{2}\zeta^{2}\epsilon^{+}$$

$\left(t_{2g}^{5}\cdot e_{g}^{2}\right)1^{2}T_{1}$:

$$^{2}T_{1}\quad \frac{1}{2}x=-\frac{\sqrt{3}}{2\sqrt{2}}\xi^{+}\eta^{2}\zeta^{2}(\epsilon^{2}-\theta^{2})-\frac{1}{2\sqrt{2}}\xi^{+}\eta^{2}\zeta^{2}(\theta^{+}\epsilon^{-}-\theta^{-}\epsilon^{+})$$

$$\frac{1}{2}y=\frac{\sqrt{3}}{2\sqrt{2}}\xi^{2}\eta^{+}\zeta^{2}(\epsilon^{2}-\theta^{2})-\frac{1}{2\sqrt{2}}\xi^{2}\eta^{+}\zeta^{2}(\theta^{+}\epsilon^{-}-\theta^{-}\epsilon^{+})$$

$$\frac{1}{2}z=\frac{1}{\sqrt{2}}\xi^{2}\eta^{2}\zeta^{+}(\theta^{+}\epsilon^{-}-\theta^{-}\epsilon^{+})$$

$\left(t_{2g}^{5}\cdot e_{g}^{2}\right)2^{2}T_{1}$:

$$^{2}T_{1}\quad \frac{1}{2}x=\frac{1}{\sqrt{6}}[\xi^{-}\eta^{2}\zeta^{2}\theta^{+}\epsilon^{+}+\xi^{+}\eta^{2}\zeta^{2}\theta^{-}\epsilon^{+}-2\xi^{+}\eta^{2}\zeta^{2}\theta^{+}\epsilon^{-}]$$

$$\frac{1}{2}y=\frac{1}{\sqrt{6}}[\xi^2\eta^-\zeta^2\theta^+\epsilon^+ + \xi^2\eta^+\zeta^2\theta^-\epsilon^+ - 2\xi^2\eta^+\zeta^2\theta^+\epsilon^-]$$

$$\frac{1}{2}z=\frac{1}{\sqrt{6}}[\xi^2\eta^2\zeta^-\theta^+\epsilon^+ + \xi^2\eta^2\zeta^+\theta^-\epsilon^+ - 2\xi^2\eta^2\zeta^+\theta^+\epsilon^-]$$

$(t_{2g}^5\cdot e_g^2)1^2T_2$:

$$^2T_2 \quad \tfrac{1}{2}\xi=-\frac{1}{2\sqrt{2}}\xi^+\eta^2\zeta^2(\epsilon^2-\theta^2)+\frac{\sqrt{3}}{2\sqrt{2}}\xi^+\eta^2\zeta^2(\theta^+\epsilon^- - \theta^-\epsilon^+)$$

$$\tfrac{1}{2}\eta=-\frac{1}{2\sqrt{2}}\xi^2\eta^+\zeta^2(\epsilon^2-\theta^2)-\frac{\sqrt{3}}{2\sqrt{2}}\xi^2\eta^+\zeta^2(\theta^+\epsilon^- - \theta^-\epsilon^+)$$

$$\tfrac{1}{2}\zeta=\frac{1}{\sqrt{2}}\xi^2\eta^2\zeta^+(\epsilon^2-\theta^2)$$

$(t_{2g}^5\cdot e_g^2)2^2T_2$:

$$^2T_2 \quad \frac{1}{2}\xi=\frac{1}{\sqrt{2}}\xi^+\eta^2\zeta^2(\theta^2+\epsilon^2)$$

$$\frac{1}{2}\eta=\frac{1}{\sqrt{2}}\xi^2\eta^+\zeta^2(\theta^2+\epsilon^2)$$

$$\frac{1}{2}\zeta=\frac{1}{\sqrt{2}}\xi^2\eta^2\zeta^+(\theta^2+\epsilon^2)$$

$(t_{2g}^5\cdot e_g^2)^4T_1$:

$$^4T_1 \quad \frac{3}{2}x=\xi^+\eta^2\zeta^2\theta^+\epsilon^+$$

$$\frac{3}{2}y=\xi^2\eta^+\zeta^2\theta^+\epsilon^+$$

$$\frac{3}{2}z=\xi^2\eta^2\zeta^+\theta^+\epsilon^+$$

$(t_{2g}^4\cdot e_g^3)^4T_1$:

$$^4T_1 \quad \frac{3}{2}x=-\frac{1}{2}\xi^2\eta^+\zeta^+\theta^+\epsilon^2+\frac{\sqrt{3}}{2}\xi^2\eta^+\zeta^+\theta^2\epsilon^+$$

$$\frac{3}{2}y=\frac{1}{2}\xi^+\eta^2\zeta^+\theta^+\epsilon^2+\frac{\sqrt{3}}{2}\xi^+\eta^2\zeta^+\theta^2\epsilon^+$$

$$\frac{3}{2}z=\xi^+\eta^+\zeta^2\theta^+\epsilon^2$$

$(t_{2g}^4\cdot e_g^3)^4T_2$:

$$^4T_2 \quad \frac{3}{2}\xi=-\frac{\sqrt{3}}{2}\xi^2\eta^+\zeta^+\theta^+\epsilon^2-\frac{1}{2}\xi^2\eta^+\zeta^+\theta^2\epsilon^+$$

$$\frac{3}{2}\eta=-\frac{\sqrt{3}}{2}\xi^+\eta^2\zeta^+\theta^+\epsilon^2+\frac{1}{2}\xi^+\eta^2\zeta^+\theta^2\epsilon^+$$

$$\frac{3}{2}\zeta=\xi^+\eta^+\zeta^2\theta^2\epsilon^+$$

Table VIII.7

Configuration d^8

$$(t_{2g}^6 \cdot e_g^2)\,^3A_{2g}:$$

$$^3A_{2g} \quad 1a_2 = \xi^2\eta^2\zeta^2\theta^+\epsilon^+$$

$$(t_{2g}^5 \cdot e_g^3)\,^3T_{1g}:$$

$$^3T_{1g} \quad 1x = -\frac{\sqrt{3}}{2}\xi^+\eta^2\zeta^2\theta^+\epsilon^2 - \frac{1}{2}\xi^+\eta^2\zeta^2\theta^2\epsilon^+$$

$$1y = \frac{\sqrt{3}}{2}\xi^2\eta^+\zeta^2\theta^+\epsilon^2 - \frac{1}{2}\xi^2\eta^+\zeta^2\theta^2\epsilon^+$$

$$1z = \xi^2\eta^2\zeta^+\theta^2\epsilon^+$$

$$(t_{2g}^5 \cdot e_g^3)\,^3T_{2g}:$$

$$^3T_{2g} \quad 1\xi = -\frac{1}{2}\xi^+\eta^2\zeta^2\theta^+\epsilon^2 + \frac{\sqrt{3}}{2}\xi^+\eta^2\zeta^2\theta^2\epsilon^+$$

$$1\eta = -\frac{1}{2}\xi^2\eta^+\zeta^2\theta^+\epsilon^2 - \frac{\sqrt{3}}{2}\xi^2\eta^+\zeta^2\theta^2\epsilon^+$$

$$1\zeta = \xi^2\eta^2\zeta^+\theta^+\epsilon^2$$

Table VIII.8

Configuration d^9

$$(t_{2g}^6 \cdot e_g^3)\,^2E_g:$$

$$^2E_g \quad \frac{1}{2}\theta = \xi^2\eta^2\zeta^2\theta^+\epsilon^2$$

$$\frac{1}{2}\epsilon = \xi^2\eta^2\zeta^2\theta^2\epsilon^+$$

$$(t_{2g}^5 \cdot e_g^4)\,^2T_2:$$

$$^2T_2 \quad \frac{1}{2}\xi = \xi^+\eta^2\zeta^2\theta^2\epsilon^2$$

$$\frac{1}{2}\eta = \xi^2\eta^+\zeta^2\theta^2\epsilon^2$$

$$\frac{1}{2}\zeta = \xi^2\eta^2\zeta^+\theta^2\epsilon^2$$

In the wavefunctions given above, a plus sign denotes an up spin and a minus sign denotes a down spin. Also ξ^2, etc., denotes $\xi^+\xi^-$, that is , two electrons are on the ξ orbital, one with up spin and the other with down spin, as already mentioned in the beginning of this appendix; ξ, η, ζ, θ, ϵ are the five wavefunctions belonging to the d orbital in an octahedral crystal field. These have already been given in Chapter 9 of the text.

Appendix IX

Spin-Orbit and Racah Parameters, the Weak-Field and Strong Field Wavefunctions, the Energy Matrices, and the Transformation Coefficients for the Palladium and Platinum Group Ions

Table IX.1

The Spin-orbit parameter ξ for $4d^n$ ions

Configuration	Ions	ζ (experimental)[a] (cm^{-1})	ζ (calculated)[b] (cm^{-1})
$4d^1$	Y^{2+}	300	312
	Zr^{3+}	500	507
	Nb^{4+}	750	
	Mo^{5+}	1030	
$4d^2$	Zr^{2+}	425	432
	Nb^{3+}	670	644
	Mo^{4+}	950	
$4d^3$	Nb^{2+}	555	560
	Mo^{3+}	800	812
	Tc^{4+}	1150	
$4d^4$	Mo^{2+}	695	717
	Tc^{3+}	990	
	Ru^{4+}	1350	
$4d^5$	Mo^{+}	630	
	Tc^{2+}	850	
	Ru^{3+}	1180	1197
	Rh^{4+}	1570	
$4d^6$	Ru^{2+}	1000	1077
	Rh^{3+}	1400	1416
$4d^7$	Rh^{2+}	1640	1664
	Pd^{3+}	1600	1529
$4d^8$	Pd^{2+}	1600	1529
	Ag^{3+}	1930	1940
$4d^9$	Ag^{2+}	1840	1794

[a]Values from the estimates of T. M. Dunn, *Trans. Faraday Soc.* **57**, 1441 (1961).

[b]Values from M. Blume, A. J. Freeman, and R. E. Watson, *Phys. Rev.* **134**, A 320 (1964).

Table IX.2

Racah parameters and $\langle r^4 \rangle$ for $4d^n$ transition metal ions

Configuration	Ion	Watson's wavefunctions[a] B (cm^{-1})	C/B	$\langle r^4 \rangle$ (atomic units)	Richardson et al. wavefunctions[b] B (cm^{-1})	C/B	$\langle r^4 \rangle$ (atomic units)
$4d^1$	Y^{2+}			59.0			54.594
	Zr^{3+}			25.328			29.389
$4d^2$	Zr^{2+}	335	3.98	37.861	333	3.96	33.142
	Nb^{3+}	391	4.04	18.600	391	4.03	17.065
	Mo^{4+}				440	4.08	10.413
$4d^3$	Nb^{2+}	364	3.99	26.982	360	3.97	25.213
	Mo^{3+}	417	4.04	14.386	416	4.03	13.468
	Tc^{4+}						
$4d^4$	Nb^{1+}	390	3.99	20.219	324	3.89	39.761
	Mo^{2+}				387	3.98	18.941
	Tc^{3+}				440	4.03	11.183
	Ru^{4+}						
$4d^5$	Mo^{1+}	360	3.94	32.985	355	3.91	28.146
	Tc^{2+}	417	4.00	15.410	414	3.99	14.745
	Ru^{3+}	466	4.05	9.169	464	4.04	8.957
	Rh^{4+}						
$4d^6$	Ru^{2+}	438	4.00	2.867	436	3.99	11.808
	Rh^{3+}	486	4.04	7.785	484	4.03	7.564
$4d^7$	Rh^{2+}	438	4.00	10.600	458	3.98	9.908
	Pd^{3+}	507	4.04	6.591	506	4.03	6.447
$4d^8$	Rh^{1+}	482	4.00	8.832	427	3.93	13.717
	Pd^{2+}	529	4.04	5.610	480	3.99	8.239
	Ag^{3+}				528	4.03	5.411
$4d^9$	Pd^{1+}	504	4.00	7.410	451	3.94	11.115
	Ag^{2+}				502	3.99	6.854

[a]R. E. Watson, *Tech. Rep.* 12 (Solid State and Molecular Theory Group, MIT), 1959; private communication, 1971. We are extremely grateful to him for supplying us the wavefunctions.

[b]J. W. Richardson, M. J. Blackman, and J. E. Ranschak, *J. Chem. Phys.* **58**, 3010 (1973).

Table IX.3

Racah parameters and $\langle r^4 \rangle$ for $5d^n$ transition metal ions

		Burns's wavefunctions[a]		
Configuration	Ion	$\langle r^4 \rangle$ (atomic units)	B (cm^{-1})	C/B
$5d^1$	Lu^{2+}	4.392		
	Hf^{3+}	3.662		
	Ta^{4+}	3.077		
	W^{5+}	2.605		
	Re^{6+}	2.219		
$5d^2$	Hf^{2+}	3.899	595	4.26
	Ta^{3+}	3.267	622	4.27
	W^{4+}	2.759	648	4.26
	Re^{5+}	2.346	675	4.27
	Os^{6+}	2.007	702	4.26
$5d^3$	Ta^{2+}	3.473	612	4.27
	W^{3+}	2.925	639	4.26
	Re^{4+}	2.481	666	4.27
	Os^{5+}	2.118	693	4.26
	Ir^{6+}	1.819	720	4.27
$5d^4$	W^{2+}	3.103	630	4.26
	Os^{4+}	2.237	683	4.26
	Ir^{5+}	1.917	710	4.27
	Pt^{6+}	1.653	737	4.26
	Re^{3+}	2.626	656	4.27
$5d^5$	Re^{2+}	2.782	647	4.27
	Os^{3+}	2.364	674	4.26
	Ir^{4+}	2.022	701	4.27
	Pt^{5+}	1.740	728	4.26
$5d^6$	Os^{2+}	2.501	665	4.26
	Ir^{3+}	2.135	691	4.27
	Pt^{4+}	1.833	718	4.26
	Au^{5+}	1.583	745	4.27
$5d^7$	Ir^{2+}	2.255	682	4.27
	Pt^{3+}	1.932	709	4.26
	Au^{4+}	1.665	736	4.27
	Hg^{5+}	1.443	764	4.22
$5d^8$	Pt^{2+}	2.038	699	4.26
	Au^{3+}	1.753	726	4.27
	Hg^{4+}	1.516	753	4.26
$5d^9$	Pt^{1+}	2.151	690	4.26
	Au^{2+}	1.847	717	4.27
	Hg^{3+}	1.594	744	4.26

[a]The $5d^n$ wavefunctions are taken from G. Burns, *J. Chem. Phys.* **41**, 1521 (1964).

Table IX.4

The zero-field spin-orbit wavefunctions and energies of a d^1 configuration

Term	Wavefunctions	Energy
$^2D_{3/2}$	$\|\frac{3}{2},\frac{3}{2}\rangle = \sqrt{\frac{4}{5}}\,\|\overset{-}{2}\rangle - \sqrt{\frac{1}{5}}\,\|\overset{+}{1}\rangle$	$-\frac{3}{2}\zeta$
	$\|\frac{3}{2},\frac{1}{2}\rangle = \sqrt{\frac{3}{5}}\,\|\overset{-}{1}\rangle - \sqrt{\frac{2}{5}}\,\|\overset{+}{0}\rangle$	
	$\|\frac{3}{2},-\frac{1}{2}\rangle = \sqrt{\frac{2}{5}}\,\|\overset{-}{0} - \sqrt{\frac{3}{5}}\,\|-\overset{-}{1}\rangle$	
	$\|\frac{3}{2},-\frac{3}{2}\rangle = \sqrt{\frac{1}{5}}\,\|-\overset{-}{1}\rangle - \sqrt{\frac{4}{5}}\,\|-\overset{+}{2}\rangle$	
$^2D_{5/2}$	$\|\frac{5}{2},\frac{5}{2}\rangle = \|\overset{+}{2}\rangle$	ζ
	$\|\frac{5}{2},\frac{3}{2}\rangle = \sqrt{\frac{4}{5}}\,\|\overset{+}{1}\rangle + \sqrt{\frac{1}{5}}\,\|\overset{-}{2}\rangle$	
	$\|\frac{5}{2},\frac{1}{2}\rangle = \sqrt{\frac{3}{5}}\,\|\overset{+}{0}\rangle + \sqrt{\frac{2}{5}}\,\|\overset{-}{1}\rangle$	
	$\|\frac{5}{2},-\frac{1}{2}\rangle = \sqrt{\frac{3}{5}}\,\|\overset{-}{0}\rangle + \sqrt{\frac{2}{5}}\,\|-\overset{+}{1}\rangle$	
	$\|\frac{5}{2},-\frac{3}{2}\rangle = \sqrt{\frac{4}{5}}\,\|-\overset{-}{1}\rangle + \sqrt{\frac{1}{5}}\,\|-\overset{-}{2}\rangle$	
	$\|\frac{5}{2},-\frac{5}{2}\rangle = \|-\overset{-}{2}\rangle$	

Table IX.5

The weak-field wavefunctions with the energies of a d^1 configuration

Term	Wavefunctions	Energy
$\Gamma_7(^2D_{5/2})$	$\Gamma_7^{1/2}=\sqrt{\frac{1}{6}}\,\lvert\frac{5}{2},\frac{5}{2}\rangle-\sqrt{\frac{5}{6}}\,\lvert\frac{5}{2},-\frac{3}{2}\rangle$	$\zeta-4Dq$
	$=\sqrt{\frac{1}{6}}\,\lvert\overset{+}{2}\rangle-\sqrt{\frac{4}{6}}\,\lvert-\overset{-}{1}\rangle-\sqrt{\frac{1}{6}}\,\lvert-\overset{+}{2}\rangle$	
	$\Gamma_7^{-1/2}=\sqrt{\frac{1}{6}}\,\lvert\frac{5}{2},-\frac{5}{2}\rangle-\sqrt{\frac{5}{6}}\,\lvert\frac{5}{2},\frac{3}{2}\rangle$	
	$=\sqrt{\frac{1}{6}}\,\lvert-\overset{-}{2}\rangle-\sqrt{\frac{4}{6}}\,\lvert\overset{+}{1}\rangle-\sqrt{\frac{1}{6}}\,\lvert\overset{-}{2}\rangle$	
$\Gamma_8(^2D_{5/2})$	$\Gamma_8^{3/2}=-\sqrt{\frac{1}{6}}\,\lvert\frac{5}{2},\frac{3}{2}\rangle-\sqrt{\frac{5}{6}}\,\lvert\frac{5}{2},-\frac{5}{2}\rangle$	$\zeta+2Dq$
	$=-\sqrt{\frac{5}{6}}\,\lvert-\overset{-}{2}\rangle-\sqrt{\frac{4}{30}}\,\lvert\overset{+}{1}\rangle-\sqrt{\frac{1}{30}}\,\lvert\overset{-}{2}\rangle$	
	$\Gamma_8^{1/2}=\lvert\frac{5}{2},\frac{1}{2}\rangle$	
	$=\sqrt{\frac{3}{5}}\,\lvert\overset{+}{0}\rangle+\sqrt{\frac{2}{5}}\,\lvert\overset{-}{1}\rangle$	
	$\Gamma_8^{-1/2}=-\lvert\frac{5}{2},-\frac{1}{2}\rangle$	
	$=-\sqrt{\frac{2}{5}}\,\lvert-\overset{+}{1}\rangle-\sqrt{\frac{3}{5}}\,\lvert\overset{-}{0}\rangle$	
	$\Gamma_8^{-3/2}=\sqrt{\frac{5}{6}}\,\lvert\frac{5}{2},\frac{5}{2}\rangle+\sqrt{\frac{1}{6}}\,\lvert\frac{5}{2},-\frac{3}{2}\rangle$	
	$=\sqrt{\frac{5}{6}}\,\lvert\overset{+}{2}\rangle+\sqrt{\frac{4}{30}}\,\lvert-\overset{-}{1}\rangle+\sqrt{\frac{1}{30}}\,\lvert-\overset{+}{2}\rangle$	

Table IX.6

The Crystal field and spin-orbit energy matrices calculated by using the wavefunctions listed in Table IX.5

(*a*) Cubic field (O_h)

Γ_8	$\Gamma_8(^2D_{5/2})$	$\Gamma_8(^2D_{3/2})$
	$\zeta+2Dq$	$-2\sqrt{6}\,Dq$
		$-\frac{3}{2}\zeta$

(*b*) Tetragonal field (D_{4h}). $DST=3Ds-5Dt$

Γ_6^T	$\Gamma_8(^2D_{5/2})$	$\Gamma_8(^2D_{3/2})$
	$\zeta+2Dq-\frac{8}{5}Ds-2Dt$	$-2\sqrt{6}\,Dq+\frac{\sqrt{6}}{5}(Ds+10Dt)$
		$-\frac{3}{2}\zeta-\frac{7}{5}Ds$

Γ_7^T	$\Gamma_7(^2D_{5/2})$	$\Gamma_8(^2D_{5/2})$	$\Gamma_8(^2D_{3/2})$
	$\zeta-4Dq+\frac{7}{3}Dt$	$(2\sqrt{5}\,/15)DST$	$\frac{2\sqrt{5}}{5\cdot\sqrt{6}}DST$
		$\zeta+2Dq+\frac{1}{15}(24Ds-Dt)$	$-2\sqrt{6}\,Dq-\frac{\sqrt{6}}{15}DST$
			$-\frac{3}{2}\zeta+\frac{7}{5}Ds$

(*c*) Trigonal field (D_3 or C_{3v}) $\left(\cos\beta=-\frac{1}{3}, \sin\beta=\frac{2\sqrt{2}}{3}\right)$

$$D\sigma\tau=\frac{1}{15}D\sigma+\frac{2}{3}D\tau$$

Γ_4^τ	$\Gamma_7(^2D_{5/2})$	$\Gamma_8(^2D_{5/2})$	$\Gamma_8(^2D_{3/2})$
	$\zeta-4Dq-\frac{14}{9}D\tau$	$-2\sqrt{5}\,(\frac{2}{5}D\sigma+\frac{1}{9}D\tau)$	$-\sqrt{30}\,D\sigma\tau$
		$\zeta+2Dq+\frac{2}{5}D\sigma-\frac{13}{9}D\tau$	$-2\sqrt{6}\,(Dq+D\sigma\tau)$
			$-\frac{3}{2}\zeta-\frac{7}{5}D\sigma$

$\Gamma_5^\tau, \Gamma_6^\tau$	$\Gamma_8(^2D_{5/2})$	$\Gamma_8(^2D_{3/2})$
	$\zeta+2Dq-\frac{2}{5}D\sigma+3D\tau$	$-2\sqrt{6}\,Dq+\frac{2}{5}e^{(3i/2)(\pi+\beta)}(3D\sigma-5D\tau)$
		$-\frac{3}{2}\zeta+\frac{7}{5}D\sigma$

Table IX.7

The strong-field wavefunctions with the energies of a d^1 configuration

Terms	Wavefunctions	Energy
$\Gamma_8^1(^2T_{2g})$	$\|\Gamma_8 \tfrac{3}{2}\rangle = \frac{i}{\sqrt{6}}\xi^+ + \frac{1}{\sqrt{6}}\eta^+ + \frac{i\sqrt{2}}{\sqrt{3}}\zeta^+$	$-4Dq - \frac{1}{2}\zeta$
	$\|\Gamma_8 \tfrac{1}{2}\rangle = -\frac{i}{\sqrt{2}}\xi^- - \frac{1}{\sqrt{2}}\eta^-$	
	$\|\Gamma_8 -\tfrac{1}{2}\rangle = \frac{i}{\sqrt{2}}\xi^+ - \frac{1}{\sqrt{2}}\eta^+$	
	$\|\Gamma_8 -\tfrac{3}{2}\rangle = -\frac{i}{\sqrt{6}}\xi^- + \frac{1}{\sqrt{6}}\eta^- + \frac{i\sqrt{2}}{\sqrt{3}}\zeta^+$	
$\Gamma_7(^2T_{2g})$	$\|\Gamma_7 \tfrac{1}{2}\rangle = -\frac{i}{\sqrt{3}}\xi^- + \frac{1}{\sqrt{3}}\eta^- - \frac{i}{\sqrt{3}}\zeta^+$	$-4Dq + \zeta$
	$\|\Gamma_7 -\tfrac{1}{2}\rangle = -\frac{i}{\sqrt{3}}\xi^+ - \frac{1}{\sqrt{3}}\eta^+ + \frac{i}{\sqrt{3}}\zeta^-$	
$\Gamma_8^u(^2E_g)$	$\|\Gamma_8' \tfrac{3}{2}\rangle = \epsilon^-$	$6Dq$
	$\|\Gamma_8' \tfrac{1}{2}\rangle = -\theta^+$	
	$\|\Gamma_8' -\tfrac{1}{2}\rangle = \theta^-$	
	$\|\Gamma_8' -\tfrac{3}{2}\rangle = -\epsilon^+$	

Table IX.8

The crystal field and spin-orbit energy matrices calculated by using the wavefunctions listed in Table IX.7

(*a*) Cubic field (O_h)

Γ_8	$\Gamma_8(^2T_{2g})$	$\Gamma_8(^2E_g)$
	$-\frac{1}{2}\zeta-4Dq$	$\sqrt{\frac{3}{2}}\,\zeta$
		$6Dq$

(*b*) Tetragonal field (D_{4h})

Γ_6^T	$\Gamma_8^{\pm 1/2}(^2T_{2g})$	$\Gamma_8^{\pm 1/2}(^2E_g)$
	$-\frac{1}{2}\zeta-4Dq-Ds+4Dt$	$\sqrt{\frac{3}{2}}\,\zeta$
		$6Dq-2Ds-6Dt$

Γ_7^T	$\Gamma_7^{\mp 1/2}(^2T_{2g})$	$\Gamma_8^{\pm 3/2}(^2T_{2g})$	$\Gamma_8^{\pm 3/2}(^2E_g)$
	$\zeta-4Dq+\frac{7}{3}Dt$	$\mp\frac{\sqrt{2}}{3}(3Ds-5Dt)$	0
		$-\frac{1}{2}\zeta-4Dq+Ds+\frac{2}{3}Dt$	$\sqrt{\frac{3}{2}}\,\zeta$
			$6Dq+2Ds-Dt$

(*c*) Trigonal field $\left(\cos\beta=-\frac{1}{3}, \sin\beta=\frac{2\sqrt{2}}{3}\right)$; $D\sigma\tau=2D\sigma-5D\tau$.

γ_4^τ	$\Gamma_7(^2T_{2g})$	$\Gamma_8(^2T_{2g})$	$\Gamma_8(^2E_g)$
	$\zeta-4Dq-\frac{14}{9}D\tau$	$\frac{\sqrt{2}}{9}(9D\sigma+20D\tau)$	$\frac{2}{3\sqrt{3}}D\sigma\tau$
		$-\frac{1}{2}\zeta-4Dq-\frac{1}{9}(9D\sigma+34D\tau)$	$\sqrt{\frac{3}{2}}\,\zeta+\frac{\sqrt{2}}{3\sqrt{3}}D\sigma\tau$
			$6Dq+\frac{7}{3}D\tau$

$\Gamma_5^\tau, \Gamma_6^\tau$	$\Gamma_7(^2T_{2g})$	$\Gamma_8(^2E_g)$
	$-\frac{1}{2}\zeta-4Dq+D\sigma+\frac{2}{3}D\tau$	$\sqrt{\frac{3}{2}}\,\zeta-\frac{\sqrt{2}}{3}e^{i\beta/2}D\sigma\tau$
		$6Dq+\frac{7}{3}D\tau$

Table IX.9

Transformation coefficients between the states of the intermediate coupling scheme and the strong-field scheme

A_1	$(\Gamma_8^a)^2$	$(\Gamma_8^b)^2$	$\Gamma_8^a\Gamma_8^b$	Γ_7^2
$t_2^2(^1A_1)$	$\sqrt{\frac{2}{3}}$	0	0	$\frac{1}{\sqrt{3}}$
$t_2^2(^3T_1)$	$-\frac{1}{\sqrt{3}}$	0	0	$\sqrt{\frac{2}{3}}$
$e^2(^1A_1)$	0	1	0	0
$t_2e(^3T_1)$	0	0	-1	0

E	$(\Gamma_8^b)^2$	$(\Gamma_8^a)^2$	$\Gamma_8^a\Gamma_8^b$	$\Gamma_7\Gamma_8^a$	$\Gamma_7\Gamma_8^b$
$e^2(^1E)$	-1	0	0	0	0
$t_2^2(^1E)$	0	$\frac{1}{\sqrt{3}}$	0	$-\sqrt{\frac{2}{3}}$	0
$t_2^2(^3T_1)$	0	$\sqrt{\frac{2}{3}}$	0	$\frac{1}{\sqrt{3}}$	0
$t_2e(^3T_1)$	0	0	$-\frac{1}{\sqrt{2}}$	0	$\frac{1}{\sqrt{2}}$
$t_2e(^3T_2)$	0	0	$\frac{1}{\sqrt{2}}$	0	$\frac{1}{\sqrt{2}}$

T_1	$\Gamma_7\Gamma_8^a$	$\Gamma_7\Gamma_8^b$	$\Gamma_8^a\Gamma_8^b$(I)	$\Gamma_8^a\Gamma_8^b$(II)
$t_2^2(^3T_1)$	1	0	0	0
$t_2e(^1T_2)$	0	$\frac{1}{\sqrt{3}}$	$-\sqrt{\frac{3}{5}}$	$-\frac{1}{\sqrt{15}}$
$t_2e(^3T_1)$	0	$\frac{1}{\sqrt{6}}$	0	$\sqrt{\frac{5}{6}}$
$t_2e(^3T_2)$	0	$-\frac{1}{\sqrt{2}}$	$-\sqrt{\frac{2}{5}}$	$\frac{1}{\sqrt{10}}$

T_2	$(\Gamma_8^b)^2$	$(\Gamma_8^a)^2$	$\Gamma_7\Gamma_8^a$	$\Gamma_8^a\Gamma_8^b$(I)	$\Gamma_8^a\Gamma_8^b$(II)	$\Gamma_7\Gamma_8^b$
$e^2(^3A_2)$	-1	0	0	0	0	0
$t_2^2(^3T_1)$	0	$-\sqrt{\frac{2}{3}}$	$-\frac{1}{\sqrt{3}}$	0	0	0
$t_2^2(^1T_2)$	0	$\frac{1}{\sqrt{3}}$	$-\sqrt{\frac{2}{3}}$	0	0	0
$t_2e(^1T_2)$	0	0	0	$\frac{1}{\sqrt{3}}$	$\frac{1}{\sqrt{3}}$	$\frac{1}{\sqrt{3}}$
$t_2e(^3T_1)$	0	0	0	0	$-\frac{1}{\sqrt{2}}$	$\frac{1}{\sqrt{2}}$
$t_2e(^3T_2)$	0	0	0	$-\sqrt{\frac{2}{3}}$	$-\frac{1}{\sqrt{6}}$	$\frac{1}{\sqrt{6}}$

Table IX.9 (continued)

A_2	$\Gamma_8^a\Gamma_8^b$
$t_2e(^3T_2)$	-1

$\Gamma_8^a=\Gamma_8(t_{2g})$
$\Gamma_8^b=\Gamma_8(e_g)$
$\Gamma_7=\Gamma_7(t_2)$

Appendix X

Proof of the Center of Gravity Rule

In this appendix we prove an important theorem, known as the center of gravity rule, that is, if a perturbation splits the energy level of a single-particle Hamiltonian, the center of gravity of the level remains unaltered.

If $\mathcal{H}$ is the total single-particle Hamiltonian, $\mathcal{H}_0$ the unperturbed part, and $\mathcal{H}_1$ the perturbation, then

$$\mathcal{H} = \mathcal{H}_0 + \mathcal{H}_1 \tag{A.1}$$

Let $\mathcal{H}_0$ belong to the symmetry group G_0. If we ignore the accidental degeneracy, then, since $\mathcal{H}_1$ removes the degeneracy of the energy levels due to $\mathcal{H}_0$, $\mathcal{H}_1$ has lower symmetry than $\mathcal{H}_0$. Now let us assume that $\mathcal{H}_1$ belongs to the symmetry group $\mathcal{G}_1$ (so that $\mathcal{H}$ will have the symmetry $\mathcal{G}_1$). Although $\mathcal{H}_1$ remains invariant under the operations of $\mathcal{G}_1$, it is no longer invariant under $\mathcal{G}_0$, and $\mathcal{G}_1$ is a subgroup of $\mathcal{G}_0$. As an operator in $\mathcal{G}_0$, $\mathcal{H}_1$ can be written as a sum of tensor operators in $\mathcal{G}_0$, that is,

$$\mathcal{H}_1 = \sum c_i h_{\gamma_i}^{\Gamma_i} \tag{A.2}$$

Let $\mathcal{E}_\Gamma$ be a degenerate level in $\mathcal{H}_0$. Since we do not consider the accidental degeneracies, the degenerate states corresponding to $\mathcal{E}_\Gamma$ span the Γth irreducible representations of $\mathcal{G}_0$, that is,

$$\mathcal{H}_0|\Gamma\alpha\rangle = \mathcal{E}_\Gamma|\Gamma\alpha\rangle \tag{A.3}$$

To find the results of perturbation $\mathcal{H}_1$ on $\mathcal{E}_\Gamma$, we have to find the roots of the matrix $\langle\Gamma\alpha'|\mathcal{H}_1|\Gamma\alpha\rangle$. The center of gravity (C.G.) is given by the trace of the matrix $\langle\Gamma\alpha'|\mathcal{H}_1|\Gamma\alpha\rangle$, that is,

$$\begin{aligned} \text{C.G.} &= \text{Tr}\langle\Gamma\alpha'|\mathcal{H}_1|\Gamma\alpha\rangle \\ &= \sum_{\alpha}\langle\Gamma\alpha|\mathcal{H}_1|\Gamma\alpha\rangle \end{aligned} \tag{A.4}$$

Using (A.2), we get

$$\text{C.G.} = \sum_{\alpha,i} c_i\langle\Gamma\alpha|h^{\Gamma_i}_{\gamma_i}|\Gamma\alpha\rangle \tag{A.5}$$

Now, with the help of the Wigner-Eckart theorem (Chapter 5), this can be written as

$$\text{C.G.} = \sum_{\alpha,i} c_i\langle\Gamma\|h^{\Gamma_i}\|\Gamma\rangle\lambda(\Gamma)^{-1/2}\langle\Gamma\alpha\Gamma_i\alpha_i;\Gamma\alpha\rangle^* \tag{A.6}$$

Using the well-known symmetry properties of the coupling coefficients, we can write

$$\langle\Gamma\alpha\Gamma_i\alpha_i;\Gamma\alpha\rangle = \pm\sqrt{\frac{\lambda(\Gamma)}{\lambda(\Gamma_i)}}\;\langle\Gamma\alpha\Gamma\alpha;\Gamma_i\gamma_i\rangle \tag{A.7}$$

The phase factor depends[2] on Γ and Γ_i. It can easily be shown that

$$\langle\Gamma\alpha\Gamma\alpha;A_1a_1\rangle = [\lambda(\Gamma)]^{-1/2} \tag{A.8}$$

where A_1 is the unit representation of G_0. Making use of (A.7) and (A.8), we can write (A.6) as

$$\text{C.G.} = \sum_i (\pm)c_i\sqrt{\frac{\lambda(\Gamma)}{\lambda(\Gamma_i)}}\;\sum_{\alpha}\langle\Gamma\alpha\Gamma\alpha;\Gamma_i\gamma_i\rangle^*\langle\Gamma\alpha\Gamma\alpha;A_1a_1\rangle\cdot\langle\Gamma\|h^{\Gamma_i}\|\Gamma\rangle \tag{A.9}$$

Now, by using the orthogonality relation between the coupling coefficients, the sum over α reduces to

$$\sum_{\alpha}\langle\Gamma\alpha\Gamma\alpha;\Gamma_i\gamma_i\rangle\langle\Gamma\alpha\Gamma\alpha;A_1a_1\rangle = \delta_{\Gamma_iA_1} \tag{A.10}$$

Since we are interested only in the parts of $\mathcal{H}_1$ that lift the degeneracy, we have $\Gamma_i \neq A_1$. This shows that the sum of the roots of the matrix $\langle\Gamma\alpha|\mathcal{H}_1|\Gamma'\alpha'\rangle$ are zero—hence the theorem.

In the proof given above we did not consider the possibility that, in the product $\Gamma\otimes\Gamma_i$, Γ may be repeated. However, in this case also it can be shown that the theorem is true. The proof is a straightforward extension of the one above.

REFERENCES

1. V. P. Desai, Ph.D. Thesis, University of Calcutta, 1976.
2. J. S. Griffith, The Irreducible Tensor Methods of Molecular Symmetry Groups, Prentice-Hall, Englewood Cliffs, N.J., 1962.

Author Index

Subject Index